A MODERN INTRODUCTION TO CLASSICAL ELECTRODYNAMICS

OXFORD MASTER SERIES IN PHYSICS

The Oxford Master Series is designed for final year undergraduate and beginning graduate students in physics and related disciplines. It has been driven by a perceived gap in the literature today. While basic undergraduate physics texts often show little or no connection with the huge explosion of research over the last two decades, more advanced and specialized texts tend to be rather daunting for students. In this series, all topics and their consequences are treated at a simple level, while pointers to recent developments are provided at various stages. The emphasis is on clear physical principles like symmetry, quantum mechanics, and electromagnetism which underlie the whole of physics. At the same time, the subjects are related to real measurements and to the experimental techniques and devices currently used by physicists in academe and industry. Books in this series are written as course books, and include ample tutorial material, examples, illustrations, revision points, and problem sets. They can likewise be used as preparation for students starting a doctorate in physics and related fields, or for recent graduates starting research in one of these fields in industry.

CONDENSED MATTER PHYSICS

1. M.T. Dove: *Structure and dynamics: an atomic view of materials*
2. J. Singleton: *Band theory and electronic properties of solids*
3. A.M. Fox: *Optical properties of solids, second edition*
4. S.J. Blundell: *Magnetism in condensed matter*
5. J.F. Annett: *Superconductivity, superfluids, and condensates*
6. R.A.L. Jones: *Soft condensed matter*
17. S. Tautz: *Surfaces of condensed matter*
18. H. Bruus: *Theoretical microfluidics*
19. C.L. Dennis, J.F. Gregg: *The art of spintronics: an introduction*
21. T.T. Heikkilä: *The physics of nanoelectronics: transport and fluctuation phenomena at low temperatures*
22. M. Geoghegan, G. Hadziioannou: *Polymer electronics*

ATOMIC, OPTICAL, AND LASER PHYSICS

7. C.J. Foot: *Atomic physics*
8. G.A. Brooker: *Modern classical optics*
9. S.M. Hooker, C.E. Webb: *Laser physics*
15. A.M. Fox: Quantum optics: *an introduction*
16. S.M. Barnett: *Quantum information*
23. P. Blood: *Quantum confined laser devices*

PARTICLE PHYSICS, ASTROPHYSICS, AND COSMOLOGY

10. D.H. Perkins: *Particle astrophysics, second edition*
11. Ta-Pei Cheng: *Relativity, gravitation and cosmology, second edition*
24. G. Barr, R. Devenish, R. Walczak, T. Weidberg: *Particle physics in the LHC era*
26. M. E. Peskin: *Concepts of elementary particle physics*

STATISTICAL, COMPUTATIONAL, AND THEORETICAL PHYSICS

12. M. Maggiore: *A modern introduction to quantum field theory*
13. W. Krauth: *Statistical mechanics: algorithms and computations*
14. J.P. Sethna: *Statistical mechanics: entropy, order parameters, and complexity, second edition*
20. S.N. Dorogovtsev: *Lectures on complex networks*
25. R. Soto: *Kinetic theory and transport phenomena*
27. M. Maggiore: *A modern introduction to classical electrodynamics*

A Modern Introduction to Classical Electrodynamics

Michele Maggiore

Département de Physique Théorique
Université de Genève

OXFORD
UNIVERSITY PRESS

OXFORD
UNIVERSITY PRESS

Great Clarendon Street, Oxford, OX2 6DP,
United Kingdom

Oxford University Press is a department of the University of Oxford.
It furthers the University's objective of excellence in research, scholarship,
and education by publishing worldwide. Oxford is a registered trade mark of
Oxford University Press in the UK and in certain other countries

Published in the United States of America by Oxford University Press
198 Madison Avenue, New York, NY 10016, United States of America

British Library Cataloguing in Publication Data

Data available

Library of Congress Control Number: 2023935108

ISBN 9780192867421
ISBN 9780192867438 (pbk.)

DOI: 10.1093/oso/9780192867421.001.0001

Printed and bound by
CPI Group (UK) Ltd, Croydon, CR0 4YY

Links to third party websites are provided by Oxford in good faith and
for information only. Oxford disclaims any responsibility for the materials
contained in any third party website referenced in this work.

A Maura, Sara, Ilaria e Lorenzo

Preface

This book evolved from the notes of a course that I teach at the University of Geneva, for undergraduate physics students. For many generations of physicists, including mine, the classic references for classical electrodynamics have been the textbook by Jackson and that by Landau and Lifschits.[1] The former is still much used, although more modern and excellent textbooks with a somewhat similar structure, such as Garg (2012) or Zangwill (2013), now exist, while the latter is by now rarely used, even as an auxiliary reference text for a course. Because of my field-theoretical background, as my notes were growing I realized that they were naturally drifting toward what looked to me as a modern version of Landau and Lifschits, and this stimulated me to expand them further into a book.

While this book is meant as a modern introduction to classical electrodynamics, it is by no means intended as a first introduction to the subject. The reader is assumed to have already had a first course on electrodynamics, at a level covered for instance by Griffiths (2017). This also implies a different structure of the presentation. In a first course of electrodynamics, it is natural to take a 'bottom-up' approach, where one starts from experimental observations in the simple settings of electrostatics and magnetostatics, and then moves toward time-dependent phenomena and electromagnetic induction, which eventually leads to generalizing the equations governing electrostatics and magnetostatics into the synthesis provided by the full Maxwell's equation. This approach is the natural one for a first introduction because, first of all, gives the correct historical perspective and shows how Maxwell's equations emerged from the unification of a large body of observations; furthermore, it also allows one to start with more elementary mathematical tools, for the benefit of the student that meets some of them for the first time, while at the same time discovering all these new and fundamental physics concepts. The price that is paid is that the approach, following the historical developments, is sometimes heuristic, and the logic of the arguments and derivations is not always tight.

For this more advanced text, I have chosen instead a 'top-down' approach. Maxwell's equations are introduced immediately (after an introductory chapter on mathematical tools) as the 'definition' of the theory, and their consequences are then systematically developed. This has the advantage of a better logical clarity. It will also allow us to always go into the 'real story', rather than presenting at first a simpler version, to be later improved.

[1] In the text we will refer to the latest editions of these books, Jackson (1998) and Landau and Lifschits (1975). However, these books went through many editions: the first edition of Jackson appeared in 1962, while that of Landau and Lifschits even dates back to 1939.

An important aspect of our presentation is that we keep distinct the discussion of electrodynamics 'in vacuum' (i.e., the computation of the electromagnetic fields generated by localized sources, in the region outside the sources) from the study of Maxwell's equations inside materials. The study of the equations 'in vacuum' reveals the underlying fundamental structure of the theory, while classical electrodynamics in material media is basically a phenomenological theory. Mixing the two treatments, because of a formal similarity among the equations, can be conceptually confusing. Until Chapter 12 we will focus uniquely on vacuum electrodynamics, while from Chapter 13 we study electrodynamics in materials.[2]

Focusing first on vacuum electrodynamics allows us to bring out the two most important structural aspects of the theory at its fundamental level, namely gauge invariance and the fact that Special Relativity is hidden in the Maxwell's equations. We will introduce immediately and in full generality the gauge potentials, and work out most of the equations and derivations of vacuum electrodynamics in terms of them. From a modern field-theoretical perspective, we know that classical electrodynamics is the prototype of a gauge theory, and the notion of gauge fields and gauge invariance is central to all modern particle physics, as well as to condensed matter theory. Similarly, after having duly derived from Maxwell's equations the most elementary results of electrostatics and magnetostatics, as well as the notions of work and electromagnetic energy and the expansion in static multipoles, we move as fast as possible to Special Relativity, introducing the covariant formalism and showing how Maxwell's equations can be reformulated in a covariant form.[3] Having in our hands the gauge potentials and the covariant formalism, most of the subsequent derivations in Chapters 8–12 are performed in terms of them, with a clear advantage in technical and conceptual clarity.

Even if this book was born from my notes for an undergraduate course, and is meant to be used for such a course, it has obviously grown well beyond the original scope, and some parts of it are quite advanced. More technical sections, or whole chapters that are more specialized, are clearly marked, so that the book can be used at different levels, from the undergraduate student, to the researcher that needs to check a textbook as a reference. Classical electrodynamics, for its richness and importance, is a subject to which one returns over and over during a scientist's career.

Finally, an important point, when writing a textbook of electrodynamics, is the choice of the system of units. In mechanics, the transformation between systems such as c.g.s. (centimeter-gram-second) and m.k.s. (meter-kilogram-second) is trivial, and just amounts to multiplicative factors. However, in electromagnetism there are further complications. This has led to two main systems of units for classical electrodynamics: the SI system, and the Gaussian system. As we will discuss in Chapter 2, the essential difference is that, for electromagnetism, the SI system beside the units of length, mass and time, introduces a fourth independent base unit of current, the ampere, while in the Gaussian system the unit

---

[2]This approach is different from, e.g. that of Jackson, or Zangwill. It is instead the same followed by Garg, and especially by Landau and Lifschits, that even separated the subjects into two different books: "The Classical Theory of Fields", Landau and Lifschits (1975), for vacuum electrodynamics, and "Electrodynamics of Continuous Media", Landau and Lifschits (1984) for electrodynamics in materials.

[3]By comparison, Jackson introduces the gauge potentials in full generality for the first time only after about 220 pages and Zangwill after about 500 pages, and their introduction is in general presented simply as a trick for simplifying the equations. However, their role is much more fundamental, since they are the basic dynamical variables in a field-theoretical treatment (which also implies that they will become the basic variables also when one moves to a quantum treatment). As for Special Relativity, Jackson introduces it only after more than 500 pages, while Zangwill relegates it to Chapter 22, after 820 pages, and Garg to Chapter 24.

of charge, and therefore of current, is derived from the three basic units of length, mass and time.

The SI system is the natural one for applications to the macroscopic world: currents are measured in amperes, voltages in volts, and so on. This makes the SI system the obvious choice for laboratory applications and in electrical engineering, and SI units are by now the almost universal standard for electrodynamics courses at the undergraduate level. The Gaussian system, on the other hand, has advantages in other contexts, and in particular leads to neater formulas when relativistic effects are important.[4]

This state of affairs has led to a rather peculiar situation. In general, undergraduate textbooks of classical electromagnetism always use SI units; in contrast, more advanced textbooks of classical electrodynamics are often split between SI and Gaussian units, and all textbooks on quantum electrodynamics and quantum field theory use Gaussian units. The difficulty of the choice is exemplified by the Jackson's textbook, that has been the 'bible' of classical electrodynamics for generations of physicists. The second edition (1975), as the first, used Gaussian units throughout. However, the third edition (1998) switched to SI units for the first 10 chapters, in recognition of the fact that almost all other undergraduate level textbooks used SI units; then, from Chapter 11 (Special Theory of Relativity) on, it goes back to Gaussian units, in recognition of the fact that they are more appropriate than SI units for relativistic phenomena.[5] Gaussian units are also the most common choice in quantum mechanics textbooks: when computing the energy level of the hydrogen atom, almost all textbooks use a Coulomb potential in Gaussian units, $-e^2/r$, rather than the SI expression $-e^2/(4\pi\epsilon_0 r)$.[6]

In this book we will use SI units, since this is nowadays the almost universal standard for an undergraduate textbook on classical electrodynamics. However, it is important to be familiar also with the Gaussian system, as a bridge toward graduate and more specialized courses. This is particularly important for the student that wishes to go into theoretical high-energy physics where, eventually, only the Gaussian system will be used. We will then discuss in Section 2.2 how to quickly translate from SI to Gaussian units, and, in Appendix A, we will provide an explicit translation in Gaussian units of the most important results and formulas of the main text.

Finally, I wish to thank Enis Belgacem, Francesco Iacovelli and Michele Mancarella, who gave the exercise sessions of the course for various years, and Stefano Foffa for useful discussions. I thank again Francesco Iacovelli for producing a very large number of figures of the book. I am grateful to Stephen Blundell for extremely useful comments on the manuscript. Last but not least, as with my previous books with OUP, I wish to thank Sonke Adlung, for his friendly and always very useful advice, as well as all the staff at OUP.

Geneva, January 2023

[4]Actually, its real virtues appear when combining electromagnetism with quantum mechanics. In this case, the reduction from four to three base units obtained with the Gaussian system can be pushed further, using a system of units where one also sets $\hbar = c = 1$, with the result that one remains with a single base unit, typically taken to be the mass unit. In quantum field theory this system is so convenient that units $\hbar = c = 1$ are called *natural units* (we will briefly mention them in Section 2.2). As a consequence, all generalizations from classical electrodynamics to quantum electrodynamics (and its extensions such as the Standard Model of electroweak and strong interactions) are nowadays uniquely discussed using the Gaussian system (or, rather, a variant of it, Heaviside–Lorentz or rationalized Gaussian units, differing just by the placing of some 4π factors, that we will also introduce in Chapter 2), supplemented by units $\hbar = c = 1$.

[5]Among the other 'old-time' classics, Landau and Lifschits (1975) used Gaussian units, while the Feynman Lectures on Physics, Feynman *et al.* (1964), used SI. The first two editions of the classic textbook by Purcell used Gaussian units, but switched to SI for the 3rd edition, Purcell and Morin (2013). Among more recent books, SI is used in Griffiths (2017), Zangwill (2013) and Tong (2015), while Gaussian units are used in Garg (2012) (with frequent translations to SI units).

[6]The most notable exception is the quantum mechanics textbook Griffiths (2004), that uses SI units, consistently with the classical electrodynamics book by the same author.

Contents

Mathematical tools

1

Classical electrodynamics requires a good familiarity with a set of mathematical tools, which will then find applications basically everywhere in physics. We find it convenient to begin by recalling some of these concepts so that, later, the understanding of the physics will not be obscured by the mathematical manipulations. We will focus here on tools that will be of more immediate use. Further mathematical tools will be discussed along the way, in the rest of the book, as they will be needed.

1.1 Vector algebra

A vector $\mathbf{a}$ has Cartesian components a_i. We use the convention that repeated indices are summed over, so,

$$\mathbf{a}\cdot\mathbf{b} = \sum_i a_i b_i \equiv a_i b_i \,, \tag{1.1}$$

where the sum runs over $i = 1, 2, 3$ in three spatial dimensions, as we will assume next, or, more generally, over $i = 1, \ldots, d$ in d spatial dimensions. We can introduce the "Kronecker delta" δ_{ij}, which is equal to 1 if $i = j$ and zero otherwise. Note that, with the convention of the sum over repeated indices, we have the identity

$$a_i = \delta_{ij} a_j \,. \tag{1.2}$$

Then, we can also rewrite

$$\mathbf{a}\cdot\mathbf{b} = \delta_{ij} a_i b_j \,. \tag{1.3}$$

From the definition it follows that, in three dimensions, $\delta_{ii} = 3$. Note that δ_{ij} are just the components of the 3×3 identity matrix I, $\delta_{ij} = (I)_{ij}$ and δ_{ii} is the trace of the 3×3 identity matrix (or, in d dimensions, δ_{ij} are the components of the $d \times d$ identity matrix, and $\delta_{ii} = d$).

When using the convention of summing over repeated indices, one must be careful not to use the same dummy index for different summations. For example, writing the sums explicitly,

$$(\mathbf{a}\cdot\mathbf{b})(\mathbf{c}\cdot\mathbf{d}) = \left(\sum_{i=1}^{3} a_i b_i\right)\left(\sum_{j=1}^{3} c_j d_j\right) \,. \tag{1.4}$$

With the convention of the sum over repeated indices, the right-hand side becomes $a_i b_i c_j d_j$. Notice that it was important here to use two different letters, i, j, for the dummy indices involved.

The vector product is given by

$$(\mathbf{a} \times \mathbf{b})_i = \epsilon_{ijk} a_j b_k \,, \tag{1.5}$$

where ϵ_{ijk} is the totally antisymmetric tensor (or the Levi–Civita tensor), defined by $\epsilon_{123} = +1$, together with the condition that it is antisymmetric under any exchange of indices. Therefore $\epsilon_{ijk} = 0$ if two indices have the same value and, e.g., $\epsilon_{213} = -\epsilon_{123} = -1$. This also implies that the tensor is cyclic, $\epsilon_{ijk} = -\epsilon_{jik} = \epsilon_{jki}$, i.e., it is unchanged under a cyclic permutation of the indices. Note that $\mathbf{a} \times \mathbf{b} = -\mathbf{b} \times \mathbf{a}$. We will see in Section 1.6 that the tensors δ_{ij} and ϵ_{ijk} play a special role in the theory of representations of the rotation group. Unit vectors are denoted by a hat; for instance, $\hat{\mathbf{x}}$, $\hat{\mathbf{y}}$, and $\hat{\mathbf{z}}$ are the unit vectors along the x, y and z axes, respectively. Note that, e.g.,

$$\hat{\mathbf{x}} \times \hat{\mathbf{y}} = \hat{\mathbf{z}} \,. \tag{1.6}$$

A very useful identity is

$$\epsilon_{ijk} \epsilon_{ilm} = \delta_{jl}\delta_{km} - \delta_{jm}\delta_{kl} \,. \tag{1.7}$$

(Prove it!) Note the structure of the indices: on the left, the index i is summed over, so it does not appear on the right-hand side. It is a dummy index, and we could have used a different name for it. For instance, the left-hand side of eq. (1.7) could be written as $\epsilon_{pjk}\epsilon_{plm}$, with a different letter p. In contrast, the indices j, k, l, m are free indices so, if they appear on the left-hand side, they must also appear on the right-hand side. Note also that, because of the cyclic property of the epsilon tensor, the left-hand side of eq. (1.7) can also be written as $\epsilon_{jki}\epsilon_{ilm}$.

Exercise 1.1 Show that

$$\mathbf{a} \cdot (\mathbf{b} \times \mathbf{c}) = \epsilon_{ijk} a_i b_j c_k \,. \tag{1.8}$$

Exercise 1.2 Using eq. (1.7), show that

$$\mathbf{a} \times (\mathbf{b} \times \mathbf{c}) = (\mathbf{a} \cdot \mathbf{c})\mathbf{b} - (\mathbf{a} \cdot \mathbf{b})\mathbf{c} \,, \tag{1.9}$$
$$(\mathbf{a} \times \mathbf{b}) \times \mathbf{c} = (\mathbf{a} \cdot \mathbf{c})\mathbf{b} - (\mathbf{b} \cdot \mathbf{c})\mathbf{a} \,. \tag{1.10}$$

Observe that the vector product is not associative: in general, $\mathbf{a} \times (\mathbf{b} \times \mathbf{c})$ is different from $(\mathbf{a} \times \mathbf{b}) \times \mathbf{c}$.

Exercise 1.3 Using eq. (1.7), show that

$$\epsilon_{ijk} \epsilon_{ijm} = 2\delta_{km} \,. \tag{1.11}$$

Double-check the result by directly identifying the combinations of indices that give a non-zero contributions to the left-hand side of eq. (1.11).

1.2   Differential operators on scalar and vector fields

We will use the notation $\partial_i = \partial/\partial x^i$ for the partial derivative with respect to the Cartesian coordinates x^i. Then, if $f(\mathbf{x})$ is a function of the spatial coordinates, its gradient ∇f is a vector field (i.e., a vector defined at each point of space) whose components, in Cartesian coordinates, are given by

$$(\nabla f)_i = \partial_i f \qquad \text{(gradient of a scalar function)}, \qquad (1.12)$$

or, in vector form,

$$\nabla f = (\partial_x f)\hat{\mathbf{x}} + (\partial_y f)\hat{\mathbf{y}} + (\partial_z f)\hat{\mathbf{z}}. \qquad (1.13)$$

The expression in polar coordinates (r, θ, ϕ) can be obtained by performing explicitly the transformation between the derivatives $\partial_x, \partial_y, \partial_z$ and the derivatives $\partial_r = \partial/\partial r, \partial_\theta = \partial/\partial \theta$, and $\partial_\phi = \partial/\partial \phi$, and expressing $\hat{\mathbf{x}}, \hat{\mathbf{y}}, \hat{\mathbf{z}}$ in terms of the unit vectors $\hat{\mathbf{r}}, \hat{\boldsymbol{\theta}}, \hat{\boldsymbol{\phi}}$ (and similarly in any other coordinate system, such as cylindrical coordinates); we will give the results in polar ad cylindric coordinates for different operators at the end of this section. Given a vector field $\mathbf{v}(\mathbf{x})$, we can form two notable quantities with the action of ∇: the divergence

$$\nabla\cdot\mathbf{v} = \partial_i v_i \qquad \text{(divergence of a vector field)}, \qquad (1.14)$$

which is a scalar field (i.e., a quantity invariant under rotations, defined at each point of space), and the curl, $\nabla \times \mathbf{v}$, which is again a vector field, with Cartesian components

$$(\nabla \times \mathbf{v})_i = \epsilon_{ijk}\partial_j v_k \qquad \text{(curl of a vector field)}. \qquad (1.15)$$

Given a function f, after forming the vector field ∇f, we can obtain again a scalar by taking the divergence of ∇f. This defines the Laplacian ∇^2, $\nabla^2 f = \nabla\cdot(\nabla f)$, or

$$\nabla^2 f = \partial_i \partial_i f = (\partial_x^2 + \partial_y^2 + \partial_z^2)f, \qquad (1.16)$$

where $\partial_x = \partial/\partial x$, etc. Similarly, we can differentiate further the divergence or the curl of a vector field. For instance, $\nabla(\nabla\cdot\mathbf{v})$ is a vector field with components

$$[\nabla(\nabla\cdot\mathbf{v})]_i = \partial_i \partial_j v_j, \qquad (1.17)$$

while $\nabla\cdot(\nabla \times \mathbf{v})$ is a scalar field, since it is the divergence of a vector field. However, from the explicit computation in components,

$$\begin{aligned} \nabla\cdot(\nabla \times \mathbf{v}) &= \partial_i(\epsilon_{ijk}\partial_j v_k) \\ &= \epsilon_{ijk}\partial_i \partial_j v_k \\ &= 0. \end{aligned} \qquad (1.18)$$

The last equality follows because $\partial_i \partial_j$ is an operator symmetric in the (i,j) indices (we always assume that the derivatives act on functions, or on vector fields, that are continuous and infinitely differentiable everywhere, so the derivatives commute, $\partial_i \partial_j = \partial_j \partial_i$), and therefore it gives zero when contracted with the antisymmetric tensor ϵ_{ijk}. Thus, *the gradient of a curl vanishes*. Similarly, the curl of a gradient vanishes:

$$(\boldsymbol{\nabla} \times \boldsymbol{\nabla} f)_i = \epsilon_{ijk} \partial_j \partial_k f = 0 \,, \tag{1.19}$$

again because $\partial_j \partial_k$ is symmetric in (j,k). The Laplacian of a vector field is defined, using Cartesian coordinates, as

$$(\boldsymbol{\nabla}^2 \mathbf{v})_i \equiv \partial_j \partial_j v_i = (\partial_x^2 + \partial_y^2 + \partial_z^2) v_i \,. \tag{1.20}$$

Exercise 1.4 Using eq. (1.7), show that

$$\boldsymbol{\nabla} \times (\boldsymbol{\nabla} \times \mathbf{v}) = \boldsymbol{\nabla}(\boldsymbol{\nabla} \cdot \mathbf{v}) - \boldsymbol{\nabla}^2 \mathbf{v} \,. \tag{1.21}$$

For future reference, we give the expression of the gradient and Laplacian of a scalar field, and of the divergence and curl of a vector field, in Cartesian coordinates (x, y, x), in polar coordinates (r, θ, ϕ), and in cylindrical coordinates (ρ, φ, z).[1,2] Denoting by $\{\hat{\mathbf{x}}, \hat{\mathbf{y}}, \hat{\mathbf{z}}\}$, $\{\hat{\mathbf{r}}, \hat{\boldsymbol{\theta}}, \hat{\boldsymbol{\phi}}\}$, and $\{\hat{\boldsymbol{\rho}}, \hat{\boldsymbol{\varphi}}, \hat{\mathbf{z}}\}$, respectively, the unit vectors in the corresponding directions, we have

$$
\begin{aligned}
\boldsymbol{\nabla} f &= (\partial_x f)\hat{\mathbf{x}} + (\partial_y f)\hat{\mathbf{y}} + (\partial_z f)\hat{\mathbf{z}} & (1.22) \\
&= (\partial_r f)\hat{\mathbf{r}} + \frac{1}{r}(\partial_\theta f)\hat{\boldsymbol{\theta}} + \frac{1}{r \sin \theta}(\partial_\phi f)\hat{\boldsymbol{\phi}} & (1.23) \\
&= (\partial_\rho f)\hat{\boldsymbol{\rho}} + \frac{1}{\rho}(\partial_\varphi f)\hat{\boldsymbol{\varphi}} + (\partial_z f)\hat{\mathbf{z}} \,, & (1.24)
\end{aligned}
$$

$$
\begin{aligned}
\boldsymbol{\nabla}^2 f &= \partial_x^2 f + \partial_y^2 f + \partial_z^2 f & (1.25) \\
&= \frac{1}{r^2}\partial_r(r^2 \partial_r f) + \frac{1}{r^2 \sin \theta}\partial_\theta(\sin \theta \partial_\theta f) + \frac{1}{r^2 \sin^2 \theta}\partial_\phi^2 f & (1.26) \\
&= \frac{1}{\rho}\partial_\rho(\rho \partial_\rho f) + \frac{1}{\rho^2}\partial_\varphi^2 f + \partial_z^2 f \,, & (1.27)
\end{aligned}
$$

$$
\begin{aligned}
\boldsymbol{\nabla} \cdot \mathbf{v} &= \partial_x v_x + \partial_y v_y + \partial_z v_z & (1.28) \\
&= \frac{1}{r^2}\partial_r(r^2 v_r) + \frac{1}{r \sin \theta}\partial_\theta(v_\theta \sin \theta) + \frac{1}{r \sin \theta}\partial_\phi v_\phi & (1.29) \\
&= \frac{1}{\rho}\partial_\rho(\rho v_\rho) + \frac{1}{\rho}\partial_\varphi v_\varphi + \partial_z v_z \,, & (1.30)
\end{aligned}
$$

$$
\begin{aligned}
\boldsymbol{\nabla} \times \mathbf{v} &= \hat{\mathbf{x}}(\partial_y v_z - \partial_z v_y) + \hat{\mathbf{y}}(\partial_z v_x - \partial_x v_z) + \hat{\mathbf{z}}(\partial_x v_y - \partial_y v_x) & (1.31) \\
&= \hat{\mathbf{r}}\frac{1}{r \sin \theta}\left[\partial_\theta(v_\phi \sin \theta) - \partial_\phi v_\theta\right] + \hat{\boldsymbol{\theta}}\left[\frac{1}{r \sin \theta}\partial_\phi v_r - \frac{1}{r}\partial_r(r v_\phi)\right] \\
&\quad + \hat{\boldsymbol{\phi}}\frac{1}{r}\left[\partial_r(r v_\theta) - \partial_\theta v_r\right] & (1.32) \\
&= \hat{\boldsymbol{\rho}}\left(\frac{1}{\rho}\partial_\varphi v_z - \partial_z v_\varphi\right) + \hat{\boldsymbol{\varphi}}\left(\partial_z v_\rho - \partial_\rho v_z\right) + \hat{\mathbf{z}}\frac{1}{\rho}\left[\partial_\rho(\rho v_\varphi) - \partial_\varphi v_\rho\right] \,. & (1.33)
\end{aligned}
$$

[1] Our convention on the polar angles is such that spherical coordinates are related to Cartesian coordinates by $x = r \sin \theta \cos \phi$, $y = r \sin \theta \sin \phi$, $z = r \cos \theta$, with $\theta \in [0, \pi]$ and $\phi \in [0, 2\pi]$. Cylindrical coordinates are related to Cartesian coordinates by $x = \rho \cos \varphi$, $y = \rho \sin \varphi$ (with $\varphi \in [0, 2\pi]$) and $z = z$. Then, as φ increases, we rotate counterclockwise with respect to the z axis. This means that $\hat{\boldsymbol{\rho}} \times \hat{\boldsymbol{\varphi}} = +\hat{\mathbf{z}}$.

[2] The Laplacian of a vector field has been defined from eq. (1.20) in terms of the Cartesian coordinates and, in this case, on each component of the vector field, it has the same form as the Laplacian acting on a scalar. This is no longer true in polar of cylindrical coordinates. In that case, $\boldsymbol{\nabla}^2 \mathbf{v}$ can be more easily obtained from eq. (1.21), using the corresponding expressions of $\boldsymbol{\nabla} \cdot \mathbf{v}$ and $\boldsymbol{\nabla} \times \mathbf{v}$.

1.3 Integration of vector fields. Gauss's and Stokes's theorems

Given a vector field $\mathbf{v}(\mathbf{x})$ and a curve C, one can define the line integral

$$\int_C d\boldsymbol{\ell} \cdot \mathbf{v}\,, \qquad (1.34)$$

by breaking the curve over infinitesimal segments, and introducing, in each segment, a vector $d\boldsymbol{\ell}$ of length $d\ell$, tangent to the curve. Notice that the line integral defined in this way is a scalar quantity. If the curve C is closed, the line integral (1.34) is known as the *circulation* of $\mathbf{v}$ around the curve. For a closed curve, we will denote the line integral by $\oint d\boldsymbol{\ell}$.

The integral over a two-dimensional surface S can be defined similarly. We split the surface in infinitesimal surface elements of area ds,[3] and we define $d\mathbf{s}$ as the vector of modulus ds, pointing in the direction perpendicular to the surface element (for a closed surface, the convention is to choose the outward normal, otherwise a choice of orientation must be made). Writing the unit vector normal to the surface as $\hat{\mathbf{n}}$, we have $d\mathbf{s} = \hat{\mathbf{n}}\,ds$. The surface integral of a vector field $\mathbf{v}(\mathbf{x})$ is then given by

$$\int_S d\mathbf{s} \cdot \mathbf{v} = \int_S ds\,(\mathbf{v}\cdot\hat{\mathbf{n}})\,. \qquad (1.35)$$

For a closed surface, this defines the *flux* of $\mathbf{v}$ through S. In the case of a closed surface, we will denote the surface integral by $\oint d\mathbf{s}$.

The fundamental theorem of calculus states that, for a function of a single variable x,

$$\int_{x_1}^{x_2} dx\,\frac{df}{dx} = f(x_2) - f(x_1)\,. \qquad (1.36)$$

This can be generalized to the line integral of a function of the three-dimensional variable $\mathbf{x}$: from the definition of the line integral (1.34) one can show (do it!) that, for a function $f(\mathbf{x})$ integrated over a curve C with endpoints $\mathbf{x}_1$ and $\mathbf{x}_2$,

$$\int_C d\boldsymbol{\ell} \cdot \boldsymbol{\nabla} f = f(\mathbf{x}_2) - f(\mathbf{x}_1)\,. \qquad (1.37)$$

Note in particular that, if C is closed, the line integral of a gradient vanishes. Stokes's theorem and Gauss's theorem are generalizations of eq. (1.37) to surfaces and to volumes, respectively. In particular, let C be a *closed* curve and let S be *any* surface that has C as its boundary (i.e., $\partial S = C$, where the notation ∂S stands for the boundary of S). Then, Stokes's theorem asserts that, for a vector field $\mathbf{v}(\mathbf{x})$ (with our usual assumptions of differentiability, that we will not repeat further),

$$\boxed{\int_S d\mathbf{s} \cdot (\boldsymbol{\nabla} \times \mathbf{v}) = \oint_C d\boldsymbol{\ell} \cdot \mathbf{v} \quad \text{(Stokes's theorem)}\,.} \qquad (1.38)$$

---

[3] When we want to stress that this is a two-dimensional surface element, we will write it as $d^2 s$. Otherwise, to simplify the notation, we write simply ds.

The orientation convention is that, if we go around the loop C in the direction of the line integral, the normal to S is obtained with the right-hand rule.

Another useful identity is obtained by setting, in Stokes's theorem, $\mathbf{v}(\mathbf{x}) = \psi(\mathbf{x})\mathbf{w}$, where $\mathbf{w}$ is a constant vector. Then, $(\mathbf{\nabla} \times \mathbf{v})_i = \epsilon_{ijk}(\partial_j \psi)w_k$ and eq. (1.38) becomes

$$w_k \int_S ds_i \epsilon_{ijk} \partial_j \psi = w_k \oint_C d\ell_k \, \psi \, . \tag{1.39}$$

Since this is valid for generic $\mathbf{w}$, we get

$$\int_S ds_i \epsilon_{ijk} \partial_j \psi = \oint_C d\ell_k \, \psi \, , \tag{1.40}$$

or, in vector notation,

$$\boxed{\int_S d\mathbf{s} \times \mathbf{\nabla} \psi = \oint_C d\boldsymbol{\ell} \, \psi \, .} \tag{1.41}$$

Yet another useful identity following from Stokes's theorem is obtained by setting $\mathbf{v}(\mathbf{x}) = \mathbf{u}(\mathbf{x}) \times \mathbf{w}$, where, again, $\mathbf{w}$ is a constant vector. Then,

$$\begin{aligned}
(\mathbf{\nabla} \times \mathbf{v})_i &= \epsilon_{ijk} \partial_j \left(\epsilon_{klm} u_l w_m \right) \\
&= \epsilon_{ijk} \epsilon_{klm} (\partial_j u_l) w_m \\
&= (\partial_j u_i) w_j - (\partial_j u_j) w_i \, ,
\end{aligned} \tag{1.42}$$

where, in the last line, we used eq. (1.7). Then eq. (1.38) gives

$$\begin{aligned}
w_k \epsilon_{ijk} \oint_C d\ell_i u_j &= \int_S ds_i \left[(\partial_j u_i) w_j - (\partial_j u_j) w_i \right] \\
&= w_k \left[\int_S ds_i \, \partial_k u_i - \int_S ds_k \, \partial_i u_i \right] \, , \tag{1.43}
\end{aligned}$$

and therefore

$$\epsilon_{ijk} \oint_C d\ell_i u_j = \int_S ds_i \, \partial_k u_i - \int_S ds_k \, \partial_i u_i \, . \tag{1.44}$$

A useful application of this formula is obtained choosing $u_i(\mathbf{x}) = x_i$. Then $\partial_k u_i = \delta_{ik}$ and $\partial_i u_i = 3$, so eq. (1.44) gives

$$\epsilon_{ijk} \oint_C d\ell_i x_j = -2 \int_S ds_k \, . \tag{1.45}$$

However, for a planar surface

$$\int_S ds_k \equiv A\hat{n}_k \, , \tag{1.46}$$

where A is the area of the surface and $\hat{\mathbf{n}}$ is the unit vector normal to it. We therefore obtain an elegant formula for the area A of a planar surface S, bounded by a curve $\mathcal{C}$,

$$A\hat{\mathbf{n}} = \frac{1}{2} \oint_{\mathcal{C}} \mathbf{x} \times d\boldsymbol{\ell} \, . \tag{1.47}$$

Gauss's theorem extends Stokes's theorem further, to integration over volumes: let V be a finite volume bounded by the surface S, i.e., $\partial V = S$. Then,

$$\int_V d^3x \, \boldsymbol{\nabla}\cdot\mathbf{v} = \int_S d\mathbf{s}\cdot\mathbf{v} \qquad \text{(Gauss's theorem)}. \qquad (1.48)$$

We will make use very often of both Gauss's and Stokes's theorems.[4]

A vector field such that $\boldsymbol{\nabla}\times\mathbf{v} = 0$ everywhere is called *irrotational*, or curl-free. We have seen in eq. (1.19) that, if $\mathbf{v}$ is the gradient of a function, $\mathbf{v} = \boldsymbol{\nabla}f$, then it is irrotational. A sort of converse of this statement holds:

Theorem for curl-free fields. Let $\mathbf{v}$ be a vector field such that $\boldsymbol{\nabla}\times\mathbf{v} = 0$ everywhere in a region V simply connected (i.e., such that every loop in V can be continuously shrunk to a point). Then, there exists a function f such that $\mathbf{v} = \boldsymbol{\nabla}f$.

A vector field $\mathbf{v}$ such that $\boldsymbol{\nabla}\cdot\mathbf{v} = 0$ is called *solenoidal*, or divergence-free. Similarly, there is a sort of inverse to eq. (1.18):

Theorem for divergence-free fields. Let $\mathbf{v}$ be a vector field such that $\boldsymbol{\nabla}\cdot\mathbf{v} = 0$ everywhere in a volume V such that every surface in V can be continuously shrunk to a point. Then, there exists a vector field $\mathbf{w}$ such that $\mathbf{v} = \boldsymbol{\nabla}\times\mathbf{w}$.

[4]Proceeding as for Stokes's theorem, if we set $\mathbf{v}(\mathbf{x}) = \psi(\mathbf{x})\mathbf{w}$, with $\mathbf{w}$ constant, we also get the useful identity

$$\int_V d^3x \, \boldsymbol{\nabla}\psi = \int_S d\mathbf{s}\,\psi, \qquad (1.49)$$

while, setting $\mathbf{v}(\mathbf{x}) = \mathbf{u}(\mathbf{x})\times\mathbf{w}$, we get

$$\int_V d^3x \, \boldsymbol{\nabla}\times\mathbf{u} = \int_S d\mathbf{s}\times\mathbf{u}. \qquad (1.50)$$

1.4   Dirac delta

The Dirac delta is an especially useful mathematical object, that appears everywhere in physics. Physically, it can be seen as the modelization of a point-like object. The Dirac delta $\delta(x)$ is not a function in the proper sense. Rather, in one dimension, it is defined from the conditions that

$$\delta(x - x_0) = 0 \qquad \text{if} \quad x \neq x_0, \qquad (1.51)$$

and that, for any function $f(x)$ regular in an integration region I that includes x_0,

$$\int_I dx \, \delta(x - x_0)f(x) = f(x_0). \qquad (1.52)$$

Note that the integral on the left-hand side vanishes if I does not include x_0 because of eq. (1.51) [and of the assumed regularity of $f(x)$]. On the other hand, again because of eq. (1.51), the integral in the left-hand side of eq. (1.52) is independent of I, as long as $x_0 \in I$. In the following we will set for definiteness $I = (-\infty, +\infty)$, so eq. (1.52) reads

$$\int_{-\infty}^{+\infty} dx \, \delta(x - x_0)f(x) = f(x_0). \qquad (1.53)$$

Observe that, applying this definition to the case of the function $f(x) = 1$, we get the normalization condition

$$\int_{-\infty}^{+\infty} dx\,\delta(x) = 1\,. \tag{1.54}$$

Since the Dirac delta vanishes at all points $x \neq x_0$, but still the integral in the left-hand side of eq. (1.53) is non-zero, it must be singular in x_0. Actually, the Dirac delta is not a proper function, but can be defined by considering a sequence of functions $\delta_n(x, x_0)$ such that, as $n \to \infty$, $\delta_n(x, x_0) \to 0$ for $x \neq x_0$, and $\delta_n(x, x_0) \to +\infty$ for $x = x_0$, while maintaining the normalization condition (1.54). As an example, one can take a sequence of gaussians centered on x_0, with smaller and smaller width,

$$\delta_n(x - x_0) = \frac{1}{\sqrt{2\pi}\,\sigma_n}\,e^{-(x-x_0)^2/(2\sigma_n^2)}\,, \tag{1.55}$$

with $\sigma_n = 1/n$. Another option could be to use

$$\delta_n(x - x_0) = \begin{cases} n & \text{for} \quad |x - x_0| < 1/(2n) \\ 0 & \text{for} \quad |x - x_0| > 1/(2n)\,. \end{cases} \tag{1.56}$$

These two sequences of functions are shown in Fig. 1.1. In both cases, the limit of $\delta_n(x - x_0)$ for $n \to \infty$ does not exists, since it diverges when $x = x_0$, and therefore does not define a proper function $\delta(x)$. However, one can generalize the notion of functions to the notion of *distributions* (or "improper functions"), which are defined from their action inside an integral, when convolved with "test" functions $f(x)$ (with suitably defined properties of regularity and, possibly, behavior at infinity). In the case of the Dirac delta, the definition in the sense of distributions is given by eq. (1.53). Using the explicit expression of the functions $\delta_n(x)$ given in eq. (1.56), for a function $f(x)$ regular near x_0, we get

$$\lim_{n \to \infty} \int_{-\infty}^{+\infty} dx\,\delta_n(x - x_0)f(x) = \lim_{n \to \infty} n \int_{x_0 - \frac{1}{2n}}^{x_0 + \frac{1}{2n}} dx\,f(x)$$

$$= \lim_{n \to \infty} n \left[\left(x_0 + \frac{1}{2n} \right) - \left(x_0 - \frac{1}{2n} \right) \right] f(x_0)$$

$$= f(x_0)\,. \tag{1.57}$$

Therefore, in the sense of distributions, i.e., after multiplying by a smooth function $f(x)$ and integrating, we have

$$\delta(x - x_0) = \lim_{n \to \infty} \delta_n(x - x_0)\,. \tag{1.58}$$

From the definition, we see that the Dirac delta only makes sense when it appears inside an integral. In physics, however, with an abuse of notation, the universal use is to treat it as if it were a normal function (and it is even called the Dirac delta "function"!), with the understanding that the relations in which it enters must be understood in the sense of distributions, i.e., multiplied by a test function and integrated.

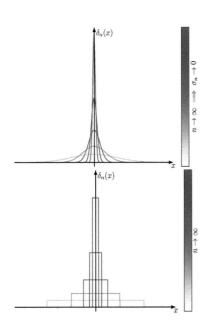

Fig. 1.1 A sequence of approximations to the Dirac δ function, using the gaussians (1.55) (top panel) or the "rectangles" (1.56) (lower panel).

From the definitions (1.55) or (1.56), setting $x_0 = 0$, we see that the Dirac delta function is even in x, $\delta(x) = \delta(-x)$. Two more useful properties are left as an exercise to the reader:

Exercise 1.5 Using the definition (1.53) show that, if a is a non-zero real number,

$$\delta(ax) = \frac{1}{|a|}\delta(x) \, . \tag{1.59}$$

Exercise 1.6 Using again eq. (1.53) show that, if $g(x)$ is a function which has only one simple zero in $x = x_0$, then

$$\delta\left(g(x)\right) = \frac{1}{|g'(x_0)|}\delta(x - x_0) \, , \tag{1.60}$$

where $g'(x) = dg/dx$. Show that, if $g(x)$ has several simple zeros at the points $x = x_i$ ($i = 1, \ldots, n$), this generalizes to

$$\delta\left(g(x)\right) = \sum_{i=1}^{n} \frac{1}{|g'(x_i)|}\delta(x - x_i) \, . \tag{1.61}$$

Exercise 1.7 Show that, in the sense of distributions,

$$x\delta(x) = 0 \, . \tag{1.62}$$

Another useful notion is the derivative of the Dirac delta, $\delta'(x)$, which, in the sense of distributions, is defined from

$$\int_{-\infty}^{+\infty} dx \, \delta'(x - x_0) f(x) \;\equiv\; -\int_{-\infty}^{+\infty} dx \, \delta(x - x_0) f'(x)$$
$$= \; -f'(x_0) \, . \tag{1.63}$$

This definition is clearly motivated by the analogy with the integration by parts of a normal function. Taking $\delta(x)$ as the limit for $n \to \infty$ (in the sense of distributions) of a series $\delta_n(x)$ of continuous and differentiable functions, such as the gaussians (1.55), one can see that $\delta'(x)$ is obtained, again in the sense of distributions, from the limit $n \to \infty$ of $\delta'_n(x)$. Indeed, for $\delta_n(x)$ differentiable (and vanishing at $x = \pm\infty$), the standard integration by parts goes through,[5] and

$$\lim_{n\to\infty} \int_{-\infty}^{+\infty} dx \, \delta'_n(x - x_0) f(x) \;=\; -\lim_{n\to\infty} \int_{-\infty}^{+\infty} dx \, \delta_n(x - x_0) f'(x)$$
$$= \; -f'(x_0) \, , \tag{1.64}$$

where the last equality follows from eq. (1.58). Therefore, in the same sense as eq. (1.58),

$$\delta'(x - x_0) = \lim_{n\to\infty} \delta'_n(x - x_0) \, . \tag{1.65}$$

Notice that $\delta'(x)$ is an odd function of x, $\delta'(-x) = -\delta'(x)$, as it is clear from its representation in terms of $\delta'_n(x)$, with $\delta_n(x)$ given by the

[5] We also assume that the test functions $f(x)$ go to zero at $\pm\infty$. In fact, here it is sufficient that they do not grow so fast to compensate the exponential decay of the gaussian $\delta_n(x)$, so that, in the integration by parts, we can discard the boundary terms at infinity.

sequence of gaussians. Higher-order derivatives of the Dirac delta are defined similarly,

$$\int_{-\infty}^{+\infty} dx \left[\frac{d^n}{dx^n}\delta(x-x_0)\right] f(x) \equiv (-1)^n \int_{-\infty}^{+\infty} dx\, \delta(x-x_0)\frac{d^n f(x)}{dx^n}$$

$$= (-1)^n \frac{d^n f(x)}{dx^n}\Big|_{x=x_0}. \qquad (1.66)$$

There is an interesting relation between the Dirac delta and the Heaviside theta function (also called simply the "theta function"), defined by

$$\theta(x) = \begin{cases} 1 & \text{for} \quad x > 0 \\ 0 & \text{for} \quad x < 0. \end{cases} \qquad (1.67)$$

Observe that, from the definition of the Dirac delta,

$$\int_{-\infty}^{x} dx'\, \delta(x') = \theta(x). \qquad (1.68)$$

Indeed, if $x < 0$, the integration region does not include the point $x' = 0$ where $\delta(x')$ is singular, so the integral vanishes. For $x > 0$, instead, we can extend the integral in eq. (1.68) up to $x' = \infty$, since anyhow $\delta(x') = 0$ for $x' > 0$, and we can then use eq. (1.54) to show that the integral is equal to one. Conversely, for a differentiable function $f(x)$ that vanishes at infinity, treating $\theta'(x)$ as a distribution and defining its derivative as we have done for the Dirac delta,

$$\int_{-\infty}^{+\infty} dx\, \theta'(x) f(x) \equiv -\int_{-\infty}^{+\infty} dx\, \theta(x) f'(x)$$

$$= -\int_{0}^{\infty} dx\, \frac{df}{dx}$$

$$= -[f(\infty) - f(0)]$$

$$= f(0). \qquad (1.69)$$

This shows that, in the sense of distributions,

$$\theta'(x) = \delta(x), \qquad (1.70)$$

which could also have been formally derived by taking the derivative of eq. (1.68).[6] This result could have also been proved using a sequence $\delta_n(x)$ of approximations to the Dirac delta, and showing that, plugging it on the left hand side of eq. (1.68), we get a continuous and differentiable approximation to the theta function.

The Dirac delta has an extremely useful integral representation. A simple way to obtain it is to use the sequence of gaussians (1.55). Then,

$$\int_{-\infty}^{+\infty} dx\, \delta_n(x) e^{-ikx} = \frac{n}{\sqrt{2\pi}} \int_{-\infty}^{+\infty} dx\, e^{-\frac{1}{2}n^2 x^2 - ikx}$$

$$= e^{-k^2/(2n^2)}, \qquad (1.71)$$

as can be proven by carrying out the integral on the right-hand side of the first line.[7] Similarly,

---

[6] Note that we have proved eq. (1.68) for $x > 0$ and for $x < 0$. Whether it also holds for $x = 0$ depends on the series $\delta_n(x)$ that we use for approximating $\delta(x)$, and on how we define $\theta(x=0)$. If we use functions $\delta_n(x)$ that are symmetric around $x = 0$, as in eqs. (1.55) or (1.56), the equality holds if we define $\theta(0) = 1/2$. However, one could use different, non-symmetric series $\delta_n(x)$, and one should then assign a different value to $\theta(0)$ for the equality to hold. In fact, the whole issue is irrelevant since the relation (1.70) holds only in the sense of distributions, i.e., after integrating, and the fact that it holds in a single point or not does not affect any integrated relation.

[7] For instance, by completing the square in the exponent one gets

$$\int_{-\infty}^{+\infty} dx\, e^{-\frac{1}{2}n^2 x^2 - ikx}$$

$$= e^{-\frac{k^2}{2n^2}} \int_{-\infty}^{+\infty} dx\, e^{-\frac{n^2}{2}(x+\frac{ik}{n^2})^2}$$

$$= e^{-\frac{k^2}{2n^2}} \int_{-\infty}^{+\infty} dx'\, e^{-\frac{n^2}{2}(x')^2}$$

$$= e^{-\frac{k^2}{2n^2}} \frac{\sqrt{2\pi}}{n}, \qquad (1.72)$$

where we introduced $x' = x + ik/n^2$. More precisely, we have actually considered a closed contour in the complex plane $z = x + iy$, composed by the x axis $y = 0$, and by the parallel line $y = ik/n^2$, and closed the contour joining these two lines at infinity. Since the integrand has no singularity inside this contour, by the Cauchy theorem the integral over the x axis is the same as the integral over the line $y = ik/n^2$, i.e., over the variable $z = x + ik/n^2$.

$$\int_{-\infty}^{+\infty} \frac{dk}{2\pi}\, e^{-\frac{k^2}{2n^2}+ikx} = \frac{n}{\sqrt{2\pi}}\, e^{-\frac{1}{2}n^2x^2}$$
$$= \delta_n(x)\,. \tag{1.73}$$

Taking the limit $n \to \infty$ of these relations we therefore get

$$\int_{-\infty}^{+\infty} dx\, \delta(x) e^{-ikx} = 1\,, \tag{1.74}$$

and its inverse relation

$$\int_{-\infty}^{+\infty} \frac{dk}{2\pi}\, e^{ikx} = \delta(x)\,. \tag{1.75}$$

Equation (1.74) could have been derived more simply from the defining property of the Dirac delta, eq. (1.53), observing that, in $x = 0$, $e^{-ikx} = 1$. Equation (1.75) was less evident, and provides a very useful integral representation of the Dirac delta.[8] In Section 1.5, we will use it to give a simple derivation of the inversion formula of the Fourier transform.

The generalization to more than one dimension is straightforward. In particular, in three spatial dimensions, we define the three-dimensional Dirac delta as

$$\delta^{(3)}(\mathbf{x}) = \delta(x)\delta(y)\delta(z)\,, \tag{1.77}$$

and this is a distribution to be multiplied by test functions $f(\mathbf{x})$ and integrated over d^3x. Then, eq. (1.53) becomes

$$\int d^3x\, \delta^{(3)}(\mathbf{x} - \mathbf{x_0}) f(\mathbf{x}) = f(\mathbf{x_0})\,, \tag{1.78}$$

while the integral representation (1.75) becomes

$$\int \frac{d^3k}{(2\pi)^3}\, e^{i\mathbf{k}\cdot\mathbf{x}} = \delta^{(3)}(\mathbf{x})\,. \tag{1.79}$$

[8] Note that, renaming the integration variable $k \to -k$, we can also write eq. (1.75) as

$$\int_{-\infty}^{+\infty} \frac{dk}{2\pi}\, e^{-ikx} = \delta(x)\,. \tag{1.76}$$

Example 1.1 Divergence of $\hat{\mathbf{r}}/r^2$. As a particularly important application of the concepts developed in Sections 1.2, 1.3, and in the present section, we perform the computation of the divergence of the vector field

$$\mathbf{v}(\mathbf{x}) = \frac{\mathbf{x}}{r^3}$$
$$= \frac{\hat{\mathbf{r}}}{r^2}\,, \tag{1.80}$$

where $r = |\mathbf{x}|$ and $\hat{\mathbf{r}}$ is the unit vector in the radial direction. In polar coordinates, $v_r = 1/r^2$, $v_\theta = v_\phi = 0$. If we use eq. (1.29), we apparently get

$$\boldsymbol{\nabla}\cdot\mathbf{v} = \frac{1}{r^2}\partial_r \left[r^2 \times \frac{1}{r^2} \right] \qquad (r \neq 0)$$
$$= 0\,. \tag{1.81}$$

Recall that the infinitesimal solid angle $d\Omega$ is defined from the transformation from Cartesian to spherical coordinates. In two dimensions, from $x = r\cos\theta$ and $y = r\sin\theta$, it follows that $dxdy = rdrd\theta$. Similarly, in three dimensions, the relation between Cartesian and spherical coordinates (as we already mentioned in Note 1 on page 4) is $x = r\sin\theta\cos\phi$, $y = r\sin\theta\sin\phi$, $z = r\cos\theta$, with $\theta \in [0, \pi]$ and $\phi \in [0, 2\pi]$. Computing the Jacobian of the transformation,

$$dxdydz = r^2\sin\theta\,drd\theta d\phi$$
$$= r^2 dr d\Omega, \qquad (1.82)$$

where

$$d\Omega = \sin\theta d\theta d\phi. \qquad (1.83)$$

Then, the total integral over the solid angle is

$$\int d\Omega = \int_0^\pi d\theta \sin\theta \int_0^{2\pi} d\phi$$
$$= 4\pi, \qquad (1.84)$$

so the total solid angle in three dimensions is 4π. Observe that we can also write $d\Omega = d\cos\theta\,d\phi$, reabsorbing the minus sign from $d\cos\theta = -\sin\theta d\theta$ into a change of the integration limits, so that

$$\int d\Omega = \int_{-1}^1 d\cos\theta \int_0^{2\pi} d\phi. \qquad (1.85)$$

However, we have stressed that this only holds for $r \neq 0$, since these manipulations become undefined at $r = 0$. To understand the behavior of $\boldsymbol{\nabla}\!\cdot\!\mathbf{v}$ in $r = 0$, we use Gauss's theorem (1.48) in a volume V given by a spherical ball of radius R. Its boundary is the sphere S^2, and on the boundary $\mathbf{v} = \hat{\mathbf{r}}/R^2$, while $d\mathbf{s} = R^2 d\Omega\,\hat{\mathbf{r}}$, where $d\Omega$ is the infinitesimal solid angle.[9] Then,

$$\begin{aligned} \int_V d^3x\,\boldsymbol{\nabla}\!\cdot\!\mathbf{v} &= \int_{S^2} d\mathbf{s}\cdot\mathbf{v} \\ &= \int_{S^2} R^2 d\Omega\,\hat{\mathbf{r}}\cdot\frac{\hat{\mathbf{r}}}{R^2} \\ &= \int_{S^2} d\Omega \\ &= 4\pi. \end{aligned} \qquad (1.86)$$

This shows that $\boldsymbol{\nabla}\!\cdot\!\mathbf{v}$ cannot be zero everywhere. Rather, since $\boldsymbol{\nabla}\!\cdot\!\mathbf{v} = 0$ for $r \neq 0$, but still its integral in d^3x over any volume V is equal to 4π, we must have

$$\boxed{\boldsymbol{\nabla}\!\cdot\!\left(\frac{\hat{\mathbf{r}}}{r^2}\right) = 4\pi\,\delta^{(3)}(\mathbf{x}).} \qquad (1.87)$$

From this, we can obtain another very useful result. Using the expression (1.23) of the gradient in polar coordinates, we get

$$\boldsymbol{\nabla}\left(\frac{1}{r}\right) = -\frac{\hat{\mathbf{r}}}{r^2}. \qquad (1.88)$$

Writing

$$\begin{aligned} \boldsymbol{\nabla}^2\left(\frac{1}{r}\right) &= \boldsymbol{\nabla}\!\cdot\!\left(\boldsymbol{\nabla}\frac{1}{r}\right) \\ &= -\boldsymbol{\nabla}\!\cdot\!\left(\frac{\hat{\mathbf{r}}}{r^2}\right), \end{aligned} \qquad (1.89)$$

we see that eq. (1.87) implies that

$$\boldsymbol{\nabla}^2\frac{1}{r} = -4\pi\delta^{(3)}(\mathbf{x}). \qquad (1.90)$$

Replacing $\mathbf{x}$ by $\mathbf{x} - \mathbf{x}'$, for $\mathbf{x}'$ generic, we then also have

$$\boxed{\boldsymbol{\nabla}^2\frac{1}{|\mathbf{x} - \mathbf{x}'|} = -4\pi\delta^{(3)}(\mathbf{x} - \mathbf{x}').} \qquad (1.91)$$

We will use this result in Section 4.1.2, when we will introduce the notion of Green's function of the Laplacian.

1.5 Fourier transform

We next recall the definition and some basic properties of the Fourier transform. For a function of one spatial variable $f(x)$, we define the Fourier transform $\tilde{f}(k)$ as[10]

$$\tilde{f}(k) = \int_{-\infty}^{+\infty} dx \, f(x) e^{-ikx} \, . \tag{1.92}$$

Observe that, if $f(x)$ is real, $\tilde{f}^*(k) = \tilde{f}(-k)$. The simplest way to invert this relation between $f(x)$ and $\tilde{f}(k)$ is to use the integral representation of the Dirac delta, eq. (1.75).[11] Multiplying eq. (1.92) by e^{ikx}, integrating over $dk/(2\pi)$, and changing the name of the integration variable to x' in the right-hand side of eq. (1.92), we get

$$
\begin{aligned}
\int_{-\infty}^{+\infty} \frac{dk}{2\pi} \tilde{f}(k) e^{ikx} &= \int_{-\infty}^{+\infty} \frac{dk}{2\pi} e^{ikx} \int_{-\infty}^{+\infty} dx' \, f(x') e^{-ikx'} \\
&= \int_{-\infty}^{+\infty} dx' \, f(x') \int_{-\infty}^{+\infty} \frac{dk}{2\pi} e^{ik(x-x')} \\
&= \int_{-\infty}^{+\infty} dx' \, f(x') \delta(x - x') \\
&= f(x) \, . \tag{1.93}
\end{aligned}
$$

Therefore, the inversion of eq. (1.92) is[12]

$$f(x) = \int_{-\infty}^{+\infty} \frac{dk}{2\pi} \tilde{f}(k) e^{ikx} \, . \tag{1.96}$$

Another useful relation is obtained considering the convolution of two functions $f(x)$ and $g(x)$, defined by

$$F(x) = \int_{-\infty}^{+\infty} dx' \, f(x') g(x - x') \, . \tag{1.97}$$

Taking the Fourier transform we get[13]

$$\tilde{F}(k) = \tilde{f}(k) \tilde{g}(k) \, . \tag{1.98}$$

Therefore, the Fourier transform of a convolution is equal to the product of the Fourier transforms. This is known as the *convolution theorem*.

Equations (1.92), (1.96), and (1.98) are easily generalized to any number of spatial dimensions. In particular, in three spatial dimensions, the Fourier transform is defined as

$$\tilde{f}(\mathbf{k}) = \int d^3x \, f(\mathbf{x}) e^{-i\mathbf{k}\cdot\mathbf{x}} \, , \tag{1.99}$$

and its inversion gives

$$f(\mathbf{x}) = \int \frac{d^3k}{(2\pi)^3} \tilde{f}(\mathbf{k}) e^{i\mathbf{k}\cdot\mathbf{x}} \, . \tag{1.100}$$

[10]More precisely, one must restrict to a space of functions such that the manipulations below are well defined. This can be obtained for instance considering $f \in L^1(\mathbb{R})$, the space of functions whose absolute value is integrable over $\mathbb{R}$.

[11]This was not the historical path. The theory of distributions, which puts the notion of Dirac delta on a sound mathematical basis, was only developed in the first half of the 20th century, while the original work of Fourier dates back to 1822.

[12]There are different conventions for the factors 2π in the definition of the Fourier transform. The one that we have used is, nowadays, the most common in physics. Another common choice is to define

$$\tilde{f}(k) = \frac{1}{\sqrt{2\pi}} \int_{-\infty}^{+\infty} dx \, f(x) e^{-ikx} \, , \tag{1.94}$$

in which case eq. (1.96) becomes

$$f(x) = \frac{1}{\sqrt{2\pi}} \int_{-\infty}^{+\infty} dk \, \tilde{f}(k) e^{ikx} \, . \tag{1.95}$$

[13]The explicit computation goes as follows:

$$
\begin{aligned}
\tilde{F}(k) &= \int_{-\infty}^{+\infty} dx \, F(x) e^{-ikx} \\
&= \int_{-\infty}^{+\infty} dx \int_{-\infty}^{+\infty} dx' f(x') \\
&\quad \times g(x - x') e^{-ik[(x-x')+x']} \\
&= \int_{-\infty}^{+\infty} dx' f(x') e^{-ikx'} \\
&\quad \times \int_{-\infty}^{+\infty} dx \, g(x - x') e^{-ik(x-x')} \, .
\end{aligned}
$$

Introducing $y = x - x'$, the last integral over dx at fixed x' is the same as an integral over dy, so

$$
\begin{aligned}
\tilde{F}(k) &= \int_{-\infty}^{+\infty} dx' f(x') e^{-ikx'} \\
&\quad \times \int_{-\infty}^{+\infty} dy \, g(y) e^{-iky} \\
&= \tilde{f}(k) \tilde{g}(k) \, .
\end{aligned}
$$

If $f(\mathbf{x})$ is real, $\tilde{f}^*(\mathbf{k}) = \tilde{f}(-\mathbf{k})$. The convolution theorem now tells us that, if

$$F(\mathbf{x}) = \int d^3x'\, f(\mathbf{x}')g(\mathbf{x}-\mathbf{x}')\,, \tag{1.101}$$

then

$$\tilde{F}(\mathbf{k}) = \tilde{f}(\mathbf{k})\tilde{g}(\mathbf{k})\,. \tag{1.102}$$

For a function of time $f(t)$ we will denote the integration variable that enters in the Fourier transform by ω, and we will also use a different sign convention, defining the Fourier transform $\tilde{f}(\omega)$ as

$$\tilde{f}(\omega) = \int dt\, f(t)e^{i\omega t}\,, \tag{1.103}$$

so that

$$f(t) = \int \frac{d\omega}{2\pi}\, \tilde{f}(\omega)e^{-i\omega t}\,. \tag{1.104}$$

In the context of Special Relativity, the advantage of using a different sign convention between the spatial and temporal Fourier transforms is that the Fourier transform of a function $f(t,\mathbf{x})$, with respect to both t and $\mathbf{x}$, can be written as

$$f(t,\mathbf{x}) = \int \frac{d\omega d^3k}{(2\pi)^4}\, \tilde{f}(\omega,\mathbf{k})e^{-i(\omega t - \mathbf{k}\cdot\mathbf{x})}\,, \tag{1.105}$$

i.e.,

$$\tilde{f}(\omega,\mathbf{k}) = \int dt d^3x\, f(t,\mathbf{x})e^{i(\omega t - \mathbf{k}\cdot\mathbf{x})}\,, \tag{1.106}$$

and, as we will see, the combination $(\omega t - \mathbf{k}\cdot\mathbf{x})$ is more natural from the point of view of Special Relativity and Lorentz invariance.

Example 1.2 In this example, we compute the three-dimensional Fourier transform of the function

$$f(r) = \frac{1}{r}\,, \tag{1.107}$$

that appears in the Coulomb potential. A direct computation leads to a problem of convergence.[14] Indeed, from eqs. (1.99), (1.82), and (1.85),

$$\begin{aligned}\tilde{f}(\mathbf{k}) &= \int d^3x\, \frac{1}{r}e^{-i\mathbf{k}\cdot\mathbf{x}} \\ &= \int_0^\infty dr\, r^2 \int_0^{2\pi} d\phi \int_{-1}^1 d\cos\theta\, \frac{1}{r}e^{-ikr\cos\theta}\,,\end{aligned} \tag{1.108}$$

where, to perform the integral, we have written d^3x in polar coordinates with $\mathbf{k}$ as polar axis, so θ is the angle between $\mathbf{k}$ and $\mathbf{x}$. The integral over $d\phi$ just gives a 2π factor, and the integral over $\cos\theta$ is also elementary,

$$\int_{-1}^1 d\alpha\, e^{-ikr\alpha} = \frac{2}{kr}\sin(kr)\,, \tag{1.109}$$

so we get

$$\tilde{f}(\mathbf{k}) = \frac{4\pi}{k^2}\int_0^\infty du\, \sin u\,, \tag{1.110}$$

[14]Indeed, this function does not belong to $L^1(\mathbb{R}^3)$, compare with Note 10 on page 13.

where $k = |\mathbf{k}|$ and $u = kr$. The integral over du, however, does not converge at $u = \infty$, and is not well defined without a prescription. We then start by computing first the Fourier transform of the function

$$V(r) = \frac{e^{-\mu r}}{r}, \tag{1.111}$$

where $\mu > 0$, and has the dimensions of the inverse of a length. This function, which corresponds to an interaction potential known as the Yukawa potential, reduces to the Coulomb potential as $\mu \to 0^+$. Performing the same passages as before gives

$$\tilde{V}(\mathbf{k}) = \frac{4\pi}{k^2} \int_0^\infty du \, \sin u \, e^{-\epsilon u}, \tag{1.112}$$

where $\epsilon = \mu/k$. For non-zero ϵ, i.e., non-zero μ, the factor $e^{-\epsilon u}$ ensures the convergence of the integral. Writing $\sin u = (e^{iu} - e^{-iu})/(2i)$, the integral is elementary,

$$
\begin{aligned}
\tilde{V}(\mathbf{k}) &= \frac{4\pi}{k^2} \frac{1}{2i} \left[\frac{1}{i - \epsilon} e^{(i-\epsilon)u} + \frac{1}{i + \epsilon} e^{-(i+\epsilon)u} \right]_0^\infty \\
&= \frac{4\pi}{k^2} \frac{1}{1 + \epsilon^2} \\
&= \frac{4\pi}{k^2 + \mu^2}.
\end{aligned}
\tag{1.113}
$$

Therefore,

$$V(\mathbf{x}) = \frac{e^{-\mu r}}{r} \quad \Longleftrightarrow \quad \tilde{V}(\mathbf{k}) = \frac{4\pi}{k^2 + \mu^2}. \tag{1.114}$$

Since the limit $\mu \to 0^+$ of these expressions is well defined, we can now define the Fourier transform of the Coulomb potential as the limit for $\mu \to 0^+$ of the Fourier transform of the Yukawa potential,[15] so

$$\boxed{f(\mathbf{x}) = \frac{1}{r} \quad \Longleftrightarrow \quad \tilde{f}(\mathbf{k}) = \frac{4\pi}{k^2}.} \tag{1.115}$$

The correctness of this limiting procedure can be checked observing that, if we rather start from $\tilde{f}(\mathbf{k}) = 4\pi/k^2$ and compute the inverse Fourier transform, we get $1/r$ without the need of regulating any divergence. Indeed, in this case we get

$$
\begin{aligned}
f(r) &= \int \frac{d^3 k}{(2\pi)^3} \frac{4\pi}{k^2} e^{i\mathbf{k} \cdot \mathbf{x}} \\
&= \frac{4\pi}{8\pi^3} 2\pi \int_0^\infty k^2 dk \int_{-1}^1 d\cos\theta \, \frac{1}{k^2} e^{ikr\cos\theta} \\
&= \frac{1}{r} \frac{2}{\pi} \int_0^\infty du \, \frac{\sin u}{u} \\
&= \frac{1}{r},
\end{aligned}
\tag{1.116}
$$

since $\int_0^\infty du \, (\sin u)/u$ converges, and has the value $\pi/2$.

[15] A note for the advanced reader. In quantum field theory, the Coulomb potential is understood as the interaction between two static charges mediated by the exchange of a massless particle, the photon. The exchange of a massive particle produces instead a Yukawa potential, with the constant μ related to the mass m of the exchanged particle by $\mu = mc/\hbar$, see e.g., Section 6.6 of Maggiore (2005). This way of regularizing the integral therefore corresponds, physically, to assigning a small mass m_γ to the photon and then taking the limit $m_\gamma \to 0$. Indeed, one way to put limits on the photon mass is to assume that the Coulomb interaction is replaced by a Yukawa potential, and obtain limits on m_γ. Currently, the strongest limit on the photon mass coming from a direct test of the Coulomb law gives $m_\gamma c^2 < 1 \times 10^{-14}$ eV. Writing $\mu = 1/r_0$, so that $r_0 = \hbar/(m_\gamma c)$, this translate into a limit $r_0 > 2 \times 10^7$ m, i.e., there is no sign of an exponential decay corresponding to a Yukawa potential up to such scales. Other limits on the mass of the photon, not based on a direct measurement of the Coulomb law, are even stronger, with the most stringent being $m_\gamma c^2 < 1 \times 10^{-18}$ eV, corresponding to $r_0 > 2 \times 10^{11}$ m, see https://pdg.lbl.gov/2021/listings/contents_listings.html. Another way to search for deviations from Coulomb's law is to look for a force proportional to $1/r^{2+\epsilon}$, and set limits on ϵ. This is a purely phenomenological parametrization and, contrary to the case of the Yukawa potential (and to statements in some textbook), it has no field-theoretical interpretation in terms of a photon mass.

1.6   Tensors and rotations

The elementary definition of a vector is based on the notion of an arrow, i.e., an object with a length and a direction. It is very useful to understand vectors in a more elaborate language, that of representations of the rotation group. This will allow us to better understand the meaning of other objects, such as tensors, and will be a useful preparation for understanding the formalism of Special Relativity, where the fundamental quantities will be given by representations of a broader group, the Lorentz group, that we will introduce later. In this section, we will give a first, simpler treatment, while in Section 1.7 we will provide a more formal group-theoretical description (that will not be necessary for the rest of the book but gives a deeper understanding).

Consider a vector $\mathbf{v}$ in two dimensions, defined as an arrow; with respect to a given system of Cartesian axes, it will have components (v_x, v_y). Consider now a counterclockwise rotation of the vector by an angle θ in the plane. After the rotation, the new components are given by[16]

$$v'_x = v_x \cos\theta - v_y \sin\theta, \qquad (1.117)$$
$$v'_y = v_x \sin\theta + v_y \cos\theta. \qquad (1.118)$$

Therefore, under a rotation by an angle θ, the transformation of $\mathbf{v}$ is given by

$$\begin{pmatrix} v_x \\ v_y \end{pmatrix} \rightarrow \begin{pmatrix} v'_x \\ v'_y \end{pmatrix} = \begin{pmatrix} \cos\theta & -\sin\theta \\ \sin\theta & \cos\theta \end{pmatrix} \begin{pmatrix} v_x \\ v_y \end{pmatrix}. \qquad (1.119)$$

Using our notation with a sum over repeated indices, we can rewrite this as

$$v_i \rightarrow v'_i = R_{ij} v_j, \qquad (1.120)$$

where the indices i, j take the two values $\{x, y\}$ (or, equivalently, $\{1, 2\}$) and $R_{ij} = R_{ij}(\theta)$ are the matrix elements of the 2×2 matrix that appears in eq. (1.119).

We can now promote eq. (1.119) to the basic relation that *defines* a vector, stating that a vector in two spatial dimensions is defined as a set of two numbers (its components) with a well-defined transformation property under rotations, expressed by eq. (1.119). This definition totally abstracts from the original notion of an arrow, and has the advantage that it can be generalized both to objects with different transformation properties and to arbitrary dimensions.

Consider first the generalization to arbitrary dimensions.[17] We begin by observing that the rotation (1.119) preserves the length of the vector, so the squared norm of the vector v (that, with our convention of summation of repeating indices, can be written as $v_i v_i$) is equal to the norm of the vector v' obtained applying a rotation to v,

$$v'_i v'_i = v_i v_i. \qquad (1.121)$$

[16] Here, we are keeping the axes of the reference frame fixed, and rotate the vector by an angle θ. Then v'_x and v'_y in eqs. (1.117) and (1.118) are the components of the rotated vector with respect to these fixed axes. This is called the "active" point of view. Equivalently, we can keep the vector fixed and rotate the axes by an angle $-\theta$. Then, v'_x and v'_y in eqs. (1.117) and (1.118) are the components of the vector v with respect to the new system of axes. This is called the "passive" point of view.

[17] In elementary physics, we are usually interested only in rotations in two or in three spatial dimensions. The generalization to arbitrary dimensions, however, is already a useful preparation for the extension to more complicated transformations, such as Lorentz transformations.

We now use this relation to define rotations and vectors, in any dimensions, as follows. We consider a set of d objects v_i, $i = 1, \ldots, d$, and a linear transformation of the form (1.120), where now R_{ij} is a $d \times d$ matrix. We define rotations in d dimensions as the transformations R_{ij} that preserve the norm of v,[18] so that eq. (1.121) (with $i = 1, \ldots, d$) holds. The set of objects v_i that transforms as in eq. (1.120) when R_{ij} is a rotation matrix are then called "vectors under rotations" or, more simply, vectors.

[18]We will qualify this more precisely at the end of this section and in Section 1.7, where we will see that "proper" rotations are obtained factoring out some discrete parity symmetry.

This definition allows us to characterize rotations as follows. Plugging eq. (1.120) into eq. (1.121) we get

$$v_i v_i = R_{ij} v_j R_{ik} v_k \,. \tag{1.122}$$

Using eq. (1.3) and renaming the dummy indices on the right-hand side as $i \to k, j \to i, k \to j$, this can be rewritten in the form

$$
\begin{aligned}
\delta_{ij} v_i v_j &= R_{ij} R_{ik} v_j v_k \\
&= R_{ki} R_{kj} v_i v_j \,.
\end{aligned}
\tag{1.123}
$$

Since this must hold for any vector v_i, it follows that

$$R_{ki} R_{kj} = \delta_{ij} \,. \tag{1.124}$$

Recall that, if R_{ij} are the matrix elements of the matrix R, the transpose matrix R^{T} has matrix elements $(R^{\mathrm{T}})_{ij} = R_{ji}$. Then eq. (1.129) reads $R^{\mathrm{T}}_{ik} R_{kj} = \delta_{ij}$ or, in matrix form

$$R^{\mathrm{T}} R = I \,, \tag{1.125}$$

where I is the identity matrix. Therefore R^{T} is the left inverse of R, and then it also holds[19]

$$R R^{\mathrm{T}} = I \,. \tag{1.126}$$

In components, this reads

$$R_{ik} R_{jk} = \delta_{ij} \,. \tag{1.127}$$

Therefore, rotation matrices satisfy

$$R R^{\mathrm{T}} = R^{\mathrm{T}} R = I \,, \tag{1.128}$$

or, in components,

$$R_{ki} R_{kj} = R_{ik} R_{jk} = \delta_{ij} \,. \tag{1.129}$$

Equation (1.128) is just the definition of orthogonal matrices. We have then found that rotations in d dimensions are given by $d \times d$ orthogonal matrices, and vectors are defined by the transformation property (1.120) under rotations.

The transformation property of vectors can be generalized to objects with different transformation properties. In particular, in any dimension

[19]This follows from the fact that, for a square matrix, the left and right inverse are the same. Indeed, let A be a matrix, and suppose that its left inverse, L, exists, so that $LA = I$. Then, multiplying by L from the right, we get $LAL = L$. Multiplying further by L^{-1} from the left, we get $AL = I$. Then, L is also the right inverse.

d we can define a tensor T_{ij} with two indices as an object that, under rotations, transforms as

$$T_{ij} \rightarrow T'_{ij} = R_{ik}R_{jl}T_{kl}, \qquad (1.130)$$

(with the indices running over two values $\{x, y\}$ in two dimensions, over $\{x, y, z\}$ in three dimensions, and more generally over d values $1, \ldots, d$ in d dimensions), with the same matrix R_{ij} that appears in the transformation of vectors. Similarly, a tensor T_{ijk} with three indices is defined as an object that, under rotations, transforms as

$$T'_{ijk} = R_{ii'}R_{jj'}R_{kk'}T_{i'j'k'}, \qquad (1.131)$$

and so on. Note that, for tensors, the connection with the notion of arrow is completely lost, and a tensor is only defined by its transformation property (1.130).

Consider now a tensor T_{ij} that, in a given reference frame, has components δ_{ij}. Normally, the numerical values of the components of a tensor change if we perform a rotation, according to eq. (1.130). In this case, however, in the new frame

$$
\begin{aligned}
T'_{ij} &= R_{ik}R_{jl}T_{kl} \\
&= R_{ik}R_{jl}\delta_{kl} \\
&= R_{ik}R_{jk} \\
&= \delta_{ij}, \qquad (1.132)
\end{aligned}
$$

where in the last equality we have used eq. (1.129). Thus, δ_{ij} is a very special tensor, whose components have the same numerical values in all frames. It is then called an *invariant tensor*. In three dimensions, the only other invariant tensor of the rotation group is ϵ_{ijk}, since it transforms as

$$\epsilon_{ijk} \rightarrow R_{ii'}R_{jj'}R_{kk'}\epsilon_{i'j'k'}. \qquad (1.133)$$

However, from the definition of the determinant of a 3×3 matrix, we can check that the right-hand side of eq. (1.133) is equal to $(\det R)\epsilon_{ijk}$. Equation (1.128), together with the fact that, for two matrices A and B, $\det(AB) = \det(A)\det(B)$, and $\det(R^{\mathrm{T}}) = \det(R)$, implies that $(\det R)^2 = 1$, so $\det R = \pm 1$. "Proper" rotations are defined as the transformations with $\det R = +1$ (while transformations with $\det R = -1$ correspond to parity transformations, that change the orientation of one axis, or of all three axes, possibly combined with proper rotations), so ϵ_{ijk} is indeed invariant under (proper) rotations.

1.7   Groups and representations

This section is more formal and can (actually, should!) be omitted at first reading.

Vectors and tensors are usually the first examples that one encounters of a much more general concept, that of representations of groups, which is ubiquitous in modern theoretical physics. Even if this will not be strictly necessary for the rest of the book, it can be interesting to expand on the previous discussion, taking a more abstract and mathematical point of view. The underlying mathematics here is that of *group theory* and group representations. Group theory is a fundamental element of the mathematical arsenal of modern theoretical physics. While, especially with hindsight, it was already implicitly present in the classical physics of the 19th century, it acquired a central role first of all because of Special Relativity and its connection with electromagnetism (in relation, in particular, to the work of Lorentz and Poincaré), that we will explore in the following chapters. The role of group theory in physics then became even more central with the advent of quantum mechanics where, for instance, concepts such as the spin of the electron cannot be really understood without it. Modern particle physics, such as the Standard Model that unifies weak and electromagnetic interactions, as well as all attempts at going beyond it, are also formulated in the language of group theory.[20]

A *group* G is a set of objects $g_1, g_2, \ldots$ (discrete or continuous) among which is defined a composition operation $g_1 \circ g_2$, such that

- if $g_1 \in G$ and $g_2 \in G$, then also $g_1 \circ g_2 \in G$.
- The composition is associative, $g_1 \circ (g_2 \circ g_3) = (g_1 \circ g_2) \circ g_3$.
- In G there is the identity element e, defined by the fact that, for each $g \in G$, $g \circ e = e \circ g = g$.[21]
- For each $g \in G$ there is an inverse element, that we denote g^{-1}, such that $g \circ g^{-1} = g^{-1} \circ g = e$.

Note that the composition operation is not necessarily commutative. If it is, the group is called commutative (or abelian), otherwise is called non-commutative (or non-abelian). Rotations form a group (in any dimensions), since they satisfy the previous axioms.

Exercise 1.8 Show that rotations in two dimensions form a commutative group, while in three dimensions they form a non-commutative group.

[Hint: take an object such as a book, lie it on a table, denote by x and y the axes on the plane of the table so that the lower edge of the book is along the x axis, and the rib of the book is along the y axis. Perform first a rotation of the book by 90° around the x axis and then a rotation of the resulting configuration by 90° around the y axis. Compare this with what happens if you first perform the rotation by 90° around the y axis and then by 90° around the x axis.]

A (linear) *representation* R of a group is an operation that assigns to a generic, abstract element g of a group a linear operator $D_R(g)$ defined on a (real or complex) vector space,

$$g \mapsto D_R(g), \tag{1.134}$$

with the property that, for all $g_1, g_2 \in G$

$$D_R(g_1) D_R(g_2) = D_R(g_1 \circ g_2). \tag{1.135}$$

[20] For an introduction to group theory in physics, see e.g., Zee (2016).

[21] The identity element is unique: in fact, assume that e_1 and e_2 are two identity elements. Then, using the fact that e_2 is an identity element, we have $e_1 \circ e_2 = e_1$. However, using the fact that e_1 is an identity element, we also have $e_1 \circ e_2 = e_2$. Therefore $e_1 = e_2$.

This condition means that the mapping preserves the group structure, i.e., that it is the same to compose the elements at the group level (with the $\circ$ operation) and then represent the resulting element $g_1 \circ g_2$, or first represent g_1 and g_2 in terms of linear operators and then compose the resulting linear operators. Note that the composition operation in $D_R(g_1)D_R(g_2)$ is the one between linear operators (such as the matrix product when the linear operators are matrices, see next), while the composition operation in $g_1 \circ g_2$ is the one at the abstract group level. Setting $g_1 = g$, $g_2 = e$ in eq. (1.135), we find that, for all $g \in G$, $D_R(g)D_R(e) = D_R(g)$; similarly, setting $g_1 = e$, $g_2 = g$, we get $D_R(e)D_R(g) = D_R(g)$. This implies that the identity element of the group e must be mapped to the identity operator I, i.e., $D_R(e) = I$. Similarly, we can show that $D_R(g^{-1}) = [D_R(g)]^{-1}$.

The vector space on which the operators D_R act is called the *basis*, or the *base space*, for the representation R and, as we have mentioned, it can be a real vector space or a complex vector space. We will be particularly interested in *matrix representations*. In this case, the base space is a vector space of finite (real or complex) dimension n, and an abstract group element g is represented by a $n \times n$ matrix $[D_R(g)]_{ij}$, with $i, j = 1, \ldots, n$. The *dimension* of the representation is defined as the dimension n of the base space.

Writing a generic element of the base space as a vector $\mathbf{v}$ with components $(v_1, \ldots, v_n)$, a group element g can be associated with a linear operator $[D_R(g)]_{ij}$ acting on the base space, and therefore to a linear transformation of the base space

$$v_i \to [D_R(g)]_{ij} v_j \,, \tag{1.136}$$

with our usual summation convention on repeated indices. The important point is that eq. (1.136) allows us to attach a physical meaning to a group element: before introducing the concept of representation, a group element g is just an abstract mathematical object defined by its composition rules with the other group members. Choosing a specific representation, instead, allows us to interpret g as a transformation acting on a vector space.[22]

[22]Apart from matrix representations, the other typical situation encountered in physics is when a group element is represented by a differential operator acting on a space of functions. Since the space of function is infinite-dimensional, this representation is infinite-dimensional.

1.7.1   Reducible and irreducible representations

The different representations of a group describe all possible ways in which objects can transform under the action of the group, for instance under rotations. However, not all possible representations describe genuinely different possibilities. Given a representation R of a group G, whose basis is a space X, and another representation R' of G, whose basis is a space X', R and R' are called *equivalent* if there is an isomorphism $S : X \to X'$ such that, for all $g \in G$,

$$D_R(g) = S^{-1} D_{R'}(g) S \,. \tag{1.137}$$

Comparing with eq. (1.136), we see that, in the case of representations of finite dimension, equivalent representations correspond to a change of basis in the vector space spanned by the v^i, obtained acting on the basis vector with a matrix S, so we do not consider them as genuinely different representations.

Furthermore, some representations can be obtained trivially, just by stacking together representations of lower dimensions, and do not really describe novel ways of transforming under the action of the group. As a simple example, consider two vectors $\mathbf{v}$ and $\mathbf{w}$, in two dimensions. Under a rotation in the plane by an angle θ, $\mathbf{v}$ transforms as in eq. (1.119), and similarly

$$
\begin{pmatrix} w_x \\ w_y \end{pmatrix} \rightarrow \begin{pmatrix} w'_x \\ w'_y \end{pmatrix} = \begin{pmatrix} \cos\theta & -\sin\theta \\ \sin\theta & \cos\theta \end{pmatrix} \begin{pmatrix} w_x \\ w_y \end{pmatrix}, \tag{1.138}
$$

which expresses the fact that, in two dimensions, vectors are representations of dimension two of the rotation group. Naively, we might think that we can find a new type of representation of dimension four, by putting together the components of $\mathbf{v}$ and $\mathbf{w}$ into a single object with components (v_x, v_y, w_x, w_y). Under rotations,

$$
\begin{pmatrix} v_x \\ v_y \\ w_x \\ w_y \end{pmatrix} \rightarrow \begin{pmatrix} v'_x \\ v'_y \\ w'_x \\ w'_y \end{pmatrix} = \begin{pmatrix} \cos\theta & -\sin\theta & 0 & 0 \\ \sin\theta & \cos\theta & 0 & 0 \\ 0 & 0 & \cos\theta & -\sin\theta \\ 0 & 0 & \sin\theta & \cos\theta \end{pmatrix} \begin{pmatrix} v_x \\ v_y \\ w_x \\ w_y \end{pmatrix}.
$$

$$\tag{1.139}$$

However, here we have not discovered a genuinely new type of representation of dimension four; we have simply stack together two vectors. Mathematically, the fact that this is not a genuinely new representation is revealed by the fact that the matrix in eq. (1.139) is block diagonal (for all rotations, so in this case for all values of θ), so that there is no rotation that mixes the components of $\mathbf{v}$ with that of $\mathbf{w}$.

This example motivates the definition of reducible and irreducible representations. A representation R is called reducible if it has an invariant subspace, i.e., if the action of any $D_R(g)$ on the vectors in the subspace gives another vector of the subspace. This corresponds to the fact that [possibly after a suitable change of basis of the form (1.137)] there is a subset of components of the basis vector that never mixes with the others, for all transformations $D_R(g)$. A representation R is called completely reducible if, for all elements g, the matrices $D_R(g)$ have a block diagonal form, or can be put in a block diagonal form with a change of basis corresponding to the equivalence relation (1.137). Conversely, a representation with no invariant subspace is called irreducible. Irreducible representations describe the genuinely different way in which physical quantities can transform under the group of transformations in questions.

To put some flesh into these rather abstract notions, we now look at the simplest irreducible representations of the rotation group.

1.7.2   The rotation group and its irreducible tensor representations

The group of rotations in d spatial dimensions can be defined as the group of linear transformations of a d-dimensional space with coordinate $(x_1, \ldots x_d)$, of the form

$$x_i \to x_i' = R_{ij} x_j \, , \qquad (1.140)$$

which leaves invariant the quadratic form

$$|\mathbf{x}|^2 = x_1^2 + \ldots + x_d^2 \, . \qquad (1.141)$$

As we already found in eq. (1.128), the corresponding condition on the matrix R_{ij} is that it must be an orthogonal matrix, $RR^{\mathrm{T}} = R^{\mathrm{T}} R = I$. The group of $d \times d$ orthogonal matrices is denoted by $O(d)$. As we already found below eq. (1.133), from $R^{\mathrm{T}} R = I$ it follows that $(\det R)^2 = 1$, so $\det R = \pm 1$. The transformations with determinant $+1$ form a subgroup, since the product of two matrices with determinant plus $+1$ is still a matrix with determinant $+1$ (which is not the case for those with determinant -1, given that the product of two matrices with determinant -1 is a matrix with determinant $+1$). The subgroup of $O(d)$ with determinant equal to $+1$ is denoted by $SO(d)$ and is identified with the proper rotation group in d dimension. As already mentioned below eq. (1.133), transformations with $\det R = -1$ rather correspond to discrete parity transformations, such as $(x \to -x, y \to y, z \to z)$ or $(x \to -x, y \to -y, z \to -z)$ or, more generally, to parity transformations combined with proper rotations. By "rotations" we will always mean the proper rotations. In particular, the rotation group in three dimensions is $SO(3)$.

The simplest irreducible representation of the rotation group, as of any other group, is the so-called trivial representation, that assign to each group element the 1×1 identity matrix, i.e., $D_R(g) = 1$ for each g. In this case, the basic properties (1.135) of a representation are obviously satisfied (note, however, that if we assign $D_R(g) = I$ with I the $n \times n$ identity matrix with dimension $n > 1$, we do not get an irreducible representation, since the identity matrix is block diagonal). Despite being trivial from the mathematical point of view, this representation is very significant physically. According to eq. (1.136), the basis for this representation is an object of dimension 1, i.e., a single quantity ϕ, that, under any rotation, remains invariant, $\phi \to \phi$. Such quantities are called *scalars* under rotation.

The next simplest representation of $SO(3)$ is the vector representation v_i. Indeed, the very definition that we have given of the group $SO(3)$, based on eq. (1.140) and the condition that it leaves invariant the quadratic form (1.141), already introduces a transformation property under rotation, in this case for the coordinates. More generally, vectors are then defined as objects that transform as in eq. (1.120), and the coordinates of three-dimensional space provide the simplest example of a vector, $\mathbf{x} = (x_1, x_2, x_3)$ [or $\mathbf{x} = (x, y, z)$; we will use the two

notations interchangeably]. Other obvious examples of vectors from elementary mechanics are the momentum **p**, or the angular momentum
L.

The vector representation can be taken to be real, since the rotation matrix R_{ij} has only real elements, and therefore transform real
vectors into real vectors. The real vector representation of $SO(3)$ is irreducible, since we cannot find a subset of the coordinates (x_1, x_2, x_3)
that, whichever rotation we perform, never mix with the others. For
instance, a rotation around the third axis mixes x_1 with x_2, a rotation
around the x_1 axis mixes x_2 with x_3, and one around the x_2 axis mixes
x_1 with x_3. Since the vector representation enters in the very definition of the group $SO(3)$, it is also called the *defining*, or *fundamental*
representation of $SO(3)$.[23]

Consider next the tensor representation (1.130). In three dimensions
a tensor T_{ij} has nine components, and eq. (1.130) states that these nine
components transform linearly among them, and therefore form a basis
for a representation of the rotation group of dimension nine. We could for
instance express the corresponding representation of rotations as 9×9
matrices, using as a basis $(T_{11}, T_{12}, \ldots, T_{33})$. We want to understand
whether this representation is reducible, i.e., if (possibly after a suitable
change of basis) there are subsets of elements of the basis that never
mix with the other elements. The best way to address the problem is to
observe that, given a generic tensor T_{ij}, we can always separate it into
its symmetric and antisymmetric parts

$$
\begin{aligned}
T_{ij} &= \frac{T_{ij}+T_{ji}}{2} + \frac{T_{ij}-T_{ji}}{2} \\
&\equiv S_{ij} + A_{ij}.
\end{aligned}
\tag{1.142}
$$

Using eq. (1.130), we can show that, under any rotation, a symmetric
tensor remains symmetric:

$$
\begin{aligned}
S'_{ij} &= R_{ik}R_{jl}S_{kl} \\
&= R_{ik}R_{jl}S_{lk} \\
&= R_{jk}R_{il}S_{kl} \\
&= S'_{ji},
\end{aligned}
\tag{1.143}
$$

where, to get the second equality, we used the symmetry of S_{kl} and, in
the third, we renamed the dummy indices $k \to l$ and $l \to k$. Similarly, an
antisymmetric tensor remains antisymmetric. Therefore, S_{ij} and A_{ij} do
not mix under rotations. Consider now the trace of the symmetric part,
$S = \delta_{ij}S^{ij}$. Using eq. (1.130) and the property (1.129) of orthogonal
matrices, we see that it is invariant under spatial rotations. It is then
convenient to separate S_{ij} as

$$
\begin{aligned}
S_{ij} &= \left(S_{ij} - \frac{1}{3}\delta_{ij}S\right) + \frac{1}{3}\delta_{ij}S \\
&\equiv S_{ij}^{\rm T} + \frac{1}{3}\delta_{ij}S.
\end{aligned}
\tag{1.144}
$$

[23]Observe that we are considering here
real representations, i.e., representations where the basis vectors are real.
One can sometimes assemble real representations of $SO(d)$ into complex
representations of lower (complex) dimension. For instance, for $SO(2)$,
eq. (1.119) defines an irreducible real
representation of (real) dimension two,
since the real quantities v_x and v_y mix
among them. However, if we form the
complex combinations $v_\pm = v_x \pm iv_y$,
we see that $v_+ \to e^{i\theta}v_+$ and $v_- \to
e^{-i\theta}v_-$ so, over complex numbers, we
have two irreducible representations of
complex dimension one. Indeed, a theorem states that all irreducible complex
representations of abelian groups, such
as $SO(2)$, are one-dimensional.
For other groups, such as the groups
$U(N)$ of $N \times N$ unitary matrices, or
$SU(N)$ if we require the determinant
to be +1, the matrix elements are complex, so, in general, one must consider
complex representations.

The term S_{ij}^{T} is traceless (as can be shown contracting it with δ_{ij} and using $\delta_{ij}\delta_{ij} = \delta_{ii} = 3$).[24] Since the trace is invariant, under rotations it remains traceless, and therefore it does not mix with the term $\delta_{ij}S$. We have therefore separated the nine components of a generic tensor T_{ij} into its symmetric traceless part S_{ij}^{T} (which has five components, corresponding to the six independent components of a symmetric 3×3 matrix, minus the condition of zero trace), its trace S (one component), and an antisymmetric tensor A_{ij} which, in three dimensions, has three independent components,

$$T_{ij} = S_{ij}^{\mathrm{T}} + \frac{1}{3}\delta_{ij}S + A_{ij}. \tag{1.145}$$

The trace $S = \delta^{ij}T_{ij}$, being invariant under rotations, corresponds to the scalar representation that we already encountered. The antisymmetric tensor might look like a new representation of the rotation group, but in fact, introducing

$$A_i = \frac{1}{2}\epsilon_{ijk}A_{jk}, \tag{1.146}$$

one sees that A_i is just a spatial vector. We have inserted a factor of $1/2$ for convenience since, for instance, in this way $A_1 = (1/2)\epsilon_{1jk}A_{jk} = (1/2)(\epsilon_{123}A_{23} + \epsilon_{132}A_{32}) = A_{23}$ [using the antisymmetry of ϵ_{ijk} and of A_{jk} with respect to (j,k)]. So, $A_{23} = A_1$, and similarly one finds $A_{12} = A_3$, $A_{31} = A_2$, so the inversion of eq. (1.146) can be written compactly as

$$A_{ij} = \epsilon_{ijk}A_k. \tag{1.147}$$

Therefore, the three independent components of an antisymmetric tensor can be rearranged into the three independent components of a vector. This means that these two representations are equivalent, in the sense of eq. (1.137), and we have not discovered a genuinely new irreducible representation of rotations.

In contrast, the traceless symmetric tensor is irreducible (since there is no further symmetry that forbids its components to mix among them, and then we can always find rotations that mix a given components of S_{ij}^{T} with any other component). We have therefore found a new irreducible representation of the rotation group, of dimension five, the traceless symmetric tensors, and eq. (1.145) can be rewritten as

$$T_{ij} = S_{ij}^{\mathrm{T}} + \frac{1}{3}\delta_{ij}S + \epsilon_{ijk}A_k, \tag{1.148}$$

in terms of the invariant tensors δ_{ij} and ϵ_{ijk}, and of the irreducible representations provided by the scalar S, the vector A_i, and the traceless symmetric tensor S_{ij}^{T}. Note how the nine independent components of T_{ij} have been shared between a scalar (one component), a vector (three components), and a traceless symmetric tensor (five components). One can work out similarly the decomposition into irreducible representations of tensors with more indices, such as T_{ijk}.

Observe that the dimension of the scalar, vector, and traceless symmetric tensor representations can all be written as $2s + 1$ for $s = 0$

(scalar), $s = 1$ (vector) and $s = 2$ (traceless symmetric tensor). Higher-order tensorial irreducible representations, obtained from tensors with more indices, give representations of dimension $2s + 1$ for all other integer values of s. The representations of the rotation group play an important role also in quantum mechanics, where the (tensorial) representations that we have found turn out to correspond to massive particles with integer spin s (in units of $\hbar$).

Finally, it is instructive to consider infinitesimal rotations, and see how they can be parametrized. A rotation matrix that differs from the identity transformation by an infinitesimal quantity can be written as

$$R_{ij} = \delta_{ij} + \omega_{ij} + \mathcal{O}(\omega^2)\,, \tag{1.149}$$

with ω_{ij} infinitesimal parameters, which describe the deviation from the identity transformation. Requiring that this satisfies eq. (1.129), to first order in ω we get

$$
\begin{aligned}
\delta_{ij} &= \left[\delta_{ik} + \omega_{ik} + \mathcal{O}(\omega^2)\right]\left[\delta_{jk} + \omega_{jk} + \mathcal{O}(\omega^2)\right] \\
&= \delta_{ij} + \omega_{ij} + \omega_{ji} + \mathcal{O}(\omega^2)\,,
\end{aligned}
\tag{1.150}
$$

and therefore, requiring that the linear order cancels, we get[25]

$$\omega_{ij} = -\omega_{ij}\,, \tag{1.151}$$

i.e., ω_{ij} is an antisymmetric tensor. An antisymmetric tensor ω_{ij} can be written in terms of a vector θ^i as in eq. (1.147), i.e., we can write $\omega_{ij} = -\epsilon_{ijk}\delta\theta_k$ (where the minus sign is a convention, chosen so that $\delta\theta^i$ corresponds to counterclockwise rotations, and the use of the notation $\delta\theta_k$ stresses that this is an infinitesimal angle), so

$$R_{ij} = \delta_{ij} - \epsilon_{ijk}\delta\theta_k\,. \tag{1.152}$$

Then, the infinitesimal form of eq. (1.140) is

$$
\begin{aligned}
x^i &\to x^i + \omega^{ij}x^j \\
&= x^i - \epsilon^{ijk}x^j\delta\theta^k\,.
\end{aligned}
\tag{1.153}
$$

As a check, we can consider a rotation around the z axis, so that $\delta\theta^1 = \delta\theta^2 = 0$ and $\delta\theta^3 = \delta\theta$. Then, eq. (1.153) gives $x \to x - (\delta\theta)y$, $y \to y + (\delta\theta)x$, $z \to z$, which is the infinitesimal form of the counterclockwise rotation around the z axis, $x \to x\cos\theta - y\sin\theta$, $y \to x\sin\theta + y\cos\theta$.

[25] Requiring the cancelation also to quadratic and higher order does not give any further constraint. This is a general property of Lie groups, (i.e., group parametrized in a continuous and differentiable manner by a set of parameters), see e.g., Section 2.1 of Maggiore (2005).

Systems of units

2

As discussed in the Preface, there are two common systems of units, SI and Gaussian, each one with its own advantages. We begin by defining them and discussing their relation, which involves some subtleties. The SI system is the most widely used in most contexts but an acquaintance with the Gaussian system can also be very useful, in particular to prepare the transition from classical electrodynamics to quantum field theory, where (for reasons that we will explain) nowadays only Gaussian units are used. Therefore, while we will use SI units in this book, we will explain how to quickly translate the results into Gaussian units and, in App. A, we will collect some of the most important formulas, written in Gaussian units.

2.1 The SI system

Since its establishment in 1960, the International System of Units, or SI (from the French *Système International*), has imposed itself as the international standard. The SI is based on seven basic units from which all other units can be derived. For our course, we will really only need four of them: the meter (the unit of length, m), the kilogram (the unit of mass, kg), the second (the unit of time, s), that together form the m.k.s. system used in classical mechanics, and the ampere (the unit of electric current, A).[1,2]

Over the years, the general trend in defining the basic SI units has been to get rid of definitions involving any man-made standard, or properties of macroscopic materials, or descriptions of measurements, and rather define them using fundamental constants of Nature, or quantum properties of matter at the atomic level. For instance, the meter was originally defined in 1793 as one ten-millionth of the distance from the equator to the North Pole along a great circle and, after other redefinitions, was eventually redefined in 1899 as the distance between two lines marked on a prototype meter bar made of an alloy of platinum and iridium and conserved in the International Bureau of Weights and Measures in Sèvres, France. Such a definition had obvious intrinsic problems of reproducibility (national meter prototype had to be fabricated and distributed), stability with time, and made reference to specific experimental conditions (e.g., the distance between the lines had to be measured at the melting point of ice). Similarly, the ancient definition of the second was based on the Earth's rotation period, so that 1 day = 24 hr = 24×3600 s. The modern definitions of the units of time and of

[1]Two more basic units are the kelvin (unit of temperature, K) and the mole (amount of substance, mol). The SI includes a seventh base unit, the candela (cd), related to luminous intensity as perceived by the human eye, mostly of interest for biology and physiology. See https://www.bipm.org/utils/common/pdf/si-brochure/SI-Brochure-9-EN.pdf for a detailed description.

[2]Note that, while the correct spelling of the last name of André-Marie Ampère (1775–1836) involves an accent, in English the unit of measure "ampere" is written without the accent. So, we will refer to Ampère's law, but the current unit is the ampere. Note also that units derived from names of persons are written with lower cases, as in ampere, coulomb, or kelvin (except when grammatical rules require an upper case letter), while their symbols are written in upper case, A for ampere, C for coulomb, K for kelvin.

[3]The exact definition is that the un-
perturbed ground state hyperfine transition frequency of an atom of Cs 133 is
$\Delta \nu = 9\,192\,631\,770$ Hz, where 1 Hz $=$
$1\,\mathrm{s}^{-1}$.

length are different. The second is defined in terms of the frequency of a specific atomic transition.[3] Given this definition of a second, the meter is now defined in term of the speed of light c, stating that, by definition,

$$c = 299\,792\,458 \,\mathrm{m/s}\,. \tag{2.1}$$

This definition is consistent with the previous definition of the meter based on a reference bar within the experimental error, but get rids of any reference to specific macroscopic objects or detailed measurement processes. In the same spirit, the kilogram is now no longer defined in terms of a reference object, but rather from the Planck constant h, stating that, by definition, $h = 6.626\,070\,15 \times 10^{-34}\,\mathrm{J\,s}$, where the unit of energy, the Joule, is related to the basic units by $\mathrm{J} = \mathrm{kg\,m^2\,s^{-2}}$. Note that, in this way, h and c are used to define the units and therefore their values are fixed by definition, and have no experimental error associated with them.[4]

For electromagnetic phenomena, the SI system proceeds by defining a base unit for electric current, the ampere (A). Since a current is a charge flowing per unit time, this induces the definition of a derived SI unit of charge, the coulomb (C), from $1\,\mathrm{A} = 1\,\mathrm{C/s}$. In 2019, the SI definition of the ampere (that, as we discuss in more detail next, was previously based on the force between two parallel wires carrying a current) has been changed, in order to relate it to fundamental constants rather than to experimental settings: now the ampere is defined from $1\,\mathrm{A} = 1\,\mathrm{C/s}$, and the coulomb is defined in terms of the electron charge $-e$, stating that, by definition,

$$e \equiv 1.602\,176\,634 \times 10^{-19}\,\mathrm{C}\,. \tag{2.4}$$

To understand the reason underlying this definition and the relation with the earlier definition, let us look at the force exerted between static charges, and at the force between parallel wires carrying currents. It is an experimental fact that two static charges, at a distance r, attract or repel each other with a force inversely proportional to the square of the distance. Even before having defined the units of electric charge, we also know that the force is proportional to the charge q_1 of the first body and to the charge q_2 of the second body, as could be shown comparing systems with a different number of individual electron charges. This gives *Coulomb's law*,

$$\mathbf{F} = k\frac{q_1 q_2}{r^2}\hat{\mathbf{r}}\,. \tag{2.5}$$

The value of the proportionality constant k depends, of course, on the units chosen for the electric charge or, vice versa, could be used to define the unit of electric charge. In the SI system, the constant k is denoted as $1/(4\pi\epsilon_0)$, so

$$\mathbf{F} = \frac{1}{4\pi\epsilon_0}\frac{q_1 q_2}{r^2}\hat{\mathbf{r}}\,. \tag{2.6}$$

The constant ϵ_0 is called the *vacuum electric permittivity*, or just the *vacuum permittivity*.

Similarly, if we take two parallel wires (approximated as infinitely long and of negligible thickness) carrying currents I_1 and I_2 and separated by a distance d, the observation shows that they attract or repel each other with a force per unit length, $dF/d\ell$, proportional to $I_1 I_2/d$. The proportionality constant in the SI system is called $\mu_0/(2\pi)$, so, for the modulus, we have

$$\left| \frac{d\mathbf{F}}{d\ell} \right| = \frac{\mu_0}{2\pi} \frac{I_1 I_2}{d} . \tag{2.7}$$

The force is attractive if the currents are parallel, and repulsive if they are antiparallel (we will see how eqs. (2.6) and (2.7) follow from Maxwell's equations in Sections 4.1.1 and 4.2.3, respectively). The constant μ_0 is called the *vacuum magnetic permeability*. As we will see, Maxwell's equations tell us that μ_0 and ϵ_0 are not independent, but are related by the exact relation

$$\epsilon_0 \mu_0 = \frac{1}{c^2} , \tag{2.8}$$

where c is the speed of light. In eq. (2.6) the value of ϵ_0 depends on the choice of units of the electric charge and in eq. (2.7) the value of μ_0 depends on the choice of units of the electric current. However, since ϵ_0 and μ_0 are related by eq. (2.8), fixing one of the two units fixes the other.

The definition of the ampere before 2019 was obtained from eq. (2.7), by stating that, if we take two long (and infinitesimally thin) parallel wires each carrying a current of 1 A and at a distance of 1 m, the force per unit length in eq. (2.7) is equal to $2 \times 10^{-7} \mathrm{N/m}$, where the Newton (N) is the derived SI unit of force. The reasons for this numerical value are historical, but the orders of magnitude are such that, for two wires carrying a current of 1 A (and at a distance d of order, say, of a few centimeters, rather than one meter), this force could be measured in a laboratory with rather simple techniques. This definition of the ampere (and therefore of the coulomb) amounts to fixing by definition μ_0 in eq. (2.7) to the value

$$\mu_0 = 4\pi \times 10^{-7} \, \frac{\mathrm{N}}{\mathrm{A}^2} , \tag{2.9}$$

so μ_0 has no observational error. Equation (2.8), together with eq. (2.1), then fixes ϵ_0,

$$\frac{1}{4\pi\epsilon_0} = (2.99792458)^2 \times 10^9 \, \frac{\mathrm{N \, m^2}}{\mathrm{C}^2} \tag{2.10}$$

$$= (8.988\ldots) \times 10^9 \, \frac{\mathrm{N \, m^2}}{\mathrm{C}^2} , \tag{2.11}$$

or[5]

$$\epsilon_0 \simeq (8.854\ldots) \times 10^{-12} \, \frac{\mathrm{C}^2}{\mathrm{N \, m^2}} , \tag{2.13}$$

again with no observational error, since both μ_0 and c are exact numbers. In contrast, the electron charge becomes a measurable quantity. For instance, we could in principle use eq. (2.6) to measure the electric force between two electrons at a given distance and, since ϵ_0 has been fixed,

[5]In terms of the fundamental SI units,

$$1 \, \frac{\mathrm{C}^2}{\mathrm{N \, m^2}} = 1 \, \frac{\mathrm{A}^2 \, \mathrm{s}^4}{\mathrm{kg \, m^3}} . \tag{2.12}$$

we would get a measurement of the electron charge. In practice, the most accurate measurement of the electron charge are obtained with other methods, such as the quantum Hall effects. In any case, the point is that, with this pre-2019 definition of the ampere, the electron charge was a measurable quantity with an observational error. For instance, the 2018 edition of the *Review of Particle Physics*,[6] that contains a standard compilation of physical quantities and of elementary particle properties, quoted the value $e = 1.602\,176\,6208(98) \times 10^{-19}$ C.

[6] https://pdg.lbl. gov/2018/reviews/ rpp2018-rev-phys-constants.pdf.

In 2019, the official SI definition of the ampere was changed and now, in the spirit of using only fundamental constants to define the units, it is based on eq. (2.4), which defines the coulomb in terms of e (and then the ampere from 1 A = 1 C/s). Therefore, now the value of the electron charge is fixed by definition and no longer has an observational error associated with it (just as the speed of light and the Planck constant). Having fixed the definition of the ampere in this way, we can now measure the force per unit length between two parallel wires at a given distance and carrying a current of, say, 1 A each. Then, from eq. (2.7) we now get a measurement of μ_0, so μ_0 becomes a measurable quantity with an observational error; its measured value, as of 2019, was consistent with the older definition $\mu_0 = 4\pi \times 10^{-7}$ N/A^2 to a relative standard uncertainty of 2.3×10^{-10}. Since eq. (2.8) remains exact, being a mathematical consequence of Maxwell's equations, once measured μ_0 one also gets ϵ_0 and the relative accuracy on ϵ_0 is the same as that on μ_0, since c has no error.

From eq. (2.4) we see that 1 C corresponds to a huge number of elementary charges, of order 10^{19}. This clearly shows how the SI units have their roots in the laboratory, since a huge number of elementary charges flowing per second is needed to produce a typical current observed in simple laboratory situations.

The definition of electromagnetic units is completed by the definition of the electric and magnetic fields. These are defined through the Lorentz force equation that, in SI units, reads

$$\mathbf{F} = q\left(\mathbf{E} + \mathbf{v} \times \mathbf{B}\right) . \tag{2.14}$$

Maxwell's equations will be discussed in detail in Chapter 3, but for completeness we write them also here, in SI units:

$$\boldsymbol{\nabla}\cdot\mathbf{E} \;=\; \frac{1}{\epsilon_0}\rho\,, \tag{2.15}$$

$$\boldsymbol{\nabla}\times\mathbf{B} \;=\; \mu_0\mathbf{j} + \mu_0\epsilon_0\frac{\partial\mathbf{E}}{\partial t}\,, \tag{2.16}$$

$$\boldsymbol{\nabla}\cdot\mathbf{B} \;=\; 0\,, \tag{2.17}$$

$$\boldsymbol{\nabla}\times\mathbf{E} \;=\; -\frac{\partial\mathbf{B}}{\partial t}\,. \tag{2.18}$$

[7] As we have seen, the fourth fundamental base unit of the SI has been chosen as the ampere rather than the coulomb. It is sometime natural to use the coulomb as the fourth base unit, as we occasionally do here and in other places in the book.

As it is clear from eq. (2.14), the SI dimensions of the electric field are N/C, or kg m/(s^{2}C).[7] We also introduce, as a derived unit, the volt (V), defined as V = J/C (where J is the Joule). The volt is therefore an energy per unit charge, so is the unit used for potentials. From $J = kg\,m^2/s^2$ it follows that, in SI units, the electric field has dimensions of V/m,

$$[\mathbf{E}] = \frac{kg\,m}{s^2\,C} = \frac{V}{m}\,. \tag{2.19}$$

Again from eq. (2.14), we see that in the SI the magnetic field has units of $N/(C\,m\,s^{-1}) = kg/(C\,s)$. This quantity is called the tesla (T), and is the derived SI unit for magnetic field:

$$[\mathbf{B}] = \frac{kg}{s^2\,A} = T\,. \tag{2.20}$$

From eq. (2.15), together with eqs. (2.10) and (2.19), we see that ρ has dimensions C/m^3, so is an electric charge per unit volume. The electric charge Q contained in a finite volume V is then given by

$$Q = \int_V d^3x\,\rho\,. \tag{2.21}$$

From eq. (2.16), using the dimensions of μ_0 given in eq. (2.9), we find that $\mathbf{j}$ has dimensions $C/(m^2\,s) = A/m^2$, so $\mathbf{j}$ is a current per unit surface. The current dI flowing through an infinitesimal surface is obtained defining $d\mathbf{s}$ as a vector whose modulus ds is the area of the surface, and whose direction is equal to the normal of the surface (with a given choice of orientation), and is then given by $dI = d\mathbf{s}\cdot\mathbf{j}$; so, the current I flowing through a finite surface S is given by

$$I = \int_S d\mathbf{s}\cdot\mathbf{j}\,. \tag{2.22}$$

2.2   Gaussian units

Gaussian units, first of all, are based on the c.g.s. system (centimeters-grams-seconds) rather than on the m.k.s. system. By itself, this would only lead to trivial conversion factors. The crucial difference, however, is in the definition of the electric charge. One starts again from eq. (2.5), but now one defines the unit of electric charge setting $k = 1$ by definition, so

$$\mathbf{F} = \frac{q_1 q_2}{r^2}\hat{\mathbf{r}}\,. \tag{2.23}$$

This is not just a rescaling of numbers compared to SI units. In the SI system, the ampere is a fourth independent base quantity, with respect to the units of lengths, time, and mass. We cannot express the ampere (and therefore also the coulomb) as a combination of positive or negative powers of meters, seconds, and kilograms. As a result, in the SI the constant $k = 1/(4\pi\epsilon_0)$ is not a pure number, but has dimensions $N\,m^2/C^2$, see eq. (2.11) (or, re-expressing it in terms of the base units, it has dimensions $kg\,m^3\,s^{-4}\,A^{-2}$). In contrast, in the Gaussian system, k is taken to be a pure number without dimensions. The unit of charge defined in this way is called the *esu* (for "electrostatic unit") of charge or, equivalently, the *statcoulomb* (statC). Since, in Gaussian units, the force is measured in dyne, with $1\,\mathrm{dyne} = 1\,\mathrm{gr\,cm\,s^{-2}}$, from eq. (2.23) we see that, in Gaussian units, the unit of charge is a derived quantity, given by

$$1\,\mathrm{statC} = 1\,\mathrm{gr^{1/2}\,cm^{3/2}\,s^{-1}}\,. \tag{2.24}$$

So, as far as electromagnetism is concerned, the SI system has four base units (kg, m, s, A) while the Gaussian system has only three (gr, cm, s).

Comparing the Coulomb law in the Gaussian and SI systems, we see that the respective definitions of electric charges are related by

$$q_{\mathrm{gau}} = \frac{q_{\mathrm{SI}}}{\sqrt{4\pi\epsilon_0}}\,. \tag{2.25}$$

Observe, once again, that the electric charge in the two systems have different dimensions, since ϵ_0 is not a pure number. We can now consider a charge such that $q_{\mathrm{SI}} = 1$ C, and ask what the corresponding value of q_{gau} is. Inserting the value (2.10) into eq. (2.25), we get

$$
\begin{aligned}
q_{\mathrm{gau}} &= 2.99792458 \times \frac{\sqrt{10^9\,\mathrm{N\,m^2}}}{\mathrm{C}} \times 1\,\mathrm{C} \\
&= 2.99792458 \times \sqrt{10^9 \times 10^5 \mathrm{dyne} \times 10^4 \mathrm{cm^2}}\,, \tag{2.26} \\
&= 2.99792458 \times 10^9\,\mathrm{statC}\,, \tag{2.27}
\end{aligned}
$$

where we have used the conversion $1\,\mathrm{N} = 10^5$ dyne and, from eq. (2.24), $1\,\mathrm{dyne\,cm^2} = 1\,\mathrm{gr\,cm^3\,s^{-2}} = 1\,\mathrm{statC^2}$. Therefore, a charge of 1 C in the SI corresponds to a charge of 2.99792458×10^9 statC in the Gaussian system. In this sense, one might be tempted to write

$$\text{``}1\,\mathrm{C} = 2.997\,924\,58 \times 10^9\,\mathrm{statC}\,.\text{''} \tag{2.28}$$

We have put the equality between quotes because, written as an equality in this form, this is wrong. The relation between C and statC is not just a simple proportionality factor, as say, in the relation $1\,\mathrm{m} = 10^2$ cm. As we have stressed previously, in Gaussian units the statC is a derived unit, that can be expressed in terms of cm, gr, and s, while in the SI system, the ampere (and therefore the coulomb) is a fourth independent base unit, independent from m, kg, and s. If one would take eq. (2.28) literally, it would be possble to transform the statC into m.k.s. units, using, from eq. (2.24), $1\,\mathrm{statC} = (10^{-3}\mathrm{kg})^{1/2}\,(10^{-2}\,\mathrm{m})^{3/2}\,\mathrm{s^{-1}}$ and then eq. (2.28) would give the coulomb in terms of m, kg, and s. This would be wrong, since the coulomb is independent from m, kg, and s. The sense in which eq. (2.28) is correct is that, as we have written above, a charge of 1 C in the SI system corresponds to a charge of 2.99792458×10^9 statC in the Gaussian system, not that 1 C is equal to 2.99792458×10^9 statC. In other words, q_{SI} and q_{gau} are quantities with different dimensions, and C and statC also have different dimensions. However, $q_{\mathrm{SI}}/\mathrm{C}$ and $q_{\mathrm{gau}}/\mathrm{statC}$ are both pure numbers, and are related by

$$\left(\frac{q_{\mathrm{gau}}}{1\,\mathrm{statC}}\right) = 2.997\,924\,58 \times 10^9 \left(\frac{q_{\mathrm{SI}}}{1\,\mathrm{C}}\right), \tag{2.29}$$

which is the real meaning of eq. (2.28).

Note also that, in the derivation of the relation expressed by eq. (2.28), we have taken eq. (2.10) as an exact relation, as in the pre-2021 definition of charge in the SI system. We can adapt this to the new SI definition of the electric charge by stating that the the coulomb is now defined by eq. (2.4) and that the relation (2.29) remains exact. This amounts to saying that, in Gaussian units, the electron charge is $-e_{\text{gau}}$, where

$$e_{\text{gau}} \equiv 1.602\,176\,634 \times 10^{-19} \times 2.99792458 \times 10^9 \,\text{statC}\,, \qquad (2.30)$$

and this relation is exact. Numerically, this gives $e_{\text{gau}} \simeq 4.803\ldots \times 10^{-10}\,\text{statC}$. Then, the constant k in eq. (2.5) becomes a measurable quantity, whose current value is consistent with $k = 1$ at the level of 10 decimal figures.

Another important difference between Gaussian and SI units is in the definition of $\mathbf{E}$ and $\mathbf{B}$, which, in Gaussian units, are now defined writing the Lorentz force equation as

$$\mathbf{F} = q \left(\mathbf{E} + \frac{\mathbf{v}}{c} \times \mathbf{B} \right)\,, \qquad (2.31)$$

where, here, $q = q_{\text{gau}}$, $\mathbf{E} = \mathbf{E}_{\text{gau}}$ and $\mathbf{B} = \mathbf{B}_{\text{gau}}$, and c is the speed of light. Then, comparing eqs. (2.14) and (2.31), and using eq. (2.25), we see that the relation between the definitions of the electric and magnetic fields in the Gaussian system and in the SI system are

$$\mathbf{E}_{\text{gau}} = \sqrt{4\pi\epsilon_0}\, \mathbf{E}_{\text{SI}}\,, \qquad (2.32)$$

$$\mathbf{B}_{\text{gau}} = c\sqrt{4\pi\epsilon_0}\, \mathbf{B}_{\text{gau}} = \sqrt{\frac{4\pi}{\mu_0}}\, \mathbf{B}_{\text{SI}}\,. \qquad (2.33)$$

Notice that, dimensionally, $[\mathbf{B}_{\text{SI}}] = [\mathbf{E}_{\text{SI}}]/[v]$, where the brackets denotes the dimensions of a quantity, and v is a velocity. In contrast, $\mathbf{E}_{\text{gau}}$ and $\mathbf{B}_{\text{gau}}$ have the same dimensions. Observe that, taking into account eq. (2.25), $(q\mathbf{E})_{\text{SI}} = (q\mathbf{E})_{\text{gau}}$. In contrast eqs. (2.25) and (2.33) imply that $(q\mathbf{B})_{\text{SI}} = (q\mathbf{B})_{\text{gau}}/c$. Equation (2.25) also implies $\mathbf{j}_{\text{gau}} = \mathbf{j}_{\text{SI}}/\sqrt{4\pi\epsilon_0}$.

Performing these replacements in eqs. (2.15)–(2.18), we get Maxwell's equations in Gaussian units,

$$\nabla \cdot \mathbf{E} = 4\pi\rho\,, \qquad (2.34)$$

$$\nabla \times \mathbf{B} - \frac{1}{c}\frac{\partial \mathbf{E}}{\partial t} = \frac{4\pi}{c}\mathbf{j}\,, \qquad (2.35)$$

$$\nabla \cdot \mathbf{B} = 0\,, \qquad (2.36)$$

$$\nabla \times \mathbf{E} + \frac{1}{c}\frac{\partial \mathbf{B}}{\partial t} = 0\,, \qquad (2.37)$$

where we did not write explicitly the subscript "gau" on $\mathbf{E}$, $\mathbf{B}$, ρ, and $\mathbf{j}$. Observe that, formally, Maxwell's equations and the Lorentz force in the SI system can be transformed into the corresponding equations of the Gaussian system with the replacements

$$\epsilon_0 \to \frac{1}{4\pi}\,, \qquad \mu_0 \to \frac{4\pi}{c^2}\,, \qquad \mathbf{E} \to \mathbf{E} \qquad \mathbf{B} \to \frac{\mathbf{B}}{c}\,, \qquad (2.38)$$

and without changing ρ and $\mathbf{j}$. As we have seen, this does not corresponds to the actual conceptual relation between the two systems; electric and magnetic fields in the two systems have different dimensions, and their correct relation is given by eqs. (2.32) and (2.33), and similarly the correct relation between electric charges is given by eq. (2.25). However, the formal replacement (2.38) is a useful trick for passing quickly from equations in the SI system to the corresponding equations in Gaussian units.

The transition from classical to quantum electrodynamics is more naturally performed in Gaussian units, and Gaussian units are the only ones that are used nowadays in quantum electrodynamics and its generalization to the Standard Model of particle physics. More precisely, in quantum field theory it is customary to use a slight modification of the Gaussian system, called *rationalized* Gaussian units (or Heaviside–Lorentz units), which differs from (unrationalized) Gaussian units just by the placing of some 4π factors. Denoting by the label "rat" the quantities in rationalized Gaussian units and, as before, by "gau" the quantities in (unrationalized) Gaussian units, the relations are

$$q_{\text{rat}} \;=\; \sqrt{4\pi}\, q_{\text{gau}}\,, \tag{2.39}$$

$$\mathbf{E}_{\text{rat}} \;=\; \frac{1}{\sqrt{4\pi}}\, \mathbf{E}_{\text{gau}}\,, \qquad \mathbf{B}_{\text{rat}} = \frac{1}{\sqrt{4\pi}}\, \mathbf{B}_{\text{gau}}\,. \tag{2.40}$$

Thus, in rationalized Gaussian units the Coulomb force reads

$$\mathbf{F} = \frac{q_1 q_2}{4\pi r^2}\, \hat{\mathbf{r}}\,, \tag{2.41}$$

which is formally the same as eq. (2.6) with $\epsilon_0 = 1$. The Lorentz force still keeps the form (2.31), because the $\sqrt{4\pi}$ factor in q cancels those in $\mathbf{E}, \mathbf{B}$, while Maxwell's equations (2.34)–(2.37) become

$$\boldsymbol{\nabla}\cdot\mathbf{E} \;=\; \rho\,, \tag{2.42}$$

$$c\boldsymbol{\nabla}\times\mathbf{B} - \frac{\partial\mathbf{E}}{\partial t} \;=\; \mathbf{j}\,, \tag{2.43}$$

$$\boldsymbol{\nabla}\cdot\mathbf{B} \;=\; 0\,, \tag{2.44}$$

$$c\boldsymbol{\nabla}\times\mathbf{E} + \frac{\partial\mathbf{B}}{\partial t} \;=\; 0\,. \tag{2.45}$$

Similarly to eq. (2.38), eqs. (2.42)–(2.45) can be obtained from the Maxwell's equations in SI units with the formal replacements

$$\epsilon_0 \to 1\,, \qquad \mu_0 \to \frac{1}{c^2}\,, \qquad \mathbf{E} \to \mathbf{E} \qquad \mathbf{B} \to \frac{\mathbf{B}}{c}\,, \tag{2.46}$$

while leaving ρ and $\mathbf{j}$ unchanged.

The underlying reason why (rationalized) Gaussian units are the standard choice in the context of quantum field theory is that they provide a first step toward the definition of a system of units defined by setting $\hbar = c = 1$, which is very convenient in the context of quantum field theory. We have seen that the Gaussian system is characterized by

The final part of this section involves notions that go beyond the scope of a course of classical electrodynamics, and is not needed in the rest of the book. It can be skipped at first reading without loss of continuity.

setting $k = 1$ in eq. (2.5), i.e., $\epsilon_0 = 1/(4\pi)$ in eq. (2.6) (exactly, as in the pre-2019 definition of the electric charge unit, or within the current experimental accuracy, with the modern definition based on the electron charge). Similarly, the rationalized Gaussian system is characterized by the choice $\epsilon_0 = 1$. Here, the crucial point is not the specific numerical value chosen for ϵ_0, but rather the fact that ϵ_0 is declared "by law" to be a pure number, without dimensions. As a consequence, as we have seen, the unit of electric charge becomes a derived unit that can be expressed in terms of the units of mass, length, and time; then, the four independent base units of the SI system relevant for electrodynamics (m, kg, s, A) reduce to only three independent units in the Gaussian system, that can be taken as (cm, gr, s). One can push this logic further and, after having defined the unit of time from the frequency of an atomic transition (see Note 3 on page 28), one can define the unit of length in terms of the speed of light stating that, by definition, $c = 1$. Here, again, the crucial point is not so much the precise numerical value assigned to c, but rather the fact that we declare the speed of light to be a dimensionless quantity. Thus, in this system of units, we have a further reduction of base units, and the unit of length is now the same as the unit of time, while any velocity is a dimensionless number. Of course, it is easy to go back to standard units; e.g., a velocity $v = 0.9$ in units $c = 1$ corresponds to $v = 0.9 \times 2.99792458 \times 10^{10}$cm/s in c.g.s. units.

At this stage, we remain with only two base units, the unit of time and the unit of mass. One can now make a further step, and also set the reduced Planck constant $\hbar = 1$, and now time becomes, dimensionally, the inverse of a mass. We are then left with a single base unit, that can be taken as the unit of mass. All other quantities are related to it in a simple way. Denoting by $[M], [L], [T]$ the dimensions of mass, length, and time, respectively, we have $[L] = [T] = [M]^{-1}$, while velocities are pure numbers, and therefore also linear momentum and energy have dimensions of mass. In these units, even the electric charge becomes a pure number. This can be seen observing that (having set $\epsilon_0 = 1$), q^2/r has dimensions of an energy (it is the Coulomb energy of a system of two charges $q_1 = q_2 = q$); however, in units $\hbar = c = 1$, $1/r$ has dimensions of mass, exactly as the energy, so q must be dimensionless. In quantum electrodynamics, an important role is played by the *fine structure* constant α. In rationalized Gaussian units (while still using "normal" c.g.s. units), it is defined as $\alpha = e^2/(4\pi\hbar c)$ (or by $\alpha = e^2/(\hbar c)$ in unrationalized Gaussian units), and it can be seen, with standard dimensional analysis, that it is a pure number. Using the value of e given in eq. (2.30), one finds that its numerical value is $\alpha \simeq 1/137$. If we then set also $\hbar = c = 1$, we see that the electron charge becomes a pure number in agreement with the previous argument, and we find that its numerical value is given by $e^2/(4\pi) \simeq 1/137$.[8]

This system of units, defined by setting ϵ_0 to a pure number (whether exactly $\epsilon_0 = 1$, or consistent with it within the current accuracy, given the definition (2.30) of the electron charge) and also $\hbar = c = 1$, might look weird at first sight, but, in fact, it is so convenient in quantum

[8]More precisely, in a quantum field theory course one learns that the electron charge depends, logarithmically, on the energy scale at which it is probed (is a "running coupling constant," in quantum fields theory jargon). The value $\alpha \simeq 1/137$ is actually the value at low energies.

[9]One can go even further and, when quantum gravity enters the game, one can also set Newton's constant $G = 1$. At this point, all quantities become dimensionless. These are called *Planck units*.

mechanics and in quantum field theory that, in this context, these units are called *natural units*.[9] Of course, one has a significant loss in the possibility of spotting mistakes in the computations by making use of dimensional analysis, since there is a much smaller variety of basic units. However, this is a minor aspect, and it is largely compensated by the gain in simplicity and physical clarity of many formulas. This subject would bring us too far from the scope of this course, but the interested reader can find an extended discussion of $\hbar = c = 1$ units, and on practical ways of using them, in Chapter 1 of Maggiore (2005).

2.3   SI or Gaussian?

After having defined the two systems, one can now try to tackle the question of which system is "better." The answer, in fact, depends on the context.

The Gaussian system is more natural from the point of view of Special Relativity. This is partly due to the definition of the magnetic field, where a factor of c is reabsorbed in **B** compared to the SI system, so that **E** and **B** have the same dimensions. This is much more natural from the point of view of Special Relativity since, as we will see in Chapter 8, in a relativistic formalism **E** and **B** enter on the same footing as the six components of an antisymmetric tensor $F^{\mu\nu}$. As a result, many equations of electrodynamics, in particular in its relativistic formulation, look more elegant and more natural in Gaussian units. On the other hand, SI units are much more natural in all laboratory situations.

The situation is quite similar to the option of using units $\hbar = c = 1$. In quantum field theory, this is the only natural choice, by now universally used, and all equations of relativistic quantum field theory look much cleaner and elegant without being "cluttered" by factors of $\hbar$ and c. Measuring speeds in units of the speed of light is the only natural option in particle physics (compare the statement "a particle has speed $v = 0.99$," in units $c = 1$, with "a particle has speed $v = 2.96795\ldots \times 10^8 \mathrm{m/s}$"). However, outside the relativistic domain, measuring speeds in units $c = 1$ might be very weird (a speed limit at $v = 60$ km/h, in units $c = 1$ would become a limit at $v = 5.559\ldots \times 10^{-8}$). So, elegance of the fundamental equations is not the only criterion, and the attempt at defining a unique "best" choice, independent of the context, is futile.

In a sense, one might argue that the Gaussian system stopped in the middle of a commendable path. The same logic that suggests to fix $\epsilon_0 = 1/(4\pi)$ (or $\epsilon_0 = 1$) also suggests to fix $c = 1$ (and, in a quantum context, $\hbar = 1$). These choices all have the effect of making the fundamental equations of the theory more transparent, getting rid of constants whose numerical value, in the SI system, is ultimately related to the human experience (e.g., the value of the speed of light in the SI system depends on the definition of the meter which, as we discussed, was originally defined as one ten-millionth of the distance from the equator to the North Pole along a great circle) and have nothing to do with Nature at its most fundamental level.

So, one can see the Gaussian system (with c kept explicit) as an intermediate choice between two cases: one might decide of not using at all the logic of reducing the number of independent units through the fundamental constants, and keep ϵ_0 and c explicit, and define the units so that they are convenient in typical laboratory situations; this gives the SI system. Or else, one could decide to use this logic to the very end, setting not only $4\pi\epsilon_0 = 1$ (or $\epsilon_0 = 1$) but also $c = 1$ and, in a quantum context, $\hbar = 1$. This leads to more transparent formulas at a fundamental level, in particular in a relativistic context and in quantum field theory.

In conclusion, the best choice of units depends on the situation. In this book, we will use SI units because of the broader contexts in which they are used. Furthermore, it is trivial to pass from SI units to the rationalized Gaussian units with $c = 1$ used in quantum field theory: from eq. (2.46), we see that one can just make the formal replacements $c \to 1$, $\epsilon_0 \to 1$, and $\mu_0 \to 1$ in all SI equations so, in practice, one can just ignore them in all the SI equations. The inverse path, from (rationalized or unrationalized) Gaussian units with $c = 1$ to SI, requires instead some extra work, with dimensional analysis required to inserted the appropriate factors of c and ϵ_0 (or μ_0).

If one rather wants to translate equations from SI to unrationalized Gaussian units, furthermore keeping c explicit, one must take care of the placing of 4π factors and of the powers of c, and many relevant formulas are collected in App. A.

Maxwell's equations

We now enter in the heart of the subject, presenting Maxwell's equations and beginning to extract their consequences. We assume that the reader already has an elementary knowledge of the basic phenomenology of electrostatic and magnetostatic phenomena, from a first introductory course of electromagnetism, and we will proceed in a more formal and advanced manner. We will first present Maxwell's equations, both in the local and the integrated form. We will then show how they imply conservation laws for the electric charge and for energy and momentum and, in the process, we will identify the energy density and momentum density of the electromagnetic field. We will then introduce the gauge potentials. While, at first, their introduction might look just a trick for simplifying the equations, in fact we will gradually discover that the gauge potentials play a fundamental role at the conceptual level. Their introduction also brings in the notion of gauge invariance, of which electromagnetism is the prototype example, and which is fundamental to all modern theoretical physics. Contrary to most other textbook treatments, we will therefore introduce them at a very early stage of our presentation. Identifying the symmetries of the theory is another basic aspects of a modern approach, and we will then discuss some of the symmetries of Maxwell's equations (leaving, however, the discovery of Lorentz symmetry and Special Relativity for Chapter 8, after we will have developed the necessary tools). In this chapter, we will keep a rather formal approach. In Chapter 4 we will then discuss several examples and applications to electrostatics and magnetostatics, both for their intrinsic importance, and to illustrate the more general concepts developed here in simple settings, making contact with more elementary treatments of classical electromagnetism.

3.1 Maxwell's equations in vector form

3.1.1 Local form of Maxwell's equations

Maxwell's equations are formulated in terms of the electric field $\mathbf{E}(t, \mathbf{x})$ and the magnetic field $\mathbf{B}(t, \mathbf{x})$. The *field* concept is among the most fundamental of modern physics, in particular in connection with Special Relativity. In classical mechanics, the dynamical variables describing a mechanical system are "generalized coordinates" of the form $q_i(t)$, with i a discrete index corresponding to the degrees of freedom of the system; these could be, for instance, the three spatial components of the position of a particle, $\{q_x(t), q_y(t), q_z(t)\}$, or the $3N$ coordinates of a system of N particles. A field can be seen as a collection of dynamical variables labeled by a continuous spatial variable $\mathbf{x}$, rather than by a discrete index i, so at each point of space we have a dynamical variable (or a set of dynamical variables, such as the three components of the electric field and the three components of the magnetic field). As we will see in due course, fields are the natural language for describing the electromagnetic interactions in a way consistent with the principles of Special Relativity.

In classical electromagnetism, the dynamics of the electric and magnetic fields is governed by Maxwell's equations, that we have already presented in Chapter 2, but rewrite here. In SI units, they read

$$\boldsymbol{\nabla}\cdot\mathbf{E} \;=\; \frac{\rho}{\epsilon_0}\,, \qquad\qquad\qquad \text{(Gauss's law),} \qquad\qquad (3.1)$$

$$\boldsymbol{\nabla}\times\mathbf{B} \;=\; \mu_0\mathbf{j} + \mu_0\epsilon_0\frac{\partial\mathbf{E}}{\partial t}\,, \qquad \text{(Ampère–Maxwell law),} \quad (3.2)$$

$$\boldsymbol{\nabla}\cdot\mathbf{B} \;=\; 0\,, \qquad\qquad\qquad\qquad\qquad\qquad\qquad\qquad (3.3)$$

$$\boldsymbol{\nabla}\times\mathbf{E} \;=\; -\frac{\partial\mathbf{B}}{\partial t}\,, \qquad\qquad \text{(Faraday's law).} \qquad (3.4)$$

As shown in Section 2.1, $\rho(t,\mathbf{x})$ has dimensions of charge per unit volume, so is an electric charge density, while $\mathbf{j}(t,\mathbf{x})$ has the dimensions of a current per unit surface, and, in this sense, we will often refer to it as the current density. These equations allow us, in principle, to solve for $\mathbf{E}$ and $\mathbf{B}$ once assigned the "source terms" ρ and $\mathbf{j}$ (and the geometric setting and boundary conditions of the problems). The motion of a non-relativistic test charge,[1] with charge q and velocity $\mathbf{v}$, in these fields is determined by the Lorentz force

$$\mathbf{F} = q\left(\mathbf{E} + \mathbf{v}\times\mathbf{B}\right)\,, \qquad\qquad (3.5)$$

where $\mathbf{E}$ and $\mathbf{B}$ are computed at the position of the particle. For a non-relativistic point particle, we have Newton's law $\mathbf{F} = d\mathbf{p}/dt$, where $\mathbf{p}$ is the momentum of the particle, and therefore

$$\frac{d\mathbf{p}}{dt} = q\left(\mathbf{E} + \mathbf{v}\times\mathbf{B}\right)\,. \qquad\qquad (3.6)$$

In a relativistic context, force, with its instantaneous character, is no longer a fundamental concept, while momentum still is, and we will see later that eq. (3.6) remains valid.[2]

We will take ϵ_0 and μ_0 as two basic constants characterizing electromagnetic phenomena. We have already met them in Section 2.1, and we will indeed show later that eqs. (2.6) and (2.7) are a consequence of eqs. (3.1)–(3.4). At this stage, it is also convenient to *define* a constant c from

$$\epsilon_0\mu_0 = \frac{1}{c^2}\,. \qquad\qquad (3.7)$$

As we saw in Section 2.1, ϵ_0 has dimensions of $\mathrm{C}^2/(\mathrm{N\,m}^2)$ while μ_0 has dimensions of $\mathrm{N\,s}^2/\mathrm{C}^2$, so c has dimensions of a velocity. The use of the letter c obviously hints to the fact that this will turn out to be the

[1] A test charge is a charged particle which is taken to move in a given external electromagnetic field. Physically, this means that we can neglect the back-action of the charge on the system that generates the external field.

[2] With a common abuse of language, we will in general refer to eq. (3.6) as the "Lorentz force equation" even in a relativistic context. A more accurate wording could be "Lorentz equation of motion," but we will not tamper with such a standard nomenclature.

speed of light, although this will only emerge later, from the study of Maxwell's equations. For the moment, we presume that we do not know this yet, and we treat it just as another constant, and we use it to rewrite Maxwell's equations as

$$\boldsymbol{\nabla}\cdot\mathbf{E} = \frac{\rho}{\epsilon_0}, \tag{3.8}$$

$$\boldsymbol{\nabla}\times\mathbf{B} - \frac{1}{c^2}\frac{\partial\mathbf{E}}{\partial t} = \mu_0\mathbf{j}, \tag{3.9}$$

$$\boldsymbol{\nabla}\cdot\mathbf{B} = 0, \tag{3.10}$$

$$\boldsymbol{\nabla}\times\mathbf{E} + \frac{\partial\mathbf{B}}{\partial t} = 0, \tag{3.11}$$

where, if one wishes, any one of the three constants ϵ_0, μ_0, and c can be completely eliminated from the equations, using eq. (3.7). This form of Maxwell's equation is useful because, as we will see in Chapter 8, the structure of the terms on the left-hand sides, that involve only electric and magnetic fields, is dictated by Lorentz invariance, while the right-hand sides of eqs. (3.8) and (3.9) show how ϵ_0 and μ_0 determine the coupling to the charge density and to the current density, respectively.

Maxwell's equations have gradually emerged in the 19th century as a means of describing and unifying a vast body of observations of electric and magnetic phenomena and, for the moment, we will take them, together with the Lorentz force equation in the form (3.6), as the basic postulates that *define* the theory of classical electromagnetism. However, after having developed the appropriate tools, we will see how they emerge very naturally (and, to a large extent, uniquely) from the requirement of constructing a theory of electric and magnetic phenomena which respects the principles of Special Relativity.[3]

At first sight, the electric and magnetic fields might look just like mathematical constructions which are convenient as an intermediate step: the charge and current density determine $\mathbf{E}$ and $\mathbf{B}$ through Maxwell's equations, and $\mathbf{E}$ and $\mathbf{B}$ determine the motion of particles, through the Lorentz force equation. As we work our way through classical electrodynamics, we will see that, in fact, $\mathbf{E}$ and $\mathbf{B}$ (or, even more precisely, other fields, the gauge potentials, from which they can be derived, and that we will introduce shortly) are the truly fundamental dynamical variables for the description of electromagnetism. We will see that field configurations carry energy, momentum and angular momentum, just like particles, and can propagate as waves in vacuum and in materials. Eventually, the fundamental nature of these fields is most clearly revealed in the context of quantum field theory, where one discovers that, in fact, all particles are described in terms of fields, and the gauge potentials provide a description of a particle, the photon, which is the mediator of the electromagnetic interaction. Even if all these developments have to wait for later chapters (and the quantum aspects belong to the domain of quantum field theory and will not be covered in this book), it is useful to already have in mind the fundamental nature of the field concept.

[3]In this context, it is interesting to observe that Einstein's 1905 paper on Special Relativity had the title "Zur Elektrodynamik bewegter Körper" (*Annalen der Physik*, **17**:891, 1905), whose English translation is "On the electrodynamics of moving bodies." The link between Special Relativity and electromagnetism was there from the start! An online edition of the English translation of Einstein's 1905 paper can be found at https://www.fourmilab.ch/etexts/einstein/specrel/www/.

3.1.2   Integrated form of Maxwell's equations

Using Gauss's or Stokes's theorems, we can obtain useful integrated forms of these equations.[4] Integrating eq. (3.8) over a volume V bounded by a closed surface $S = \partial V$ and using Gauss theorem (1.48), we get

$$\oint_S d\mathbf{s} \cdot \mathbf{E}(t, \mathbf{x}) = \frac{1}{\epsilon_0} Q_V(t) \,, \tag{3.12}$$

where $Q_V(t)$ is the electric charge inside the volume V,

$$Q_V(t) = \int_V d^3x \, \rho(t, \mathbf{x}) \,. \tag{3.13}$$

We can rewrite eq. (3.12) as

$$\Phi_E(t) = \frac{1}{\epsilon_0} Q_V(t) \,, \tag{3.14}$$

where

$$\Phi_E(t) \equiv \oint_S d\mathbf{s} \cdot \mathbf{E}(t, \mathbf{x}) \tag{3.15}$$

is the flux of $\mathbf{E}$ through the closed surface $S = \partial V$. Note that, since the coordinates $\mathbf{x}$ are restricted to be on the fixed boundary S of V, and one integrates over this boundary, the integral on the left-hand side of eq. (3.15) is a function of the time coordinate only (and, implicitly, on the choice of volume V, and therefore on its boundary ∂V; however, to keep the notation lighter, we omit the label V in Φ_E). A similar integration of eq. (3.10) over the volume V gives

$$\oint_S d\mathbf{s} \cdot \mathbf{B}(t, \mathbf{x}) = 0 \,, \tag{3.16}$$

i.e.,

$$\Phi_B(t) = 0 \,, \tag{3.17}$$

where

$$\Phi_B(t) \equiv \oint_S d\mathbf{s} \cdot \mathbf{B}(t, \mathbf{x}) \tag{3.18}$$

is the flux of the magnetic field through the closed surface $S = \partial V$. The flux of the magnetic field through any closed surface vanishes, because of the absence of magnetic charges.

Integrating eq. (3.9) over a surface S with boundary $\mathcal{C} = \partial S$ and using Stokes's theorem (1.38), we get the integrated form of the Ampère–Maxwell law,

$$\oint_{\mathcal{C}} d\boldsymbol{\ell} \cdot \mathbf{B}(t, \mathbf{x}) = \mu_0 I(t) + \frac{1}{c^2} \frac{d\Phi_E(t)}{dt} \,, \tag{3.19}$$

where

$$I(t) = \int_S d\mathbf{s} \cdot \mathbf{j}(t, \mathbf{x}) \tag{3.20}$$

is the current passing through the surface S [as we already saw in eq. (2.22)] and Φ_E is the electric flux passing through S. Note that the integral on the left-hand side of eq. (3.19) is the circulation of $\mathbf{B}$ around the closed curve $\mathcal{C}$. Finally, integrating eq. (3.11) again over a closed surface S with boundary $\mathcal{C} = \partial S$ and using Stokes's theorem, we get the integrated form of Faraday's law,

$$\oint_{\mathcal{C}} d\boldsymbol{\ell} \cdot \mathbf{E}(t, \mathbf{x}) = -\frac{d\Phi_B(t)}{dt}, \tag{3.21}$$

where Φ_B is the magnetic flux passing through S.[5] The line integral of $\mathbf{E}$ along the curve $\mathcal{C}$ is called the *electromotive force*, or *emf*.[6] This result implies that, when the magnetic flux enclosed by a loop made by a conducting wire changes with time, a voltage is induced around the loop, and therefore a current appears in the wire, and in this form it was originally discovered by Faraday. The effect can be induced on a fixed loop by a time-dependent magnetic field but can also take place just moving the orientation of the loop with respect to a static magnetic field. We will discuss this in more detail in Section 4.3.

[5]Observe that, in eq. (3.17), Φ_B was the flux through a *closed* surface, while in eq. (3.21) it is the flux through a surface which is not closed, but rather has a boundary $\mathcal{C}$.

[6]The name is historical and not well chosen given that it is not a force, but rather the line integral of a force per unit charge, i.e., a potential; since $\mathbf{E}$ has dimensions of V/m, see eq. (2.19), the electromotive force is measured in volts.

3.2   Conservation laws

We next show that Maxwell's equations imply a conservation law for the electric charge, as well as a conservation law for energy and momentum. In the process, we will also be able to identify the energy and momentum carried by the electromagnetic field. In general, just as in classical mechanics, conservation laws are a consequence of the invariance of the system under some transformation. For instance, in classical mechanics energy conservation is a consequence of invariance under time translations and momentum conservation is a consequence of invariance under spatial translations. We will see in Section 8.7.3 how this relation between symmetries and conservation laws generalizes to classical field theory and, in particular, to the electromagnetic field. In this section we will rather see how these conservation laws emerge from simple manipulations of Maxwell's equations.

3.2.1   Conservation of the electric charge

A first immediate consequence of Maxwell's equations is a conservation law for the electric charge. Taking the time derivative of eq. (3.8) and combining it with the divergence of eq. (3.9) [and recalling that the divergence of a curl is zero, see eq. (1.18)], we get

$$\frac{\partial \rho}{\partial t} + \boldsymbol{\nabla} \cdot \mathbf{j} = 0. \tag{3.22}$$

This is a *continuity equation*. Since it is a mathematical consequence of eqs. (3.8) and (3.9), we see that, in Maxwell's equations, the charge density ρ and the current density $\mathbf{j}$ cannot be chosen arbitrarily but must respect eq. (3.22). The physical meaning of this equation can be understood better by integrating it over a finite volume V, with boundary $S = \partial V$, and using Gauss's theorem (1.48). This gives

$$\frac{d}{dt}\int_V d^3x\,\rho(t,\mathbf{x}) = -\int_{\partial V} d\mathbf{s}\cdot\mathbf{j}\,. \tag{3.23}$$

On the left-hand side we have the total electric charge Q_V inside the volume V, see eq. (3.13), so eq. (3.23) states that

$$\frac{dQ_V(t)}{dt} = -\int_{\partial V} d\mathbf{s}\cdot\mathbf{j}\,, \tag{3.24}$$

i.e., that the variation of the charge inside a volume V is due to the flux of electric current going through its boundary, i.e., to the charges escaping the volume or entering it. If no charge escapes or enters, Q_V is conserved. Inside the volume V, charges are neither created nor destroyed.[7]

We will often consider the charge and current density generated by an ensemble of point-like particles. The charge density of a point-like particle with charge q_a, on a trajectory $\mathbf{x}_a(t)$, is

$$\rho_a(t,\mathbf{x}) = q_a\delta^{(3)}[\mathbf{x} - \mathbf{x}_a(t)]\,, \tag{3.25}$$

and its contribution to the current density is obtained multiplying this by its velocity $\mathbf{v}_a(t)$,[8]

$$\begin{align} \mathbf{j}_a(t,\mathbf{x}) &= \rho_a(t,\mathbf{x})\mathbf{v}_a(t) \tag{3.26} \\ &= q_a\mathbf{v}_a(t)\delta^{(3)}[\mathbf{x}-\mathbf{x}_a(t)]\,. \tag{3.27} \end{align}$$

Then, for a collection of charges labeled by an index a,

$$\begin{align} \rho(t,\mathbf{x}) &= \sum_a \rho_a(t,\mathbf{x})\,, \tag{3.28} \\ \mathbf{j}(t,\mathbf{x}) &= \sum_a \mathbf{j}_a(t,\mathbf{x})\,. \tag{3.29} \end{align}$$

We can check that these expressions indeed satisfy the continuity equation (3.22), observing that

$$\begin{align} \frac{\partial\rho_a}{\partial t} &= q_a\frac{\partial}{\partial t}\delta^{(3)}[\mathbf{x}-\mathbf{x}_a(t)] \\ &= q_a\frac{dx_a^i(t)}{dt}\frac{\partial}{\partial x_a^i}\delta^{(3)}[\mathbf{x}-\mathbf{x}_a(t)] \\ &= -q_a v_a^i(t)\frac{\partial}{\partial x^i}\delta^{(3)}[\mathbf{x}-\mathbf{x}_a(t)] \\ &= -\frac{\partial}{\partial x^i}\left\{q_a v_a^i(t)\delta^{(3)}[\mathbf{x}-\mathbf{x}_a(t)]\right\} \\ &= -\boldsymbol{\nabla}\cdot\mathbf{j}_a\,. \tag{3.30} \end{align}$$

We will see in eqs. (13.35)–(13.37) how the expressions for the current density and the continuity equation generalize from point particles to fluids.

[7]As a historical note, the original Ampère law was simply $\boldsymbol{\nabla}\times\mathbf{B}=\mu_0\mathbf{j}$, to be contrasted with the full Ampère–Maxwell law (3.2). If we take its divergence, we find $\boldsymbol{\nabla}\cdot\mathbf{j}=0$, rather than the continuity equation (3.22). In 1862, Maxwell, with a heuristic mechanical reasoning, postulated the presence of the extra term $\mu_0\epsilon_0\partial\mathbf{E}/\partial t$ in eq. (3.2). Since the term $\epsilon_0\partial\mathbf{E}/\partial t$ formally adds up to $\mathbf{j}$, it was identified with a form of current, and was called the "displacement current." We now understand that this term rather belongs to the left-hand side of Maxwell's equations, as we have written in eq. (3.9), being also present in vacuum, where there are no charges and currents, and it is crucial to obtain current conservation in the form (3.22). However, the idea of current conservation was not available to Maxwell, since the electron was yet to be discovered, and he did not associate electric currents with charges in motion. See Zangwill (2013), Section 2.2.5, and references therein, for a historical discussion.

[8]To obtain the current density, consider a beam of charged particles, with charge density ρ and velocity $\mathbf{v}$, and take a surface of area dA transverse to $\mathbf{v}$. In a time dt, the charges that pass through dA have filled a volume $dV = dA\times v dt$. The electric charge that has gone through the surface is therefore $dQ = \rho dA v dt$. The current dI flowing through the surface dA is then given by $dI = dQ/dt = \rho v dA$ and the current density $\mathbf{j}$, i.e., the current per unit surface, has modulus $j = dI/dA = \rho v$. Since, as a vector, it has the same direction of $\mathbf{v}$, then $\mathbf{j} = \rho\mathbf{v}$.

3.2.2 Energy, momentum, and angular momentum of the electromagnetic field

Energy density and energy flux

We next show that, from Maxwell's equations, we can obtain conservation equations that allow us to associate energy and momentum to the electromagnetic field. To identify the expression for the energy of the electromagnetic field we take the scalar product of eq. (3.11) with $\mathbf{B}$, and of eq. (3.9) with $\mathbf{E}$. Subtracting them, we get

$$[\mathbf{B}\cdot(\nabla \times \mathbf{E}) - \mathbf{E}\cdot(\nabla \times \mathbf{B})] + \frac{1}{2}\frac{\partial}{\partial t}\left(\frac{1}{c^2}\mathbf{E}^2 + \mathbf{B}^2\right) = -\mu_0\mathbf{E}\cdot\mathbf{j}. \quad (3.31)$$

The term in brackets can be rewritten using[9]

$$\mathbf{B}\cdot(\nabla \times \mathbf{E}) - \mathbf{E}\cdot(\nabla \times \mathbf{B}) = \nabla\cdot(\mathbf{E} \times \mathbf{B}). \quad (3.32)$$

Therefore, dividing by μ_0, eq. (3.31) becomes

$$\frac{\partial}{\partial t}\frac{\left(\epsilon_0\mathbf{E}^2 + \mu_0^{-1}\mathbf{B}^2\right)}{2} = -\frac{1}{\mu_0}\nabla\cdot(\mathbf{E} \times \mathbf{B}) - \mathbf{E}\cdot\mathbf{j}, \quad (3.33)$$

where we used eq. (3.7). Let us define the *Poynting vector*[10]

$$\boxed{\mathbf{S} = \mu_0^{-1}\mathbf{E} \times \mathbf{B}.} \quad (3.34)$$

Integrating eq. (3.33) over a finite volume V with boundary ∂V and using Gauss's theorem, we get

$$\boxed{\frac{d}{dt}\int_V d^3x \frac{\left(\epsilon_0\mathbf{E}^2 + \mu_0^{-1}\mathbf{B}^2\right)}{2} + \int_V d^3x\,\mathbf{E}\cdot\mathbf{j} = -\int_{\partial V} d\mathbf{s}\cdot\mathbf{S},}$$

$$(3.35)$$

which is known as *Poynting's theorem*. In the absence of external charges, $\mathbf{j} = 0$, this equation is already in the form of a conservation equation: on the left-hand side we have the time derivative of a quantity with the dimensions of an energy, and on the right-hand side a flux coming out or in from the boundary of the volume. So, this already suggests that the energy density of the electromagnetic field is given by $\left(\epsilon_0\mathbf{E}^2 + \mu_0^{-1}\mathbf{B}^2\right)/2$ and the energy flux is given by the Poynting vector. However, if we set $\mathbf{j} = 0$, we could multiply both sides of eq. (3.35) by an arbitrary number α, so this argument at most tells us that the energy density is of the form $\alpha\left(\epsilon_0\mathbf{E}^2 + \mu_0^{-1}\mathbf{B}^2\right)/2$ and the energy flux is $\alpha\mathbf{S}$, for some α. To confirm this interpretation as an energy conservation law, and to fix the normalization factor, we must better understand the term $\mathbf{E}\cdot\mathbf{j}$. To this purpose, we consider a collection of non-relativistic point-like charges q_a, $a = 1,\ldots,N$ inside the volume V, with positions $\mathbf{x}_a(t)$ and velocities $\mathbf{v}_a(t)$. We could perform the computation for generic relativistic

[9]The explicit steps are as follows:

$$B_i\epsilon_{ijk}\partial_j E_k - E_i\epsilon_{ijk}\partial_j B_k$$
$$= \epsilon_{ijk}\left(B_i\partial_j E_k - E_i\partial_j B_k\right)$$
$$= -\epsilon_{ijk}\left(B_k\partial_j E_i + E_i\partial_j B_k\right)$$
$$= -\partial_j(\epsilon_{ijk}E_i B_k)$$
$$= +\partial_j(\epsilon_{ikj}E_i B_k)$$
$$= \nabla\cdot(\mathbf{E} \times \mathbf{B}),$$

where, in the second equality, we have used the antisymmetry of ϵ_{ijk} to write $\epsilon_{ijk}B_i\partial_j E_k = -\epsilon_{ijk}B_k\partial_j E_i$.

[10]Named after John Henry Poynting, who derived this conservation equation in 1884.

particles (see next) but, to understand that this is indeed an energy conservation equation, and to fix the normalization factor α of the energy density, it is sufficient to limit ourselves to non-relativistic particles. Using eq. (3.27)

$$\int_V d^3x\, \mathbf{E}(t,\mathbf{x})\cdot\mathbf{j}(t,\mathbf{x}) = \sum_a q_a \mathbf{v}_a(t)\cdot\mathbf{E}[t,\mathbf{x}_a(t)], \qquad (3.36)$$

where the sum over a runs over all charges inside V. We now use the Lorentz force equation (3.6) that, for a non-relativistic particle located in $\mathbf{x}_a(t)$, with velocity $\mathbf{v}_a(t)$, charge q_a, and mass m_a, becomes

$$m_a \frac{d\mathbf{v}_a}{dt} = q_a \left\{ \mathbf{E}[t,\mathbf{x}_a(t)] + \mathbf{v}_a(t) \times \mathbf{B}[t,\mathbf{x}_a(t)] \right\}. \qquad (3.37)$$

Multiplying both sides of this equation by $\mathbf{v}_a(t)$ and using $\mathbf{v}_a\cdot(\mathbf{v}_a\times\mathbf{B}) = 0$, we get

$$
\begin{aligned}
q_a \mathbf{v}_a(t)\cdot\mathbf{E}[t,\mathbf{x}_a(t)] &= m_a \mathbf{v}_a(t)\cdot\frac{d\mathbf{v}_a}{dt} \\
&= \frac{d}{dt}\left(\frac{1}{2}m_a \mathbf{v}_a^2 \right). \qquad (3.38)
\end{aligned}
$$

Thus,

$$
\begin{aligned}
\int_V d^3x\, \mathbf{E}(t,\mathbf{x})\cdot\mathbf{j}(t,\mathbf{x}) &= \frac{d}{dt}\sum_a \frac{1}{2}m_a \mathbf{v}_a^2 \\
&= \frac{d}{dt}\mathcal{E}_{\text{kin}}, \qquad (3.39)
\end{aligned}
$$

where $\mathcal{E}_{\text{kin}}$ is the total kinetic energy of the system. Therefore eq. (3.35) can be rewritten as

$$\frac{d}{dt}\int_V d^3x\, \frac{(\epsilon_0 \mathbf{E}^2 + \mu_0^{-1}\mathbf{B}^2)}{2} + \frac{d}{dt}\mathcal{E}_{\text{kin}} = -\int_{\partial V} d\mathbf{s}\cdot\mathbf{S}. \qquad (3.40)$$

This shows that, indeed, the energy of the electromagnetic field in a volume V is

$$\mathcal{E}_{\text{em}} = \int_V d^3x\, \frac{(\epsilon_0 \mathbf{E}^2 + \mu_0^{-1}\mathbf{B}^2)}{2}, \qquad (3.41)$$

so eq. (3.40) can be written as

$$\frac{d}{dt}(\mathcal{E}_{\text{em}} + \mathcal{E}_{\text{kin}}) = -\int_{\partial V} d\mathbf{s}\cdot\mathbf{S}. \qquad (3.42)$$

The energy density of the electromagnetic fields, that we denote by $u(t,\mathbf{x})$, can then be identified with the integrand in eq. (3.41),

$$u = \frac{1}{2}\left(\epsilon_0 \mathbf{E}^2 + \mu_0^{-1}\mathbf{B}^2 \right), \qquad (3.43)$$

(where, for notational simplicity, we did not write explicitly the arguments $(t, \mathbf{x})$ in u, $\mathbf{E}$ and $\mathbf{B}$). Equivalently, eliminating μ_0 with eq. (3.7),

$$u = \frac{\epsilon_0}{2} \left(\mathbf{E}^2 + c^2 \mathbf{B}^2 \right) . \tag{3.44}$$

The Poynting vector (3.34) gives, instead, the energy flux across a surface. In terms of u and $\mathbf{S}$, eq. (3.33) takes the form[11]

$$\boxed{\frac{\partial u}{\partial t} + \boldsymbol{\nabla} \cdot \mathbf{S} = -\mathbf{E} \cdot \mathbf{j} .} \tag{3.47}$$

When $\mathbf{j} = 0$, as in a region of space where there are no charged particles, this is a local conservation equation of the same form as eq. (3.22) for the electric charge. However, for generic $\mathbf{j}$, the right-hand side is non-vanishing. While the electric charge inside a volume can only change if charges flow in or out from the boundary, the energy of the electromagnetic field inside a volume can be exchanged with the mechanical energy of the particles, because of the work made by the electric field on the charged particles, so the integrated conservation law rather has the form (3.42), and the local conservation law has the form (3.47), with a term $-\mathbf{E} \cdot \mathbf{j}$, in general non-vanishing, on the right-hand side. Also note that, as a by-product of this derivation, eq. (3.39) shows that

$$\frac{dW_E}{dt} = \int_V d^3 x \, \mathbf{E}(t, \mathbf{x}) \cdot \mathbf{j}(t, \mathbf{x}) \tag{3.48}$$

is the rate at which the electric field performs work on a system of charges and currents (while the magnetic field performs no work, since $\mathbf{v}_a \cdot (\mathbf{v}_a \times \mathbf{B}) = 0$).

Our use of the non-relativistic limit for the point-like charges was fully sufficient to understand that eq. (3.35) is an equation for energy conservation, and to fix the overall coefficient in the energy density of the electromagnetic field. However, it is not difficult, and instructive, to also perform the computation for fully relativistic particles.[12] This can be done if we anticipate that the Lorentz force equation for a relativistic particle has the form (3.6), where, for a relativistic particle of mass m_a and velocity $\mathbf{v}_a$, the momentum is

$$\mathbf{p}_a = \frac{m_a \mathbf{v}_a}{\sqrt{1 - (v_a^2 / c^2)}} , \tag{3.49}$$

while the energy is

$$\mathcal{E}_a = c \sqrt{m_a^2 c^2 + p_a^2} . \tag{3.50}$$

(We will prove these results in Sections 7.4.3 and 8.6.1.) Inserting eq. (3.49) into eq. (3.50) we can also write

$$\mathcal{E}_a = \frac{m_a c^2}{\sqrt{1 - (v_a^2 / c^2)}} . \tag{3.51}$$

Equation (3.49) can be inverted to give

$$\mathbf{v}_a = \frac{c \, \mathbf{p}_a}{\sqrt{m_a^2 c^2 + p_a^2}} . \tag{3.52}$$

[11]Observe that the energy density and the energy flux are not uniquely fixed by the integrated conservation equation (3.35). Indeed, defining

$$u' = u - \boldsymbol{\nabla} \cdot \mathbf{v}, \tag{3.45}$$
$$\mathbf{S}' = \mathbf{S} + \partial_t \mathbf{v} + \boldsymbol{\nabla} \times \mathbf{w}, \tag{3.46}$$

where u and $\mathbf{S}$ are still given by eqs. (3.34) and (3.43), and $\mathbf{v}$ and $\mathbf{w}$ are arbitrary vector fields, the local conservation equation (3.47) is still satisfied. The vector field $\mathbf{v}$ reshuffles the relative contributions of the integral over the volume V and that over the boundary ∂V, while $\mathbf{w}$ adds a term to the energy flux whose surface integral vanishes, since its divergence vanishes (and vice versa, on a topologically trivial space any vector field with vanishing divergence can be written as $\boldsymbol{\nabla} \times \mathbf{w}$, see the theorem for divergence-free fields on page 7). The expressions for the energy density and the energy flux obtained from eqs. (3.34) and (3.43), however, also emerge naturally in a relativistic formalism, as we will see in Section 8.4, where they become part of an energy-momentum tensor [although, in general, some freedom also exists in the definition of the energy-momentum tensor of a field, see Sections 3.2.1 and 3.5.3 of Maggiore (2007)]. Eventually, at the theoretical level, the best argument for the uniqueness of the expressions (3.34) and (3.43) comes from General Relativity, where the expression of the energy-momentum tensor is uniquely defined. So, for instance, the energy density u in eq. (3.43) is the quantity that couples to the gravitational field in the way required for a local energy density.

[12]We follow the derivation in Garg (2012).

Then, multiplying the Lorentz equation (3.6) by $\mathbf{v}_a$, we get

$$
\begin{aligned}
q_a \mathbf{v}_a \cdot \mathbf{E}[t, \mathbf{x}_a(t)] &= \mathbf{v}_a \cdot \frac{d\mathbf{p}_a}{dt} \\
&= \frac{c\,\mathbf{p}_a}{\sqrt{m_a^2 c^2 + p_a^2}} \cdot \frac{d\mathbf{p}_a}{dt} \\
&= \frac{c}{2\sqrt{m_a^2 c^2 + p_a^2}} \frac{dp_a^2}{dt} \, .
\end{aligned}
\tag{3.53}
$$

On the other hand, from eq. (3.50), the term in the last equality is just the same as $d\mathcal{E}_a/dt$. Thus,

$$
\begin{aligned}
\int_V d^3x\, \mathbf{E}(t, \mathbf{x}) \cdot \mathbf{j}(t, \mathbf{x}) &= \sum_a q_a \mathbf{v}_a \cdot \mathbf{E}[t, \mathbf{x}_a(t)] \\
&= \frac{d\mathcal{E}_{\text{kin.rel.}}}{dt} \, ,
\end{aligned}
\tag{3.54}
$$

where $\mathcal{E}_{\text{kin.rel.}} = \sum_a \mathcal{E}_a$ is the total relativistic kinetic energy of the particles, and eq. (3.40) holds in the form

$$
\frac{d}{dt} \int_V d^3x\, \frac{(\epsilon_0 \mathbf{E}^2 + \mu_0^{-1} \mathbf{B}^2)}{2} + \frac{d\mathcal{E}_{\text{kin.rel.}}}{dt} = -\int_{\partial V} d\mathbf{s} \cdot \mathbf{S} \, .
\tag{3.55}
$$

Momentum and momentum flux

Beside energy, the electromagnetic field also carries momentum. To identify its expression, we can proceed similarly to what we have done for energy conservation, using Maxwell's equations to obtain a conservation law in which the part depending on the sources can be identified with the time derivative of their mechanical momentum (just as, in eq. (3.40), we found a conservation equation in which a term was the time derivative of the mechanical energy of the sources). To this purpose, we consider the quantity[13]

$$
\mathbf{g}(t, \mathbf{x}) = \epsilon_0\, \mathbf{E}(t, \mathbf{x}) \times \mathbf{B}(t, \mathbf{x}) \, ,
\tag{3.56}
$$

which is related to the Poynting vector (3.34) by

$$
\mathbf{g}(t, \mathbf{x}) = \frac{1}{c^2} \mathbf{S}(t, \mathbf{x}) \, .
\tag{3.57}
$$

To prove that eq. (3.56) indeed represents the momentum of the electromagnetic field, we take its time derivative,

$$
\frac{\partial g_i}{\partial t} = \epsilon_0\, \epsilon_{ijk} \left(\frac{\partial E_j}{\partial t} B_k + E_j \frac{\partial B_k}{\partial t} \right) \, .
\tag{3.58}
$$

Using Maxwell's equations (3.9, 3.11) to compute $\partial E_j/\partial t$ and $\partial B_k/\partial t$, we get

$$
\frac{\partial g_i}{\partial t} = \epsilon_0 E_j(\partial_j E_i - \partial_i E_j) + \epsilon_0 c^2 B_j(\partial_j B_i - \partial_i B_j) - (\mathbf{j} \times \mathbf{B})_i \, ,
\tag{3.59}
$$

[13] A quantity proportional to $\mathbf{E} \times \mathbf{B}$ is a natural candidate for the momentum density of the electromagnetic field, since momentum density is a spatial vector, and $\mathbf{E} \times \mathbf{B}$ is the only vector that we can form with $\mathbf{E}$ and $\mathbf{B}$. Furthermore, we will see in Section 3.4 that $\mathbf{B}$ is a pseudovector under parity, and then $\mathbf{E} \times \mathbf{B}$ is a true vector under parity, just as the momentum density.

where, as usual, $\partial_j = \partial/\partial x^j$. We now use the identity

$$
\begin{aligned}
E_j(\partial_j E_i - \partial_i E_j) &= -\partial_j\left(\frac{1}{2}E^2\delta_{ij} - E_i E_j\right) - E_i \boldsymbol{\nabla}\!\cdot\!\mathbf{E} \\
&= -\partial_j\left(\frac{1}{2}E^2\delta_{ij} - E_i E_j\right) - \frac{1}{\epsilon_0}\rho E_i \,, \quad (3.60)
\end{aligned}
$$

where, in the second line, we used Gauss's law (3.8). Similarly,

$$
B_j(\partial_j B_i - \partial_i B_j) = -\partial_j\left(\frac{1}{2}B^2\delta_{ij} - B_i B_j\right)\,, \qquad (3.61)
$$

since, in this case $\boldsymbol{\nabla}\!\cdot\!\mathbf{B} = 0$. Then, eq. (3.58) becomes

$$
\boxed{\frac{\partial g_i}{\partial t} = -\partial_j T_{ij} - (\rho\mathbf{E} + \mathbf{j}\times\mathbf{B})_i \,,} \qquad (3.62)
$$

where

$$
T_{ij} = \epsilon_0\left[\left(\frac{1}{2}E^2\delta_{ij} - E_i E_j\right) + c^2\left(\frac{1}{2}B^2\delta_{ij} - B_i B_j\right)\right]\,, \qquad (3.63)
$$

or, equivalently,

$$
\boxed{T_{ij} = \epsilon_0\left(\frac{1}{2}E^2\delta_{ij} - E_i E_j\right) + \mu_0^{-1}\left(\frac{1}{2}B^2\delta_{ij} - B_i B_j\right)\,.}
$$

$$(3.64)$$

This tensor (or, depending on conventions, its sign-reversed), is called the *Maxwell stress tensor*.[14]

Integrating eq. (3.62) over a volume V that includes all charges and currents, and using eq. (3.66), we get

$$
\left(\frac{d\mathbf{P}_{\text{em}}}{dt}\right)_i + \int_V d^3x\,(\rho\mathbf{E} + \mathbf{j}\times\mathbf{B})_i = -\int_{\partial V} ds\,\hat{n}_j T_{ij}\,, \qquad (3.65)
$$

where we have defined

$$
\begin{aligned}
\mathbf{P}_{\text{em}}(t) &= \int_V d^3x\,\mathbf{g}(t,\mathbf{x}) \qquad\qquad\qquad\qquad (3.66) \\
&= \epsilon_0\int_V d^3x\,(\mathbf{E}\times\mathbf{B})(t,\mathbf{x})\,. \qquad\quad (3.67)
\end{aligned}
$$

We now observe that the Lorentz force exerted on an infinitesimal volume d^3x, which contains a charge distribution $\rho(t,\mathbf{x})$ with a velocity field $\mathbf{v}(t,\mathbf{x})$, is obtained writing $dq = \rho d^3x$ in eq. (3.6), and $\rho(t,\mathbf{x})\mathbf{v}(t,\mathbf{x}) = \mathbf{j}(t,\mathbf{x})$ is the current density at the point $\mathbf{x}$ and time t, see Note 8 on page 44. Therefore, the Lorentz force equation on an extended distribution of charges and currents can be written as

$$
\boxed{\frac{d\mathbf{P}_{\text{mech}}}{dt} = \int_V d^3x\,(\rho\mathbf{E} + \mathbf{j}\times\mathbf{B})\,,} \qquad (3.68)
$$

[14]There are different conventions on the sign used in the definition of the Maxwell stress tensor. For instance, Landau and Lifschits (1975) and Garg (2012) define the Maxwell stress tensor as the expression that appears in eq. (3.64), as we do, while in Jackson (1998) and Griffiths (2017) it is defined as minus the tensor given in eq. (3.64). An advantage of using the tensor T^{ij} defined in eq. (3.63) or eq. (3.64) rather than its negative (independently of which it is called the "Maxwell stress tensor") is that, from eq. (3.62), T_{ij} is the momentum flux, rather than its negative, so momentum conservation takes the same form as energy conservation, eq. (3.40). A related advantage emerges in the context of the relativistic formulation of electrodynamics: as we will see in Section 8.4, the energy density of the electromagnetic field will turn out to be the (00) component of a Lorentz tensor $T^{\mu\nu}$, and T^{ij}, as defined from eq. (3.63), will turn out to be equal to the $\mu = i, \nu = j$ component of $T^{\mu\nu}$, rather than its negative.

where $\mathbf{P}_{\mathrm{mech}}$ is the mechanical momentum of the extended source distribution. Therefore, eq. (3.65) can be written as

$$\frac{d}{dt}\left(\mathbf{P}_{\mathrm{em}}+\mathbf{P}_{\mathrm{mech}}\right)_i = -\int_{\partial V} ds\,\hat{n}_j T_{ij}\,. \qquad (3.69)$$

This has the required form of a conservation equation, and shows that $\mathbf{P}_{\mathrm{em}}$, defined in eq. (3.67), is indeed the momentum of the electromagnetic field (so $\mathbf{g}(t,\mathbf{x})$ is the momentum density), while $\hat{n}_j T_{ij}$ is the flow of the i-th component of the momentum of the electromagnetic field through a surface normal to $\hat{\mathbf{n}}$.

We will confirm this result in Section 8.4 with a covariant computation (which will first require the development of the covariant formulation of electrodynamics),[15] and in Section 8.7.3, with a computation that will make use of the machinery of classical field theory and Noether's theorem, which is more involved, but will make it clear that the conservation law (3.69) is related to the invariance under spatial translations, confirming the interpretation of $\mathbf{P}_{\mathrm{em}}$ as the momentum of the electromagnetic field.

Equation (3.69) is valid in full generality, for a relativistic source. In the non-relativistic limit, $d\mathbf{P}_{\mathrm{mech}}/dt$ becomes the same as the mechanical force $\mathbf{F}$ exerted on a system of charges and currents, localized in a volume V. Therefore, eq. (3.69) can be written as

$$\begin{aligned} F_i &= -\frac{1}{c^2}\int_V d^3x\,\frac{\partial S_i}{\partial t} - \int_{\partial V} ds\,\hat{n}_j T_{ij} && (3.70)\\ &= -\int_V d^3x\left[\frac{1}{c^2}\frac{\partial S_i}{\partial t} + \partial_j T_{ij}\right], && (3.71) \end{aligned}$$

where, in the second line, we have "undone" Gauss's theorem to write back the surface integral as a volume integral. Therefore, the electromagnetic field exerts a force per unit volume $f_i \equiv dF_i/dV$, given by

$$f_i = -\frac{1}{c^2}\frac{\partial S_i}{\partial t} - \partial_j T_{ij}\,. \qquad (3.72)$$

Whenever $\partial S_i/\partial t = 0$, the force in eq. (3.71) can be written as a surface integral. This happens in particular in electrostatics, where $\mathbf{E} = 0$, or in magnetostatics, where $\mathbf{B} = 0$, so in both cases, $\mathbf{S} = 0$. We will discuss some applications of these results in Section 4.1.7.

The previous result shows that the momentum density of the electromagnetic field, $\mathbf{g}$, is related to the energy flux, which is given by the Poynting vector $\mathbf{S}$, by eq. (3.57). It is interesting to understand this relation in the following way. Consider a beam of relativistic particles propagating with velocity v along a given direction, all with energy $\mathcal{E}$ and momentum p. According to eqs. (3.49) and (3.51),

$$p = \frac{v\mathcal{E}}{c^2}\,. \qquad (3.73)$$

Let $dN = ndAdt$ be the number of particles of the beam that cross a transverse area dA in a time dt. Since each particle carries an energy $\mathcal{E}$,

[15] The expression "covariant," referred to an equation, means that the left- and right-hand sides of the equation transform in the same way under the given transformation. This is a generalization of the notion of invariance, in which the left- and right-hand sides do not change. For instance, an equation such as $\mathbf{F} = m\mathbf{a}$ is covariant under rotations, since both the left- and right-hand sides transform as vectors under rotations. Therefore, if the equation holds in a reference frame, it also holds in a rotated frame. When we use the word "covariant" without further specification, we will refer to the covariance under the transformation of the Lorentz group (spatial rotations and boosts), that we will introduce in Chapter 7.

the corresponding energy flux (energy per unit area and unit time) is $n\mathcal{E}$ while, since each particle carries a momentum p, the total momentum carried by these particles is $pdN = pndAdt$. On the other hand, given that their velocity is v, in a time dt they have filled a volume $dV = dA \times vdt$ so their momentum density is

$$\frac{pndAdt}{dAvdt} = \frac{pn}{v}. \tag{3.74}$$

Inserting p from eq. (3.73), we therefore find that the momentum density is $n\mathcal{E}/c^2$, so it is equal to $1/c^2$ times the energy flux. So, eq. (3.57) is the relation that should be expected for a collection of particles. This may come as a surprise. In the classical treatment of electromagnetism, there is no notion of a particle associated with the momentum and the energy flux of the electromagnetic field. This is actually a hint of the fact that, at the quantum level, a notion of particle, the photon, will emerge.

Angular momentum

One can similarly show that the electromagnetic field carries an angular momentum

$$\mathbf{J}_{\mathrm{em}} = \int d^3x\, \mathbf{x} \times \mathbf{g}(t, \mathbf{x}), \tag{3.75}$$

where $\mathbf{g}(t, \mathbf{x})$ is the momentum density, i.e., $\mathbf{x} \times \mathbf{g}$ is the angular momentum density of the electromagnetic field.[16] Using the explicit expression (3.56),

$$\mathbf{J}_{\mathrm{em}} = \epsilon_0 \int d^3x\, \mathbf{x} \times (\mathbf{E} \times \mathbf{B}). \tag{3.76}$$

Taking the time derivative and performing manipulations analogous to those performed previously for momentum conservation, gives angular momentum conservation in the form

$$\frac{d}{dt}\left(\mathbf{J}_{\mathrm{em}} + \mathbf{J}_{\mathrm{mech}}\right)_i = -\int_{\partial V} d^2s\, \hat{n}_j M_{ij}, \tag{3.77}$$

where the flux of angular momentum is

$$M_{ij} = \epsilon_{ikl} x_k T_{lj}, \tag{3.78}$$

(where T_{ij} is the Maxwell's stree tensor), and

$$\frac{d\mathbf{J}_{\mathrm{mech}}}{dt} = \int_V d^3x\, \mathbf{x} \times (\rho\mathbf{E} + \mathbf{j} \times \mathbf{B}). \tag{3.79}$$

We will rederive this result explicitly in Section 8.7.3, using the formalism of Noether's theorem. The fact that the electromagnetic field carries energy, momentum, and angular momentum, just as a mechanical system, shows that, already at the classical level, $\mathbf{E}$ and $\mathbf{B}$ must be considered as real physical entities (in the same sense in which we think of particles as real physical entities), and are not just useful mathematical constructions.

[16]In analogy with the notation used in quantum mechanics, for the angular momentum of the electromagnetic field we prefer to use the notation $\mathbf{J}_{\mathrm{em}}$, rather than $\mathbf{L}_{\mathrm{em}}$. Indeed, as we will show in Section 8.7.3 using the formalism of classical field theory and Noether's theorem, the expression given in eq. (3.75) can be rewritten as the sum of two terms that, at quantum level, correspond to the orbital angular momentum and to the spin part. We will then denote the mechanical angular momentum of the particles by $\mathbf{J}_{\mathrm{mech}}$.

3.3   Gauge potentials and gauge invariance

We will now rewrite Maxwell's equations introducing a scalar potential ϕ and a vector potential $\mathbf{A}$, that we will call collectively the "gauge potentials." At the level of classical electromagnetism, this might look at first just like a useful mathematical trick for rewriting the equations in a simpler form. However, gauge potentials are much more fundamental. One realizes this, already at the level of classical electrodynamics, when expressing the theory in a Lagrangian formalism, as we will do in Section 8.7.2. There, one discovers that they play the role that "generalized coordinates" have in the description of classical mechanical systems. The Lagrangian formalism is the starting point for the quantization of a field theory, so the gauge potentials also have the role of the basic fields in terms of which the quantization procedure is carried out. Classical electrodynamics is also the simplest example of a *gauge theory*, i.e., a theory built with gauge fields, of the type that we will introduce below, and with an invariance under "gauge transformations," that again will be introduced next. The generalization of these concepts plays a crucial role in modern physics, particularly in particle physics and in condensed matter. Even if the extension to more general gauge theories, as well as all aspects related to quantization, go beyond the scope of this book, it is useful to be aware of them to already have a correct perspective.[17]

[17] See Maggiore (2005) for an introduction to quantum field theory and quantum electrodynamics with a conceptual unity with this book.

At the mathematical level Maxwell's equations, in the form (3.8)–(3.11), are two vector equations (three components each) and two scalar equations, for a total of eight equations, for six fields E_i and B_i. It is therefore clear that there must be some degree of interdependence among them, otherwise in general they would not admit solutions. The introduction of the gauge potential will show, first of all, how to reduce them to four equations for the four fields $(\phi, \mathbf{A})$. Furthermore, we will see that, in terms of these variables, there is an extra symmetry (gauge symmetry), that allows us to further reduce the number of independent fields and equations.

Let us begin with the Maxwell equation (3.10), $\boldsymbol{\nabla}\!\cdot\!\mathbf{B} = 0$. Because of the theorem for divergence-free fields given on page 7 (and taking into account that we work in $\mathbb{R}^3$, so that every surface in V can be continuously shrunk to a point), there exists a vector field $\mathbf{A}(t, \mathbf{x})$ such that

$$\boxed{\mathbf{B} = \boldsymbol{\nabla} \times \mathbf{A}\,.}\tag{3.80}$$

The field $\mathbf{A}$ is called the *vector potential*. Consider next Faraday's law (3.11). In terms of $\mathbf{A}$, it can be rewritten as

$$\boldsymbol{\nabla} \times \left(\mathbf{E} + \frac{\partial \mathbf{A}}{\partial t} \right) = 0\,.\tag{3.81}$$

Using now the theorem for curl-free fields, see again page 7, we conclude that there exists a function $\phi(t, \mathbf{x})$ such that

$$\mathbf{E} + \frac{\partial \mathbf{A}}{\partial t} = -\boldsymbol{\nabla}\phi\,,\tag{3.82}$$

(where the minus sign is a convention in the definition of ϕ), and therefore

$$\mathbf{E} = -\boldsymbol{\nabla}\phi - \frac{\partial \mathbf{A}}{\partial t} \, . \tag{3.83}$$

Thus, in terms of the scalar and vector potentials, the two Maxwell's equations that do not depend on the sources, eqs. (3.10) and (3.11), are automatically satisfied, because of the identities $\boldsymbol{\nabla} \cdot (\boldsymbol{\nabla} \times \mathbf{A}) = 0$ and $\boldsymbol{\nabla} \times (\boldsymbol{\nabla}\phi) = 0$. We can now insert eqs. (3.80) and (3.83) into the two remaining Maxwell's equations, eqs. (3.8) and (3.9). This gives

$$\nabla^2\phi + \frac{\partial}{\partial t}(\boldsymbol{\nabla} \cdot \mathbf{A}) = -\frac{\rho}{\epsilon_0} \, , \tag{3.84}$$

and

$$\nabla^2\mathbf{A} - \frac{1}{c^2}\frac{\partial^2 \mathbf{A}}{\partial t^2} - \boldsymbol{\nabla}\left(\boldsymbol{\nabla} \cdot \mathbf{A} + \frac{1}{c^2}\frac{\partial\phi}{\partial t}\right) = -\mu_0\mathbf{j} \, , \tag{3.85}$$

where we have used eq. (1.21). Thus, Maxwell's equations are completely equivalent to eqs. (3.84) and (3.85), i.e., to four equations for the four fields $(\phi, \mathbf{A})$. The solutions for $\mathbf{E}$ and $\mathbf{B}$ can then be obtained using the definitions (3.80) and (3.83). The electromagnetic field is therefore completely specified by four functions, the scalar field ϕ and the three components of the vector field $\mathbf{A}$, rather than by six quantities (the three components of $\mathbf{E}$ and the three components of $\mathbf{B}$).

In fact, even this description in terms of gauge potentials is redundant. Indeed, consider the simultaneous transformation of ϕ and $\mathbf{A}$ given by

$$\begin{aligned} \mathbf{A} \to \mathbf{A}' &= \mathbf{A} - \boldsymbol{\nabla}\theta \, , \\ \phi \to \phi' &= \phi + \frac{\partial\theta}{\partial t} \, , \end{aligned} \tag{3.86}$$

where θ is an arbitrary function.[18] Since $\boldsymbol{\nabla} \times (\boldsymbol{\nabla}\theta) = 0$, this transformation does not affect $\mathbf{B}$. Similarly, the transformation of ϕ is chosen so as to cancel the transformation of $\mathbf{A}$ in eq. (3.83), so $\mathbf{E}$ is also unchanged. The physical, observable, quantities are the electric and magnetic fields. The potentials $(\phi, \mathbf{A})$ and $(\phi', \mathbf{A}')$ are therefore physically equivalent, since they describe the same electric and magnetic fields. The transformation (3.86) is called a *gauge transformations*. The fields $\mathbf{E}$ and $\mathbf{B}$ do not change under this transformation and are therefore examples of *gauge-invariant* quantities. Maxwell's equations (3.8)–(3.11) are obviously gauge invariant, since they depend only on $\mathbf{E}$ and $\mathbf{B}$.

Since θ can be chosen arbitrarily without changing the physics, we can choose it so that eqs. (3.84) and (3.85), written in terms of the new

[18]We always assume that functions such as $\theta(t, \mathbf{x})$ are continuous and infinitely differentiable. In particular, subsequent derivatives applied to ϕ, $\mathbf{A}$, or θ commute, e.g., $\partial_i\partial_j\theta = \partial_j\partial_i\theta$. We will not repeat these conditions further in the following.

gauge potentials $(\phi', \mathbf{A}')$, are simpler. A convenient choice is obtained observing that, from eq. (3.86)

$$\boldsymbol{\nabla} \cdot \mathbf{A}' + \frac{1}{c^2} \frac{\partial \phi'}{\partial t} = \left(\boldsymbol{\nabla} \cdot \mathbf{A} + \frac{1}{c^2} \frac{\partial \phi}{\partial t} \right) - \Box \theta \,, \qquad (3.87)$$

where

$$\Box = -\frac{1}{c^2} \frac{\partial^2}{\partial t^2} + \boldsymbol{\nabla}^2 \qquad (3.88)$$

is called the *d'Alembertian* operator (or, colloquially, the "Box" operator). The d'Alembertian is invertible (we will perform its inversion explicitly in Section 10.1, using the method of Green's functions). Therefore, an equation of the form $\Box \theta = f$ always admits solutions [with suitable boundary conditions at spatial infinity for $f(t, \mathbf{x})$] so, for any given value of the initial gauge potentials ϕ and $\mathbf{A}$, we can choose θ such that the left-hand side of eq. (3.87) vanishes. Omitting hereafter, for notational simplicity, the prime on the transformed gauge fields, we have reached the gauge

$$\boxed{\boldsymbol{\nabla} \cdot \mathbf{A} + \frac{1}{c^2} \frac{\partial \phi}{\partial t} = 0 \,.} \qquad (3.89)$$

[19]This gauge was first introduced by L. V. Lorenz in 1867. It is often misspelled as "Lorentz gauge" (with an extra "t"), after H. A. Lorentz (the person after whom the Lorentz transformations are named; in 1867, however, he was just 14 years old...). This "misprint" has only been widely recognized in relatively recent times, thanks to Jackson and Okun (2001).

This is called the *Lorenz* gauge.[19] In this gauge, eqs. (3.84) and (3.85) become

$$\boxed{\Box \phi = -\frac{\rho}{\epsilon_0} \,,} \qquad (3.90)$$

and

$$\boxed{\Box \mathbf{A} = -\mu_0 \, \mathbf{j} \,,} \qquad (3.91)$$

and therefore have the form of wave equations, that we will study in detail in Chapter 9.

Another convenient gauge choice is

$$\boxed{\boldsymbol{\nabla} \cdot \mathbf{A} = 0 \,,} \qquad (3.92)$$

which can always be reached because the Laplacian is invertible (again, assuming suitable boundary conditions at spatial infinity). This is called the *Coulomb gauge*. In this gauge, eqs. (3.84) and (3.85) become

$$\boxed{\nabla^2 \phi = -\frac{\rho}{\epsilon_0} \,,} \qquad (3.93)$$

and

$$\boxed{\Box \mathbf{A} = -\mu_0 \mathbf{j} + \frac{1}{c^2} \boldsymbol{\nabla} \frac{\partial \phi}{\partial t} \,.} \qquad (3.94)$$

As we will discuss in Section 4.1.1, for static sources eq. (3.93) is just *Poisson's equation*, the basic equation of electrostatics and, for a point-like charge, it gives rise to the Coulomb potential (this is the origin of the name of this choice of gauge). For a generic charge distribution, we will see in Section 4.1.2 how eq. (3.93) can be solved for ϕ by inverting the Laplacian. The solution for ϕ can then be inserted in eq. (3.94) to solve for $\mathbf{A}$. Note, however, that in this gauge the equations for ϕ and for $\mathbf{A}$ are very different. The equation for ϕ involves a Laplacian, and the solution vanishes if the source term $\rho = 0$ (again, with vanishing boundary conditions at infinity). In contrast, the equation for $\mathbf{A}$ involves the d'Alembertian. As we will see in Chapter 9, even when the source term vanishes, $\mathbf{j} = 0$, and $\phi = 0$, this equation has non-vanishing solutions in the form of plane waves.

3.4 Symmetries of Maxwell's equations

We will now examine some of the symmetries of Maxwell's equations. First of all, Maxwell's equations are obviously invariant under spatial or temporal translations. There is no preferred origin of time or of space in the equations, and time and space derivatives are invariant under translations, e.g., $\partial/[\partial(t + t_0)] = \partial/\partial t$, for any constant t_0. As we know from classical mechanics, in a mechanical system invariance under time translations implies energy conservation, and invariance under spatial translations implies momentum conservation. When we will develop a field theoretical approach in Section 8.7, we will see how this translates in the conservation equations for energy and momentum that we have found in Section 3.2.

Maxwell's equations are also clearly invariant under spatial rotations: in eqs. (3.8) and (3.10) both the left- and right-hand sides are scalars under rotations, so if the equations hold in a reference frame, they also hold in a rotated frame. Similarly, in eqs. (3.9) and (3.11) both the left- and right-hand sides are vectors under rotations so, again, they transform in the same manner under rotations and, if they hold in a reference frame, they hold in a rotated frame. Just as in classical mechanics, this will translate into the conservation of angular momentum.

Another important symmetry of Maxwell's equations is parity, which is related to reflection of the axes. We see by inspection that, if we transform the spatial coordinates as $\mathbf{x} \to -\mathbf{x}$, while at the same time we transform the fields and the sources as

$$\mathbf{E}(t, \mathbf{x}) \to -\mathbf{E}(t, -\mathbf{x}), \qquad \mathbf{B}(t, \mathbf{x}) \to \mathbf{B}(t, -\mathbf{x}), \qquad (3.95)$$

and

$$\rho(t, \mathbf{x}) \to \rho(t, -\mathbf{x}), \qquad \mathbf{j}(t, \mathbf{x}) \to -\mathbf{j}(t, -\mathbf{x}), \qquad (3.96)$$

the transformed Maxwell's equations, rewritten in terms of the transformed variable $\mathbf{x}' = -\mathbf{x}$, are the same as the original ones. Note that a space-time event P, that has coordinates $(t, \mathbf{x})$ in the original frame, has coordinates $(t, -\mathbf{x})$ in the transformed frame. The change in the arguments in the previous transformations therefore simply reflects this change of label of the given space-time point P. The non-trivial aspect is the sign in front of the various quantities: the transformation of $\mathbf{E}$ is that of a *polar* (or "normal," or "true") vector, while that of $\mathbf{B}$ is that of an *axial* vector (or "pseudovector"). Similarly, $\rho(t, \mathbf{x})$ is a scalar under parity (while a pseudoscalar is defined as a quantity that, under rotations, is a scalar, but picks an overall minus sign under parity) and $\mathbf{j}$ is a polar vector. Finally, comparing the definitions (3.80) and (3.83) with eq. (3.95), we see that, under parity, ϕ is a scalar field (rather than a pseudoscalar) and $\mathbf{A}$ is a polar vector field.

The fact that electrodynamics is invariant under parity might seem, naively, an obvious fact. Shouldn't the law of physics be the same under a change of the orientation of the three axes, or under reflection in a mirror? This was, somewhat implicitly, the point of view until the 1950s, when it was discovered that there is another interaction, the weak interaction, that, in fact, is not invariant under parity. So, the invariance of electromagnetism under parity is a non-trivial fact.

Another symmetry of Maxwell's equations is time-reversal $t \to -t$. In this case, the equations are invariant if we transform

$$\mathbf{E}(t, \mathbf{x}) \to \mathbf{E}(-t, \mathbf{x}) \,, \qquad \mathbf{B}(t, \mathbf{x}) \to -\mathbf{B}(-t, \mathbf{x}) \,, \qquad (3.97)$$

while we transform the sources as

$$\rho(t, \mathbf{x}) \to \rho(-t, \mathbf{x}) \,, \qquad \mathbf{j}(t, \mathbf{x}) \to -\mathbf{j}(-t, \mathbf{x}) \,. \qquad (3.98)$$

These are the natural transformation properties under time reversal: a current is proportional to the velocity of the charges that produce it, so it changes sign if we reverse the direction of time. Magnetic fields, which are generated by currents, therefore must also change the sign. Electric charges and electric fields, instead, do not reverse the sign. Just as with parity, time-reversal is an invariance of electromagnetic interactions, but is violated by weak interactions.

Actually, Maxwell's equations have a much larger symmetry which is not readily apparent from the form (3.8)–(3.11) in which we have written them, which is based on the spatial vectors $\mathbf{E}$ and $\mathbf{B}$. This symmetry is the covariance under Lorentz transformations, i.e., the transformations of Special Relativity. We will discover this symmetry in Chapter 8, after having developed a more convenient formalism in Chapter 7.

Finally, in the absence of sources, i.e., when $\rho = 0$ and $\mathbf{j} = 0$, Maxwell's equations also have a *duality symmetry*

$$\mathbf{E} \to c\mathbf{B} \,, \qquad \mathbf{B} \to -\frac{\mathbf{E}}{c} \,. \qquad (3.99)$$

This symmetry is broken by the source terms.

Elementary applications of Maxwell's equations

4

Historically, the mathematical descriptions of phenomena in the domains of electrostatics, magnetostatics, and electromagnetic induction provided the building blocks from which Maxwell's equations were eventually inferred, with a "bottom-up" approach in which equations valid in specific settings were generalized and unified. Such a bottom-up approach, beside being important for providing a historical perspective, is also appropriate for a first elementary introduction to electromagnetism. In this more advanced course, we take instead a "top-down" approach in which, after having presented the full Maxwell's equations in the previous chapter, we systematically develop their consequences starting, in this chapter, with the most elementary applications. This, of course, does not respect the actual historical development, but has the advantage of allowing for a logically clear and streamlined presentation. In the main text of the chapter we discuss a selection of important results, while several other applications are discussed in a long Solved Problems section, at the end of the chapter.

4.1 Electrostatics

In terms of the electric field, the fundamental equations of electrostatics are obtained from eqs. (3.8) and (3.11), setting the time derivative in eq. (3.11) to zero, so they are

$$\boldsymbol{\nabla}\cdot\mathbf{E} \;=\; \frac{\rho}{\epsilon_0}, \tag{4.1}$$

$$\boldsymbol{\nabla}\times\mathbf{E} \;=\; 0. \tag{4.2}$$

It is convenient to use eq. (4.2) to introduce the field ϕ from $\mathbf{E} = -\boldsymbol{\nabla}\phi$ (making use of the theorem for curl-free fields, see page 7), so that eq. (4.2) is automatically satisfied, and eq. (4.1) becomes Poisson's equation

$$\boxed{\nabla^2\phi = -\frac{\rho}{\epsilon_0}\,.} \tag{4.3}$$

Of course, this could have also been obtained setting the time derivatives in eqs. (3.83) and (3.84) to zero. We will use eq. (4.3) as our basic equation to be solved in electrostatics.

4.1.1   Coulomb's law

As a first application we will show how, in electrostatics, Maxwell's equations imply Coulomb's law. In Section 4.1.2 we will solve eq. (4.3) for a generic distribution of charges. Here, we limit ourselves to a point-like charge q located at the origin, so that

$$\rho(\mathbf{x}) = q\delta^{(3)}(\mathbf{x}) \,. \tag{4.4}$$

Then, eq. (4.3) becomes

$$\nabla^2\phi = -\frac{q}{\epsilon_0}\delta^{(3)}(\mathbf{x}) \,. \tag{4.5}$$

A solution can be found immediately using eq. (1.90), that gives

$$\phi(\mathbf{x}) = \frac{q}{4\pi\epsilon_0}\frac{1}{r} \,, \tag{4.6}$$

which, as expected, is the Coulomb potential of a point-like charge. This solution is unique once one imposes the boundary condition that ϕ vanishes as $r \to \infty$ (in Section 4.1.5 we will provide a proof of the uniqueness of the solution to electrostatic problems). Using the expression (1.23) of the gradient in polar coordinates, the electric field $\mathbf{E} = -\nabla\phi$ generated by a charge q at the origin is therefore

$$\mathbf{E} = \frac{1}{4\pi\epsilon_0}\frac{q}{r^2}\hat{\mathbf{r}} \,. \tag{4.7}$$

From the Lorentz force equation (3.5), the force $\mathbf{F}_2$ exerted on a charge q_2 located in $\mathbf{x}_2 = \mathbf{r}$ by a charge q_1 located at $\mathbf{x}_1 = 0$ is $\mathbf{F}_2 = q_2\mathbf{E}_1$, so

$$\mathbf{F}_2 = \frac{1}{4\pi\epsilon_0}\frac{q_1 q_2}{r^2}\hat{\mathbf{r}} \,. \tag{4.8}$$

This confirms that the constant ϵ_0 that appears in the Maxwell equation (3.1) is the same that appears in the Coulomb force (2.6).

The result (4.6) is immediately generalized to the case where the source term is given by a set of N point particles with charges q_a and position $\mathbf{x}_a$, so that

$$\rho(\mathbf{x}) = \sum_{a=1}^{N} q_a\,\delta^{(3)}(\mathbf{x}-\mathbf{x}_a) \,. \tag{4.9}$$

Then

$$\nabla^2\phi = -\frac{1}{\epsilon_0}\sum_{a} q_a\,\delta^{(3)}(\mathbf{x}-\mathbf{x}_a) \,, \tag{4.10}$$

whose solution is[1]

$$\phi(\mathbf{x}) = \frac{1}{4\pi\epsilon_0}\sum_{a=1}^{N}\frac{q_a}{|\mathbf{x}-\mathbf{x}_a|} \,. \tag{4.11}$$

[1]This is an example of the *superposition principle*. Since Poisson's equation (4.3) is linear, both in ϕ and in the source term ρ, if ρ is taken to be a sum of terms $\rho = \sum_a \rho_a$, the solution for ϕ (that, as we will see, is unique) is $\phi = \sum_a \phi_a$, where ϕ_a is the solution when the source term is ρ_a. This linearity is not just a property of electrostatics but extends to the full Maxwell's equation. As we see from eqs. (3.8)–(3.11), Maxwell's equations are linear, both in the fields $\mathbf{E}$, $\mathbf{B}$, and in the sources $\rho, \mathbf{j}$. As a consequence, suppose that, when the charge and current densities are given by some functions $\rho_1, \mathbf{j}_1$, Maxwell's equations have a solution $\mathbf{E}_1$, $\mathbf{B}_1$ (we omit the arguments $(t, \mathbf{x})$ for notational simplicity) and that, when the sources are given by $\rho_2, \mathbf{j}_2$, they have a solution $\mathbf{E}_2$, $\mathbf{B}_2$. Then, from the linearity of the equations it follows that, when the charge and current densities are given by $\rho = \rho_1 + \rho_2$ and $\mathbf{j} = \mathbf{j}_1 + \mathbf{j}_2$, Maxwell's equations have a solution $\mathbf{E} = \mathbf{E}_1 + \mathbf{E}_2$, $\mathbf{B} = \mathbf{B}_1 + \mathbf{B}_2$.

4.1.2 Electric field from a generic static charge density

We now consider an electric charge density $\rho(\mathbf{x})$, independent of time and localized in space (i.e., non-vanishing only inside a finite volume), but otherwise generic, while we set the electric current $\mathbf{j} = 0$. Note that this choice is consistent with the continuity equation (3.22). It is convenient to work in the Coulomb gauge (3.92), where Maxwell's equations take the form (3.93, 3.94). Since $\rho(\mathbf{x})$ is independent of time, also $\phi(\mathbf{x})$, derived from eq. (3.93), is independent of time.[2] Then, eq. (3.94) becomes simply $\Box \mathbf{A} = 0$. Non-vanishing solutions of this equation describe plane waves, and will be discussed in Chapter 9. Here, we are interested in the solution sourced by the static charge density, so we set $\mathbf{A} = 0$, which is the trivial solution of $\Box \mathbf{A} = 0$ (and is obviously consistent with the condition $\nabla \cdot \mathbf{A} = 0$ that defines the Coulomb gauge).

An equation such as (3.93) can be conveniently solved with the method of Green's functions. The Green's function of the Laplacian operator, $G(\mathbf{x}, \mathbf{x}')$, is defined as the solution of the equation

$$\nabla_{\mathbf{x}}^2 G(\mathbf{x}, \mathbf{x}') = \delta^{(3)}(\mathbf{x} - \mathbf{x}'), \qquad (4.12)$$

[where the subscript $\mathbf{x}$ in the Laplacian indicates that $\nabla_{\mathbf{x}}^2$ acts on the $\mathbf{x}$ variable of $G(\mathbf{x}, \mathbf{x}')$]. Since the right-hand side depends only on the difference $\mathbf{x} - \mathbf{x}'$, we can already anticipate that, in fact, $G(\mathbf{x}, \mathbf{x}')$ will only depend on $\mathbf{x}, \mathbf{x}'$ through the combination $\mathbf{x} - \mathbf{x}'$, and we will therefore write it as $G(\mathbf{x} - \mathbf{x}')$.[3] Once found a solution of eq. (4.12), the corresponding solution of eq. (3.93) is given by[4]

$$\phi(\mathbf{x}) = -\frac{1}{\epsilon_0} \int d^3 x' \, G(\mathbf{x} - \mathbf{x}') \rho(\mathbf{x}'). \qquad (4.13)$$

In fact, taking the Laplacian of both sides,

$$
\begin{aligned}
\nabla^2 \phi(\mathbf{x}) &= -\frac{1}{\epsilon_0} \int d^3 x' \, [\nabla_{\mathbf{x}}^2 G(\mathbf{x} - \mathbf{x}')] \rho(\mathbf{x}') \\
&= -\frac{1}{\epsilon_0} \int d^3 x' \, \delta^{(3)}(\mathbf{x} - \mathbf{x}') \rho(\mathbf{x}') \\
&= -\frac{\rho(\mathbf{x})}{\epsilon_0}. \qquad (4.14)
\end{aligned}
$$

The Green's function method can be applied to more general linear differential equations, where a differential operator (here the Laplacian), acting on a function, must be equal to a given source term. Its advantage is that it allows us to separate the problem of solving a differential equation such as (3.93) into two steps. First, one searches for the solution of the Green's function. This part of the problem is independent of the source term [here $\rho(\mathbf{x})$] and for several operators, such as the Laplacian, can be solved exactly. Then, one remains with the computation of the integral in (4.13). For a generic source term it will not be possible to perform it exactly, but the integral might be amenable to useful analytic approximations (or direct numerical evaluation).

[2] One could always add to the solution of the inhomogeneous equation $\nabla^2 \phi = -\rho/\epsilon_0$, a solution of the homogeneous equation $\nabla^2 \phi = 0$, which could be taken to be time dependent. However, for a localized distribution of charges we impose the boundary condition that $\phi = 0$ as $r \to \infty$, and this fixes to zero the solution of the homogeneous equation, see Section 4.1.5.

[3] Actually, setting for instance $\mathbf{x}' = 0$, we see that eq. (4.12) is invariant under rotations of the variable $\mathbf{x}$ around the origin (or, for $\mathbf{x}'$ generic, is invariant under rotations of $\mathbf{x} - \mathbf{x}'$), so we can also anticipate that $G(\mathbf{x} - \mathbf{x}')$ (that, as we will see later, is unique) will actually be a function only of the modulus $|\mathbf{x} - \mathbf{x}'|$.

[4] As long as the integral converges at infinity, which, as we will see below, is the case for a localized distribution of charges.

The first step for solving the differential equation (3.93) is therefore to find the Green's function of the Laplacian.[5] This can be done using eq. (1.91), which shows that the Green's function of the Laplacian is

$$G(\mathbf{x} - \mathbf{x}') = -\frac{1}{4\pi |\mathbf{x} - \mathbf{x}'|}. \tag{4.15}$$

As anticipated, this is actually a function only of $|\mathbf{x} - \mathbf{x}'|$. Then, from eq. (4.13), the general solution for the scalar potential generated by a localized charge density is

$$\phi(\mathbf{x}) = \frac{1}{4\pi\epsilon_0} \int d^3x' \frac{\rho(\mathbf{x}')}{|\mathbf{x} - \mathbf{x}'|}. \tag{4.16}$$

Observe that the condition that the charge density is localized, i.e., that the function $\rho(\mathbf{x})$ has compact support, ensures the convergence of the integral at infinity. For a set of point-like particles, inserting eq. (4.9) into eq. (4.16), we recover eq. (4.11). In electrostatics, the electric field is obtained from

$$\mathbf{E}(\mathbf{x}) = -\boldsymbol{\nabla}\phi, \tag{4.17}$$

since $\mathbf{A} = 0$ in eq. (3.83). Therefore,

$$
\begin{aligned}
E_i(\mathbf{x}) &= -\frac{1}{4\pi\epsilon_0}\partial_i \int d^3x' \frac{\rho(\mathbf{x}')}{|\mathbf{x} - \mathbf{x}'|} \\
&= -\frac{1}{4\pi\epsilon_0} \int d^3x' \, \rho(\mathbf{x}')\partial_i \frac{1}{|\mathbf{x} - \mathbf{x}'|} \\
&= \frac{1}{4\pi\epsilon_0} \int d^3x' \, \rho(\mathbf{x}')\frac{x_i - x_i'}{|\mathbf{x} - \mathbf{x}'|^3},
\end{aligned}
\tag{4.18}
$$

[6]This can be shown as follows:
$$
\begin{aligned}
\partial_i \frac{1}{|\mathbf{x} - \mathbf{x}'|} &= -\frac{1}{|\mathbf{x} - \mathbf{x}'|^2}\partial_i|\mathbf{x} - \mathbf{x}'| \\
&= -\frac{1}{2|\mathbf{x} - \mathbf{x}'|^3}\partial_i|\mathbf{x} - \mathbf{x}'|^2 \\
&= -\frac{1}{2|\mathbf{x} - \mathbf{x}'|^3} \\
&\quad \times \partial_i(x_j x_j - 2x_j x_j' + x_j' x_j') \\
&= -\frac{1}{2|\mathbf{x} - \mathbf{x}'|^3}(2x_i - 2x_i') \\
&= -\frac{x_i - x_i'}{|\mathbf{x} - \mathbf{x}'|^3}.
\end{aligned}
$$

where we used[6]

$$\partial_i \frac{1}{|\mathbf{x} - \mathbf{x}'|} = -\frac{x_i - x_i'}{|\mathbf{x} - \mathbf{x}'|^3}. \tag{4.19}$$

Therefore, in vector form,

$$\mathbf{E}(\mathbf{x}) = \frac{1}{4\pi\epsilon_0} \int d^3x' \, \rho(\mathbf{x}')\frac{\mathbf{x} - \mathbf{x}'}{|\mathbf{x} - \mathbf{x}'|^3}. \tag{4.20}$$

Another useful variant of eq. (4.20) is obtained starting from the second line in eq. (4.18), and writing

$$
\begin{aligned}
E_i(\mathbf{x}) &= -\frac{1}{4\pi\epsilon_0} \int d^3x' \, \rho(\mathbf{x}')\frac{\partial}{\partial x_i} \frac{1}{|\mathbf{x} - \mathbf{x}'|} \\
&= +\frac{1}{4\pi\epsilon_0} \int d^3x' \, \rho(\mathbf{x}')\frac{\partial}{\partial x_i'} \frac{1}{|\mathbf{x} - \mathbf{x}'|} \\
&= -\frac{1}{4\pi\epsilon_0} \int d^3x' \, \frac{\partial \rho(\mathbf{x}')}{\partial x_i'} \frac{1}{|\mathbf{x} - \mathbf{x}'|},
\end{aligned}
\tag{4.21}
$$

where, in the last line, we integrated by parts, and used the fact that $\rho(\mathbf{x})$ is localized to discard the boundary term. Then,

$$\mathbf{E}(\mathbf{x}) = -\frac{1}{4\pi\epsilon_0} \int d^3x' \, \frac{\nabla_{\mathbf{x}'}\rho(\mathbf{x}')}{|\mathbf{x} - \mathbf{x}'|} \,. \tag{4.22}$$

We can then compute the force exerted between two distributions of charges, $\rho_1(\mathbf{x})$ and $\rho_2(\mathbf{x})$, that we take as localized in two non-overlapping volumes, denoted by V_1 and V_2, respectively. The force $d\mathbf{F}_1$ exerted on the infinitesimal charge

$$dq_1 = d^3x \, \rho_1(\mathbf{x}) \,, \tag{4.23}$$

contained in the infinitesimal volume d^3x, by the electric field $\mathbf{E}_2(\mathbf{x})$ generated by a charge distribution $\rho_2(\mathbf{x})$, is given by

$$d\mathbf{F}_1 = dq_1 \mathbf{E}_2(\mathbf{x}) \,. \tag{4.24}$$

Therefore, integrating over the volume of the first charge,

$$\mathbf{F}_1 = \int d^3x \, \rho_1(\mathbf{x})\mathbf{E}_2(\mathbf{x}) \,. \tag{4.25}$$

Using eq. (4.20), this can be written as

$$\mathbf{F}_1 = \frac{1}{4\pi\epsilon_0} \int d^3x \, d^3x' \, \rho_1(\mathbf{x})\rho_2(\mathbf{x}') \frac{\mathbf{x} - \mathbf{x}'}{|\mathbf{x} - \mathbf{x}'|^3} \,. \tag{4.26}$$

Note that, since ρ_1 vanishes outside V_1 and ρ_2 vanishes outside V_2, we could actually extend the integrations over d^3x and d^3x' to all of space.[7]

The force $\mathbf{F}_2$ exerted on the current density ρ_2 by the charge density ρ_1 is obtained exchanging $1 \leftrightarrow 2$ in eq. (4.26). Then, after also exchanging the names of the integration variables $\mathbf{x} \leftrightarrow \mathbf{x}'$, we get

$$\mathbf{F}_2 = \frac{1}{4\pi\epsilon_0} \int d^3x \, d^3x' \, \rho_1(\mathbf{x})\rho_2(\mathbf{x}') \frac{\mathbf{x}' - \mathbf{x}}{|\mathbf{x} - \mathbf{x}'|^3} \,, \tag{4.27}$$

and see that $\mathbf{F}_2 = -\mathbf{F}_1$, so the force satisfies Newton's third law. If we apply this to two point charges, setting

$$\rho_1(\mathbf{x}) = q_1 \delta^{(3)}(\mathbf{x}) \tag{4.28}$$

and

$$\rho_2(\mathbf{x}') = q_2 \delta^{(3)}(\mathbf{x}' - \mathbf{r}) \,, \tag{4.29}$$

eq. (4.27) gives back, of course, Coulomb's law (4.8).

In Section 5.5, we will discuss how to obtain this force from a continuous generalization of the Coulomb potential, and the relation of such a potential to the energy stored in the electric field generated by $\rho_1(\mathbf{x})$ and $\rho_2(\mathbf{x})$.

[7]Note also that, thanks to the condition that the two charge densities are non overlapping, there is no problem of convergence of the integral as $|\mathbf{x} - \mathbf{x}'| \to 0$, since there is a minimum distance d between the localization region of the two distributions, so that $\rho_1(\mathbf{x})\rho_2(\mathbf{x}') = 0$ if $|\mathbf{x} - \mathbf{x}'| < d$.

4.1.3   Scalar gauge potential and electrostatic potential

In the elementary treatment of electrostatics, one starts from eq. (3.4) with a vanishing magnetic field, $\boldsymbol{\nabla}\times\mathbf{E} = 0$, and uses the theorem for curl-free vectors, see page 7, to introduce the electrostatic potential $\varphi(\mathbf{x})$ from $\mathbf{E} = -\boldsymbol{\nabla}\varphi$. Comparing with eq. (3.83), we see that the scalar gauge potential $\phi(t, \mathbf{x})$ is the generalization of the electrostatic potential $\varphi(\mathbf{x})$ to the general setting of time-dependent electric and magnetic fields, and reduces to it when we can neglect all time dependences. We will then use the notation $\phi(\mathbf{x})$ even in the context of electrostatics. Using eq. (1.37), the equation

$$\mathbf{E}(\mathbf{x}) = -\boldsymbol{\nabla}\phi(\mathbf{x}), \tag{4.30}$$

can be integrated to give

$$\phi(\mathbf{x}) - \phi(\mathbf{x}_0) = -\int_{\mathbf{x}_0}^{\mathbf{x}} d\mathbf{x}' \cdot \mathbf{E}(\mathbf{x}'), \tag{4.31}$$

where the line integral is carried out over an arbitrary path $\mathcal{C}$ connecting an arbitrary initial point $\mathbf{x}_0$ to the point $\mathbf{x}$. The integral in eq. (4.31) is independent of the path $\mathcal{C}$ that connects $\mathbf{x}_0$ and $\mathbf{x}$: the difference between the integral computed on a path $\mathcal{C}_1$ and that on a path $\mathcal{C}_2$, both with endpoints in $\mathbf{x}_0$ and $\mathbf{x}$, is in fact the same as the integral over the closed loop $\mathcal{C}_1 - \mathcal{C}_2$ (where we denote by $\mathcal{C}_1 - \mathcal{C}_2$ the loop where first we go from $\mathbf{x}_0$ to $\mathbf{x}$ following $\mathcal{C}_1$, and then back to $\mathbf{x}_0$ following $\mathcal{C}_2$ in the opposite sense). As already discussed after eq. (1.37), for a function $\mathbf{E}$ that can be written as a gradient, the line integral over a closed loop vanishes, so the integral in eq. (4.31) is independent of the path.

Consider now a particle with charge q_a, moving on a trajectory $\mathbf{x}_a(t)$ under the action of an electromagnetic field. We take the particle as non-relativistic, so we can use the Lorentz force equation in the form (3.5). In eq. (4.31), for the path $\mathcal{C}$ we use the actual trajectory $\mathbf{x}_a(t)$ of the particle, and we denote its velocity by $\mathbf{v}_a(t) = d\mathbf{x}_a/dt$. Since $\mathbf{v}_a \cdot (\mathbf{v}_a \times \mathbf{B}) = 0$, eq. (3.5) implies that $\mathbf{v}_a \cdot \mathbf{F}(\mathbf{x}_a) = q_a \mathbf{v}_a \cdot \mathbf{E}(\mathbf{x}_a)$ or, equivalently, $d\mathbf{x}_a \cdot \mathbf{F}(\mathbf{x}_a) = q_a d\mathbf{x}_a \cdot \mathbf{E}(\mathbf{x}_a)$. Then, eq. (4.31) can be written as

$$q_a \left[\phi(\mathbf{x}) - \phi(\mathbf{x}_0)\right] = -\int_{\mathbf{x}_0}^{\mathbf{x}} \mathbf{F}(\mathbf{x}') \cdot d\mathbf{x}'. \tag{4.32}$$

On the right-hand side, we recognize minus the work made by the Lorentz force on the particle. The fact that, as we have seen, the integral is independent of the path $\mathcal{C}$ connecting $\mathbf{x}_0$ and $\mathbf{x}$ means, in the language of mechanics, that the force $\mathbf{F}$, given in our case by the Lorentz force, is conservative.

The work made by an external agent against the electric field to move a particle from $\mathbf{x}_0$ to $\mathbf{x}$ is equal to minus the work made by the Lorentz force,[8] so

$$\begin{aligned} W_{\text{ext}} &= -\int_{\mathbf{x}_0}^{\mathbf{x}} \mathbf{F}(\mathbf{x}') \cdot d\mathbf{x}' \\ &= q_a \left[\phi(\mathbf{x}) - \phi(\mathbf{x}_0)\right]. \end{aligned} \tag{4.33}$$

[8] Compare, for instance, with the prototype mechanical example, where the positive work $W_{\text{ext}} = mgh$ made by an external agent to lift a mass m from $z = 0$ to $z = h$ against the gravitational force $\mathbf{F} = -mg\hat{\mathbf{z}}$ is minus the work $W_{\text{grav}} = \int \mathbf{F} \cdot d\mathbf{x} = \int_0^h (-mg\hat{\mathbf{z}}) \cdot (dz\,\hat{\mathbf{z}}) = -mgh$, made by the gravitational field.

4.1.4   Instability of a system of static charges

We now prove *Earnshaw's theorem*, which states that, in a finite region R, free of charges, the electrostatic potential $\phi(\mathbf{x})$ takes its maximum and its minimum on the boundary ∂R. This theorem can then be used to show that (in classical electrodynamics) a system of static charges interacting only electromagnetically cannot be in a state of mechanical equilibrium.

The proof of the theorem is as follows. Suppose, to the contrary, that $\phi(\mathbf{x})$ has a minimum at a point $\mathbf{x}$ in the interior of R. We can then construct an infinitesimal volume V, which encloses $\mathbf{x}$ and is still inside R, and is therefore charge-free. Let ∂V be the boundary of V. If $\phi(\mathbf{x})$ has a minimum at $\mathbf{x}$, its gradient is such that, for any vector $\mathbf{v}$, $\mathbf{v}\cdot\boldsymbol{\nabla}\phi$ is strictly positive in $\mathbf{x}$, and remains positive at an infinitesimal distance from $\mathbf{x}$. Then, on any point of ∂V we should have $\hat{\mathbf{n}}\cdot\boldsymbol{\nabla}\phi > 0$, where $\hat{\mathbf{n}}$ is the unit normal to ∂V at that point, and therefore we should have

$$\int_{\partial V} d^2 s\,\hat{\mathbf{n}}\cdot\boldsymbol{\nabla}\phi > 0. \tag{4.34}$$

However, this is not possible since

$$\begin{aligned} \int_{\partial V} d^2 s\,\hat{\mathbf{n}}\cdot\boldsymbol{\nabla}\phi &= -\int_{\partial V} d^2 s\,\hat{\mathbf{n}}\cdot\mathbf{E} \\ &= -\int_V d^3 x\,\boldsymbol{\nabla}\cdot\mathbf{E}, \end{aligned} \tag{4.35}$$

and this vanishes, since $\boldsymbol{\nabla}\cdot\mathbf{E} = 0$ in a charge-free region. Similarly, one shows that there can be no maximum. Actually, even if we have stated the theorem in the language of electrodynamics, using the electric field, the theorem states, more generally, that any harmonic function, i.e., any function ϕ that satisfies $\boldsymbol{\nabla}^2\phi = 0$ in a region R, has its minima and maxima on the boundary ∂R. In fact, from Gauss's theorem

$$\begin{aligned} \int_{\partial V} d^2 s\,\hat{\mathbf{n}}\cdot\boldsymbol{\nabla}\phi &= \int_V d^3 x\,\boldsymbol{\nabla}\cdot(\boldsymbol{\nabla}\phi) \\ &= \int_V d^3 x\,\boldsymbol{\nabla}^2\phi, \end{aligned} \tag{4.36}$$

and this vanishes if $\boldsymbol{\nabla}^2\phi = 0$ in V.

An important consequence of Earnshaw's theorem is that a set of isolated charges cannot be in a state of stable equilibrium under the action of electrostatic forces only. Indeed, consider the electrostatic potential $\phi(\mathbf{x})$ generated by a given distribution of charges, localized in a volume V, and imagine placing a test charge q_a at some position $\mathbf{x}_a$ inside V, where there was no other charge. A stable equilibrium situation is then only obtained if $\phi(\mathbf{x})$ has a minimum (or a maximum, depending on the sign of q_a) at $\mathbf{x} = \mathbf{x}_a$. However, as we have seen, this is not possible. Therefore, any point-like charge inserted into a pre-existing electrostatic potential cannot be in equilibrium.[9] Another important application of Earnshaw's theorem will be discussed in Section 4.1.6, where we will see that it implies that the electric field inside a hollow conductor vanishes.

[9]This shows that a classical model of matter, based only on point-like electrons and nuclei interacting with static electromagnetic interactions, cannot be stable. As we will see in Problem 10.2, the same is true also beyond the static limit, and a model of an atom made of a classical electron rotating around a positively charged nucleus decays on a very short time scale because of the emission of electromagnetic radiation. This is an intrinsic limitation of classical electrodynamics, and hints to the fact that, at some microscopic scale, the classical description must be replaced by quantum mechanics.

4.1.5   Uniqueness of the solution of electrostatic problems

We now ask to what extent the solution of a problem of electrostatics, determined by eq. (4.3) together with the geometry of the system and the boundary conditions, is unique. Let us assume that there are two solutions $\phi_1(\mathbf{x})$ and $\phi_2(\mathbf{x})$ of eq. (4.3). Then, the difference $\psi(\mathbf{x}) = \phi_1(\mathbf{x}) - \phi_2(\mathbf{x})$ satisfies

$$\nabla^2 \psi = 0. \qquad (4.37)$$

Let V be a volume with boundary ∂V. We use the identity[10]

$$\int_V d^3x\, \boldsymbol{\nabla}\psi \cdot \boldsymbol{\nabla}\psi = \int_V d^3x \left[\boldsymbol{\nabla}\cdot(\psi \boldsymbol{\nabla}\psi) - \psi\nabla^2\psi \right]. \qquad (4.40)$$

Then, using eq. (4.37) for the second term on the right-hand side, and Gauss's theorem for the first, we have

$$\int_V d^3x\, |\boldsymbol{\nabla}\psi|^2 = \int_{\partial V} d\mathbf{s}\cdot(\psi\boldsymbol{\nabla}\psi). \qquad (4.41)$$

Consider first the situation in which the source term in eq. (4.3) is localized, and the space is just a large volume in three-dimensional space, with no inner boundaries, and a boundary ∂V at large distances from the sources. If the distribution of charges is localized, we can take as volume V a sphere of radius R enclosing all charges, so

$$\int_{\partial V} d\mathbf{s}\cdot\psi\boldsymbol{\nabla}\psi = R^2\int d\Omega\,\hat{\mathbf{r}}\cdot\psi\boldsymbol{\nabla}\psi, \qquad (4.42)$$

where we used the fact that, for a sphere, $d\mathbf{s} = R^2 d\Omega\,\hat{\mathbf{r}}$. At sufficiently large distances any solution $\phi(\mathbf{x})$ of eq. (4.3) decreases with distances at least as $1/r$, so $\boldsymbol{\nabla}\phi$ decreases at least as $1/r^2$.[11] Therefore, on the surface of the sphere, $\psi\boldsymbol{\nabla}\psi$ is of order $1/R^3$ or smaller. Then, taking the limit $R \to \infty$, the right-hand side of eq. (4.41) vanishes, and eq. (4.41) gives

$$\int d^3x\, |\boldsymbol{\nabla}\psi|^2 = 0, \qquad (4.43)$$

where the integral is now over all three-dimensional space. Since $|\boldsymbol{\nabla}\psi|^2$ is a non-negative quantity, this can only be satisfied if $\boldsymbol{\nabla}\psi = 0$ everywhere, so ψ must be a constant. This shows that, if $\phi_1(\mathbf{x})$ and $\phi_2(\mathbf{x})$ are two distinct solutions of eq. (4.3), they can differ at most by a constant. A constant addition to a potential is irrelevant, since it does not affect the electric field, and can be fixed simply by imposing that the solution of eq. (4.3) vanishes at infinity. Therefore, the solution of eq. (4.3) for a localized distribution of charges, in a space with no inner boundaries, is unique.

This argument can be easily generalized to the situation in which we consider a space that, rather than being the whole $\mathbb{R}^3$, has one or several inner boundaries S_i, that could correspond, for instance, to surfaces of

---

[10]This follows trivially, expanding $\boldsymbol{\nabla}\cdot(\psi\boldsymbol{\nabla}\psi)$ on the right-hand side. More generally, given two functions ϕ and ψ, we have the identity

$$\int_V d^3x\, (\boldsymbol{\nabla}\phi\cdot\boldsymbol{\nabla}\psi + \phi\nabla^2\psi)$$
$$= \int_V d^3x\, \boldsymbol{\nabla}\cdot(\phi\boldsymbol{\nabla}\psi)$$

and therefore, upon use of Gauss's theorem,

$$\int_V d^3x\, (\boldsymbol{\nabla}\phi\cdot\boldsymbol{\nabla}\psi + \phi\nabla^2\psi)$$
$$= \int_{\partial V} d\mathbf{s}\cdot(\phi\boldsymbol{\nabla}\psi), \qquad (4.38)$$

which is called *Green's first identity*. Rewriting this exchanging ϕ with ψ we have

$$\int_V d^3x\, (\boldsymbol{\nabla}\psi\cdot\boldsymbol{\nabla}\phi + \psi\nabla^2\phi)$$
$$= \int_{\partial V} d\mathbf{s}\cdot(\psi\boldsymbol{\nabla}\phi),$$

and, subtracting this from eq. (4.38) we get

$$\int_V d^3x\, (\phi\nabla^2\psi - \psi\nabla^2\phi)$$
$$= \int_{\partial V} d\mathbf{s}\cdot(\phi\boldsymbol{\nabla}\psi - \psi\boldsymbol{\nabla}\phi), \qquad (4.39)$$

which is called *Green's second identity*.

[11]We will formalize this more precisely in Chapter 6, where we will see that, for a distribution of charges, the Coulomb potential $\phi \propto 1/r$ is the first term in a multipole expansion. If the total charge of the distribution vanishes, the $1/r$ term is absent and $\phi(\mathbf{x})$ goes to zero faster that $1/r$.

material bodies (we will study the case where these bodies are perfect conductors in Section 4.1.6). In this case, eq. (4.41) becomes

$$\int_V d^3x \, |\boldsymbol{\nabla}\psi|^2 = \sum_i \int_{S_i} d\mathbf{s} \cdot \psi \boldsymbol{\nabla}\psi \, . \tag{4.44}$$

To have a well-defined problem we must assign boundary conditions for $\phi_1(\mathbf{x})$ and $\phi_2(\mathbf{x})$ on the inner boundaries S_i, which will induce corresponding boundary conditions on ψ. There are two natural boundary conditions on a surface:

- *Dirichlet boundary conditions*: in this case we fix the value of the potential ϕ on the surfaces, so we set $\phi_1(\mathbf{x}) = \phi_2(\mathbf{x}) = f_i(x)$ on the surface S_i. This implies $\psi(x) = 0$ on each surface S_i, so the right-hand side of eq. (4.44) vanishes. Then, we find again eq. (4.43), which implies that $\boldsymbol{\nabla}\psi = 0$ and therefore ψ is a constant.

- *Neumann boundary conditions*: in this case we require that, for any solution ϕ of eq. (4.3), the component of $\boldsymbol{\nabla}\phi$ normal to the surface vanishes, so $\hat{\mathbf{n}}\cdot\boldsymbol{\nabla}\phi_{1,2} = 0$ and therefore also $\hat{\mathbf{n}}\cdot\boldsymbol{\nabla}\psi = 0$. Since $d\mathbf{s} = ds\,\hat{\mathbf{n}}$, we find again that the right-hand side of eq. (4.44) vanishes and that ψ is a constant.

Then, also in these cases, the solution for ϕ is unique, apart from an irrelevant constant.

A related result is that (on a topologically trivial space, such as $\mathbb{R}^3$) a vector field $\mathbf{V}(\mathbf{x})$ is uniquely determined by its divergence and its curl, modulo the gradient of a function ψ that satisfies $\boldsymbol{\nabla}^2\psi = 0$. In fact, consider the equations

$$\boldsymbol{\nabla}\cdot\mathbf{V} = f(\mathbf{x})\,, \qquad \boldsymbol{\nabla}\times\mathbf{V} = \mathbf{u}(\mathbf{x})\,, \tag{4.45}$$

with $f(\mathbf{x})$ and $\mathbf{u}(\mathbf{x})$ given. Let $\mathbf{V}_1(\mathbf{x})$ and $\mathbf{V}_2(\mathbf{x})$ be two solutions of these equations. Then the vector field $\mathbf{w}(\mathbf{x}) = \mathbf{V}_2(\mathbf{x}) - \mathbf{V}_1(\mathbf{x})$ satisfies

$$\boldsymbol{\nabla}\cdot\mathbf{w} = 0\,, \qquad \boldsymbol{\nabla}\times\mathbf{w} = 0\,. \tag{4.46}$$

From the theorem for curl-free fields given on page 7, $\boldsymbol{\nabla}\times\mathbf{w} = 0$ implies that (on $\mathbb{R}^3$) we can write $\mathbf{w} = \boldsymbol{\nabla}\psi$. Then, $\boldsymbol{\nabla}\cdot\mathbf{w} = 0$ becomes $\boldsymbol{\nabla}^2\psi = 0$. Therefore,

$$\mathbf{V}_2(\mathbf{x}) = \mathbf{V}_1(\mathbf{x}) + \boldsymbol{\nabla}\psi\,, \qquad \text{with} \quad \boldsymbol{\nabla}^2\psi = 0\,. \tag{4.47}$$

If, furthermore, the boundary conditions of the problem are such that ψ goes to zero at infinity, then the only solution of $\boldsymbol{\nabla}^2\psi = 0$ is $\psi = 0$, and $\mathbf{V}_2(\mathbf{x}) = \mathbf{V}_1(\mathbf{x})$.[12]

In electrostatics, we have $\boldsymbol{\nabla}\cdot\mathbf{E} = -\rho(\mathbf{x})/\epsilon_0$ and $\boldsymbol{\nabla}\times\mathbf{E} = 0$, so, in $\mathbb{R}^3$, $\rho(\mathbf{x})$ uniquely determines the solution for $\mathbf{E}(\mathbf{x})$, modulo the gradient of a function that satisfies $\boldsymbol{\nabla}^2\psi = 0$. For a localized distribution of charge, the argument in eqs. (4.37)–(4.43) then shows that $\boldsymbol{\nabla}\psi = 0$, so we find again that $\mathbf{E}$ is uniquely determined by $\rho(\mathbf{x})$. In this form, however, the theorem extends also to magnetostatics (with a localized distribution of currents), which, as we will discuss in more detail later, is governed by the equations $\boldsymbol{\nabla}\cdot\mathbf{B} = 0$ and $\boldsymbol{\nabla}\times\mathbf{B} = \mu_0\mathbf{j}$, so both the divergence and the curl of $\mathbf{B}$ are given.

[12]In Solved Problem 11.1, we will show explicitly how to compute $\mathbf{V}$ in terms of its curl and its divergence, see eq. (11.174). In eqs. (13.52) and (13.56), we will then use these results to give the corresponding solutions for $\mathbf{E}$ and $\mathbf{B}$ in material media, in full generality.

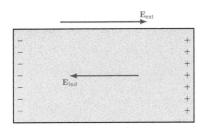

Fig. 4.1 A conductor in an external electric field $\mathbf{E}_{\text{ext}}$. Under its action, the free electrons move to the surface and create an induced electric field $\mathbf{E}_{\text{ind}}$, that screens $\mathbf{E}_{\text{ext}}$.

[13] Note that this only holds for a static situation in which a conductor, subject only to a static external electric field, has rearranged its surface charges and reached an equilibrium situation, where the external field is screened. If, instead, we use a battery to keep a potential difference between two points of a conducting wire (which amounts to continually removing the surface charge imbalance), a steady current will flow.

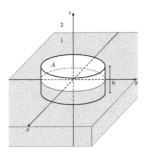

Fig. 4.2 An infinitesimal cylinder across the boundary between a conductor (shaded part, marked as medium 1) and the vacuum (medium 2).

4.1.6   Electrostatics of conductors

Consider now a conductor, with volume V and boundary $S = \partial V$. A fundamental property of conductors is that, in a static situation, the electric field inside them must vanish. Indeed, suppose that we apply an external time-independent electric field. In a conductor, the electrons are free to move, and, because of their negative charge, they will move in the direction opposite to the applied electric field. They will then eventually accumulate on some parts of the surface, which will therefore be negatively charged, and deplete other parts of the surface. The latter will therefore be overall positively charged because, there, the positive charge of the ions (which stay fixed) is no longer fully compensated by the electrons, see Fig. 4.1. This charge imbalance creates an induced electric field in the direction opposite to the applied electric field. The process will continue until the external field is completely screened and the total electric field vanishes (we are assuming here an ideal conductor, with an infinite supply of free charges; in normal situations this is a very good approximation for a metal). From Gauss's law $\boldsymbol{\nabla}\cdot\mathbf{E} = \rho/\epsilon_0$, $\mathbf{E} = 0$ implies $\rho = 0$. Therefore, at equilibrium, in the interior of the conductor positive and negative charges balance perfectly, and any charge imbalance is on the surface of the conductor.[13]

Another consequence of this screening process is that the surface of the conductor is an *equipotential surface*, i.e., the potential ϕ has a constant value on the surface. This follows from eq. (4.31), taking $\mathbf{x}_0$ and $\mathbf{x}$ to be two points on the surface of the conductor, and $\mathcal{C}$ any path that connects them passing through the interior of the conductor, where $\mathbf{E} = 0$. The fact that ϕ is constant on the surface also means that, on the surface, the components of $\boldsymbol{\nabla}\phi$ parallel to the surface vanish. The electric field at the surface of the conductor is therefore perpendicular to the surface. Again, physically, what happens is that, as long as there is a component of $\mathbf{E}$ tangential to the surface, the surface charges move in such a way as to screen it, until an equilibrium configuration with zero tangential electric field is reached. The component of $\mathbf{E}$ perpendicular to the surface of the conductor can be computed using the integrated Gauss's law, that we rewrite here as

$$\int_V d^3x\, \boldsymbol{\nabla}\cdot\mathbf{E} = \frac{1}{\epsilon_0}\int_V d^3x\, \rho\,. \qquad (4.48)$$

We take as volume V the cylinder shown in Fig. 4.2, which straddles across the boundary between the conductor and the vacuum. We take the z axis along the height of the cylinder, so

$$\int_V d^3x = \int_A dx\,dy \int_{-h/2}^{+h/2} dz\,. \qquad (4.49)$$

In the limit $h \to 0$,

$$\int_{-h/2}^{+h/2} dz\, \rho = \sigma \qquad (4.50)$$

is the surface charge density. We take A sufficiently small, so that σ can be taken to be spatially constant over A. Then,

$$\int_A dxdy \int_{-h/2}^{+h/2} dz\, \rho = A\sigma\,. \tag{4.51}$$

On the left-hand side of eq. (4.48), in the limit $h \to 0$ we have

$$
\begin{aligned}
\int_V d^3x\, \boldsymbol{\nabla}\cdot\mathbf{E} &= \int_A dxdy \int_{-h/2}^{+h/2} dz\, \partial_z E_z \\
&\quad + \int_{-h/2}^{+h/2} dz \int_A dxdy\,(\partial_x E_x + \partial_y E_y) \\
&= A[E_z(2) - E_z(1)] + \mathcal{O}(h)\,, \tag{4.52}
\end{aligned}
$$

where $E_z(2) \equiv E_z$ is the value of E_z as we approach the conductor from the medium 2, here taken to be just the vacuum, and $E_z(1)$ is the value when we approach the boundary from the medium 1. In the case where the latter is a conductor, $E_z(1) = 0$. Therefore, sending $h \to 0$ with A sufficiently small so that σ is constant over it, but still its linear size much larger than h, we find that, on the surface,

$$E_z = \frac{\sigma}{\epsilon_0}\,. \tag{4.53}$$

For a boundary with a normal $\hat{\mathbf{n}}$ (pointing outward from the conductor) in a generic direction, rather than along $\hat{\mathbf{z}}$, we then have

$$\boxed{\hat{\mathbf{n}}\cdot\mathbf{E} = \frac{\sigma}{\epsilon_0}\,.} \tag{4.54}$$

We can now use the fact that on the surface of conductors ϕ is constant, combined with the results of Section 4.1.5, to investigate the uniqueness of the solution of electrostatic problems in the presence of conductors. Let ϕ_1 and ϕ_2 be two solutions of $\nabla^2\phi = -\rho/\epsilon_0$, and define $\psi = \phi_1 - \phi_2$. According to the discussion in Section 4.1.5, we want to understand under what conditions the integral on the right-hand side of eq. (4.44) vanishes. We write $\mathbf{E}_1 = -\boldsymbol{\nabla}\phi_1$ and $\mathbf{E}_2 = -\boldsymbol{\nabla}\phi_2$, and we consider for simplicity the case of a single conductor with surface S, so that the volume V that includes the conductor has S as its inner boundary. Then

$$
\begin{aligned}
\int_S d\mathbf{s}\cdot\psi\boldsymbol{\nabla}\psi &= -(\phi_1 - \phi_2)_{|S} \int_S d\mathbf{s}\cdot(\mathbf{E}_1 - \mathbf{E}_2) \\
&= -(\phi_1 - \phi_2)_{|S} \int_V d^3x\, \boldsymbol{\nabla}\cdot(\mathbf{E}_1 - \mathbf{E}_2) \\
&= -(\phi_1 - \phi_2)_{|S}\frac{1}{\epsilon_0} \int_V d^3x\,(\rho_1 - \rho_2) \\
&= -(\phi_1 - \phi_2)_{|S}\frac{Q_1 - Q_2}{\epsilon_0}\,, \tag{4.55}
\end{aligned}
$$

where, in the first line, we used the fact that ϕ_1 and ϕ_2 are constants on S and can therefore be carried out of the integral, and we then used Gauss's theorem (1.48) and Gauss's law (3.1). From this, we see that there are two ways of making the left-hand side of eq. (4.55) vanish: (1) we can fix the value of the potential ϕ on the surface, so that $(\phi_1 - \phi_2)|_S = 0$. This is a Dirichlet boundary condition; (2) we rather fix the total charge on the conductor, so that $Q_1 = Q_2$. As we see from the previous steps, this is an integrated form of a Neumann boundary condition, since it corresponds to fixing the surface integral of the component of $\nabla \phi$ normal to the surface (rather than assigning its value at each point of the surface). In both cases, the left-hand side of eq. (4.55) vanishes and, by the argument in Section 4.1.5, the solution for ϕ is then unique.

Finally, consider a hollow conductor, i.e., a conductor which, in its interior, has a cavity region R, where no charges are present. In this region, the potential satisfies $\nabla^2 \phi = 0$. Furthermore, at the boundary ∂R between the cavity and the conductor, we have $\phi = \phi_0$, with ϕ_0 a constant, since we have seen that the surface of a conductor must be an equipotential surface (and the argument holds independently of whether this is an external or an internal boundary surface). However, as we discussed in Section 4.1.4, Earnshaw's theorem states that ϕ cannot have minima or maxima inside a charge-free region R. A function that is constant on the boundary ∂R and has no maxima or minima inside is necessarily constant everywhere in R. Therefore, in all R, $\phi = \phi_0$, and $\mathbf{E} = -\nabla \phi = 0$. This is a remarkable result, that shows that, no matter the geometry of the inner cavity, the surface charges distribute themselves on the boundary in such a way that they screen any electric field in its interior. This is the principle at the basis of *Faraday's cage*.

Note, however, that the argument no longer goes through if there are charges inside the cavity, since in this case $\nabla^2 \phi \neq 0$ inside. Therefore, the surface charges on the inner and outer surfaces of a hollow conductor distribute themselves so as to cancel the electric field generated by charges outside the conductor, but do not cancel the electric field generated by charges in the internal cavity.

4.1.7 Electrostatic forces from surface integrals

We now elaborate on eq. (3.69) to show that, in electrostatics, the equations of motion of charged particles (or of extended bodies) can be written in terms of surface integrals. We consider a set of extended bodies (or point charges) and, for the integration volume that enters in eq. (3.68), we choose a volume V_a that includes only the a-th body, so that $\mathbf{P}_{\text{mech}}$ is the momentum $\mathbf{p}_a$ of the a-th extended body. In electrostatics $\mathbf{B} = 0$, and therefore $\mathbf{P}_{\text{em}}$ vanishes, see eq. (3.67). Then, eq. (3.69) becomes

$$\left(\frac{d\mathbf{p}_a}{dt} \right)_i = - \int_{\partial V_a} d^2 s \, \hat{n}_j T_{ij} \,. \tag{4.56}$$

Note that the derivation of eq. (3.69) was completely general and did not assume that the particles are non-relativistic. When they are non-relativistic, we can also use Newton's law $d\mathbf{p}_a/dt = \mathbf{F}_a$, where $\mathbf{F}_a$ is the force acting on the a-th body, and eq. (4.56) can be rewritten as

$$(\mathbf{F}_a)_i = -\int_{\partial V_a} d^2s\, \hat{n}_j T_{ij}\,. \qquad (4.57)$$

It is remarkable that, in electrostatics, the total electric force acting over a body can be computed as an integral over any surface enclosing it (so, in particular, over its boundary surface), without apparently knowing the forces on the individual volume elements inside the body.[14] In electrostatics, the Maxwell stress tensor (3.63) reduces to

$$
\begin{aligned}
T_{ij} &= -\epsilon_0 \left(E_i E_j - \frac{1}{2} E^2 \delta_{ij} \right) && (4.58) \\
&= -\epsilon_0 \left(\partial_i \phi \partial_j \phi - \frac{1}{2} \delta_{ij} \partial_k \phi \partial_k \phi \right), && (4.59)
\end{aligned}
$$

where we used eq. (4.30). Using eq. (4.58), we can rewrite eq. (4.56) as

$$\boxed{\frac{d\mathbf{p}_a}{dt} = \epsilon_0 \int_{\partial V_a} d^2s \left[(\mathbf{E}\cdot\hat{\mathbf{n}})\mathbf{E} - \frac{1}{2} E^2 \hat{\mathbf{n}} \right].} \qquad (4.60)$$

This expresses the force on a charge q_a as an integral of the total electric field (including that generated by the charge q_a itself) over a surface enclosing the charge. This is, apparently, quite different from the standard expression $\mathbf{F} = q_a \mathbf{E}'(\mathbf{x}_a)$, where we denote here by $\mathbf{E}'(\mathbf{x}_a)$ the field generated by all other charges, *except* q_a, evaluated at the position $\mathbf{x}_a$ of the charge q_a.

As an application, it is instructive to see how eq. (4.60) reproduces the Coulomb force between two point charges. We consider for simplicity two equal charges $q_1 = q$ and $q_2 = q$ (we take $q > 0$) at a distance $2d$, and we set their positions at $\mathbf{x}_1 = (0, 0, +d)$ and $\mathbf{x}_2 = (0, 0, -d)$, respectively. To compute the force on the first charge we take, as the volume V_1 that encloses it, a hemisphere of radius R in the $z \geq 0$ region, i.e., the volume defined by the conditions $x^2 + y^2 + z^2 \leq R$ and $z \geq 0$, see Fig. 4.3, and we send $R \to \infty$. The boundary ∂V_1 is then given by the union of the (x, y) plane and the surface of the hemisphere at infinity. The electric field to be used in eq. (4.60) is the total electric field created by the two charges, since this is the quantity that enters in the energy-momentum tensor (4.58). As $R \to \infty$, $\mathbf{E}$ is of order $1/R^2$. So, on the surface of the hemisphere, the term in bracket in eq. (4.60) is of order $1/R^4$, while $d^2s = R^2 d\Omega$. Therefore, as $R \to \infty$, the contribution from the surface of the hemisphere at infinity vanishes, and only the contribution from the (x, y) plane matters. This can be computed as follows. Consider first the electric field $\mathbf{E}_1(x, y)$ produced by the charge q_1 in a point $\mathbf{x} = (x, y, 0)$ of the plane. The squared distance between the charge q_1 located at $\mathbf{x}_1 = (0, 0, +d)$ and the point $\mathbf{x} = (x, y, 0)$ is $x^2 + y^2 + d^2$, so the modulus of $\mathbf{E}_1(x, y)$ is given by

$$|\mathbf{E}_1(x, y)| = \frac{q}{4\pi\epsilon_0} \frac{1}{x^2 + y^2 + d^2}\,. \qquad (4.61)$$

[14] The underlying reason is that, in electrostatics, the knowledge of the field on the boundary uniquely determines the field everywhere, as we have seen in Section 4.1.5, so, in fact, the information on the field inside the body is implicitly there.

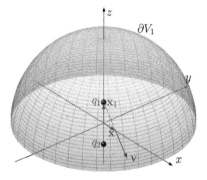

Fig. 4.3 The hemisphere surrounding the charge q_1. The vector $\mathbf{v}$ indicates the direction of the electric field generated by the charge q_1 at the point with coordinates $(x, y, 0)$.

For positive q, $\mathbf{E}_1(x, y)$ points in the direction of the vector $\mathbf{v}$ given by $\mathbf{x}_1 + \mathbf{v} = \mathbf{x}$, see again Fig. 4.3, so $\mathbf{v} = \mathbf{x} - \mathbf{x}_1 = (x, y, -d)$. The unit vector in this direction is therefore

$$\hat{\mathbf{v}} = \frac{1}{(x^2 + y^2 + d^2)^{1/2}} (x, y, -d),\qquad(4.62)$$

and therefore

$$\mathbf{E}_1(x, y) = \frac{q}{4\pi\epsilon_0} \frac{1}{(x^2 + y^2 + d^2)^{3/2}} (x, y, -d).\qquad(4.63)$$

The field $\mathbf{E}_2(x, y)$ generated by the second charge at the point $\mathbf{x} = (x, y, 0)$ is simply obtained by replacing $d \to -d$ in the previous expression. Therefore, the total electric field $\mathbf{E}(x, y) = \mathbf{E}_1(x, y) + \mathbf{E}_2(x, y)$ is given by

$$\mathbf{E}(x, y) = \frac{2q}{4\pi\epsilon_0} \frac{1}{(x^2 + y^2 + d^2)^{3/2}} (x, y, 0).\qquad(4.64)$$

Note that, on the (x, y) plane, the electric field has no $\hat{\mathbf{z}}$ component, as was clear from the symmetry of the problem, since we have taken $q_1 = q_2$. We now introduce polar coordinates in the (x, y) plane, $x = \rho\cos\phi$, $y = \rho\sin\phi$. Then, in eq. (4.60), $d^2s = \rho d\rho d\phi$ and, for the volume V_1, the outer normal to the plane is $\hat{\mathbf{n}} = -\hat{\mathbf{z}}$. Therefore, in eq. (4.60), the term $\mathbf{E}\cdot\hat{\mathbf{n}}$ vanishes and, for the force $\mathbf{F}_1 = d\mathbf{p}_1/dt$ exerted on the first particle, we get[15]

$$\begin{aligned}\mathbf{F}_1 &= \frac{1}{2}\epsilon_0\left(\frac{2q}{4\pi\epsilon_0}\right)^2 \hat{\mathbf{z}} \int_0^\infty \rho d\rho \int_0^{2\pi} d\phi \frac{\rho^2}{(\rho^2 + d^2)^3}\\[2mm] &= \hat{\mathbf{z}}\frac{q^2}{4\pi\epsilon_0}\frac{1}{(2d)^2}.\end{aligned}\qquad(4.66)$$

This correctly gives the Coulomb force on the first charge, due to the second charge at distance $2d$. In particular, a force in the $+\hat{\mathbf{z}}$ direction on the first charge, which is located at $(0, 0, d)$, exerted by the charge at $(0, 0, -d)$, corresponds to a repulsive force along the line joining the two charges, which is the correct result given that we took charges with the same sign.

It might appear that we have killed an ant with a hammer, given the long computation performed just for getting back the Coulomb force. However, conceptually it is quite interesting to see how the force on a charge can be obtained without using the electric field at the position of the charge itself, but rather using the total electric field on a surface surrounding it, which could even be chosen at a large distance from it. Note also that, in the usual computation of the force $\mathbf{F}_1$ exerted on a charge q_1 by a charge q_2, we have $\mathbf{F}_1 = q_1\mathbf{E}_2(\mathbf{x}_1)$, where $\mathbf{E}_2(\mathbf{x}_1)$ is the electric field generated in $\mathbf{x}_1$ by the charge q_2 (or, in general, by all other charges present, except q_1 itself). In contrast, in the computation of the force $\mathbf{F}_1$ from the Maxwell stress tensor enters the total electric field $\mathbf{E}$ generated by all charges, including q_1.[16]

---

[15]To compute the integral over ρ we introduce $u = \rho^2/d^2$ and we use

$$\int_0^\infty du\, \frac{u}{(u + 1)^3} = \frac{1}{2}.\qquad(4.65)$$

[16]A note for the advanced reader. This way of writing the equations of motions in term of surface integrals can also be performed in Newtonian gravity, with the gravitational potential taking the role of the scalar potential ϕ in eq. (4.59) (and ϵ_0 replaced by $1/(4\pi G)$, where G is Newton's constant), see eqs. (5.221)–(5.224) of Maggiore (2007). This admits an elegant extension to General Relativity, developed in a classic work by Einstein, Infeld, and Hoffmann in 1938. In General Relativity, this can provide significant advantages because the dynamics can then be written in a way that only involves fields at large distances from the source, i.e., the weak-field regime, independently of the internal structure of the sources, where, in General Relativity, in particular for black holes and neutron stars, complicated non-linear effects might take place.

4.2 Magnetostatics

We now turn to magnetostatics, i.e., situations that involve only static magnetic fields. In this case the Ampère–Maxwell law (3.2) reduces to Ampère's law (see Note 7 on page 44),

$$\boxed{\nabla \times \mathbf{B} = \mu_0 \mathbf{j}\,,} \qquad (4.67)$$

and this, together with eq. (3.3), that we repeat here,

$$\boxed{\nabla \cdot \mathbf{B} = 0\,,} \qquad (4.68)$$

determines the magnetic field. Observe that, in magnetostatics, only the combination $\mu_0 = 1/(\epsilon_0 c^2)$ enters. Equation (4.67) implies

$$\nabla \cdot \mathbf{j} = 0\,, \qquad (4.69)$$

as we see by taking the divergence of both sides. This is consistent with the conservation equation (3.22) if we set the net electric charge density $\rho = 0$ or, more generally, if $\partial\rho/\partial t = 0$. The integrated form of the Ampère–Maxwell law, eq. (3.19), reduces to

$$\oint_{\mathcal{C}} d\boldsymbol{\ell} \cdot \mathbf{B}(\mathbf{x}) = \mu_0 I\,, \qquad (4.70)$$

where I is the current through any surface bounded by $\mathcal{C}$, while the integrated form of eq. (4.68) is still given by eq. (3.16), which we repeat here,

$$\oint_{S} d\mathbf{s} \cdot \mathbf{B}(t, \mathbf{x}) = 0\,. \qquad (4.71)$$

4.2.1 Magnetic field of an infinite straight wire

As a first application, we compute the magnetic field generated by an infinite straight wire carrying a steady current I. The problem of computing the magnetic field produced by a generic static distribution of currents will be studied, in full generality, in Section 4.2.2. We set the wire along the z direction, and we use cylindrical coordinates (ρ, φ, z), as shown in Fig. 4.4. Since the problem is invariant under translations along the z axis and rotations around the wire, $\mathbf{B}(\rho, \varphi, z)$ must be of the form

$$\mathbf{B}(\rho, \varphi, z) = B_\rho(\rho)\hat{\boldsymbol{\rho}} + B_\varphi(\rho)\hat{\boldsymbol{\varphi}} + B_z(\rho)\hat{\mathbf{z}}\,, \qquad (4.72)$$

where B_ρ, B_φ, and B_z are independent of z and φ. Writing $\mathbf{j} = j_z(\rho)\hat{\mathbf{z}}$,[17] and using eqs. (1.30) and (1.33) for the divergence and curl in cylindrical coordinates, as well as the fact that $\mathbf{B}$ is independent of φ and z, Ampère's law (4.67) becomes

$$-\hat{\boldsymbol{\varphi}}\partial_\rho B_z + \hat{\mathbf{z}}\frac{1}{\rho}\partial_\rho(\rho B_\varphi) = \mu_0 j_z(\rho)\hat{\mathbf{z}}\,, \qquad (4.73)$$

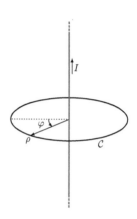

Fig. 4.4 The cylindrical coordinate system centered on the wire (with the wire along the z axis), and the loop $\mathcal{C}$ used to apply the integrated Ampère's law.

[17] Here, we consider the limit of an infinitely thin wire, so we can write the current density as $j_z(\mathbf{x}) = I\delta(x)\delta(y)$, and the proportionality constant is the current I, fixed by $I = \int dx dy\, j_z$. However, later, it will be useful to model it using a function $j_z(\rho)$ that vanishes for $\rho > a$, with a the transverse size of the wire, that we will eventually send to zero.

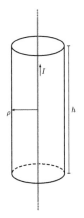

Fig. 4.5 The cylinder used in the integrated Maxwell equation (4.71).

[18]In fact, it is not even necessary to use this information; even if B_z were non-zero, invariance under translation along the z direction would imply that B_z is the same at $z = \pm h/2$, so the flux entering from the lower face would cancel against that coming out from the upper face.

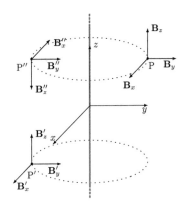

Fig. 4.6 The geometry of the symmetry argument based on parity and subsequent 180° rotations, discussed in the text.

while eq. (4.68) becomes

$$\frac{1}{\rho}\partial_\rho(\rho B_\rho) = 0 \,. \qquad (4.74)$$

From eq. (4.73), we can immediately conclude that $\partial_\rho B_z = 0$ and therefore B_z is a constant. As a boundary condition, for physical reasons we require that, at infinite distance from the wire, i.e., at $\rho \to \infty$, B_z vanishes. Then, this constant is actually zero, so $B_z = 0$ everywhere.

Next, we can prove that the radial component B_ρ also vanishes. This can be seen more easily from the integrated Maxwell equation (4.71), taking as surface S the boundary of a cylinder of height h and radius $\rho > 0$, whose axis coincides with the wire, and with faces at $z = \pm h/2$, see Fig. 4.5. Equation (4.71) states that the magnetic flux through the boundary S of the cylinder vanishes. The flux through the faces of the cylinder at $z = \pm h/2$ vanishes because $B_z = 0$,[18] while the flux through the lateral surface of the cylinder is $2\pi\rho h B_\rho(\rho)$. Since this must vanish, we get $B_\rho(\rho) = 0$ (for $\rho \neq 0$, i.e., outside the wire, that we have taken here as infinitesimally thin).

The vanishing of B_z and B_ρ can actually be understood using a symmetry argument based on parity. The argument, however, has some subtleties, so it is instructive to go through it in some detail. Consider the parity transformation ("Π"), $\mathbf{x} \to -\mathbf{x}$. Since $\mathbf{j}$ is a true vector (rather than a pseudovector), it changes sign under parity, $\mathbf{j}(\mathbf{x}) \to -\mathbf{j}(-\mathbf{x})$. In our case $\mathbf{j}(\mathbf{x})$ is proportional to a Dirac delta in the transverse plane and is independent of z, so the change $\mathbf{x} \to -\mathbf{x}$ in its argument has no effect, and simply $\mathbf{j} \to -\mathbf{j}$. So, if before parity $\mathbf{j}$ was in the $+\hat{\mathbf{z}}$ direction, after the parity transformation it points in the $-\hat{\mathbf{z}}$ direction. Therefore, the geometry of this problem is not invariant under parity alone. However, we can combine this with a rotation by 180° around the y axis, that we denote by R_y, that sends the current back toward the positive $\hat{\mathbf{z}}$ direction. Thus, the combined transformation $R_y\Pi$ is a symmetry transformation of the system.

Consider now how the magnetic field $\mathbf{B}$, at a generic point $\mathbf{x}$ is space, transforms under this combined operation. We found in eq. (3.95) that $\mathbf{B}$ is a pseudovector under parity, i.e., for a static field, $\mathbf{B}(\mathbf{x}) \to +\mathbf{B}(-\mathbf{x})$. Note that its argument changes from $\mathbf{x}$ to $-\mathbf{x}$, i.e., the parity operation sends the point $P = (x, y, z)$ into the antipodal point $P' = (-x, -y, -z)$. We assume, without loss of generality, that the point P has coordinates $(x = 0, y, z)$ and we consider, for definiteness, a magnetic field that, in P, is the sum of three vectors $\mathbf{B}_x$, $\mathbf{B}_y$, and $\mathbf{B}_z$ pointing, respectively, toward the positive x, y, and z axes, as shown in Fig. 4.6. Note that, having set $x = 0$, a vector in the positive x direction corresponds to a clockwise azimuthal component, and a vector in the positive y direction to an outward radial component. After the parity operation, the three vector components of the magnetic field $\mathbf{B}'$ at P' will be as shown in the figure, i.e. $\mathbf{B}'_x$, $\mathbf{B}'_y$, and $\mathbf{B}'_z$ would still point in the same directions as before the transformation, since $\mathbf{B} \to +\mathbf{B}$; however, now $\mathbf{B}'_y$ corresponds to an inward radial vector, and $\mathbf{B}'_x$ to a counterclockwise azimuthal vector.

After the subsequent R_y rotation, the point P$'$ is transformed into the point P$''$ in Fig. 4.6 and, because of the 180° rotation in the (x,z) plane, $\mathbf{B}''_x = -\mathbf{B}'_x$ and $\mathbf{B}''_z = -\mathbf{B}'_z$ while $\mathbf{B}''_y = \mathbf{B}'_y$. The three vector components of the magnetic field at P$''$ will therefore be as shown in the figure. We can finally perform a 180° rotation around the z axis, R_z, that brings the point P$''$ back onto the initial point P. Thus, in the end, after the combined $R_z R_y \Pi$ transformation, eventually $\mathbf{x} \to \mathbf{x}$, while

$$B_\rho(\mathbf{x}) \to -B_\rho(\mathbf{x}), \quad B_\varphi(\mathbf{x}) \to +B_\varphi(\mathbf{x}), \quad B_z(\mathbf{x}) \to -B_z(\mathbf{x}). \quad (4.75)$$

Thus, we have found a transformation that leaves the system invariant, and such that, under it, both $B_\rho(\mathbf{x})$ and $B_z(\mathbf{x})$ change sign; one is therefore tempted to conclude that they must vanish. While, as we have seen from the explicit computation, this conclusion is indeed true, the correct logic still requires some more steps. Indeed, one can already be perplexed by the fact that, with this argument based on symmetries, it seems that we do not even need to impose the boundary condition $B_z = 0$ at infinity, that was necessary in the derivation from eq. (4.73). After all, symmetry arguments can be an elegant way of extracting the consequences of the equations of a theory, but do not contain more information than the equations themselves. In fact, the correct chain of reasoning here is as follows. First of all, it is useful to stress that, behind any use of arguments based on parity invariance, stands the fact that Maxwell's equations are indeed invariant under parity, as we discussed in Section 3.4.[19] Second, the invariance of the geometry of the problem and of the equations governing a theory under a transformation, is not yet enough to guarantee that the corresponding solutions will also be invariant under this transformation. A necessary condition is that even the boundary conditions will be chosen to be invariant under the symmetry transformation. Maxwell's equations are invariant under parity and under rotations; still, a solution of Maxwell's equations invariant under this combination of parity and subsequent 180° rotations, can only emerge if also the boundary conditions are invariant under it. In our problem, a natural boundary condition on B_z is imposed at $\rho \to \infty$, and the only such boundary condition that respects this symmetry is $B_z(\rho = \infty) = 0$, which is indeed the condition that we imposed, for physical reasons, when searching the solution of eq. (4.73). In principle, we could rather decide to impose a boundary condition $B_z(\rho = \infty) = B_\infty \neq 0$; this would break the symmetry (in particular, the parity trasnformation). The corresponding solution for $B_z(\rho)$, which in this case would simply be a constant equal to the boundary value, $B_z(\rho) = B_\infty$, would not be invariant under parity. Still, at the mathematical level, it would be a perfectly well-defined solution of Maxwell's equations. For B_ρ, instead, the correct boundary condition is $B_\rho(\rho = 0) = 0$, since a radial field that does not vanish at $\rho = 0$ would be mathematically ill-defined. Again, this boundary condition respects the parity and rotation symmetries.

Last but not least, in general, even when the geometry of the problem, the equations of the theory, and the boundary conditions respect a symmetry, this does not yet necessarily imply that the solution re-

[19] As we stressed there, despite its apparently obvious geometric nature, invariance under parity is not guaranteed a priori. Electrodynamics happens to respect it, but, for example, another fundamental interaction, the weak interaction, is not invariant under parity; we do not perceive this in the macroscopic world just because the range of weak interactions is limited to subnuclear scales.

[21] More accurately, they are covariant, i.e., the left- and right-hand sides of the equations transform in the same way. See Note 15 on page 50.

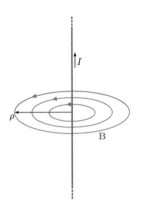

Fig. 4.7 Some field lines of the solution for the magnetic field.

[22] It is also interesting to find the corresponding solution for the vector potential. Using the expression (1.33) for the curl in cylindrical coordinate, one can see directly that the magnetic field (4.79) can be obtained from

$$\mathbf{A}(\mathbf{x}) = A_z(\rho)\hat{\mathbf{z}}\,, \qquad (4.80)$$

with

$$A_z(\rho) = -\frac{\mu_0 I}{2\pi}\,\ln\rho\,. \qquad (4.81)$$

This gauge potential satisfies the Coulomb gauge condition, since $\nabla\cdot\mathbf{A} = \partial_z A_z(\rho) = 0$. Note that $\mathbf{A}$ grows without bounds at large distances from the wire. This is an artifact of having taken a current distribution that is not localized, and rather extends from $z = -\infty$ to $z = +\infty$. We will see in Section 6.2 that, for a localized current distribution, $\mathbf{A}$ vanishes at infinity.

spects it. This only happens if the solution is unique. The other option is that there is a family of solutions, that transform into each other under this symmetry transformation.[20] For instance, in our case, using only symmetry arguments, one could not exclude the possibility that, even once imposed the boundary condition $B_z(\rho = \infty) = 0$, there could be two solutions, one with $B_z(\rho) > 0$ and one with $B_z(\rho) < 0$, that both approach zero at large ρ, and that transform into each other under parity. In our case this does not happen because, as we have seen at the end of Section 4.1.5, in magnetostatics, once assigned the boundary conditions, the solution is unique. So, the complete logic of the symmetry argument is: (1) Maxwell's equations are invariant under parity and under rotations.[21] (2) The problem is invariant under a combination of parity and 180° rotations. (3) For physical reasons, as boundary condition on B_z we choose $B_z(\rho = \infty) = 0$, while on B_ρ we impose $B_\rho(\rho = 0) = 0$ for mathematical consistency; these boundary conditions are invariant under the combined parity and 180° rotation transformations. (4) In magnetostatics, once given the boundary conditions, the solution is unique. Then, we can finally conclude that the solution must be invariant under this combined parity plus rotation transformation and, since under this transformation B_z and B_ρ change sign, they must vanish.

In conclusion, either from the explicit computation using the integrated Maxwell's equations, or from the symmetry argument, we find that in cylindrical coordinates the only non-vanishing component of the magnetic field is $B_\varphi(\rho)$, and eq. (4.72) becomes

$$\mathbf{B}(\rho, \varphi, z) = B_\varphi(\rho)\hat{\boldsymbol{\varphi}}\,. \qquad (4.76)$$

The function $B_\varphi(\rho)$ can now be determined using the integrated form of Ampère's law, eq. (4.70). Taking as curve $\mathcal{C}$ a circle of radius ρ in the plane transverse to the wire and centered on the wire, as shown in Fig. 4.4, in eq. (4.70) we have $d\boldsymbol{\ell} = \rho d\varphi\,\hat{\boldsymbol{\varphi}}$, so

$$\rho \int_0^{2\pi} d\varphi\, B_\varphi(\rho) = \mu_0 I\,, \qquad (4.77)$$

and therefore

$$B_\varphi(\rho) = \mu_0 \frac{I}{2\pi\rho}\,. \qquad (4.78)$$

In conclusion,

$$\boxed{\mathbf{B}(\rho, \varphi, z) = \mu_0 \frac{I}{2\pi\rho}\,\hat{\boldsymbol{\varphi}}\,.} \qquad (4.79)$$

Some field lines of this solution in a transverse plane are shown in Fig. 4.7.[22]

It is instructive to see how the same result can be obtained from a direct integration of eqs. (4.73) and (4.74), since some subtleties appear. Equation (4.74) tells us that $\rho B_\rho(\rho) = \alpha$, with α a constant, so

$B_\rho(\rho) = \alpha/\rho$. Setting $\alpha = 0$ we get back the result $B_\rho(\rho) = 0$ found previously. However, it seems that any field of the form $B_\rho(\rho)\hat{\boldsymbol{\rho}} = \alpha\hat{\boldsymbol{\rho}}/\rho$, with arbitrary α, would give an acceptable solution of eq. (4.74). This, however, is not the case, because eq. (4.74) becomes singular at $\rho = 0$, and more care is needed. To compute the divergence of a vector field that, in cylindrical coordinates, has the form $\mathbf{v}(\rho, \varphi, z) = \hat{\boldsymbol{\rho}}/\rho$, we proceed similarly to what we did in eqs. (1.86) and (1.87) for a field that, in polar coordinates, had the form $\hat{\mathbf{r}}/r^2$. In the present case, we take as volume V the same cylinder used in Fig. 4.5, with radius ρ and height h. Its lateral surface has a surface element $d\mathbf{s} = dz\rho d\varphi\hat{\boldsymbol{\rho}}$. Then, using Gauss's theorem

$$
\begin{aligned}
\int_V d^3x\, \boldsymbol{\nabla}\cdot\mathbf{v} &= \int_{\partial V} d\mathbf{s}\cdot\mathbf{v} \\
&= \int_{-h/2}^{h/2} dz \int_0^{2\pi} \rho d\varphi\hat{\boldsymbol{\rho}}\cdot\frac{\hat{\boldsymbol{\rho}}}{\rho} \\
&= 2\pi h\,.
\end{aligned}
\tag{4.82}
$$

We now write d^3x, on the left-hand side of eq. (4.82), as $d^3x = dz d^2\mathbf{x}_\perp$, where $\mathbf{x}_\perp = x\hat{\mathbf{x}}+y\hat{\mathbf{y}}$ is a two-dimensional vector spanning the transverse plane. Then, the left-hand side of eq. (4.82) can also be written as

$$
\begin{aligned}
\int_V d^3x\, \boldsymbol{\nabla}\cdot\mathbf{v} &= \int_{-h/2}^{h/2} dz \int_{|\mathbf{x}_\perp|<\rho} d^2\mathbf{x}_\perp\, \boldsymbol{\nabla}\cdot\mathbf{v} \\
&= h \int_{|\mathbf{x}_\perp|<\rho} d^2\mathbf{x}_\perp\, \boldsymbol{\nabla}\cdot\mathbf{v}\,,
\end{aligned}
\tag{4.83}
$$

where the integral over dz is trivial because, for the vector field $\mathbf{v}(\rho, \varphi, z) = \hat{\boldsymbol{\rho}}/\rho$ that we are considering, $\boldsymbol{\nabla}\cdot\mathbf{v}$ is independent of z. Since this holds for arbitrary $\rho > 0$, and therefore also for ρ infinitesimally small, comparison of eqs. (4.82) and (4.83) shows that

$$
\boxed{\boldsymbol{\nabla}\cdot\mathbf{v} = 2\pi\delta^{(2)}(\mathbf{x}_\perp)\,, \qquad \text{where} \quad \mathbf{v}(\rho, \varphi, z) = \frac{\hat{\boldsymbol{\rho}}}{\rho}\,.}
\tag{4.84}
$$

Therefore, a magnetic field of the form $\mathbf{B}(\rho, \varphi, z) = \alpha\hat{\boldsymbol{\rho}}/\rho$ would satisfy

$$
\boldsymbol{\nabla}\cdot\mathbf{B} = 2\pi\alpha\,\delta^{(2)}(\mathbf{x}_\perp)\,,
\tag{4.85}
$$

so it is not a solution of $\boldsymbol{\nabla}\cdot\mathbf{B} = 0$, unless $\alpha = 0$.

Finally, we can determine B_φ from a direct integration of eq. (4.73), whose $\hat{\mathbf{z}}$ component is

$$
\partial_\rho(\rho B_\varphi) = \mu_0 j_z(\rho)\rho\,.
\tag{4.86}
$$

Rather than taking an infinitesimally thick wire, it is simpler here to take a model for a circular wire of radius a, with j_z constant for $\rho \leq a$ and $j_z = 0$ for $\rho > a$. Then, integrating eq. (4.86) from $\rho = 0$ to

[23]It should be appreciated how, in this problem, the use of the integrated Maxwell's equations provided a simpler and more elegant way of obtaining the solution for B_ρ and B_φ. In particular, in the direct integration of eq. (4.73), we had to deal with the subtle issue of the spurious solution $B_\rho(\rho) = \alpha/\rho$ with non-vanishing α that, as we have seen, generates a Dirac delta on the z axis in $\boldsymbol{\nabla}\!\cdot\!\mathbf{B}$, and is therefore not acceptable. In the integrated form of the Ampère law, in contrast, we only dealt with the field at a surface of the cylinder used in the integrated form of the equation, and the spurious solution never appeared. Notice also that, for computing B_φ from the direct integration of eq. (4.73), we had to make a choice for the functional form of $\mathbf{j}$ inside the wire, that we just took to be a constant for $\rho < a$. The fact that the radius a of the wire eventually disappeared from the right-hand side of eq. (4.89) is an indication of the fact that the result, outside the wire, is independent of the modelization used; however, the integrated form of the Ampère law makes this apparent, since no modelization of the wire was ever needed.

[24]As in the electrostatic case, we could add an arbitrary time-dependent solution of the equation $\Box\mathbf{A} = 0$ that, as we will see in Chapter 9, describes plane waves. Here, we only consider the solution sourced by the current.

[25]Recall that, in Cartesian coordinates, the Laplacian of a vector field has the same form as the Laplacian on a scalar field on each component, so we have $\boldsymbol{\nabla}^2 A^i = -\mu_0 j^i$, with $\boldsymbol{\nabla}^2 = \partial_x^2 + \partial_y^2 + \partial_z^2$ just as on scalars. Correspondingly, also the Green's function for each component is the same as in the scalar case. As discussed in Note 2 on page 4, this is no longer true in polar or cylindrical coordinates. However, working in Cartesian coordinates suffices to obtain the Green's function and solve eq. (4.91). Then, since the solution is written in a vector form, it holds in any coordinate system.

$\rho = a$ (with the boundary condition $B_\varphi(\rho = 0) = 0$, necessary because a non-vanishing $\hat{\boldsymbol{\varphi}}$ component would be singular at $\rho = 0$) we get

$$B_\varphi(\rho) = \frac{\mu_0 j_z \rho}{2}, \qquad (\rho \le a), \qquad (4.87)$$

so

$$B_\varphi(\rho = a) = \frac{\mu_0 I}{2\pi a}, \qquad (4.88)$$

where $I = j_z \pi a^2$ is the current flowing through the wire (recall that $\mathbf{j}$ is a current per unit surface). Outside the wire, i.e., for $\rho > a$, $j_z = 0$, and eq. (4.86) gives $B_\varphi \propto 1/\rho$. The proportionality constant is obtained requiring continuity at $\rho = a$, which then gives

$$B_\varphi(\rho) = \frac{\mu_0 I}{2\pi\rho}, \qquad (\rho \ge a), \qquad (4.89)$$

so we have recovered eq. (4.78) outside the wire (and we can now also send $a \to 0$ to describe an infinitesimally thin wire carrying a fixed current I).[23]

4.2.2   Magnetic field of a static current density

We now compute the magnetic field produced by a current density $\mathbf{j}(\mathbf{x})$, that we take localized in space and time-independent, but otherwise arbitrary, and we also set $\rho = 0$ for the total charge density. According to eq. (3.22), this implies

$$\boldsymbol{\nabla}\!\cdot\!\mathbf{j} = 0. \qquad (4.90)$$

This situation is realized, for instance, if we have a wire forming a loop, with a steady current flowing through it. Within the wire, the positive charge density of the ions compensates for the negative charge density of the flowing electrons, so there is no net electric charge density. However, the ions do not have a bulk motion, while the electrons do, thereby creating a net electric current. The condition $\boldsymbol{\nabla}\!\cdot\!\mathbf{j} = 0$ expresses the fact that electrons are neither created nor destroyed inside the wire and, in each infinitesimal volume inside the wire, bounded by surfaces S_1 and S_2 in the transverse directions and of length dl in the longitudinal direction, the flow of electrons entering through S_1 is compensated by the electrons flowing out through S_2.

It is convenient to work in terms of the gauge potentials, in the Coulomb gauge. Since $\rho = 0$, the solution of eq. (3.93) is $\phi = 0$ and, looking for a static solution $\mathbf{A}(\mathbf{x})$,[24] eq. (3.94) reduces to

$$\boxed{\boldsymbol{\nabla}^2 \mathbf{A} = -\mu_0 \mathbf{j}.} \qquad (4.91)$$

The solution is obtained again from the Green's function (4.15) of the Laplacian, and is given by[25]

$$\boxed{\mathbf{A}(\mathbf{x}) = \frac{\mu_0}{4\pi} \int d^3 x' \, \frac{\mathbf{j}(\mathbf{x}')}{|\mathbf{x} - \mathbf{x}'|},} \qquad (4.92)$$

as long as the integral converges at infinity. This is the case, in particular, with our assumption that $\mathbf{j}(\mathbf{x})$ is localized in space. As a check, observe that this solution was found by writing Maxwell's equations in the Coulomb gauge $\nabla \cdot \mathbf{A} = 0$, and therefore must satisfy this gauge condition. This indeed holds, thanks to eq. (4.90). In fact,

$$
\begin{aligned}
\nabla_{\mathbf{x}} \cdot \mathbf{A}(\mathbf{x}) &= \frac{\mu_0}{4\pi} \int d^3 x' \, \mathbf{j}(\mathbf{x}') \cdot \nabla_{\mathbf{x}} \frac{1}{|\mathbf{x} - \mathbf{x}'|} \\
&= -\frac{\mu_0}{4\pi} \int d^3 x' \, \mathbf{j}(\mathbf{x}') \cdot \nabla_{\mathbf{x}'} \frac{1}{|\mathbf{x} - \mathbf{x}'|} \\
&= \frac{\mu_0}{4\pi} \int d^3 x' \, [\nabla_{\mathbf{x}'} \cdot \mathbf{j}(\mathbf{x}')] \frac{1}{|\mathbf{x} - \mathbf{x}'|} \\
&= 0, \quad\quad\quad\quad\quad\quad\quad\quad\quad\quad\quad\quad (4.93)
\end{aligned}
$$

where we added to the ∇ operator a label $\mathbf{x}$ or $\mathbf{x}'$ to stress on which variable it acts. We then integrated $\nabla_{\mathbf{x}'}$ by parts (setting to zero the boundary terms, by the assumption that $\mathbf{j}(\mathbf{x})$ is localized), and we finally used eq. (4.90). Then, from $\mathbf{B} = \nabla \times \mathbf{A}$,

$$
\mathbf{B}(\mathbf{x}) = \frac{\mu_0}{4\pi} \nabla \times \left(\int d^3 x' \, \frac{\mathbf{j}(\mathbf{x}')}{|\mathbf{x} - \mathbf{x}'|} \right), \quad\quad (4.94)
$$

or, equivalently,[26]

$$
\boxed{ \mathbf{B}(\mathbf{x}) = \frac{\mu_0}{4\pi} \int d^3 x' \, \frac{\mathbf{j}(\mathbf{x}') \times (\mathbf{x} - \mathbf{x}')}{|\mathbf{x} - \mathbf{x}'|^3} . } \quad\quad (4.95)
$$

As an application, consider the situation in which the current is carried by a thin wire that forms a closed loop. We idealize the thin wire as a loop $\mathcal{C}$ of zero thickness and, at a generic point $\mathbf{x} \in \mathcal{C}$, we denote by $d\boldsymbol{\ell}$ the vector tangent to the loop (after having chosen one of the two possible directions for running along the loop $\mathcal{C}$), and of infinitesimal length $d\ell$, see Fig. 4.8. We also denote by $\hat{\mathbf{e}}_{d\ell}$ the unit vector in the direction of $d\boldsymbol{\ell}$, so that

$$
d\boldsymbol{\ell} = d\ell \, \hat{\mathbf{e}}_{d\ell} . \quad\quad (4.96)
$$

We denote by $\mathbf{x}_\perp$ the two dimensional Cartesian coordinates orthogonal to $d\boldsymbol{\ell}$ at the point $\mathbf{x}$ so, for instance, if in $\mathbf{x}$ we have $d\boldsymbol{\ell} = d\ell \, \hat{\mathbf{x}}$, then $\mathbf{x}_\perp = y\hat{\mathbf{y}} + z\hat{\mathbf{z}}$. It is convenient to introduce a one-dimensional coordinate ℓ that parametrizes the position of a point on the loop $\mathcal{C}$, as follows. We arbitrarily choose a point P in $\mathcal{C}$, and we assign to it the value $\ell = 0$. This corresponds to a choice of origin for this coordinate. Given another point $P' \in \mathcal{C}$, we assign it the coordinate ℓ given by

$$
\ell = \int_P^{P'} d\ell , \quad\quad (4.97)
$$

where the integral is a line integral along $\mathcal{C}$. We have therefore constructed a convenient coordinate system $(\ell, \mathbf{x}_\perp)$, useful along the loop and in its immediate neighborhood.[27] The idealization of zero thick-

[26] Explicitly,

$$
\begin{aligned}
B_i(\mathbf{x}) &= \frac{\mu_0}{4\pi} \epsilon_{ijk} \partial_j \int d^3 x' \, \frac{j_k(\mathbf{x}')}{|\mathbf{x} - \mathbf{x}'|} \\
&= \frac{\mu_0}{4\pi} \epsilon_{ijk} \int d^3 x' \, j_k(\mathbf{x}') \, \partial_j \frac{1}{|\mathbf{x} - \mathbf{x}'|} \\
&= -\frac{\mu_0}{4\pi} \epsilon_{ijk} \int d^3 x' \, j_k(\mathbf{x}') \frac{x_j - x_j'}{|\mathbf{x} - \mathbf{x}'|^3} \\
&= +\frac{\mu_0}{4\pi} \epsilon_{ijk} \int d^3 x' \, j_j(\mathbf{x}') \frac{x_k - x_k'}{|\mathbf{x} - \mathbf{x}'|^3} ,
\end{aligned}
$$

where we used eq. (4.19).

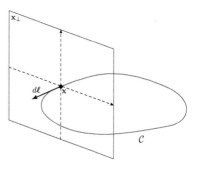

Fig. 4.8 A graphical illustration of the definitions of $d\boldsymbol{\ell}$ and $\mathbf{x}_\perp$, for a loop $\mathcal{C}$.

[27] Actually, in general this coordinate system is well defined (in the sense that it is in one-to-one correspondence with the $\mathbf{x}$ coordinates of the three-dimensional space in which $\mathcal{C}$ is embedded), only in a sufficiently small transverse region around the loop. Indeed, for a closed loop, starting from a point P on the loop with coordinates $(\ell = \ell_1, \mathbf{x}_\perp = 0)$, and moving in the transverse direction, for some value $\mathbf{x}_\perp^*$ of the coordinates $\mathbf{x}_\perp$ we would eventually reach another point P' of the loop, corresponding to a value ℓ_2 of the ℓ coordinate. Then, the coordinate values $(\ell = \ell_1, \mathbf{x}_\perp = \mathbf{x}_\perp^*)$ and $(\ell = \ell_2, \mathbf{x}_\perp = 0)$ would describe the same point in three-dimensional space. However, in the following, we only use this coordinate system in an infinitesimal neighborhood of the loop, so the problem does not arise.

ness means that the current density $\mathbf{j}(\mathbf{x}) = \mathbf{j}(\ell, \mathbf{x}_\perp)$ is a two-dimensional Dirac delta in the transverse directions, i.e.,

$$\mathbf{j}(\ell, \mathbf{x}_\perp) = I(\ell)\delta^{(2)}(\mathbf{x}_\perp)\hat{\mathbf{e}}_{d\ell}, \qquad (4.98)$$

where $I(\ell, \mathbf{x}_\perp = 0) \equiv I(\ell)$ could a priori still be a function of ℓ (while the multiplication by $\delta^{(2)}(\mathbf{x}_\perp)$ allows us to set to zero its argument $\mathbf{x}_\perp$). However, the condition $\mathbf{\nabla} \cdot \mathbf{j} = 0$ implies that $I(\ell)$ is actually independent also of ℓ. For instance, if at the point $\mathbf{x}$ we orient the axes so that $d\boldsymbol{\ell} = dx\,\hat{\mathbf{x}}$, in $\mathbf{x}$ we have $\mathbf{j} = (j_x, 0, 0)$ and current conservation becomes $\partial_x j_x = 0$. Since in this case $d\ell = dx$, this condition is equivalent to

$$\frac{dI(\ell)}{d\ell} = 0, \qquad (4.99)$$

and this holds on a generic point on the loop, i.e., for ℓ generic. Therefore, for an infinitesimally thin wire

$$\mathbf{j}(\ell, \mathbf{x}_\perp) = I\delta^{(2)}(\mathbf{x}_\perp)\hat{\mathbf{e}}_{d\ell}, \qquad (4.100)$$

with I a constant. Recall that $\mathbf{j}(\mathbf{x})$ is a current per unit surface. Since

$$\int d^2\mathbf{x}_\perp \, \mathbf{j}(\ell, \mathbf{x}_\perp) = I\hat{\mathbf{e}}_{d\ell}, \qquad (4.101)$$

I is just the current flowing in the wire. Its independence on ℓ means that, in a single closed loop, the current is the same at all points of the loop. Sufficiently close to a point on the loop labeled, in three-space, by the coordinates $\mathbf{x}$, the variables $\mathbf{x}_\perp$ and ℓ form an orthogonal system of coordinates, so we have $d^3x = d^2\mathbf{x}_\perp d\ell$. Therefore

$$\begin{aligned}
\mathbf{j}(\mathbf{x})d^3x &= I\delta^{(2)}(\mathbf{x}_\perp)\hat{\mathbf{e}}_{d\ell}\, d^2\mathbf{x}_\perp d\ell \\
&= I\delta^{(2)}(\mathbf{x}_\perp)\, d^2\mathbf{x}_\perp d\boldsymbol{\ell}, \qquad (4.102)
\end{aligned}$$

where we used eq. (4.96). Inside an integral, we can first carry out the integration over $d^2\mathbf{x}_\perp$ with the help of the Dirac delta; this amounts to replacing

$$\boxed{\mathbf{j}(\mathbf{x})d^3x \to I d\boldsymbol{\ell},} \qquad (4.103)$$

while setting $\mathbf{x}_\perp = 0$ in any occurrence of $\mathbf{x}_\perp$ in the integrand. Equations (4.92) and (4.95) can then be written as loop integrals, as

$$\boxed{\mathbf{A}(\mathbf{x}) = \frac{\mu_0 I}{4\pi}\oint_C d\boldsymbol{\ell}\,\frac{1}{|\mathbf{x} - \mathbf{x}(\ell)|},} \qquad (4.104)$$

and

$$\boxed{\mathbf{B}(\mathbf{x}) = \frac{\mu_0 I}{4\pi}\oint_C \frac{d\boldsymbol{\ell}\times[\mathbf{x} - \mathbf{x}(\ell)]}{|\mathbf{x} - \mathbf{x}(\ell)|^3},} \qquad (4.105)$$

respectively, where $\mathbf{x}$ is the generic point in space where we compute the field, and $\mathbf{x}(\ell)$ is the coordinate in three-dimensional space of the point of the loop parametrized by ℓ. Equation (4.105) is called the *Biot–Savart law*.[28]

---

[28]Sometime, the name "Biot–Savart law" is given directly to the more general expression (4.95).

4.2.3 Force of a magnetic field on a wire and between two parallel wires

We now compute the force exerted by an external magnetic field $\mathbf{B}$ on a wire currying a current, and we will then use this result to compute the force between two parallel wires. In Section 4.2.4, we will compute the force between two arbitrary current distributions.

Consider a line element $d\boldsymbol{\ell} = d\ell\,\hat{\mathbf{e}}_{d\ell}$ of the wire, see eq. (4.96). We denote by n_e the number of electrons per unit length in the wire, all taken to have a velocity $\mathbf{v} = -v\hat{\mathbf{e}}_{d\ell}$. Given that the electron charge is $q = -e < 0$, we have $q\mathbf{v} = ev\hat{\mathbf{e}}_{d\ell}$, so the current flows in the direction $+\hat{\mathbf{e}}_{d\ell}$.[29] The Lorentz force (3.5) on a single electron, due to the external magnetic field, is given by $\mathbf{F} = (-e)(-v\hat{\mathbf{e}}_{d\ell})\times\mathbf{B} = ev\hat{\mathbf{e}}_{d\ell}\times\mathbf{B}$. The total charge contained in a line element $d\ell$ is $(-e)n_e d\ell$, so the force on it is $d\mathbf{F} = (-en_e d\ell)(-v\hat{\mathbf{e}}_{d\ell}) \times \mathbf{B} = (en_e v)\,d\ell\,\hat{\mathbf{e}}_{d\ell} \times \mathbf{B}$. In a time dt, the total charge passing through a transverse surface of the wire is $dQ = en_e(vdt)$, so the current $I = dQ/dt$ flowing in the wire is $I = en_e v$. Therefore

$$\boxed{d\mathbf{F} = Id\boldsymbol{\ell} \times \mathbf{B}\,,} \qquad (4.106)$$

or, equivalently,

$$\frac{d\mathbf{F}}{d\ell} = I\,\hat{\mathbf{e}}_{d\ell} \times \mathbf{B}\,, \qquad (4.107)$$

and the total force on the loop is[30]

$$\boxed{\mathbf{F} = I \oint_C d\boldsymbol{\ell} \times \mathbf{B}\,.} \qquad (4.109)$$

More generally, if we have a current density $\mathbf{j}(\mathbf{x})$ not necessarily confined to a one-dimensional wire, we can repeat the same argument with $\mathbf{j}(\mathbf{x})d^3x$ replacing $Id\boldsymbol{\ell}$ [compare with eq. (4.103)], and we get[31]

$$\boxed{\mathbf{F} = \int d^3x\, \mathbf{j}(\mathbf{x})\times\mathbf{B}(\mathbf{x})\,.} \qquad (4.110)$$

We next compute the force between two infinite straight parallel wires carrying steady currents I_1 and I_2. We consider first the case where the two currents are in the same direction, that we take to be the $+\hat{\mathbf{z}}$ direction, and we denote by d the distance between the two wires. We denote by $d\mathbf{F}_2/dl$ the force per unit length on the second wire (which we take parallel to the z axis and located at a distance d from it) due to the magnetic field of the first wire (again parallel to the z axis and located in $\mathbf{x}_\perp = 0$). From eq. (4.107),

$$\frac{d\mathbf{F}_2}{d\ell} = I_2\hat{\mathbf{z}} \times \mathbf{B}_1\,, \qquad (4.111)$$

[29]This is not how currents really flow in metals. As we will discuss in Section 14.4, individual electrons undergo collisions on microscopic distance scales, and $\mathbf{v}$ should really be understood as a macroscopic drift velocity of the ensemble of electrons, not as the velocity of individual electrons. However, once understood in this average sense, this derivation of the force on a wire is correct.

[30]Observe that, if $\mathbf{B}(\mathbf{x}) = \mathbf{B}_0$ is spatially constant, the total force on a closed wire vanishes,

$$\mathbf{F} = I\left(\oint_C d\boldsymbol{\ell}\right) \times \mathbf{B}_0 = 0\,, \qquad (4.108)$$

since $\oint_C d\boldsymbol{\ell} = 0$.

[31]Actually, we already knew this result from the derivation of momentum conservation in eq. (3.68). Rather than deriving eq. (4.109) from the Lorentz force on the individual electrons, we could have equivalently started from eq. (3.68) to write eq. (4.110) and then, specializing to a wire using eq. (4.103), we get eq. (4.109).

where $\mathbf{B}_1$ is the magnetic field generated by the first wire. Using eq. (4.79) gives

$$
\begin{aligned}
\frac{d\mathbf{F}_2}{d\ell} &= I_2\hat{\mathbf{z}} \times \left(\mu_0 \frac{I_1}{2\pi d} \hat{\boldsymbol{\varphi}} \right) \\
&= -\frac{\mu_0}{2\pi} \frac{I_1 I_2}{d} \hat{\boldsymbol{\rho}} , \quad (4.112)
\end{aligned}
$$

[32] Note that, in order for $\{\hat{\boldsymbol{\rho}}, \hat{\boldsymbol{\varphi}}, \hat{\mathbf{z}}\}$ to be a right-handed frame, we must have $\hat{\boldsymbol{\rho}} \times \hat{\boldsymbol{\varphi}} = +\hat{\mathbf{z}}$, so $\hat{\mathbf{z}} \times \hat{\boldsymbol{\varphi}} = -\hat{\boldsymbol{\rho}}$. This convention has already been implicitly used when we derived the solution (4.79), since it is implied by the expressions (1.30) and (1.33) of the divergence and curl in cylindrical coordinates, see Note 1 on page 4.

where we used $\hat{\mathbf{z}} \times \hat{\boldsymbol{\varphi}} = -\hat{\boldsymbol{\rho}}$.[32] We have therefore recovered eq. (2.7), that we anticipated in Section 2.1. This shows that the vacuum magnetic permeability μ_0, that was originally defined from eq. (2.7), is indeed the same as the constant μ_0 that appears in the Ampère law (4.67). Note that, if the currents are parallel, the force is attractive. The case of anti-parallel currents can be obtained reversing the sign of $I_1\hat{\mathbf{z}}$ in the previous derivation, and therefore the force becomes repulsive.

4.2.4   Force between generic static current distributions

We can now compute the magnetic force between two arbitrary static current densities $\mathbf{j}_1(\mathbf{x})$ and $\mathbf{j}_2(\mathbf{x})$. We set again $\rho_1 = \rho_2 = 0$, and we take the currents to be separately conserved,[33] so that

[33] This is the case, in particular, if the two current densities are spatially non-overlapping, as, for instance, in the case where they are confined to two different wires.

$$
\boldsymbol{\nabla} \cdot \mathbf{j}_1 = \boldsymbol{\nabla} \cdot \mathbf{j}_2 = 0 . \quad (4.113)
$$

From eq. (4.110), the force $\mathbf{F}_1$ exerted on the current density $\mathbf{j}_1$ by the magnetic field generated by the current density $\mathbf{j}_2$ is

$$
\mathbf{F}_1 = \int d^3x\, \mathbf{j}_1(\mathbf{x}) \times \mathbf{B}_2(\mathbf{x}) . \quad (4.114)
$$

Using eq. (4.95),

$$
\mathbf{B}_2(\mathbf{x}) = \frac{\mu_0}{4\pi} \int d^3x'\, \frac{\mathbf{j}_2(\mathbf{x}') \times (\mathbf{x} - \mathbf{x}')}{|\mathbf{x} - \mathbf{x}'|^3} , \quad (4.115)
$$

and therefore

$$
\begin{aligned}
\mathbf{F}_1 &= \frac{\mu_0}{4\pi} \int d^3x\, d^3x'\, \frac{\mathbf{j}_1(\mathbf{x}) \times [\mathbf{j}_2(\mathbf{x}') \times (\mathbf{x} - \mathbf{x}')]}{|\mathbf{x} - \mathbf{x}'|^3} \quad (4.116) \\
&= \frac{\mu_0}{4\pi} \int d^3x\, d^3x'\, \frac{[\mathbf{j}_1(\mathbf{x}) \cdot (\mathbf{x} - \mathbf{x}')]\, \mathbf{j}_2(\mathbf{x}') - [\mathbf{j}_1(\mathbf{x}) \cdot \mathbf{j}_2(\mathbf{x}')]\, (\mathbf{x} - \mathbf{x}')}{|\mathbf{x} - \mathbf{x}'|^3} ,
\end{aligned}
$$

where we used eq. (1.9) to expand the triple vector product. We now observe that

$$
\begin{aligned}
\int d^3x\, \frac{\mathbf{j}_1(\mathbf{x}) \cdot (\mathbf{x} - \mathbf{x}')}{|\mathbf{x} - \mathbf{x}'|^3} &= -\int d^3x\, \mathbf{j}_1(\mathbf{x}) \cdot \boldsymbol{\nabla}_{\mathbf{x}} \left(\frac{1}{|\mathbf{x} - \mathbf{x}'|} \right) \\
&= +\int d^3x\, [\boldsymbol{\nabla}_{\mathbf{x}} \cdot \mathbf{j}_1(\mathbf{x})] \frac{1}{|\mathbf{x} - \mathbf{x}'|} \\
&= 0 , \quad (4.117)
\end{aligned}
$$

where we used eq. (4.19), we integrated by parts discarding the boundary term (which is valid if $\mathbf{j}_1(\mathbf{x})$ is localized, or, more generally, if it decreases faster than $1/|\mathbf{x}|$ as $|\mathbf{x}| \to \infty$), and we finally used current conservation in the form (4.113). Therefore, eq. (4.116) simplifies to

$$\mathbf{F}_1 = -\frac{\mu_0}{4\pi} \int d^3x\, d^3x'\, \mathbf{j}_1(\mathbf{x}) \cdot \mathbf{j}_2(\mathbf{x}')\, \frac{\mathbf{x} - \mathbf{x}'}{|\mathbf{x} - \mathbf{x}'|^3}\,. \qquad (4.118)$$

Equation (4.118) is the analogue of eq. (4.26) for the magnetostatic case. For two current loops $\mathcal{C}_1$ and $\mathcal{C}_2$, using eq. (4.103), we get

$$\mathbf{F}_1 = -\frac{\mu_0 I_1 I_2}{4\pi} \oint_{\mathcal{C}_1} \oint_{\mathcal{C}_2} d\boldsymbol{\ell}_1 \cdot d\boldsymbol{\ell}_2\, \frac{\mathbf{x}(\ell_1) - \mathbf{x}(\ell_2)}{|\mathbf{x}(\ell_1) - \mathbf{x}(\ell_2)|^3}\,, \qquad (4.119)$$

where $\mathbf{x}(\ell_1)$ is the spatial coordinate of the points of the loop $\mathcal{C}_1$ labeled by the coordinate ℓ_1 along the loop [defined as in eq. (4.97)], and $\mathbf{x}(\ell_2)$ is the spatial coordinate of the points of the loop $\mathcal{C}_2$ labeled by the coordinate ℓ_2.

Similarly to the electric case [see the discussion following eq. (4.26)], the force $\mathbf{F}_2$ exerted on the current density $\mathbf{j}_2$ by the current density $\mathbf{j}_1$ is obtained exchanging $1 \leftrightarrow 2$ in the previous expression,

$$\mathbf{F}_2 = -\frac{\mu_0}{4\pi} \int d^3x\, d^3x'\, \mathbf{j}_2(\mathbf{x}) \cdot \mathbf{j}_1(\mathbf{x}')\, \frac{\mathbf{x} - \mathbf{x}'}{|\mathbf{x} - \mathbf{x}'|^3}\,, \qquad (4.120)$$

and, renaming $\mathbf{x} \leftrightarrow \mathbf{x}'$, we see that $\mathbf{F}_2 = -\mathbf{F}_1$, so the magnetostatic force satisfies Newton's third law. We also observe that, if we set $\mathbf{j}_2(\mathbf{x}) = \mathbf{j}_1(\mathbf{x})$, the integral vanishes because it becomes odd under $\mathbf{x} \leftrightarrow \mathbf{x}'$. This shows that a current distribution does not exert a force on itself.

To make contact with the setting of Section 4.2.3, we now choose $\mathbf{j}_1(\mathbf{x}') = I_1 \delta^{(2)}(\mathbf{x}'_\perp)\hat{\mathbf{z}}$ and $\mathbf{j}_2(\mathbf{x}) = I_2 \delta^{(2)}(\mathbf{x}_\perp - \mathbf{d}_\perp)\hat{\mathbf{z}}$ [compare this with eq. (4.100)], where $\mathbf{d}_\perp$ is a vector of modulus d in the transverse (x, y) plane. This corresponds to two infinite straight wires with parallel currents, separated by a distance $d = |\mathbf{d}_\perp|$ in the transverse plane. For definiteness, we take $\mathbf{d}_\perp$ along the x axis, so that $\mathbf{d}_\perp = (d, 0)$. Writing $d^3x = d^2\mathbf{x}_\perp dz$ and $d^3x' = d^2\mathbf{x}'_\perp dz'$, eq. (4.120) gives

$$\frac{d\mathbf{F}_2}{dz} = -\frac{\mu_0 I_1 I_2}{4\pi} \int_{-\infty}^{\infty} dz'\, \frac{\mathbf{x}_2(z) - \mathbf{x}_1(z')}{|\mathbf{x}_1(z') - \mathbf{x}_2(z)|^3}\,, \qquad (4.121)$$

where, having performed the integration over $d^2\mathbf{x}_\perp$ and $d^2\mathbf{x}'_\perp$ with the help of the Dirac deltas $\delta^{(2)}(\mathbf{x}'_\perp)$ and $\delta^{(2)}(\mathbf{x}_\perp - \mathbf{d}_\perp)$, we have $\mathbf{x}_2(z) = (d, 0, z)$ and $\mathbf{x}_1(z') = (0, 0, z')$. Carrying out the integral,[34] we get

$$\frac{d\mathbf{F}_2}{dz} = -\frac{\mu_0}{2\pi} \frac{I_1 I_2}{d}\, \hat{\mathbf{x}}\,. \qquad (4.122)$$

Since we have chosen $\hat{\boldsymbol{\rho}} = \hat{\mathbf{x}}$, where, in general, $\hat{\boldsymbol{\rho}}$ is the unit vector in the radial direction of the transverse plane, from the wire 1 sitting at

---

[34]Explicitly, we write $\mathbf{x}_2(z) - \mathbf{x}_1(z') = (d, 0, z - z')$, and therefore

$$\left(\frac{d\mathbf{F}_2}{dz}\right)_i = -\frac{\mu_0 I_1 I_2}{4\pi}$$
$$\times \int_{-\infty}^{\infty} dz'\, \frac{(d, 0, z - z')_i}{[d^2 + (z - z')^2]^{3/2}}\,.$$

Passing to the integration variable $u = (z - z')/d$, we see that the third component in this vector expression vanishes because the integrand is odd in u, while, for the first component, we get

$$\int_{-\infty}^{\infty} dz'\, \frac{d}{[d^2 + (z - z')^2]^{3/2}}$$
$$= \frac{1}{d} \int_{-\infty}^{\infty} du\, \frac{1}{[1 + u^2]^{3/2}}$$
$$= \frac{2}{d}\,.$$

$\mathbf{x}_\perp = 0$ toward the wire 2 (and $dz = d\ell$, since the wires are parallel to the z axis), we see that we have recovered eq. (4.112).

It is interesting to observe that the results for the electrostatic and the magnetostatic forces, eqs. (4.26) and (4.118), are completely analogous, except for the overall sign, which is such that the electric force between charges of the same sign is repulsive, while the magnetic force between parallel currents is attractive. The similar structure of the integrals comes from the fact that in both cases we had to solve a Laplace equation, for the scalar or the vector potential. The relative minus sign can be traced to the different nature (scalar or vector, respectively) of the scalar and vector potentials, resulting in different tensor structures of the indices involving the $\boldsymbol{\nabla}$ operator, see in particular the minus sign coming from the triple vector product in eq. (4.116).[35]

[35] The difference between the behavior of the Coulomb force between two point charges, which decreases as $1/r^2$, and the magnetic force between two infinite straight parallel wires, which decreases with distance only as $1/d$, is due to the different structure of the sources, which is a three-dimensional Dirac delta for a point-like charge, but just a two-dimensional Dirac delta in the transverse plane, extending infinitely in the longitudinal direction, for an infinite straight wire. Note, however, that when the current densities are localized, as for two well separated wire loops, the factor $\mathbf{j}_1(\mathbf{x}) \cdot \mathbf{j}_2(\mathbf{x}')$ in eq. (4.118) gets both positive and negative contributions, depending on the relative orientations of different portions of the wires, and therefore there are partial cancellations. As a result, the magnetic force decreases faster than the Coulomb force between two charge densities $\rho_1(\mathbf{x})$ and $\rho_2(\mathbf{x}')$ with fixed signs (and is very sensitive to the relative orientation of the loops). We will come back to this when we study the expansion in electric and magnetic multipoles, in Chapter 6.

4.2.5   Magnetic forces from surface integrals

We now show that, similarly to the situation discussed in Section 4.1.7 for electrostatics, the force exerted by a static magnetic field on a localized current distribution can be written as a surface integral, on a surface enclosing the source.

We start from eq. (4.110). We observe that, in that equation, $\mathbf{B}$ was an external magnetic field acting on a current $\mathbf{j}$. Below eq. (4.120) we have found, however, that the force generated by a current distribution on itself vanishes. We can therefore extend eq. (4.110) to a generic current $\mathbf{j}$, possibly made of several disjoint contributions, and we can take $\mathbf{B}$ as the total magnetic field generated by $\mathbf{j}$. We can then use Ampère's law (4.67), so that

$$\mathbf{F} = \frac{1}{\mu_0} \int d^3x \, (\boldsymbol{\nabla} \times \mathbf{B}) \times \mathbf{B} \,. \tag{4.123}$$

Using the identity (1.7), we have

$$[(\boldsymbol{\nabla} \times \mathbf{B}) \times \mathbf{B}]_i = B_k \partial_k B_i - (\partial_i B_k) B_k \,. \tag{4.124}$$

Using $\boldsymbol{\nabla} \cdot \mathbf{B} = 0$, we can rewrite $B_k \partial_k B_i = \partial_k (B_k B_i)$, so

$$[(\boldsymbol{\nabla} \times \mathbf{B}) \times \mathbf{B}]_i = \partial_k \left(B_i B_k - \frac{1}{2} \delta_{ik} B^2 \right) \,. \tag{4.125}$$

Therefore, if V is any volume such that the $\mathbf{j}$ vanishes outside it,

$$\begin{aligned} F_i &= \frac{1}{\mu_0} \int_V d^3x \, \partial_k \left(B_i B_k - \frac{1}{2} \delta_{ik} B^2 \right) \\ &= \frac{1}{\mu_0} \int_{\partial V} d^2s \, n_k \left(B_i B_k - \frac{1}{2} \delta_{ik} B^2 \right) \,, \end{aligned} \tag{4.126}$$

or, in vector form,

$$\boxed{\mathbf{F} = \frac{1}{\mu_0} \int_{\partial V} d^2s \left[(\mathbf{B} \cdot \hat{\mathbf{n}}) \mathbf{B} - \frac{1}{2} B^2 \hat{\mathbf{n}} \right] \,,} \tag{4.127}$$

which is the magnetic analog of eq. (4.60). The force is then expressed in terms of the magnetic part of Maxwell's stress tensor (3.64).

4.3 Electromagnetic induction

We now take a closer look to Faraday's law, in the integrated form (3.21). This law governs the induction phenomena on which are based ac generators, transformers, etc., and therefore has a fundamental role in electrical engineering and all the related technology. Here we will limit to two basic examples that put in evidence the basic principles.

4.3.1 Time-varying magnetic field and Lenz's law

Equation (3.21) tells us that, if the flux of the magnetic field through a surface S, with boundary $\mathcal{C} = \partial S$, changes in time, this induces a circulation of the electric field along the closed curve $\mathcal{C}$. Therefore, a time-varying magnetic field generates an electric field, which is called the "induced" electric field. We are interested, in particular, in the case in which $\mathcal{C}$ is the loop made by a conducting wire. As the simplest example, we consider a wire in the (x, y) plane, at rest in our reference frame, and a magnetic field $\mathbf{B}_{\text{ext}}$ in the $\hat{\mathbf{z}}$ direction, see Fig. 4.9, and we increase the magnetic field with time. Then, there will be an induced electric field in the wire. This will generate a current in the wire (the "induced current") and, in turn, a wire carrying a current generates a magnetic field (the "induced magnetic field"). Since we will eventually use the approximation of magnetostatics to compute the induced magnetic field, the following computation is only valid if the change in time of the external magnetic field is quasi-adiabatic.

Lenz's law states that the induced magnetic field has the direction that opposes the change of the flux of the external magnetic field. To see how this comes out from eq. (3.21), recall that the flux Φ_B is defined choosing an orientation for the normal of the surface S; for the surface bounded by the wire in Fig. 4.9, let us choose its normal, for instance, in the positive $\hat{\mathbf{z}}$ direction. From Stokes's theorem, this choice then fixes the direction of the line element $d\boldsymbol{\ell}$ in eq. (3.21), according to the "right-hand rule": if we close the right hand along the direction of integration of the loop $\mathcal{C}$, the thumb must point in the direction of the normal to the surface. Therefore, choosing the $+\hat{\mathbf{z}}$ direction for the normal of the surface implies that the line integral on the left-hand side of eq. (3.21) runs counterclockwise. In our setting, we have chosen $\mathbf{B}_{\text{ext}} = B_{\text{ext}}(t)\hat{\mathbf{z}}$ with $B_{\text{ext}}(t) > 0$ and $dB_{\text{ext}}/dt > 0$, and, having chosen the normal $\hat{\mathbf{n}}$ to S equal to $+\hat{\mathbf{z}}$, so we have $\Phi_{B_{\text{ext}}} > 0$ and $d\Phi_{B_{\text{ext}}}/dt > 0$. Therefore, from eq. (3.21), along the wire the electric field must point in the clockwise direction, so that $d\boldsymbol{\ell}\cdot\mathbf{E} < 0$, in order to compensate for the minus sign on the right-hand side.[36]

This electric field generates in the wire a current $\mathbf{j}$ in the direction of $\mathbf{E}$,[37] which therefore also runs clockwise, as shown in the figure. In turn, this current generates a magnetic field (the "induced magnetic field") $\mathbf{B}_{\text{ind}}$. From eq. (4.105), one can see that the induced magnetic field circulates around the wire, in the direction shown in Fig. 4.9; at a sufficiently small distance from the wire, where it can be approximated as

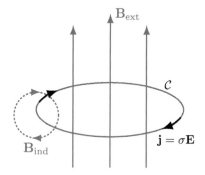

Fig. 4.9 The induced current and induced magnetic field created by increasing the flux of an external magnetic field through the closed loop $\mathcal{C}$.

[36] Had we chosen $-\hat{\mathbf{z}}$ as the direction of the normal to S, we would now have Φ_B negative and increasing in absolute value, so $d\Phi_B/dt < 0$. Then, overall, the right-hand side of eq. (3.21) would now be positive. On the other hand, with this choice of the normal, the line integral in Stokes's theorem would run clockwise. Since we would now need $d\boldsymbol{\ell}\cdot\mathbf{E} > 0$ to match the sign on the right-hand side, we would still conclude that $\mathbf{E}$ points in the clockwise direction. Of course, the direction of the induced electric field does not depend on our arbitrary choice of the direction of the normal to the surface S.

[37] As we will discuss in Section 13.6.2, for a simple conducting wire the current is given by Ohm's law, $\mathbf{j} = \sigma\mathbf{E}$, where the positive constant σ is the conductivity of the material.

straight, this can also be seen more simply using the result for a straight wire found in Section 4.2.1, see Fig. 4.7. So, inside the wire, $\mathbf{B}_{\text{ind}}$ is in the direction opposite to the external field. It therefore generates a flux that goes in the direction of opposing the increase of the flux of the external magnetic field, which is the content of Lenz's law.

Note that Lenz's law states that the induced magnetic field opposes the change of the flux, not the flux itself. For instance, if we repeat the same argument as before starting from a given magnetic field $\mathbf{B}_{\text{ext}} = B_{\text{ext}}\hat{\mathbf{z}}$, with $B_{\text{ext}} > 0$, and we decrease B_{ext} toward zero, we find that the induced current will flow counterclockwise, so, in the direction that tries to restore the original value of the field inside the loop.

4.3.2 Induction on moving loops

We next consider the case in which a wire, which makes a closed loop, moves with respect to a magnetic field. It takes no extra effort to consider the most general case in which the magnetic field also changes with time, so we will include both effects. The situation is depicted in Fig. 4.10, where we show the position of a loop $\mathcal{C}$, representing a closed wire, at time t and at time $t + \delta t$, and two corresponding surfaces, $S(t)$ and $S(t + \delta t)$, that have $\mathcal{C}(t)$ and $\mathcal{C}(t + \delta t)$, respectively, as boundaries. We do not need to assume that the loop moves rigidly. The motion of the loop is determined by giving, at each point of $\mathcal{C}(t)$, the corresponding velocity $\mathbf{v}(t)$. The position of that point at time $t + \delta t$ is then obtained by adding to it the vector $\mathbf{v}(t)\delta t$, as shown in the figure, and $\mathbf{v}(t)$ is allowed to change from point to point of the loop (we should then write $\mathbf{v}[t, \mathbf{x}(\ell)]$ where ℓ is the curvilinear coordinate along the loop, see eq. (4.97), but we will keep the notation simple).

The difference between the magnetic flux going through $S(t + \delta t)$ and that going through $S(t)$ is given by

$$\delta\Phi_B(t) = \int_{S(t+\delta t)} d\mathbf{s} \cdot \mathbf{B}(t+\delta t, \mathbf{x}) - \int_{S(t)} d\mathbf{s} \cdot \mathbf{B}(t, \mathbf{x}). \qquad (4.128)$$

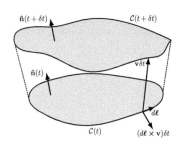

Fig. 4.10 A closed loop evolving in time, and two surfaces that have $\mathcal{C}(t)$ and $\mathcal{C}(t + \delta t)$, respectively, as boundaries.

Working to first order in the infinitesimal quantity δt, we can manipulate this as

$$\begin{aligned}
\delta\Phi_B(t) &= \int_{S(t+\delta t)} d\mathbf{s} \cdot [\mathbf{B}(t+\delta t, \mathbf{x}) - \mathbf{B}(t, \mathbf{x})] \qquad (4.129)\\
&\quad + \int_{S(t+\delta t)} d\mathbf{s} \cdot \mathbf{B}(t, \mathbf{x}) - \int_{S(t)} d\mathbf{s} \cdot \mathbf{B}(t, \mathbf{x})\\
&= \delta t \int_{S(t)} d\mathbf{s} \cdot \frac{\partial\mathbf{B}}{\partial t} + \int_{S(t+\delta t)} d\mathbf{s} \cdot \mathbf{B}(t, \mathbf{x}) - \int_{S(t)} d\mathbf{s} \cdot \mathbf{B}(t, \mathbf{x}),
\end{aligned}$$

where we first added and subtracted the same quantity $\int_{S(t+\delta t)} d\mathbf{s} \cdot \mathbf{B}(t, \mathbf{x})$ and then, in the first integral of the last line, to first order in δt we could replace $S(t+\delta t)$ by $S(t)$, since this terms already has a factor δt in front. We now consider the closed volume V bounded by the surfaces $\mathcal{C}(t)$ and $\mathcal{C}(t+\delta t)$, and by the lateral cylindrical region S_L swept by the loop when it evolves from $\mathcal{C}(t)$ to $\mathcal{C}(t + \delta t)$. Observe that, on $S(t + \delta t)$, the outer normal to ∂V is the same as the normal $\hat{\mathbf{n}}(t + \delta t)$ of $\mathcal{C}(t + \delta t)$, while, on $S(t)$, the outer normal to ∂V is the same as $-\hat{\mathbf{n}}(t)$, see Fig. 4.10. Using the fact that $\boldsymbol{\nabla}\cdot\mathbf{B} = 0$, together with Gauss's theorem, we have

$$
\begin{aligned}
0 &= \int_V d^3x \, \boldsymbol{\nabla}\cdot\mathbf{B} \qquad\qquad\qquad\qquad\qquad\quad (4.130) \\
&= \int_{S(t+\delta t)} d\mathbf{s}\cdot \mathbf{B}(t,\mathbf{x}) - \int_{S(t)} d\mathbf{s}\cdot\mathbf{B}(t,\mathbf{x}) + \int_{S_L} d\mathbf{s}\cdot \mathbf{B}(t,\mathbf{x}) ,
\end{aligned}
$$

where the minus sign in front of the second term is due to the fact that the outer normal of ∂V on $S(t)$ is minus the normal $\hat{\mathbf{n}}(t)$ shown in Fig. 4.10. We now observe, again from Fig. 4.10, that the surface element of the lateral surface S_L is given by

$$
d\mathbf{s} = d\boldsymbol{\ell} \times (\mathbf{v}\delta t) , \qquad\qquad\qquad (4.131)
$$

where $d\boldsymbol{\ell}$ is the line element on $\mathcal{C}(t)$. Therefore, eq. (4.130) gives

$$
\begin{aligned}
\int_{S(t+\delta t)} d\mathbf{s}\cdot \mathbf{B}(t,\mathbf{x}) - \int_{S(t)} d\mathbf{s}\cdot\mathbf{B}(t,\mathbf{x}) &= -\delta t \oint_{\mathcal{C}} (d\boldsymbol{\ell}\times\mathbf{v})\cdot \mathbf{B}(t,\mathbf{x}) \\
&= -\delta t \oint_{\mathcal{C}(t)} d\boldsymbol{\ell} \cdot (\mathbf{v}\times\mathbf{B}) , \quad (4.132)
\end{aligned}
$$

where, more explicitly, $\mathbf{v} = \mathbf{v}[t,\mathbf{x}(\ell)]$ and $\mathbf{B} = \mathbf{B}[t,\mathbf{x}(\ell)]$. Plugging this into eq. (4.129) and taking the limit $\delta t \to 0$, we get

$$
\frac{d\Phi_B}{dt} = \int_{S(t)} d\mathbf{s}\cdot\frac{\partial\mathbf{B}}{\partial t} - \oint_{\mathcal{C}(t)} d\boldsymbol{\ell}\cdot(\mathbf{v}\times\mathbf{B}) . \qquad (4.133)
$$

Finally, in the first integral we express $\partial\mathbf{B}/\partial t$ in terms of $\boldsymbol{\nabla}\times\mathbf{E}$ using Faraday's law (3.4) and we use Stokes's theorem (1.38). This gives the final expression

$$
\boxed{\frac{d\Phi_B}{dt} = - \oint_{\mathcal{C}(t)} d\boldsymbol{\ell}\cdot(\mathbf{E} + \mathbf{v}\times\mathbf{B}) .} \qquad (4.134)
$$

Equation (4.134) shows that, for a moving loop, the electromotive force $\mathcal{E}_{\text{emf}}$, that, for a static loop, we have already introduced after eq. (3.21), can be written as

$$
\mathcal{E}_{\text{emf}} = \oint_{\mathcal{C}(t)} d\boldsymbol{\ell}\cdot(\mathbf{E} + \mathbf{v}\times\mathbf{B}) , \qquad (4.135)
$$

and is made of two terms: the first is the electric field induced by the time derivative of the magnetic field, and the second (the "motional emf") is due to the motion of the loop in the magnetic field. Note that the latter gives an electromotive force even in a static magnetic field.

Observe that, on the right-hand side of eq. (4.134) we have obtained the combination of electric and magnetic fields, $\mathbf{E} + \mathbf{v}\times\mathbf{B}$, that enters in the Lorentz force (3.5).[38]

---

[38]This, eventually, is dictated by the underlying Lorentz covariance of Maxwell's equations. As we will see in Section 8.6.1, the combination $\mathbf{E}+\mathbf{v}\times\mathbf{B}$ is the spatial component of a four-vector.

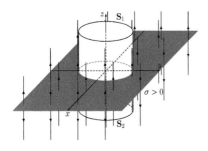

Fig. 4.11 The electric field generated by a charged plane, and the cylinder through which we compute the flux of the electric field. The arrows denote the lines of the electric field.

[39] As we have already seen in Section 4.2.1, there is a subtle limitation to this use of symmetry arguments, which is related to the phenomenon of *spontaneous symmetry breaking* and is especially important in particle physics and in condensed matter. In general, the symmetries of the problem (i.e., the symmetries of the equations that govern the problem, including its geometry and boundary conditions) are the same as the symmetries of the solution of these equations only when the solution is unique. More generally, one can have a family of solutions, that are not invariant under the symmetry transformation of the problem (in our example of an infinite plane, rotations around the z axis), but are transformed into each other by such transformations. A classic example is given by a ferromagnet: the fundamental equations governing the interaction between elementary dipole moments in a ferromagnet are invariant under the full group of rotations in three dimensions. However, in its magnetized phase, in a ferromagnet the dipole moments are aligned along one specific direction, randomly chosen, so this solution is not invariant under rotations. The symmetry under rotations is now reflected in the fact that the ferromagnet can align itself along any direction, so there is a family of solutions, related to each other by rotations. However, the solution to electrostatic problems is unique, as we have shown in Section 4.1.5, so in our case this use of symmetry principles is correct.

4.4   Solved problems

In this section we collect, in the form of Solved Problems, a number of other rather classic applications of electrostatics and magnetostatics.

Problem 4.1. Electric field of an infinite charged plane

The integrated form of Maxwell's equations, presented in Section 3.1.2, is particularly useful when the geometry of the problem has a high degree of symmetry. As an example, consider the electric field generated by a static charge density, distributed uniformly on a plane, idealized to be infinite in extent and with zero thickness. Let σ be the charge per unit surface on the plane. We orient the axes so that the charged plane coincides with the (x, y) plane, as in Fig. 4.11. For symmetry reasons, the electric field must then point in the $\pm \hat{\mathbf{z}}$ directions and its modulus cannot depend on x, y.[39] Taking for definiteness a charge density $\sigma > 0$, so that the lines of electric field go out from the plane, we have

$$\mathbf{E} = \pm E(z) \hat{\mathbf{z}}, \tag{4.136}$$

where the plus sign holds for $z > 0$ and the minus sign for $z < 0$, and $E(z) = |\mathbf{E}(z)|$. To compute $E(z)$, we consider a volume V bounded by the cylinder shown in Fig. 4.11, with end-faces $\mathbf{S}_1 = S\hat{\mathbf{z}}$ and $\mathbf{S}_2 = S \times (-\hat{\mathbf{z}})$, both of area S, located at $\pm z$, respectively, and we use Gauss's law in the integrated form (3.12). The charge inside the cylinder is $Q = \sigma S$. The fluxes through $\mathbf{S}_1$ and $\mathbf{S}_2$ are equal and add up, since, at $\mathbf{S}_1$, both $\mathbf{E}$ and the outer normal to the surface are in the $\hat{\mathbf{z}}$ direction, while at $\mathbf{S}_2$ they are both in the $-\hat{\mathbf{z}}$ direction, so $\mathbf{E}(z){\cdot}\mathbf{S}_1 = \mathbf{E}(-z){\cdot}\mathbf{S}_2 = E(z)S$, while there is no flux from the lateral surface of the cylinder. Then, eq. (3.12) gives

$$2SE(z) = \frac{\sigma S}{\epsilon_0}, \tag{4.137}$$

so, in the end, the modulus of the electric field is independent even of z, and

$$\mathbf{E} = \pm \frac{\sigma}{2\epsilon_0} \hat{\mathbf{z}}. \tag{4.138}$$

It is instructive to repeat the computation by performing explicitly the integration over the electric fields generated by the infinitesimal surface elements of the plane. We choose the coordinates so that the charged plane corresponds to $z = 0$, and we compute the electric field at a given point P with $z > 0$. We set the origin of the reference frame so that the coordinates of this point are $(0, 0, z)$ and, in the (x, y) plane, we use polar coordinates (ρ, φ). We compute first the electric field generated in P by the charges in an infinitesimal ring, lying in the charged plane, and with radial coordinate between ρ and $\rho + d\rho$, while $0 \leq \varphi < 2\pi$. Integrating over φ, the total contribution to the components of $\mathbf{E}$ parallel to the plane, E_x and E_y, vanishes, since the contribution from an infinitesimal surface $\rho d\rho d\varphi$ at a given value of φ is canceled by the contribution from the infinitesimal surface on the other side of the ring, at $\varphi + \pi$, so $E_x = E_y = 0$. This explicitly confirms the result obtained previously from symmetry arguments. Writing $E_z = |\mathbf{E}| \cos \theta$, all the points in the ring, i.e., with the same value of ρ, contributes in the same way to the modulus, as well as to $\cos \theta$,

$$|\mathbf{E}| = \frac{1}{4\pi\epsilon_0} \frac{\sigma \rho d\rho d\varphi}{\rho^2 + z^2}, \qquad \cos \theta = \frac{z}{(\rho^2 + z^2)^{1/2}}. \tag{4.139}$$

Therefore,

$$E_z = \frac{\sigma}{4\pi\epsilon_0} \, 2\pi z \int_0^\infty d\rho \, \frac{\rho}{(\rho^2 + z^2)^{3/2}} \, . \qquad (4.140)$$

Passing to the integration variable $u = \rho/z$, we see that z cancels and

$$E_z = \frac{\sigma}{2\epsilon_0} \int_0^\infty du \, \frac{u}{(u^2 + 1)^{3/2}} \, . \qquad (4.141)$$

The integral is elementary and equal to 1, so we recover eq. (4.138).

We can now determine the corresponding gauge potentials ϕ and $\mathbf{A}$. In a problem of electrostatics the magnetic field vanishes, and we can set $\mathbf{A} = 0$. Then, eq. (3.83) gives $\mathbf{E} = -\nabla\phi$. In this case, again for the symmetry of the problem, ϕ cannot depend on (x, y), so $\phi = \phi(z)$ and $\nabla\phi = (d\phi/dz)\hat{\mathbf{z}}$. Then, from eq. (4.138),

$$\frac{d\phi}{dz} = \mp\frac{\sigma}{2\epsilon_0} \, , \qquad (4.142)$$

so

$$\phi(z) = \phi_0 - \frac{\sigma}{2\epsilon_0}|z| \, , \qquad (4.143)$$

where ϕ_0 is an arbitrary integration constant. The use of the integrated form of Maxwell's equations (in this case, just the integrated form of Gauss's law), for a problem with a high degree of symmetry, has allowed us to very quickly solve directly in terms of the electric field. It is instructive to compare this with the direct integration of Poisson's equation (4.3). In our idealized setting of a charged plane with zero thickness, we have $\rho(\mathbf{x}) = \sigma\delta(z)$. Since $\rho(\mathbf{x})$ only depends on z, we search for a solution $\phi = \phi(z)$ and eq. (4.3) becomes

$$\frac{d^2\phi}{dz^2} = -\frac{\sigma}{\epsilon_0}\delta(z) \, . \qquad (4.144)$$

From eq. (1.70) we see that the most general solution for the $d\phi/dz$ is of the form

$$\frac{d\phi}{dz} = -\frac{\sigma}{\epsilon_0}\left[\theta(z) + a\right] \, , \qquad (4.145)$$

with a a (dimensionless) integration constant. Quite commonly, in problems of electrostatics, integration constants are fixed by requiring that the electric field vanishes at infinite distance from the sources. Here, however, this is not possible, since we have considered an infinite plane and, with such an idealization, there is no guarantee that the electric field will vanish at $z \to \pm\infty$. In fact, we already know from the solution (4.138) that this will not be the case. Rather, the symmetry of the problem requires that the modulus $|d\phi/dz|$ must be invariant under the parity transformation $z \to -z$ (we have in fact already used this condition when solving the problem using the integrated Gauss's law). The function $\theta(z)$ is equal to 0 for $z < 0$ and to 1 for $z > 0$, so its absolute value is not an even function, but this is easily fixed by choosing the integration constant $a = -1/2$, since $\theta(z) - 1/2$ is equal to $-1/2$ for $z < 0$ and $+1/2$ for $z > 0$, so $|\theta(z) - 1/2|$ is an even function. Then, eq. (4.145) becomes

$$\frac{d\phi}{dz} = -\frac{\sigma}{\epsilon_0}\left[\theta(z) - \frac{1}{2}\right] \, , \qquad (4.146)$$

which agrees with eq. (4.142). Observe also that, using the identity

$$\theta(z) + \theta(-z) = 1 \, , \qquad (4.147)$$

we can also rewrite eq. (4.146) as

$$\frac{d\phi}{dz} = -\frac{\sigma}{2\epsilon_0}\left[\theta(z) - \theta(-z)\right] \, . \qquad (4.148)$$

Problem 4.2. Electric field of a spherical charge distribution

Consider now the electric field generated by a spherical charge distribution of radius d. In this case, by symmetry, the electric field will be in the radial direction and its modulus will only depend on r, so $\mathbf{E} = E(r)\hat{\mathbf{r}}$. We take a spherical surface S of radius $r > d$ and use eq. (3.12). Writing $d\mathbf{s} = r^2 d\Omega\,\hat{\mathbf{r}}$ we get

$$
\begin{aligned}
\Phi_E(t) &= \int_S r^2 d\Omega\,\hat{\mathbf{r}} \cdot E(r)\hat{\mathbf{r}} \\
&= 4\pi r^2 E(r),
\end{aligned}
\tag{4.149}
$$

and therefore, from eq. (3.12),

$$
E(r) = \frac{1}{4\pi\epsilon_0}\frac{Q}{r^2}, \qquad (r > d),
\tag{4.150}
$$

where Q is the total charge of the sphere. This is the famous result (which is due to Newton, who found it for the gravitational case, but is valid for any force proportional to $1/r^2$) that the field outside a uniformly charged sphere is the same as if all the charge (or, in the gravitational case, all the mass) of the sphere were concentrated at its center.

If, instead, we take $r < d$, Φ_E is still given by eq. (4.149), but, on the right-hand side of eq. (3.12) the only contribution comes from the charge at $r < d$. In particular, if all the charge is on the surface of the sphere, the electric field at any point inside the sphere is zero! To understand how this comes out from the cancelation among the contribution of different charges, consider the setting of Fig. 4.12 in which, for graphical clarity, we only show a section of the sphere. We want to compute the electric field at a generic point P inside the sphere by summing over the contributions from the charges on the surface of the sphere. Consider first the charges in the infinitesimal region dS_1 which, with respect to the point P, subtends a solid angle $d\Omega$ and is at a distance r_1. Then, $dS_1 = r_1^2 d\Omega$ and, if the surface charge on the sphere is σ (which is a constant, independent of the polar angles θ, ϕ, because of the assumption of spherical symmetry), the charge in dS_1 is $\sigma r_1^2 d\Omega$. Taking $\sigma > 0$, we see from the figure that the electric field produced in P from the charge in dS_1 is directed radially and inward, toward the center O of the sphere. Therefore, it generates an electric field

$$
\begin{aligned}
d\mathbf{E}_1 &= \frac{1}{4\pi\epsilon_0}\frac{\sigma r_1^2 d\Omega}{r_1^2}(-\hat{\mathbf{r}}) \\
&= -\frac{1}{4\pi\epsilon_0}\sigma d\Omega\,\hat{\mathbf{r}}.
\end{aligned}
\tag{4.151}
$$

The crucial point is that r_1^2 canceled between numerator and denominator. Consider now the contribution of the antipodal surface dS_2, subtending the same solid angle $d\Omega$. Now, as we see from the figure, the contribution to the electric field is in the $+\hat{\mathbf{r}}$ direction. We denote by r_2 the distance of dS_2 from P. Then

$$
\begin{aligned}
d\mathbf{E}_2 &= \frac{1}{4\pi\epsilon_0}\frac{\sigma r_2^2 d\Omega}{r_2^2}(+\hat{\mathbf{r}}) \\
&= +\frac{1}{4\pi\epsilon_0}\sigma d\Omega\,\hat{\mathbf{r}}.
\end{aligned}
\tag{4.152}
$$

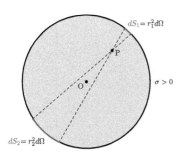

Fig. 4.12 The geometry for computing the contribution of the electric field at the point P, due to the charges on two antipodal infinitesimal regions on the surface of the sphere.

This is equal and opposite to the contribution from dS_1. Therefore, when integrating over the whole sphere, the contribution of each surface element is canceled by that of its antipodal surface elements, and we recover the result that $\mathbf{E} = 0$ inside the sphere.

Problem 4.3. Parallel-plate capacitor

We next consider a parallel-plate capacitor, made of two parallel and oppositely charged infinite planes, separated by a distance d along the z axis, with surface densities $\sigma > 0$ at $z = 0$ and $-\sigma$ at $z = d$, as shown in Fig. 4.13. The electric field can be computed using the superposition principle, that we discussed in Section 4.1.1. The electric field of the parallel-plate capacitor is then immediately obtained from the result of Problem 4.1, just by summing the electric fields produced by the two plates. Taking into account that the modulus of the electric field of an infinite plane is independent of z and its direction changes sign on the two sides, and that we have taken the opposite signs for the surface charge densities on the two planes, we see that, for $z < 0$ and for $z > d$, the fields produced by the two planes cancel out, while inside the capacitor they add up, so

$$\mathbf{E} = \frac{\sigma}{\epsilon_0}\hat{\mathbf{z}}, \qquad (0 < z < d). \qquad (4.153)$$

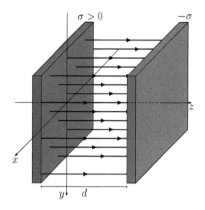

Fig. 4.13 The electric field generated by a parallel plate capacitor.

The potential ϕ is therefore constant for $z < 0$ and $z > d$, while, inside, it satisfies

$$\frac{d\phi}{dz} = -\frac{\sigma}{\epsilon_0}, \qquad (4.154)$$

so

$$\phi(z) = \phi_0 - \frac{\sigma}{\epsilon_0}z. \qquad (4.155)$$

The absolute value of the difference in the potential between the charged plates is therefore

$$V \equiv |\phi(d) - \phi(0)| = \frac{\sigma d}{\epsilon_0}. \qquad (4.156)$$

If, rather than taking an infinite extent in the (x, y) plane, we take the charged plane to have a finite area A, the total charge Q on each plate is, in absolute value, $Q = \sigma A$. We take A sufficiently large so that the effects near the finite boundary can be neglected, and the previous computation of the electric field still goes through, except near the boundaries. For a generic capacitor the *capacitance* C is defined by

$$C = \frac{Q}{V}, \qquad (4.157)$$

where Q is the charge of the positively charged plate, so $Q > 0$. Since V is the absolute value of the potential difference between the two conductors, C is positive by definition. Then, from eq. (4.156) we find that, for a parallel-plate capacitor,

$$C = \frac{\epsilon_0 A}{d}. \qquad (4.158)$$

From eq. (4.157), we see that capacitances are naturally measured in coulombs per volts, C/V, and this derived SI unit is called the farad (F), in honor of Michael Faraday. In terms of the base units, $1\,\mathrm{F} = 1\,\mathrm{C}^2\mathrm{s}^2/(\mathrm{kg\,m}^2)$.[40] The farad, however, turns out to be an unreasonably large unit, in practice. Typical values for capacitors are of order picofarad ($\mathrm{pF} = 10^{-12}\,\mathrm{F}$) to microfarad ($\mu\mathrm{F} = 10^{-6}\,\mathrm{F}$). Note also that the combination $\mathrm{C}^2/(\mathrm{N\,m}^2)$, that gives the units of ϵ_0, is the same as farads/meter; we see this, for instance, from the

[40]As mentioned in Note 7 on page 30, we sometimes use the coulomb instead of the ampere as a basic unit.

fact that $C \times V = N \times m$, since, dimensionally, they are both energies, and therefore

$$
\begin{aligned}
\frac{C^2}{N\,m^2} &= C\left(\frac{Nm}{V}\right)\left(\frac{1}{N\,m^2}\right) \\
&= \frac{C}{Vm} \\
&= \frac{F}{m}\,.
\end{aligned} \tag{4.159}
$$

We could also have seen this directly from eq. (4.158). Therefore, eq. (2.13) can be rewritten as

$$
\epsilon_0 \simeq 8.854\ldots \times 10^{-12}\,\frac{F}{m}\,. \tag{4.160}
$$

From this we see why the picofarad is a more appropriate unit than the farad. For a plane parallel capacitor filled with vacuum, as in the example that we have considered, taking for instance $A \sim 1\,cm^2$ and $d \sim 1\,mm$, from eq. (4.158) we get C of order of a pF.

The force exerted by a plate on the other is attractive, given that the two plates are oppositely charged. Its modulus can be obtained from eq. (4.138), which gives the electric field generated by one plate, multiplying it by the absolute value Q of the charge on the other plate, so

$$
F = \frac{Q\sigma}{2\epsilon_0}\,. \tag{4.161}
$$

We can re-express it in terms of the total electric field E between the two plates, given in eq. (4.153), which is twice as large as the field generated by each plate, so

$$
F = \frac{1}{2}QE\,. \tag{4.162}
$$

Typical circuits contain a large number of capacitors. The linearity of Maxwell's equation implies that the total charges Q_a on the capacitors (where $a = 1,\ldots,N$ labels the capacitor) and the potentials ϕ_a on their surfaces are related linearly,

$$
Q_a = \sum_{b=1}^{N} C_{ab}\phi_b\,. \tag{4.163}
$$

The coefficients C_{ab} depend only on the geometry of the system, i.e., the shape of the conductors and their relative arrangement. They are called the *coefficients of capacitance*, and form the *capacitance matrix* C. The off-diagonal elements C_{ab}, with $a \neq b$, are called the mutual (or cross) capacitances, while the diagonal elements C_{aa} are the (self) capacitances. Their values are not the same as the value of the capacitance of the a-th conductors in the absence of all other. Indeed, even when all ϕ_b with $b \neq a$ are set to zero (i.e., the conductors are "grounded"), there are charges on their surfaces that influence the potential on the surface of the a-th conductor. In Problem 5.2, we will prove some useful properties obeyed by the coefficients C_{ab}.

The inverse matrix C^{-1} is often denoted by P, so

$$
\phi_a = \sum_{b=1}^{N} P_{ab}Q_b\,, \tag{4.164}
$$

and P_{ab} are called the *coefficients of potential*.

Problem 4.4. Spherical capacitor

We now consider a spherical capacitor, made of two concentric spherical shells of radius a and b, with $a > b$, and charges $-Q$ and Q, respectively. Again, by symmetry, the electric field is radial, $\mathbf{E} = E(r)\hat{\mathbf{r}}$. Applying the integrated form of Gauss's law, eq. (3.12), to a spherical volume with a radius $r < b$ we find that the flux is zero because there is no charge inside the volume, and therefore $\mathbf{E} = 0$ at $r < b$. Similarly, at $r > a$, the charges of the two spherical shells compensate each other and the total charge is $Q - Q = 0$, so again eq. (3.12) gives $\mathbf{E} = 0$. The field between the two shells can be computed using eq. (3.12) where the volume V used in the integration is now a sphere of radius r, with $b < r < a$. The total charge inside this volume is just the charge Q of the inner spherical shell, while the external spherical shell has no effect so, as in eq. (4.150),

$$E(r) = \frac{1}{4\pi\epsilon_0} \frac{Q}{r^2}, \qquad (b < r < a). \qquad (4.165)$$

The potential difference between the two shells is therefore, in absolute value,

$$
\begin{aligned}
V &= \int_b^a E(r)\,dr \\
&= \frac{Q}{4\pi\epsilon_0} \left(\frac{1}{b} - \frac{1}{a} \right), \qquad (4.166)
\end{aligned}
$$

and the capacitance is

$$C = \epsilon_0 \frac{4\pi ab}{a - b}. \qquad (4.167)$$

Comparing with the result (4.158), we see that we still have at denominator the distance $d = a - b$ between the two elements of the capacitor, while the area factor A in eq. (4.158) is replaced by $4\pi ab$ in eq. (4.167). In the limit $b \to a$, $4\pi ab$ tends to the area A of the sphere, consistently with the fact that, if the radii of curvature of the spheres are much larger than their distance $d = a - b$, locally, the geometry is the same as that of a parallel plane capacitor.

Problem 4.5. Electrostatic energy of an ionic crystal

We now discuss the electrostatic energy of an ionic crystal, such as NaCl, where the positively charged Na^+ ions and the negatively charged Cl^- ions are arranged in a cubic lattice, with lattice spacing a (so that each ion has six nearest neighbors with opposite charge, at a distance a). To compute it, we can select one given ion, say of Na^+, and compute its interaction energy with all other Na^+ and Cl^- ions. This gives the interaction energy per unit ion, u. Since the computation performed choosing any of the $N/2$ positive ions or any of the $N/2$ negative ions gives the same result, the total energy potential energy is then obtained multiplying this result by N, and then dividing by 2, since in this way each pair has been counted twice, so $U = uN/2$.

We set the origin of the reference frame on the chosen ion. The positions of the ions, in a regular cubic lattice of spacing a, are given by $\mathbf{x} = a\hat{\mathbf{n}} = a(n_x, n_y, n_z)$ with n_x, n_y, n_z integers running from zero to infinity, so their distance from the origin is $a(n_x^2 + n_y^2 + n_z^2)^{1/2}$. The charge of the ion at the position $a\hat{\mathbf{n}}$ is $(-1)^{n_x + n_y + n_z} e$.[41] Therefore, the potential energy per unit ion

[41]This can be checked setting at first $n_y = n_z = 0$ and moving along the x axis. At $n_x = 0$ we have our chosen Na^+ ion, with charge $+e$; for n_x even we find again positively charges ions, and for n_x odd negatively charged ions. In the line defined by, say, $n_y = 1, n_z = 0$, the situation is inverted. Just on top of our chosen ion, at $n_x = 0$, we find a negative charge, at $n_x = 1$ a positive charge, and so on.

of an ionic crystal is given by

$$u = \frac{1}{4\pi\epsilon_0} \frac{e^2}{a} \sum_{\hat{n} \neq 0} \frac{(-1)^{n_x + n_y + n_z}}{(n_x^2 + n_y^2 + n_z^2)^{1/2}} . \qquad (4.168)$$

The sum in eq. (4.170) is called the *Madelung's sum*. Here, however, we encounter a problem. This series is not absolutely convergent, i.e., its convergence (and its finite value in the case that it converges) depends on the order in which we sum the terms. For instance, if we would first sum up all the terms with $n_x + n_y + n_z$ even, we would get a divergent result, since a sum of terms, all with the same sign, and that goes asymptotically as $\sum_n 1/n$, diverges. This mathematical problem reflects a physical ambiguity. The sum can be organized moving outward from our chosen ion, and including in the sum the interaction with the ions which belong to larger and larger volumes, with our chosen ion near their center. One possibility is to choose these volumes such that they all contain a zero net charge. Another option is to choose them so that, near their boundaries, they are deformed so as to include only ions of a certain type, e.g., positively charged. Choosing different sequences of volumes, that include or exclude ions at given positions on the boundary, we can obtain any desired distribution of surface charges on this sequence of volumes. Such a distribution of surface charges can, in general, have a finite or even a divergent interaction energy with the chosen ion, even when the boundary is at a very large distance r from it, since the contribution to the sum from the ions at a large distance r decreases as $1/r$, but, taking for instance the extreme case of a constant surface charge (i.e., the situation in which the boundary is deformed so as to include only ions of a given sign), the total charge on the surface grows as r^2; in this case, the interaction of the surface charges with the chosen ions diverges as $r \to \infty$. With different organizations of the sequence of volumes used in the sum, corresponding to different surface charge distributions, we can obtain different results, finite or divergent; this is the physical reason why the result depends on how the sum in eq. (4.168) is organized. For computing the electrostatic energy of a crystal, we are interested in the situation in which there is no surface charge; therefore, the sum must be organized through a series of electrically neutral subsequent volumes. Using a series of expanding and electrically neutral cubes, one can show that the sum converges, and can be computed numerically.[42] The result can be written as

$$u = -\frac{A}{4\pi\epsilon_0} \frac{e^2}{a} , \qquad (4.169)$$

where $A \simeq 1.7476$ is the *Madelung's constant* for a cubic lattice, such as that of NaCl. Equivalently, using $U = uN/2$, the total potential energy of a crystal with N ions is

$$U = -N \frac{A}{8\pi\epsilon_0} \frac{e^2}{a} . \qquad (4.170)$$

We can compare the value of A obtained from this procedure with that obtained starting from a given Na ion and computing its interaction energy with the 6 nearest neighbor Cl ions at distance a, the 12 Na ions at distance $a\sqrt{2}$ and the 8 Cl ions at distance $a\sqrt{3}$ (note that this configuration is not electrically neutral). This gives

$$u = \frac{1}{4\pi\epsilon_0} \frac{e^2}{a} \left[-6 + \frac{12}{\sqrt{2}} - \frac{8}{\sqrt{3}} + \dots \right] , \qquad (4.171)$$

and therefore in this approximation $A = 6 - (12/\sqrt{2}) + (8/\sqrt{3}) + \dots \simeq 2.13$, to be compared with the correct value $A \simeq 1.7476$.

---

[42] Actually, there are further mathematical subtleties here. If one uses a series of expanding cubes the sum converges while, if one uses a series of expanding spheres, it can be proven that the sum diverges. In fact, the problem is quite interesting and challenging from a mathematical point of view. The rigorous definition of the sum in eq. (4.168) is obtained from the analytic continuation to $s = 1$ of the series

$$\sum_{\hat{n} \neq 0} \frac{(-1)^{n_x + n_y + n_z}}{(n_x^2 + n_y^2 + n_z^2)^{s/2}}$$

which is absolutely convergent for $\mathrm{Re}(s) > 1$, and the sum over expanding cubes converges to the correct result; see the discussion in Section 3 of Bailey, Borwein, Kapoor, and Weisstein (2006), https://www.davidhbailey.com//dhbpapers/tenproblems.pdf.

Two aspects of eq. (4.170) are noteworthy. First, the sum gives a negative result for U. This means that the attraction between opposite charges wins over the repulsion between charges of the same sign, resulting in a bound state. This is a welcome result, since it goes in the direction of explaining the cohesion of a crystal. However, we can now ask what fixes the value of the lattice spacing a. From eq. (4.169), the smaller is a, the more negative is the potential energy, and the state with the largest binding energy is obtained for $a \to 0^+$. Therefore, if eq. (4.169) was the end of the story, the crystal could lower its energy by decreasing a indefinitely and would therefore collapse. This result, however, is not unexpected: we have already found in Section 4.1.4 that no set of electric charges can be in electrostatic equilibrium. Once again, at atomic scales quantum mechanics must come to the rescue to ensure the stability of matter.[43]

Indeed, quantum effects induce an effective repulsion between the core electrons of the Na and Cl ions, that counterbalances the tendency of the electrostatic potential to induce the collapse of the lattice. Without entering into details beyond the scope of this course, all we need to know here is that, phenomenologically, this quantum repulsion can be described by a potential per unit ion of the form B/r^n, for some positive constants B and a power index $n > 1$ (with $n \simeq 8$ for NaCl).[44] The equilibrium value of the lattice spacing is then determined minimizing, with respect to r, the total potential energy per unit ion,

$$u(r) = -\frac{A}{4\pi\epsilon_0}\frac{e^2}{r} + \frac{B}{r^n}\,. \qquad (4.172)$$

Requiring $du/dr = 0$ fixes the equilibrium value of r, which we denote by a, from the relation

$$B = \frac{A}{4\pi\epsilon_0}\frac{e^2 a^{n-1}}{n}\,, \qquad (4.173)$$

so the lattice spacing a is fixed in terms of A, B, and n. If we use this relation to eliminate B from eq. (4.172), we find that the potential per unit ion is

$$u = -\frac{A}{4\pi\epsilon_0}\frac{e^2}{a}\left(1 - \frac{1}{n}\right). \qquad (4.174)$$

The term $1/n$ therefore gives a quantum correction to the binding energy. Using the numerical values of e and $1/(4\pi\epsilon_0)$ from eqs. (2.4) and (2.11), together with the NaCl values $A \simeq 1.7476$, $n \simeq 8$ and $a \simeq 0.281$ nm, gives a dissociation energy per unit molecule (i.e., the energy per unit pair of NaCl that must be provided to dissociate the crystal) $u_{\mathrm{diss}} \equiv -u \simeq 1.255 \times 10^{-18}$ J. The energy needed to dissociate a mole of NaCl is then

$$
\begin{aligned}
U_{\mathrm{diss}} &= u_{\mathrm{diss}}N_A \\
&\simeq 756\,\mathrm{kJ/mol}\,, \qquad (4.175)
\end{aligned}
$$

where N_A is the Avogadro number (2.3). Using the conversion factor $1\,\mathrm{kJ} \simeq 0.239\,\mathrm{kcal}$, we can also rewrite it as $U_{\mathrm{diss}} \simeq 181\,\mathrm{kcal/mol}$. Alternatively, a convenient unit for atomic physics is the electronvolt (eV), defined by the exact relation[45]

$$1\,\mathrm{eV} = 1.602\,176\,634 \times 10^{-19}\,\mathrm{J}\,. \qquad (4.176)$$

Then, the dissociation energy per unit pair of NaCl can be written as

$$u_{\mathrm{diss}} \simeq 7.84\,\mathrm{eV}\,. \qquad (4.177)$$

[43]Actually, the result that, for this problem, no stable result is possible in classical electrodynamics, can already be derived just with dimensional analysis. The electrostatic energy per ion pair must be proportional to the product of their charges and therefore, in SI units, to $e^2/(4\pi\epsilon_0)$. The only way to obtain an energy from this quantity is to divide it by a length, and in this problem we have only one length-scale available, the lattice spacing a. Therefore, the result must necessarily be of the form (4.169), for some numerical constant A, positive or negative. If $A > 0$, as is actually the case, the energy is minimized for $a \to 0^+$, and the lattice collapse. If one had found $A < 0$, u would have been given by a positive constant times $1/a$, and this is minimized for $a \to \infty$, so in this case repulsion wins and the lattice "explodes," with its ions dispersing at infinite distance from each other. Quantum mechanics is able to solve the problem because it has at its disposal another fundamental constant, the Planck constant $\hbar$, and with it, together with the mass m of a particle (or, here, of an ion), we can form the combination $\hbar/(mc)$, which has dimensions of length. Therefore, we have another length-scale at our disposal, which allows for more complicated functional forms of the potential, such as that in eq. (4.172) (note, in fact, that B there is not a pure number, contrary to A).

[44]For the reader with some elementary knowledge of quantum mechanics, the mechanism that stabilizes the system is actually the Pauli exclusion principle: when two ions get too close, the core electrons of an ion begin to feel the presence of the core electrons of the other ion and the Pauli principle (more precisely, the antisymmetrization of the wave functions of these electrons) provides an effective repulsion between them.

[45]Note the relation with the absolute value of the electron charge, e, given in eq. (2.4): the eV is the energy that a charge e acquires by going through a potential difference of 1 V.

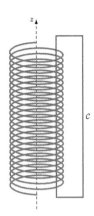

Fig. 4.14 A solenoid, made by a wire winding along a cylinder, and the contour $\mathcal{C}$ used for the integrated Ampère law.

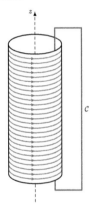

Fig. 4.15 As in Fig. 4.14, for a wire that winds very tightly around the cylinder, so that $\mathbf{j} = j\hat{\boldsymbol{\varphi}}$.

Problem 4.6. Magnetic field of a straight solenoid

We next compute the magnetic field of an infinite straight solenoid, i.e., of a wire carrying a current, that winds around an infinitely long cylinder, as in Fig. 4.14. We proceed similarly to the computation for an infinite straight wire in Section 4.2.1. We use again cylindrical coordinates (ρ, φ, z). We assume that the wire winds very tightly around the cylinder, so that it practically defines a surface current density $\mathbf{j} = j\hat{\boldsymbol{\varphi}}$, see Fig. 4.15. Then, the problem is invariant under translations along the z axis and rotations around the cylinder, and therefore $\mathbf{B}(\rho, \varphi, z)$ must again be of the form (4.72), as was the case for an infinitely long straight wire.

The radial component B_ρ vanishes, by essentially the same argument used in Section 4.2.1: we denote the radius of the solenoid by a and we consider a cylinder of radius $R > a$, such as that in Fig. 4.5, which encloses the solenoid (rather than enclosing a wire, as was the case in Fig. 4.5). The flux through the surface of the cylinder must vanish, by eq. (4.71). Since B_z is the same at the faces of the cylinder at $z = \pm h/2$ (because of the invariance of the problem under translations along z), the flux from the lower face cancels that from the upper face, and then the flux from the lateral surface of the cylinder must also vanish, and this implies $B_\rho = 0$. This remains true if we put this cylindrical volume inside the solenoid, i.e., if we take $R < a$. Therefore, B_ρ vanishes both inside and outside the solenoid.

Similarly, taking a loop $\mathcal{C}$ in the plane transverse to the cylinder, such as that in Fig. 4.4 (where, again, now the loop encloses the solenoid rather than a wire) and of radius $\rho > a$, the left-hand side of eq. (4.70) is $2\pi\rho B_\varphi$, while the right-hand side vanishes, because for the (tightly winding) solenoid the current is uniquely in the $\hat{\boldsymbol{\varphi}}$ direction, and has no $\hat{\mathbf{z}}$ component, so there is no current flowing through a surface bounded by $\mathcal{C}$; the same holds for $\rho < a$, and therefore $B_\varphi = 0$. This means that, in this problem,

$$\mathbf{B}(\rho, \varphi, z) = B_z(\rho)\hat{\mathbf{z}}. \tag{4.178}$$

This could have also been shown using symmetry arguments, similarly to what we did in Section 4.2.1 for an infinite wire: we begin by considering a parity transformation $\mathbf{x} \to -\mathbf{x}$, denoted by Π. Under it, a point on the surface of the cylinder, say at $\mathbf{x} = (0, a, z)$ (where the axes are oriented as in Fig. 4.6), is sent into the antipodal point $\mathbf{x}' = (0, -a, -z)$, which is still on the surface of the cylinder. Under this transformation, $\mathbf{j}$ transforms as $\mathbf{j}(\mathbf{x}) \to -\mathbf{j}(\mathbf{x}')$. So, if at the point $\mathbf{x}$ the current was flowing inward with respect to the plane of the page in Fig. 4.15, after the transformation the current at $\mathbf{x}'$ is flowing outward from the plane of the page. This is indeed how the current at the antipodal point actually flows in Fig. 4.15, so the configuration in Fig. 4.15 is invariant under parity. We next study how $\mathbf{B}(\mathbf{x})$ transforms, for a generic starting point P with coordinates $\mathbf{x}$. After this parity transformation, the situation is exactly the same as for the transformation from the point P to P$'$ in Fig. 4.6. Similarly to what we did in the discussion of Fig. 4.6, we then perform further symmetry transformations that bring back the point P$'$ onto P. In this case, a convenient choice is to perform first a rotation by 180° around the z axis, that we denote by R_z, that brings P$'$ to a point P$''$ with coordinates $\mathbf{x}'' = (0, a, -z)$, and finally a translation T_z along the z axis that brings it back to P. Under R_z, B_z is invariant while B_x and B_y change sign, while under the translation all components are invariant. So, in the end, the combined transformation $T_z R_z \Pi$ is a symmetry transformation of the system and, under it, $B_x(\mathbf{x}) \to -B_x(\mathbf{x})$, $B_y(\mathbf{x}) \to -B_y(\mathbf{x})$, and $B_z(\mathbf{x}) \to +B_z(\mathbf{x})$. Therefore, by the same chain of arguments discussed in Section 4.2.1 (including a choice of boundary conditions that respect this symmetry), both B_x and B_y (and therefore B_ρ and B_φ) must vanish and, in this setting, only B_z is non-zero, confirming the result that we found by a direct use of the integrated Maxwell's equations.[46]

So, either from a direct use of Maxwell's equations, or from symmetry arguments, the magnetic field of an infinite straight solenoid has the form (4.178). We then plug it into Ampère's law (4.67). Using the curl in cylindrical coordinates, eq. (1.33), we get

$$\partial_\rho B_z = -\mu_0 j. \tag{4.179}$$

Outside the solenoid, i.e., for $\rho > a$, we have no current, so $\mathbf{j} = 0$ and $\partial_\rho B_z = 0$. The boundary condition $B_z = 0$ at $\rho \to \infty$ then fixes $B_z = 0$. Inside the solenoid, at $\rho < a$, again $\mathbf{j} = 0$ and B_z is constant. However, we cannot appeal to continuity across the solenoid to fix this constant to zero, because of the singular current density that is present there. Rather, we take a loop $\mathcal{C}$ such as that in Fig. 4.15, and we apply the integrated Ampère-law (4.70),

$$\oint_{\mathcal{C}} d\boldsymbol{\ell} \cdot \mathbf{B}(\mathbf{x}) = \mu_0 I_S, \tag{4.180}$$

where we denote by I_S the total current flowing through the surface S bounded by $\mathcal{C}$. If we denote by I the current carried by the wire, by n the number of loops of the wire per unit length, and by L the vertical length of the loop $\mathcal{C}$, we have $I_S = nLI$. The portions of $\mathcal{C}$ in the radial direction do not contribute to the integral in eq. (4.180) because there $d\boldsymbol{\ell} \cdot \mathbf{B} = d\rho\,\hat{\boldsymbol{\rho}} \cdot \mathbf{B} = d\rho B_\rho$, and we have seen that $B_\rho = 0$ everywhere. Similarly, the portion of the loop in the $\hat{\mathbf{z}}$ direction outside the solenoid does not contribute because, there, $B_z = 0$. The only contribution therefore comes from the part of $\mathcal{C}$ along the $\hat{\mathbf{z}}$ direction and inside the solenoid, and there

$$\int d\boldsymbol{\ell} \cdot \mathbf{B}(\mathbf{x}) = \int_{-L/2}^{L/2} dz\, B_z(\rho)$$
$$= L B_z(\rho). \tag{4.181}$$

Equation (4.180) therefore gives $LB_z(\rho) = \mu_0 nLI$, so B_z is actually also independent of ρ and given by $B_z = \mu_0 nI$. In conclusion, in an infinite cylindrical solenoid the magnetic field is non-vanishing only inside and, there, it is oriented along the axis of the solenoid and is spatially uniform,

$$\boxed{\mathbf{B} = \mu_0 nI\hat{\mathbf{z}}.} \tag{4.182}$$

Problem 4.7. Energy dissipation in a conducting wire

We now consider a resistive wire, i.e., a wire where the relation between the current density and an applied electric field is given by Ohm's law,[47]

$$\mathbf{j} = \sigma\mathbf{E}, \tag{4.183}$$

where σ (not to be confused with a surface charge density) is a proportionality constant called the *conductivity*. We will discuss Ohm's law and its microscopic justification (as well as its frequency-dependent generalization) in Section 14.4;

[46]We could reach the same conclusion by a combination of time reversal and rotations. The transformation of $\mathbf{j}$ and $\mathbf{B}$ under time reversal was given in eqs. (3.97) and (3.98). For time independent fields, this reduces to $\mathbf{B}(\mathbf{x}) \to -\mathbf{B}(\mathbf{x})$ and $\mathbf{j}(\mathbf{x}) \to -\mathbf{j}(\mathbf{x})$ and corresponds to the fact that eqs. (4.67) and (4.68) are invariant under a transformation that flips simultaneously the signs of $\mathbf{B}$ and $\mathbf{j}$, without touching the spatial coordinates. We can now observe that, under a rotation by 180° around the x axis (denoted again by R_x), the solenoid in Fig. 4.15 is geometrically unchanged, except that now $\mathbf{j}$ runs clockwise rather than counterclockwise. This, however, can be compensated by a time reversal transformation T. Therefore, TR_x is a symmetry of the system and, proceeding similarly to what we did previously, one finds that, under this transformation (followed by a translation along the z axis to get back to the original point P with coordinate $\mathbf{x}$), $B_x(\mathbf{x}) \to -B_x(\mathbf{x})$, while $B_y(\mathbf{x}) \to B_y(\mathbf{x})$ and $B_z(\mathbf{x}) \to B_z(\mathbf{x})$. Then, with the by now usual chain of arguments, $B_x = 0$. In the same way, using the transformation TR_y, one finds $B_y = 0$. So, again, we find that only B_z is non-zero.

[47]In Section 4.1.6 we proved that, inside a conductor, $\mathbf{E} = 0$. This, however, was shown for static charges at equilibrium, i.e., when a conductor, isolated and subject only to a static external electric field, has rearranged its surface charges and reached an equilibrium situation, where the external field is screened and $\mathbf{j} = 0$. Here, we are interested precisely in the opposite situation where, using for instance a battery, a potential difference is kept among two points of a conducting wire, driving a steady current inside it.

[48]This, of course, can be seen also as a consequence of Gauss's law, given that the overall charge density inside the wire is zero. In the wire, the ions are fixed while the electrons drift with the current. However, consider a microscopic volume $dV = A\,dl$, where A is the transverse size of the wire and dl a line element along the wire (here dV is small compared to macroscopic scales, but still sufficiently large to allow us to perform an average over many ions and electrons, as necessary to define "macroscopic" quantities such as ρ and $\mathbf{j}$, smoothing out fluctuations on atomic scale; we will discuss these averaging procedures in detail in Chapter 13). The positive charge of the ions present in the volume dV is always compensated by the negative charges of the electrons that, at the given time, happens to be in dV. The individual electrons will continuously flow out of dV from one side but, in a steady current, they will be replaced by an equal flux of electrons that enter dV from the opposite side.

[49]Instead of the conductivity σ, it is common to use the *resistivity* $\rho = 1/\sigma$ (again, not to be confused with a charge density!) Then, for a thin wire,

$$R = \frac{\rho d}{A}. \qquad (4.187)$$

We see that, dimensionally, ρ is the same as a resistance times a length and is normally given in the derived SI units of $\Omega\,\mathrm{m}$. For instance, for copper at $20°$, $\rho \simeq 1.68 \times 10^{-8}\,\Omega\,\mathrm{m}$. Conductivities are then given in $(\Omega\,\mathrm{m})^{-1}$. In the SI system, the inverse of the ohm is called the *siemens* (S), so $1\ \Omega^{-1} = 1$ S. In the SI system, conductivities are then measured in siemens per meter, S/m. In terms of the fundamental SI units,

$$1\,\mathrm{S} = 1\,\frac{\mathrm{A}^2\,\mathrm{s}^3}{\mathrm{kg}\,\mathrm{m}^2}. \qquad (4.188)$$

for the moment, we just take it as a simple phenomenological relation which, observationally, turns out to have a very broad range of applicability. Note that, for a steady current, in the continuity equation (3.22) we have $\partial\rho/\partial t = 0$, and therefore, $\boldsymbol{\nabla}\!\cdot\!\mathbf{j} = 0$. From eq. (4.183), it then follows that also

$$\boldsymbol{\nabla}\!\cdot\!\mathbf{E} = 0, \qquad (4.184)$$

so, even if, in this case, $\mathbf{E} \neq 0$ inside a conductor, still its divergence is zero.[48] From the discussion in eqs. (4.96)–(4.99) we know that $\boldsymbol{\nabla}\!\cdot\!\mathbf{j} = 0$ implies that $\mathbf{j}$ is uniform along the wire and therefore the same holds for $\mathbf{E}$.

If we apply a potential difference V_{ab} between two points a and b of the wire at a distance L apart, this will induce a current I, and the resistance R_{ab} between these two points is defined by $V_{ab} = R_{ab}I$, or (leaving henceforth implicit the labels a, b)

$$V = RI, \qquad (4.185)$$

which is the most elementary version of Ohm's law. The resistance R is determined by the conductivity σ, the geometry of the wire, and the distance between the points a, b. For example, taking for simplicity a wire of cross-sectional area A, with j uniform inside it, we have $I = jA$. On the other hand, we have seen that the electric field is uniform along the wire so, if it produces a potential difference V among two points a, b at distance d, its modulus is $E = V/d$. Then, multiplying eq. (4.183) by A, we get $I = \sigma AE = (\sigma A/d)V$ which shows that, in this geometry,

$$R = \frac{d}{\sigma A}. \qquad (4.186)$$

From eq. (4.185), R has dimensions of volts per ampere, V/A. This derived SI unit is called the *ohm* (Ω).[49]

As we have mentioned before, in the absence on an external agent such as a battery, the current in the conductor would quickly set to the equilibrium value $\mathbf{j} = 0$. The mechanism that damps any initial current is given by the collisions of the electrons, accelerated by the external field, against the fixed ions, so that the kinetic energy of the electrons is dissipated into heat ("Joule heating"). We will study this mechanism with a simple model in Section 14.4. In any case, to keep a steady current going, we need to supply continuously energy to the system, for instance with a battery. The corresponding work per unit time can be computed observing that, if we take two points a and b on the wire, with potential difference V, the work done to transfer a charge dq from a to b is $dW = V\,dq$. In a steady current I, a charge $dq = I\,dt$ is transferred in a time dt, so the required work is $dW = VI\,dt$. Then, using $V = RI$ (where, again, $V = V_{ab}$ and $R = R_{ab}$ are, respectively, the potential and the resistance between the points a and b but, according to standard use, we suppress the labels a, b), we get

$$\frac{dW}{dt} = VI = RI^2 = \frac{V^2}{R}. \qquad (4.189)$$

This is the energy per unit time that must be supplied to keep a steady current going, to compensate the energy dissipated into heat in the wire.

It is instructive to compare this result with the flow of electromagnetic energy into the wire, obtained from the Poynting vector. Observe that, outside the wire, $\mathbf{E} = 0$ and the Poynting vector vanishes. Inside, however, both $\mathbf{E}$ and $\mathbf{B}$ are non-vanishing. In particular, the flow of energy into the wire can be computed from the Poynting vector at the wire surface. We take the wire as circular, of radius a. Using cylindrical coordinates with the z axis along the wire, the magnetic field is given by eq. (4.79),

$$\mathbf{B}(\rho = a) = \frac{\mu_0 I}{2\pi a} \hat{\boldsymbol{\varphi}}, \qquad (4.190)$$

where $I = j\pi a^2$. The electric field at the wire surface is obtained from eq. (4.183) with $\mathbf{j} = j\hat{\mathbf{z}}$ and $j = I/(\pi a^2)$, so

$$\mathbf{E}(\rho = a) = \frac{I}{\sigma \pi a^2} \hat{\mathbf{z}}. \qquad (4.191)$$

Therefore, using eq. (3.34) together with $\hat{\mathbf{z}} \times \hat{\boldsymbol{\varphi}} = -\hat{\boldsymbol{\rho}}$, we get[50]

$$\mathbf{S} = -\frac{I^2}{2\pi^2 a^3 \sigma} \hat{\boldsymbol{\rho}}, \qquad (4.194)$$

Note that the energy flux points toward the wire, so the electromagnetic field at the exterior of the wire feeds energy into the wire. The energy per unit time that flows into a portion of wire of length L is obtained from the right-hand side of Poynting theorem (3.35), taking as volume V a cylinder of length L along the z axis and radius a. Its lateral surface element is $dz\, a d\varphi\, \hat{\boldsymbol{\rho}}$, so

$$
\begin{aligned}
-\int_{\partial V} d\mathbf{s} \cdot \mathbf{S} &= \frac{I^2}{2\pi^2 a^3 \sigma} \int_0^L dz \int_0^{2\pi} a d\varphi \\
&= \frac{I^2 L}{\sigma A},
\end{aligned}
\qquad (4.195)
$$

where $A = \pi a^2$. From eq. (4.186), the resistance between two points at distance L is $R = L/(\sigma A)$, so we see that

$$-\int_{\partial V} d\mathbf{s} \cdot \mathbf{S} = RI^2. \qquad (4.196)$$

Therefore, the flow of energy from the electromagnetic field into the wire balances the losses due to dissipation in the wire, providing the continuous inflow of energy necessary to maintain a steady current. Note that the flow of energy is in the radial direction $-\hat{\boldsymbol{\rho}}$, even if eventually the flow of the current is in $\hat{\mathbf{z}}$ direction.

As discussed in Note 11 on page 47, one can find different expressions for the energy density and Poynting vector, that give the same integrated energy conservation law. In particular for a static problem, where $\mathbf{E} = -\boldsymbol{\nabla}\phi$ and $\boldsymbol{\nabla} \times \mathbf{B} = \mu_0 \mathbf{j}$, we can rewrite the Poynting vector as[51]

$$\mathbf{S} = \phi \mathbf{j}. \qquad (4.197)$$

So, in this case the energy that flows into a portion of wire of length L, obtained as before taking as volume V a cylinder of length L along the z axis and radius a, now appears to enters from the lower face of the cylinder and flow out from the upper face (with a net difference between incoming and outgoing flow due to the fact that the potential ϕ in eq. (4.197) grows linearly, in absolute value, along the wire), rather than from the lateral faces of the cylinder, as in eq. (4.194).

In any case, the important notion is that the energy is transferred to the wire by the surrounding electromagnetic field, and energy conservation, as derived from Maxwell's equations, is fundamentally an integrated relation.

---

[50]This can be rewritten as

$$\mathbf{S} = -\frac{\sigma}{2} E^2(\rho = a) a \hat{\boldsymbol{\rho}}. \qquad (4.192)$$

Observe that, if j is uniform across the wire, the same argument, taking as volume V a cylinder of length L along the z axis and radius $\rho < a$, shows that, inside the wire,

$$\mathbf{S} = -\frac{\sigma}{2} E^2(\rho) \boldsymbol{\rho}, \qquad (4.193)$$

where $\boldsymbol{\rho} = \rho \hat{\boldsymbol{\rho}}$.

[51]Explicitly,

$$
\begin{aligned}
\mathbf{S} &= \frac{1}{\mu_0} \mathbf{E} \times \mathbf{B} \\
&= -\frac{1}{\mu_0} (\boldsymbol{\nabla}\phi) \times \mathbf{B} \\
&= -\frac{1}{\mu_0} [\boldsymbol{\nabla} \times (\phi \mathbf{B}) - \phi \boldsymbol{\nabla} \times \mathbf{B}] \\
&= -\frac{1}{\mu_0} [\boldsymbol{\nabla} \times (\phi \mathbf{B}) - \phi \mu_0 \mathbf{j}] \\
&= \phi \mathbf{j} - \frac{1}{\mu_0} \boldsymbol{\nabla} \times (\phi \mathbf{B}).
\end{aligned}
$$

The term $\boldsymbol{\nabla} \times (\phi \mathbf{B})$ does not contribute to the integrated conservation equation, since its integral over a closed boundary ∂V vanishes. It corresponds to the freedom of adding to $\mathbf{S}$ a term $\boldsymbol{\nabla} \times \mathbf{w}$, for an arbitrary vector field $\mathbf{w}$ (see Note 11 on page 47). We can therefore drop it, and use eq. (4.197) for the Poynting vector.

Problem 4.8. Inductance of a circuit

Consider a set of closed loops $\mathcal{C}_a$, $(a = 1, \ldots, N)$, carrying currents I_a. These currents generate a magnetic field, and we denote the flux of the total magnetic field through the a-th loop by $\Phi_{B,a}$,

$$\Phi_{B,a} = \int_{S_a} d\mathbf{s} \cdot \mathbf{B} \,, \tag{4.198}$$

where S_a is any surface having $\mathcal{C}_a$ as the boundary. The linearity of Maxwell's equation implies that the relations between the currents and the flux is linear, i.e.,

$$\Phi_{B,a} = \sum_{a=1}^{N} L_{ab} I_b \,. \tag{4.199}$$

The coefficients L_{ab}, with $a \neq b$, are called the *mutual inductances*, while the diagonal element $L_a \equiv L_{aa}$ is the self-inductance (or, simply, the inductance) of the a-th loop. In Section 5.3, we will prove explicitly eq. (4.199), and we will show how to compute the mutual inductances in terms of the shapes and positions of the loops. From eq. (4.199), we see that the dimension of L_{ab}, in SI units, are the same as Tm^2/A. Comparing eqs. (2.19) and (2.20) we see that this is the same as Vs/A. This derived SI unit for inductance is called the *henry* (H).

If we have just a single loop, we can omit the index a and write, more simply,

$$\Phi_B = LI \,. \tag{4.200}$$

As an application, consider the situation in which we have a single loop; the current in the loop is initially zero and the loop is connected to a battery, which is suddenly switched on at time $t = 0$. We denote by V_0 the electromotive force (emf) provided by the battery. This emf drives a current in the loop, so $I(t)$ raises with time. According to eq. (4.200), the time evolution of the current induces a time evolution of the flux Φ_B. In turn, according to Faraday's law (3.21), this induces an electromotive force $-d\Phi_B/dt$. Then the total electromotive force in the loop is

$$\mathcal{E}_{\mathrm{emf}} = V_0 - L\frac{dI}{dt} \,. \tag{4.201}$$

The relative minus sign is the content of Lenz's law, as we have seen in Section 4.3.1, and is such that the induced electromotive force opposes the change in the flux produced by the external source. If the loop has a resistance R, then Ohm's law (4.185) states that $\mathcal{E}_{\mathrm{emf}} = RI$. Therefore, the equation that governs the time evolution of the current, after the battery is switched on, is

$$V_0 - L\frac{dI}{dt} = RI \,, \tag{4.202}$$

i.e.,

$$\tau\frac{dI}{dt} + I = \frac{V_0}{R} \,, \tag{4.203}$$

where $\tau = L/R$. The solution, with the initial condition $I(t = 0) = 0$, is

$$I(t) = \frac{V_0}{R}\left(1 - e^{-t/\tau}\right) \,. \tag{4.204}$$

Therefore, τ is the timescale on which the current raises to its asymptotic value V_0/R. For a given value of R, the larger is L, the slower is the raise of the current.

Electromagnetic energy

As we have seen in Section 3.2.2, the electromagnetic field carries energy. In this chapter, we will see that the energy stored in a static electric field is equal to the work that has been done to assemble the configuration of charges that generates it (although, for point charges, this requires to deal with some "self-energy" divergences) and, similarly, the energy stored in a static magnetic field is equal to the work needed to produce the configuration of currents that generates it. We will also show how to rewrite the electromagnetic energy in different useful forms, in electrostatics and magnetostatics.

Another important aspect is the relation between the electromagnetic energy and the "mechanical potentials" U, from which (for non-relativistic systems) the electromagnetic force acting on the sources can be derived, as $\mathbf{F} = -\boldsymbol{\nabla} U$. We will see that in electrostatics, in particular when dealing with a set of conductors, to compute the forces acting on them we must distinguish between a mechanical potential at fixed charges, and a mechanical potential computed keeping fixed the electrostatic potentials on their surfaces. A similar issue arises in magnetostatics.

5.1 Work and energy in electrostatics

In this section, we evaluate the energy of a system of point-like charges q_a, at given positions $\mathbf{x}_a$, using the mechanical definition of energy of a system as the work that should be done, by an external agent, to build the desired configuration. We will then consider the generalization to continuous charge distributions. In Section 5.2, we will compare with the energy stored in the corresponding electromagnetic field.

We consider first a system of two charges, q_1 and q_2, and we compute the work done by an external agent to bring the second charge from infinity to a final position $\mathbf{x}_2$, in the potential generated by the first charge, which is located at a fixed position $\mathbf{x}_1$.[1] To compute the work done, we assume that the position of the first charge is nailed down in $\mathbf{x}_1$, so that it does not move. Then the work is given by eq. (4.33), where, from eq. (4.6), the potential generated by the first charge is

$$\phi(\mathbf{x}) = \frac{1}{4\pi\epsilon_0} \frac{q_1}{|\mathbf{x} - \mathbf{x}_1|}. \qquad (5.1)$$

If the charge comes from an initial position $\mathbf{x}_0$ with $|\mathbf{x}_0| \to \infty$, we have

[1] Observe that we are interest in building a final *static* configuration of charges by bringing them into the desired position from infinity. We can imagine doing it very slowly, so that the velocities of the particles during the whole process are infinitesimally small, and we can then apply the non-relativistic notions of force and work to obtain the exact energy required to build the final field configuration.

$\phi(\mathbf{x}_0) = 0$, so,

$$
\begin{aligned}
W_{\text{ext}}^{(2)} &= q_2 \phi(\mathbf{x}_2) \\
&= \frac{1}{4\pi\epsilon_0} \frac{q_1 q_2}{|\mathbf{x}_2 - \mathbf{x}_1|},
\end{aligned}
\tag{5.2}
$$

where the superscript (2) reminds us that this is the work made to bring the charge q_2 at the specified position.[2]

We now keep the positions $\mathbf{x}_1$ and $\mathbf{x}_2$ of these two charges fixed, and we compute the work needed to bring a third charge, q_3, from infinity to a given position $\mathbf{x}_3$. We therefore use again eq. (4.33), where now $\phi(\mathbf{x})$ is the potential generated by the first two charges,[3]

$$
\phi(\mathbf{x}) = \frac{1}{4\pi\epsilon_0} \left[\frac{q_1}{|\mathbf{x} - \mathbf{x}_1|} + \frac{q_2}{|\mathbf{x} - \mathbf{x}_2|} \right].
\tag{5.3}
$$

The work needed to bring the charge q_3 at the specified position is therefore

$$
W_{\text{ext}}^{(3)} = \frac{1}{4\pi\epsilon_0} \left[\frac{q_1 q_3}{|\mathbf{x} - \mathbf{x}_1|} + \frac{q_2 q_3}{|\mathbf{x} - \mathbf{x}_2|} \right].
\tag{5.4}
$$

The total work done to build the configuration made of these three charges is $W_{\text{ext}}^{(2)} + W_{\text{ext}}^{(3)}$, so,

$$
W_{\text{ext}} = \frac{1}{4\pi\epsilon_0} \left[\frac{q_1 q_2}{|\mathbf{x}_2 - \mathbf{x}_1|} + \frac{q_1 q_3}{|\mathbf{x}_3 - \mathbf{x}_1|} + \frac{q_2 q_3}{|\mathbf{x}_3 - \mathbf{x}_2|} \right].
\tag{5.5}
$$

Note that the expression is fully symmetrical with respect to the charges, independently of the order in which they are brought from infinity to their final position; as we have seen in Section 4.1.3, it is also independent of the path chosen to bring the charges in their final positions, and is therefore a function of the final configuration only.

It is clear that we can now proceed iteratively and, for a system of N charges,

$$
W_{\text{ext}} = \frac{1}{4\pi\epsilon_0} \sum_{a=1}^{N} \sum_{b>a}^{N} \frac{q_a q_b}{|\mathbf{x}_a - \mathbf{x}_b|}.
\tag{5.6}
$$

Since the energy of a (non-relativistic) system is the same as the work made by an external agent to build it, the electrostatic energy $\mathcal{E}_E$ of a static system of point charges is

$$
\boxed{(\mathcal{E}_E)_{\text{p.p.}} = \frac{1}{8\pi\epsilon_0} \sum_{a=1}^{N} \sum_{b \neq a}^{N} \frac{q_a q_b}{|\mathbf{x}_a - \mathbf{x}_b|},}
\tag{5.7}
$$

where the subscript "p.p." stands for "point particles", and we included in the sum over b both $b > a$ and $b < a$, so that each pair is counted twice, and we compensated this dividing by two. From eq. (4.11) we see that the electrostatic potential felt by the a-th charge because of the interaction with all other charges is

$$
\phi_a(\mathbf{x}_1, \dots \mathbf{x}_n) \equiv \frac{1}{4\pi\epsilon_0} \sum_{b \neq a}^{N} \frac{q_b}{|\mathbf{x}_a - \mathbf{x}_b|},
\tag{5.8}
$$

[2] As a check of the sign, we observe that the work needed to bring closer to each other two charges of the same sign must be positive, to overcome their repulsion, and this is correctly reproduced by eq. (5.2). Also note that, as discussed after eq. (4.31), the result is independent of the path used to bring the charge q_2 from infinity to the position $\mathbf{x}_2$.

[3] It is often remarked that we are using here the *superposition principle*, that states the the electric and magnetic fields generated by an ensemble of charges with pre-assigned positions or trajectories is the (vector) sum of the fields generated by the individual charges. However, as we discussed in Section 4.1.1, in classical electrodynamics this is not a separate principle, but just a consequence of the linearity of Maxwell's equations. At the quantum level, there can be effects that generate non-linearities, which will manifest at microscopic scales or for sufficiently strong fields.

so eq. (5.7) can also be written as

$$\boxed{(\mathcal{E}_E)_{\text{p.p.}} = \frac{1}{2} \sum_{a=1}^{N} q_a \phi_a \,.} \tag{5.9}$$

The formal generalization of eqs. (5.7) and (5.9) to a continuous charge distribution $\rho(\mathbf{x})$ are

$$\mathcal{E}_E = \frac{1}{8\pi\epsilon_0} \int d^3x\, d^3x'\, \frac{\rho(\mathbf{x})\rho(\mathbf{x}')}{|\mathbf{x}-\mathbf{x}'|} \,, \tag{5.10}$$

and

$$\mathcal{E}_E = \frac{1}{2} \int d^3x\, \rho(\mathbf{x})\phi(\mathbf{x}) \,, \tag{5.11}$$

respectively. We note, however, that in the continuous formulation the condition $a \neq b$ that appears in eq. (5.7) is absent. We will discuss this point in great detail in Section 5.2.2.

Consider now the situation where we have two continuous charge distributions $\rho_1(\mathbf{x})$ and $\rho_2(\mathbf{x})$, localized in two non-overlapping volumes V_1 and V_2, respectively. Setting $\rho(\mathbf{x}) = \rho_1(\mathbf{x}) + \rho_2(\mathbf{x})$ in eq. (5.10), we get $\mathcal{E}_E = \mathcal{E}_E^{(1)} + \mathcal{E}_E^{(2)} + \mathcal{E}_E^{\text{int}}$ where

$$\mathcal{E}_E^{(a)} = \frac{1}{8\pi\epsilon_0} \int d^3x\, d^3x'\, \frac{\rho_a(\mathbf{x})\rho_a(\mathbf{x}')}{|\mathbf{x}-\mathbf{x}'|} \,, \tag{5.12}$$

(with $a = 1, 2$), can be seen as the electromagnetic "self-energy" of the a-th charge distribution, while

$$\mathcal{E}_E^{\text{int}} = \frac{1}{4\pi\epsilon_0} \int d^3x\, d^3x'\, \frac{\rho_1(\mathbf{x})\rho_2(\mathbf{x}')}{|\mathbf{x}-\mathbf{x}'|} \tag{5.13}$$

is the interaction energy between the two charge distributions.

5.2   Energy stored in a static electric field

The energy stored in a static electric field can be obtained from the general results of Section 3.2.2. We will see how it can be rewritten in different useful forms. We will then take the limit of point-like charges, and compare with the work needed to build the corresponding charge configuration, that we computed in Section 5.1. This computation is instructive, also because it allows us to illustrate some subtleties in the passage from a continuous charge distribution to a set of point-like charges.

5.2.1   Continuous charge distribution

We start from the expression (3.41) for the energy density of the electromagnetic field. In the context of electrostatics, $\mathbf{B} = 0$ and $\mathbf{E}$ depends only on $\mathbf{x}$. Furthermore, $\mathbf{E}$ can be written in terms of ϕ as in eq. (4.30).

Then the electrostatic energy in a volume V, sufficiently large to include all the charges under consideration, is

$$
\begin{aligned}
\mathcal{E}_E &= \frac{\epsilon_0}{2} \int_V d^3x\, \mathbf{E}^2(\mathbf{x}) \\
&= \frac{\epsilon_0}{2} \int_V d^3x\, \boldsymbol{\nabla}\phi \cdot \boldsymbol{\nabla}\phi \\
&= \frac{\epsilon_0}{2} \int_V d^3x\, \left[\boldsymbol{\nabla}\cdot(\phi\boldsymbol{\nabla}\phi) - \phi\boldsymbol{\nabla}^2\phi \right].
\end{aligned}
\tag{5.14}
$$

As we already saw in the discussion of eqs. (4.41) and (4.42), for a localized charge distribution, the term $\boldsymbol{\nabla}\cdot(\phi\boldsymbol{\nabla}\phi)$ gives a boundary term on the surface ∂V, that vanishes when we send the integration volume to infinity. Furthermore, in electrostatics ϕ obeys Poisson's equation (4.3). Then, we get

$$
\mathcal{E}_E = \frac{1}{2} \int d^3x\, \rho(\mathbf{x})\phi(\mathbf{x}),
\tag{5.15}
$$

where we have sent $V \to \infty$, so the integration is now over all of space, in order to eliminate the boundary term. The solution of Poisson's equation (4.3) for a generic charge distribution is given by eq. (4.16). Inserting it into eq. (5.15) we get

$$
\mathcal{E}_E = \frac{1}{8\pi\epsilon_0} \int d^3x\, d^3x'\, \frac{\rho(\mathbf{x})\rho(\mathbf{x}')}{|\mathbf{x} - \mathbf{x}'|},
\tag{5.16}
$$

which allows us to express the energy stored in the electric field, in the electrostatic limit, in terms of the charge distribution that generates it.

These results agree with eqs. (5.10) and (5.11), that were obtained computing the work performed to assemble a configuration of point charges and performing a "naive" generalization to a continuous distribution. As we mentioned, a subtle point is that, in the point-particle formulation, ϕ_a is the electrostatic potential generated by all charges *except* the a-th charge, see eq. (5.8). In the continuous formulation, a condition equivalent to $b \neq a$ in eq. (5.8) is absent. We will examine this point in Section 5.2.2. Barring a clarification of this point, we have then found that the energy of a static charge configuration, defined as the work done by an external agent to assemble it, is the same as the energy stored in the electric field that this charge configuration generates.

A first comment on these results is that the previous manipulations raise a question on the uniqueness of the expression for the energy density. From eq. (3.41), we would naturally conclude that the energy density of the electromagnetic field, in the electrostatic limit, is

$$
u(\mathbf{x}) = \frac{\epsilon_0}{2}|\mathbf{E}(\mathbf{x})|^2,
\tag{5.17}
$$

as we wrote indeed in eq. (3.43). In contrast, eq. (5.15) might suggest that we identify the energy density with

$$
u(\mathbf{x}) \overset{?}{=} \frac{1}{2}\rho(\mathbf{x})\phi(\mathbf{x}).
\tag{5.18}
$$

Locally, the two expressions are very different. In particular, $u(\mathbf{x})$ in eq. (5.18) is localized on the position of the charges, and vanishes in regions where $\rho = 0$ (so, in particular, for a set of point-like charges it would be a sum of Dirac delta functions), while the expression in eq. (5.17) is non-vanishing even in charge-free regions. Also note that $|\mathbf{E}(\mathbf{x})|^2$ is definite positive, while $\rho(\mathbf{x})\phi(\mathbf{x})$ is not. The correct interpretation is that the energy density of the electromagnetic field is given by eq. (3.43) and so, for electrostatics, by eq. (5.17). Indeed, the derivation leading to eq. (3.43) was completely general, independently of the specific form of the charge and current distributions in Maxwell's equations. In contrast, eq. (5.15) only holds in electrostatics. Basically, what we have done has been to use the equations of electrostatics to rewrite the integral over $|\mathbf{E}(\mathbf{x})|^2$ as an integral over a different integrand that gives the same result. This, however, was specific to electrostatics, and even to the situation when the integral is over all of space rather than over a finite volume (otherwise we cannot discard the boundary term), while eq. (3.43) is much more general. For instance, we will see in Chapter 9 that we can associate an energy density even to electromagnetic waves propagating in vacuum, in which case the source terms are just vanishing. Therefore, no special general meaning should be attached to $\rho(\mathbf{x})\phi(\mathbf{x})/2$, apart from being a function whose spatial integral over $\mathbb{R}^3$, in electrostatics, happens to coincide with the spatial integral of $\epsilon_0|\mathbf{E}(\mathbf{x})|^2/2$. The general expression for the energy density of the electromagnetic field is given by eq. (3.43).[4]

A second comment is that the integrand in eq. (5.16) becomes singular when $|\mathbf{x}-\mathbf{x}'| \to 0$. We must therefore understand under what conditions the integral converges, since only in that case eq. (5.16) provides a well-defined expression for the energy of a static charge distribution. The issue is clearer in Fourier space. According to the definition (1.100), we write

$$\rho(\mathbf{x}) = \int \frac{d^3k}{(2\pi)^3}\, \tilde{\rho}(\mathbf{k})e^{i\mathbf{k}\cdot\mathbf{x}}, \qquad (5.19)$$

and we use eq. (1.115) for the Fourier transform of $1/|\mathbf{x}|$, which allows us to write

$$\frac{1}{|\mathbf{x}-\mathbf{x}'|} = \int \frac{d^3k}{(2\pi)^3}\, \frac{4\pi}{k^2} e^{i\mathbf{k}\cdot(\mathbf{x}-\mathbf{x}')}, \qquad (5.20)$$

where $k = |\mathbf{k}|$. Then eq. (5.16) can be rewritten as[5]

$$\boxed{\ \mathcal{E}_E = \frac{1}{2\epsilon_0} \int \frac{d^3k}{(2\pi)^3}\, \frac{\tilde{\rho}(-\mathbf{k})\tilde{\rho}(\mathbf{k})}{k^2}.\ } \qquad (5.21)$$

The possible divergence of the integral in eq. (5.16) as $|\mathbf{x}-\mathbf{x}'| \to 0$ translates into a possible divergence of eq. (5.21) at large k. This is to be expected from the general properties of the Fourier transform since, in a relation such as (1.100), the term $\tilde{f}(\mathbf{k})e^{i\mathbf{k}\cdot\mathbf{x}}$ describes features of the function $f(\mathbf{x})$ at scales $|\mathbf{x}| \sim 1/|\mathbf{k}|$, so the Fourier modes with large $\mathbf{k}$ describe the short-distance behavior. Writing $d^3k = k^2 dk d\Omega$, eq. (5.21)

[4]As we have discussed in Note 11 on page 47, there is some potential ambiguity even for this expression of the energy density, corresponding to the possibility of adding to it a term $\nabla\cdot\mathbf{v}$, as in eq. (3.45). As we mentioned in Note 11, this, however, can be set to zero appealing to the Lorentz covariance of the energy-momentum tensor in Special Relativity, or even better, using General Relativity to identify the energy density that couples to the gravitational field, which is indeed given by eq. (3.43).

[5]The explicit computation goes as follows:

$$\int d^3x d^3x'\, \frac{\rho(\mathbf{x})\rho(\mathbf{x}')}{|\mathbf{x}-\mathbf{x}'|}$$

$$= \int d^3x d^3x' \int \frac{d^3k_1}{(2\pi)^3}\frac{d^3k_2}{(2\pi)^3}\frac{d^3k_3}{(2\pi)^3}$$

$$\times \tilde{\rho}(\mathbf{k}_1)e^{i\mathbf{k}_1\cdot\mathbf{x}}\, \frac{4\pi}{k_2^2}e^{i\mathbf{k}_2\cdot(\mathbf{x}-\mathbf{x}')}\tilde{\rho}(\mathbf{k}_3)e^{i\mathbf{k}_3\cdot\mathbf{x}'}$$

$$= 4\pi \int \frac{d^3k_1}{(2\pi)^3}\frac{d^3k_2}{(2\pi)^3}\frac{d^3k_3}{(2\pi)^3}\frac{\tilde{\rho}(\mathbf{k}_1)\tilde{\rho}(\mathbf{k}_3)}{k_2^2}$$

$$\times \int d^3x\, e^{i(\mathbf{k}_1+\mathbf{k}_2)\cdot\mathbf{x}} \int d^3x'\, e^{i(\mathbf{k}_3-\mathbf{k}_2)\cdot\mathbf{x}'}$$

$$= 4\pi \int \frac{d^3k_1}{(2\pi)^3}\frac{d^3k_2}{(2\pi)^3}\frac{d^3k_3}{(2\pi)^3}\frac{\tilde{\rho}(\mathbf{k}_1)\tilde{\rho}(\mathbf{k}_3)}{k_2^2}$$

$$\times (2\pi)^3\delta^{(3)}(\mathbf{k}_1+\mathbf{k}_2)(2\pi)^3\delta^{(3)}(\mathbf{k}_3-\mathbf{k}_2)$$

$$= 4\pi \int \frac{d^3k_1}{(2\pi)^3}\, \frac{\tilde{\rho}(-\mathbf{k}_1)\tilde{\rho}(\mathbf{k}_1)}{k_1^2}.$$

Renaming $\mathbf{k}_1 = \mathbf{k}$ we get eq. (5.21).

becomes

$$\mathcal{E}_E = \frac{1}{2\epsilon_0 \, (2\pi)^3} \int_0^\infty dk \int d\Omega \, \tilde{\rho}(-\mathbf{k})\tilde{\rho}(\mathbf{k}) \,. \tag{5.22}$$

The condition for convergence is therefore that the Fourier modes $\tilde{\rho}(\mathbf{k})$ go to zero as $k \to \infty$ sufficiently fast, so that this integral converge. A sufficient condition, for this, is that $\tilde{\rho}(\mathbf{k})$ goes to zero faster that $1/k^{1/2}$, so $\tilde{\rho}(-\mathbf{k})\tilde{\rho}(\mathbf{k})$ goes to zero faster than $1/k$.[6] Therefore if, in this sense, $\rho(\mathbf{x})$ is sufficiently smooth on short scales, the integral in eq. (5.21) [and, therefore, that in eq. (5.16)], converges and provides a well-defined expression for the energy generated by a static charge distribution.

5.2.2   The point-like limit and particle self-energies

The most notable exception to the smooth behavior defined above is obtained for an ensemble of point-like charges, described by Dirac deltas. Consider a set of charged particles with charges q_a and fixed positions $\mathbf{x}_a$, with $a = 1, \ldots, N$. In this case,

$$\rho(\mathbf{x}) = \sum_{a=1}^N q_a \delta^{(3)}(\mathbf{x} - \mathbf{x}_a) \,, \tag{5.23}$$

and, from eq. (1.99),

$$\tilde{\rho}(\mathbf{k}) = \sum_{a=1}^N q_a \, e^{-i\mathbf{k}\cdot\mathbf{x}_a} \,. \tag{5.24}$$

We see that, for point-like charges, the Fourier modes $\tilde{\rho}(\mathbf{k})$ do not even go to zero as $k \to \infty$, and the integrals in eqs. (5.16) and (5.21) diverge. Indeed, if we naively insert eq. (5.23) into eq. (5.16), we get

$$\mathcal{E}_E \stackrel{?}{=} \frac{1}{8\pi\epsilon_0} \sum_{a=1}^N \sum_{b=1}^N \frac{q_a q_b}{|\mathbf{x}_a - \mathbf{x}_b|} \,. \tag{5.25}$$

The question mark stresses that what we are doing is not correct, since $\tilde{\rho}(\mathbf{k})$ in this case does not satisfy the convergence condition; as a result, the right-hand side is divergent, because of the contribution of the terms with $a = b$.

To understand the meaning of this apparently nonsensical result, consider first the terms in eq. (5.25) with $b \neq a$. These are just the terms that we denoted as $(\mathcal{E}_E)_{\text{p.p.}}$ in eq. (5.7). As we have seen in Section 5.1, this is equal to the work that one must perform to assemble this distribution of charges, starting from a set of charges at infinite distances from each other, and therefore agrees with the definition of energy of the system as the work needed to build the given configuration.

So, the terms with $a \neq b$ in eq. (5.25) give precisely the result expected for a system of point-like particles with Coulomb interaction, and the trouble comes from the terms where $a = b$. This is a sort of "self-energy" term, and it is interesting to understand the origin, and the cure, of this divergence in some detail.[7] The problem is in the assumption of exactly

---

[6]This condition is sufficient, but it is not necessary. A term in $\tilde{\rho}(-\mathbf{k})\tilde{\rho}(\mathbf{k})$ that goes to zero more slowly could still give a vanishing contribution when integrated over the solid angle.

[7]This part can be skipped at first reading. The bottomline of the following discussion is that these divergent self-energy terms, once regularized (i.e., suitably treated from the mathematical point of view), are just constants, and can be discarded.
The reader with experience in quantum field theory, on the other hand, will notice that, even if we are in a purely classical context, we have patterned our discussion in analogy to the procedure of regularization and renormalization in quantum field theory. In Section 12.3.1, we will expand on this approach, and we will compare it to attempts to deal with these problems by building a classical model of an extended electron.

point-like particles, and its resolution comes from the realization that
the notion of an exactly point-like particle is a mathematical idealization that is useful in many situations, but this is one case in which the
idealization leads us astray. When we write the charge density as in
eq. (5.23), we are implicitly assuming that we know the structure of
an elementary particle down to infinitely small length scales or, equivalently, eq. (5.24) assumes that we know the Fourier modes of its charge
distribution up to infinitely large values of $\mathbf{k}$. However, the structure of
elementary particles can only be discussed in the framework of quantum
mechanics or, in fact, quantum field theory. Within the context of a
classical treatment, the best that one can do is to acknowledge one's
ignorance of physics at sufficiently small scales, where quantum effects
enter into play. If we denote by ℓ the length-scale below which a classical
treatment of elementary particles breaks down, this amounts to saying
that we do now know the Fourier modes of the charge distribution for
modes with $|\mathbf{k}| \gtrsim \mathcal{O}(1/\ell)$. We should then put a cutoff on the integrals
over d^3k, integrating only over the Fourier modes with $|\mathbf{k}|$ smaller than
some cutoff value Λ of order $1/\ell$, and deferring the proper treatment
of higher values of $|\mathbf{k}|$ to quantum mechanics and quantum field the-
ory. Then eq. (5.23), that, using the integral representation (1.79) of
the Dirac delta, can be written as

$$\rho(\mathbf{x}) = \sum_{a=1}^{N} q_a \int \frac{d^3k}{(2\pi)^3} e^{i\mathbf{k}\cdot(\mathbf{x}-\mathbf{x}_a)} , \tag{5.26}$$

should be replaced by

$$\rho(\mathbf{x}) = \sum_{a=1}^{N} q_a \int_{|\mathbf{k}|<\Lambda} \frac{d^3k}{(2\pi)^3} e^{i\mathbf{k}\cdot(\mathbf{x}-\mathbf{x}_a)} . \tag{5.27}$$

The resulting expression is a smoothed approximation to the Dirac delta,
of the type of the functions used in Section 1.4 to approximate the Dirac
delta (which is actually a distribution) with a sequence of "normal"
functions, and with a smoothing length-scale $\ell \sim 1/\Lambda$. Equation (5.27)
is formally equivalent to setting to zero all Fourier modes of $\tilde{\rho}(\mathbf{k})$ with
$|\mathbf{k}| > \Lambda$ and therefore has the effect of restricting also the integral in
eq. (5.21) to $|\mathbf{k}| < \Lambda$. Then eq. (5.21) becomes

$$(\mathcal{E}_E)_{\text{reg}} = \frac{1}{2\epsilon_0} \int_{|\mathbf{k}|<\Lambda} \frac{d^3k}{(2\pi)^3} \frac{\tilde{\rho}(-\mathbf{k})\tilde{\rho}(\mathbf{k})}{k^2} , \tag{5.28}$$

and is now finite. The subscript "reg" stands for "regularized" and
means that now this expression is, at least, mathematically well defined.
We can then insert eq. (5.24), which is still valid for the modes with
$|\mathbf{k}| < \Lambda$, into eq. (5.28). This gives

$$(\mathcal{E}_E)_{\text{reg}} = \frac{1}{2\epsilon_0} \sum_{a=1}^{N}\sum_{b=1}^{N} q_a q_b \int_{|\mathbf{k}|<\Lambda} \frac{d^3k}{(2\pi)^3} \frac{e^{i\mathbf{k}\cdot(\mathbf{x}_a-\mathbf{x}_b)}}{k^2} \tag{5.29}$$

$$= \frac{1}{\epsilon_0} \sum_{a=1}^{N}\sum_{b>a}^{N} q_a q_b \int_{|\mathbf{k}|<\Lambda} \frac{d^3k}{(2\pi)^3} \frac{e^{i\mathbf{k}\cdot(\mathbf{x}_a-\mathbf{x}_b)}}{k^2} + \frac{1}{2\epsilon_0} \sum_{a=1}^{N} q_a^2 \int_{|\mathbf{k}|<\Lambda} \frac{d^3k}{(2\pi)^3} \frac{1}{k^2} .$$

[8]See Section 12.3.1 for a more accurate formalization of this subtraction, within the logic of renormalization. It would also be tempting to interpret the self-energy term as the origin of the rest mass of a charged particle. Then, for an electron, one could be tempted to fix ℓ from $[1/(4\pi\epsilon_0)](e^2/\ell) = m_e c^2$, which would give

$$\ell = \frac{1}{4\pi\epsilon_0} \frac{e^2}{m_e c^2} \equiv r_0 \,. \quad (5.32)$$

We will meet r_0 again in Section 16.2, where we will see that it enters the scattering of electromagnetic waves off free electrons and is called the "classical electron radius." The idea of an electromagnetic origin of the mass of elementary particles is suggestive but, once again, classical field theory is not the right framework for addressing this type of questions. In particular, quantum mechanics introduces another length-scale associated with the electron, called the Compton radius,

$$r_C \equiv \frac{\hbar}{m_e c} \,, \quad (5.33)$$

(where $\hbar$ is Planck's constant) that we will also meet in Section 16.2. The relation between r_C and r_0 becomes clearer introducing the *fine structure constant*

$$\alpha = \frac{1}{4\pi\epsilon_0} \frac{e^2}{\hbar c} \,. \quad (5.34)$$

Dimensionally, this quantity is a pure number, and has the value $\alpha \simeq 1/137$, so $\alpha \ll 1$. In terms of α, the relation between r_0 and r_C reads $r_0 = \alpha r_C$, so quantum mechanics enters into play at a scale $r_C = r_0/\alpha \gg r_0$. Therefore, one should already stop the classical treatment at a scale $\ell \sim r_C$. Then, the self-energy term in eq. (5.31) is of order $\alpha m_e c^2$, so should rather be seen as a small electromagnetic correction to the rest-mass of the electron. Once again, within our classical discussion we cannot push these reasonings too far, but the latter interpretation is closer to the actual treatment in quantum electrodynamics, where one starts with a "bare" mass term for the electron, and radiative effects, computed as an expansion in powers of the small parameter α, give corrections to it producing a "renormalized" mass, which is then identified with the observed mass; see e.g., Maggiore (2005) for an introductory textbook.

The terms with $a \neq b$ have a finite limit for $\Lambda \to \infty$. This means that they are insensitive to the details of the structure of elementary particles at short distances, and for these terms we can take the limit $\Lambda \to \infty$. In this limit, using eq. (1.115), we get back eq. (5.7). The integral in the second term is easily computed analytically: passing in polar coordinates

$$\frac{1}{(2\pi)^3} \int_0^\Lambda k^2 dk \int d\Omega \frac{1}{k^2} = \frac{\Lambda}{2\pi^2} \,, \quad (5.30)$$

and diverges if we remove the cutoff, i.e., if we send $\Lambda \to \infty$. The result become nicer defining $\ell = \pi/\Lambda$ (note that ℓ has dimensions of length). Then, apart from terms that vanish as $\ell \to 0$,

$$(\mathcal{E}_E)_{\rm reg} = \frac{1}{4\pi\epsilon_0} \sum_{a=1}^N \sum_{b>a}^N \frac{q_a q_b}{|\mathbf{x}_a - \mathbf{x}_b|} + \frac{1}{4\pi\epsilon_0} \sum_{a=1}^N \frac{q_a^2}{\ell} \,. \quad (5.31)$$

The last term can be interpreted as a "Coulomb self-energy" term, associated with individual charges, rather than to their interaction. This interpretation should not be taken too literally, first of all because, again, classical electrodynamics is not the proper framework for studying the nature of elementary particles. Observe, furthermore, that the exact numerical value of the self-energy term depends on the details of the regularization procedure. For instance, instead of putting a sharp cutoff setting to zero all modes with $|\mathbf{k}| > \Lambda$, we could have used a smooth cutoff that suppresses them gradually, and the precise numbers would have been different, so this term reflects the details of our mathematical regularization procedure, rather than a direct physical property. The important points, however, are: (1) this term is divergent if we remove the cutoff, sending $\ell \to 0$. For this reason, it was necessary to "regularize" eq. (5.21) before applying it to point-like particles. (2) This "self-energy" term, for given ℓ, is just a numerical constants. The energy of a static system of charges can be redefined as the difference between its value in a given configuration and the value when all charges are at an infinite distance among them, since this represent the work that should be done to assemble the charges in the given spatial configuration; the charges themselves are taken as given, and we must not perform work to create them. The constant self-energy term then cancels in this difference and can therefore simply be discarded. We then get back the expression (5.7) of the Coulomb energy of a system of charges interacting among them.[8]

In summary, eqs. (5.16) or (5.21), which have been derived from the general formula (3.41) for the energy of the electromagnetic field, give the energy density of a localized and static charge distribution $\rho(\mathbf{x})$, as long as the latter is sufficiently smooth on short scales. Basically, the exception to this behavior is given by point-like particles. In this case, one must be aware of the fact that the notion of point-like particle is just a mathematical idealization and, after a suitable regularization procedure and the subtraction of a constant term (that would diverge when removing the cutoff), one obtains the expected result (5.7) for the Coulomb potential energy of a set of charges.

5.2.3   Energy of charges in an external electric field

Another important quantity is the electrostatic energy of a system of charges in an external electric field. Eventually, the external field will itself be generated by a set of charges. So, we can model the situation considering two sets of spatially separated charges, localized into two non-overlapping volumes V_1 and V_2. Then, using a discrete formulation first, the sum over a in the equations of Section 5.1 splits into a sum over the charges in V_1 and those in V_2, and eq. (5.7) becomes

$$(\mathcal{E}_E)_{\text{p.p.}} = \left(\sum_{a,b\in V_1, b\neq a} +2\sum_{a\in V_1}\sum_{b\in V_2} + \sum_{a,b\in V_2, b\neq a} \right) \frac{q_a q_b}{8\pi\epsilon_0|\mathbf{x}_a - \mathbf{x}_b|} . \quad (5.35)$$

The first term gives the interaction energy among the charges in V_1, while the third term gives the interaction energy among the charges in V_2. The middle term is the interaction between the two groups.[9] Then, the electrostatic interaction energy between the two groups of particles is given by

$$\frac{1}{4\pi\epsilon_0} \sum_{a\in V_1}\sum_{b\in V_2} \frac{q_a q_b}{|\mathbf{x}_a - \mathbf{x}_b|} . \quad (5.36)$$

We next observe, from eq. (4.11), that

$$\phi_{\text{ext}}(\mathbf{x}) \equiv \frac{1}{4\pi\epsilon_0} \sum_{b\in V_2} \frac{q_b}{|\mathbf{x} - \mathbf{x}_b|} , \quad (5.37)$$

is the potential generated by the charges in V_2 at the point $\mathbf{x}$. The label "ext" stresses that we consider this as an "external potential," from the point of view of the charges in V_1.[10] Correspondingly, we denote the expression in (5.36) as $(\mathcal{E}_E)_{\text{ext}}$, to stress that this is the electrostatic energy of a set of particles (the particles in V_1) due to the interaction with a given external field. Then, combining with eq. (5.37), we get

$$\boxed{(\mathcal{E}_E)_{\text{ext}} = \sum_{a\in V_1} q_a \phi_{\text{ext}}(\mathbf{x}_a) .} \quad (5.38)$$

[Note the absence of the factor $1/2$, compared to eq. (5.9)]. The continuous versions of these results are obtained from eq. (5.13), considering the charge density $\rho_2(\mathbf{x}) \equiv \rho_{\text{ext}}(\mathbf{x})$ as an external source,[11] and therefore are given by

$$\boxed{(\mathcal{E}_E)_{\text{ext}} = \int d^3x\, \rho(\mathbf{x})\phi_{\text{ext}}(\mathbf{x}) ,} \quad (5.39)$$

where we denoted $\rho_1(\mathbf{x})$ simply as $\rho(\mathbf{x})$, and

$$\phi_{\text{ext}}(\mathbf{x}) = \frac{1}{4\pi\epsilon_0} \int d^3x' \frac{\rho_{\text{ext}}(\mathbf{x}')}{|\mathbf{x} - \mathbf{x}'|} , \quad (5.40)$$

is the scalar potential generated by the second charge distribution, which can be seen as an external potential from the point of view of the first charge distribution.

[9] Note that, there, we could omit the condition $b\neq a$ since in this sum a and b run over different groups of particles, so this condition is automatically satisfied.

[10] This distinction is useful in particularly when the back-reaction of the charges in V_1 on those in V_2 is small (for instance, the charges in V_1 are a small group of elementary charges, while those in V_2 form a macroscopic object, e.g., a capacitor, creating a macroscopic electric field), so we can consider the external field as given, independently of the motion of the particles in V_1.

[11] We will then label again the corresponding energy as $(\mathcal{E}_E)_{\text{ext}}$, to stress that it is the energy of $\rho_1(\mathbf{x})$ in an external field, while in eq. (5.13) it was labeled $\mathcal{E}_E^{\text{int}}$, to stress that it is the interaction energy between $\rho_1(\mathbf{x})$ and $\rho_2(\mathbf{x})$.

5.2.4   Energy of a system of conductors

Consider a set of extended charge distributions $\rho_a(\mathbf{x})$, labeled by an index $a = 1, \ldots, N$, each localized in a volume V_a, non-overlapping with the other volumes. Then $\rho(\mathbf{x}) = \sum_{a=1}^{N} \rho_a(\mathbf{x})$, and the integral in eq. (5.15) splits into a sum over non-overlapping volumes,

$$\mathcal{E}_E = \frac{1}{2} \sum_{a=1}^{N} \int_{V_a} d^3x \, \rho_a(\mathbf{x}) \phi(\mathbf{x}) \,. \tag{5.41}$$

In the special case in which these N bodies are conductors, this expression simplifies considerably. Recall from Section 4.1.6 that, for conductors at equilibrium, the charge density is zero inside the conductor, and non-vanishing only on its surface. Furthermore, we found that, on the surface, the electrostatic potential is constant. Denoting by ϕ_a the constant value of $\phi(\mathbf{x})$ on the surface of the a-th conductor, eq. (5.41) becomes

$$\mathcal{E}_E = \frac{1}{2} \sum_{a=1}^{N} \phi_a \int_{V_a} d^3x \, \rho_a(\mathbf{x}) \,, \tag{5.42}$$

where we could extract $\phi(\mathbf{x})$ from the integrals, since $\rho_a(\mathbf{x})$ is proportional to a two-dimensional Dirac delta on the surface of the a-th conductor, and there $\phi(\mathbf{x}) = \phi_a$. The remaining integral is the total charge Q_a of the a-th conductor, so we see that the energy of an ensemble of conductors can be expressed as

$$\boxed{\mathcal{E}_E = \frac{1}{2} \sum_{a=1}^{N} Q_a \phi_a \,,} \tag{5.43}$$

which is analogous to eq. (5.9) for point particles. As we discussed in Problem 4.3, for a set of conductors the charges Q_a and the potential ϕ_a are linearly related, through eq. (4.163) or the inverse relation (4.164). Therefore, eq. (5.43) can be rewritten in terms of the potentials ϕ_a and the coefficients of capacitance C_{ab} as

$$\mathcal{E}_E = \frac{1}{2} \sum_{a,b=1}^{N} C_{ab} \phi_a \phi_b \,, \tag{5.44}$$

or in terms of the charges Q_a and the coefficients of potential P_{ab} as

$$\mathcal{E}_E = \frac{1}{2} \sum_{a,b=1}^{N} P_{ab} Q_a Q_b \,. \tag{5.45}$$

In Problem 5.1, we will check this expression for a single capacitor, computing the work need to assemble its charge configuration.

The relation between the energy of a system of conductors, and the mechanical potentials from which we can derive the forces acting on them is somewhat subtle and depends on whether the conductors are kept at fixed charge or at fixed electrostatic potential. We will examine this in Section 5.5.1.

5.3 Work and energy in magnetostatics

We now discuss the corresponding issues in the context of magneto-statics. We begin by computing the work necessary to build a static magnetic field configuration. The source of the magnetic field is a current distribution $\mathbf{j}(\mathbf{x})$, and we must therefore ask what work is necessary to build up such a current distribution, starting from zero.

For the sake of the argument, imagine an external agent that "grabs" an electron at rest, and starts to accelerate it, moving it in a circular orbit, until it reaches a given velocity corresponding to a given current. Concretely, a current will rather be produced by accelerating a large number of electrons, and the force that accelerates them will be an external electric field, but we do not need to specify this for the following argument. If we neglect for a moment any effect related to electromagnetism, we would conclude that the work done by the external force goes into the kinetic energy of the electron; or, in the more realistic setting of an electric wire, it goes partly in the kinetic energy of the electrons and partly into dissipation in the wire. Once we include Maxwell's equations in our considerations, we realize that the final configuration, with a given final current, generates a magnetic field. We already know that a magnetic field carries energy, so energy conservation must now take a different form: to conserve energy, the work that has been done by the external force during the process of building the final configuration must be equal to the kinetic energy of the final electron (or the kinetic energy of the electrons in a wire plus the energy lost to dissipation, in a more realistic setting) plus the energy of the final magnetic field. In other words, the work done by the external force to accelerate a particle to a given velocity must be different when the particle is charged from when it is electrically neutral, and the difference must be precisely the work done to create the corresponding magnetic field.

The reason why the work is different for a charged and for neutral particle is that the charged particle interacts with the electromagnetic field that is being built. The magnetic field, by itself, does no work on the charged particles on which it acts, given that $(\mathbf{v} \times \mathbf{B}) \cdot (\mathbf{v}dt) = 0$. However, as we raise the current, the magnetic field that it generates also raises. Therefore, during the transient period needed to reach the final static field configuration starting from zero initial field, $\partial \mathbf{B}/\partial t$ is non-zero. According to Faraday's law (3.4), a non-zero value of $\partial \mathbf{B}/\partial t$ generates an electric field (the "induced" electric field). The induced electric field performs work on the charged particles that make up the current and, as we have seen in Section 4.3.1, for a closed loop it acts to oppose the increase of the magnetic flux (Lenz's law). So, we have to supply extra external work to counterbalance the work made by this induced electric field during the transient period needed to reach the desired final value of the magnetic field, even if this final configuration is static. One might try to circumvent the problem by raising the current very slowly, so that $\partial \mathbf{B}/\partial t$ is very small, and, at any given time, the effect might seem negligible. However, in this case it would take a very long time to reach the desired final value of $\mathbf{j}$ and, as the following computation will make clear, the two effects compensate.

The work per unit time made by an electric field on a current distribution $\mathbf{j}(\mathbf{x})$ has already been computed in eq. (3.48). The work made by the external agent is the negative of this, so, at a time t during the transient period, when the current has a value $\mathbf{j}(t, \mathbf{x})$, we have

$$\frac{dW_{\text{ext}}}{dt} = -\int d^3x \, \mathbf{E}(t, \mathbf{x}) \cdot \mathbf{j}(t, \mathbf{x}) \, , \qquad (5.46)$$

where $\mathbf{E}(t, \mathbf{x})$ is the induced electric field at that time. In a magneto-static setting, we want to raise the current very slowly; then the Ampère–Maxwell law (3.2) reduces to Ampère's law, and

$$\mathbf{j}(\mathbf{x}) = \frac{1}{\mu_0} \boldsymbol{\nabla} \times \mathbf{B} \, . \qquad (5.47)$$

In other words, while we must take into account $\partial \mathbf{B}/\partial t$ during the transient period, otherwise we miss the leading contribution to W_{ext}, we can assume that the time derivative of the induced electric field is sufficiently small to be neglected. Then eq. (5.46) becomes

$$\begin{aligned}
\frac{dW_{\text{ext}}}{dt} &= -\frac{1}{\mu_0} \int d^3x \, \mathbf{E} \cdot (\boldsymbol{\nabla} \times \mathbf{B}) \\
&= -\frac{1}{\mu_0} \int d^3x \, \mathbf{B} \cdot (\boldsymbol{\nabla} \times \mathbf{E}) \\
&= \frac{1}{\mu_0} \int d^3x \, \mathbf{B} \cdot \frac{\partial \mathbf{B}}{\partial t} \\
&= \frac{1}{2\mu_0} \frac{d}{dt} \int d^3x \, |\mathbf{B}(t, \mathbf{x})|^2 \, ,
\end{aligned} \qquad (5.48)$$

where, in the first line, we have integrated by parts, so that $E_i(\epsilon_{ijk} \partial_j B_k)$ gives $-(\epsilon_{ijk} \partial_j E_i) B_k$, which is the same as $+(\epsilon_{ijk} \partial_i E_j) B_k = \mathbf{B} \cdot (\boldsymbol{\nabla} \times \mathbf{E})$, and in the second line we used Faraday's law (3.4). Therefore, integrating with respect to time, with the initial condition that $W_{\text{ext}} = 0$ at the initial time when $\mathbf{B} = 0$, we find that the work needed to obtain a given final static field $\mathbf{B}(\mathbf{x})$ is

$$W_{\text{ext}} = \frac{1}{2\mu_0} \int d^3x \, |\mathbf{B}|^2 \, , \qquad (5.49)$$

in full agreement with the general result (3.41) with $\mathbf{E} = 0$. Therefore, the energy associated with a static magnetic field $\mathbf{B}$, defined as the work needed to build the configuration of current that generates it, is

$$\boxed{\mathcal{E}_B = \frac{1}{2\mu_0} \int d^3x \, |\mathbf{B}(\mathbf{x})|^2 \, ,} \qquad (5.50)$$

and agrees with the expression of the energy stored in the final magnetic field obtained from eq. (3.41).

5.4   Energy stored in a static magnetic field

In this section, in analogy with the treatment in Section 5.2 for electrostatics, we rewrite the energy (5.50) in some other useful forms, in terms of the vector potential $\mathbf{A}(\mathbf{x})$ corresponding to the final magnetic field and of the final current distribution $\mathbf{j}(\mathbf{x})$ that generates it, and we will also write the corresponding expression when the current distribution corresponds to a set of closed loops.[12] We write

$$
\begin{aligned}
\mathcal{E}_B &= \frac{1}{2\mu_0} \int d^3x\, \mathbf{B}\cdot\mathbf{B} \\
&= \frac{1}{2\mu_0} \int d^3x\, \mathbf{B}\cdot(\boldsymbol{\nabla}\times\mathbf{A}) \\
&= \frac{1}{2\mu_0} \int d^3x\, (\boldsymbol{\nabla}\times\mathbf{B})\cdot\mathbf{A} \\
&= \frac{1}{2} \int d^3x\, \mathbf{j}\cdot\mathbf{A}\,,
\end{aligned}
\tag{5.51}
$$

where in the second line we integrated by parts [similarly to the passages made after eq. (5.48)] and we then used eq. (5.47). Note that the last passage is again specific to magnetostatics. Therefore, in magnetostatics we can write the energy as

$$
\boxed{\mathcal{E}_B = \frac{1}{2} \int d^3x\, \mathbf{j}(\mathbf{x})\cdot\mathbf{A}(\mathbf{x})\,.}
\tag{5.52}
$$

This is the analog of eq. (5.15) in electrostatics.[13] Another useful form is obtained using the solution for $\mathbf{A}(\mathbf{x})$ in terms of the current distribution given by eq. (4.92),[14] which gives

$$
\boxed{\mathcal{E}_B = \frac{\mu_0}{8\pi} \int d^3x\, d^3x'\, \frac{\mathbf{j}(\mathbf{x})\cdot\mathbf{j}(\mathbf{x}')}{|\mathbf{x}-\mathbf{x}'|}\,.}
\tag{5.53}
$$

This is the analog for magnetostatics of eq. (5.16) in electrostatics. Similarly to eq. (5.21), we can take the Fourier transform of eq. (5.53), and write

$$
\mathcal{E}_B = \frac{\mu_0}{2} \int \frac{d^3k}{(2\pi)^3}\, \frac{\tilde{\mathbf{j}}(-\mathbf{k})\cdot\tilde{\mathbf{j}}(\mathbf{k})}{k^2}\,.
\tag{5.54}
$$

As in eq. (5.21), the convergence of the integral is assured if $\tilde{\mathbf{j}}(\mathbf{k})$ goes to zero faster that $1/k^{1/2}$.

If $\mathbf{j}(\mathbf{x}) = \mathbf{j}_1(\mathbf{x}) + \mathbf{j}_2(\mathbf{x})$, where $\mathbf{j}_1(\mathbf{x})$ and $\mathbf{j}_2(\mathbf{x})$ are localized in two non-overlapping volumes V_1 and V_2, eq. (5.53) becomes

$$
\mathcal{E}_B = \mathcal{E}_B^{(1)} + \mathcal{E}_B^{(2)} + \mathcal{E}_B^{\mathrm{int}}\,,
\tag{5.55}
$$

where

$$
\mathcal{E}_B^{(a)} = \frac{\mu_0}{8\pi} \int d^3x\, d^3x'\, \frac{\mathbf{j}^{(a)}(\mathbf{x})\cdot\mathbf{j}^{(a)}(\mathbf{x}')}{|\mathbf{x}-\mathbf{x}'|}\,,
\tag{5.56}
$$

---

[12]Note that, in the electrostatic case, we considered separately a discrete and a continuous distribution of charges. In magnetostatics, currents are necessarily extended, so we work directly in the continuous formalism. Alternatively, one could consider the interaction between two closed loops, with a size small compared to their distance. We will discuss this setting in the context of the multipole expansion in Section 6.3, where we will see that this corresponds to the magnetic dipole interaction. Since the equivalent of the electric charge (the electric "monopole," in the language of multipole expansion) does not exist for the magnetic source, the analogy with electrostatics is clearer working directly in terms of extended distributions.

[13]Observe that this expression is gauge invariant, as it should be for an energy. This is evident from the original expression (5.50), because it only involve $\mathbf{B}$, which is gauge invariant. Equation (5.52), in contrast, is written in terms of $\mathbf{A}$, which is not gauge invariant. However, under the gauge transformation $\mathbf{A} \to \mathbf{A} - \boldsymbol{\nabla}\theta$,

$$
\int d^3x\, \mathbf{j}\cdot\mathbf{A} \to \int d^3x\, \mathbf{j}\cdot(\mathbf{A} - \boldsymbol{\nabla}\theta)\,,
$$

and the extra term vanishes after integrating by parts and using $\boldsymbol{\nabla}\cdot\mathbf{j} = 0$, which is valid in magnetostatics, see eq. (4.90).

[14]This solution for $\mathbf{A}$ was obtained in the Coulomb gauge but, as shown in the previous note, eq. (5.52) is anyhow gauge invariant.

(with $a = 1, 2$), can be seen as the "self-energy" of the a-th current distribution, while

$$\mathcal{E}_B^{\text{int}} = \frac{\mu_0}{4\pi} \int d^3x d^3x' \frac{\mathbf{j}_1(\mathbf{x}) \cdot \mathbf{j}_2(\mathbf{x}')}{|\mathbf{x} - \mathbf{x}'|} \, , \tag{5.57}$$

is the interaction energy between the two current distributions.[15]

When the back-action of $\mathbf{j}_1$ on $\mathbf{j}_2$ is negligible, from the point of view of the first current distribution it can be useful to see the second current distribution as generating a given "external field." We can then rewrite eq. (5.57) as

$$(\mathcal{E}_B)_{\text{ext}} = \int d^3x \, \mathbf{j}_1(\mathbf{x}) \cdot \mathbf{A}_{\text{ext}}(\mathbf{x}) \, , \tag{5.58}$$

where $\mathbf{A}_2(\mathbf{x}) \equiv \mathbf{A}_{\text{ext}}(\mathbf{x})$ is the vector potential generated by $\mathbf{j}_2(\mathbf{x})$, according to eq. (4.92). Equations (5.52) and (5.58) can be compared to eqs. (5.15) and (5.39) for the electrostatic case.

While in electrostatic the most elementary source is provided by a point charge, in magnetostatics the simplest source is a closed loop carrying a current. We therefore discuss how the previous general expressions simplify for a collection of loops. We consider a set of closed loops $\mathcal{C}_a$, $(a = 1, \ldots, N)$, carrying currents I_a. From eq. (4.103),

$$\mathbf{j}(\mathbf{x}) d^3x = \sum_{a=1}^{N} \mathbf{j}_a(\mathbf{x}) d^3x$$

$$\rightarrow \sum_{a=1}^{N} I_a d\boldsymbol{\ell}_a \, . \tag{5.59}$$

Then, using Stokes's theorem (1.38), we can transform eq. (5.52) as

$$\begin{aligned}
\mathcal{E}_B &= \frac{1}{2} \sum_{a=1}^{N} I_a \oint_{\mathcal{C}_a} d\boldsymbol{\ell}_a \cdot \mathbf{A} \\
&= \frac{1}{2} \sum_{a=1}^{N} I_a \int_{S_a} d\mathbf{s} \cdot (\boldsymbol{\nabla} \times \mathbf{A}) \\
&= \frac{1}{2} \sum_{a=1}^{N} I_a \int_{S_a} d\mathbf{s} \cdot \mathbf{B} \, ,
\end{aligned} \tag{5.60}$$

where S_a is any surface having $\mathcal{C}_a$ as the boundary and $\mathbf{B}$ is the total magnetic field generated by all the loops. The integral on the right-hand side is just the magnetic flux of $\mathbf{B}$ through S_a, that we denote as $\Phi_{B,a}$, so

$$\mathcal{E}_B = \frac{1}{2} \sum_{a=1}^{N} I_a \Phi_{B,a} \, , \tag{5.61}$$

[15]Note that, as in eq. (5.13), which is the analogous result for electrostatics, the integral over d^3x takes contributions only when $\mathbf{x}$ is inside the volume V_1 where $\mathbf{j}_1(\mathbf{x})$ is non-vanishing, and the integral over d^3x' takes contributions only when $\mathbf{x}'$ is inside the volume V_2 where $\mathbf{j}_2(\mathbf{x}')$ is non-vanishing. Therefore, having taken V_1 and V_2 non-overlapping, $|\mathbf{x} - \mathbf{x}'|$ is always above a minimum value and there is no problem of convergence in the integral in eq. (5.57).

which can be compared with eq. (5.43) for a system of conductors. For a closed loop $\mathcal{C}$ carrying a current I, placed in an external magnetic field $\mathbf{B}_{\text{ext}}$, proceeding as in eq. (5.60) we see that eq. (5.58) can be rewritten as

$$(\mathcal{E}_B)_{\text{ext}} = I\Phi_{B_{\text{ext}}}, \tag{5.62}$$

where we denoted the magnetic flux of B_{ext} through $\mathcal{C}$ by $\Phi_{B_{\text{ext}}}$.

We now observe that the knowledge of the position and shape of the loops, and of their currents, uniquely determines the magnetic field everywhere, and therefore the fluxes through the loops. As we have already mentioned in Problem 4.8, the linearity of Maxwell's equation implies that these relations are linear, so that

$$\Phi_{B,a} = \sum_{b=1}^{N} L_{ab}I_b. \tag{5.63}$$

The coefficients L_{ab} are called the *mutual inductances*, while the diagonal terms $L_{aa} \equiv L_a$ are called the self-inductances and depend on the shapes and positions $\mathbf{x}_a$ of all loops. Equation (5.63) is the analog of eq. (4.164) in electrostatics. The derived SI units for the inductance is the *henry* (H).

The mutual inductance between two loops $\mathcal{C}_a$ and $\mathcal{C}_b$ (with $a \neq b$) can be written explicitly, in terms of the geometry of the loops, computing the contribution to the flux through loop a produced by the current I_b. Denoting this as $(\Phi_B)_{a,b}$, and using the same manipulations as in eq. (5.60) in reverse order,

$$
\begin{aligned}
(\Phi_B)_{a,b} &= \int_{S_a} d\mathbf{s} \cdot \mathbf{B}_b \\
&= \int_{S_a} d\mathbf{s} \cdot (\boldsymbol{\nabla} \times \mathbf{A}_b) \\
&= \oint_{\mathcal{C}_a} d\boldsymbol{\ell}_a \cdot \mathbf{A}_b[\mathbf{x}(\ell_a)],
\end{aligned}
\tag{5.64}
$$

where $\mathbf{B}_b$ and $\mathbf{A}_b$ are the magnetic field and, respectively, the vector potential, generated by the loop b, and we wrote explicitly that the argument of $\mathbf{A}_b$ is the value of $\mathbf{x}$ that corresponds to the coordinate ℓ_a on the loop, $\mathbf{x}(\ell_a)$. Using eq. (4.104), we get

$$(\Phi_B)_{a,b} = \frac{\mu_0}{4\pi} I_b \oint_{\mathcal{C}_a} \oint_{\mathcal{C}_b} \frac{d\boldsymbol{\ell}_a \cdot d\boldsymbol{\ell}_b}{|\mathbf{x}(\ell_a) - \mathbf{x}(\ell_b)|}. \tag{5.65}$$

This shows that a relation of the form (5.63) indeed holds, and provides an explicit expression for the coefficients L_{ab},

$$L_{ab} = \frac{\mu_0}{4\pi} \oint_{\mathcal{C}_a} \oint_{\mathcal{C}_b} \frac{d\boldsymbol{\ell}_a \cdot d\boldsymbol{\ell}_b}{|\mathbf{x}(\ell_a) - \mathbf{x}(\ell_b)|}. \tag{5.66}$$

From this explicit expression we see that $L_{ab} = L_{ba}$. Using eq. (5.63), eq. (5.61) can be rewritten as

$$\mathcal{E}_B = \frac{1}{2} \sum_{a=1}^{N} L_{ab} I_a I_b \,. \tag{5.67}$$

This is analogous to eq. (5.44) for conductors. Therefore the a-th loop has a self-energy

$$(\mathcal{E}_B)_a = \frac{1}{2} L_a I_a^2 \,, \tag{5.68}$$

while, using $L_{ab} = L_{ba}$, the interaction energy between loop a and loop b is

$$(\mathcal{E}_B)_{ab} = L_{ab} I_a I_b \,, \tag{5.69}$$

or, using the explicit expression (5.66)

$$(\mathcal{E}_B)_{ab} = \frac{\mu_0}{4\pi} I_a I_b \oint_{C_a} \oint_{C_b} \frac{d\boldsymbol{\ell}_a \cdot d\boldsymbol{\ell}_b}{|\mathbf{x}(\ell_a) - \mathbf{x}(\ell_b)|} \,. \tag{5.70}$$

From eq. (5.68), we see that the self-inductance of a loop must be a positive quantity and provides a measure of how much energy should be given to a circuit to raise its current to a given value. The larger L_a, the larger is the energy that must be supplied to reach a given value of I, just as, in the expression $\mathcal{E}_{\text{kin}} = (1/2)mv^2$ of the kinetic energy of a non-relativistic particle, the larger the mass, the larger is the energy that must be supplied to reach a given velocity. More generally, the condition that the quadratic form (5.67) must be positive imposes the constraints $L_a > 0$ and

$$L_a L_b > L_{ab}^2 \,. \tag{5.71}$$

5.5   Forces and mechanical potentials

This section is quite long and technical, and should be skipped at first reading.

In non-relativistic mechanics, conservative forces acting on a system can be obtained from mechanical potential functions, that we generically denote by U, defined by the fact that the force can be written as $\mathbf{F} = -\boldsymbol{\nabla} U$. Similar relations hold for the forces acting on charge and current distributions in the full Maxwell theory.[16] However, one must be aware of some subtleties: a correct treatment is analogous to that of thermodynamical potentials, where one must specify which, among the variables that determine the state of the system, are kept constant, and which are varied, and different mechanical potentials, related by Legendre transforms, are appropriate to different situations.

Let us begin with the most obvious example, which is the Coulomb potential. We consider a system of N point charges, $q_1, \ldots, q_N$, located at positions $\mathbf{x}_1, \ldots, \mathbf{x}_N$, respectively. The force $\mathbf{F}_a$ acting on the a-th charge is obtained summing the contribution from all other charges, which are given individually by Coulomb's law (4.8), and is therefore

[16] It should be stressed that force, with its instantaneous character, is an intrinsically non-relativistic notion that has no place in a full relativistic theory, and the same holds for the "instantaneous" mechanical potential U (in the various forms that it can take, that will be described in the following). Therefore, the whole discussion in this section is only valid for non-relativistic sources. In Section 12.2.2, we will see how the effective dynamics of a system of charges can be written in terms of instantaneous interactions, order by order in v/c.

$$\mathbf{F}_a = \frac{1}{4\pi\epsilon_0} \sum_{b \neq a}^{N} q_a q_b \frac{\mathbf{x}_a - \mathbf{x}_b}{|\mathbf{x}_a - \mathbf{x}_b|^3} \,. \tag{5.72}$$

Using eq. (4.19), we see that this force can be written as

$$\mathbf{F}_a = -\frac{\partial U_{\text{Coul}}}{\partial \mathbf{x}_a}, \tag{5.73}$$

where[17]

$$U_{\text{Coul}}(\mathbf{x}_1, \ldots, \mathbf{x}_N) = \frac{1}{8\pi\epsilon_0} \sum_{a=1}^{N} \sum_{b \neq a}^{N} \frac{q_a q_b}{|\mathbf{x}_a - \mathbf{x}_b|}. \tag{5.74}$$

This expression defines the Coulomb potential. Comparing with eq. (5.7), we see that the Coulomb potential is the same as the energy of a system of static charges (which is the same as their interaction energy, given that the kinetic energy vanishes for static charges). Thus, eqs. (5.73), (5.74), and (5.7) reveal that the mechanical potential from which we can derive the force on a static charge due to all other static charges is equal to the energy of the whole configuration of static charges. To mitigate the feeling that we are elaborating on the obvious, let us anticipate that we will find in Section 5.5.1 that this result only holds because we have implicitly assumed that the charges on the individual bodies are conserved. For a set of point particles this is a somewhat obvious assumption; however, in other contexts other options are possible; in particular, for a set of conductors, one could rather fix the value of the electrostatic potentials on their surfaces, and in that case we will find that the equality between the relevant mechanical potential and the energy of a static configuration does not hold. We will meet similar situations in magnetostatics.

First, however, we examine the generalization to a continuous charge distribution. The forces between two charge distributions $\rho_1(\mathbf{x})$ and $\rho_2(\mathbf{x})$, localized in two non-overlapping volumes V_1 and V_2, have already been computed in eqs. (4.26) and (4.27), while the continuous generalization of the Coulomb interaction potential (5.74) is given by[18]

$$U_E^{\text{int}} \equiv \frac{1}{4\pi\epsilon_0} \int d^3x\, d^3x' \frac{\rho_1(\mathbf{x})\rho_2(\mathbf{x}')}{|\mathbf{x} - \mathbf{x}'|}. \tag{5.75}$$

We now want to understand how to get the forces in eqs. (4.26) and (4.27) by taking gradients of this potential. At first, one might even wonder where the variables are with respect to which we could take derivatives of U_E^{int}, since in eq. (5.75) $\mathbf{x}$ and $\mathbf{x}'$ are integrated over and therefore do not appear as free variables on the left-hand side. Actually, the required variables are hidden in $\rho_1(\mathbf{x})$ and $\rho_2(\mathbf{x}')$, through the fact that these charge distributions are localized in the volumes V_1 and V_2, respectively. We take V_1 and V_2 to be volumes with a fixed shape and orientation, so their position in space can be given by specifying the coordinates $\mathbf{x}_1$ and $\mathbf{x}_2$ of just one of their points, such as, for instance, their geometrical centers. For instance, if V_1 and V_2 are spheres of given radii R_1 and R_2, we can take $\mathbf{x}_1$ and $\mathbf{x}_2$ as the respective centers of these spheres. A change in $\mathbf{x}_1$ corresponds to a rigid displacement of the first charge distribution, and similarly a change in $\mathbf{x}_2$ corresponds to a rigid displacement of the second charge distribution. Let us denote by $\rho_1^{(0)}(\mathbf{x})$

[17]When computing $\partial/\partial\mathbf{x}_a$, one must be careful to use a different dummy index in the double summation, writing e.g.,

$$\frac{\partial U_{\text{Coul}}}{\partial \mathbf{x}_a} = \frac{1}{8\pi\epsilon_0} \frac{\partial}{\partial \mathbf{x}_a} \sum_{c=1}^{N} \sum_{b \neq c}^{N} \frac{q_c q_b}{|\mathbf{x}_c - \mathbf{x}_b|}.$$

Then, there are two equal contributions from $c = a$ and from $b = a$, that transform the factor $1/(8\pi\epsilon_0)$ in the potential into $1/(4\pi\epsilon_0)$ in the force.

[18]The overall factor $1/(4\pi\epsilon_0)$ in eq. (5.75) differs by a factor of 2 from the factor $1/(8\pi\epsilon_0)$ in eq. (5.74); this comes from the (by now, usual) fact that, setting e.g., $N = 2$, in eq. (5.74) we have $\sum_{a=1}^{2}\sum_{b=1}^{2}$ with the condition $b \neq a$, so there are two equal contributions to the sum, from $(a = 1, b = 2)$ and from $(a = 2, b = 1)$. The result is then the same as that obtained from eq. (5.75) when we set $\rho_1(\mathbf{x}) = q_1\delta^{(3)}(\mathbf{x} - \mathbf{x}_1)$ and $\rho_2(\mathbf{x}') = q_2\delta^{(3)}(\mathbf{x}' - \mathbf{x}_2)$.

the first charge density distribution in the reference frame where $\mathbf{x}_1 = 0$. Then, in a translated reference frame where the coordinate specifying the position of V_1 has the generic value $\mathbf{x}_1$, we have

$$\rho_1(\mathbf{x}) = \rho_1^{(0)}(\mathbf{x} - \mathbf{x}_1). \tag{5.76}$$

Similarly, denoting by $\rho_2^{(0)}(\mathbf{x})$ the second charge density distribution in the reference frame where $\mathbf{x}_2 = 0$, in a generic frame we have

$$\rho_2(\mathbf{x}) = \rho_2^{(0)}(\mathbf{x} - \mathbf{x}_2). \tag{5.77}$$

For instance, for a point charge q_1 located at the origin, $\rho_1^{(0)}(\mathbf{x}) = q_1 \delta^{(3)}(\mathbf{x})$, and then $\rho_1(\mathbf{x}) = q_1 \delta^{(3)}(\mathbf{x} - \mathbf{x}_1)$ is the charge density in the frame where the charge is located in the position $\mathbf{x}_1$. So, $\mathbf{x}_1$ and $\mathbf{x}_2$ are the "hidden" variables on which the potential depends. The use of the densities $\rho_{1,2}^{(0)}(\mathbf{x})$ has the advantage of explicitly bringing out the dependence on $\mathbf{x}_1$ and $\mathbf{x}_2$, which is otherwise implicit in $\rho_{1,2}(\mathbf{x})$, and eqs. (4.26), (4.27), and (5.75) can be written, more explicitly, as

$$\mathbf{F}_1(\mathbf{x}_1, \mathbf{x}_2) = \frac{1}{4\pi\epsilon_0} \int d^3x\, d^3x'\, \rho_1^{(0)}(\mathbf{x} - \mathbf{x}_1) \rho_2^{(0)}(\mathbf{x}' - \mathbf{x}_2) \frac{\mathbf{x} - \mathbf{x}'}{|\mathbf{x} - \mathbf{x}'|^3}, \tag{5.78}$$

$$\mathbf{F}_2(\mathbf{x}_1, \mathbf{x}_2) = \frac{1}{4\pi\epsilon_0} \int d^3x\, d^3x'\, \rho_1^{(0)}(\mathbf{x} - \mathbf{x}_1) \rho_2^{(0)}(\mathbf{x}' - \mathbf{x}_2) \frac{\mathbf{x}' - \mathbf{x}}{|\mathbf{x} - \mathbf{x}'|^3}, \tag{5.79}$$

and

$$U_E^{\rm int}(\mathbf{x}_1, \mathbf{x}_2) = \frac{1}{4\pi\epsilon_0} \int d^3x\, d^3x'\, \frac{\rho_1^{(0)}(\mathbf{x} - \mathbf{x}_1) \rho_2^{(0)}(\mathbf{x}' - \mathbf{x}_2)}{|\mathbf{x} - \mathbf{x}'|}. \tag{5.80}$$

We can now check that

$$\mathbf{F}_1(\mathbf{x}_1, \mathbf{x}_2) = -\frac{\partial U_E^{\rm int}}{\partial \mathbf{x}_1}, \tag{5.81}$$

and, similarly, $\mathbf{F}_2(\mathbf{x}_1, \mathbf{x}_2) = -\partial U_E^{\rm int}/\partial \mathbf{x}_2$, where it is understood that the partial derivative with respect to $\mathbf{x}_1$ is taken at $\mathbf{x}_2$ constant, and vice versa.[19] Also note, for future reference, that the derivatives have been taken keeping fixed the total charges in the volumes V_1 and V_2.

Comparing eq. (5.75) with eq. (5.13) we see that, also in the continuous case, the mechanical potential from which the forces acting on the charges can be derived is the same as the energy that these charges have in a static configuration, i.e., their interaction energy. If, instead of just two charge distributions ρ_1 and ρ_2, we have N of them, labeled as $\rho_a(\mathbf{x})$ and localized in volumes V_a, identified by coordinates $\mathbf{x}_a$, eq. (5.75) is generalized by summing over all possible pairs,

$$U_E^{\rm int}(\mathbf{x}_1, \dots, \mathbf{x}_N) = \frac{1}{8\pi\epsilon_0} \sum_{a=1}^{N} \sum_{b\neq a}^{N} \int d^3x\, d^3x'\, \frac{\rho_a(\mathbf{x})\rho_b(\mathbf{x}')}{|\mathbf{x} - \mathbf{x}'|}, \tag{5.82}$$

where, as usual, we have counted each pair in the sum twice, and compensated for this by an extra factor $1/2$; again, the dependence on $\mathbf{x}_a$ is hidden in $\rho_a(\mathbf{x})$, and can be shown explicitly using instead $\rho_a^{(0)}(\mathbf{x} - \mathbf{x}_a)$.

[19]The explicit computation is as follows:

$$-(4\pi\epsilon_0)\frac{\partial U_E^{\rm int}}{\partial \mathbf{x}_1}$$

$$= -\int d^3x\, d^3x'\, \frac{\partial \rho_1^{(0)}(\mathbf{x} - \mathbf{x}_1)}{\partial \mathbf{x}_1}$$

$$\times \frac{1}{|\mathbf{x} - \mathbf{x}'|}\rho_2^{(0)}(\mathbf{x}' - \mathbf{x}_2)$$

$$= \int d^3x\, d^3x'\, \frac{\partial \rho_1^{(0)}(\mathbf{x} - \mathbf{x}_1)}{\partial \mathbf{x}}$$

$$\times \frac{1}{|\mathbf{x} - \mathbf{x}'|}\rho_2^{(0)}(\mathbf{x}' - \mathbf{x}_2)$$

$$= -\int d^3x\, d^3x'\, \rho_1^{(0)}(\mathbf{x} - \mathbf{x}_1)$$

$$\times \left(\frac{\partial}{\partial \mathbf{x}}\frac{1}{|\mathbf{x} - \mathbf{x}'|}\right)\rho_2^{(0)}(\mathbf{x}' - \mathbf{x}_2)$$

$$= \int d^3x\, d^3x'\, \rho_1^{(0)}(\mathbf{x} - \mathbf{x}_1)$$

$$\times \frac{\mathbf{x} - \mathbf{x}'}{|\mathbf{x} - \mathbf{x}'|^3}\rho_2^{(0)}(\mathbf{x}' - \mathbf{x}_2),$$

where we used $\partial\rho_1^{(0)}(\mathbf{x} - \mathbf{x}_1)/\partial \mathbf{x}_1 = -\partial\rho_1^{(0)}(\mathbf{x} - \mathbf{x}_1)\partial\mathbf{x}$, we then integrated $\partial/\partial\mathbf{x}$ by parts (discarding the boundary term since ρ_1 is localized), and we finally used eq. (4.19).

It is useful to define

$$U_E[\rho] \equiv \frac{1}{8\pi\epsilon_0} \int d^3x d^3x' \, \frac{\rho(\mathbf{x})\rho(\mathbf{x}')}{|\mathbf{x} - \mathbf{x}'|} \,. \tag{5.83}$$

Similarly to eqs. (5.12) and (5.13), we now write $\rho(\mathbf{x}) = \rho_1(\mathbf{x}) + \rho_2(\mathbf{x})$, with $\rho_1(\mathbf{x})$ and $\rho_2(\mathbf{x})$ localized to volumes V_1 and V_2, respectively, identified by the positions $\mathbf{x}_1$ and $\mathbf{x}_2$ and, correspondingly, we use the notation $U_E(\mathbf{x}_1, \mathbf{x}_2)$ for $U_E[\rho]$. Then, we get

$$U_E(\mathbf{x}_1, \mathbf{x}_2) = U_E^{(1)} + U_E^{(2)} + U_E^{\text{int}}(\mathbf{x}_1, \mathbf{x}_2) \,, \tag{5.84}$$

where

$$U_E^{(a)} = \frac{1}{8\pi\epsilon_0} \int d^3x d^3x' \, \frac{\rho_a(\mathbf{x})\rho_a(\mathbf{x}')}{|\mathbf{x} - \mathbf{x}'|} \,, \tag{5.85}$$

while $U_E^{\text{int}}(\mathbf{x}_1, \mathbf{x}_2)$ is given by eq. (5.75). In terms of the densities $\rho^{(0)}$,

$$U_E^{\text{int}}(\mathbf{x}_1, \mathbf{x}_2) = \frac{1}{4\pi\epsilon_0} \int d^3x d^3x' \, \frac{\rho^{(0)}(\mathbf{x} - \mathbf{x}_1)\rho^{(0)}(\mathbf{x}' - \mathbf{x}_2)}{|\mathbf{x} - \mathbf{x}'|} \,, \tag{5.86}$$

while

$$U_E^{(a)} = \frac{1}{8\pi\epsilon_0} \int d^3x d^3x' \, \frac{\rho^{(0)}(\mathbf{x} - \mathbf{x}_a)\rho^{(0)}(\mathbf{x}' - \mathbf{x}_a)}{|\mathbf{x} - \mathbf{x}'|} \,. \tag{5.87}$$

We now observe that, shifting simultaneously the integration variables $\mathbf{x} \to \mathbf{x} + \mathbf{x}_a$ and $\mathbf{x}' \to \mathbf{x}' + \mathbf{x}_a$ we can eliminate the dependence on $\mathbf{x}_a$ in eq. (5.87), since this eliminates the dependence on $\mathbf{x}_a$ from both $\rho^{(0)}(\mathbf{x} - \mathbf{x}_a)$ and $\rho^{(0)}(\mathbf{x}' - \mathbf{x}_a)$, while $|\mathbf{x} - \mathbf{x}'|$ is invariant under this shift. So, in the end, for each a, $U_E^{(a)}$ is independent of $\mathbf{x}_a$. This is physically clear, since it means that the self-energy of the a-th charge distribution is independent of where the volume V_a is located, i.e., is invariant under translations. Therefore, we can use $U_E(\mathbf{x}_1, \mathbf{x}_2)$ as a mechanical potential function in place of $U_E^{\text{int}}(\mathbf{x}_1, \mathbf{x}_2)$, i.e.,

$$\mathbf{F}_1(\mathbf{x}_1, \mathbf{x}_2) = -\frac{\partial U_E}{\partial \mathbf{x}_1} \,, \tag{5.88}$$

and, similarly, $\mathbf{F}_2(\mathbf{x}_1, \mathbf{x}_2) = -\partial U_E / \partial \mathbf{x}_2$. The form (5.83) of the mechanical potential is convenient because it is also valid for N charged objects with densities $\rho_a(\mathbf{x})$ ($a = 1, \ldots, N$), localized in non-overlapping volumes V_a identified by the position $\mathbf{x}_a$, simply setting $\rho(\mathbf{x}) = \sum_a \rho_a(\mathbf{x})$. Then, eq. (5.83) becomes

$$U_E(\mathbf{x}_1, \ldots, \mathbf{x}_N) = \frac{1}{8\pi\epsilon_0} \sum_{a,b=1}^{N} \int d^3x d^3x' \, \frac{\rho_a(\mathbf{x})\rho_b(\mathbf{x}')}{|\mathbf{x} - \mathbf{x}'|} \,. \tag{5.89}$$

The terms of the sum with $a = b$ are self-energy terms that, as we have seen, are independents of all coordinates $\mathbf{x}_a$ and do not contribute when taking the gradients, while the terms with $a \neq b$ give back eq. (5.82). We can also use eq. (5.89) for point charges, with the understanding that the self-energy terms are regularized as in eq. (5.31), and then are again irrelevant constants.

Comparison of eqs. (5.83) and (5.16) shows that the mechanical potential U_E needed to compute the forces is the same as the total interaction energy $\mathcal{E}_E$ of the system a static charge distributions,

$$\boxed{U_E = \mathcal{E}_E \,.} \tag{5.90}$$

Given the equality between eqs. (5.15) and (5.16), we can also rewrite U_E in the form

$$U_E = \frac{1}{2} \int d^3 x \, \rho(\mathbf{x}) \phi(\mathbf{x}) \,, \tag{5.91}$$

where $\phi(\mathbf{x})$ is the total electrostatic potential generated by $\rho(\mathbf{x})$.

5.5.1 Mechanical potentials for conductors

In the previous section, we have considered a system of N charged bodies, with charge densities $\rho_a(\mathbf{x})$ localized in non-overlapping volumes V_a, so that the total charge density is $\rho(\mathbf{x}) = \sum_a \rho_a(\mathbf{x})$. We have identified the position of the volume V_a by a coordinate $\mathbf{x}_a$ that describes rigid displacements of the volume (corresponding, e.g., to the geometrical center of the body). The total charge of each extended body is

$$Q_a = \int d^3 x \, \rho_a(\mathbf{x}) \,. \tag{5.92}$$

Since $\rho_a(\mathbf{x})$ vanishes outside the volume V_a, it is actually not necessary to restrict explicitly the integral in $d^3 x$ to the volume V_a. We have then found that the force acting on the a-th charge can be written as

$$\mathbf{F}_a = -\frac{\partial}{\partial \mathbf{x}_a} U_E(Q_1, \ldots Q_N; \mathbf{x}_1, \ldots, \mathbf{x}_N) \,, \tag{5.93}$$

where U_E can be written in the equivalent forms

$$
\begin{aligned}
U_E(Q_1, \ldots Q_N; \mathbf{x}_1, \ldots, \mathbf{x}_N) &= \frac{1}{8\pi\epsilon_0} \int d^3 x \, d^3 x' \, \frac{\rho(\mathbf{x})\rho(\mathbf{x}')}{|\mathbf{x} - \mathbf{x}'|} \\
&= \frac{1}{8\pi\epsilon_0} \sum_{a,b=1}^{N} \int d^3 x \, d^3 x' \, \frac{\rho_a(\mathbf{x})\rho_b(\mathbf{x}')}{|\mathbf{x} - \mathbf{x}'|} \\
&= \frac{1}{8\pi\epsilon_0} \sum_{a,b=1}^{N} \int d^3 x \, d^3 x' \, \frac{\rho_a^{(0)}(\mathbf{x} - \mathbf{x}_a)\rho_b^{(0)}(\mathbf{x}' - \mathbf{x}_b)}{|\mathbf{x} - \mathbf{x}'|} \,.
\end{aligned} \tag{5.94}
$$

[20] For extended bodies, in principle, one should also include the variables that define their shapes and orientation in space. Such variables could be added trivially to the steps that we will perform in the following. To keep the notation simpler, we neglect these variables, assuming for instance that the bodies are spherical, of fixed radii. Each volume V_a can then be identified uniquely by the position $\mathbf{x}_a$ of its center. The treatment discussed here can then be generalized to the inclusion of thermodynamic variables such as the temperature, see Chapter 4 of Landau and Lifschits (1984).

Compared to the previous section, we have slightly changed the notation, also writing explicitly the dependence of U_E on the charges Q_a.[20] This different notation reflects a broader of point of view, inspired by the treatment of thermodynamic potentials in statistical physics. First of all, we have already observed in Section 5.1 that the work needed to build a configuration of charges is independent of the order in which we bring the charges from infinity, and on the path that we follow to bring them to their final positions. In the language of thermodynamics, this means that U_E in eq. (5.94) is a function of state, i.e., depends only on the variables $Q = \{Q_1, \ldots Q_N\}$ and $\mathbf{x} = \{\mathbf{x}_1, \ldots, \mathbf{x}_N\}$ that identify the final state, and not on how we reached it. A second point to stress is that the definition of functions of state depends on which quantities are kept constant during the process of building the given configuration. In the case where we built a configuration of point charges bringing them from infinity, it was implicit that the charges of the individual particles were kept constant. For a set of elementary charges this was an obvious assumptions, since the electric charge associated with each elementary particle is a conserved quantity. However, now it is useful to keep this in mind and, in eq. (5.94), we stressed this by writing explicitly $Q_1, \ldots Q_N$ among the arguments of U_E.

We now restrict to the case of extended conductors. In this case, starting from eq. (5.91) and performing the same steps as in eqs. (5.41)–(5.43), we can write

$$U_E = \frac{1}{2} \sum_{a=1}^{N} Q_a \phi_a \, . \tag{5.95}$$

A more accurate notation is

$$U_E(Q_1, \ldots Q_N; \mathbf{x}_1, \ldots, \mathbf{x}_N) = \frac{1}{2} \sum_{a=1}^{N} Q_a \phi_a(Q_1, \ldots Q_N; \mathbf{x}_1, \ldots, \mathbf{x}_N) \, , \tag{5.96}$$

which stresses that, for given values of the charges Q_a, the value of the electrostatic potential at the surface of the a-th conductor, ϕ_a, depends on the positions $\mathbf{x}_1, \ldots, \mathbf{x}_N$ of all conductors (as identified by their centers, see Note 20 on page 118), as well as on their charges.

We now want to compute the partial derivative of U_E with respect to Q_a, for a given a, while keeping fixed all other charges, and all the variables $\mathbf{x}_1, \ldots, \mathbf{x}_N$. Starting from eq. (5.96) would require us to compute the derivative of ϕ_a with respect to Q_a. Actually, it is simpler to use U_E in the form given in the second line of eq. (5.94), and compute how U_E changes under

$$\rho_a \to \rho_a + \delta\rho_a \, . \tag{5.97}$$

We denote this variation by δ_a, so that $\delta_a[\rho_b(\mathbf{x})] = \delta_{ab}\delta\rho_a(\mathbf{x})$, where δ_{ab} is the Kronecker symbol. Then (taking care of changing the name of the dummy summation indices),

$$
\begin{aligned}
\delta_a U_E &= \frac{1}{8\pi\epsilon_0} \sum_{b,c=1}^{N} \int d^3x \, d^3x' \, \frac{\delta_a[\rho_c(\mathbf{x})\rho_b(\mathbf{x}')]}{|\mathbf{x} - \mathbf{x}'|} \\
&= \frac{1}{8\pi\epsilon_0} \sum_{b,c=1}^{N} \int d^3x \, d^3x' \, \frac{[\delta_a\rho_c(\mathbf{x})]\rho_b(\mathbf{x}') + \rho_c(\mathbf{x})[\delta_a\rho_b(\mathbf{x}')]}{|\mathbf{x} - \mathbf{x}'|} \\
&= \frac{1}{4\pi\epsilon_0} \sum_{b=1}^{N} \int d^3x \, \delta\rho_a(\mathbf{x}) \int d^3x' \, \frac{\rho_b(\mathbf{x}')}{|\mathbf{x} - \mathbf{x}'|} \, ,
\end{aligned}
\tag{5.98}
$$

where the second variation in the second line gives the same contribution as the first, after renaming $\mathbf{x} \leftrightarrow \mathbf{x}'$ and $b \leftrightarrow c$. We now observe that

$$\frac{1}{4\pi\epsilon_0} \sum_{b=1}^{N} \int d^3x' \, \frac{\rho_b(\mathbf{x}')}{|\mathbf{x} - \mathbf{x}'|} = \phi(\mathbf{x}) \qquad (5.99)$$

is the total electrostatic potential generated by the charge densities. Therefore

$$\delta_a U_E = \int d^3x \, \delta\rho_a(\mathbf{x}) \, \phi(x) \,. \qquad (5.100)$$

These manipulations were valid for generic charge distributions. We now specialize to conductors using again the fact that, for conductors at equilibrium, the charge density is zero inside the conductor, and non-vanishing only on its surface, and that the electrostatic potential on the surface is constant. Denoting again by ϕ_a this constant value of $\phi(\mathbf{x})$ on the surface of the a-th conductor, and using the fact that $\delta\rho_a(\mathbf{x})$ is proportional to a two-dimensional Dirac delta on the surface of the a-th conductor, eq. (5.100) becomes

$$
\begin{aligned}
\delta_a U_E &= \phi_a \int d^3x \, \delta\rho_a(\mathbf{x}) \\
&= \phi_a \delta Q_a \,,
\end{aligned}
\qquad (5.101)
$$

where δQ_a is the variation of the charge Q_a of the a-th conductor, induced by the variation (5.97) of its charge density. This shows that, for a set of conductors,

$$\frac{\partial}{\partial Q_a} U_E(Q_1, \dots Q_N; \mathbf{x}_1, \dots, \mathbf{x}_N) = \phi_a \,, \qquad (5.102)$$

where ϕ_a is the total electrostatic potential on the surface of the a-th conductor. Therefore, for conductors we can write

$$
\begin{aligned}
dU_E &= \sum_{a=1}^{N} \left(\frac{\partial U_E}{\partial Q_a} \right)_{Q',\mathbf{x}} dQ_a + \sum_{a=1}^{N} \left(\frac{\partial U_E}{\partial \mathbf{x}_a} \right)_{Q,\mathbf{x}'} \cdot d\mathbf{x}_a \\
&= \sum_{a=1}^{N} \left(\phi_a dQ_a - \mathbf{F}_a \cdot d\mathbf{x}_a \right) \,,
\end{aligned}
\qquad (5.103)
$$

where the subscripts in the partial derivatives indicate the variables that are kept constant when taking the partial derivative: the subscript $\{Q', \mathbf{x}\}$ in $\partial U_E / \partial Q_a$ means that we keep fixed all the charges except Q_a, and all the $\mathbf{x}$, while the subscript $\{Q, \mathbf{x}'\}$ in $\partial U_E / \partial \mathbf{x}_a$ means that we keep fixed all the charges, and all the $\mathbf{x}$ except $\mathbf{x}_a$. In this more precise notation, we can rewrite eqs. (5.93) and (5.102) as

$$\phi_a = \left(\frac{\partial U_E}{\partial Q_a} \right)_{Q',\mathbf{x}} \,, \qquad (5.104)$$

$$\mathbf{F}_a = -\left(\frac{\partial U_E}{\partial \mathbf{x}_a} \right)_{Q,\mathbf{x}'} \,. \qquad (5.105)$$

Therefore, we can see U_E as a "mechanical potential function at fixed charge," in the sense that, from it, we can obtain the mechanical forces by taking (minus) the gradient with respect to $\mathbf{x}_a$, at fixed charges.

For a set of point-like particles, working at fixed charges is the only natural option, so, in Section 5.1 we computed the work needed to assemble a given configuration of point charges, bringing them one by one to the desired final location, while keeping their charges constant. For conductors, however, there are two natural options. The first is to keep a conductor at fixed charge as we move it from infinity to the desired location. This is automatically obtained if the conductor is isolated, so that the charge on it is conserved. Note that, if a conductor is isolated, a second conductor that is moved toward it induces a change in the electrostatic potential of its surface, so Q_a is constant while ϕ_a is not.

Another possibility is to keep the electrostatic potential at its surface, ϕ_a, at a fixed value. For this, the conductor cannot be isolated. For instance, it might "grounded," i.e., connected by a thin conducting wire to the Earth, so that the potential at its surface is at the same value as the potential of the Earth, taken to be the reference value $\phi = 0$; or, more generally, it could be connected to some external source such as a battery, that keeps its surface to the desired value of ϕ_a. The external source, such as a battery or the Earth, exchanges charges with the conductor connected to it, in order to re-equilibrate the effect of the external disturbances and keep the electrostatic potential on its surface constant. In this case, the work done to reach the final configuration must be computed at constant ϕ_a, rather than at constant Q_a.

We therefore want to define another potential function $\hat{U}_E$, which is the appropriate one when we keep fixed the electrostatic potential, rather than the electric charge. The relevant tool here is the *Legendre transform*, which is the standard technique used in similar contexts in classical mechanics and in thermodynamics. Suppose that we have a function of two variables $U(Q, x)$, so that

$$dU = \left(\frac{\partial U}{\partial Q}\right)_x dQ + \left(\frac{\partial U}{\partial x}\right)_Q dx, \tag{5.106}$$

and we define

$$\phi(Q, x) = \left(\frac{\partial U}{\partial Q}\right)_x. \tag{5.107}$$

We assume that this relation can be inverted, so as to give $Q = Q(\phi, x)$. Then, one defines the Legendre transform $\hat{U}(\phi, x)$ as

$$\hat{U}(\phi, x) \equiv U(Q, x) - Q\phi, \tag{5.108}$$

where, on the right-hand side, $Q = Q(\phi, x)$. This definition is chosen so that

$$
\begin{aligned}
d\hat{U} &= dU - Qd\phi - \phi dQ \\
&= \left(\frac{\partial U}{\partial Q}\right)_x dQ + \left(\frac{\partial U}{\partial x}\right)_Q dx - Qd\phi - \phi dQ. \tag{5.109}
\end{aligned}
$$

Using eq. (5.107), the terms proportional to dQ cancel, and we get

$$d\hat{U} = -Qd\phi + \left(\frac{\partial U}{\partial x}\right)_Q dx. \tag{5.110}$$

We see that $\hat{U}$ is a function of ϕ and x, such that

$$\left(\frac{\partial \hat{U}}{\partial \phi}\right)_x = -Q, \tag{5.111}$$

$$\left(\frac{\partial \hat{U}}{\partial x}\right)_\phi = \left(\frac{\partial U}{\partial x}\right)_Q. \tag{5.112}$$

In this context, ϕ and Q are called "conjugate variables." We now apply this procedure to a system of conductors. If all of them are kept at constant potential, we perform the Legendre transform with respect to all variables Q_a, defining

$$\hat{U}_E(\phi_1, \dots \phi_N; \mathbf{x}_1, \dots, \mathbf{x}_N) = U_E(Q_1, \dots Q_N; \mathbf{x}_1, \dots, \mathbf{x}_N) - \sum_{a=1}^{N} Q_a \phi_a, \tag{5.113}$$

where, on the right-hand side, the charges Q_a are expressed as functions of the electrostatic potential ϕ_a.[21] For conductors, this relation is linear and invertible and is expressed by eq. (4.163). Note that the coefficients of capacitance C_{ab} depend on the positions $\mathbf{x}_1, \dots, \mathbf{x}_N$ of all conductors.

Similarly to eq. (5.110), using eq. (5.103) we get

$$\begin{aligned} d\hat{U}_E &= dU_E - \sum_{a=1}^{N} \phi_a dQ_a - \sum_{a=1}^{N} Q_a d\phi_a \\ &= \sum_{a=1}^{N} \left(-Q_a d\phi_a - \mathbf{F}_a \cdot d\mathbf{x}_a\right), \end{aligned} \tag{5.114}$$

and therefore

$$Q_a = -\left(\frac{\partial \hat{U}_E}{\partial \phi_a}\right)_{\phi', \mathbf{x}}, \tag{5.115}$$

and

$$\mathbf{F}_a = -\left(\frac{\partial \hat{U}_E}{\partial \mathbf{x}_a}\right)_{\phi, \mathbf{x}'}. \tag{5.116}$$

The knowledge of $\hat{U}_E$ therefore allows us to obtain the total charge on the surface of the a-th conductor, using eq. (5.115), and the force exerted on it, using eq. (5.116).

We now recall that the potential U_E can be written in the simple form (5.95), where $\phi_a = \phi_a(Q_1, \dots Q_N; \mathbf{x}_1, \dots, \mathbf{x}_N)$ are given, in terms of the charges $Q_1, \dots Q_N$ and of the positions $\mathbf{x}_1, \dots, \mathbf{x}_N$, by the inversion of eq. (4.163),

$$\phi_a = \sum_{b=1}^{N} P_{ab} Q_b, \tag{5.117}$$

[21] If some conductors are kept at fixed electrostatic potential and some are isolated, so are at fixed charge, we perform the Legendre transform only for those that are at fixed electrostatic potential, obtaining a function $\hat{U}_E$ of the variables $\phi_1, \dots, \phi_n$, $Q_1, \dots, Q_m$, and $\mathbf{x}_1, \dots, \mathbf{x}_N$, where $i = 1, \dots, n$ labels the conductors at fixed electrostatic potential and $j = 1, \dots, m$ those at fixed charge (with $n + m = N$). For notational simplicity, we just consider the case where all conductors are at fixed electrostatic potential.

as in eq. (4.164). Note, again, that the dependence on $\mathbf{x}_1, \ldots, \mathbf{x}_N$ enters through the coefficients C_{ab}, or through the coefficients P_{ab} of the inverse matrix. Then, from eq. (5.113), performing the Legendre transform on U_E has the effect of just flipping the sign,

$$\hat{U}_E(\phi_1, \ldots \phi_N; \mathbf{x}_1, \ldots, \mathbf{x}_N) = -\frac{1}{2} \sum_{a=1}^{N} Q_a \phi_a, \qquad (5.118)$$

where now the ϕ_a are taken as the independent variables, and the Q_a are expressed in terms of them, and of the $\mathbf{x}_a$, using eq. (4.163). In other words,

$$\hat{U}_E(\phi_1, \ldots \phi_N; \mathbf{x}_1, \ldots, \mathbf{x}_N) = -U_E(Q_1, \ldots Q_N; \mathbf{x}_1, \ldots, \mathbf{x}_N), \qquad (5.119)$$

where, on the right-hand side, the charges $Q_1, \ldots Q_N$ must be expressed as functions of $\phi_1, \ldots \phi_N$ and $\mathbf{x}_1, \ldots, \mathbf{x}_N$; or, equivalently,

$$U_E(Q_1, \ldots Q_N; \mathbf{x}_1, \ldots, \mathbf{x}_N) = -\hat{U}_E(\phi_1, \ldots \phi_N; \mathbf{x}_1, \ldots, \mathbf{x}_N), \qquad (5.120)$$

where, on the right-hand side, the electrostatic potentials $\phi_1, \ldots \phi_N$ must be expressed as functions of $Q_1, \ldots Q_N$ and $\mathbf{x}_1, \ldots, \mathbf{x}_N$. Combining this with eq. (5.90), we see that

$$\boxed{\hat{U}_E = -\mathcal{E}_E.} \qquad (5.121)$$

In conclusion, if the charges on the conductors are conserved, the appropriate potential function to use is U_E. Using eq. (5.96), together with eq. (5.117), we can write

$$U_E(Q_1, \ldots Q_N; \mathbf{x}_1, \ldots, \mathbf{x}_N) = \frac{1}{2} \sum_{a,b=1}^{N} P_{ab}(\mathbf{x}_1, \ldots, \mathbf{x}_N) Q_a Q_b. \qquad (5.122)$$

The force acting on the a-th conductor can then be computed using eq. (5.105), which gives

$$\mathbf{F}_a = -\frac{1}{2} \sum_{b,c=1}^{N} \left(\frac{\partial P_{bc}}{\partial \mathbf{x}_a} \right)_{\mathbf{x}'} Q_b Q_c. \qquad (5.123)$$

If, instead, we fix the values of the electrostatic potentials ϕ_a on the conductors, the appropriate potential function is $\hat{U}_E$, given in eq. (5.118). Using eq. (4.163), we can write it as

$$\hat{U}_E(\phi_1, \ldots \phi_N; \mathbf{x}_1, \ldots, \mathbf{x}_N) = -\frac{1}{2} \sum_{a,b=1}^{N} C_{ab}(\mathbf{x}_1, \ldots, \mathbf{x}_N) \phi_a \phi_b. \qquad (5.124)$$

Equation (5.116) then gives the force on the a-th conductor in terms of the electrostatic potentials $\phi_1, \ldots \phi_N$ assigned on the surfaces of the conductors,

$$\mathbf{F}_a = \frac{1}{2} \sum_{b,c=1}^{N} \left(\frac{\partial C_{bc}}{\partial \mathbf{x}_a} \right)_{\mathbf{x}'} \phi_b \phi_c. \qquad (5.125)$$

Note, however, that the force acting on a conductor is an instantaneous property of the system, independent of whether the system was assembled keeping the charges fixed or the electrostatic potentials, so, in the end, eqs. (5.123) and (5.125) must give the same numerical result, in one case expressed in terms of the charges, and in the other in terms of the electrostatic potentials generated by these charges in the given configuration of conductors. This can be explicitly shown recalling, from eqs. (4.163) and (4.164) that, denoting by C the matrix with matrix elements C_{ab}, and by P the matrix with matrix elements P_{ab}, we have $P = C^{-1}$. We denote by Q the (column) vector with components Q_a and by ϕ the (column) vector with components ϕ_a. Then, in matrix form, eq. (4.163) reads

$$Q = C\phi, \tag{5.126}$$

while eq. (5.123) reads

$$\mathbf{F}_a = -\frac{1}{2} Q^T \left(\frac{\partial C^{-1}}{\partial \mathbf{x}_a} \right) Q, \tag{5.127}$$

where Q^T is the transpose vector, i.e., the vector written as a row rather than as a column. From eq. (5.126), we also have $Q^T = \phi^T C^T$, where C^T is the transpose matrix. However, C is a symmetric matrix, $C_{ab} = C_{ba}$ so, $C = C^T$ (we will prove this in Problem 5.2) and therefore $Q^T = \phi^T C$. We also use $\partial_x C^{-1} = -C^{-1}(\partial_x C)C^{-1}$.[22] Then,

[22] This follows from the fact that $C^{-1}C$ is equal to the identity matrix I, and therefore

$$
\begin{aligned}
0 &= \partial_x (C^{-1}C) \\
&= (\partial_x C^{-1})C + C^{-1}(\partial_x C).
\end{aligned}
$$

Multiplying by C^{-1} from the right, we get $\partial_x C^{-1} = -C^{-1}(\partial_x C)C^{-1}$.

$$
\begin{aligned}
\mathbf{F}_a &= -\frac{1}{2} Q^T \left(\frac{\partial C^{-1}}{\partial \mathbf{x}_a} \right) Q \\
&= \frac{1}{2} (\phi^T C) \left(C^{-1} \frac{\partial C}{\partial \mathbf{x}_a} C^{-1} \right) C\phi \\
&= \frac{1}{2} \phi^T \left(\frac{\partial C}{\partial \mathbf{x}_a} \right) \phi,
\end{aligned} \tag{5.128}
$$

which is just eq. (5.125), written in matrix form.

5.5.2 Mechanical potentials in magnetostatics

The force between two static current distributions was computed in Section 4.2.4, see eq. (4.118). We now want to find a mechanical potential function from which this magnetic force can be obtained. The procedure is analogous to that presented in Section 5.5.1 for the electrostatic case. Similarly to the discussion there, we consider two current distributions $\mathbf{j}_1(\mathbf{x})$ and $\mathbf{j}_2(\mathbf{x})$, localized in the non-overlapping volumes V_1 and V_2, respectively (and, therefore, separately conserved). We use a coordinate $\mathbf{x}_1$ to label rigid displacements of V_1, and $\mathbf{x}_2$ for V_2. For instance, for two circular loops of currents, $\mathbf{x}_1$ and $\mathbf{x}_2$ could be the coordinates of the centers of the two loops. We then denote by $\mathbf{j}_1^{(0)}(\mathbf{x})$ the first current density, in the reference frame where $\mathbf{x}_1 = 0$, and similarly by $\mathbf{j}_2^{(0)}(\mathbf{x})$ the second current density, in the reference frame where $\mathbf{x}_2 = 0$. Then, in a frame where the centers of the two loops have generic coordinates

$\mathbf{x}_1$ and $\mathbf{x}_2$, respectively, we have

$$\mathbf{j}_1(\mathbf{x}) = \mathbf{j}_1^{(0)}(\mathbf{x} - \mathbf{x}_1)\,, \tag{5.129}$$

and

$$\mathbf{j}_2(\mathbf{x}) = \mathbf{j}_2^{(0)}(\mathbf{x} - \mathbf{x}_2)\,. \tag{5.130}$$

In this notation, the forces (4.118) and (4.120) read

$$\mathbf{F}_1(\mathbf{x}_1, \mathbf{x}_2) = -\frac{\mu_0}{4\pi} \int d^3x\, d^3x'\, \mathbf{j}_1^{(0)}(\mathbf{x} - \mathbf{x}_1) {\cdot} \mathbf{j}_2^{(0)}(\mathbf{x}' - \mathbf{x}_2) \frac{\mathbf{x} - \mathbf{x}'}{|\mathbf{x} - \mathbf{x}'|^3}\,, \tag{5.131}$$

and [after renaming $\mathbf{x} \leftrightarrow \mathbf{x}'$ in eq. (4.120)]

$$\mathbf{F}_2(\mathbf{x}_1, \mathbf{x}_2) = -\frac{\mu_0}{4\pi} \int d^3x\, d^3x'\, \mathbf{j}_1^{(0)}(\mathbf{x} - \mathbf{x}_1) {\cdot} \mathbf{j}_2^{(0)}(\mathbf{x}' - \mathbf{x}_2) \frac{\mathbf{x}' - \mathbf{x}}{|\mathbf{x} - \mathbf{x}'|^3}\,, \tag{5.132}$$

i.e., $\mathbf{F}_2(\mathbf{x}_1, \mathbf{x}_2) = -\mathbf{F}_1(\mathbf{x}_1, \mathbf{x}_2)$. Proceeding exactly as in Note 19 on page 116, we see that the desired potential function is

$$\hat{U}_B^{\rm int}(\mathbf{x}_1, \mathbf{x}_2) = -\frac{\mu_0}{4\pi} \int d^3x\, d^3x'\, \frac{\mathbf{j}_1^{(0)}(\mathbf{x} - \mathbf{x}_1) {\cdot} \mathbf{j}_2^{(0)}(\mathbf{x}' - \mathbf{x}_2)}{|\mathbf{x} - \mathbf{x}'|}\,, \tag{5.133}$$

since the magnetic forces (5.131) and (5.132) can be obtained from

$$\mathbf{F}_1(\mathbf{x}_1, \mathbf{x}_2) = -\left(\frac{\partial \hat{U}_B^{\rm int}}{\partial \mathbf{x}_1}\right)_{\mathbf{x}_2, \mathbf{j}}\,, \tag{5.134}$$

and $\mathbf{F}_2(\mathbf{x}_1, \mathbf{x}_2) = -\left(\partial \hat{U}_B^{\rm int}/\partial \mathbf{x}_2\right)_{\mathbf{x}_1, \mathbf{j}}$. Rewriting eq. (5.133) in terms of $\mathbf{j}_1(\mathbf{x})$ and $\mathbf{j}_2(\mathbf{x})$, we have

$$\hat{U}_B^{\rm int}(\mathbf{x}_1, \mathbf{x}_2) = -\frac{\mu_0}{4\pi} \int d^3x\, d^3x'\, \frac{\mathbf{j}_1(\mathbf{x}) {\cdot} \mathbf{j}_2(\mathbf{x}')}{|\mathbf{x} - \mathbf{x}'|}\,, \tag{5.135}$$

where the dependence on $\mathbf{x}_1$ and $\mathbf{x}_2$ is now implicit in $\mathbf{j}_1(\mathbf{x})$ and $\mathbf{j}_2(\mathbf{x}')$, respectively, just as in the discussion in Section 5.5 for the electrostatic case. The reason why for this function we used the notation $\hat{U}_B^{\rm int}$, rather than just $U_B^{\rm int}$, will become clear soon.

Exactly as we did in eqs. (5.83)–(5.89) for the electrostatic case, it is convenient to introduce a function

$$\hat{U}_B[\mathbf{j}] = -\frac{\mu_0}{8\pi} \int d^3x\, d^3x'\, \frac{\mathbf{j}(\mathbf{x}) {\cdot} \mathbf{j}(\mathbf{x}')}{|\mathbf{x} - \mathbf{x}'|}\,. \tag{5.136}$$

If $\mathbf{j}(\mathbf{x}) = \mathbf{j}_1(\mathbf{x}) + \mathbf{j}_2(\mathbf{x})$, where $\mathbf{j}_1(\mathbf{x})$ and $\mathbf{j}_2(\mathbf{x})$ are localized in two non-overlapping volumes V_1 and V_2 identified by coordinates $\mathbf{x}_1$ and $\mathbf{x}_2$, respectively, eq. (5.136) becomes

$$\hat{U}_B(\mathbf{x}_1, \mathbf{x}_2) = \hat{U}_B^{(1)} + \hat{U}_B^{(2)} + \hat{U}_B^{\rm int}(\mathbf{x}_1, \mathbf{x}_2)\,, \tag{5.137}$$

where

$$\hat{U}_B^{(a)} = -\frac{\mu_0}{8\pi} \int d^3x\, d^3x'\, \frac{\mathbf{j}^{(a)}(\mathbf{x}) \cdot \mathbf{j}^{(a)}(\mathbf{x}')}{|\mathbf{x} - \mathbf{x}'|}\,, \tag{5.138}$$

(with $a = 1, 2$), while the interaction term $\hat{U}_B^{\text{int}}$ is given by eq. (5.135). Just as in the discussion following eq. (5.87), the self-energy terms $\hat{U}_B^{(1)}$ and $\hat{U}_B^{(2)}$ do not depend on $\mathbf{x}_1$ and $\mathbf{x}_2$, so we can use $\hat{U}_B(\mathbf{x}_1, \mathbf{x}_2)$ as the mechanical potential, in place of $\hat{U}_B^{\text{int}}(\mathbf{x}_1, \mathbf{x}_2)$. As in the electrostatic case, the form (5.136) of the mechanical potential is also valid for N charged objects with current densities $\mathbf{j}_a(\mathbf{x})$ $(a = 1, \ldots, N)$, localized in non-overlapping volumes V_a identified by the position $\mathbf{x}_a$, simply setting $\mathbf{j}(\mathbf{x}) = \sum_a \mathbf{j}_a(\mathbf{x})$.

From eqs. (5.53) and (5.136), we get

$$\hat{U}_B = -\mathcal{E}_B . \tag{5.139}$$

Comparing this to the electrostatic case, where eqs. (5.90) and (5.121) hold, we see that $\hat{U}_B$ is really the magnetostatic analog of $\hat{U}_E$, rather than of U_E, which is the reason why we used the notation $\hat{U}_B$ when we introduced it in eqs. (5.133) and (5.136).

The reason is that, while the charge on an isolated conductor moving in an electric field is conserved, the current of an isolated loop moving in a magnetic field is not. If we want to build a configuration of charges at given positions, all the work that we do goes into the mechanical potential energy, and no work is needed to keep the charges at the initial value. Therefore, in this case, the energy of the system, defined as the work needed to assemble it, is the same as the mechanical potential energy. As we saw in Section 5.5.1, this is not the case when we assemble a configuration of conductors with given values of the electrostatic potentials ϕ_a at their surfaces, since the latter are not conserved quantity, and we must also provide some extra "electric" work, e.g., through batteries that keep the electrostatic potentials constants. The magnetostatic case at fixed currents that we are considering here is analogous to the electrostatic case at fixed ϕ_a, rather than at fixed charges. To assemble a configuration of loops at given positions and with given currents, it is not sufficient to take a set of loops with the desired currents, at infinite distance from each other, and compute the mechanical work necessary to bring them to the desired final positions, as we did in Section 5.1 for a set of charges. We must also provide, along the way, further electric work to maintain the currents fixed, connecting the loops to batteries. Otherwise, as we move a loop in the magnetic field created by the others, an electric field is induced in the loop, as we saw in Section 4.3.2, and the corresponding "motional" electromotive force changes the value of the current in the loop. The net effect is that the final energy $\mathcal{E}_B$ of the system has the form $\mathcal{E}_B = \hat{U}_B + W_e$, where W_e is the required electric work. Since we have found that $\hat{U}_B = -\mathcal{E}_B$, we conclude that the required electric work is $W_e = 2\mathcal{E}_B$. The same holds for the electrostatic case at fixed ϕ_a.

We now consider a set of current loops, and we ask what the magnetic analog of the potential U_E is. For a set of loops, with currents I_a and identified by the position $\mathbf{x}_a$ (given, e.g., by the position of their centers, for circular loops), using eqs. (5.61) and (5.139), we have

$$\hat{U}_B(I_1, \ldots, I_N; \mathbf{x}_1, \ldots, \mathbf{x}_N) = -\frac{1}{2} \sum_{a=1}^{N} I_a \Phi_{B,a} \,. \tag{5.140}$$

Here, the magnetic fluxes $\Phi_{B,a}$ are expressed in terms of the currents through eq. (5.63).[23] Consider now the function of the fluxes defined by

$$U_B(\Phi_{B,1}, \ldots, \Phi_{B,N}; \mathbf{x}_1, \ldots, \mathbf{x}_N) = \frac{1}{2} \sum_{a=1}^{N} I_a \Phi_{B,a} \,, \tag{5.142}$$

where now the currents are expressed in terms of the fluxes by inverting eq. (5.63). It is clear that $\hat{U}_B$ is just the Legendre transform of U_B, with the currents I_a and the magnetic fluxes $\Phi_{B,a}$ playing the role of conjugate variables since, from the explicit expressions given in eqs. (5.140) and (5.142), we have

$$\begin{aligned} \hat{U}_B(I_1, \ldots, I_N; \mathbf{x}_1, \ldots, \mathbf{x}_N) \;=\;& U_B(\Phi_{B,1}, \ldots, \Phi_{B,N}; \mathbf{x}_1, \ldots, \mathbf{x}_N) \\ &- \sum_{a=1}^{N} I_a \Phi_{B,a} \,. \end{aligned} \tag{5.143}$$

Then, eq. (5.139) implies

$$U_B = \mathcal{E}_B \,. \tag{5.144}$$

We see that the situation is exactly the same as in the electrostatic case, with the replacements $U_E \leftrightarrow U_B$, $\hat{U}_E \leftrightarrow \hat{U}_B$, $\phi_a \leftrightarrow I_a$ and $Q_a \leftrightarrow \Phi_{B,a}$. Then, proceeding exactly as we did in Section 5.5.1 for the electrostatic case, we now find that

$$\Phi_{B,a} = -\left(\frac{\partial \hat{U}_B}{\partial I_a} \right)_{I',\mathbf{x}} , \tag{5.145}$$

and

$$\mathbf{F}_a = -\left(\frac{\partial \hat{U}_B}{\partial \mathbf{x}_a} \right)_{I,\mathbf{x}'} , \tag{5.146}$$

which are the analogues of eqs. (5.115) and (5.116). Therefore, the mechanical potential $\hat{U}_B$ can be used to obtain the force on the a-th loop for a system of loops at fixed currents, taking minus its gradient, while (minus) its derivative with respect to the current I_a, with all other currents fixed, gives the magnetic flux though the a-th loop. Using eq. (5.63), we can express $\hat{U}_B$ directly in terms of the currents, as

$$\hat{U}_B(I_1, \ldots I_N; \mathbf{x}_1, \ldots, \mathbf{x}_N) = -\frac{1}{2} \sum_{a,b=1}^{N} L_{ab}(\mathbf{x}_1, \ldots, \mathbf{x}_N) I_a I_b \,, \tag{5.147}$$

and therefore the force on the a-th loop, in a set of loops at fixed currents, is

$$\mathbf{F}_a = \frac{1}{2} \sum_{b,c=1}^{N} \left(\frac{\partial L_{bc}}{\partial \mathbf{x}_a} \right)_{\mathbf{x}'} I_b I_c \,, \tag{5.148}$$

[23] Similarly to the electrostatic case, the dependence on the positions $\mathbf{x}_1, \ldots, \mathbf{x}_N$ enter through the fact that, for given currents, the fluxes depend on the currents and on the positions of the loops; in particular, the dependence on the positions is carried by the mutual inductances L_{ab}, so eq. (5.63), more precisely, reads

$$\begin{aligned} \Phi_{B,a}&(I_1, \ldots, I_N; \mathbf{x}_1, \ldots, \mathbf{x}_N) \\ &= \sum_{b=1}^{N} L_{ab}(\mathbf{x}_1, \ldots, \mathbf{x}_N) I_b \,. \end{aligned} \tag{5.141}$$

to be compared with eq. (5.125) for the case of conductors at fixed electrostatic potentials. The potential U_B satisfies instead

$$I_a = \left(\frac{\partial U_B}{\partial \Phi_{B,a}}\right)_{\Phi'_B, \mathbf{x}}, \tag{5.149}$$

$$\mathbf{F}_a = -\left(\frac{\partial U_B}{\partial \mathbf{x}_a}\right)_{\Phi_B, \mathbf{x}'}, \tag{5.150}$$

which are the analogues of eqs. (5.149) and (5.150). Denoting by L the matrix with elements L_{ab} and by L^{-1} the inverse matrix with matrix elements L^{-1}_{ab}, we have

$$I_a = \sum_{b=1}^{N} L^{-1}_{ab} \Phi_{B,b}, \tag{5.151}$$

$$U_B(\Phi_{B,1}, \dots \Phi_{B,N}; \mathbf{x}_1, \dots, \mathbf{x}_N) = \frac{1}{2} \sum_{a,b=1}^{N} L^{-1}_{ab}(\mathbf{x}_1, \dots, \mathbf{x}_N) \Phi_{B,a} \Phi_{B,b}, \tag{5.152}$$

and

$$\mathbf{F}_a = -\frac{1}{2} \sum_{b,c=1}^{N} \left(\frac{\partial L^{-1}_{bc}}{\partial \mathbf{x}_a}\right)_{\mathbf{x}'} \Phi_{B,b} \Phi_{B,c}. \tag{5.153}$$

Exactly as we did in eqs. (5.126)–(5.128), we can then prove that the force computed using eq. (5.148) is the same as that computed using eq. (5.153), in one case expressed in terms of the currents, and in the other in terms of the magnetic fluxes generated by these currents in the given configuration of loops.

5.6   Solved problems

Problem 5.1. Energy stored in a capacitor

In this chapter we have shown, in full generality, that the energy stored in the electric field of a system with given static charges is the same as the work that has been done (at fixed charge) to assemble the corresponding charge configuration.

It is instructive to check this explicitly for a single capacitor made of two plates, computing on the one hand the work made to charge it, and on the other hand the electromagnetic energy stored in it. We consider an initial configuration in which the two plates are both uncharged, and we transfer elementary charges from one plate to the other so that, at the end, the two plates have charges Q and $-Q$, respectively (we take $Q > 0$).[24] Let q and $-q$ be the value of the charges on the two plates at some stage of this charging process. From eq. (4.157), for a capacitor of capacitance C, when the charge on the positive plate is q, the potential difference between the plates has a value $V = q/C$. We now imagine to transport an infinitesimal positive charge dq from the negatively charged to the positively charged plate (more realistically, we would rather move the negatively charged electrons in the

[24] This is analogous to the assembling of a static charge configuration by carrying elementary charged particles from infinity to the desired final position, studied in Section 5.1. We are therefore working a fixed charge in the sense that no extra work is needed to ensure that the charges of the elementary particles stay fixed.

opposite direction, but the mathematics is the same), so that the charges on the two plates become $-(q + dq)$ and $q + dq$, respectively. The work made by an external agent to carry this charge across a potential difference V is $dW_{\text{ext}} = V\,dq$, so, using $V = q/C$,

$$
\begin{aligned}
W_{\text{ext}} &= \int_0^Q dq\, V \\
&= \int_0^Q dq\, \frac{q}{C} \\
&= \frac{Q^2}{2C}\,.
\end{aligned}
\tag{5.154}
$$

Using $Q = CV$, this can be rewritten in the equivalent forms $W_{\text{ext}} = (1/2)CV^2$ or $W_{\text{ext}} = (1/2)QV$. We can compare this result with that obtained from the general expression (3.41) for the energy of the electromagnetic field. For a parallel-plate capacitor, $\mathbf{E}$ is given by eq. (4.153) (while $\mathbf{B} = 0$). Then eq. (3.41) gives

$$
\begin{aligned}
\mathcal{E}_{\text{em}} &= \frac{\epsilon_0}{2} \int_V d^3x\, \mathbf{E}^2 \\
&= \frac{\epsilon_0}{2} \left(\frac{\sigma}{\epsilon_0} \right)^2 Ad \\
&= \frac{Q^2}{2C}\,,
\end{aligned}
\tag{5.155}
$$

where we used $Q = \sigma A$ and $C = \epsilon_0 A/d$, from eq. (4.158). Comparing this with eq. (5.154), we therefore see that the work done to assemble the static charge configuration of a parallel-plate capacitor is equal to the energy stored in its final electric field, in agreement with the general result. Note that we could have obtained eq. (5.155) without even using the explicit expressions for the electric field inside the parallel-plate capacitor and for its capacitance C, but just using eq. (5.44) or eq. (5.45), that give, in full generality, the energy stored in the electric field of a system of capacitors. For the single capacitor that we are considering, the matrix C_{ab} in eq. (5.44) becomes a single number C and the inverse matrix P_{ab} becomes the single number $1/C$, so eq. (5.45) reduces to eq. (5.155).

We can make the same check for a spherical capacitor. Inserting eq. (4.165) into eq. (3.41),

$$
\begin{aligned}
\mathcal{E}_{\text{em}} &= \frac{\epsilon_0}{2} \int_V d^3x\, \mathbf{E}^2 \\
&= \frac{\epsilon_0}{2} 4\pi \int_b^a r^2 dr\, \frac{Q^2}{(4\pi\epsilon_0)^2}\frac{1}{r^4} \\
&= \frac{1}{2}\frac{Q^2}{4\pi\epsilon_0} \left(\frac{1}{b} - \frac{1}{a} \right) \\
&= \frac{1}{2}QV\,,
\end{aligned}
\tag{5.156}
$$

where, in the last line, we used eq. (4.166). Using $V = Q/C$, we get again $\mathcal{E}_{\text{em}} = Q^2/(2C)$, in agreement with eq. (5.154) (which was obtained independently of the plate's geometry).

Problem 5.2. Green's reciprocity relation and properties of mutual capacitances

Consider a charge distributions $\rho_1(\mathbf{x})$. The potential that it generates is

$$\phi_1(\mathbf{x}) = \frac{1}{4\pi\epsilon_0} \int d^3x' \, \frac{\rho_1(\mathbf{x}')}{|\mathbf{x} - \mathbf{x}'|} \, . \tag{5.157}$$

Similarly, another charge distributions $\rho_2(\mathbf{x})$ generates a potential

$$\phi_2(\mathbf{x}) = \frac{1}{4\pi\epsilon_0} \int d^3x' \, \frac{\rho_2(\mathbf{x}')}{|\mathbf{x} - \mathbf{x}'|} \, . \tag{5.158}$$

Then,

$$\int d^3x \rho_1(\mathbf{x})\phi_2(\mathbf{x}) = \frac{1}{4\pi\epsilon_0} \int d^3x d^3x' \, \frac{\rho_1(\mathbf{x})\rho_2(\mathbf{x}')}{|\mathbf{x} - \mathbf{x}'|} \, . \tag{5.159}$$

On the other hand,

$$\int d^3x \rho_2(\mathbf{x})\phi_1(\mathbf{x}) = \frac{1}{4\pi\epsilon_0} \int d^3x d^3x' \, \frac{\rho_2(\mathbf{x})\rho_1(\mathbf{x}')}{|\mathbf{x} - \mathbf{x}'|} \, , \tag{5.160}$$

which, upon renaming $\mathbf{x} \leftrightarrow \mathbf{x}'$, is the same as the right-hand side of eq. (5.159). Therefore, we have the identity

$$\int d^3x \, \rho_1(\mathbf{x})\phi_2(\mathbf{x}) = \int d^3x \, \rho_2(\mathbf{x})\phi_1(\mathbf{x}) \, , \tag{5.161}$$

which is valid for arbitrary charge densities $\rho_1(\mathbf{x})$ and $\rho_2(\mathbf{x})$, and the potentials $\phi_1(\mathbf{x})$ and $\phi_2(\mathbf{x})$ that they generate. Equation (5.161) is known as *Green's reciprocity relation*. A discrete version of this identity is obtained considering, as first charge distribution, a set of conductors with charges $Q_1, \ldots, Q_N$. We denote by $\phi_1, \ldots, \phi_N$ the corresponding values of the electrostatic potentials on their surface. As the second charge distribution, we take the same set of conductors, in the same positions, but with charges $Q'_1, \ldots, Q'_N$, and we denote by $\phi'_1, \ldots, \phi'_N$ the corresponding values of the electrostatic potential on their surface. Then, $\rho_1(\mathbf{x}) = \sum_a \rho_{1,a}(\mathbf{x})$, where $\rho_{1,a}$ is non-vanishing only on the surface of the a-conductor, where $\phi_2(\mathbf{x})$ has the constant value ϕ'_a, and therefore,

$$\begin{aligned}
\int d^3x \, \rho_1(\mathbf{x})\phi_2(\mathbf{x}) &= \sum_{a=1}^{N} \phi'_a \int d^3x \, \rho_{1,a}(\mathbf{x}) \\
&= \sum_{a=1}^{N} Q_a \phi'_a \, . \tag{5.162}
\end{aligned}$$

Similarly,

$$\begin{aligned}
\int d^3x \, \rho_2(\mathbf{x})\phi_1(\mathbf{x}) &= \sum_{a=1}^{N} \phi_a \int d^3x \, \rho_{2,a}(\mathbf{x}) \\
&= \sum_{a=1}^{N} Q'_a \phi_a \, . \tag{5.163}
\end{aligned}$$

Therefore, for a set of conductors, Green's reciprocity relation becomes

$$\sum_{a=1}^{N} Q_a \phi'_a = \sum_{a=1}^{N} Q'_a \phi_a \, . \tag{5.164}$$

From eq. (4.163), we have $Q_a = \sum_{b=1}^{N} C_{ab}\phi_b$ and $Q'_a = \sum_{b=1}^{N} C_{ab}\phi'_b$, with the same coefficients C_{ab}, since these depend only on the geometry of the system, i.e., on the positions and shapes of the conductors, that we have taken to be the same in the two cases. Then eq. (5.164) gives

$$\sum_{a,b=1}^{N} C_{ab}\phi_b\phi'_a = \sum_{a,b=1}^{N} C_{ab}\phi'_b\phi_a , \qquad (5.165)$$

or, renaming $a \leftrightarrow b$ in the second sum,

$$\sum_{a,b=1}^{N} C_{ab}\phi'_a\phi_b = \sum_{a,b=1}^{N} C_{ba}\phi'_a\phi_b . \qquad (5.166)$$

Since this is valid for arbitrary choices of the charge configurations Q and Q', and therefore for arbitrary values of ϕ and ϕ', we conclude that

$$C_{ab} = C_{ba} , \qquad (5.167)$$

i.e., the capacitance matrix is symmetric. We can repeat exactly the same argument in magnetostatics. In this case we get

$$\int d^3x\, \mathbf{j}_1(\mathbf{x})\cdot\mathbf{A}_2(\mathbf{x}) = \int d^3x\, \mathbf{j}_2(\mathbf{x})\cdot\mathbf{A}_1(\mathbf{x}) . \qquad (5.168)$$

We can then apply it to a given configuration of loops. Let $I_1, \ldots, I_N$ be a set of values of the of currents in the loops, and $\Phi_{B,1}, \ldots, \Phi_{B,N}$ the corresponding magnetic fluxes; and let $I'_1, \ldots, I'_N$ be another set of values of the currents of the same loops, and $\Phi'_{B,1}, \ldots, \Phi'_{B,N}$ the corresponding magnetic fluxes. Then,

$$\sum_{a=1}^{N} I_a\Phi'_{B,a} = \sum_{a=1}^{N} I'_a\Phi_{B,a} . \qquad (5.169)$$

From this, we can prove in the same way as before that $L_{ab} = L_{ba}$. In this case, we already knew this from the explicit expression (5.66), while for C_{ab} there is not an equally simple general expression.

Problem 5.3. Energy stored in a wire loop

We now consider the work that should be done to create a current I in a loop. This is a special case of the discussion in Section 5.3 and allows us to illustrate that general analysis in a simpler setting. As in the discussion of Section 5.3, we raise the current in the loop from zero to the final value I. Even if the final value is constant in time, during the transient period necessary to reach the final value, I is a function of time, $I = I(t)$. As a consequence, according to eq. (4.200), the magnetic flux Φ_B through the loop also changes in time,

$$\frac{d\Phi_B}{dt} = L\frac{dI}{dt} , \qquad (5.170)$$

where L is the self-inductance of the loop. As already discussed after eq. (4.200), this creates a "back-emf"

$$\mathcal{E}_{\text{emf}}^{\text{back}} = -\frac{d\Phi_B}{dt} , \qquad (5.171)$$

that opposes the increase in the current. An external agent must therefore provide work against this back-emf. From the definition of emf as the line integral of $\mathbf{E}$ around the circuit, the work needed to push a charge dq through a single trip around the loop is

$$
\begin{aligned}
dW_{\text{ext}} &= -\mathcal{E}_{\text{emf}}^{\text{back}}\, dq \\
&= L\frac{dI}{dt}\, dq \,.
\end{aligned}
\tag{5.172}
$$

On the other hand, if we have a current I in the loop, the charge that moves around the circuit in a time dt is $dq = I\,dt$, where I is the current I that flows in the wire. Therefore,

$$
\frac{dW_{\text{ext}}}{dt} = LI\frac{dI}{dt} \,.
\tag{5.173}
$$

With the initial condition $W_{\text{ext}} = 0$ when $I = 0$, this integrates to

$$
W_{\text{ext}} = \frac{1}{2}LI^2 \,,
\tag{5.174}
$$

or, using eq. (4.200),

$$
W_{\text{ext}} = \frac{1}{2}I\Phi_B \,.
\tag{5.175}
$$

This is the work done to create the desired final steady current I in the circuit. As we see comparing with eq. (5.68), this is indeed the same as the energy $\mathcal{E}_B$ stored in the magnetic field generated by the current I, that we obtained starting from the general definition

$$
\mathcal{E}_B = \frac{1}{2\mu_0} \int d^3x\, |\mathbf{B}|^2 \,.
\tag{5.176}
$$

It is important to appreciate that this energy, which is stored in the magnetic field and is associated with the self-induction L, is fully recoverable, for instance, turning the current back to zero. This should be contrasted with the energy dissipated in the wire because of its resistance R, which is irretrievably lost to Joule heating.

Multipole expansion for static fields

When a source is localized in a volume V of typical linear size d, and we are only interested in the field that it generates at distances $r \gg d$, we expect, physically, that only the gross features of the source distribution will be important, rather than all its fine details. The tool that allows us to formalize this intuition is the multipole expansion. If the source is static, the multipole expansion is an expansion in powers of d/r. When the source is type-dependent a new parameter enters into play, which is the typical frequency ω of the source, and this defines a new length-scale $\lambdabar \equiv c/\omega$. In this case the expansion is more complex, and even when $r \gg d$, still depends on the relative values of r and $\lambdabar$. In this chapter we study the multipole expansion for static sources, both in electrostatics and in magnetostatics. This will allow us to introduce the *static multipoles* of the source. The time-dependent situation, that will lead to a multipole expansion of the electromagnetic radiation in terms of *radiative multipoles*, is more complicated and will be studied in Section 11.2, after we will have developed the formalism for dealing with time-dependent fields and electromagnetic radiation.

6.1 Electric multipoles

In eq. (4.16) we found the solution for the scalar gauge potential $\phi(\mathbf{x})$ in the presence of a charge distribution $\rho(\mathbf{x})$, taken to be static and lo-calized (or, at least, decreasing sufficiently rapidly at large distances, so that the integral in eq. (4.16) converges), but otherwise arbitrary. This solution is still given in terms of an integral, that cannot be computed analytically for a generic $\rho(\mathbf{x})$. We now assume that the source is local-ized in a region of typical linear size d, and we study the solution for $\phi(\mathbf{x})$, and the corresponding electric field, in the limit $r \gg d$.

We begin by observing that, in eq. (4.16), $\mathbf{x}'$ is an integration variable that in principle runs over all the three-dimensional space $\mathbb{R}^3$. However, since we have assumed that $\rho(\mathbf{x}') = 0$ for $|\mathbf{x}'| > d$, in practice the integral in eq. (4.16) only runs over the values of $\mathbf{x}'$ with $|\mathbf{x}'| < d$. In the limit $r \gg d$ we can therefore expand $|\mathbf{x} - \mathbf{x}'|$ for $|\mathbf{x}'| \ll |\mathbf{x}|$, i.e., in powers of the small parameter d/r. Before studying the full expansion, let us consider the first two non-trivial terms. These can be obtained defining

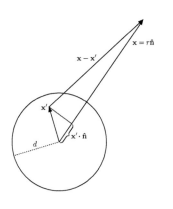

$\hat{\mathbf{n}}$ as the unit vector in the direction of $\mathbf{x}$, so that $\mathbf{x} = r\hat{\mathbf{n}}$, and writing

$$
\begin{aligned}
|\mathbf{x} - \mathbf{x}'|^2 &= r^2 - 2r\hat{\mathbf{n}}\cdot\mathbf{x}' + |\mathbf{x}'|^2 \\
&= r^2\left[1 - \frac{2\hat{\mathbf{n}}\cdot\mathbf{x}'}{r} + \mathcal{O}\left(\frac{d^2}{r^2}\right)\right],
\end{aligned}
\tag{6.1}
$$

so that

$$
\begin{aligned}
|\mathbf{x} - \mathbf{x}'| &= r\left[1 - \frac{\hat{\mathbf{n}}\cdot\mathbf{x}'}{r} + \mathcal{O}\left(\frac{d^2}{r^2}\right)\right] \tag{6.2} \\
&= r - \hat{\mathbf{n}}\cdot\mathbf{x}' + \mathcal{O}\left(\frac{d^2}{r}\right). \tag{6.3}
\end{aligned}
$$

The first two terms of this expansion have a simple graphical interpretation shown in Fig. 6.1. From eq. (6.2) we find that

$$
\frac{1}{|\mathbf{x} - \mathbf{x}'|} = \frac{1}{r}\left[1 + \frac{\hat{\mathbf{n}}\cdot\mathbf{x}'}{r} + \mathcal{O}\left(\frac{d^2}{r^2}\right)\right].
\tag{6.4}
$$

Fig. 6.1 A graphical illustration of the relation given in eq. (6.3).

Inserting this into eq. (4.16) we get

$$
\phi(\mathbf{x}) = \frac{1}{4\pi\epsilon_0}\frac{1}{r}\int d^3x'\, \rho(\mathbf{x}')\left[1 + \frac{\hat{\mathbf{n}}\cdot\mathbf{x}'}{r} + \mathcal{O}\left(\frac{d^2}{r^2}\right)\right].
\tag{6.5}
$$

These are the first two terms of the so-called *multipole expansion*.[1] The first term, that in this context is also called the "monopole" term, is just the Coulomb potential (4.6),

[1] Observe that r is the distance of the point $\mathbf{x}$ from an origin of the reference frame, that we have chosen as an arbitrary point inside the charge distribution. We will later discuss the dependence of the multipole expansion from the choice of origin.

$$
\phi_{\text{monopole}}(r) = \frac{1}{4\pi\epsilon_0}\frac{q}{r},
\tag{6.6}
$$

where

$$
q = \int d^3x\, \rho(\mathbf{x}),
\tag{6.7}
$$

is the total electric charge of the system. We define the dipole moment of the electric charge distribution, or, simply, the "*electric dipole*," as

$$
\boxed{\mathbf{d} = \int d^3x\, \rho(\mathbf{x})\mathbf{x}.}
\tag{6.8}
$$

Then, the second term in eq. (6.5) can be written as

$$
\boxed{\phi_{\text{dipole}}(\mathbf{x}) = \frac{1}{4\pi\epsilon_0}\frac{\mathbf{d}\cdot\hat{\mathbf{n}}}{r^2},}
\tag{6.9}
$$

or, equivalently,

$$
\phi_{\text{dipole}}(\mathbf{x}) = \frac{1}{4\pi\epsilon_0}\frac{\mathbf{d}\cdot\mathbf{x}}{r^3}.
\tag{6.10}
$$

Observe that, while ϕ_{monopole} depends only on r, $\phi_{\text{dipole}}(\mathbf{x})$ depends both on r and on $\hat{\mathbf{n}}$. The corresponding electric fields are obtained from

$\mathbf{E} = -\boldsymbol{\nabla}\phi$, since $\mathbf{A} = 0$. Of course, $\phi_{\text{monopole}}(r)$ generates the Coulomb electric field $\mathbf{E}_{\text{Coulomb}} = q\hat{\mathbf{r}}/(4\pi\epsilon_0 r^2)$. The electric field generated by the dipole is

$$
\begin{aligned}
(\mathbf{E}_{\text{dipole}})_i &= -\frac{1}{4\pi\epsilon_0}d_j\partial_i\left(\frac{x_j}{r^3}\right) \\
&= \frac{1}{4\pi\epsilon_0}\frac{1}{r^3}\left(3n_i n_j - \delta_{ij}\right)d_j\,, \quad (6.11)
\end{aligned}
$$

where we used eq. (6.10) and, to compute ∂_i on a function of r, we used $\partial_i f(r) = (\partial_i r)df/dr$, and

$$
\begin{aligned}
\partial_i r &= \partial_i(x_j x_j)^{1/2} \\
&= \frac{x_i}{r} \\
&= n_i\,. \quad (6.12)
\end{aligned}
$$

We can rewrite eq. (6.11) in vector form, as

$$
\boxed{\mathbf{E}_{\text{dipole}} = \frac{1}{4\pi\epsilon_0}\frac{3(\mathbf{d}\cdot\hat{\mathbf{n}})\hat{\mathbf{n}} - \mathbf{d}}{r^3}\,.} \quad (6.13)
$$

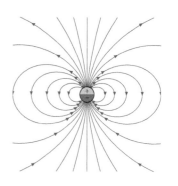

Fig. 6.2 The field lines of an electric dipole.

Fig. 6.2 shows the field lines of the dipole electric field. We now work out the next term in the expansion. To systematically carry out the multipole expansion to higher orders it is convenient to write the expansion of $1/|\mathbf{x} - \mathbf{x}'|$, for $|\mathbf{x}'|$ small compared to $|\mathbf{x}|$, in the form

$$
\frac{1}{|\mathbf{x} - \mathbf{x}'|} = \frac{1}{r} - x_i'\partial_i\frac{1}{r} + \frac{1}{2}x_i'x_j'\partial_i\partial_j\frac{1}{r} + \dots\,. \quad (6.14)
$$

where, again, $r = |\mathbf{x}|$ and ∂_i is the derivative with respect to x^i. The corresponding expansion of $\phi(\mathbf{x})$ in eq. (4.16) is

$$
\begin{aligned}
4\pi\epsilon_0\,\phi(x) &= \frac{1}{r}\int d^3x'\,\rho(\mathbf{x}') - \left(\partial_i\frac{1}{r}\right)\int d^3x'\,\rho(\mathbf{x}')x_i' \\
&\quad + \frac{1}{2}\left(\partial_i\partial_j\frac{1}{r}\right)\int d^3x'\,\rho(\mathbf{x}')x_i'x_j' + \dots\,. \quad (6.15)
\end{aligned}
$$

The first two terms give again the monopole and dipole terms computed previously. The third term can be further transformed observing that we are interested in the field at distances r much larger than the size of the region d where the source is localized, so, in particular, for $r \neq 0$. In this case, from eq. (1.90), $\boldsymbol{\nabla}^2(1/r) = 0$. Then, in the third term of eq. (6.15), we can replace $x_i'x_j'$ with the traceless combination $x_i'x_j' - (1/3)\delta_{ij}|\mathbf{x}'|^2$, since

$$
x_i'x_j'\partial_i\partial_j\frac{1}{r} = \left(x_i'x_j' - \frac{1}{3}\delta_{ij}|\mathbf{x}'|^2\right)\partial_i\partial_j\frac{1}{r} + \frac{1}{3}|\mathbf{x}'|^2\delta_{ij}\partial_i\partial_j\frac{1}{r}\,, \quad (6.16)
$$

and, in the last term on the right-hand side of eq. (6.16), we can use $\delta_{ij}\partial_i\partial_j(1/r) = \boldsymbol{\nabla}^2(1/r) = 0$. We then define

$$
q_{ij} = \int d^3x\,\rho(\mathbf{x})\left(x_i x_j - \frac{1}{3}\delta_{ij}|\mathbf{x}|^2\right)\,, \quad (6.17)
$$

which is called the "reduced quadrupole moment" of the charge distribution. The advantage of removing the trace part in q_{ij} is that, as we mentioned in Section 1.7.2 (and we showed explicitly for a tensor T_{ij} with two indices), symmetric trace-less tensors (or "STF" tensors, for "symmetric trace-free") provide irreducible representations of the rotation group. Working with the irreducible representations, we are directly working with the fundamental "building blocks" of the representation theory, so it is conceptually clearer, and this also typically brings technical simplifications in various computations.

While the normalization in (6.17) is the most natural from the point of view of STF tensors, for historical reasons the name "quadrupole moment" is usually reserved for the quantity $Q_{ij} \equiv 3q_{ij}$,[2]

$$ Q_{ij} = \int d^3x \, \rho(\mathbf{x}) \left(3x_i x_j - \delta_{ij}|\mathbf{x}|^2 \right) . \tag{6.18} $$

The corresponding term in the potential is given by

$$
\begin{aligned}
4\pi\epsilon_0 \, \phi_{\text{quadrupole}}(\mathbf{x}) &= \frac{1}{2} q_{ij} \partial_i \partial_j \frac{1}{r} \\
&= -\frac{1}{2r^3}(\delta_{ij} - 3n_i n_j) q_{ij} \\
&= \frac{3}{2r^3} n_i n_j q_{ij} \\
&= \frac{1}{2r^3} n_i n_j Q_{ij} .
\end{aligned}
\tag{6.19}
$$

Putting together the terms up to the quadrupole, we therefore have

$$ \phi(\mathbf{x}) = \frac{1}{4\pi\epsilon_0} \left[\frac{q}{r} + \frac{n_i d_i}{r^2} + \frac{n_i n_j Q_{ij}}{2r^3} + \dots \right] . \tag{6.20} $$

[2]Actually, there are different conventions in the literature. Our definition agrees with Jackson (1998) (and with Landau and Lifschits (1975), that, however, denotes it as D_{ij}) while, for instance, Griffiths (2017) defines the quadrupole moment as

$$ Q_{ij} = \frac{1}{2} \int d^3x \, \rho(\mathbf{x}) \left(3x_i x_j - \delta_{ij}|\mathbf{x}|^2 \right), $$

while Zangwill (2013) defines it as

$$ Q_{ij} = \frac{1}{2} \int d^3x \, \rho(\mathbf{x}) \, x_i x_j . $$

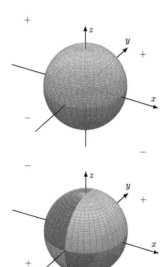

Fig. 6.3 An example of a distribution of charges leading to a dipole moment (upper panel) and to a quadrupole moment (lower panel).

In Fig. 6.3 we show an example of a distribution of charges on the surface of a sphere, leading to an electric dipole moment (upper panel) or to an electric quadrupole moment (lower panel). The two different colors describe an excess of positive and negative charges, respectively. In all cases, the net charge is zero. In the configuration in the upper panel there is a positive net charge in the $z > 0$ hemisphere and a negative net charge in the $z < 0$ hemisphere. The charge distribution $\rho(\mathbf{x})$ has been taken to be odd under $z \to -z$. Then, the integrand $\rho(\mathbf{x})x_i$ in eq. (6.8) is even only when the index $i = z$, while for $i = x$ or $i = y$ the integration over $d^3\mathbf{x}$ gives zero (we do not put the prime here over the integration variable). This leads to a dipole moment along the z axis, while d_x and d_y vanish. In the distribution shown in the lower panel, in contrast, $\rho(\mathbf{x}) = \rho(-\mathbf{x})$. Then, the dipole vanishes since the integrand $\rho(\mathbf{x})\mathbf{x}$ is odd under $\mathbf{x} \to -\mathbf{x}$. However, this configuration has a non-vanishing quadrupole moment and, in this example, only Q_{xz} and Q_{zx} are non-zero. This can be understood observing that the charge distribution in the lower panel of Fig. 6.3 is odd under the parity transformation $\{x \to$

$-x, y \to y, z \to z\}$ and also under the parity transformation $\{x \to x, y \to y, z \to -z\}$. Therefore, the integral of $\rho(\mathbf{x}')|\mathbf{x}'|^2$ in eq. (6.18) vanishes, because its integrand is odd under any of these parity transformations. The term $\rho(\mathbf{x})x_i x_j$ is even under both, and therefore has a non-vanishing integral, only when $i = x, j = z$ (or $i = z, j = x$), so, for the charge distribution shown in the right panel of Fig. 6.3, only the $Q_{xz} = Q_{zx}$ component of the symmetric traceless tensor Q_{ij} is non-vanishing.

It should be observed that the expression for the multipole moments depends on the choice of the origin of the reference frame. If we perform a translation of the origin of the reference frame by a vector $-\mathbf{s}$, a point P that before the transformation had the coordinate $\mathbf{x}$ will have coordinate $\mathbf{x}' = \mathbf{x} + \mathbf{s}$ in the new frame, i.e., the coordinates transform as

$$\mathbf{x} \to \mathbf{x}' = \mathbf{x} + \mathbf{s}. \tag{6.21}$$

Under this transformation the charge density transforms as a "scalar under translations,"[3]

$$\rho(\mathbf{x}) \to \rho'(\mathbf{x}') = \rho(\mathbf{x}). \tag{6.22}$$

Since also d^3x is invariant under translation, $d^3x' = d^3x$, the total charge q of the system is independent of the choice of the origin of the reference frame, as of course we expect. However, the dipole moment transforms as

$$
\begin{aligned}
\mathbf{d} \quad &\to \quad \int d^3x'\, \rho'(\mathbf{x}')\mathbf{x}' \\
&= \quad \int d^3x\, \rho(\mathbf{x})\,(\mathbf{x} + \mathbf{s}) \\
&= \quad \mathbf{d} + q\mathbf{s}, \tag{6.23}
\end{aligned}
$$

where q is the total charge. Therefore, the electric dipole moment is invariant under shifts of the origin only for a system with zero total charge. Similarly, from eq. (6.18),

$$Q_{ij} \to Q_{ij} + q\left(3s_i s_j - \delta_{ij}|\mathbf{s}|^2\right) + 3(d_i s_j + d_j s_i) - 2\delta_{ij}(\mathbf{s}{\cdot}\mathbf{d}), \tag{6.24}$$

and therefore it is invariant only if both q and $\mathbf{d}$ vanish. These transformations are, indeed, precisely those required so that the full expansion (6.5), when carried out to all orders, is independent of the choice of the origin. This is as it should be, since the original expression that we are expanding, eq. (4.16), makes no reference to a choice of origin: as we have seen d^3x' and $\rho(\mathbf{x}')$ are invariant under translations, and also $\mathbf{x} - \mathbf{x}'$ is invariant under translation. However, the latter property is apparently lost in the expansion (6.4), at least as long as we truncate it to a finite order.

To see how the independence from the choice of the origin is recovered, thanks to the transformations of the multipole moments such as those in eqs. (6.23) and (6.24) for the dipole and quadrupole, we proceed as follows. Under the transformation (6.21), the coordinate of the point P

[3]This expresses the fact that the numerical value of the charge density at a given point P is the same independently of where we put the origin of a reference frame. In a frame with a given origin, the point P will be labeled by a coordinate $\mathbf{x}$, while in a frame with a different origin it will be labeled by a coordinate $\mathbf{x}'$ and, numerically, $\mathbf{x}' \neq \mathbf{x}$. However, under this change of reference frame, the *functional form* of the charge density changes, from $\rho(\mathbf{x})$ to a new function $\rho'(\mathbf{x})$, such that the *numerical value* of ρ' in $\mathbf{x}'$ is the same as the numerical value of ρ in $\mathbf{x}$, so that, in the end, the numerical value at the point P remains the same, independently of how P is labeled.

in which we are computing the field changes from $\mathbf{x}$ to $\mathbf{x}+\mathbf{s}$. Therefore, using eq. (6.4) and working to first order in $\mathbf{s}$ only, the term q/r in eq. (6.20) transforms as

$$\frac{q}{r} \rightarrow \frac{q}{|\mathbf{x}+\mathbf{s}|}$$
$$= \frac{q}{r} - qs_i \frac{n_i}{r^2} + \dots \, . \tag{6.25}$$

This extra term $-qs_i n_i/r^2$ is precisely canceled by the transformation (6.23) of the dipole, in the term $n_i d_i/r^2$ in eq. (6.20). To second order in $\mathbf{s}$, there will be a contribution from the q/r term in eq. (6.20), due to the expansion of eq. (6.25) to second order in $\mathbf{s}$, as well as a further contribution from the term $n_i d_i/r^2$, due to the expansion of $1/r^2$ to first order in $\mathbf{s}$. These are precisely the contributions that are canceled by the transformation (6.24) of the quadrupole. This continues to all orders, with cancellations among terms of different orders in the multipole expansion. In this way, the full expansion at all orders is independent of the choice of origin. Note, however, that any truncation of the expansion to finite order (e.g., restricting to the monopole and dipole) will retain a dependence on the choice of origin. In practice, as long as we compute the field at a distance r from the localized charge distribution, much larger than the linear size d of the distribution itself, any choice of origin inside the localization volume of the charge will give basically equivalent results, with differences that disappear quickly as we go to larger distances (or include higher multipoles in the truncation). From the mathematical point of view, nothing forbids us from performing the expansion (6.5) choosing an origin outside, or even very far from the localization region of the charge; however, since in eq. (6.4) we are expanding in powers of $\mathbf{x}'$, taken to be small with respect to $\mathbf{x}$, it is clearly convenient to use a choice of origin such that, when $\rho(\mathbf{x}')$ in eq. (6.5) is non-zero, $\mathbf{x}'$ is as small as possible, which is obtained by choosing the origin somewhere inside the charge distribution (for instance, in the "center-of-charge," the equivalent of the center of mass for the charge density). Otherwise, the consequence would be that a truncation of the expansion to any finite order would be a worse approximation to the exact result.

The expansion in STF tensors is easily generalized to arbitrary order. Equation (6.14) becomes

$$\frac{1}{|\mathbf{x}-\mathbf{x}'|} = \sum_{l=0}^{\infty} \frac{(-1)^l}{l!} x'_{i_1} \dots x'_{i_l} \partial_{i_1} \dots \partial_{i_l} \frac{1}{r} \, . \tag{6.26}$$

Again, using $\mathbf{\nabla}^2(1/r) = 0$ for $r \neq 0$, we can remove the traces with respect to any pair of indices. We then define

$$q_{i_1 \dots i_l} = \int d^3 x' \, x'_{\langle i_1} \dots x'_{i_l \rangle} \, \rho(\mathbf{x}') \, , \tag{6.27}$$

where the brackets in $x'_{\langle i_1} \dots x'_{i_l \rangle}$ mean that we must take the trace-free part of the symmetric tensor $x'_{i_1} \dots x'_{i_l}$ i.e., we must remove the traces with respect to all pair of indices. For instance, for the tensor with three indices, we can replace $x_i x_j x_k$ with

$$x_{\langle i}x_j x_{k\rangle} = x_i x_j x_k - \frac{1}{3}|\mathbf{x}|^2\left(\delta_{ij}x_k + \delta_{ik}x_j + \delta_{jk}x_i\right).\tag{6.28}$$

The coefficients $q_{i_1...i_l}$ are called the STF multipole moments of the charge distribution. Then, inserting eq. (6.26) into eq. (4.16), we get

$$\phi(\mathbf{x}) = \frac{1}{4\pi\epsilon_0}\sum_{l=0}^{\infty}\frac{(-1)^l}{l!}q_{i_1...i_l}\partial_{i_1}\ldots\partial_{i_l}\left(\frac{1}{r}\right).\tag{6.29}$$

The terms $l=0$, $l=1$, and $l=2$ in eq. (6.29) are just the monopole, dipole, and quadrupole terms that we found previously.[4]

[4]The expansion in STF tensors can be rewritten as an expansion in spherical harmonics. We will not elaborate on this here. The interested reader can find a discussion of the relation between these two expansions, as well as the extension to vector spherical harmonics (relevant in electromagnetism) and tensor spherical harmonics (relevant for the gravitational field) in Section 3.5.2 of Maggiore (2007).

6.2   Magnetic multipoles

We now consider the multipole expansion in magnetostatics. We start from eq. (4.92), and we consider the situation in which the current $\mathbf{j}(\mathbf{x}')$ is localized in a region with $|\mathbf{x}'| < d$. We then compute $\mathbf{A}(\mathbf{x})$ at distances $r \gg d$, where, as before, $r \equiv |\mathbf{x}|$. We limit ourselves to the first non-trivial term that, as we will see, is the magnetic dipole term. The generic expansion to all orders in STF tensors can be performed similarly to what we have done in the electric case, but one rarely encounters situations where magnetic multipoles higher than the dipole play a role.

We therefore insert the expansion (6.4) into eq. (4.92) [technically, it is slightly simpler to perform the multipole expansion of $\mathbf{A}(\mathbf{x})$ and then obtain the corresponding expansion for $\mathbf{B}(\mathbf{x})$ from $\mathbf{B} = \nabla\times\mathbf{A}$, rather than performing the corresponding expansion directly in eq. (4.95)]. This gives

$$\mathbf{A}(\mathbf{x}) = \frac{\mu_0}{4\pi}\frac{1}{r}\int d^3x'\,\mathbf{j}(\mathbf{x}')\left[1 + \frac{\hat{\mathbf{n}}\cdot\mathbf{x}'}{r} + \mathcal{O}\left(\frac{d^2}{r^2}\right)\right].\tag{6.30}$$

The first term in the expansion actually vanishes, as a consequence of current conservation. This can be shown using the identity $\partial_j x_i = \delta_{ij}$ to rewrite, trivially, $j_i(\mathbf{x}) = j_j(\mathbf{x})\partial_j x_i$. Then

$$\begin{aligned}\int d^3\mathbf{x}\,j_i(\mathbf{x}) &= \int d^3\mathbf{x}\,j_j(\mathbf{x})\partial_j x_i\\ &= -\int d^3\mathbf{x}\,x_i\partial_j j_j(\mathbf{x})\\ &= 0,\end{aligned}\tag{6.31}$$

where in the second line we integrated by parts (setting the boundary terms to zero, because $\mathbf{j}(\mathbf{x})$ is localized) and we used current conservation in the form (4.90), valid for magnetostatics. The leading term is then given by the second term, which we denote as $\mathbf{A}_{\text{dipole}}$. In components,

$$A_{i,\text{dipole}}(\mathbf{x}) = \frac{\mu_0}{4\pi}\frac{x_j}{r^3}\int d^3x'\,j_i(\mathbf{x}')x_j',\tag{6.32}$$

where we wrote n_j as x_j/r. To manipulate this expression, we proceed similarly to the previous computation, integrating by parts and using current conservation,

$$
\begin{aligned}
\int d^3x\, j_i(\mathbf{x})x_j &= \int d^3x\, (\partial_k x_i)\, j_k(\mathbf{x})x_j \\
&= -\int d^3x\, x_i \partial_k [j_k(\mathbf{x})x_j] \\
&= -\int d^3x\, x_i j_k(\mathbf{x})\partial_k x_j \\
&= -\int d^3x\, j_j(\mathbf{x})x_i .
\end{aligned}
\tag{6.33}
$$

This shows that the integral is antisymmetric in the (i,j) indices, and therefore

$$
\begin{aligned}
\int d^3x\, j_i(\mathbf{x})x_j &= \frac{1}{2}\int d^3x\, [j_i(\mathbf{x})x_j - j_j(\mathbf{x})x_i] \\
&= -\frac{1}{2}\epsilon_{ijk}\int d^3x\, [\mathbf{x}\times\mathbf{j}(\mathbf{x})]_k ,
\end{aligned}
\tag{6.34}
$$

where the last identity can be verified writing $\epsilon_{ijk}(\mathbf{x}\times\mathbf{j})_k = \epsilon_{ijk}\epsilon_{klm}x_l j_m$ and using eq. (1.7). We define the *magnetic dipole moment* of the current distribution (or, more simply, the "magnetic moment"), as[5]

$$
\boxed{\mathbf{m} = \frac{1}{2}\int d^3x\, \mathbf{x}\times\mathbf{j}(\mathbf{x}) .}
\tag{6.36}
$$

Therefore

$$
\int d^3x\, j_i(\mathbf{x})x_j = -\epsilon_{ijk}m_k ,
\tag{6.37}
$$

and eq. (6.32) becomes

$$
\boxed{\mathbf{A}_{\text{dipole}}(\mathbf{x}) = \frac{\mu_0}{4\pi}\frac{\mathbf{m}\times\mathbf{x}}{r^3} .}
\tag{6.38}
$$

Taking the curl (for $r \neq 0$, given that we are computing the magnetic field generated by a localized source, in an expansion for $r \gg d$), we get

$$
(\mathbf{B}_{\text{dipole}})_i = \frac{\mu_0}{4\pi}\frac{1}{r^3}(3n_i n_j - \delta_{ij})m_j ,
\tag{6.39}
$$

or, in vector form,

$$
\boxed{\mathbf{B}_{\text{dipole}} = \frac{\mu_0}{4\pi}\frac{3(\mathbf{m}\cdot\hat{\mathbf{n}})\hat{\mathbf{n}} - \mathbf{m}}{r^3} ,}
\tag{6.40}
$$

to be compared with eq. (6.13) for the electric field generated by an electric dipole. This is the leading term for the magnetic field generated by

[5]Observe that, under a change of the origin used to define the multipole moment, corresponding to changing $\mathbf{x} \to \mathbf{x}' = \mathbf{x} + \mathbf{s}$, we have $\mathbf{j}(\mathbf{x}) \to \mathbf{j}'(\mathbf{x}') = \mathbf{j}(\mathbf{x})$, i.e., the current density is invariant under translations, and $d^3x' = d^3x$, and therefore

$$
m_i \to m_i + \frac{1}{2}\epsilon_{ijk}s_j\int d^3x\, j_k .
\tag{6.35}
$$

However, because of eq. (6.31), for a localized current density the extra term vanishes, and therefore, in magnetostatics (i.e., when $\boldsymbol{\nabla}\cdot\mathbf{j} = 0$ holds, so that also eq. (6.31) holds) the magnetic dipole moment is independent of the choice of the origin. This is due to the fact that, in the expansion in magnetic multipoles, there is no monopole term, i.e., no magnetic charge, so the situation is analogous to eq. (6.23) for the electric dipole, when $q = 0$. For magnetic multipoles, the dependence on the choice of the origin starts from the magnetic quadrupole (if the magnetic dipole is non-vanishing, otherwise it starts at even higher orders).

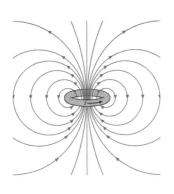

Fig. 6.4 The field lines of a magnetic dipole, represented here by a closed loop carrying a current.

a localized current, at large distances. Fig. 6.4 shows the corresponding lines of the magnetic field, to be compared with Fig. 6.2 for the electric field from the electric dipole.

A typical situations in which a magnetic dipole moment appears is given by a loop $\mathcal{C}$ carrying a current. Using eq. (4.103), eq. (6.36) becomes

$$\mathbf{m} = \frac{I}{2} \oint_{\mathcal{C}} \mathbf{x} \times d\boldsymbol{\ell}. \tag{6.41}$$

For a loop lying in a plane, the integral in this expression is just twice the area A of the loop times the normal $\hat{\mathbf{n}}$ to the surface, as we proved in eq. (1.47) using Stokes's theorem,[6] so

$$\mathbf{m} = IA\,\hat{\mathbf{n}}. \tag{6.42}$$

Another important case is given by a non-relativistic charged particle with charge q_a, mass m_a and velocity $\mathbf{v}_a$. The corresponding current is given by eq. (3.27) which, inserted into eq. (6.36), gives[7]

$$\mathbf{m} = \frac{q_a}{2m_a} \mathbf{L}_a, \tag{6.43}$$

where $\mathbf{L}_a = m_a \mathbf{x}_a \times \mathbf{v}_a$ is the angular momentum of a non-relativistic particle.[8]

6.3 Point-like electric or magnetic dipoles

Equation (6.13) gives the electric field generated by an electric dipole, at distances r much larger than the size d of the source, since we obtained it performing an expansion in the small parameter d/r. Consider now as source a point-like electric dipole $\mathbf{d}$ located at the origin (which could be, for instance, a classical modelization of a microscopic object with an intrinsic dipole moment). In this case the size d of the source vanishes, and an expansion in powers of d/r gives the exact result for all values of r except $r = 0$, so eq. (6.13) is exact for all $r \neq 0$. In principle, however, there could still be a Dirac delta singularity at the origin. To test for the presence of a Dirac delta in the electric field $\mathbf{E}_{\text{dipole}}$ generated by a point dipole, we integrate it over a volume V with boundary ∂V,

$$\int_V d^3x \, (E_{\text{dipole}})_i = -\int_V d^3x \, \partial_i \phi_{\text{dipole}}$$
$$= -\int_{\partial V} d^2s \, n_i \, \phi_{\text{dipole}}, \tag{6.44}$$

where $d\mathbf{s} = d^2s\,\hat{\mathbf{n}}$ is the surface element on the boundary ∂V, and $\hat{\mathbf{n}}$ is the outer normal of the boundary. We now insert here the expression (6.9) for ϕ_{dipole}, and we take V to be a sphere of radius R, so that $d^2s = R^2 d\Omega$. Then,

$$\int_V d^3x \, (E_{\text{dipole}})_i = -\frac{1}{4\pi\epsilon_0} \int R^2 d\Omega \, n_i \frac{n_j d_j}{R^2}$$
$$= -\frac{1}{\epsilon_0} d_j \int \frac{d\Omega}{4\pi} n_i n_j. \tag{6.45}$$

[6]This can also be shown, in a less elegant but possibly more direct manner, just by explicitly computing the integral for a square loop, setting the origin for instance in the center of the square and computing the separate contributions of the four sides. The result for a generic planar loop is then obtained by filling it with infinitesimal square loops.

[7]Note, however, that in this case the current also depends on time and satisfies the full continuity equation rather than $\boldsymbol{\nabla}\cdot\mathbf{j} = 0$, see eq. (3.30). Therefore, this example does not belong to the domain of magnetostatics and in this case eq. (6.31) does not hold. As a consequence, it is no longer true that the magnetic dipole is independent of the choice of the origin (compare with Note 5 on page 140). Indeed, we see from eq. (6.43) that, in this case, the magnetic dipole is given in terms of the orbital angular momentum, which is another quantity that depends on the choice of origin.

[8]In quantum mechanics, particles carry an intrinsic angular moment called spin, usually denoted by $\mathbf{S}$. One would then be tempted to assume that, beside having a magnetic moment associated with their angular momentum $\mathbf{L}$, which indeed even in quantum mechanics is given by eq. (6.43), particles should also carry an intrinsic magnetic moment $\mathbf{m} = [q_a/(2m_a)]\,\mathbf{S}$ associated with spin. However, spin is a purely quantum concept, and the previous classical derivation does not go through. It turns out that the actual intrinsic magnetic moment of a charged particle is rather given by $\mathbf{m} = g_a[q_a/(2m_a)]\,\mathbf{S}_a$, where g_a (usually written simply as g, when it is clear to which particle it refers) is a number that depends on the type of particle. For an elementary particle of spin 1/2, such as the electron, in a first approximation (given by the Dirac equation) $g = 2$, and quantum field theory gives corrections to this result that can be computed as an expansion in powers of the fine structure constant α, see e.g., Section 3.6 and Solved Problem 7.2 of Maggiore (2005).

The remaining angular integral could in principle be performed component by component, writing $d\Omega = d\cos\theta d\phi$ and using the explicit expression of the unit normal vector in polar coordinates,

$$\hat{\mathbf{n}} = (\sin\theta\cos\phi, \sin\theta\sin\phi, \cos\theta). \tag{6.46}$$

There is, however, a much more clever way of computing an integral such as

$$I_{ij} \equiv \int \frac{d\Omega}{4\pi}\, n_i n_j\,, \tag{6.47}$$

based on the observation that $n_i n_j$ is a tensor under spatial rotation. After integration over the whole sphere with the measure $d\Omega$, which is invariant under rotations, the result must still be a tensor. However, after having integrated over all directions, there is no longer a preferred direction in space, so no vector on which the result could depend. Since the result is a symmetric tensor with two indices, the only possibility is that it is proportional to δ_{ij}, with some proportionality constant κ,

$$\int \frac{d\Omega}{4\pi}\, n_i n_j = \kappa\delta_{ij}\,. \tag{6.48}$$

Taking the trace of both sides, we get $\int d\Omega/4\pi = 3\kappa$, and therefore $\kappa = 1/3$. Thus, without even computing a single integral, we get the identity

$$\int \frac{d\Omega}{4\pi}\, n_i n_j = \frac{1}{3}\delta_{ij}\,. \tag{6.49}$$

Applying this to eq. (6.45), we get

$$\int_V d^3x\, (E_{\text{dipole}})_i = -\frac{1}{3\epsilon_0}\, d_i\,. \tag{6.50}$$

This result, for a point dipole, is exact, because, to compute it, we only used the expression for ϕ_{dipole} on a surface at $r = R > 0$, see eq. (6.44). In the case of a point-like source that we are considering, as long as $r \neq 0$ the exact result for the electric field is given by eq. (6.11). Allowing for a possibility of a Dirac delta at the origin, the most general form of the electric field generated by a point-like electric dipole is

$$(E_{\text{dipole}})_i = \frac{1}{4\pi\epsilon_0}\left[\frac{1}{r^3}(3n_in_j - \delta_{ij})d_j + \kappa_i^E\delta^{(3)}(\mathbf{x})\right]. \tag{6.51}$$

The constant vector κ_i^E can be fixed inserting eq. (6.51) into eq. (6.50). The first term in bracket in eq. (6.51) does not contribute to the integral. In fact,

$$\int_V d^3x\, \frac{1}{r^3}(3n_in_j - \delta_{ij})d_j = d_j\int_0^R dr\, r^2\frac{1}{r^3}\int d\Omega(3n_in_j - \delta_{ij})\,, \tag{6.52}$$

and the angular integral vanishes because of eq. (6.49).[9] Then, eq. (6.50) gives $\kappa_i^E = -(4\pi/3)d_i$. The conclusion is that the electric field generated

---

[9]Actually, one should also observe that the radial integral diverges in $r = 0$. A more correct procedure is to regularize it by integrating from $r = \epsilon$ to $r = R$, and in the end taking the limit for $\epsilon \to 0^+$. However, since the angular integral vanishes identically, this limit is indeed zero.

by a point-like electric dipole $\mathbf{d}$ at the origin of the coordinate system is given by

$$\mathbf{E} = \frac{1}{4\pi\epsilon_0}\left[\frac{3(\mathbf{d}\cdot\hat{\mathbf{n}})\hat{\mathbf{n}} - \mathbf{d}}{r^3} - \frac{4\pi}{3}\,\mathbf{d}\,\delta^{(3)}(\mathbf{x})\right]\,. \tag{6.53}$$

We can proceed in exactly the same way to compute the magnetic field generated by a point-like magnetic dipole $\mathbf{m}$ (which could be, for instance, a classical modelization of an elementary particle with an intrinsic magnetic moment, see Note 8 on page 141). Again, in the limit in which the size d of the source vanishes, an expansion in powers of d/r gives the exact result for all values of $r \neq 0$, so eqs. (6.38) and (6.40) are exact for all $r \neq 0$. Just as we did for the electric field, to test for the presence of a Dirac delta in $\mathbf{B}$ we integrate it over a volume V with boundary ∂V,

$$\int_V d^3x\, B_i(\mathbf{x}) = \epsilon_{ijk}\int_V d^3x\, \partial_j A_k(\mathbf{x})$$

$$= \epsilon_{ijk}\int_{\partial V} d^2s\, n_j A_k(\mathbf{x})\,. \tag{6.54}$$

We insert here the expression (6.38) for $\mathbf{A}$, and we take V to be a sphere of radius R, so that $d^2s = R^2 d\Omega$. Then

$$\int_V d^3x\, B_i(\mathbf{x}) = \frac{\mu_0}{4\pi}\,\epsilon_{ijk}\epsilon_{klm}m_l\int R^2 d\Omega\, n_j\frac{x_m}{R^3}$$

$$= \mu_0\,(\delta_{il}\delta_{jm} - \delta_{im}\delta_{jl})m_l\int\frac{d\Omega}{4\pi}\,n_j n_m\,, \tag{6.55}$$

where we made use of eq. (1.7) and of the fact that, on the surface of the sphere, $x_m = Rn_m$. The remaining angular integral is performed using eq. (6.49), so we eventually get

$$\int_V d^3x\, B_i(\mathbf{x}) = \mu_0\,(\delta_{il}\delta_{jm} - \delta_{im}\delta_{jl})m_l\frac{1}{3}\delta_{jm}$$

$$= \frac{2\mu_0}{3}m_i\,. \tag{6.56}$$

In the case of a point-like source that we are considering, for $r \neq 0$, the exact result for the magnetic field is given by eq. (6.39). Allowing for a possibility of a Dirac delta at the origin, the most general form of the magnetic field generated by a point-like magnetic dipole is therefore

$$B_i = \frac{\mu_0}{4\pi}\left[\frac{1}{r^3}\,(3n_i n_j - \delta_{ij})m_j + \kappa_i^B\delta^{(3)}(\mathbf{x})\right]\,, \tag{6.57}$$

and the constant vector κ_i^B can be fixed by comparing with eq. (6.56). The first term in bracket in eq. (6.57) has the same angular dependence as in the electric case and so does not contribute to the integral. Then, eq. (6.55) gives $\kappa_i^B = (8\pi/3)m_i$. The conclusion is that the magnetic field generated by a point-like magnetic dipole $\mathbf{m}$ is given by[10]

$$\mathbf{B} = \frac{\mu_0}{4\pi}\left[\frac{3(\mathbf{m}\cdot\hat{\mathbf{n}})\hat{\mathbf{n}} - \mathbf{m}}{r^3} + \frac{8\pi}{3}\,\mathbf{m}\,\delta^{(3)}(\mathbf{x})\right]\,. \tag{6.58}$$

[10]The term proportional to the Dirac delta in eq. (6.58) has an important application in quantum mechanics, where it contributes to the hyperfine splitting of the hydrogen atom.

6.4 Multipole expansion of interaction potentials

We now study the interaction potentials associated with electric and magnetic multipoles. We consider electric multipoles in an external static electric field, as well as the interaction between the multipoles of two charge distributions. We will then repeat the analysis for magnetic multipoles.

6.4.1 Electric multipoles in external field

As we found in eq. (5.90), in electrostatics the mechanical potential at fixed charges, U_E, is the same as the energy $\mathcal{E}_E$ stored in the electric field. We will then write the following equations in terms of U_E, which is the quantity that enters directly when computing the forces to which a given multipole moment is subject, or the forces between multipoles of different charge distributions, but it can be kept in mind that the same equations hold in terms of the energy $\mathcal{E}_E$ stored in the electric field.

We consider first a charge distribution $\rho(\mathbf{x})$ in an external electric field. We assume that $\rho(\mathbf{x})$ is localized inside a region V that can be enclosed in a sphere of radius d. We also assume that the external potential is generated by charges localized in a region V', at a distance $r \gg d$ from V (so, in particular, there is no overlap between V and V'). The condition $r \gg d$ implies that the external electrostatic potential $\phi_{\text{ext}}(\mathbf{x})$ varies slowly across V. We then choose an origin inside V, so that the multipole moments are defined with respect to that origin, and we expand $\phi_{\text{ext}}(\mathbf{x})$ in a Taylor series around that origin. Denoting by $(U_E)_{\text{ext}}$ the mechanical potential at fixed charge in an external field, we have $(U_E)_{\text{ext}} = (\mathcal{E}_E)_{\text{ext}}$ and the Taylor expansion of $\phi_{\text{ext}}(\mathbf{x})$ in eq. (5.39) gives

$$
\begin{aligned}
(U_E)_{\text{ext}} &= \int_V d^3x\, \rho(\mathbf{x}) \left[\phi_{\text{ext}}(0) + x^i \partial_i \phi_{\text{ext}}(0) + \frac{1}{2} x^i x^j \partial_i \partial_j \phi_{\text{ext}}(0) + \dots \right] \\
&= \phi_{\text{ext}}(0) \int_V d^3x\, \rho(\mathbf{x}) + \partial_i \phi_{\text{ext}}(0) \int_V d^3x\, \rho(\mathbf{x}) x^i \\
&\quad + \frac{1}{2} \partial_i \partial_j \phi_{\text{ext}}(0) \int_V d^3x\, \rho(\mathbf{x}) x^i x^j + \dots \\
&= \phi_{\text{ext}}(0) q - E_{i,\text{ext}}(0) d^i - \frac{1}{2} \partial_i E_{j,\text{ext}}(0) \int_V d^3x\, \rho(\mathbf{x}) x^i x^j + \dots, \quad (6.59)
\end{aligned}
$$

where $E_{i,\text{ext}} = -\partial_i \phi_{\text{ext}}$ is the external electric field. The first term is the "monopole interaction" energy, i.e., the interaction energy of the total charge q of the system with the external potential. The second term is the mechanical potential energy associated with an electric dipole in an external electric field, when the position of the volume V is identified by the value $\mathbf{x} = 0$ of the coordinate of one of its points (and the multipoles are defined with respect to that point). If we perform a rigid displacement of the charge density, in the fixed external field, the position $\mathbf{x} = 0$ is replaced by a generic position $\mathbf{x}$ (with respect to which we still define the multipoles), and eq. (6.59) becomes

$$(U_E)_{\text{ext}}(\mathbf{x}) = q\phi_{\text{ext}}(\mathbf{x}) - d^i E_{i,\text{ext}}(\mathbf{x}) - \frac{1}{2}\partial_i E_{j,\text{ext}}(\mathbf{x})\int_V d^3x'\,\rho(\mathbf{x}')x'^i x'^j + \dots . \tag{6.60}$$

The term associated with the dipole defines the potential $(U_E)_{\text{dipole}}(\mathbf{x})$,

$$\boxed{(U_E)_{\text{dipole}}(\mathbf{x}) = -\mathbf{d}\cdot\mathbf{E}_{\text{ext}}(\mathbf{x}) .} \tag{6.61}$$

According to eq. (5.88), the mechanical force exerted on the dipole is therefore[11]

$$F_k = d_i\partial_k E_{i,\text{ext}} . \tag{6.62}$$

Since, in electrostatics, $\boldsymbol{\nabla}\times\mathbf{E} = 0$, we have $\partial_k E_{i,\text{ext}} = \partial_i E_{k,\text{ext}}$, and eq. (6.62) can also be written as

$$F_k = d_i\partial_i E_{k,\text{ext}} , \tag{6.63}$$

or, in vector notation,

$$\mathbf{F} = (\mathbf{d}\cdot\boldsymbol{\nabla})\mathbf{E}_{\text{ext}} . \tag{6.64}$$

We can also use $(U_E)_{\text{dipole}}$ to find the torque acting on a dipole. Consider a rotation of the dipole by $\delta\boldsymbol{\theta}$ (where, writing $\delta\boldsymbol{\theta} = \delta\theta\,\hat{\mathbf{n}}$, $\hat{\mathbf{n}}$ defines the direction of the axis around which we perform a rotation and $\delta\theta$ the rotation angle) around the origin O with respect to which the multipoles are defined. Under this rotation the electric dipole moment changes by

$$\delta\mathbf{d} = \delta\boldsymbol{\theta}\times\mathbf{d} , \tag{6.65}$$

which is the transformation of a vector under infinitesimal rotations, see eq. (1.153). The corresponding change in $(U_E)_{\text{dipole}}$ is

$$\begin{aligned}
\delta(U_E)_{\text{dipole}} &= -(\delta\mathbf{d})\cdot\mathbf{E}_{\text{ext}} \\
&= -(\delta\boldsymbol{\theta}\times\mathbf{d})\cdot\mathbf{E}_{\text{ext}} \\
&= -(\mathbf{d}\times\mathbf{E}_{\text{ext}})\cdot\delta\boldsymbol{\theta} .
\end{aligned} \tag{6.66}$$

Just as the force is obtained from a potential from eq. (5.88), i.e., from $\delta U = -\mathbf{F}\cdot\delta\mathbf{x}$, the torque $\mathbf{N}$ is obtained from[12]

$$\delta U = -\mathbf{N}\cdot\delta\boldsymbol{\theta} . \tag{6.67}$$

Therefore, the torque on a dipole in an external electric field is

$$\mathbf{N} = \mathbf{d}\times\mathbf{E}_{\text{ext}} . \tag{6.68}$$

This force tends to align the dipole so that it is parallel to the external electric field, so that the potential (6.61) is minimized.

Note that this is the torque around the center of the dipole, i.e., in the frame where the dipole center has the position $\mathbf{x} = 0$. In a frame with a different origin of the axes, in which the dipole center has a generic

[11]Note that the dipole moment $\mathbf{d}$ of the charge distribution does not change under rigid translations of the charge distribution, since it is always defined with respect to the new, "translated" point chosen to identify the position of the volume V. The dependence on $\mathbf{x}$ enters only through a possible spatial dependence of the external electric field. If the external field is uniform, the force vanishes.

[12]Note that, to define the torque as in eq. (6.66), we compute how $(U_E)_{\text{dipole}}$ changes when we rotate the dipole with respect to a fixed electric field. Given that $(U_E)_{\text{dipole}}$ is a scalar, if we would simply rotate the reference frame, transforming both $\mathbf{d}$ and $\mathbf{E}_{\text{ext}}$ accordingly, $(U_E)_{\text{dipole}}$ would not change and $\delta(U_E)_{\text{dipole}} = 0$. This is of course the same that we do when we define the force on an object from $\delta U = -\mathbf{F}\cdot d\mathbf{x}$, where we consider the change in the potential when the position of the object changes, with respect to a fixed external field.

position $\mathbf{x}$, in addition to this there will be a torque exerted by the force $\mathbf{F}$ in eq. (6.64), that will make the dipole rotate around the new origin, so the total torque is

$$\mathbf{N} = \mathbf{d} \times \mathbf{E}_{\text{ext}} + \mathbf{x} \times \left[(\mathbf{d} \cdot \boldsymbol{\nabla}) \mathbf{E}_{\text{ext}} \right] . \tag{6.69}$$

The second term vanishes if the external electric field is uniform, while the first is present even for a uniform field.

The next term in eq. (6.59) involves the electric quadrupole. Actually, the reduced quadrupole defined in eq. (6.17) also involves a term proportional to δ_{ij}. However, $\mathbf{E}_{\text{ext}}$ is defined as the electric generated by the external charges ρ_{ext}, and therefore satisfies $\boldsymbol{\nabla} \cdot \mathbf{E}_{\text{ext}} = \rho_{\text{ext}}/\epsilon_0$. Since the external charges are localized in the volume V', which has no overlap with V, inside V we have $\boldsymbol{\nabla} \cdot \mathbf{E}_{\text{ext}} = 0$; then δ_{ij}, when contracted with $\partial_i E_{j,\text{ext}}$, gives zero. Therefore, we are free to add to the term $x^i x^j$ in eq. (6.59), the term proportional to δ_{ij} that completes the definition (6.17) of the reduces quadrupole moment. Also taking into account the factor of 3 in the relation between the reduced quadrupole moment q_{ij} and the quadrupole moment Q_{ij}, see eq. (6.18), we see that the energy associated with an electric quadrupole in an external electric field is

$$\boxed{ (U_E)_{\text{quadr}}(\mathbf{x}) = -\frac{1}{6} Q_{ij} \partial_i E_{j,\text{ext}}(\mathbf{x}) . } \tag{6.70}$$

The force exerted on the quadrupole by the external electric field is obtained from $F_k = -\partial_k (U_E)_{\text{quadr}}$, so

$$F_k = \frac{1}{6} Q_{ij} \partial_i \partial_k E_{j,\text{ext}} . \tag{6.71}$$

[13] We now need to use the fact that, under rotations, a tensor Q_{ij} with two indices transforms as

$$Q_{ij} \to R_{ik} R_{jl} Q_{kl} ,$$

see eq. (1.130). For an infinitesimal rotation, we write the rotation matrix R_{ij} as in eq. (1.152). This gives (taking into account that $Q_{ij} = Q_{ji}$)

$$Q_{ij} \to Q_{ij} - (\epsilon_{ilm} Q_{lj} + \epsilon_{jlm} Q_{li}) \delta\theta_m .$$

Then, taking again into account that, for a static external field, $\partial_i E_{j,\text{ext}} = \partial_j E_{i,\text{ext}}$, eq. (6.70) gives

$$\delta(U_E)_{\text{quadr}} = \frac{1}{3} \epsilon_{ilm} Q_{lj} \partial_j E_{i,\text{ext}} \delta\theta_m .$$

Therefore, the quadrupole contribution to the torque is

$$N_m = -\frac{1}{3} \epsilon_{ilm} Q_{lj} \partial_j E_{i,\text{ext}} , \tag{6.73}$$

which, after renaming the indices as $m \to i$, $i \to k$, $l \to j$ and $j \to l$, gives eq. (6.74).

Again, using the fact that, for static fields, $\partial_k E_{j,\text{ext}} = \partial_j E_{k,\text{ext}}$, we can rewrite this as

$$\mathbf{F} = \frac{1}{6} Q_{ij} \partial_i \partial_j \mathbf{E}_{\text{ext}} . \tag{6.72}$$

Similarly, for the total torque on a quadrupole with respect to its origin, we get [13]

$$N_i = \frac{1}{3} \epsilon_{ijk} Q_{jl} \partial_l E_{k,\text{ext}} . \tag{6.74}$$

If we denote by $\mathbf{Q} \cdot \boldsymbol{\nabla}$ the vector differential operator whose j-th component is $Q_{jl} \partial_l$, we can write this in a more compact form as

$$\mathbf{N} = \frac{1}{3} (\mathbf{Q} \cdot \boldsymbol{\nabla}) \times \mathbf{E}_{\text{ext}} . \tag{6.75}$$

Again, in a frame where the origin used to define the quadrupole moment is at a generic position $\mathbf{x}$, rather than at $\mathbf{x} = 0$, we must add to this the term $\mathbf{x} \times \mathbf{F}$, where now $\mathbf{F}$ is given by eq. (6.72), and therefore

$$\mathbf{N} = \frac{1}{3} (\mathbf{Q} \cdot \boldsymbol{\nabla}) \times \mathbf{E}_{\text{ext}} + \frac{1}{6} \mathbf{x} \times (Q_{ij} \partial_i \partial_j \mathbf{E}_{\text{ext}}) . \tag{6.76}$$

6.4.2 Interaction between the electric multipoles of two charge distributions

A description in terms of interaction of the multipole moments of a charge distribution with a given external electric field is particularly appropriate when the external field is generated by a macroscopic object, and we study its interaction with a microscopic charge distribution. If, instead, we have two localized charge distributions of similar size, e.g., two molecules, interacting among them, a symmetric treatment of the two systems can be more appropriate. In this case, we start from eq. (5.75), that we rewrite here

$$U_E^{int} = \frac{1}{4\pi\epsilon_0} \int_V d^3x \int_{V'} d^3x' \frac{\rho_1(\mathbf{x})\rho_2(\mathbf{x}')}{|\mathbf{x} - \mathbf{x}'|} . \tag{6.77}$$

We choose the origin O of the reference frame at some point inside the volume V, see Fig. 6.5; we will use this origin to define the multipole moments of the charge distribution $\rho_1(\mathbf{x})$ (recall the discussion of eqs. (6.21)–(6.25) on the dependence of multipole moments from the choice of origin). We denote by O' a fixed point inside the volume V', and we will use this origin to define the multipole moments of the charge distribution $\rho_2(\mathbf{x})$. We denote the vector from O to O' by $\mathbf{r}$. A point inside the volume V can be labeled by a vector $\mathbf{x}$ starting from the origin O. Similarly, a point inside V' can be labeled by a vector $\mathbf{y}$ starting from O'. With respect to the origin O, the coordinate $\mathbf{x}'$ of the latter point is given by $\mathbf{x}' = \mathbf{r} + \mathbf{y}$. Then, eq. (6.77) can be rewritten as

$$U_E^{int}(\mathbf{r}) = \frac{1}{4\pi\epsilon_0} \int_V d^3x \int_{V'} d^3x' \frac{\rho_1(\mathbf{x})\rho_2(\mathbf{x}')}{|\mathbf{r} - (\mathbf{x} - \mathbf{y})|} , \tag{6.78}$$

where $\mathbf{r}$ is fixed and $\mathbf{y}$ is a function of the integration variable $\mathbf{x}'$, given by $\mathbf{y} = \mathbf{x}' - \mathbf{r}$. We write it in this form, however, because we can now expand the denominator in the limit $|\mathbf{x}| \ll r$, $|\mathbf{y}| \ll r$ (where $r = |\mathbf{r}|$), corresponding to the fact that we are interested in the limit in which the linear sizes of the volumes V and V' are much smaller than r. We then expand the denominator to second order, as in eq. (6.14),

$$\frac{1}{|\mathbf{r} - (\mathbf{x} - \mathbf{y})|} = \frac{1}{r} + \frac{(x^i - y^i)r_i}{r^3} \tag{6.79}$$

$$+ \frac{1}{2r^5}(x^i - y^i)(x^j - y^j)(3r_ir_j - \delta_{ij}r^2) + \dots .$$

Then, collecting the various terms,

$$(4\pi\epsilon_0)\,U_E^{int}(\mathbf{r}) = \frac{1}{r}\int_V d^3x\,\rho_1(\mathbf{x})\int_{V'}d^3x'\,\rho_2(\mathbf{x}') \tag{6.80}$$

$$+ \frac{r_i}{r^3}\left[\int_V d^3x\,\rho_1(\mathbf{x})x^i\int_{V'}d^3x'\,\rho_2(\mathbf{x}') - \int_V d^3x\,\rho_1(\mathbf{x})\int_{V'}d^3x'\,\rho_2(\mathbf{x}')y^i\right]$$

$$+ \frac{3r_ir_j - r^2\delta_{ij}}{2r^5}\int_V d^3x\int_{V'}d^3x'\,\rho_1(\mathbf{x})\rho_2(\mathbf{x}')(x^ix^j - x^iy^j - y^ix^j + y^iy^j) .$$

Fig. 6.5 Two charge distributions localized on non-overlapping regions, and the coordinates and origins described in the text.

In the first line, we see that the integrals give the total charges q_1 and q_2 of the two charge distribution. In the second line, we recognize the electric dipole moment of the first charge distribution,

$$d_1^i = \int_V d^3x\, \rho_1(\mathbf{x}) x^i \,, \qquad (6.81)$$

defined with respect to the origin O, which is the natural choice for this charge distribution. Similarly,

$$
\begin{aligned}
d_2^i &= \int_{V'} d^3x'\, \rho_2(\mathbf{x}') y^i \\
&= \int_{V'} d^3x'\, \rho_2(\mathbf{x}')(\mathbf{x}' - \mathbf{r})^i \,,
\end{aligned} \qquad (6.82)
$$

is the dipole moment of the second charge distribution, now defined with respect to the origin O', which is the natural definition for this charge distribution.[14] In the last line of eq. (6.80), the quantity

$$\int_V d^3x\, \rho_1(\mathbf{x}) x^i x^j \qquad (6.84)$$

is the reduced quadrupole moment of the charge distribution $\rho_1(\mathbf{x})$, again with respect to the origin O, except that the term proportional to δ_{ij} in eq. (6.17) is missing. However, this expression is contracted with $3r_ir_j - r^2\delta_{ij}$, and, using $\delta_{ij}\delta_{ij} = 3$

$$(3r_ir_j - r^2\delta_{ij})\delta_{ij} = 0 \,, \qquad (6.85)$$

so we can add for free the missing term proportional to δ_{ij} and reconstruct the full expression for the reduced quadrupole moment q_1^{ij}. In the same way,

$$\int_{V'} d^3x'\, \rho_2(\mathbf{x}') y^i y^j \,, \qquad (6.86)$$

is the reduced quadrupole moment of the charge distribution $\rho_2(\mathbf{x}')$, with respect to its natural origin O' (again, apart from a term proportional to δ_{ij}, which anyhow gives zero upon contraction). So, putting everything together, and writing $q^{ij} = Q^{ij}/3$, we get

$$
\begin{aligned}
(4\pi\epsilon_0)\, U_E^{\text{int}}(\mathbf{r}) &= \frac{q_1 q_2}{r} + \frac{1}{r^3}(q_2\mathbf{d}_1\cdot\mathbf{r} - q_1\mathbf{d}_2\cdot\mathbf{r}) \\
&+ \frac{3r_ir_j - r^2\delta_{ij}}{2r^5}\left[\frac{1}{3}\left(q_1 Q_2^{ij} + q_2 Q_1^{ij}\right) - \left(d_1^i d_2^j + d_1^j d_2^i\right)\right] + \dots \,.
\end{aligned} \qquad (6.87)
$$

The first term, proportional to $q_1 q_2/r$, is a "monopole-monopole" term, i.e., the Coulomb interaction between the total electric charges of the two charge distributions. If both q_1 and q_2 are non-zero, it is the dominant term at large r. The second term, proportional to $(q_2\mathbf{d}_1\cdot\mathbf{r} - q_1\mathbf{d}_2\cdot\mathbf{r})$, gives the interaction between the charge of a distribution and the electric dipole of the other, and is therefore a "monopole-dipole" term. Note that, as all other terms, it is symmetric under the exchange of the two

[14]Equivalently, introducing the charge distribution $\rho_2^{(0)}(\mathbf{x})$ as in eq. (5.77), we have $\rho_2(\mathbf{x}') = \rho_2^{(0)}(\mathbf{y})$ and, since $d^3x' = d^3y$, eq. (6.82) reads

$$d_2^i = \int_{V'} d^3y\, \rho_2^{(0)}(\mathbf{y}) y^i \,. \qquad (6.83)$$

charge distribution, $1 \leftrightarrow 2$ (observing that, under such an exchange, $\mathbf{r} \to -\mathbf{r}$). In the second line we have a "monopole-quadrupole" term, and a "dipole-dipole" term. As long as q_1 and q_2 are non-zero, at large distances, where this expansion is valid, the monopole-monopole term, which is of order $1/r$, dominates over the monopole-dipole term, which is of order $1/r^2$, and this in turn dominates over the monopole-quadrupole and dipole-dipole terms, which are of order $1/r^3$. However, if both localized charge distributions have an overall zero charge, $q_1 = q_2 = 0$, the dominant term becomes the dipole-dipole interaction,

$$(U_E)_{\text{dipole-dipole}} = \frac{1}{4\pi\epsilon_0} \frac{r^2 \delta_{ij} - 3r_i r_j}{2r^5} \left(d_1^i d_2^j + d_1^j d_2^i \right) . \tag{6.88}$$

Performing the contraction of indices and using $\hat{\mathbf{r}} = \mathbf{r}/r$, this gives

$$\boxed{(U_E)_{\text{dipole-dipole}} = \frac{1}{4\pi\epsilon_0} \frac{\mathbf{d}_1 \cdot \mathbf{d}_2 - 3(\mathbf{d}_1 \cdot \hat{\mathbf{r}})(\mathbf{d}_2 \cdot \hat{\mathbf{r}})}{r^3} .} \tag{6.89}$$

Observe that the interaction between dipoles can be attractive or repulsive, depending on the relative orientation of the dipoles and on their relative direction with respect to the vector $\mathbf{r}$ joining them. For instance, if the dipoles are orthogonal to $\hat{\mathbf{r}}$, so that $\mathbf{d}_1 \cdot \hat{\mathbf{r}} = \mathbf{d}_2 \cdot \hat{\mathbf{r}} = 0$, the interaction is repulsive when the dipoles are parallel and attractive when they are antiparallel. If, instead, $\mathbf{d}_1$ and $\mathbf{d}_2$ are aligned or antialigned with $\hat{\mathbf{r}}$, the interaction is attractive when the dipoles are parallel and repulsive when they are antiparallel.

Equation (6.89) could have also been obtained more simply, considering the interaction between the electric dipole of the second charge distribution with the external electric field created by the first. According to eq. (6.61), this is

$$(U_E)_{\text{dipole}} = -\mathbf{d}_2 \cdot \mathbf{E}_1 . \tag{6.90}$$

We then substitute the electric field generated by the electric dipole of the first charge distribution at the position O', that, according to eq. (6.13), is given by

$$\mathbf{E}_1 = \frac{1}{4\pi\epsilon_0} \frac{3(\mathbf{d}_1 \cdot \hat{\mathbf{r}})\hat{\mathbf{r}} - \mathbf{d}_1}{r^3} , \tag{6.91}$$

and we get again eq. (6.89). The expansion (6.87), however, provides the cleanest way of understanding the structure of the expansion and computing all terms systematically. These results have been obtained from an expansion at large distances, compared to the size of the charge distributions. For the interaction among two point-like electric dipoles, we must also add the Dirac delta in eq. (6.53), and therefore

$$(U_E)_{\text{dipole-dipole}} = \frac{1}{4\pi\epsilon_0} \frac{\mathbf{d}_1 \cdot \mathbf{d}_2 - 3(\mathbf{d}_1 \cdot \hat{\mathbf{r}})(\mathbf{d}_2 \cdot \hat{\mathbf{r}})}{r^3} + \frac{1}{3\epsilon_0} \mathbf{d}_1 \cdot \mathbf{d}_2 \delta^{(3)}(\mathbf{r}) , \tag{6.92}$$

where $\mathbf{r} = \mathbf{x}_2 - \mathbf{x}_1$ is the relative distance between the two point-like electric dipoles.

6.4.3   Interactions of magnetic multipoles

We next turn to the interactions involving the magnetic multipole moments. To compute the mechanical forces, such as those due for instance to the interaction of a magnetic dipole with an external field, or between two magnetic dipoles, the most convenient quantity is the mechanical potential $\hat{U}_B$ introduced in eq. (5.136), since from it we can obtain the mechanical forces by taking spatial derivatives at fixed currents, as in eq. (5.146), and therefore keeping the magnetic moments fixed. However, one should keep in mind that $\hat{U}_B = -\mathcal{E}_B$, see eq. (5.139), so the corresponding formulas for the magnetic energy have the opposite sign. In the following, we will work in terms of $\hat{U}_B$ rather than $\mathcal{E}_B$.

We limit ourselves to the magnetic dipole term, since higher-order magnetic multipoles are rarely encountered in practical applications. We proceed similarly to what we did for the electric dipole. We now start from eq. (5.135), that we write in the form

$$\hat{U}_B^{\rm int} = -\int d^3x\, \mathbf{j}_1(\mathbf{x})\mathbf{A}_2(\mathbf{x})\,, \tag{6.93}$$

where, from eq. (4.92)

$$\mathbf{A}_2(\mathbf{x}) = \frac{\mu_0}{4\pi}\int d^3x'\, \frac{\mathbf{j}_2(\mathbf{x}')}{|\mathbf{x}-\mathbf{x}'|}\,. \tag{6.94}$$

To stress that we consider the current distribution $\mathbf{j}_2$ as an external source from the point of view of the current distribution $\mathbf{j}_1$, we change notation writing $\mathbf{j}_1(\mathbf{x}) = \mathbf{j}(\mathbf{x})$, $\mathbf{j}_2(\mathbf{x}) = \mathbf{j}_{\rm ext}(\mathbf{x})$, $\mathbf{A}_2(\mathbf{x}) = \mathbf{A}_{\rm ext}(\mathbf{x})$ and we denote the interaction potential $\hat{U}_B^{\rm int}$ as $(\hat{U}_B)_{\rm ext}$, in analogy with the notation used in eq. (6.59) for the electrostatic case. Then

$$(\hat{U}_B)_{\rm ext} = -\int d^3x\, \mathbf{j}(\mathbf{x})\mathbf{A}_{\rm ext}(\mathbf{x})\,. \tag{6.95}$$

We assume that $\mathbf{j}(\mathbf{x})$ is localized in a finite volume V, and that $\mathbf{A}_{\rm ext}(\mathbf{x})$ varies slowly across V. Similarly to what we did in eq. (6.59), we choose an origin inside V, that we use to define the multipole moments,[15] and we expand $\mathbf{A}_{\rm ext}(\mathbf{x})$ around that origin. Then, to first order, eq. (6.95) becomes

$$(\hat{U}_B)_{\rm ext} = -A_{i,{\rm ext}}(0)\int d^3x\, j_i(\mathbf{x}) - \partial_k A_{i,{\rm ext}}(0)\int d^3x\, j_i(\mathbf{x})x_k + \ldots\,. \tag{6.96}$$

The first term vanishes because of eq. (6.31), while the second term is transformed using eq. (6.37). This gives

$$\begin{aligned}(\hat{U}_B)_{\rm ext} &= \epsilon_{ikl}\partial_k A_{i,{\rm ext}}(0)m_l + \ldots\\ &= -\epsilon_{ikl}\partial_i A_{k,{\rm ext}}(0)m_l + \ldots\\ &= -B_l(0)m_l + \ldots\,,\end{aligned} \tag{6.97}$$

and the first term in this expansion defines the interaction of a magnetic dipole with the external magnetic field, $(\hat{U}_B)_{\rm dipole}$. As in the electrostatic case, we perform a rigid displacement of the current density, in

[15] Although, as discussed in Note 5 on page 140, in magnetostatics the magnetic dipole is independent of the choice of the origin, and the dependence only starts from the magnetic quadrupole, that we will not include here.

the fixed external field, so that the position $\mathbf{x} = 0$ is replaced by a generic position $\mathbf{x}$, with respect to which we still define the multipoles. Therefore, in vector form,

$$(\hat{U}_B)_{\text{dipole}}(\mathbf{x}) = -\mathbf{m}\cdot\mathbf{B}_{\text{ext}}(\mathbf{x})\,.$$

(6.98)

Equation (6.98) can be compared to eq. (6.61) for the electric dipole. The computation of the force and torque on a magnetic dipole is then completely analogous to that performed for the electric dipole, taking into account that $\hat{U}_B$ (at fixed currents, and therefore at fixed magnetic moment) plays the role that U_E plays in electrostatics (at fixed charges). The force exerted on a magnetic dipole by an external magnetic field is obtained from eq. (5.146) and is

$$F_k = m_i \partial_k B_{i,\text{ext}}\,,$$

(6.99)

to be compared to eq. (6.62). If the current $\mathbf{j}_{\text{ext}}$ sourcing the external magnetic field has no overlap with the current distribution $\mathbf{j}$ that gives rise to the magnetic dipole $\mathbf{m}$ on which we are computing the force, then, from $\nabla\times\mathbf{B}_{\text{ext}} = \mu_0\mathbf{j}_{\text{ext}}$ it follows that $\nabla\times\mathbf{B}_{\text{ext}} = 0$ in the region where we compute the force, and eq. (6.99) can also be written as $F_k = m_i \partial_i B_{k,\text{ext}}$ or, in vector notation,

$$\mathbf{F} = (\mathbf{m}\cdot\nabla)\mathbf{B}_{\text{ext}}\,,$$

(6.100)

to be compared to eq. (6.64). The torque acting on a magnetic dipole at the origin, due to an external magnetic field, is obtained exactly as in the derivation of eq. (6.68), and is

$$\mathbf{N} = \mathbf{m}\times\mathbf{B}_{\text{ext}}\,,$$

(6.101)

and tends to orient the magnetic dipole so that it aligns with the external magnetic field, thereby minimizing $(\hat{U}_B)_{\text{dipole}}$. For a magnetic dipole located in a generic point $\mathbf{x}$, we must add to this torque the term $\mathbf{x}\times\mathbf{F}$, where $\mathbf{F}$ is given by eq. (6.99) [which, when $\nabla\times\mathbf{B}_{\text{ext}} = 0$ in the region under consideration, can also be written as in eq. (6.100)], to be compared to eq. (6.69).

Finally, we consider the interaction between the magnetic multipoles of two different current densities, similarly to the discussion in Section 6.4.2 for the electric case. However, in this case the "monopole" term is absent, and we are not interested in going beyond the magnetic dipole, since higher-order magnetic multipoles rarely appear. Therefore, the only interaction term that we need is the dipole-dipole term. Rather than performing the full expansion of $\hat{U}_B^{\text{int}}$ similarly to what we have done for U_E^{int} in eqs. (6.80)–(6.87), it is then simpler to proceed as in eqs. (6.90) and (6.91): we write

$$(\hat{U}_B)_{\text{dipole}} = -\mathbf{m}_2\cdot\mathbf{B}_1\,,$$

(6.102)

for the interaction potential between the magnetic dipole $\mathbf{m}_2$ of the second current distribution and the magnetic field $\mathbf{B}_1$ generated by the magnetic moment of the first current distribution, and, for the latter, we use eq. (6.40),

$$\mathbf{B}_1 = \frac{\mu_0}{4\pi} \frac{3(\mathbf{m}_1 \cdot \hat{\mathbf{n}})\hat{\mathbf{n}} - \mathbf{m}_1}{r^3}\,. \tag{6.103}$$

We then obtain

$$(\hat{U}_B)_{\text{dipole}-\text{dipole}} = \frac{\mu_0}{4\pi} \frac{\mathbf{m}_1 \cdot \mathbf{m}_2 - 3(\mathbf{m}_1 \cdot \hat{\mathbf{n}})(\mathbf{m}_2 \cdot \hat{\mathbf{n}})}{r^3}\,, \tag{6.104}$$

to be compared to eq. (6.89). Once again, for the interaction between two point-like magnetic dipoles, we must also add the Dirac delta in eq. (6.58), so that

$$(\hat{U}_B)_{\text{dipole}-\text{dipole}} = \frac{\mu_0}{4\pi} \frac{\mathbf{m}_1 \cdot \mathbf{m}_2 - 3(\mathbf{m}_1 \cdot \hat{\mathbf{n}})(\mathbf{m}_2 \cdot \hat{\mathbf{n}})}{r^3} - \frac{2\mu_0}{3} \mathbf{m}_1 \cdot \mathbf{m}_2\, \delta^{(3)}(\mathbf{r})\,, \tag{6.105}$$

where $\mathbf{r} = \mathbf{x}_2 - \mathbf{x}_1$ is the relative distance between the two point-like magnetic dipoles, to be compared to eq. (6.92).

6.5 Solved problems

Problem 6.1. Larmor precession

The torque gives the rate of change of the angular momentum $\mathbf{L}$, as

$$\mathbf{N} = \frac{d\mathbf{L}}{dt}\,. \tag{6.106}$$

As an application, recall from eq. (6.43) that, for a non-relativistic particle with charge q_a, mass m_a and angular momentum $\mathbf{L}$, the magnetic dipole moment is $\mathbf{m} = (q_a/2m_a)\,\mathbf{L}$. We write, more generally,

$$\mathbf{m} = \gamma_a \mathbf{L}\,, \tag{6.107}$$

(so that this equation applies also to the spin angular momentum, for which the proportionality constant is different, see Note 8 on page 141).[16] Equations (6.101) and (6.106) then give

[16]The subscript a in γ_a is unconventional, but helps to avoid confusion with the Lorentz γ factor, that will instead appear in the formula for the frequency at which the position a charged particle rotates in a magnetic field, see eq. (8.201) in Solved Problem 8.4.

$$\frac{d\mathbf{L}}{dt} = \gamma_a \mathbf{L} \times \mathbf{B}_{\text{ext}}\,, \tag{6.108}$$

which give the evolution of the angular momentum (or, equivalently, of the magnetic moment) in an external magnetic field. Performing the scalar product of both sides of eq. (6.108) with $\mathbf{L}$ we get

$$\mathbf{L} \cdot \frac{d\mathbf{L}}{dt} = 0\,, \tag{6.109}$$

since $\mathbf{L} \cdot (\mathbf{L} \times \mathbf{B}_{\text{ext}}) = 0$. This can be rewritten as

$$\frac{1}{2} \frac{d(\mathbf{L} \cdot \mathbf{L})}{dt} = 0\,, \tag{6.110}$$

and therefore the modulus $|\mathbf{L}|$ is constant. Similarly, multiplying eq. (6.108) by $\mathbf{B}_{\text{ext}}$ and using $\mathbf{B}_{\text{ext}} \cdot (\mathbf{L} \times \mathbf{B}_{\text{ext}}) = 0$, we get $\mathbf{B}_{\text{ext}} \cdot d\mathbf{L}/dt = 0$. Therefore, also the component of **L** parallel to **B** is conserved, and only the components of **L** orthogonal to **B** change. Setting $\mathbf{B}_{\text{ext}} = B\hat{\mathbf{z}}$, eq. (6.108) gives

$$\frac{dL_x}{dt} = -\omega_L L_y \,, \tag{6.111}$$

$$\frac{dL_y}{dt} = \omega_L L_x \,, \tag{6.112}$$

$$\frac{dL_z}{dt} = 0 \,, \tag{6.113}$$

where

$$\omega_L = -\gamma_a B \tag{6.114}$$

is called the *Larmor frequency*. When $\gamma_a = q_a/2m_a$, as for a classical particle of charge q_a and mass m_a,

$$\omega_L = -\frac{q_a B}{2m_a} \,. \tag{6.115}$$

The minus sign in the definition of ω_L is inserted so that, for electrons, where $q_a = -e < 0$, we have $\omega_L > 0$. The solution of eqs. (6.111) and (6.112) is

$$L_x = L_\perp \cos(\omega_L t + \varphi) \,, \tag{6.116}$$

$$L_y = L_\perp \sin(\omega_L t + \varphi) \,, \tag{6.117}$$

where $L_\perp = (L_x^2 + L_y^2)^{1/2}$ is the constant value of the projection of **L** on the (x, y) plane, and φ is a phase (that can be reabsorbed into a choice of the origin for t). So, the vector $\mathbf{L}_\perp = L_x\hat{\mathbf{x}} + L_y\hat{\mathbf{y}}$ rotates in the (x, y) plane at the Larmor frequency (counterclockwise, if $\omega_L > 0$), while L_z stays constant. This behavior is known as the *Larmor precession*. The factor γ_a, when Larmor precession is applied to the spin of an elementary particle with mass m_a and charge q_a, must be written as

$$\gamma_a = \frac{g_a q_a}{2m_a} \,, \tag{6.118}$$

where g_a is the g-factor of the particle (see again Note 8 on page 141). The constant γ_a is called the *gyromagnetic ratio* of the particle.[17]

[17]Note that the name "gyromagnetic ratio" is also often used instead for the dimensionless g-factor.

Special Relativity

7

We now introduce the basic postulates and the formalism of Special Relativity. Special Relativity is one of the pillars of modern physics. Our presentation will be tuned toward understanding how Special Relativity is hidden into Maxwell's equations, and in building a "covariant" formalism that makes this symmetry explicit.

7.1 The postulates

To introduce the postulates of Special Relativity, we first need to define *inertial frames*. These are reference frames defined by the condition that, in these frames, a body on which no external force acts moves with constant speed $\mathbf{v}$ (constant both in modulus and in direction). The special theory of relativity, as formulated by Einstein in 1905, is based on two postulates:

(1) *Principle of relativity:* the laws of nature are the same in all inertial frames.

(2) *Constancy of the speed of light:* the speed of light has the same value in all inertial frames.

Newtonian physics also has a relativity principle, that we now call *Galilean Relativity*, that again states that the laws of Newtonian physics are the same in all coordinate systems moving at uniform speed relative to one another.[1] Given two reference frames K, with coordinates $(t, \mathbf{x})$,

---

[1] From Galileo's 1632 book *Dialogue Concerning the Two Chief World Systems*, (Second Day), translated by S. Drake, University of California Press, 1953 (taken from https://en.wikipedia.org/wiki/Galileo\%27s_ship.) "Shut yourself up with some friend in the main cabin below decks on some large ship, and have with you there some flies, butterflies, and other small flying animals. Have a large bowl of water with some fish in it; hang up a bottle that empties drop by drop into a wide vessel beneath it. With the ship standing still, observe carefully how the little animals fly with equal speed to all sides of the cabin. The fish swim indifferently in all directions; the drops fall into the vessel beneath; and, in throwing something to your friend, you need throw it no more strongly in one direction than another, the distances being equal; jumping with your feet together, you pass equal spaces in every direction. When you have observed all these things carefully (though doubtless when the ship is standing still everything must happen in this way), have the ship proceed with any speed you like, so long as the motion is uniform and not fluctuating this way and that. You will discover not the least change in all the effects named, nor could you tell from any of them whether the ship was moving or standing still. In jumping, you will pass on the floor the same spaces as before, nor will you make larger jumps toward the stern than toward the prow even though the ship is moving quite rapidly, despite the

and K', with coordinates $(t', \mathbf{x}')$, such that the origin of K' moves with respect of the origin of K with speed $-\mathbf{v}_0$, in Galilean Relativity the space and time coordinates of the two frames are related by

$$t' = t, \qquad \mathbf{x}' = \mathbf{x} + \mathbf{v}_0 t \qquad (7.1)$$

(with a suitable choice for the origin in space and time, such that $t = 0$ corresponds to $t' = 0$ and, at $t = 0$, the point $\mathbf{x} = 0$ corresponds to $\mathbf{x}' = 0$). The laws of Newtonian mechanics are invariant under these transformations. Note that, in Newtonian mechanics, time is absolute, and is the same in all reference frames, i.e., for all observers.[2] In Galilean Relativity, from eq. (7.1), if a particle in the frame K moves along the trajectory $\mathbf{x}(t)$, so that its velocity is $\mathbf{v}(t) = d\mathbf{x}(t)/dt$, in the frame K' it will move on the trajectory $\mathbf{x}'(t) = \mathbf{x}(t) + \mathbf{v}_0 t$ and will therefore have a velocity $\mathbf{v}'(t) = d\mathbf{x}'(t)/dt$, such that

$$\mathbf{v}'(t) = \mathbf{v}(t) + \mathbf{v}_0 . \qquad (7.2)$$

Thus, according to Galilean Relativity, if in the frame K the speed of a light beam traveling along the $\hat{\mathbf{x}}$ axis is $\mathbf{v} = c\hat{\mathbf{x}}$, and the frame K' is related to K by a velocity transformation (7.1) along the x axis, i.e., $\hat{\mathbf{v}}_0 = v_0 \hat{\mathbf{x}}$, then in the frame K' the light beam should travel at the speed $(c + v_0)\hat{\mathbf{x}}$. Thus, the second postulate of Special Relativity marks the difference with Galilean Relativity and, as we will see more formally in the following, implies the end of the absolute notion of time.

We now develop the mathematical consequences of the two postulates of Special Relativity. Consider two inertial frames: K, with coordinates $\mathbf{x} = (x, y, z)$, and K', with coordinates $\mathbf{x}' = (x', y', z')$. We denote by t time measured in the K frame and by t' that in the K' frame. We do not assume a priori $t' = t$. The correct relation will emerge from the two basic postulates. Suppose that, in the frame K, a flash of light is emitted a time t_1 at the position (x_1, y_1, z_1) and is subsequently absorbed at time t_2 at the position (x_2, y_2, z_2). The fact that light moves at the speed c

[2] Of course, different observers can use different origins for time, so that, in general $t' = t + t_0$. This reflects another invariance of Newtonian mechanics, invariance under time translations, which is related to energy conservation. When one says that, in Newtonian mechanics, time is absolute, one means that *time differences* are the same for all observers in relative uniform motion.

fact that during the time that you are in the air the floor under you will be going in a direction opposite to your jump. In throwing something to your companion, you will need no more force to get it to him whether he is in the direction of the bow or the stern, with yourself situated opposite. The droplets will fall as before into the vessel beneath without dropping toward the stern, although while the drops are in the air the ship runs many spans. The fish in their water will swim toward the front of their bowl with no more effort than toward the back, and will go with equal ease to bait placed anywhere around the edges of the bowl. Finally the butterflies and flies will continue their flights indifferently toward every side, nor will it ever happen that they are concentrated toward the stern, as if tired out from keeping up with the course of the ship, from which they will have been separated during long intervals by keeping themselves in the air. And if smoke is made by burning some incense, it will be seen going up in the form of a little cloud, remaining still and moving no more toward one side than the other. The cause of all these correspondences of effects is the fact that the ship's motion is common to all the things contained in it, and to the air also. That is why I said you should be below decks; for if this took place above in the open air, which would not follow the course of the ship, more or less noticeable differences would be seen in some of the effects noted."

implies that

$$(x_1 - x_2)^2 + (y_1 - y_2)^2 + (z_1 - z_2)^2 - c^2(t_1 - t_2)^2 = 0 \,. \qquad (7.3)$$

In the frame K' light will be emitted a time t_1' at the position (x_1', y_1', z_1') and absorbed at time t_2' at the position (x_2', y_2', z_2'). Since, according to the second postulate, also in K' light propagates with the speed c, we have

$$(x_1' - x_2')^2 + (y_1' - y_2')^2 + (z_1' - z_2')^2 - c^2(t_1' - t_2')^2 = 0 \,. \qquad (7.4)$$

We define the *interval* s^2 between the two events as

$$\boxed{s^2 = -c^2(t_1 - t_2)^2 + (x_1 - x_2)^2 + (y_1 - y_2)^2 + (z_1 - z_2)^2 \,.}$$

$$(7.5)$$

We have therefore found that the interval between two events related by light propagation is zero, in all inertial frames. Note that the interval between two arbitrary events in general will not be zero: for example, for events along the path of a particle moving in straight line at a speed $v < c$ we have

$$(x_1 - x_2)^2 + (y_1 - y_2)^2 + (z_1 - z_2)^2 = v^2(t_1 - t_2)^2 \,, \qquad (7.6)$$

and therefore

$$
\begin{aligned}
s^2 &= (x_1 - x_2)^2 + (y_1 - y_2)^2 + (z_1 - z_2)^2 - c^2(t_1 - t_2)^2 \\
&= (v^2 - c^2)(t_1 - t_2)^2 < 0 \,.
\end{aligned}
\qquad (7.7)
$$

We can distinguish three cases:

(1) Light-like interval: $s^2 = 0$, as for the flash of light discussed above.

(2) Time-like interval: $s^2 < 0$, as is eq. (7.7). Such events correspond to the motion of particles traveling at $v < c$. This is, in particular, the case for two events happening at the same point in space, at succesive values of time, i.e., at spatial separation $\Delta \mathbf{x} = 0$, and $t_2 \neq t_1$.

(3) Space-like interval: $s^2 > 0$. This is, for instance, the case of two events such that $t_1 = t_2$ but $\mathbf{x}_1 \neq \mathbf{x}_2$. Such events cannot be joined by the trajectory of a particle moving with speed $v \leq c$. We say that they are causally disconnected, because, as we will discuss in Section 7.2.2, the first event cannot influence the second event, and vice versa.

The relation between the space-time coordinates $(t, \mathbf{x})$ of K and the space-time coordinates $(t', \mathbf{x}')$ of K' must therefore be such that, when the interval between two events is zero in K, it must also be zero in K'. We now show that, in fact, this relation must be such that, even for non-zero intervals, the interval must be the same in the two frames.[3]

[3]We follow here Section 2 of the old classic Landau and Lifschits (1975).

To this purpose, it is convenient to work with infinitesimal intervals. In the K frame, the interval between an event at (t, x, y, z) and an event at $(t + dt, x + dx, y + dy, z + dz)$ is

$$ds^2 = -c^2 dt^2 + dx^2 + dy^2 + dz^2 \,. \qquad (7.8)$$

In the frame K', the two events will have coordinates (t', x', y', z') and $(t' + dt', x' + dx', y' + dy', z' + dz')$, and the interval between them is $ds'^2 = -c^2 dt'^2 + dx'^2 + dy'^2 + dz'^2$. Since ds^2 and ds'^2 are infinitesimals of the same order, we must have

$$ds'^2 = a \, ds^2 \,, \qquad (7.9)$$

for some coefficient a. Because of the invariance under spatial and temporal translations (i.e., of the fact that there is no privileged position in space nor a privileged origin of time) the coefficient a cannot depend on the value (t, x, y, z) of the first event that enters in ds^2 (nor of the coordinates of the second event, that, furthermore, only differ infinitesimally from the first), and therefore can only depend on the relative velocity $\mathbf{v}$ between the two frames K and K'. Furthermore, because of the invariance under rotations (i.e., the isotropy of space) it can actually depend only on the modulus $v = |\mathbf{v}|$. Consider now three reference frames K_1, K_2, K_3 and denote by $\mathbf{v}_{12}$ the relative velocity of K_2 with respect to K_1, by $\mathbf{v}_{13}$ the relative velocity of K_3 with respect to K_1, and by $\mathbf{v}_{23}$ the relative velocity of K_3 with respect to K_2. Similarly, we denote by ds_1^2, ds_2^2 and ds_3^2 the respective intervals. From eq. (7.9) we have

$$ds_2^2 = a(v_{12}) ds_1^2 \,, \qquad ds_3^2 = a(v_{13}) ds_1^2 \,, \qquad ds_3^2 = a(v_{23}) ds_2^2 \,. \quad (7.10)$$

Combining these expressions we get

$$a(v_{13}) = a(v_{12}) a(v_{23}) \,. \qquad (7.11)$$

However, $v_{13} = |\mathbf{v}_{13}|$ depends not only on $v_{12} = |\mathbf{v}_{12}|$ and on $v_{23} = |\mathbf{v}_{23}|$, but also on the angle between the vectors $\mathbf{v}_{12}$ and $\mathbf{v}_{23}$.[4] This angle does not appear on the right-hand side of eq. (7.11) and therefore the only possible solution of eq. (7.11) is that a does not depend on the velocity at all and is just a constant. Then, eq. (7.11) reduces to $a^2 = a$, which has the solutions $a = 0, 1$. The solution $a = 0$ is clearly not acceptable, so we get $a = 1$. Thus, the relation between the coordinates $(t, \mathbf{x})$ of K and the coordinates $(t', \mathbf{x}')$ of K' must be such that, for all events (light-like, space-like, or time-like),

$$ds'^2 = ds^2 \,. \qquad (7.12)$$

[4] As we will see below, in the limit of velocities small with respect to the speed of light we recover the composition of velocities of Special Relativity, $\mathbf{v}_{13} = \mathbf{v}_{12} + \mathbf{v}_{23}$. However, we do not need to use this relation (that, as we will see, is not valid for generic velocities), but only the fact that v_{13} depends on the angle between the vectors $\mathbf{v}_{12}$ and $\mathbf{v}_{23}$, independently of the specific form of this dependence.

From the equality of the infinitesimal intervals also follows the equality of the finite intervals, so $s^2 = s'^2$. In conclusion, *from the two postulates it follows that the laws of Nature must be invariant under the transformations that leave invariant the interval (7.5) between two events.*

7.2   Space and time in Special Relativity

7.2.1   Lorentz transformations

We now identify the set of transformations that leave invariant the interval (7.5), which, without loss of generality, we can write as

$$s^2 = -c^2 t^2 + x^2 + y^2 + z^2 , \tag{7.13}$$

having set $t' = 0$ and $\mathbf{x}' = 0$. The set of transformations that leaves this expression invariant forms a group. As we saw in Section 1.7, the group that leaves the quadratic form $x^2 + y^2 + z^2$ invariant is the rotation group in three dimensions, $SO(3)$ (apart from the parity transformations). Similarly, the group of transformations that leaves invariant the quadratic form (7.13) is called the *Lorentz group*,[5] and the corresponding transformation are called *Lorentz transformations*. First of all, we see that rotations of the spatial coordinates, i.e., transformations of the form

$$t \to t' = t , \qquad x_i \to x_i' = R_{ij} x_j , \tag{7.14}$$

where R is the rotation matrix that we introduced in Section 1.6, leave the interval (7.13) invariant, since they do not touch time and they transform the spatial coordinates in such a way that $x^2 + y^2 + z^2$ is invariant. In three dimensions, the most general rotation can be expressed as a combination of a rotation around the z axis, i.e., in the (x, y) plane, a rotation around the x axis (i.e., in the (y, z) plane) and a rotation around the y axis, so in the (x, z) plane. For instance, a rotation around the z axis has the form

$$
\begin{aligned}
x \to x' &= x \cos\theta - y \sin\theta , & (7.15) \\
y \to y' &= x \sin\theta + y \cos\theta . & (7.16)
\end{aligned}
$$

Since rotations form a group, they are a subgroup of the Lorentz group. It is convenient to introduce $x^0 \equiv ct$, which has dimensions of length, just as the x^i, and to define the *four-vector* x^μ, with components (x^0, x, y, z) (or, for uniformity of notation, (x^0, x^1, x^2, x^3), so the "Lorentz index" μ takes the values $\{0, 1, 2, 3\}$). Just as we did for vectors, we will actually define four-vectors in terms of their transformations under the action of the Lorentz group. Let us begin by observing that, under a rotation in the (x, y) plane, $x^\mu \to x'^\mu = (x'^0, x', y', z')$, where

$$
\begin{pmatrix} x'^0 \\ x' \\ y' \\ z' \end{pmatrix}
=
\begin{pmatrix}
1 & 0 & 0 & 0 \\
0 & \cos\theta & -\sin\theta & 0 \\
0 & \sin\theta & \cos\theta & 0 \\
0 & 0 & 0 & 1
\end{pmatrix}
\begin{pmatrix} x^0 \\ x \\ y \\ z \end{pmatrix}
\tag{7.17}
$$

We can similarly write all other rotations so, denoting by Λ the 4×4 matrix of Lorentz transformations, and by R a generic 3×3 matrix describing a rotation, rotations are a special case of Lorentz transforma-

[5] We will further refine the definition of the Lorentz group later, by eliminating the discrete parity transformations.

tions, of the form

$$\Lambda = \left(\begin{array}{c|ccc} 1 & 0 & 0 & 0 \\ \hline 0 & & & \\ 0 & & R & \\ 0 & & & \end{array}\right) .$$
(7.18)

By analogy with rotations in a plane, it is also easy to find other transformations that leave the interval (7.13) invariant. We can consider for instance a transformation that does not act on y and z, and that leaves $(x^0)^2 - x^2$ invariant. This has the form of a "hyperbolic rotation"

$$\begin{aligned} x^0 \to x'^0 &= x^0 \cosh \zeta + x \sinh \zeta , \\ x \to x' &= x^0 \sinh \zeta + x \cosh \zeta , \end{aligned}$$
(7.19)

where ζ ranges in the interval $-\infty < \zeta < +\infty$ and (especially in a particle physics context) is called the *rapidity*. In matrix form

$$\begin{pmatrix} x'^0 \\ x' \\ y \\ z \end{pmatrix} = \begin{pmatrix} \cosh \zeta & \sinh \zeta & 0 & 0 \\ \sinh \zeta & \cosh \zeta & 0 & 0 \\ 0 & 0 & 1 & 0 \\ 0 & 0 & 0 & 1 \end{pmatrix} \begin{pmatrix} x^0 \\ x \\ y \\ z \end{pmatrix} .$$
(7.20)

A transformation of the form (7.19) is called a *Lorentz boost* along the x axis. We can similarly perform a hyperbolic rotation in the (t, y) and in the (t, z) planes. Thus, we have found six independent transformations that leaves the quadratic form (7.13) invariant, corresponding to three rotations and three hyperbolic rotations. We will see below that this exhausts the set of (proper) Lorentz transformations. First, let us understand the physical meaning of eq. (7.19). We introduce v_0 from

$$v_0 = c \tanh \zeta .$$
(7.21)

Since $-\infty < \zeta < +\infty$, we have $-c < v_0 < c$. Then eq. (7.19) can be rewritten as

$$\begin{aligned} x^0 \to x'^0 &= \gamma(v_0) \left(x^0 + \frac{v_0}{c} x \right) , \end{aligned}$$
(7.22)

$$\begin{aligned} x \to x' &= \gamma(v_0) \left(x + \frac{v_0}{c} x^0 \right) , \end{aligned}$$
(7.23)

or, using $t = x^0/c$ instead of x^0,

$$\begin{aligned} t \to t' &= \gamma(v_0) \left(t + \frac{v_0}{c^2} x \right) , \end{aligned}$$
(7.24)

$$\begin{aligned} x \to x' &= \gamma(v_0) \left(x + v_0 t \right) , \end{aligned}$$
(7.25)

where we have introduced the "gamma factor"[6]

$$\gamma(v) = \frac{1}{\sqrt{1 - (v/c)^2}} .$$
(7.26)

[6]Typically, when there will be no possibility of confusion, we will denote $\gamma(v)$ simply by γ.

We see that, in the limit $v_0/c \to 0$, eqs. (7.24) and (7.25) reduce to the transformation (7.1) of Galilean Relativity! Thus, from the point of view of Special Relativity, the apparent validity of Galilean Relativity in Newtonian physics, and in everyday experience, is due to the fact that we usually deal with velocities which are very small compared to the speed of light.

From eqs. (7.24) and (7.25) we can obtain the corresponding composition of velocities. Consider a particle that, with respect to an observer that uses coordinates (t, x, y, z), moves with velocity $\mathbf{v} = (v_x, v_y, v_z)$. In this frame, in a time dt its coordinates will change by an amount $d\mathbf{x} = \mathbf{v}dt$, i.e., $dx = v_x dt$, $dy = v_y dt$, and $dz = v_z dt$. Then, with respect to an observer that uses coordinates (t', x', y', z'), with the x' axis parallel to the x axis, it moves by an amount dx' in a time dt', where

$$dt' = \gamma(v_0) \left(dt + \frac{v_0}{c^2} dx \right), \tag{7.27}$$

$$dx' = \gamma(v_0)(dx + v_0 dt), \tag{7.28}$$

while $dy' = dy$ and $dz' = dz$. From this, using $\mathbf{v}' = d\mathbf{x}'/dt'$ and $\mathbf{v} = d\mathbf{x}/dt$, we get

$$v_x' = \frac{v_x + v_0}{1 + \frac{v_x v_0}{c^2}}, \tag{7.29}$$

$$v_y' = \frac{v_y}{\gamma(v_0) \left(1 + \frac{v_x v_0}{c^2} \right)}, \tag{7.30}$$

$$v_z' = \frac{v_z}{\gamma(v_0) \left(1 + \frac{v_x v_0}{c^2} \right)}. \tag{7.31}$$

In the limit $c \to \infty$ (i.e., c much larger than all other velocities in the equations) we recover the Galilean composition of velocities, eq. (7.2). However, for generic velocities the composition is different. In particular, in the limiting case of a particle moving with the speed of light along the x axis, $v_x = c$, $v_y = v_z = 0$, and a velocity transformation of parameter v_0 again along the x axis, we get $v_x' = c$, $v_y' = v_z' = 0$, independently of the value of v_0! We have therefore recovered the fact that the speed of light is the same in all inertial frames, which was our starting point. Notice also that (unless $v_y = v_z = 0$), even the transverse components of the velocity change when performing a Lorentz boost, contrary to what happens in the Galilean transformation. This is due to the fact that we are also transforming the time variable.

7.2.2 Causality and simultaneity

Physically, the fact that Nature is invariant under Lorentz transformations, rather than under the Galilean transformations of everyday experience, introduces a revolution in our notions of space and time. This is seen in a particularly stunning way in the change of the concept of causality, as well as in the notion of simultaneity of events, as illustrated in Fig. 7.1. In this plot, on the vertical axis we display $x^0 \equiv ct$ and on the horizontal axis one spatial coordinate, say x, while y and z are suppressed for graphical reasons. In this plot, light rays travel at $45°$, corresponding to the fact that the interval between two events connected

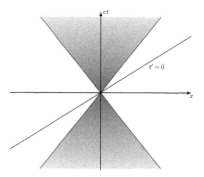

Fig. 7.1 The past and future light cones of the observer located at the origin (boundaries of the gray shaded areas). The white regions are causally disconnected from the observer at the origin. For the observer using the (t, x) coordinates, the x axis corresponds to simultaneous events, all characterized by $t = 0$. For the boosted observer K', the simultaneous events correspond to the line labeled $t' = 0$.

by a light ray has $s^2 = 0$, i.e., $|\Delta \mathbf{x}| = \pm c\Delta t$. We see that an observer located at the origin $\mathbf{x} = 0$ can receive signals, traveling at a speed less or equal than the speed of light, only from the $t < 0$ part of the shaded region in the figure; its boundary is called the *past light cone* of that observer (if one makes the plot in three dimensions, with one more spatial coordinate, it is indeed a cone with the tip at the origin, see Fig. 7.2). The region where this observer can send signals is the shaded region of Fig. 7.1 with $t > 0$, or the corresponding region in Fig. 7.2; its boundary is called the *future light cone*. The white regions to the left and to the right in Fig. 7.1 are *causally disconnected* from the observer at the origin: the events that fall in these region cannot be influenced by anything that happens at $(t = 0, \mathbf{x} = 0)$; and, vice versa, nothing that happens there can influence the events at $(t = 0, \mathbf{x} = 0)$. Only events in or inside its past light cone can influence the events at $(t = 0, \mathbf{x} = 0)$ and, conversely, what happens in $(t = 0, \mathbf{x} = 0)$ can only influence the events in, or inside, its future light cone.

Related to this, the notion of simultaneity of the events is also relative to the observer considered. In the reference frame K, that uses coordinates (t, x), the events with the same value of t are simultaneous. In Fig. 7.1, simultaneous events are along lines parallel to the x axis, and the events that take place at $t = 0$ are those along the x axis in the figure. However, for the observer K', that uses coordinate (t', x') related to (t, x) by the Lorentz boost (7.24–7.25), the events are simultaneous if they have the same value of t'. For instance, the events with $t' = 0$ correspond, according to eq. (7.24), to the events on the straight line $ct = -(v_0/c)x$. Since the boost parameter v_0 is in the range $-c < v_0 < c$, these are lines comprised between $ct = -x$ and $ct = +x$ (with the limiting lines excluded), i.e., contained in the white regions causally disconnected from the origin; an example is given by the line shown in the figure. So, simultaneity is no longer an "absolute" concept, but is relative to the observer.

We see from Fig. 7.1 that, whenever two events are causally disconnected, we can find a boosted reference frame such that, in this frame, the two events become simultaneous, since we can always find a straight line $ct = ax$, with $-1 < a < 1$, that joins the origin with a point in the white region. Conversely, when two points are causally connected, i.e., one is on (or inside) the past or in the future light cone of the other, this is not possible. This is graphically clear from the figure and can also be seen more formally as follows. Consider two events that, in the frame K, have coordinates $(t_1, \mathbf{x}_1)$ and $(t_2, \mathbf{x}_2)$; let $(t'_1, \mathbf{x}'_1)$ and $(t'_2, \mathbf{x}'_2)$ be their coordinates in the boosted frame K'. If the events are simultaneous in K', we have $t'_1 = t'_2$ and therefore the interval $s'^2_{12} = -(t'_1 - t'_2)^2 + (\mathbf{x}'_1 - \mathbf{x}'_2)^2 > 0$. However, the interval is invariant under Lorentz transformations, so we must also have $s^2_{12} > 0$. This is just the condition that (setting the first event at the origin) the second event is in the white, causally disconnected, area of Fig. 7.1.

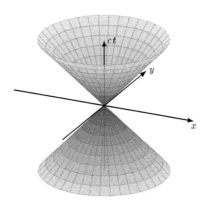

Fig. 7.2 A three-dimensional rendering of the past and future light-cones.

7.2.3 Proper time and time dilatation

We next define the *proper time* τ of an observer (or, e.g., of a particle). Suppose that, with respect to a given inertial observer K, a second observer O moves with velocity $\mathbf{v}(t)$, with $v(t) = |\mathbf{v}(t)|$ strictly smaller than c. We do not need to assume that $\mathbf{v}$ is constant, i.e., the frame moving with O need not be an inertial frame. We want to understand the relation between the time t measured by the clock of the inertial observer K, and the time τ measured by a clock moving with O. To this purpose, we consider an inertial reference frame K' such that, as some time t, O and K' have the same velocity, i.e., the frame K' is (instantaneously) comoving with O. We can imagine that at time t the observer O emits a first signal and at time $t + dt$ it emits a second signal. Each of these signals marks an event, and we can compute the infinitesimal interval between these two events. In the frame K, during the time interval dt, the observer O has moved by $d\mathbf{x} = \mathbf{v}(t)dt$. Therefore, the corresponding interval between the two events is

$$ds^2 = -c^2 dt^2 + d\mathbf{x}^2 = -c^2 dt^2 [1 - v^2(t)/c^2] \,, \qquad (7.32)$$

where $v(t) = |\mathbf{v}(t)|$. In the inertial frame K', in contrast, the observer O is instantaneously at rest so, to linear order in dt, $d\mathbf{x} = 0$. Then, calling $d\tau$ the time interval measured by a clock carried by the observer O, to lowest order in the infinitesimal quantity dt the interval measured in the inertial frame K' is

$$ds^2 = -c^2 d\tau^2 \,. \qquad (7.33)$$

Since the intervals measured in two inertial frames must be the same, we get

$$
\begin{aligned}
d\tau &= dt \sqrt{1 - \frac{v^2(t)}{c^2}} \\
&= \frac{dt}{\gamma(v)} \,,
\end{aligned}
\qquad (7.34)
$$

where $\gamma(v)$ was defined in eq. (7.26). This relation can be integrated (which, physically, means that we are using a succession of comoving inertial frames) so that, choosing the origin of times so that $t = t_0$ corresponds to $\tau = \tau_0$, we get

$$\tau(t) - \tau_0 = \int_{t_0}^{t} dt' \sqrt{1 - \frac{v^2(t')}{c^2}} \,. \qquad (7.35)$$

The quantity τ is called the *proper time* of the observer O. It is the time measured by the clock carried by this observer. Note that, since $\sqrt{1 - v^2/c^2}$ is always smaller than one, $d\tau$ is always smaller than dt. From the point of view of the observer K, the clock carried by O goes slower. This is the famous phenomenon of *time dilatation* of Special Relativity.

This apparently leads to a paradox. Suppose that O actually moves with constant velocity v, so that now also the frame moving with O is an inertial frame, that coincides with K' at all times. Then, from the point of view of an observer in K, the clock carried by the inertial observer in K' goes slower but, exactly by the same reasoning, the inertial observer K' will rather find that the clock carried by K goes slower!

In fact, there is no logical contradiction with this and, rather, this apparent paradox is at the core of the notion of "Relativity." This can be understood by specifying more carefully what should be done, operationally, to compare the two clocks. Suppose that, at some initial time, the clocks in K and that in K' are together at the same point in space and are both set to the same initial value of time, say $t = t' = 0$. To determine what the clock in K' measures at a subsequent time, compared to a clock in K, we need a second clock, at rest with respect to K and located in a second position, that the observer in K must have previously synchronized with the first clock. That is, these two clocks belonging to K have been first carried to the same place, where it has been checked that they both read the same time, and then one has been brought to the second position.[7] When the clock carried by the observer in K' will have reached the position of this second clock, a comparison can be performed. However, now the situation is no longer symmetric between the two frames. We are comparing one clock in K' with *two* clocks in K. The clock that goes slower is the one that is compared with two clocks of the other frame.[8]

We can also consider a more symmetric situation, in which each observer prepares two clocks, synchronizing them in his/her frame, and uses them to compare with a clock of the other frame. Again, each observer will find that the other observer's clock goes slower; the clock that goes slower is always the one that is checked against two clocks of the other frame. Observe also that, for the observer K', the two clocks in K are *not* synchronized! We indeed see from Fig. 7.1 that, if in K a clock is at $(t = 0, x = 0)$ and another is at $(t = 0, x = x_0)$, for some $x_0 \neq 0$ (and therefore in K the two clocks are synchronized, since they both register the same value of time, in this case $t = 0$), from the point of view of a boosted observer K' they will not be synchronized. For K', synchronized clocks are those that, in this space-time diagram, can be found on a line such as the $t' = 0$ line shown in Fig. 7.1. So, for instance, the inertial observer K could use two clocks, synchronize them (from his point of view), place them at two different positions, and use them to compare with one clock of K', and he would find that the clock of K' goes slower. The observer K' could do exactly the same, preparing two clocks synchronized in her frame, and use them to check a clock of K. Again, she would find that the clock in K goes slower. The observer K would attribute this different result to the fact that K' had made a mistake: from his point of view, the two clocks used by K' were not correctly synchronized. The observer K' would reach the same conclusion: the fault was in the fact that the clocks in K were not correctly synchronized! In fact, both observers were right, and simply there is no "absolute" notion of which clock goes slower. The fact that a moving clock goes slower is a correct statement, relative to the (two) clocks of

[7]To be precise, this second clock, that at the beginning is at the same position as the first, and with the same zero velocity with respect to K, must have been brought to the second position very gently, i.e., giving to it a negligible acceleration at the beginning, and a negligible deceleration to eventually stop it in the final position. This is in principle always possible, at the level of these "gedanken" experiments. Acceleration and deceleration indeed affect the reading of a clock, as one learns in General Relativity.

[8]Alternatively, the clock in K' might invert its motion and come back to meet the clock in K again; this however introduces extra complications due to the corresponding phase of acceleration, so K' is no longer inertial. In the context of General Relativity, this produces another apparent paradox called the Twin Paradox, on which we will not dwell here.

the observer that sees that clock in motion. This is one instance of the fact that some statement that, in Newtonian physics, have an absolute validity (a clock either goes slower than another or it does not), in Special Relativity can have a validity only *relative* to some observer (hence, the name "Relativity" given to the theory).

7.2.4   Lorentz contraction

In a similar way we can prove that, given two inertial observers K and K', the length of a rigid rod depends on the velocity at which the observer sees it moving. Consider first the frame K, where the rod is at rest along the x axis. If we call ℓ its length in this frame, the coordinates of its two endpoints can be taken, respectively, as $(x_1, 0, 0)$ and $(x_2, 0, 0)$, with $x_2 - x_1 = \ell$. Inverting eqs. (7.24) and (7.25), the relation between the coordinates (t, x) in K and the coordinates (t', x') in K' is

$$t = \gamma(v_0)\left(t' - \frac{v_0}{c^2}x'\right),\tag{7.36}$$

$$x = \gamma(v_0)\left(x' - v_0 t'\right).\tag{7.37}$$

It is straightforward to explicitly check that this provides the inversion of eqs. (7.24) and (7.25), but in fact the result can be obtained much more simply by reversing the sign of v_0. In the frame K', the bar moves with velocity v_0 along the x axis, and, at a given time t', the position of its end-points will be $(x_1', 0, 0)$ and $(x_2', 0, 0)$, respectively. The observer in K' will define the length of the bar as the difference in the position of its end-points, $x_2' - x_1'$, *measured at the same value of her time variable* t'. From eq. (7.37), we have

$$x_2' = \frac{1}{\gamma(v_0)}x_2 + v_0 t',\tag{7.38}$$

$$x_1' = \frac{1}{\gamma(v_0)}x_1 + v_0 t',\tag{7.39}$$

and therefore

$$x_2' - x_1' = \frac{1}{\gamma(v_0)}(x_2' - x_1').\tag{7.40}$$

Therefore, the length $\ell' = (x_2' - x_1')$, measured in a frame where the rod moves with velocity v_0, is related to the length ℓ in the frame where the rod is at rest, by

$$\ell' = \frac{\ell}{\gamma(v_0)}$$

$$= \ell\sqrt{1 - \frac{v_0^2}{c^2}}.\tag{7.41}$$

This is the *Lorentz contraction* of lengths. Note that the contraction only takes place in the direction of motion. The coordinates of the transverse directions, for a Lorentz boost along the x axis, satisfy $y' = y$ and $z' = z$, so transverse directions are not affected.

7.3 The mathematics of the Lorentz group

7.3.1 Four-vectors and Lorentz tensors

We now introduce a covariant formalism, that will make the transformation properties of the various quantities under Lorentz transformations explicit. In the case of rotations, in Section 1.7.2 we have defined the rotation group as the group of linear transformations (1.140), in a space with d spatial dimensions, which leave the quadratic form (1.141) invariant. We have seen that this implies that R is an orthogonal matrix. Vectors were then defined as objects that transform according to the "fundamental" representation, $v_i \to v_i' = R_{ij}v_j$, i.e., with the same matrix R_{ij} used in the definition itself of the group. It was also useful to introduce a "metric tensor" δ_{ij}, so that the scalar product between two vectors is given by $\mathbf{v} \cdot \mathbf{w} = \delta_{ij}v_iw_j$, so in particular the squared norm of a vector $\mathbf{v}$ is $|\mathbf{v}|^2 = \delta_{ij}v_iv_j$.

Similarly, after eq. (7.13) we have defined the *Lorentz group* as the group of linear transformation of a four-dimensional space, with coordinates (x^0, x^1, x^2, x^3), which leaves invariant the quadratic form

$$s^2 = -(x^0)^2 + (x^1)^2 + (x^2)^2 + (x^3)^2 \,. \tag{7.42}$$

Generalizing eq. (1.140), we now write such linear transformations with the notation

$$x^\mu \to x'^\mu = \Lambda^\mu{}_\nu x^\nu \,, \tag{7.43}$$

where the "Lorentz index" μ takes the values $0, 1, 2, 3$ and, again, the sum over repeated indices is understood. Note, however, that, contrary to the case of the rotation group, we are now careful about the positioning on the indices, and the sum is always performed by contracting an upper and a lower index. The reason for this convention will become apparent in the following.

Equation (7.43) is the transformation law that is used to define the Lorentz group and therefore, just as for vectors in the case of rotations, can also be used to introduce the "fundamental" representation of the Lorentz group, that we call the four-vector representation: four-vectors are defined as any set of four quantities (V^0, V^1, V^2, V^3), [or, with an equivalent notation, (V^0, V^x, V^y, V^z)], collectively denoted as V^μ, that, under Lorentz transformations, transform linearly among them, according to[9]

$$\boxed{V^\mu \to V'^\mu = \Lambda^\mu{}_\nu V^\nu \,.} \tag{7.44}$$

[9]More precisely, as we will see below, this transformation property defines "contravariant" four vectors, to be distinguished from "covariant" four-vectors that will be introduced in Section 7.3.2.

For instance, for a Lorentz boost along the x axis with velocity v_0, we saw in eqs. (7.22) and (7.23) that

$$
\begin{aligned}
x^0 \to x'^0 &= \gamma(v_0)(x^0 + \beta x) \,, && (7.45)\\
x \to x' &= \gamma(v_0)(x + \beta x^0) \,. && (7.46)
\end{aligned}
$$

where, for the parameter of the transformation, we have introduced the notation $\boldsymbol{\beta} = \mathbf{v}_0/c$ and $\beta = |\boldsymbol{\beta}|$. Then, for a generic four-vector V^μ,

$$
\begin{aligned}
V^0 \to V'^0 &= \gamma(v_0)\left(V^0 + \beta V^x\right), & (7.47) \\
V^x \to V'^x &= \gamma(v_0)\left(V^x + \beta V^0\right), & (7.48)
\end{aligned}
$$

(where, of course, the notation V^0 for the $\mu = 0$ component of the four-vector V^μ should not be confused with the velocity v_0 of the transformation), while $V'^y = V^y$ and $V'^z = V^z$. For a boost in a generic direction, we can write[10]

$$
\begin{aligned}
V^0 &\to V'^0 = \gamma(v_0)\left(V^0 + \boldsymbol{\beta}\!\cdot\!\mathbf{V}\right), & (7.49) \\
V_\parallel &\to V'_\parallel = \gamma(v_0)\left(V_\parallel + \beta V^0\right), & (7.50) \\
\mathbf{V}_\perp &\to \mathbf{V}'_\perp = \mathbf{V}_\perp, & (7.51)
\end{aligned}
$$

where we have split $\mathbf{V} = (V^x, V^y, V^z)$ into its components parallel and perpendicular to $\boldsymbol{\beta}$,

$$
\mathbf{V} = V_\parallel \hat{\boldsymbol{\beta}} + \mathbf{V}_\perp. \tag{7.52}
$$

The four-vector representation has dimension four and is irreducible since, with rotations, we can mix among them all the spatial components V^i, while with boosts we can mix V^0 with any of the spatial components.

Similarly to what we have done for the rotation group, after having defined the four-vector representation, we can proceed to define tensor representations of the Lorentz group (that we will call "Lorentz tensors," or, when the context is clear, simply "tensors"). For instance, a tensor $T^{\mu\nu}$ with two upper indices (or a "contravariant" tensor) is defined as an object that, under Lorentz transformations, changes as

$$
T^{\mu\nu} \to T'^{\mu\nu} = \Lambda^\mu{}_\rho \Lambda^\nu{}_\sigma T^{\rho\sigma}, \tag{7.53}
$$

and similarly for tensors with three or more Lorentz indices.

We next introduce the Minkowski metric[11]

$$
\boxed{\eta_{\mu\nu} = \mathrm{diag}(-1,1,1,1).} \tag{7.54}
$$

This plays a role analogous to the metric δ_{ij} in the case of rotations, in the sense that it allows us to define the scalar product between two four-vectors V^μ and W^μ, as

$$
(VW) \equiv \eta_{\mu\nu} V^\mu W^\nu, \tag{7.55}
$$

so that the squared norm of a four-vector V^μ is $V^2 = \eta_{\mu\nu} V^\mu V^\nu$. Notice that this scalar product is not positive definite, and V^2 can be positive, negative, or zero. The infinitesimal interval (7.8) can then be rewritten as

$$
\boxed{ds^2 = \eta_{\mu\nu} dx^\mu dx^\nu.} \tag{7.56}
$$

[10] In Section 7.3.5 we will show that the spatial components (V^x, V^y, V^z) of a four-vector V^μ transform as a vector under rotations, so we already use the notation $\mathbf{V} = (V^x, V^y, V^z)$.

[11] At first, we introduce $\eta_{\mu\nu}$ as a fixed matrix, given by eq. (7.54) in all reference frames. More precisely, we will see in Section 7.3.3 that it is actually an invariant tensor of the Lorentz group, i.e., a tensor that keeps the same numerical value in all frames related by Lorentz transformations, similarly to the metric δ_{ij} for rotations, as we saw in eq. (1.132).

In eq. (1.129) we saw that the condition that rotations must preserve the quadratic form (1.141) restricts R_{ij} to orthogonal matrices. We now derive the analogous condition on $\Lambda^\mu_{\ \nu}$. Writing $x'^\mu = \Lambda^\mu_{\ \rho} x^\rho$ and requiring that

$$\eta_{\mu\nu} x'^\mu x'^\nu = \eta_{\mu\nu} x^\mu x^\nu \,, \tag{7.57}$$

we get

$$\eta_{\mu\nu} (\Lambda^\mu_{\ \rho} x^\rho)(\Lambda^\nu_{\ \sigma} x^\sigma) = \eta_{\mu\nu} x^\mu x^\nu \,. \tag{7.58}$$

By renaming the dummy indices $\mu \to \rho, \nu \to \sigma$ on the right-hand side, and rearranging the factors, we get

$$\eta_{\mu\nu} \Lambda^\mu_{\ \rho} \Lambda^\nu_{\ \sigma} x^\rho x^\sigma = \eta_{\rho\sigma} x^\rho x^\sigma \,. \tag{7.59}$$

Since this must hold for x generic, we must have

$$\boxed{\eta_{\mu\nu} \Lambda^\mu_{\ \rho} \Lambda^\nu_{\ \sigma} = \eta_{\rho\sigma} \,.} \tag{7.60}$$

This is the analogous of the condition (1.129) for the rotation group.

In matrix notation, eq. (7.60) can be rewritten as

$$\Lambda^T \eta \Lambda = \eta \,, \tag{7.61}$$

where $(\Lambda^T)_\rho^{\ \mu} = \Lambda^\mu_{\ \rho}$ is the transpose matrix. Taking the determinant of both sides, we get $(\det \Lambda)^2 = 1$, and therefore $\det \Lambda = \pm 1$. Transformations with $\det \Lambda = -1$ can always be written as the product of a transformation with $\det \Lambda = 1$ and of a discrete transformation that reverses the sign of an odd number of coordinates, e.g., a parity transformation $(t, x, y, z) \to (t, -x, -y, -z)$, or a reflection around a single spatial axis, such as $(t, x, y, z) \to (t, -x, y, z)$, or a time-reversal transformation, $(t, x, y, z) \to (-t, x, y, z)$. Transformations with $\det \Lambda = +1$ are called *proper Lorentz transformations*.[12]

7.3.2 Contravariant and covariant quantities

From the metric $\eta_{\mu\nu}$ and a contravariant four-vector V^μ, we can form a set of four quantities V_μ with lower index, defined by

$$V_\mu \equiv \eta_{\mu\nu} V^\nu \,, \tag{7.62}$$

called a *covariant* four-vector. Explicitly,

$$V_0 = -V^0 \,, \qquad V_i = +V^i \,. \tag{7.63}$$

It is also convenient to define a matrix $\eta^{\mu\nu}$, with both upper indices, whose numerical values are still the same as for $\eta_{\mu\nu}$, i.e.,

$$\eta^{\mu\nu} = \text{diag}(-1, 1, 1, 1) \,. \tag{7.64}$$

With our convention that Lorentz indices are summed over by contracting an upper and a lower index, we can use $\eta^{\mu\nu}$ to invert eq. (7.62), writing

$$V^\mu = \eta^{\mu\nu} V_\nu \,, \tag{7.65}$$

[12]More generally, in the group theory language developed in Section 1.7, the group of transformations of a space with coordinates $(y_1, \ldots y_m, x_1, \ldots x_n)$, which leaves invariant the quadratic form

$$s^2 = -(y_1^2 + \ldots + y_m^2) + (x_1^2 + \ldots + x_n^2) \,,$$

is called the orthogonal group $O(n, m)$ [or, equivalently, $O(m, n)$], and it reduces to $O(n)$ if $m = 0$. Thus, in three spatial dimensions, the group that leaves invariant the quadratic form $s^2 = -(x^0)^2 + x^2 + y^2 + z^2$ is called $O(3, 1)$. However, in $O(n, m)$ there are both transformations with determinant $+1$ and with determinant -1. The transformations with determinant $+1$ form a subgroup, which is denoted by $SO(n, m)$. Thus, the "proper" Lorentz group, which is defined by eliminating the discrete parity transformations, is actually $SO(3, 1)$. When we will refer to the Lorentz group, we will henceforth always mean the proper Lorentz group $SO(3, 1)$, similarly to our restriction from $O(3)$ to $SO(3)$ for rotations. This can be generalized to arbitrary spatial dimensions. Just as $SO(d)$ is the group of (proper) rotations in a four-dimensional space, $SO(d, 1)$ is the Lorentz group in a space-time with d spatial dimensions and a time-like coordinate.

since this gives $V^0 = -V_0$ and $V^i = +V_i$, which is the (obvious) inversion of eq. (7.63). So, $\eta_{\mu\nu}$ can be used to lower the index of a contravariant four-vector, obtaining a covariant one; and, vice versa, $\eta^{\mu\nu}$ can be used to raise the index of a covariant four-vector, obtaining a contravariant four-vector.

Consider now the combination $\eta_{\mu\rho}\eta^{\rho\nu}$. Numerically, this is just the identity matrix. We denote it by δ_μ^ν,

$$
\begin{aligned}
\delta_\mu^\nu &= \eta_{\mu\rho}\eta^{\rho\nu} \\
&= \text{diag}(1,1,1,1),
\end{aligned}
\tag{7.66}
$$

where the position of the indices on δ_μ^ν (one upper and one lower) matches the position in $\eta_{\mu\rho}\eta^{\rho\nu}$. Observe that we have obvious identities such as $V^\mu = \delta_\nu^\mu V^\nu$.

In terms of a covariant and a contravariant four-vector, the scalar product (7.55) can then be rewritten as

$$
(VW) = V_\mu W^\mu = V^\mu W_\mu \,.
\tag{7.67}
$$

Explicitly,

$$
V_\mu W^\mu = V_0 W^0 + V_1 W^1 + V_2 W^2 + V_3 W^3 \,,
\tag{7.68}
$$

so, using a covariant and a contravariant four-vector, the scalar product takes a Euclidean form, with all plus signs.

By definition, a contravariant four-vector V^μ is an object that transforms as in eq. (7.44). From eq. (7.62), it then follows that the corresponding covariant four-vector transforms as

$$
\begin{aligned}
V_\mu &= \eta_{\mu\sigma} V^\sigma \\
&\rightarrow \eta_{\mu\sigma}\Lambda^\sigma{}_\rho V^\rho \\
&= \eta_{\mu\sigma}\Lambda^\sigma{}_\rho \eta^{\rho\nu} V_\nu \,.
\end{aligned}
\tag{7.69}
$$

and therefore

$$
V_\mu \rightarrow V_\mu' = \Lambda_\mu{}^\nu V_\nu \,,
\tag{7.70}
$$

where

$$
\Lambda_\mu{}^\nu = \eta_{\mu\sigma}\Lambda^\sigma{}_\rho \eta^{\rho\nu} \,.
\tag{7.71}
$$

The matrices $\Lambda_\mu{}^\nu$ and $\Lambda^\mu{}_\nu$ are different: because of the $\eta_{\mu\nu}$ involved in the transformation, some of their matrix elements differ by a minus sign. In particular, $\Lambda^0{}_0 = \Lambda_0{}^0$, $\Lambda^0{}_i = -\Lambda_0{}^i$, $\Lambda^i{}_0 = -\Lambda_i{}^0$, and $\Lambda^i{}_j = \Lambda_i{}^j$. However, physically V^μ and V_μ represents the same quantity in a different notation and, in the language of representation theory, they correspond to equivalent representations, related as in eq. (1.137), with $\eta_{\mu\nu}$ playing the role of the matrix S.[13]

Similarly, we can use $\eta_{\mu\nu}$ to lower one or more indices of a tensor. For instance, given a contravariant tensor with two indices $T^{\mu\nu}$, defined by the fact that it transforms as in eq. (7.53), we can define a covariant tensor $T_{\mu\nu}$ as

$$
T_{\mu\nu} = \eta_{\mu\rho}\eta_{\nu\sigma} T^{\rho\sigma} \,.
\tag{7.72}
$$

[13]More abstractly, one can define contravariant four-vectors as object V^μ that transform as in eq. (7.44), and covariant four-vectors as objects W_μ that transform as

$$
W_\mu \rightarrow W_\mu' = \Lambda_\mu{}^\nu W_\nu \,,
$$

with $\Lambda_\mu{}^\nu$ defined by eq. (7.71). At this point, however, one would discover that, given any contravariant four-vector V^μ, the quantity $V_\nu \equiv \eta_{\mu\nu} V^\nu$ is a covariant four-vector and vice versa, given any covariant four-vector W_μ, the quantity $W^\mu \equiv \eta^{\mu\nu} W_\nu$ is a contravariant four-vector. Therefore, the spaces of covariant and contravariant four-vectors are in one-to-one correspondence, so we do not lose generality by defining contravariant four-vectors starting from covariant four-vectors and lowering their indices. The same holds for the covariant and contravariant tensors that we now introduce.

Proceeding as in eq. (7.69) we see that it transforms as

$$T_{\mu\nu} \rightarrow T'_{\mu\nu} = \Lambda_\mu{}^\rho \Lambda_\nu{}^\sigma T_{\rho\sigma} \,. \tag{7.73}$$

We can also define tensors with mixed covariant and contravariant indices. For instance, defining

$$T^\mu{}_\nu = \eta_{\nu\sigma} T^{\mu\sigma} \,, \tag{7.74}$$

we find that it transforms as

$$T^\mu{}_\nu \rightarrow \Lambda^\mu{}_\rho \Lambda_\nu{}^\sigma T^\rho{}_\sigma \,. \tag{7.75}$$

We can proceed in the same way for tensors with three or more indices.

For later use, we observe that, in terms of $\Lambda_\mu{}^\nu$, the condition (7.60) can be written as[14]

$$\boxed{\eta_{\rho\sigma} \Lambda_\mu{}^\rho \Lambda_\nu{}^\sigma = \eta_{\mu\nu} \,.} \tag{7.80}$$

This is similar to eq. (7.60), except that now, on the left-hand side, the contraction of the two indices of η is made with the second indices of each of the two Λ matrices, rather than with the first indices.

7.3.3   Invariant tensors of the Lorentz group

The notation $\eta_{\mu\nu}$ for the metric (7.54), with two lower Lorentz indices, implies that it is a covariant tensor. However, $\eta_{\mu\nu}$ is a special type of contravariant tensor, that retains the same numerical value of its components in all frames connected by Lorentz transformations, i.e., is an invariant tensor of the Lorentz group [just as we found that δ_{ij} is an invariant tensor of the rotation group, see eq. (1.132)]. Indeed, consider a covariant tensor $T_{\mu\nu}$ whose components, in a given frame, are given numerically by $T_{\mu\nu} = \eta_{\mu\nu} = \mathrm{diag}(-1, 1, 1, 1)$. After a Lorentz transformation, this tensor becomes $T'_{\mu\nu} = \Lambda_\mu{}^\rho \Lambda_\nu{}^\sigma \eta_{\rho\sigma}$. However, because of the defining property of the Lorentz group, written in the form (7.80), the right-hand side of this equation is just $\eta_{\mu\nu}$ again, so $T'_{\mu\nu} = \eta_{\mu\nu}$. Thus, $\eta_{\mu\nu}$ is an invariant tensor with two lower indices.

Similarly, using eq. (7.79), we see that $\eta^{\mu\nu}$ is an invariant tensor with two upper indices. The same holds for δ^μ_ν that, being constructed from $\eta_{\mu\rho}$ and $\eta^{\rho\nu}$ as in eq. (7.66), is also an invariant tensor. It is important, however, to understand that the identity matrix is an invariant tensor *only* if we define it with an upper and a lower index. In this way it transforms so that, if it is equal to $\mathrm{diag}(1, 1, 1, 1)$ in a frame, it remains equal to $\mathrm{diag}(1, 1, 1, 1)$ in any other frame related by a Lorentz transformation. We have derived this from the fact that it is constructed with $\eta_{\mu\rho}$ and $\eta^{\rho\nu}$, and we proved that the latter are invariant tensors, but we can also show it directly from its transformation property

$$\delta^\mu_\nu \rightarrow \Lambda^\mu{}_\rho \Lambda_\nu{}^\sigma \delta^\rho_\sigma \,, \tag{7.81}$$

---

[14]To get eq. (7.80) we multiply both sides of eq. (7.60) by $\eta^{\sigma\alpha} \Lambda^\beta{}_\alpha$. This gives

$$\eta_{\mu\nu} \Lambda^\mu{}_\rho \left(\Lambda^\nu{}_\sigma \eta^{\sigma\alpha} \Lambda^\beta{}_\alpha \right) = \eta_{\rho\sigma} \eta^{\sigma\alpha} \Lambda^\beta{}_\alpha \,. \tag{7.76}$$

The right-hand side can be rewritten as

$$\begin{aligned} \eta_{\rho\sigma} \eta^{\sigma\alpha} \Lambda^\beta{}_\alpha &= \delta^\alpha_\rho \Lambda^\beta{}_\alpha \\ &= \Lambda^\beta{}_\rho \\ &= \eta_{\mu\nu} \eta^{\nu\beta} \Lambda^\mu{}_\rho \,, \end{aligned}$$

where we used $\eta_{\mu\nu} \eta^{\nu\beta} = \delta^\beta_\mu$. Therefore, writing $\eta_{\mu\nu} \Lambda^\mu{}_\rho = \Lambda_{\nu\rho}$ on both sides of eq. (7.76),

$$\Lambda_{\nu\rho} \left(\Lambda^\nu{}_\sigma \eta^{\sigma\alpha} \Lambda^\beta{}_\alpha \right) = \Lambda_{\nu\rho} \eta^{\nu\beta} \,. \tag{7.77}$$

Since $\Lambda_{\nu\rho}$ is an invertible matrix, we can factorize it out from this equation, and we get

$$\Lambda^\nu{}_\sigma \eta^{\sigma\alpha} \Lambda^\beta{}_\alpha = \eta^{\nu\beta} \,, \tag{7.78}$$

or, renaming the indices as $\sigma \rightarrow \rho$, $\alpha \rightarrow \sigma$, $\nu \rightarrow \mu$ and $\beta \rightarrow \nu$,

$$\eta^{\rho\sigma} \Lambda^\mu{}_\rho \Lambda^\nu{}_\sigma = \eta^{\mu\nu} \,. \tag{7.79}$$

Lowering the μ, ν indices on both sides, and inverting the upper/lower position of the contracted ρ, σ indices, we finally obtain eq. (7.80).

and using eq. (7.60) (with the indices properly raised and lowered). Note that the positioning of the Lorentz indices in the two Λ factors in eq. (7.81) is the one appropriate to the transformation of a tensor with one upper and one lower index. If, in contrast, we consider a tensor with two lower indices $T_{\mu\nu}$, whose numerical value in a reference frame happens to be $\mathrm{diag}(1,1,1,1)$, and we perform a Lorentz transformation transforming it as in (7.73), as appropriate for a tensor with two lower indices, in the new frame $T_{\mu\nu}$ will have different numerical values, and will no longer be of the form $\mathrm{diag}(1,1,1,1)$. So, for instance, it makes no sense to define an identity matrix with two lower indices, "$\delta_{\mu\nu} = \mathrm{diag}(1,1,1,1)$." Such an object is not an invariant tensor, and the numerical assignment $\mathrm{diag}(1,1,1,1)$ could only hold in one Lorentz frame (and in those related to it by a spatial rotation) but would change as soon as we perform a boost. Notice that, lowering the upper index of δ^{μ}_{ν}, we get $\eta_{\mu\nu}$,

$$\eta_{\mu\rho}\delta^{\rho}_{\nu} = \eta_{\mu\nu} \, , \tag{7.82}$$

and similarly raising an index of $\eta_{\mu\nu}$ we get δ^{μ}_{ν},

$$\eta^{\mu\rho}\eta_{\rho\nu} = \delta^{\mu}_{\nu} \, . \tag{7.83}$$

Just as there is no meaning in writing "$\delta_{\mu\nu} = \mathrm{diag}(1,1,1,1)$," there is no meaning in writing "$\eta^{\mu}{}_{\nu} = \mathrm{diag}(-1,1,1,1)$." Again, a tensor of the form $\mathrm{diag}(-1,1,1,1)$ maintains the same numerical value in any frame only if it is transformed according to the transformation law of a tensor with two lower indices (or with two upper indices), not with one upper and one lower index.

The only other invariant tensor of the Lorentz group (apart from all possible lowering of its indices, see below) is the totally antisymmetric tensor $\epsilon^{\mu\nu\rho\sigma}$. This tensor vanishes if two indices take the same value, satisfies $\epsilon^{0123} = +1$, and changes sign under permutations of any two indices; so, for instance, repeatedly switching the position of the 0 index, $\epsilon^{1023} = -1$, $\epsilon^{1203} = +1$, and $\epsilon^{1230} = -1$ so, in this case, it changes sign under a cyclic permutation. Note, however, that starting from $\epsilon^{1230} = -1$ and making three jumps for the index 1, we get $\epsilon^{2301} = +1$, so in this case a cyclic permutation of 0123 gives again $+1$ instead of -1. Therefore, the tensor $\epsilon^{\mu\nu\rho\sigma}$ is neither cyclic nor anti-cyclic (in contrast, for the rotation group in three dimensions, ϵ^{ijk} is cyclic, since it is again antisymmetric, but it has only three indices, so, e.g. $\epsilon^{123} = -\epsilon^{213} = +\epsilon^{231}$). Observe that $\epsilon^{0ijk} = \epsilon^{ijk}$.

The fact that $\epsilon^{\mu\nu\rho\sigma}$ is an invariant tensor follows from the fact that, from the definition of the determinant of a 4×4 matrix,

$$\Lambda^{\mu}{}_{\mu'}\Lambda^{\nu}{}_{\nu'}\Lambda^{\rho}{}_{\rho'}\Lambda^{\sigma}{}_{\sigma'}\epsilon^{\mu'\nu'\rho'\sigma'} = (\mathrm{det}\Lambda)\epsilon^{\mu\nu\rho\sigma} \, , \tag{7.84}$$

and, for the (proper) Lorentz group, $\mathrm{det}\,\Lambda = 1$. Combining $\epsilon^{\mu\nu\rho\sigma}$ with the metric tensor we can lower some of its indices. In particular, $\epsilon_{\mu\nu\rho\sigma}$ is still totally antisymmetric, while mixed combinations such as $\epsilon_{\mu}{}^{\nu\rho\sigma} = \eta_{\mu\mu'}\epsilon^{\mu'\nu\rho\sigma}$ are not.

7.3.4 Infinitesimal Lorentz transformations

With the formalism that we have developed, we can now compute how many independent parameters there are in a Lorentz transformation. This is conveniently done restricting to infinitesimal transformations. For the Lorentz group, the identity transformation is given by $\Lambda^\mu{}_\nu = \delta^\mu_\nu$, since, on any vector V^μ, we have in this case $V^\mu \to \Lambda^\mu{}_\nu V^\nu = \delta^\mu_\nu V^\nu = V^\mu$. A transformation infinitesimally close to the identity can then be written as

$$\Lambda^\mu{}_\nu = \delta^\mu_\nu + \omega^\mu{}_\nu \,, \tag{7.85}$$

where $\omega^\mu{}_\nu$ are infinitesimal of first order, that describe the deviation of $\Lambda^\mu{}_\nu$ from the identity transformation. Note the positioning of the indices in $\Lambda^\mu{}_\nu$ and in $\omega^\mu{}_\nu$, with the lower index in the second position; as we will see in a moment, it is important to keep track of it, since it will turn out that the matrix $\omega_{\mu\nu}$ is not symmetric. Plugging this into eq. (7.60), neglecting terms quadratic in the infinitesimal quantity $\omega^\mu{}_\nu$, and raising and lowering the Lorentz indices according to the rules discussed in this section, we get[15]

$$\omega_{\mu\nu} = -\omega_{\nu\mu} \,. \tag{7.86}$$

An antisymmetric 4×4 matrix has six independent elements, so the Lorentz group has six parameters. These are the three angles and the three rapidities, corresponding to the three independent rotations and the three independent boosts that we found by inspection in Section 7.2. Thus, the angles and rapidities associated with rotations and boosts, respectively, exhaust the parameters associated with Lorentz transformations.

7.3.5 Decomposition of a Lorentz tensor under rotations

Since four-vectors and four-tensors have well defined transformation properties under the Lorentz group, in particular they also have well defined transformation properties under spatial rotations, since these are a subgroup of the Lorentz group. In this subsection we explore this connection in more detail.

Rotations are a particular case of Lorentz transformations, with $\Lambda^\mu{}_\nu$ of the form (7.18) so, in components,

$$\Lambda^0{}_0 = 1 \,, \qquad \Lambda^i{}_0 = \Lambda^0{}_i = 0 \,, \qquad \Lambda^i{}_j = R^i{}_j \,, \tag{7.87}$$

where $R^i{}_j$ is the rotation matrix. Note that we now keep one index upper and one lower also on $R^i{}_j$. However, for the rotation group the spatial indices could be raised and lowered with the Kronecker delta, that can be written as δ^i_j or as δ_{ij}, and we could keep all indices lower (or upper), as indeed we have done in Section 1.6.

Consider first the Lorentz transformation of a four-vector V^μ, given by eq. (7.44). Since the matrix Λ in eq. (7.18) is in a block-diagonal

---

[15]Explicitly,

$$\eta_{\rho\sigma} = \eta_{\mu\nu}(\delta^\mu_\rho + \omega^\mu{}_\rho)(\delta^\nu_\sigma + \omega^\nu{}_\sigma)$$
$$= \eta_{\rho\sigma} + \eta_{\mu\nu}\delta^\mu_\rho \omega^\nu{}_\sigma + \eta_{\mu\nu}\delta^\nu_\sigma \omega^\mu{}_\rho$$
$$+ \mathcal{O}(\omega^2) \,.$$

Therefore, to linear order in ω,

$$\begin{aligned} 0 &= \eta_{\mu\nu}\delta^\mu_\rho \omega^\nu{}_\sigma + \eta_{\mu\nu}\delta^\nu_\sigma \omega^\mu{}_\rho \\ &= \eta_{\rho\nu}\omega^\nu{}_\sigma + \eta_{\mu\sigma}\omega^\mu{}_\rho \\ &= \omega_{\rho\sigma} + \omega_{\sigma\rho} \,. \end{aligned}$$

Observe that this generalizes to the Lorentz group the result that we found for spatial rotations in eq. (1.151).

form, under rotations V^0 does not mix with V^i, and

$$V^0 \to V^0 , \qquad V^i \to R^i{}_j V^j . \qquad (7.88)$$

Thus, under rotations, V^0 is a scalar while V^i is a vector. According to the discussion in Section 1.7.1, the fact that V^0 and V^i never mix under rotations is expressed in the language of group theory by saying that V^μ provides a reducible representation of the rotation group; in other words, it is made by separate "building blocks" (here V^0 and V^i) that do not mix among them under any rotation. However, under boosts V^0 and V^i mix, so four-vectors are an irreducible representation of the full Lorentz group.

Let us now consider the transformation of a tensor $T^{\mu\nu}$ under rotations. From eqs. (7.53) and (7.87), we find that

$$T^{00} \to \Lambda^0{}_\rho \Lambda^0{}_\sigma T^{\rho\sigma} = T^{00} , \qquad (7.89)$$

since, for rotations, $\Lambda^0{}_i = 0$ and $\Lambda^0{}_0 = 1$. This means that T^{00} is a scalar under rotations. Similarly,

$$T^{0i} \to \Lambda^0{}_\rho \Lambda^i{}_\sigma T^{\rho\sigma} = R^i{}_j T^{0j} , \qquad (7.90)$$

which is the transformation law of a spatial vector (and the same for T^{i0}), while

$$T^{ij} \to \Lambda^i{}_\rho \Lambda^j{}_\sigma T^{\rho\sigma} = R^i{}_k R^j{}_l T^{kl} , \qquad (7.91)$$

and therefore is a spatial tensor. Recalling that a spatial tensor T^{ij} further decomposes into irreducible representations of the rotation group as in eq. (1.148) we see that, from the point of view of spatial rotations, the 16 components of a Lorentz tensor $T^{\mu\nu}$ decompose into two scalars $[T^{00}$, and $S = \delta_{ij} T^{ij}$, see eq. (1.148)], three vectors (T^{0i}, T^{i0} and A^i), and a traceless symmetric tensor $S^{\rm T}_{ij}$. The counting of degrees of freedom of course matches, $4 \times 4 = 1 + 1 + 3 + 3 + 3 + 5$.

Observe also that the trace of $T^{\mu\nu}$ in the four-dimensional sense, $T = \eta_{\mu\nu} T^{\mu\nu}$, is a Lorentz invariant quantity, and therefore is invariant (i.e., a scalar) also under rotations. Writing $T = \eta_{00} T^{00} + \delta_{ij} T^{ij}$, we see that T is related to the two scalars under rotations that we have found above, T^{00} and S, by $T = -T^{00} + S$. Note that T^{00} and S are scalars under rotations but are not Lorentz scalars. For instance, T^{00} is the (00) component of a Lorentz tensor. Only their combination $-T^{00} + S$ is a Lorentz scalar.

7.3.6 Covariant transformations of fields

In classical electrodynamics the fundamental variables are fields, i.e., dynamical quantities that depend not only on time, as the typical variables $q_i(t)$ of an elementary mechanical system, but also on space. For instance, the scalar potential $\phi(t, \mathbf{x})$, the vector potential $\mathbf{A}(t, \mathbf{x})$, or the electric and magnetic fields $\mathbf{E}(t, \mathbf{x})$ and $\mathbf{B}(t, \mathbf{x})$, are all functions of time and space. Under a rotation, or under a Lorentz transformation, they therefore transform both because their arguments transform, and because of their intrinsic scalar, vector, or tensor nature. To understand these transformation properties, let us consider first spatial rotations (in which case, we can suppress for simplicity the time dependence). The simplest example of a transformation is that of a scalar field. Consider for instance the temperature $T(\mathbf{x})$ as a function of the position. The numerical values of the coordinates x_i of a point P depend on how we have chosen the reference frame. If we rotate our reference frame, they will change according to $x_i \to x_i' = R_{ij}x_j$ (for rotations we use here the simpler convention of keeping all spatial indices lower and summing over repeated lower indices). However, the temperature at the point P is the same, independently of how we choose to orient the axes of the reference frame, i.e., independently of the labels x_i that we choose to assign to the point P. This means that, when $\mathbf{x} \to \mathbf{x}'$, the function $T(\mathbf{x})$ must change as

$$T(\mathbf{x}) \to T'(\mathbf{x}') = T(\mathbf{x}) \,. \tag{7.92}$$

This relation expresses the fact that T will become a new function T' of the new coordinate $\mathbf{x}'$, and the functional form of T' must be such that, on the new label $\mathbf{x}'$ that we have given to the point P, it has the same numerical value that the old function T had on the old label $\mathbf{x}$ of P. In other words, the functional form will adapt itself to the change of the argument, so that, in the end, the temperature at a point P is the same, independently of how we have chosen to orient the axes of our reference frame.[16] Equation (7.92) can be rewritten as $T'(R\mathbf{x}) = T(\mathbf{x})$, where $R\mathbf{x}$ denotes the vector with components $R_{ij}x_j$; then, replacing the generic point $\mathbf{x}$ by $R^{-1}\mathbf{x}$, we can also rewrite it as

$$T'(\mathbf{x}) = T(R^{-1}\mathbf{x}) \,. \tag{7.93}$$

This defines the transformation of a *scalar field* (in this case, scalar under spatial rotations). If, in contrast, we consider a vector field, such as for instance the electric field $\mathbf{E}(\mathbf{x})$, when the label $\mathbf{x}$ of the point P becomes $\mathbf{x}'$, with $x_i' = R_{ij}x_j$, the vector itself (seen as an abstract geometric object, e.g., an arrow starting from P with a given length and direction) will not change, but now we must refer its components E_i to the new axes. Thus, under a rotation $x_i \to R_{ij}x_j$, they will change as

$$E_i(\mathbf{x}) \to E_i'(\mathbf{x}') = R_{ij}E_j(\mathbf{x}) \,. \tag{7.94}$$

This can be rewritten also as $\mathbf{E}'(R\mathbf{x}) = R\mathbf{E}(\mathbf{x})$, or

$$\mathbf{E}'(\mathbf{x}) = R\mathbf{E}(R^{-1}\mathbf{x}) \,. \tag{7.95}$$

This is the transformation law of a vector field.[17] Similarly, a tensor field transforms as

$$T_{ij}(\mathbf{x}) \to T_{ij}'(\mathbf{x}') = R_{ik}R_{jl}T_{kl}(\mathbf{x}) \,. \tag{7.96}$$

We now consider a scalar function $f(\mathbf{x})$, and we study how its gradient transform under rotations. If $x_i \to x_i' = R_{ij}x_j$ and $f(\mathbf{x}) \to f'(\mathbf{x}') =$

[16]This is completely analogous to the discussion of scalars under translations, see eq. (6.22) and Note 3 on page 137.

[17]As a check, consider a vector field $\mathbf{E}(\mathbf{x}) = \mathbf{x}$. This is a purely radial vector field. Then eq. (7.95) gives $\mathbf{E}'(\mathbf{x}) = R(R^{-1}\mathbf{x}) = \mathbf{x}$. Indeed, a purely radial field is rotationally invariant and does not change under rotations. The same holds if we rather write $\mathbf{E}(\mathbf{x}) = c\mathbf{x}$, with c a constant (needed to provide the correct dimensions to $\mathbf{E}$), since $R(c\mathbf{x}) = cR\mathbf{x}$, or even if we take $c = c(r)$ where $r = |\mathbf{x}|$.

$f(\mathbf{x})$, we have

$$\frac{\partial f(\mathbf{x})}{\partial x_i} \to \frac{\partial f'(\mathbf{x}')}{\partial x_i'} = \frac{\partial x_j}{\partial x_i'}\frac{\partial f(\mathbf{x})}{\partial x_j}. \tag{7.97}$$

For orthogonal matrices, the inversion of $x_i' = R_{ij}x_j$ gives $x_j = R_{ij}x_i'$,[18] and therefore $\partial x_j/\partial x_i' = R_{ij}$, so we finally obtain

$$\frac{\partial f(\mathbf{x})}{\partial x^i} \to R_{ij}\frac{\partial f(\mathbf{x})}{\partial x^j}. \tag{7.98}$$

Comparing this to eq. (7.94) we see that the gradient of a scalar function transforms as a vector field. A useful notation that makes this result more explicit is[19]

$$\partial_i \equiv \frac{\partial}{\partial x^i}, \tag{7.99}$$

so eq. (7.98) reads

$$\partial_i f(\mathbf{x}) \to R_{ij}\partial_j f(\mathbf{x}). \tag{7.100}$$

Thus, the index i in ∂_i behaves as a vector index. In the same way we can prove, for instance, that $\partial_i\partial_j f(\mathbf{x})$ is a tensor field with two indices (symmetric, since the derivative commutes, assuming as always smooth functions) or that, if $v_i(\mathbf{x})$ is a vector field under rotations, then $\partial_i v_j(\mathbf{x})$ is also a tensor field under rotation, and so on.

The generalization of these manipulations to the Lorentz group is straightforward. A field $\phi(x)$ [where we use the notation x to denote collectively $(t, \mathbf{x})$] is a scalar under Lorentz transformations if, under $x^\mu \to x'^\mu = \Lambda^\mu{}_\nu x^\nu$, it transforms as

$$\phi(x) \to \phi'(x') = \phi(x). \tag{7.101}$$

Similarly, a (contravariant) four-vector field $V^\mu(x)$ is defined as a field that transforms as

$$V^\mu(x) \to V'^\mu(x') = \Lambda^\mu{}_\nu V^\nu(x), \tag{7.102}$$

while a covariant vector field transforms as

$$V_\mu(x) \to V_\mu'(x') = \Lambda_\mu{}^\nu V_\nu(x), \tag{7.103}$$

and similarly for tensor fields. We define

$$\partial_\mu \equiv \frac{\partial}{\partial x^\mu}. \tag{7.104}$$

Using eq. (7.60), we can check that the inversion of $x'^\mu = \Lambda^\mu{}_\nu x^\nu$ is

$$x^\nu = \Lambda_\mu{}^\nu x'^\mu. \tag{7.105}$$

Then, if $\phi(x)$ is a Lorentz scalar,

$$\begin{aligned}
\frac{\partial\phi(x)}{\partial x^\mu} &\to \frac{\partial\phi(x')}{\partial x'^\mu} \\
&= \frac{\partial x^\nu}{\partial x'^\mu}\frac{\partial\phi(x)}{\partial x^\nu} \\
&= \Lambda_\mu{}^\nu\frac{\partial\phi(x)}{\partial x^\nu},
\end{aligned} \tag{7.106}$$

[18]This is seen most easily writing, in matrix form, $\mathbf{x}' = R\mathbf{x}$. The inversion is then $\mathbf{x} = R^{-1}\mathbf{x}'$. For orthogonal matrices $R^{-1} = R^{\mathrm{T}}$, see eq. (1.128), so we get $\mathbf{x} = R^{\mathrm{T}}\mathbf{x}'$. In components, this gives $x_j = R^{\mathrm{T}}_{ji}x_i'$ and, by definition of transpose matrix, $R^{\mathrm{T}}_{ji} = R_{ij}$. Otherwise, working in components, we multiply $x_i' = R_{ij}x_j$ by R_{ik}, to get $R_{ik}x_i' = R_{ik}R_{ij}x_j$ and use eq. (1.129) in the form $R_{ik}R_{ij} = \delta_{kj}$. This gives $R_{ik}x_i' = x_k$ and we then rename $k \to j$.

[19]For spatial indices, the upper/lower position is irrelevant. However, we will see in eq. (7.107) that, for Lorentz indices, the derivative with respect to a quantity with upper index gives a quantity with lower index, so in the final result (7.99) we already use the positioning of the indices appropriate for the extension to the relativistic context.

so

$$\partial_\mu \phi(x) \to \Lambda_\mu{}^\nu \partial_\nu \phi(x)\,, \tag{7.107}$$

which shows that $\partial_\mu \phi(x)$ is a covariant four-vector field. Note that ∂_μ, which is the derivative with respect of x^μ where μ is an upper index, produces a four-vector with lower index.[20] We can also define

$$\partial^\mu \equiv \frac{\partial}{\partial x_\mu}\,, \tag{7.109}$$

and, from $x_\mu = \eta_{\mu\nu} x^\nu$, we can easily prove that $\partial^\mu = \eta^{\mu\nu} \partial_\nu$ and $\partial_\mu = \eta_{\mu\nu} \partial^\nu$, so the index μ in ∂_μ or ∂^μ can be treated as a normal Lorentz index, raised and lowered with $\eta_{\mu\nu}$ and $\eta^{\mu\nu}$. Note that

$$\partial_\mu x^\nu = \delta_\mu^\nu\,, \tag{7.110}$$

while

$$\partial^\mu x^\nu = \eta^{\mu\nu}\,, \tag{7.111}$$

as can be seen by contracting both sides of $\partial_\rho x^\nu = \delta_\rho^\nu$ with $\eta^{\mu\rho}$.

We can similarly work out the transformations of other quantities involving ∂_μ. Consider for instance a tensor field $T^{\mu\nu}(x)$ and form the quantity $\partial_\mu T^{\mu\nu}(x)$. Under Lorentz transformations

$$\begin{aligned}
\frac{\partial T^{\mu\nu}(x)}{\partial x^\mu} &\to \frac{\partial T'^{\mu\nu}(x')}{\partial x'^\mu} \\
&= \frac{\partial x^\alpha}{\partial x'^\mu} \frac{\partial}{\partial x^\alpha} [\Lambda^\mu{}_\rho \Lambda^\nu{}_\sigma T^{\rho\sigma}(x)] \\
&= \Lambda_\mu{}^\alpha \Lambda^\mu{}_\rho \Lambda^\nu{}_\sigma \frac{\partial}{\partial x^\alpha} T^{\rho\sigma}(x)\,,
\end{aligned} \tag{7.112}$$

where we used eq. (7.105) to compute $\partial x^\alpha / \partial x'^\mu$. Using eq. (7.60), we get

$$\Lambda_\mu{}^\alpha \Lambda^\mu{}_\rho = \delta_\rho^\alpha\,, \tag{7.113}$$

and therefore

$$\partial_\mu T^{\mu\nu}(x) \to \Lambda^\nu{}_\sigma \partial_\rho T^{\rho\sigma}(x)\,. \tag{7.114}$$

In terms of $T^\nu(x) \equiv \partial_\mu T^{\mu\nu}(x)$, this reads $T^\nu(x) \to \Lambda^\nu{}_\sigma T^\sigma(x)$, which is the transformation law of a four-vector field. Therefore $\partial_\mu T^{\mu\nu}(x)$ is a four-vector field. From these examples, it is clear that the transformation properties of quantities obtained acting with ∂_μ on tensor fields can be read from the remaining free Lorentz indices. So, as we have seen explicitly, if $\phi(x)$ is a scalar field, $\partial_\mu \phi$ is a four-vector field; similarly, we can show that $\partial_\mu \partial_\nu \phi$ is a Lorentz tensor with two covariant indices, etc. If $V^\mu(x)$ is a four-vector field, manipulations analogous to those performed above show that $\partial_\mu V^\mu(x)$ is a scalar field, while, for instance, $\partial_\mu V_\nu(x)$ is a Lorentz tensor field with two covariant indices; given a Lorentz tensor field $T^{\mu\nu}(x)$, $\partial_\mu T^{\mu\nu}(x)$ is a four-vector field, as we have seen explicitly, and similarly $\partial_\mu \partial_\nu T^{\mu\nu}(x)$ is a scalar field, $\partial_\rho T^{\mu\nu}(x)$ is a Lorentz tensor field with one covariant and two contravariant indices, and so on.

[20]This can also be checked observing, for instance, that if we act with ∂_μ on the scalar field configuration $\phi(x) = v_\nu x^\nu$, with v_ν some four-vector independent of x, we get

$$\partial_\mu (v_\nu x^\nu) = v_\nu \frac{\partial x^\nu}{\partial x^\mu} = v_\nu \delta_\mu^\nu = v_\mu\,, \tag{7.108}$$

so the result is indeed a four-vector with lower index. The fact that $\partial x^\nu / \partial x^\mu = \delta_\mu^\nu$ follows from the fact that this derivative is one if $\nu = \mu$ and is zero otherwise, and it is a tensor, with an upper index ν inherited from x^ν. It is therefore given by δ_μ^ν.

7.3.7 More general lessons

We conclude this section by remarking that the two postulates of Special Relativity emerged from the extraordinary physical intuition of Einstein. From the modern point of view, largely stimulated by Special Relativity, as well as by the developments in the theory of fundamental interactions and quantum field theory, one of the fundamental questions is always what are the symmetries of a given theory. Special Relativity is a statement about the symmetries of Nature at the most fundamental and elementary level, namely, the symmetries of space and time. In the mathematical language that we have developed in this section, the postulates of Special Relativity are equivalent to saying that, as far as coordinate transformations are concerned, the symmetry group of Nature is given by the Lorentz group $SO(3,1)$, rather than just by its rotation subgroup $SO(3)$ [together with Galilean velocity transformations (7.2)]. In fact, with one more decade of deep thinking, Einstein went even (much) further and realized that, when gravitation enters the game, the symmetry transformations are much larger and include all coordinate diffeomorphisms. This, however, is the subject of another fascinating chapter of physics, General Relativity. Modern particle physics is also very largely based on the identification of symmetry groups, in this case at the level of the dynamics; this will be the subject of a quantum field theory course. For the purpose of the present course on classical electrodynamics, we can just stress that the revolution initiated by Special Relativity has gradually permeated basically all of theoretical physics, not only because of its specific concepts, but also for bringing the notion of symmetry groups to the forefront of the modern understanding of Nature.

7.4 Relativistic particle kinematics

7.4.1 Covariant description of particle trajectories

In Newtonian mechanics, the motion of a particle is described by giving the evolution of the three spatial coordinates as a function of time, $x^i(t)$. In our relativistic setting, for a given inertial observer K that uses coordinates $(t, \mathbf{x})$, this amounts to giving the spatial components of x^μ as a function of t or, equivalently, of $x^0 = ct$. In Special Relativity this is not a natural choice, since it obscures the fundamental Lorentz covariance of the equations, by separating artificially the x^0 coordinate from the three spatial coordinates x^i, that together form the four-vector x^μ. Furthermore, the use of time t as a way to parametrize the trajectory is also not natural from the relativistic point of view, since time is not a Lorentz-invariant quantity. However, we have seen in Section 7.2.3 that for a massive particle, moving at a speed v strictly smaller than c, we can introduce its proper time τ, as in eq. (7.33). Since, for motions with $v < c$, we have $ds^2 < 0$, $d\tau = \sqrt{-ds^2}/c$ is a real quantity. Since $d\tau$ is defined in terms of the interval ds^2, it is clearly a Lorentz-invariant quantity, and any inertial observer will agree on its value (apart from an arbitrary choice of the origin of time). For the inertial observer K that uses coordinates $(t, \mathbf{x})$ and sees the particle moving with velocity $v(t)$, the relation between the particle's proper time τ and his/her time t is given by eq. (7.35). This relation is one-to-one since, from eq. (7.34), we see that $d\tau/dt = 1/\gamma > 0$, and can be inverted to obtain $t = t(\tau)$ or, equivalently,

$$x^0 = x^0(\tau). \tag{7.115}$$

Since t can be expressed as a function of τ, instead of using $x^i(t)$ the observer K can use

$$x^i(\tau) \equiv x^i[t(\tau)]. \tag{7.116}$$

As a result, the trajectory of the particle is now described by the set of four functions $\{x^0(\tau), x^i(\tau)\}$ or, in four-vector form, $x^\mu(\tau)$. In this way, the motion of a particle in space-time is described in an explicit covariant manner, through a four-vector x^μ, which is a function of a Lorentz-invariant quantity τ. From the point of view of Lorentz covariance, this is much more natural than using $x^i(t)$, in which we have a three-vector x^i, function of a quantity t, or of x^0, which is the temporal component of a four-vector. In other words, rather than describing the motion of a particle throughout space with a function $\mathbf{x}(t)$, as in Newtonian mechanics, we prefer to use a parametric form $\mathbf{x} = \mathbf{x}(\tau)$ and $t = t(\tau)$. In principle, one could invert the latter to obtain τ as a function of t, $\tau = \tau(t)$, and plug this back into $\mathbf{x} = \mathbf{x}(\tau)$ to get back the more usual description in terms of $\mathbf{x}(t)$. However, the description in terms of $x^\mu(\tau)$ has the advantage of being explicitly Lorentz covariant. The functions $x^\mu(\tau)$ define the so-called particle *world-line*.

Given a trajectory $x^\mu(\tau)$, in an infinitesimal interval $d\tau$ of proper time $x^\mu(\tau)$ changes by an amount $dx^\mu(\tau) = (dx^0(\tau), d\mathbf{x}(\tau))$. The interval ds^2 separating the events $x^\mu(\tau)$ and $x^\mu(\tau) + dx^\mu(\tau)$ is

$$ds^2 = -[dx^0(\tau)]^2 + [d\mathbf{x}(\tau)]^2. \tag{7.117}$$

From the definition (7.33) of proper time, we then have

$$[dx^0(\tau)]^2 - [d\mathbf{x}(\tau)]^2 = c^2 d\tau^2, \tag{7.118}$$

or, in a more explicitly covariant form,

$$\eta_{\mu\nu} dx^\mu(\tau) dx^\nu(\tau) = -c^2 d\tau^2. \tag{7.119}$$

From $x^\mu(\tau)$ we can form the *four-velocity* u^μ, defined as

$$u^\mu(\tau) = \frac{dx^\mu(\tau)}{d\tau}. \tag{7.120}$$

Since x^μ is a four-vector, and τ is Lorentz-invariant, $u^\mu(\tau)$ is a four-vector. From eq. (7.119) we immediately get

$$u^2 \equiv \eta_{\mu\nu} u^\mu u^\nu = -c^2. \tag{7.121}$$

Consider an inertial frame K, with coordinates $(t, \mathbf{x})$, where the particle moves with velocity $\mathbf{v}(t)$. Using $d\tau = dt/\gamma$ from eq. (7.34), $x^0 = ct$, and $dx^i/dt = v^i$, we get

$$u^0 = \gamma \frac{dx^0}{dt} = \gamma c, \tag{7.122}$$

$$u^i = \gamma \frac{dx^i}{dt} = \gamma v^i, \tag{7.123}$$

from which one can check that eq. (7.121) indeed holds.

7.4.2   Action of a free relativistic particle

We now introduce a relativistic generalization of the action principle. This will be useful as a first example of a relativistic action principle, and also provides a clean conceptual way of defining energy and momentum for a free relativistic particle. Recall that, in classical mechanics, for a particle described by coordinates $\mathbf{q}(t)$, with components $q_i(t)$, the Lagrangian is a functional of the coordinates and their time derivatives, $L[\mathbf{q}, \dot{\mathbf{q}}]$, and the corresponding action is

$$S = \int dt \, L[\mathbf{q}, \dot{\mathbf{q}}] \,. \tag{7.124}$$

The conjugate momentum is defined by

$$p_i = \frac{\delta L}{\delta \dot{q}_i} \,. \tag{7.125}$$

The Hamiltonian is then defined as

$$H[\mathbf{q}, \mathbf{p}] = \dot{\mathbf{q}} \cdot \mathbf{p} - L[\mathbf{q}, \dot{\mathbf{q}}] \,, \tag{7.126}$$

where $\dot{\mathbf{q}}$ is expressed in terms of $\mathbf{p}$ (and possibly of $\mathbf{q}$) by inverting eq. (7.125). Writing $-H[\mathbf{q}, \mathbf{p}] = L[\mathbf{q}, \dot{\mathbf{q}}] - \dot{\mathbf{q}} \cdot \mathbf{p}$, and comparing it to the definition of Legendre transform in eq. (5.108) and the discussion following it, we see that $-H$ is the Legendre transform of L, and $\mathbf{q}$ and $\mathbf{p}$ are conjugate variables, in the sense of the Legendre transform.

The simplest example is provided by a particle of mass m in an external potential $V(\mathbf{q})$. The Lagrangian is

$$L = \frac{1}{2} m \dot{\mathbf{q}}^2 - V(\mathbf{q}) \,, \tag{7.127}$$

where $\dot{\mathbf{q}} = \mathbf{v}$ is the velocity of the particle. The momentum is then $p_i = m\dot{q}_i$, whose inversion is just $\dot{q}_i = p_i/m$, and we get

$$H = \frac{\mathbf{p}^2}{2m} + V(\mathbf{q}) \,. \tag{7.128}$$

For a free relativistic particle the action must be Lorentz invariant and must therefore be obtained from the integration of a Lorentz-invariant first-order differential. The only available Lorentz-invariant quantity (assuming the particle to be point-like and without internal degrees of freedom, such as spin) is the proper time of the particle, so we must have

$$S_{\text{free}} = -\alpha \int d\tau \, , \tag{7.129}$$

for some constant α.[21] Using $d\tau = dt/\gamma$, we can also rewrite this as

$$S_{\text{free}} = -\alpha \int dt \sqrt{1 - \frac{v^2}{c^2}} \, , \tag{7.130}$$

so the corresponding Lagrangian is

$$L = -\alpha \sqrt{1 - \frac{v^2}{c^2}} \, . \tag{7.131}$$

Expanding the square root to order v^2/c^2 we get

$$L = -\alpha + \alpha \frac{v^2}{2c^2} + \mathcal{O}\left(\frac{v^4}{c^4}\right) . \tag{7.132}$$

Comparison with the non-relativistic limit shows that $\alpha = mc^2$, so that, apart from a constant term that has no effect on the equations of motion, we get the non-relativistic Lagrangian of a free particle, $L = (1/2)mv^2$. Therefore, the action of a free relativistic particle is

$$\boxed{S_{\text{free}} = -mc^2 \int d\tau \, ,} \tag{7.133}$$

or, equivalently,

$$S_{\text{free}} = -mc^2 \int dt \sqrt{1 - \frac{v^2}{c^2}} \, . \tag{7.134}$$

7.4.3   Relativistic energy and momentum

From the relativistic Lagrangian, we can get the relativistic momentum using eq. (7.125),

$$\begin{aligned} p_i &= \frac{\delta}{\delta v_i}\left[-mc^2\sqrt{1 - \frac{v^2}{c^2}}\right] \\ &= \gamma m v_i \, , \end{aligned} \tag{7.135}$$

and the energy

$$\begin{aligned} E &= \mathbf{p}\cdot\mathbf{v} - L \\ &= \gamma mc^2 \, . \end{aligned} \tag{7.136}$$

In conclusion, we have obtained the relativistic expression for the energy and momentum of a particle,

$$\boxed{E = \gamma mc^2 = \frac{mc^2}{\sqrt{1 - (v^2/c^2)}} \, ,} \tag{7.137}$$

[21] This kind of reasoning, based on the most general structure that respects some symmetry principle, is quite typical of modern field theory. In more detail, we wish to construct the action for a free relativistic point-like particle. The fact that the particle has no inner structure means that the only variables that we have at our disposal are its proper time τ and the four-vector $x^\mu(\tau)$ that describes its trajectory. Requiring invariance under spatial and temporal translations implies that the action must be invariant under the transformation $x^\mu(\tau) \to x^\mu(\tau) + a^\mu$, where a^μ is an arbitrary constant four-vector. This means that any dependence of the action on $x^\mu(\tau)$ can only enter through $dx^\mu(\tau)/d\tau$, i.e., u^μ. The most general action could then have the form $S = \int d\tau f(u^\mu)$, for some scalar function f. To construct a scalar out of u^μ we need to contract its Lorentz index with another four-vector (or, more generally, to consider $u^\mu u^\nu$ and contract its two Lorentz indices with a tensor with two indices, or to consider $u^\mu u^\nu u^\rho$ and contract it with a tensor with three indices, and so on). Since we are considering a free particle, there is no external four-vector field (such as the gauge potential A^μ) that could be used here, and the contraction of u_μ with itself gives a constant, because of eq. (7.121). Therefore, f must be a constant, independent of u_μ. Note that this state of affairs changes completely if we assume that the particle has an internal structure. In this case, new degrees of freedom would enter the action; for instance, a massive particle could have a spin $\mathbf{s}$, that in a covariant setting can also be described by a four-vector s^μ, defined by the fact that, in the rest frame of the particle, $s^\mu = (0, \mathbf{s})$. The action then becomes more complicated, and in general contains an infinite number of possible terms that can be organized in order of importance, in a sense very similar to the multipole expansion discussed in Chapter 6. This is the logic behind the use of effective actions in modern field theory.

and

$$\mathbf{p} = \gamma m \mathbf{v} = \frac{m\mathbf{v}}{\sqrt{1 - (v^2/c^2)}} \,. \tag{7.138}$$

Equation (7.137) shows that, even for $v = 0$, a particle still has an energy

$$E = mc^2 \,, \tag{7.139}$$

associated with its mass (with no doubt, the most famous formula of physics for the general public!). Expanding to second order in v^2/c^2, the next term gives the Newtonian expression for the kinetic energy,

$$E = mc^2 + \frac{1}{2}mv^2 + \mathcal{O}\left(\frac{v^4}{c^4}\right) \,. \tag{7.140}$$

Comparing eqs. (7.137) and (7.138) to eqs. (7.122) and (7.123) we find that

$$E/c = mu^0 \,, \tag{7.141}$$
$$p^i = mu^i \,. \tag{7.142}$$

Therefore, defining

$$p^\mu \equiv mu^\mu \,, \tag{7.143}$$

we have

$$p^\mu = (E/c, \mathbf{p}) \,. \tag{7.144}$$

Since we have already proven that u^μ is a four-vector, this shows that E/c and $\mathbf{p}$ form a four-vector. We will refer to p^μ as the four-momentum of the particle. Note, from eq. (7.121), that

$$\begin{aligned} p^2 &\equiv \eta_{\mu\nu} p^\mu p^\nu \\ &= -m^2 c^2 \,, \end{aligned} \tag{7.145}$$

or, explicitly,[22]

$$E^2 = m^2 c^4 + |\mathbf{p}|^2 c^2 \,. \tag{7.146}$$

For physical reasons we only keep the positive root of this equation,[23] so

$$E = \sqrt{m^2 c^4 + |\mathbf{p}|^2 c^2} \,, \tag{7.147}$$

which is the dispersion relation for relativistic particles, that we already anticipated in eq. (3.50). Equations (7.137) and (7.138) can also be combined to give

$$\mathbf{p} = \frac{E\mathbf{v}}{c^2} \,. \tag{7.148}$$

From eq. (7.147), if $m = 0$ we have

$$E = |\mathbf{p}|c \,, \qquad (m = 0) \,. \tag{7.149}$$

[22] A comment on notation. When energy appears in equations where also the electric field appears, as was the case for instance in eq. (3.54), to avoid confusion we use $\mathcal{E}$ for the energy and $\mathbf{E}$ for the electric field (so $E = |\mathbf{E}|$ denotes the modulus of the electric field). In a context such as here, where only energy is involved, we use the more common notation E for the energy.

[23] A really satisfying understanding of the negative root only comes from quantum field theory. See e.g., Section 4.1 of Maggiore (2005).

Inserting this in eq. (7.148) and taking the modulus, we get $|\mathbf{v}| = c$. Massless particles always travel at the speed of light and, vice versa, if $|\mathbf{v}| = c$ then eqs. (7.147) and (7.148) give $m = 0$.

Consider now a particle with energy E, subject to an external interaction that changes its four-momentum. Applying a time derivative to both sides of $p_\mu p^\mu = -m^2 c^2$ it follows that

$$
\begin{aligned}
0 &= p_\mu \frac{dp^\mu}{dt} \\
&= -\frac{E}{c^2}\frac{dE}{dt} + \mathbf{p} \cdot \frac{d\mathbf{p}}{dt} \, .
\end{aligned}
\tag{7.150}
$$

Therefore, using eq. (7.148), we see that, in a full relativistic setting, we still have

$$
\boxed{\frac{dE}{dt} = \mathbf{v} \cdot \frac{d\mathbf{p}}{dt} \, .}
\tag{7.151}
$$

In the non-relativistic limit, $d\mathbf{p}/dt$ is equal to the force $\mathbf{F}$ acting on the particle, and we recover the result that $dE/dt = \mathbf{v} \cdot \mathbf{F}$, i.e., dE/dt is equal to the work per unit time performed by the external force.

The fact that p^μ is a four-vector immediately tells us how energy and momentum change under Lorentz transformations. They will simply transform as any other four-vector; so, for instance, if we make a boost with a velocity v_0 along the x axis, we can read the transformation from eqs. (7.22) and (7.23) with x^0 replaced by E/c and x replaced by p_x, which gives

$$
\begin{aligned}
E' &= \gamma(v_0)(E + v_0 p_x) \,, \tag{7.152} \\
p'_x &= \gamma(v_0)\left(p_x + \frac{v_0}{c^2} E\right) \,, \tag{7.153}
\end{aligned}
$$

while $p'_y = p_y$ and $p'_z = p_z$. In particular, if before the boost the particle is at rest, after the boost we will have $E = \gamma(v_0)mc^2$ and $p_L = \gamma(v_0)mv_0$, where we eliminated the prime from the boosted quantity and we used the more general notation p_L for the component of the momentum in the direction of the boost, i.e., in the longitudinal direction. Eliminating v_0 in terms of the rapidity ζ from eq. (7.21), we have $\gamma(v_0) = \cosh\zeta$, so $E = mc^2 \cosh\zeta$ and $p_L = mc \sinh\zeta$. Then, we have

$$
\frac{(E/c) + p_L}{(E/c) - p_L} = e^{2\zeta} \,,
\tag{7.154}
$$

and therefore[24]

$$
\zeta = \frac{1}{2} \log\left[\frac{(E/c) + p_L}{(E/c) - p_L}\right] \,.
\tag{7.156}
$$

[24] This expression for the rapidity is useful in experimental particle physics. Note that the rapidity, that we denoted by ζ, is also often denoted by y in this particle physics context, while the letter η is instead reserved to the 'pseudo-rapidity', defined as

$$
\eta = \frac{1}{2} \log\left(\frac{|\mathbf{p}| + p_L}{|\mathbf{p}| - p_L}\right) \,.
\tag{7.155}
$$

For an ultra-relativistic particle, $E \simeq |\mathbf{p}|c$, and the pseudo-rapidity becomes the same as the rapidity.

A useful property of the rapidity is that, if we consider a particle characterized by a value ζ of the rapidity, i.e., with $E = mc^2 \cosh\zeta$ and $p_L = mc \sinh\zeta$, and we perform a boost with rapidity ζ_0 along the longitudinal direction, using the analogous of eq. (7.19) we get

$$
\begin{aligned}
E'/c &= (E/c)\cosh\zeta_0 + p_L \sinh\zeta_0 \\
&= mc(\cosh\zeta \cosh\zeta_0 + \sinh\zeta \sinh\zeta_0) \\
&= mc \cosh(\zeta + \zeta_0) \,,
\end{aligned}
$$

and, similarly,

$$
p'_L = mc \sinh(\zeta + \zeta_0) \,.
$$

In other words, under boosts in the longitudinal directions, ζ transforms simply as $\zeta \to \zeta + \zeta_0$ (which is completely analogous to the composition of angles for subsequent rotations around the same axis). Therefore, $d\zeta$ is invariant under longitudinal boosts.

Covariant formulation of electrodynamics

We now use the formalism developed in the previous chapter to rewrite Maxwell's equations in a form that will make explicit their covariance under Lorentz transformation. We find it convenient to start by studying the source term, i.e., the charge and current density. After having established that they can be assembled into a four-vector, we will then see how to write Maxwell's equations in a covariant form.

8.1 The four-vector current

We begin by studying the Lorentz transformation properties of the charge density ρ and of the current density $\mathbf{j}$. For a single point-like charged particle, the charge density and the current density have been given in eqs. (3.25) and (3.27). If we denote by $\mathbf{x}(t)$ the trajectory of a particle with charge q, and by $\mathbf{x}$ a generic point in space, we can rewrite these expressions as

$$\rho(x) = q\delta^{(3)}[\mathbf{x} - \mathbf{x}(t)], \tag{8.1}$$

$$\mathbf{j}(x) = q\mathbf{v}(t)\delta^{(3)}[\mathbf{x} - \mathbf{x}(t)], \tag{8.2}$$

where $x = (ct, \mathbf{x})$ and $\mathbf{v}(t) = d\mathbf{x}(t)/dt$. A simple way of understanding their properties under Lorentz transformations is to describe the trajectory using the four-vector $x^\mu(\tau)$, defined as in Section 7.4.1, instead of $\mathbf{x}(t)$. Indeed, consider the quantity

$$j^\mu(x) \equiv q \int cd\tau' \, u^\mu(\tau')\delta^{(4)}[x - x(\tau')], \tag{8.3}$$

where the factor of c is inserted for later convenience. Note that $u^\mu(\tau) = dx^\mu(\tau)/d\tau$ is a four-vector and proper time τ is a Lorentz scalar. Furthermore, by definition,

$$\int d^4x \, \delta^{(4)}(x) = 1. \tag{8.4}$$

Under a Lorentz transformation $x^\mu \rightarrow x'^\mu = \Lambda^\mu{}_\nu x^\nu$ we have

$$d^4x \rightarrow d^4x' = (\det \Lambda)\, d^4x. \tag{8.5}$$

Since $\det \Lambda = 1$, d^4x is Lorentz invariant. From eq. (8.4), then, also $\delta^{(4)}(x)$ is Lorentz invariant.[1] It then follows that $j^\mu(x)$, defined by

[1] Equivalently, under a generic linear transformation $x^\mu \rightarrow x'^\mu = \Lambda^\mu{}_\nu x^\nu$ we have

$$\delta^{(4)}(x) \rightarrow \frac{1}{\det \Lambda}\delta^{(4)}(x), \tag{8.6}$$

and, for a Lorentz transformation, $\det \Lambda = 1$.

eq. (8.3), is a four-vector, since it is obtained multiplying the four-vector $u^\mu(\tau')$ by the Lorentz invariant quantity $\delta^{(4)}(x)$ and integrating over the Lorentz invariant differential $d\tau'$. The integral over $d\tau'$ in eq. (8.3) can be computed explicitly using eq. (1.61), that we rewrite with the notation

$$\delta[f(\tau')] = \sum_i \frac{1}{|f'(\tau_i)|} \delta(\tau' - \tau_i),\qquad(8.7)$$

where τ_i are the simple zeros of the function $f(\tau')$. We use this identity for $f(\tau') = ct - x^0(\tau')$ and we denote by τ the value of τ' such that $ct = x^0(\tau'),$[2] which is nothing but the proper time of the particle considered. Then eq. (8.3) can be rewritten as

$$
\begin{aligned}
j^\mu(x) &= q \int cd\tau'\, u^\mu(\tau')\delta^{(3)}[\mathbf{x} - \mathbf{x}(\tau')]\delta[ct - x^0(\tau')] \\
&= q \int cd\tau'\, u^\mu(\tau')\delta^{(3)}[\mathbf{x} - \mathbf{x}(\tau')] \frac{1}{|dx^0/d\tau|}\delta(\tau' - \tau) \\
&= qc\frac{u^\mu(\tau)}{u^0(\tau)}\delta^{(3)}[\mathbf{x} - \mathbf{x}(\tau)],\qquad(8.8)
\end{aligned}
$$

where we used $u^0 = dx^0/d\tau$. Recalling, from eqs. (7.122) and (7.123), that $u^0 = \gamma c$ and $u^i = \gamma v^i$, and comparing with eqs. (8.1) and (8.2), we see that

$$j^\mu = (c\rho, \mathbf{j}) .\qquad(8.9)$$

Therefore, the charge density ρ (times c) and the current density $\mathbf{j}$, defined by eqs. (8.1) and (8.2), are the temporal and spatial components, respectively, of the contravariant four-vector j^μ defined by eq. (8.3). Since any distribution of charges and currents can be obtained by superposition of individual charges, eq. (8.9) is completely general. If we lower the Lorentz index, according to our metric signature $\eta_{\mu\nu} = (-, +, +, +)$, we get $j_\mu = (-c\rho, \mathbf{j})$.

In terms of j^μ, the conservation equation (3.22) can be rewritten in an explicitly Lorentz covariant form as

$$\boxed{\partial_\mu j^\mu = 0 .}\qquad(8.10)$$

Also observe that, in the covariant language, the total charge Q over all of space is obtained from a volume integral of the $\mu = 0$ component of the four-vector j^μ,

$$\boxed{Q = \frac{1}{c}\int d^3x\, j^0(t, \mathbf{x}) .}\qquad(8.11)$$

Since the right-hand side is integrated over $\mathbf{x}$, but not over t, a priori the left-hand side could have been a function of time. However, as we saw in Section 3.2.1, current conservation implies that the charge Q is conserved in the sense of eq. (3.24), so, in particular, $dQ/dt = 0$ if the integral in eq. (8.11) is over all space, and we set the boundary condition that $\mathbf{j}$ has compact support, or anyhow vanishes sufficiently fast at infinity.[3]

---

[2]The solution exists and is unique because, for any acceptable physical trajectory (e.g., a trajectory that do not go back and forth in time) the parametrization of the trajectory in terms of proper time is in one-to-one correspondence with that in terms of "coordinate time" t. Furthermore, $dx^0/d\tau > 0$, i.e., t and τ increase in the same direction.

[3]The charge given in eq. (8.11), beside being conserved, is also a Lorentz scalar. This is not readily apparent from eq. (8.11) because the integration is only over the spatial variables d^3x, and the integrand, j^0 is the temporal component of a four-vector. The explicit proof is somewhat technical, and we give it in Solved Problem 8.1, where we also provide an explicitly Lorentz-invariant expression for it.

8.2   The four-vector potential A_μ and the $F_{\mu\nu}$ tensor

We next consider the scalar potential $\phi(x)$ and the vector potential $\mathbf{A}(x)$. We combine them into a contravariant four-vector field $A^\mu(x)$,

$$A^\mu = (\phi/c, \mathbf{A}),\qquad(8.12)$$

and we will show in Section 8.3 that, with this assignment, i.e., with this definition of how ϕ and $\mathbf{A}$ behave under Lorentz transformations, Maxwell's equations are Lorentz covariant, so that, if they hold in a frame, they also hold in a Lorentz-transformed frame. For rotations, this is already explicit from the vector form (3.8)–(3.11). The new result will be that they are covariant under the larger Lorentz group, i.e., that Special Relativity is an underlying symmetry of electromagnetism.

First, it is useful to observe that the assignment of ϕ and $\mathbf{A}$ to the four-vector field A^μ is consistent with gauge invariance, and in fact allows us to write the gauge transformation (3.86) compactly. This can be shown observing first that, lowering the Lorentz index,

$$A_\mu = (-\phi/c, \mathbf{A}).\qquad(8.13)$$

Then eq. (3.86) takes the simple and elegant form

$$\boxed{A_\mu \to A'_\mu = A_\mu - \partial_\mu \theta.}\qquad(8.14)$$

Taking θ to be a Lorentz scalar function, $\partial_\mu \theta$ is a four-vector field, and the gauge transformation preserves the fact that A_μ is a covariant four-vector field. For the contravariant field A^μ, raising the indices in eq. (8.14), we have $A^\mu \to A^\mu - \partial^\mu \theta$.

We next introduce the Lorentz tensor field

$$\boxed{F^{\mu\nu} = \partial^\mu A^\nu - \partial^\nu A^\mu.}\qquad(8.15)$$

This tensor is antisymmetric and therefore has six independent components. We can work them out explicitly as follows. Consider first F^{0i},

$$\begin{aligned}
F^{0i} &= \partial^0 A^i - \partial^i A^0 \\
&= \frac{1}{c}\left(-\partial_t A^i - \partial^i \phi\right).
\end{aligned}\qquad(8.16)$$

Comparing with eq. (3.83) (and recalling that $\partial^i = \partial_i$ is the i-th component of $\boldsymbol{\nabla}$) we discover that F^{0i} is just the i-th component of the electric field (divided by c),[4]

$$\boxed{F^{0i} = \frac{1}{c}E^i,}\qquad(8.17)$$

[4]The extra factors of c or $1/c$ in several of the subsequent formulas, as well as the relative factor of c between the temporal and spatial components of A^μ in eq. (8.12), are absent in Gaussian units, see App. A, and this is among the reasons why Gaussian units provide nicer expressions for relativistic equations.

or $E_i/c = -F_{0i} = F_{i0}$. Similarly (keeping for simplicity all indices lower, since spatial indices are raised and lowered with δ_{ij}),

$$
\begin{aligned}
\epsilon_{ijk}F_{jk} &= \epsilon_{ijk}(\partial_j A_k - \partial_k A_j) \\
&= 2\epsilon_{ijk}\partial_j A_k \,,
\end{aligned}
\tag{8.18}
$$

where in the last equality we used the antisymmetry of ϵ_{ijk}. However, $\epsilon_{ijk}\partial_j A_k$ is just $(\boldsymbol{\nabla} \times \mathbf{A})_i$. Then, comparison with eq. (3.80) shows that

$$
B_i = \frac{1}{2}\epsilon_{ijk}F_{jk} \,,
\tag{8.19}
$$

which can be inverted to give

$$
\boxed{F_{ij} = \epsilon_{ijk}B_k \,.}
\tag{8.20}
$$

Thus, the six independent components of $F_{\mu\nu}$ are given by the three components of $\mathbf{E}$ and the three components of $\mathbf{B}$. Explicitly, as a matrix,[5]

$$
F^{\mu\nu} = \begin{pmatrix} 0 & E_1/c & E_2/c & E_3/c \\ -E_1/c & 0 & B_3 & -B_2 \\ -E_2/c & -B_3 & 0 & B_1 \\ -E_3/c & B_2 & -B_1 & 0 \end{pmatrix}.
\tag{8.21}
$$

Under a gauge transformation (8.14),

$$
\begin{aligned}
F_{\mu\nu} &\rightarrow F_{\mu\nu} - (\partial_\mu\partial_\nu\theta - \partial_\nu\partial_\mu\theta) \\
&= F_{\mu\nu} \,,
\end{aligned}
\tag{8.22}
$$

(we always consider transformations such that the function θ is infinitely differentiable, and therefore on it the derivatives commute), so $F_{\mu\nu}$ is gauge invariant. Indeed, we have already seen in Section 3.3 that the electric field and the magnetic field are gauge invariant.

8.3 Covariant form of Maxwell's equations

We are now ready to write Maxwell's equations in covariant form. Consider the equation

$$
\boxed{\partial_\mu F^{\mu\nu} = -\mu_0 j^\nu \,,}
\tag{8.23}
$$

where j^ν is given by eq. (8.9).[6] Note that eq. (8.23) is a four-vector equation, since the index ν is free. For $\nu = 0$, $\partial_\mu F^{\mu 0}$ is the same as $\partial_i F^{i0}$, since $F^{00} = 0$. Then, using $F^{i0} = -E^i/c$ and $j^0 = c\rho$, we see that eq. (8.23) with $\nu = 0$ is the same as

$$
\boldsymbol{\nabla}\cdot\mathbf{E} = \mu_0 c^2 \rho \,.
\tag{8.24}
$$

[5]Note that the signs in the matrix elements depend on our choice of metric $\eta_{\mu\nu} = (-,+,+,+)$. To compare with a text that uses $\eta_{\mu\nu} = (+,-,-,-)$ [such as Jackson (1998)], one must observe that A^μ is always defined in terms of ϕ and $\mathbf{A}$ from eq. (8.12), but now $A_\mu = \eta_{\mu\nu}A^\nu$ differs from ours by an overall minus sign (while $\partial_\mu = \partial/\partial x^\mu$ is unchanged because it involves x^μ with an upper index); therefore also $F_{\mu\nu} = \partial_\mu A_\nu - \partial_\nu A_\mu$ has the opposite sign, and the same holds for $F^{\mu\nu}$, since we need two η factors for raising its indices, $F^{\mu\nu} = \eta^{\mu\rho}\eta^{\nu\sigma}F_{\rho\sigma}$.

[6]As remarked in Note 5, if one rather uses the signature $\eta_{\mu\nu} = (-,+,+,+)$ for the metric, opposite to ours, $F^{\mu\nu}$ also differs from our definition by an overall sign, and then eq. (8.23) is replaced by $\partial_\mu F^{\mu\nu} = +\mu_0 j^\nu$.

For $\nu = i$, instead,

$$
\begin{aligned}
\partial_\mu F^{\mu i} &= \partial_0 F^{0i} + \partial_j F^{ji} \\
&= \frac{1}{c^2} \partial_t E^i - \partial_j \epsilon^{ijk} B^k \\
&= \left[\frac{1}{c^2} \partial_t \mathbf{E} - \mathbf{\nabla} \times \mathbf{B} \right]^i .
\end{aligned} \tag{8.25}
$$

Therefore, eq. (8.23) with $\nu = i$ reduces to the vector equation

$$
\mathbf{\nabla} \times \mathbf{B} - \frac{1}{c^2} \frac{\partial \mathbf{E}}{\partial t} = \mu_0 \mathbf{j} . \tag{8.26}
$$

Note that eqs. (8.24) and (8.26) are written in terms of only two parameters: μ_0, that appears in eq. (8.23), and c, which physically represents the speed of light and, formally, entered these equations through the definition of the Lorentz group [defined by the condition that Lorentz transformations leave invariant the quadratic form (7.13)] and, from there, entered all other formulas of this chapter through $x^\mu = (ct, \mathbf{x})$, $\partial_0 = (1/c)\partial_t$, and so on.

We see that eqs. (8.24) and (8.26) are the same as eqs. (3.1) and (3.2) once we make the identification

$$
\epsilon_0 \mu_0 = \frac{1}{c^2} . \tag{8.27}
$$

The quantity that we formally denoted by c in eq. (3.7) is therefore the same as the speed of light that we are using here, as the notation in eq. (3.7) already anticipated.[7]

In terms of A^μ, eq. (8.23) reads

$$
\Box A^\nu - \partial^\nu (\partial_\mu A^\mu) = -\mu_0 j^\nu , \tag{8.28}
$$

which puts together eqs. (3.84) and (3.85) into a single four-vector equation.[8] Also observe that the Lorenz gauge condition (3.89) takes the compact form

$$
\partial_\mu A^\mu = 0 . \tag{8.29}
$$

Therefore, in the Lorenz gauge, eq. (8.28) becomes a simple wave equation,

$$
\Box A^\mu = -\mu_0 j^\mu , \tag{8.30}
$$

which, again, unifies the scalar and vector equations (3.90) and (3.91) into a single four-vector equation.

Let us now turn to the second pair of Maxwell's equations, eqs. (3.10) and (3.11). These equations do not involve the sources and we have

[7]We will confirm the interpretation of $1/\sqrt{\epsilon_0 \mu_0}$ as the speed of light in Chapter 9, where we will see that it is indeed the velocity at which electromagnetic waves travel in vacuum.

[8]The equation with $\nu = i$ reduces indeed trivially to eq. (3.85). Equation (8.28) with $\nu = 0$ can be written as

$$
\Box(\phi/c) - \partial^0 [\partial_0(\phi/c) + \mathbf{\nabla}\cdot\mathbf{A}] = -\mu_0 c \rho .
$$

Using $\Box - \partial^0 \partial_0 = \mathbf{\nabla}^2$, $\partial^0 = -\partial_0 = -(1/c)\partial_t$ and $\mu_0 c^2 = 1/\epsilon_0$ we get back eq. (3.84).

seen that, once the electric and magnetic fields are written in terms of the gauge potentials, they reduce to mathematical identities. In the covariant formalism this comes out as follows. We define the tensor

$$\tilde{F}^{\mu\nu} = \frac{1}{2}\epsilon^{\mu\nu\rho\sigma}F_{\rho\sigma}. \tag{8.31}$$

This is called the tensor *dual* to $F_{\mu\nu}$. Note that, since $\epsilon^{\mu\nu\rho\sigma}$ is totally antisymmetric, we can also rewrite it as

$$\tilde{F}^{\mu\nu} = \frac{1}{2}\epsilon^{\mu\nu\rho\sigma}(\partial_\rho A_\sigma - \partial_\sigma A_\rho) = \epsilon^{\mu\nu\rho\sigma}\partial_\rho A_\sigma. \tag{8.32}$$

Again because of the antisymmetry of $\epsilon^{\mu\nu\rho\sigma}$, we have the mathematical identity $\epsilon^{\mu\nu\rho\sigma}\partial_\mu\partial_\rho = 0$ (as an operator acting on any infinitely differentiable function, on which ∂_μ and ∂_ρ commute), since $\partial_\mu\partial_\rho$ is symmetric in (μ, ρ) and its contraction with the antisymmetric tensor $\epsilon^{\mu\nu\rho\sigma}$ vanishes. Therefore,

$$\partial_\mu\tilde{F}^{\mu\nu} = 0. \tag{8.33}$$

Proceeding as before (and observing that $\epsilon^{0\nu\rho\sigma}$ is different from zero only when ν, ρ, σ are all spatial indices, and ϵ^{0ijk} is equal to the three-dimensional tensor ϵ^{ijk}) one can check that, for $\nu = 0$, eq. (8.33) gives eq. (3.10), while for $\nu = i$ it gives eq. (3.11).

Equations (8.23) and (8.33) are therefore Maxwell's equations in covariant form. From these expressions, it is explicit that the underlying symmetry of Maxwell's equations is given by the Lorentz group, i.e., spatial rotations and Lorentz boosts, since both sides of eq. (8.23) transform as a four-vector, while eq. (8.33) sets a four-vector to zero, which is of course also a condition that, if it holds in a reference frame, holds in any other frame related by a Lorentz transformation. The covariant formalism therefore unveils a symmetry that, in the original formulation (3.8–3.11), was actually already present, but was not readily visible.

8.4   Energy-momentum tensor of the electromagnetic field

We can now use the covariant formalism to discuss the energy and momentum of the electromagnetic field. Let us consider the tensor

$$T^{\mu\nu} = -\frac{1}{\mu_0}\left(F^{\mu\rho}F_\rho{}^\nu + \frac{1}{4}\eta^{\mu\nu}F^{\rho\sigma}F_{\rho\sigma}\right). \tag{8.34}$$

This is called the *energy-momentum tensor* of the electromagnetic field, for reasons that we will now explain. The (00) component can be written as

$$\begin{aligned}
T^{00} &= -\frac{1}{\mu_0}\left[F^{0i}F_i{}^0 - \frac{1}{4}\left(2F^{0i}F_{0i} + F^{ij}F_{ij}\right)\right] \\
&= -\frac{1}{\mu_0}\left(\frac{1}{2}F^{0i}F_{0i} - \frac{1}{4}F^{ij}F_{ij}\right), \tag{8.35}
\end{aligned}$$

since $F_i{}^0 = -F_{i0} = F_{0i}$. Using eqs. (8.17) and (8.20), as well as eq. (8.27), we get

$$T^{00} = \frac{1}{2}\left(\epsilon_0 \mathbf{E}^2 + \mu_0^{-1} \mathbf{B}^2\right) . \tag{8.36}$$

Comparing with eq. (3.43), we see that T^{00} is the energy density u of the electromagnetic field. Similarly, setting $\mu = 0$ and $\nu = i$ in eq. (8.34), we find that

$$T^{0i} = \frac{1}{c}S^i , \tag{8.37}$$

where $\mathbf{S}$ is the Poynting vector (3.34). Given that $T^{\mu\nu}$ is symmetric, we also have $T^{i0} = T^{0i}$. Finally, setting $\mu = i$ and $\nu = j$ in eq. (8.34), and using again eq. (8.27), we get

$$T^{ij} = \epsilon_0 \left(\frac{1}{2}E^2\delta_{ij} - E_iE_j\right) + \mu_0^{-1}\left(\frac{1}{2}B^2\delta_{ij} - B_iB_j\right) , \tag{8.38}$$

which is the same as eq. (3.64). We therefore see that the energy density u of the electromagnetic field given in eq. (3.43), $1/c$ times the Poynting vector S^i given in eq. (3.34), and the Maxwell stress tensor given in eq. (3.64), form together a Lorentz tensor $T^{\mu\nu}$.[9]

The conservation equations discussed separately in Section 3.2.2 can then be derived, in a unified manner, in terms of the Lorentz tensor $T^{\mu\nu}$. Applying ∂_μ to both sides of eq. (8.34), and using the equation of motion (8.23), we find[10]

$$\boxed{\partial_\nu T^{\mu\nu} = -F^{\mu\nu}j_\nu .} \tag{8.39}$$

In particular, setting $\mu = 0$, we have

$$\partial_0 T^{00} + \partial_i T^{0i} = -F^{0i}j_i . \tag{8.40}$$

Using $\partial_0 = (1/c)\partial_t$, $T^{00} = u$, $T^{i0} = T^{0i} = (1/c)S^i$ and $F^{0i} = E^i/c$, we can rewrite this as

$$\partial_t u + \boldsymbol{\nabla}\cdot\mathbf{S} = -\mathbf{E}\cdot\mathbf{j} , \tag{8.41}$$

which is the same as eq. (3.47). Energy conservation is therefore the $\nu = 0$ component of eq. (8.39).

Similarly, recalling that $j_0 = -c\rho$, the $\mu = i$ component of eq. (8.39) gives

$$\begin{aligned}
\partial_0 T^{0i} + \partial_j T^{ij} &= -F^{i0}j_0 - F^{ik}j_k \\
&= -\rho E^i - \epsilon_{ikl}j_kB_l \\
&= -(\rho\mathbf{E} + \mathbf{j}\times\mathbf{B})_i .
\end{aligned} \tag{8.42}$$

Writing again $\partial_0 = (1/c)\partial_t$, and observing, from eqs. (3.57) and (8.37), that

$$T^{0i}(x) = cg^i(x) , \tag{8.43}$$

we get

$$\frac{\partial g_i}{\partial t} + \partial_j T_{ij} = -(\rho\mathbf{E} + \mathbf{j}\times\mathbf{B})_i , \tag{8.44}$$

[9]It is also interesting to observe that $T^{\mu\nu}$ is traceless. In fact, using $\eta_{\mu\nu}\eta^{\mu\nu} = 4$ in eq. (8.34),

$$\begin{aligned}
\eta_{\mu\nu}T^{\mu\nu} &= -\frac{1}{\mu_0}\left(F^{\mu\rho}F_{\rho\mu} + F^{\rho\sigma}F_{\rho\sigma}\right) \\
&= \frac{1}{\mu_0}\left(F^{\mu\rho}F_{\mu\rho} - F^{\rho\sigma}F_{\rho\sigma}\right) \\
&= 0.
\end{aligned}$$

In the quantum theory, this is related to the fact that the photon is a massless particle.

[10]The explicit computation is as follows. Applying ∂_μ to both sides of eq. (8.34), we get

$$\begin{aligned}
-\mu_0\partial_\mu T^{\mu\nu} &= (\partial_\mu F^{\mu\rho})F_\rho{}^\nu \\
&+ F^{\mu\rho}\partial_\mu F_\rho{}^\nu + \frac{1}{2}(\partial^\nu F^{\rho\sigma})F_{\rho\sigma} .
\end{aligned}$$

Using the equation of motion (8.23), the term on the right-hand side, in the first line, can be rewritten as

$$(\partial_\mu F^{\mu\rho})F_\rho{}^\nu = -\mu_0 j^\rho F_\rho{}^\nu .$$

The sum of the terms in the second line, instead, vanishes. In fact, we can rewrite it as

$$\begin{aligned}
&F_{\mu\rho}\partial^\mu F^{\rho\nu} + (\partial^\nu\partial^\rho A^\sigma)F_{\rho\sigma} \\
&= F_{\sigma\rho}\partial^\sigma F^{\rho\nu} + (\partial^\nu\partial^\rho A^\sigma)F_{\rho\sigma} \\
&= F_{\rho\sigma}[\partial^\nu\partial^\rho A^\sigma - \partial^\sigma(\partial^\rho A^\nu - \partial^\nu A^\rho)] \\
&= F_{\rho\sigma}[\partial^\nu(\partial^\rho A^\sigma + \partial^\sigma A^\rho) - \partial^\rho\partial^\sigma A^\nu] .
\end{aligned}$$

We now observe that $F_{\rho\sigma}$ is antisymmetric in (ρ, σ), while the expression in brackets is symmetric. Therefore, the contraction vanishes. We are then left with

$$\partial_\mu T^{\mu\nu} = j_\mu F^{\mu\nu} .$$

Renaming the indices $\mu \leftrightarrow \nu$ and using $T^{\mu\nu} = T^{\nu\mu}$ and $F^{\mu\nu} = -F^{\nu\mu}$, we finally obtain eq. (8.39).

which is the same as eq. (3.62). This confirms the result found in eq. (3.67) with a non-covariant formalism. Note that the scalar and vector conservation equations (3.47) and (3.62) are now unified into the single covariant equation (8.39).

Since T^{00} is the energy density of the electromagnetic field, and T^{0i}/c is the momentum density, the energy and momentum carried by the electromagnetic field are given by

$$\mathcal{E}_{\rm em} = \int d^3x\, T^{00}\,, \tag{8.45}$$

$$P^i_{\rm em} = \frac{1}{c} \int d^3x\, T^{0i}\,, \tag{8.46}$$

where the integral extends over all of space (or over a finite volume V, if we are interested in the energy and momentum of the electromagnetic field in that volume). As can be expected by analogy with the result (7.144) valid for point particles, the quantity

$$P^\nu_{\rm em} = \left(\mathcal{E}_{\rm em}/c, \mathbf{P}^i_{\rm em}\right) \tag{8.47}$$

$$= \frac{1}{c} \int d^3x\, T^{0\nu}\,, \tag{8.48}$$

is a four-vector. The proof is analogous to the one used to show that the charge Q given in eq. (8.11) is a Lorentz scalar, and we give it explicitly in Problem 8.2.

8.5   Lorentz transformations of electric and magnetic fields

The results of Sections 8.2 and 8.3 show that, from the point of view of Lorentz transformations, the vectors $\mathbf{E}$ and $\mathbf{B}$ are not the spatial component of a four-vector, as one could have naively guessed. There is no "E^0" that combines with $\mathbf{E}$ to form a four-vector, and similarly for $\mathbf{B}$. Rather, Maxwell's equations become Lorentz covariant when we assemble $\mathbf{E}$ and $\mathbf{B}$ together into a different representation of the Lorentz group, the antisymmetric tensor $F^{\mu\nu}$.

From this, we can immediately derive how $\mathbf{E}$ and $\mathbf{B}$ transform under Lorentz transformations. Under a Lorentz transformation

$$x^\mu \to x'^\mu = \Lambda^\mu{}_\nu x^\nu\,, \tag{8.49}$$

we have

$$F_{\mu\nu}(x) \to F'_{\mu\nu}(x') = \Lambda_\mu{}^\rho \Lambda_\nu{}^\sigma F_{\rho\sigma}(x)\,, \tag{8.50}$$

or, in matrix form,

$$F(x) \to F'(x') = \Lambda F(x) \Lambda^T\,. \tag{8.51}$$

Let K be an inertial frame where the electric and magnetic field are $\mathbf{E}(\mathbf{x})$ and $\mathbf{B}(\mathbf{x})$, and let K' be the reference frame of an observer that moves with constant velocity $\mathbf{v} = v\hat{\mathbf{x}}$ with respect to K. The coordinates of the two frames are then related by a boost along the x axis, as in eqs. (7.22) and (7.23), with $v_0 = -v$ (since, if $v > 0$, K' moves in the direction of the positive x axis with respect K, so a particle at rest in K moves toward the negative x axis in K'). Then, $\Lambda^\mu{}_\nu$ is given by the matrix in eq. (7.20), with $\tanh \zeta = -v/c$. Performing the matrix product explicitly in eq. (8.51) we get, for the electric and magnetic field seen by the K' observer,

$$E_1' = E_1 , \qquad E_2' = \gamma(E_2 - vB_3) , \qquad E_3' = \gamma(E_3 + vB_2) , \qquad (8.52)$$

and

$$B_1' = B_1 , \qquad B_2' = \gamma\left(B_2 + \frac{v}{c^2}E_3\right) , \qquad B_3' = \gamma\left(B_3 - \frac{v}{c^2}E_2\right) , \qquad (8.53)$$

where, for notational simplicity, we have not explicitly written the argument x in $\mathbf{E}(x), \mathbf{B}(x)$ and the argument x' in $\mathbf{E}'(x'), \mathbf{B}'(x')$. These expressions can be rewritten in a rotationally invariant form (which therefore holds for boosts in a generic direction) denoting by $\mathbf{E}_\parallel$ the component of $\mathbf{E}$ parallel to the boost axis (so, for a boost along the x axis, $\mathbf{E}_\parallel = E_x\hat{\mathbf{x}}$) and by $\mathbf{E}_\perp$ the component transverse to the boost axis (so, for a boost along the x axis, $\mathbf{E}_\perp = E_y\hat{\mathbf{y}} + E_z\hat{\mathbf{z}}$), and similarly for $\mathbf{B}$. Then eqs. (8.52) and (8.53) can be written as[11]

$$\mathbf{E}_\parallel' = \mathbf{E}_\parallel , \qquad \mathbf{E}_\perp' = \gamma(\mathbf{E}_\perp + \mathbf{v} \times \mathbf{B}_\perp) , \qquad (8.57)$$

and

$$\mathbf{B}_\parallel' = \mathbf{B}_\parallel , \qquad \mathbf{B}_\perp' = \gamma(\mathbf{B}_\perp - \mathbf{v} \times \mathbf{E}_\perp/c^2) . \qquad (8.58)$$

Consider, for instance, a frame where we have an electric field along the z axis, $\mathbf{E} = E\hat{\mathbf{z}}$, while $\mathbf{B} = 0$; an observer that moves with velocity $\mathbf{v} = v\hat{\mathbf{x}}$ with respect to this frame will see an electric field $\mathbf{E} = \gamma E\hat{\mathbf{z}}$ and a non-vanishing magnetic field $\mathbf{B} = \gamma(v/c^2)E\hat{\mathbf{y}}$. The latter result is at first very surprising. Boosting an electric field we generate a magnetic field! This is at the core of the fact that Maxwell's equations unify electric and magnetic phenomena into electromagnetism, and $\mathbf{E}$ and $\mathbf{B}$ are deeply interrelated, to the extent that, with a boost, we can generate a magnetic field from an electric field (and, conversely, we can generate an electric field from a magnetic field). At the mathematical level, this is expressed by the fact that $\mathbf{E}$ and $\mathbf{B}$ do not belong to two separate representations of the Lorentz group but, rather, together fill the component of an antisymmetric tensor $F_{\mu\nu}$, and therefore mix under Lorentz boosts.

Physically, this can also be understood from the fact that the charge density and the current density form a four-vector, eq. (8.9). Suppose that in the frame K we have just an electric charge at rest at the origin. This will generate a radial electric field, but no magnetic field. In the boosted frame K', however, the charge is in motion with velocity $\mathbf{v}$. It therefore produces a current $\mathbf{j}$, that generates a magnetic field perpendicular to its direction of motion.

[11]Equivalently, we can write $\mathbf{v} \times \mathbf{E}_\perp$ as $\mathbf{v} \times \mathbf{E}$, and $\mathbf{v} \times \mathbf{B}_\perp$ as $\mathbf{v} \times \mathbf{B}$, given that $\mathbf{v} \times \mathbf{E}_\parallel = \mathbf{v} \times \mathbf{B}_\parallel = 0$. Another expression equivalent to eqs. (8.57) and (8.58) is

$$\mathbf{E}' = \gamma(\mathbf{E} + \mathbf{v} \times \mathbf{B}) - (\gamma - 1)\hat{\mathbf{v}}(\hat{\mathbf{v}} \cdot \mathbf{E}) , \qquad (8.54)$$

and

$$\mathbf{B}' = \gamma(\mathbf{B} - \mathbf{v} \times \mathbf{E}/c^2) - (\gamma - 1)\hat{\mathbf{v}}(\hat{\mathbf{v}} \cdot \mathbf{B}) , \qquad (8.55)$$

where $\boldsymbol{\beta} = \mathbf{v}/c$. The equivalence of eqs. (8.57) and (8.54) can be shown writing $\mathbf{E} = \mathbf{E}_\perp + \mathbf{E}_\parallel$, $\mathbf{E}' = \mathbf{E}_\perp' + \mathbf{E}_\parallel'$, and observing that $\mathbf{E}_\parallel$ is the projection of $\mathbf{E}$ in the direction of the velocity, so $\mathbf{E}_\parallel = (\mathbf{E} \cdot \hat{\mathbf{v}})\hat{\mathbf{v}}$, while $\mathbf{v} \times \mathbf{B}$ is transverse to $\mathbf{v}$ so, in eq. (8.54), it only contributes to $\mathbf{E}_\perp$. Then eq. (8.54) gives

$$\mathbf{E}_\perp' = \gamma(\mathbf{E}_\perp + \mathbf{v} \times \mathbf{B}) , \qquad (8.56)$$

and $\mathbf{v} \times \mathbf{B} = \mathbf{v} \times \mathbf{B}_\perp$, so we recover the second equation in eq. (8.57). For the parallel component, writing $\mathbf{E}_\parallel = E_\parallel \hat{\mathbf{v}}$, $\mathbf{E}_\parallel' = E_\parallel' \hat{\mathbf{v}}$, and using $\hat{\mathbf{v}} \cdot \mathbf{E} = E_\parallel$, eq. (8.54) gives

$$E_\parallel' = \gamma E_\parallel - (\gamma - 1)E_\parallel ,$$

so $E_\parallel' = E_\parallel$. Equation (8.54) is therefore equivalent to eq. (8.57). One proceeds in the same way to show the equivalence of eqs. (8.55) and (8.58). Observe that the second term on the right-hand side of eq. (8.54) can be rewritten using the identity

$$(\gamma - 1)\hat{\mathbf{v}}(\hat{\mathbf{v}} \cdot \mathbf{E}) = \frac{\gamma^2}{\gamma + 1}\boldsymbol{\beta}(\boldsymbol{\beta} \cdot \mathbf{E}) ,$$

where $\boldsymbol{\beta} = \mathbf{v}/c$. This identity can be shown writing $\gamma^2 = 1/(1 - \beta^2)$, which gives $\beta^2 = (\gamma^2 - 1)/\gamma^2$. One can similarly rewrite the second term on the right-hand side of eq. (8.55).

Exercise 8.1 Using eqs. (8.52) and (8.53), show that $\mathbf{E}^2 - c^2\mathbf{B}^2$ and $\mathbf{E}\cdot\mathbf{B}$ are Lorentz invariant. Try to find an explicit Lorentz-invariant expression for $\mathbf{E}^2 - c^2\mathbf{B}^2$.

Exercise 8.2 Compute the Lorentz-invariant quantities $F_{\mu\nu}F^{\mu\nu}$ and $F_{\mu\nu}\tilde{F}^{\mu\nu}$ in terms of $\mathbf{E}$ and $\mathbf{B}$.

Exercise 8.3 Without using the covariant formalism, and rather working with $\mathbf{E}$ and $\mathbf{B}$, show that, under the transformation (8.52, 8.53), the full set of Maxwell's equations (3.8–3.11), with the source terms set to zero, is invariant (in the sense that, if the original fields satisfied Maxwell's equations, the transformed fields also satisfy them). Is each Maxwell equation separately invariant? Repeat the exercise including the source terms, taking into account that, according to eq. (8.9), $c\rho$ and $\mathbf{j}$ transform as a four-vector.

Exercise 8.4 As an example of the usefulness of a covariant formalism, consider a set of equations, involving two two-dimensional vectors $\mathbf{a} = (a_x, a_y)$ and $\mathbf{b} = (b_x, b_y)$, and two more quantities A, B, given by

$$B\mathbf{a} = A\mathbf{b}, \qquad \epsilon_{\alpha\beta}a_\alpha b_\beta = 0, \qquad (8.59)$$

where the indices α, β take the values $1, 2$ (or, equivalently, x, y) and $\epsilon_{\alpha\beta}$ is the antisymmetric tensor in two dimensions, $\epsilon_{12} = -\epsilon_{21} = 1$, $\epsilon_{11} = \epsilon_{22} = 0$. Show that this system of equations is invariant under rotations in the (x, y) plane, under which $\mathbf{a}$ and $\mathbf{b}$ transform as two-dimensional vectors, and A and B as scalars. Can you find a formalism that explicitly shows that this system of equations actually has a much larger symmetry, which was not apparent from eq. (8.59)? Can you draw the analogy with our discovery of Lorentz invariance in Maxwell's equations?

8.6 Relativistic formulation of the particle-field interaction

8.6.1 Covariant form of the Lorentz force equation

The final step, to obtain a fully covariant formulation of electrodynamics, is to write also the interaction of a particle with the electromagnetic field in a relativistic form. We therefore investigate whether there is a covariant expression, whose spatial components give the vector equation (3.6). To this purpose, it is useful to rewrite eq. (3.6) in terms of proper time, using $d\tau = dt/\gamma$. We also recall, from eqs. (7.122) and (7.123), that $u^0 = \gamma c$, $u^i = \gamma v^i$. Equation (3.6) can then be rewritten as

$$\begin{aligned} \frac{d\mathbf{p}}{d\tau} &= q\gamma\left(\mathbf{E} + \mathbf{v} \times \mathbf{B}\right) \\ &= q\left(\frac{u^0}{c}\mathbf{E} + \mathbf{u} \times \mathbf{B}\right), \end{aligned} \qquad (8.60)$$

or, in components

$$
\begin{aligned}
\frac{dp^i}{d\tau} &= q\left(-u_0\frac{E^i}{c} + \epsilon_{ijk}u_jB_k\right) \\
&= q\left(F^{i0}u_0 + F^{ij}u_j\right),
\end{aligned}
\tag{8.61}
$$

where we used eqs. (8.17) and (8.20), together with $F^{0i} = -F^{i0}$ and $F^{ij} = F_{ij}$. Recalling, from eq. (7.144), that $p^\mu = (\mathcal{E}/c, p^i)$ (where here we denote the energy of the particle by $\mathcal{E}$, since we reserve the notation E to the modulus of the electric field $\mathbf{E}$), we recognize this as the $\mu = i$ component of the equation

$$
\frac{dp^\mu}{d\tau} = qF^{\mu\nu}u_\nu\,,
\tag{8.62}
$$

or, equivalently, using eq. (7.143),

$$
\frac{dp^\mu}{d\tau} = \frac{q}{m}F^{\mu\nu}p_\nu\,.
\tag{8.63}
$$

Therefore, eq. (3.6), with $\mathbf{p}$ interpreted as the full relativistic momentum,

$$
\mathbf{p} = \gamma(v)m\mathbf{v}\,,
\tag{8.64}
$$

is the spatial component of a covariant equation, and therefore is also valid relativistically. Explicitly, in terms of the velocity, the spatial component of the relativistic Lorentz "force" equations (8.62) is[12]

$$
\frac{d}{dt}\left[\gamma(v)m\mathbf{v}\right] = q\left(\mathbf{E} + \mathbf{v}\times\mathbf{B}\right).
\tag{8.65}
$$

This covariantization also carries with it the $\mu = 0$ component, that must correspond to another equation that could have been derived with the non-covariant formalism. Indeed, using again $d\tau = dt/\gamma$ and $u_i = \gamma v_i$, the $\mu = 0$ component of eq. (8.62) reads

$$
\frac{d\mathcal{E}}{dt} = q\,\mathbf{E}\cdot\mathbf{v}\,.
\tag{8.66}
$$

In fact, this is just eq. (7.151), that in our present notation reads $d\mathcal{E}/dt = \mathbf{v}\cdot d\mathbf{p}/dt$, with $d\mathbf{p}/dt$ expressed through eq. (3.6). Note that only the electric field contributes to the work. This is a consequence of the fact that the contribution to $d\mathbf{p}/dt$ from the magnetic field, $q\mathbf{v}\times\mathbf{B}$, is orthogonal to $\mathbf{v}$, and therefore does not contribute to $\mathbf{v}\cdot d\mathbf{p}/dt$.

This derivation shows that the Lorentz force equation, when written in the form (3.6) with $\mathbf{p}$ given in terms of the velocity as in eq. (8.64), is not just the low-velocity limit of some fully relativistic expression, but is in fact already the spatial component of a four-vector equation, and is therefore correct even at relativistic velocities, as the covariant formalism makes explicit.

[12] As mentioned in Note 2 on page 40, the use of the word "force" here is an abuse of language, since force is associated with the Newtonian concept of instantaneous action, in which the force between two particles depends on their instantaneous positions. What makes eq. (8.65) fully consistent with the principles of relativity is that the force on a particle, at time t and position $\mathbf{x}$, is expressed in terms of the electric and magnetic fields at the same value of t and $\mathbf{x}$ but, as we will see in Chapter 10, these are determined in terms of the "retarded position" of the source, i.e., the position that the source had at an earlier time, consistent with the propagation of signals at the speed of light. A more appropriate name for eq. (8.65) could be "the relativistic equation of motion for a particle in an external electromagnetic field." We will, however, often use the common expression "Lorentz force" even in the relativistic setting.

8.6.2   The interaction action of a point particle

An alternative derivation of eq. (8.62), more in line with typical reasoning used in modern theoretical physics, is as follows. We saw in eq. (7.133) that the action of a free (point-like) particle is given by

$$S_{\text{free}} = -mc^2 \int d\tau \,. \tag{8.67}$$

Let us understand the form of the action that describes the interaction of this particle with the electromagnetic field A^μ. Just as we did for deriving the free action (8.67), we use symmetry principles, observing that, in order to respect the Lorentz covariance that we have discovered in Maxwell's equations, also the interaction action must be Lorentz invariant. The variation of a Lorentz-invariant action with respect to a four-vector dynamical variable such as $x^\mu(\tau)$ will then produce Lorentz-covariant equations of motion.

The integration variable in the interaction action must therefore be again $d\tau$, which is the only Lorentz-invariant generalization of the integral over dt that appears in the non-relativistic action of a particle. The interaction action must also involve the gauge potential $A^\mu(x)$, computed on the trajectory $x^\mu(\tau)$ of the particle. To obtain a Lorentz-invariant action we therefore need something to saturate the Lorentz index of $A^\mu(x)$, and, for a point particle without an internal structure, the only other four-vector that we have at our disposal is its four-velocity $u^\mu(\tau)$. The simplest possibility is then a linear coupling, proportional to $u_\mu(\tau)A^\mu[x(\tau)]$. The interaction must also be proportional to the charge q of the particle: if $q = 0$, there is no interaction between the particle and the electromagnetic field. We are therefore led to postulating an interaction action

$$S_{\text{int}} = q \int d\tau \, u_\mu(\tau)A^\mu[x(\tau)] \,, \tag{8.68}$$

apart from an overall numerical multiplicative constant.[13]

The total action $S = S_{\text{free}} + S_{\text{int}}$ of a point-like particle in an external electromagnetic field can therefore be written as

$$S = -mc^2 \int d\tau + q \int d\tau \, u_\mu(\tau)A^\mu[x(\tau)] \,. \tag{8.69}$$

We can now express $d\tau$ in terms of dt using $d\tau = dt/\gamma$ and, correspondingly, we use t and $\mathbf{x}(t)$ instead of $x^\mu(\tau)$ in the arguments of ϕ and $\mathbf{A}$. Using eqs. (7.122) and (7.123), as well as eq. (8.12), we get

$$S = \int dt \left[-mc^2 \sqrt{1 - \frac{v^2(t)}{c^2}} - q\phi[t, \mathbf{x}(t)] + q\mathbf{v}\cdot\mathbf{A}[t, \mathbf{x}(t)] \right] \,. \tag{8.70}$$

Then, the Lagrangian is

$$L[\mathbf{x}(t), \mathbf{v}(t)] = -mc^2 \sqrt{1 - \frac{v^2(t)}{c^2}} - q\phi[t, \mathbf{x}(t)] + q\mathbf{v}\cdot\mathbf{A}[t, \mathbf{x}(t)] \,. \tag{8.71}$$

[13] At this stage, a multiplicative constant could simply be reabsorbed into q, rather than already assuming that q is the electric charge. However, we will see below that eq. (8.68) is indeed the action whose equation of motion gives the Lorentz force equation, with q identified with the electric charge without extra multiplicative factors.

It is also interesting to observe that eq. (8.68) can be rewritten as

$$S_{\text{int}} = q \int d\tau \, \frac{dx^\mu(\tau)}{d\tau} \int d^4x \, \delta^{(4)}[x - x(\tau)] A_\mu(x) \,. \qquad (8.72)$$

Exchanging the integrals over d^4x and over $d\tau$, this can be rewritten as

$$S_{\text{int}} = \frac{1}{c} \int d^4x \, j^\mu(x) A_\mu(x) \,, \qquad (8.73)$$

where $j^\mu(x)$ is the current defined in eq. (8.3). Observe that, in this expression, the argument x in $j^\mu(x)$ and in $A_\mu(x)$ is the generic space-time point, while in eq. (8.71) the spatial argument was evaluated on the position $\mathbf{x}(t)$ of the particle at time t.

Since a generic current $j^\mu(x)$ can always be thought of as generated by a superposition of point charges, eq. (8.73) can be taken as the general form of the interaction of an arbitrary four-current density $j^\mu(x)$ with an external electromagnetic field.[14] More explicitly, using eqs. (8.9) and (8.13), in terms of $\rho(t, \mathbf{x})$, $\mathbf{j}(t, \mathbf{x})$, $\phi(t, \mathbf{x})$ and $\mathbf{A}(t, \mathbf{x})$, we have

$$S_{\text{int}} = \int dt d^3x \, [-\rho(t, \mathbf{x})\phi(t, \mathbf{x}) + \mathbf{j}(t, \mathbf{x}) \cdot \mathbf{A}(t, \mathbf{x})] \,, \qquad (8.74)$$

where we also used $dx^0 = cdt$.

The action (8.73) is gauge-invariant, as long as the current j^μ is conserved. Indeed, under the gauge transformation (8.14),

$$
\begin{aligned}
S_{\text{int}} \quad &\rightarrow \quad S_{\text{int}} - \frac{1}{c} \int d^4x \, j^\mu(t, \mathbf{x}) \partial_\mu \theta \\
&= \quad S_{\text{int}} + \frac{1}{c} \int d^4x \, \theta \partial_\mu j^\mu(t, \mathbf{x}) \\
&= \quad S_{\text{int}} \,.
\end{aligned}
\qquad (8.75)
$$

Note that, in the integration by parts, we dropped the boundary terms. This can be justified, restricting to gauge functions $\theta(x)$ that go to zero sufficiently fast at infinity (or to a localized current j^μ). We see that gauge invariance requires current conservation.[15]

The equation of motion can now be obtained from the Lagrangian variational principle: for a system with coordinates $q_i(t)$,

$$\frac{d}{dt} \frac{\delta L}{\delta \dot{q}_i} - \frac{\delta L}{\delta q_i} = 0 \,. \qquad (8.76)$$

We use the Lagrangian in the form (8.71), where the role of $q_i(t)$ is played by $x_i(t)$, while $\dot{q}_i = v_i$. We compute the derivatives explicitly. First of all,

$$
\begin{aligned}
\frac{\delta L}{\delta v_i} &= -mc^2 \frac{1}{2}\left(1 - \frac{v^2}{c^2}\right)^{-1/2}\left(\frac{-2v^i}{c^2}\right) + qA^i(t, \mathbf{x}) \\
&= \gamma m v^i + qA^i(t, \mathbf{x}) \\
&= p^i + qA^i(t, \mathbf{x}) \,.
\end{aligned}
\qquad (8.77)
$$

[14] Observe that we are considering here the electromagnetic field as a given external field. We will include the dynamics of the electromagnetic field at the level of the action in Section 8.7, see in particular eq. (8.129).

[15] This is a point with important implications in quantum field theory, on which we will not elaborate further here. See e.g., Chapter 7 of Maggiore (2005).

Note that, by definition, the derivative of the Lagrangian with respect to the velocity is the momentum conjugate to the coordinate x^i. Therefore, in the presence of an electromagnetic field, the momentum conjugate to the coordinate x^i, that we denote by P^i, is not just the same as the momentum $p^i = \gamma m v^i$ of a free particle but, rather, is given by

$$\mathbf{P} = \mathbf{p} + q\mathbf{A} \,. \tag{8.78}$$

We next compute

$$
\begin{aligned}
\frac{d}{dt}\left\{p^i + qA^i[t, \mathbf{x}(t)]\right\} &= \frac{dp^i}{dt} + q\left[\frac{\partial A^i(t, x)}{\partial t} + \frac{dx^j(t)}{dt}\partial_j A^i(t, x)\right] \\
&= \frac{dp^i}{dt} + q\left[\frac{\partial A^i(t, x)}{\partial t} + v^j \partial_j A^i(t, x)\right] . \tag{8.79}
\end{aligned}
$$

Note that, for spatial indices, we do not need to be careful about the upper/lower positioning, and we can equally well keep all of them lower and sum over repeated lower indices. We finally compute the variation with respect to x_i:

$$\frac{\delta L}{\delta x_i} = q v_j \partial_i A_j - q \partial_i \phi \,. \tag{8.80}$$

(Recall that in the variation performed to obtain the equations of motion, q_i and $\dot{q}_i$ are taken as independent variables). We now write

$$
\begin{aligned}
v_j \partial_i A_j &= v_j(\partial_i A_j - \partial_j A_i) + v_j \partial_j A_i \\
&= v_j \epsilon_{ijk} B_k + v_j \partial_j A_i \,, \tag{8.81}
\end{aligned}
$$

and putting everything together, we get

$$
\begin{aligned}
0 &= \frac{dp^i}{dt} + q\left(\frac{\partial A^i}{\partial t} + v^j \partial_j A^i\right) - q(\epsilon_{ijk} v_j B_k + v^j \partial_j A^i) + q \partial_i \phi \\
&= \frac{dp^i}{dt} + q\left(\frac{\partial A^i}{\partial t} - \epsilon_{ijk} v_j B_k\right) + q \partial_i \phi \,, \tag{8.82}
\end{aligned}
$$

where we observed that the two terms proportional to $v^j \partial_j A^i$ canceled among them. Therefore, the equation of motion derived from the Lagrangian (8.71) is

$$\frac{d\mathbf{p}}{dt} = -q\left(\boldsymbol{\nabla}\phi + \frac{\partial \mathbf{A}}{\partial t}\right) + q\mathbf{v} \times \mathbf{B} \,, \tag{8.83}$$

where $\mathbf{p} = \gamma(v)m\mathbf{v}$. Using eq. (3.83), we see that this is just the Lorentz force equation (3.6). The fact that we have derived it from a fully covariant action shows again that it is indeed the spatial component a fully relativistic equation.

Finally, we can obtain the Hamiltonian from

$$H(\mathbf{P}, \mathbf{x}) = \mathbf{P} \cdot \mathbf{v} - L \,, \tag{8.84}$$

where the Lagrangian L is given by eq. (8.71) and $\mathbf{v}$ must be written in terms of $\mathbf{P}$ and $\mathbf{A}$ using eqs. (7.138) and (8.78). The inversion of eq. (7.138) gives

$$\mathbf{v} = c\,\frac{\mathbf{p}}{\sqrt{p^2 + m^2 c^2}} \,, \tag{8.85}$$

and therefore, in terms of $\mathbf{P}$ and $\mathbf{x}$,

$$\mathbf{v} = c\,\frac{\mathbf{P} - q\mathbf{A}}{\sqrt{(\mathbf{P} - q\mathbf{A})^2 + m^2 c^2}}\,. \qquad (8.86)$$

Inserting this into eq. (8.84) we get

$$H(\mathbf{P}, \mathbf{x}) = c\,\sqrt{(\mathbf{P} - q\mathbf{A})^2 + m^2 c^2} + q\phi\,, \qquad (8.87)$$

where the potentials $\mathbf{A}(t, \mathbf{x})$ and $\phi(t, \mathbf{x})$ must be computed on the position $\mathbf{x} = \mathbf{x}(t)$ of the particle.

8.7   Field-theoretical approach to classical electrodynamics

The Lagrangian formulation of classical mechanics is particularly useful for understanding the relation between symmetries and conservation laws and is also the first step toward the Hamiltonian formulation. Beside their intrinsic elegance, the Lagrangian and Hamiltonian formalisms are also the natural bridge between the classical and the quantum theory. Maxwell's equations also admit a Lagrangian formulation which, again, allows us to better understand the formal structure of the theory and the relation between symmetries and conservation laws, and is a prototype of a classical field theory. Furthermore, even if we will not develop these aspects in this book, this field-theoretical formulation is also the starting point for the quantization of the theory, leading to quantum electrodynamics. In this section, we will develop such a field-theoretical approach to classical electrodynamics. The subject is advanced, and here we will limit ourselves to presenting briefly the main results, referring the reader to Maggiore (2005) (see in particular chapters 2 and 3) for more detailed discussions and derivations.

This section is more advanced and should be skipped at first reading

8.7.1   Euler–Lagrange equations of relativistic fields

Elementary classical mechanics deals with systems with a finite number of degrees of freedom. These are described by (generalized) coordinates $q_i(t)$, where the index i labels all degrees of freedom of the system. A typical example is a system of N particles in three dimensions, in which case the index i takes the values $1, \ldots, 3N$. Classical field theory generalizes this to systems with a *continuous* set of degrees of freedom. In the previous chapters, we have already made frequent use of the notion of field. In the context of electrodynamics, the most obvious examples are the electric field $\mathbf{E}(t, \mathbf{x})$ and the magnetic field $\mathbf{B}(t, \mathbf{x})$, or the gauge potentials $\phi(t, \mathbf{x})$ and $\mathbf{A}(t, \mathbf{x})$. We can here regard the variable $\mathbf{x}$ as a continuous generalization of the index i in $q_i(t)$, so that we have a dynamical degree of freedom, i.e., a function of time, associated with each point of space.

In a non-relativistic context, it is convenient to organize the dynamical variables according to their transformation properties under spatial rotations. For instance, the $3N$ degrees of freedom $q_i(t)$ of a system of N particles in three dimensions, with $i = 1, \ldots, 3N$, are obviously organized into a set of N vectors $\mathbf{q}_a$, where $a = 1, \ldots, N$ labels the particle and, for each a, $\mathbf{q}_a = (q_x, q_y, q_z)_a$ is a spatial vector. In the case of fields, we have already discussed their transformation properties under rotations in Section 7.3.6. For instance, we saw that the temperature is an example of a field scalar under rotations, defined by the fact that it transforms as in eq. (7.92). Another example, more pertinent to classical electrodynamics, is the scalar gauge potential $\phi(t, \mathbf{x})$. In eq. (7.92), as in all equations relative to transformations under rotations, we suppressed the time variable t, since we were only interested in the behavior under spatial rotations, that do not affect time. More generally, we can write the transformation property under rotations of a scalar field, such as $\phi(t, \mathbf{x})$, as

$$\phi(t, \mathbf{x}) \rightarrow \phi'(t, \mathbf{x}') = \phi(t, \mathbf{x}). \tag{8.88}$$

The transformation property under rotations of a vector field, such as the electric field, is given by eq. (7.94) or, also re-instating the time dependence,

$$E_i(t, \mathbf{x}) \rightarrow E_i'(t, \mathbf{x}') = R_{ij} E_j(t, \mathbf{x}), \tag{8.89}$$

and similarly for any other vector field.

In a relativistic context, we are interested in the transformation properties under Lorentz transformations, that we discussed in Section 7.3.6. In particular, a field scalar under Lorentz transformations (that, when the context is clear, we simply call a scalar field) transforms as in eq. (7.101), that we rewrite here,

$$\phi(x) \rightarrow \phi'(x') = \phi(x). \tag{8.90}$$

For rotations, $x' = (t, \mathbf{x}')$ and we get back eq. (7.92). A contravariant four-vector field $V^\mu(x)$ transforms as in eq. (7.102) so, in particular, the gauge field A^μ transforms as

$$A^\mu(x) \rightarrow A'^\mu(x') = \Lambda^\mu{}_\nu A^\nu(x). \tag{8.91}$$

Similarly, a tensor field such as $F^{\mu\nu}$ transforms as

$$F^{\mu\nu}(x) \rightarrow F'^{\mu\nu}(x') = \Lambda^\mu{}_\rho \Lambda^\nu{}_\sigma F^{\rho\sigma}(x). \tag{8.92}$$

Having introduced the basic variables of a field theory, the next step is to define their dynamics. This can be done by extending to fields the Lagrangian formalism of classical mechanics, and gives rise to the subject of classical field theory. To this purpose, we begin by considering a non-relativistic system with generalized coordinates $q_i(t)$. The Lagrangian is a function of the coordinates q_i and their time derivatives $\dot{q}_i$, that we denote collectively as $q(t)$ and $\dot{q}(t)$, respectively, i.e., $L = L[q(t), \dot{q}(t)]$.[16]

[16]More precisely, the Lagrangian is a *functional* of $q(t)$ and $\dot{q}(t)$, i.e., an object that depends on the functions $q(t)$ and $\dot{q}(t)$, rather than just on a finite number of variables. The corresponding derivatives should really be defined as functional derivatives at the level of integrated quantities (such as the action), with the rule

$$\frac{\delta q(t')}{\delta q(t)} = \delta(t - t'), \tag{8.93}$$

and the standard composition rules for derivative, so that, for instance,

$$\frac{\delta}{\delta q(t)} \int dt' q(t') j(t')$$
$$= \int dt' \frac{\delta q(t')}{\delta q(t)} j(t')$$
$$= \int dt' \delta(t - t') j(t')$$
$$= j(t). \tag{8.94}$$

In practice, this is equivalent to naively differentiating at the level of the Lagrangian, with formal rules as

$$\frac{\delta}{\delta q(t)} q(t') j(t') \rightarrow j(t). \tag{8.95}$$

In the following, we will perform the manipulations leading to the equations of motion in this form.

The action is defined as (compare to eq. (7.124), where we considered the case of a single particle)

$$S = \int dt \, L[q, \dot{q}] \, . \qquad (8.96)$$

The action principle is obtained considering the action integrated between fixed initial and final times, t_i and t_f, and studying its variation under a change of the trajectory $q_i(t) \to q_i(t) + \delta q_i(t)$, with $\delta q_i(t) = 0$ at $t = t_i$ and at $t = t_f$, i.e., we study how S varies if we change $q_i(t)$ while keeping it fixed at the initial and final times. If $q_i(t) \to q_i(t) + \delta q_i(t)$, then $\dot{q}_i(t) \to \dot{q}_i(t) + d/dt[\delta q_i(t)]$, i.e., $\delta \dot{q}_i = d/dt[\delta q_i(t)]$. Therefore, the variation of the action is

$$
\begin{aligned}
\delta S &= \int_{t_i}^{t_f} dt \sum_i \left[\frac{\delta L}{\delta q_i} \delta q_i + \frac{\delta L}{\delta \dot{q}_i} \delta \dot{q}_i \right] \\
&= \sum_i \int_{t_i}^{t_f} dt \left[\frac{\delta L}{\delta q_i} \delta q_i + \frac{\delta L}{\delta \dot{q}_i} \frac{d}{dt} \delta q_i \right] \\
&= \sum_i \int_{t_i}^{t_f} dt \left[\frac{\delta L}{\delta q_i} - \frac{d}{dt} \frac{\delta L}{\delta \dot{q}_i} \right] \delta q_i \, , \qquad (8.97)
\end{aligned}
$$

where, in the last line, we integrated d/dt by parts and used the fact that, since $\delta q_i(t) = 0$ at $t = t_i$ and $t = t_f$, the boundary term vanishes. The action principle states that the classical trajectory is such that, under such a variation of the $q_i(t)$, $\delta S = 0$. Since the variations δq_i are taken to be independent, each of the terms in the sum over i in eq. (8.97) must vanish independently and, since this must happen for an arbitrary variation $\delta q_i(t)$, we must have

$$\frac{\delta L}{\delta q_i} - \frac{d}{dt} \frac{\delta L}{\delta \dot{q}_i} = 0 \, , \qquad (8.98)$$

for each value of i. These are the equations of motion (or Euler–Lagrange equations) of the system. We now want to generalize the action principle from a system described by a finite number of mechanical variables $q_i(t)$, to fields and we want to construct a Lorentz-covariant field theory. To begin, we consider a single (Lorentz) scalar field $\phi(x)$. For a non-relativistic system with generalized coordinates $q_i(t)$, we have seen that the Lagrangian is a function of q_i and $\partial_t q_i$. However, if we want to construct a Lorentz-covariant formalism, the Lagrangian of a scalar field cannot be a function of ϕ and $\partial_t \phi$, since $\partial_t \phi$ is not a covariant quantity. Rather, we must use $\partial_\mu \phi$ which, as we saw in eq. (7.107), is a four-vector field. Therefore, the Lagrangian of a scalar field will depend on ϕ and $\partial_\mu \phi$. Note, however, that these quantities depend not only on t but also on $\mathbf{x}$. As mentioned previously, we can think of $\mathbf{x}$ as an "index" that labels the dynamical variables of the theory, corresponding to the index i of the variables $q_i(t)$ in classical mechanics, except that now this label is continuous rather than discrete: we have a dynamical variable associated with each point in space. To understand how to deal with this label in the continuous case, consider first the simple case of a set of free non-relativistic particles, with masses m_i and coordinates $q_i(t)$. The Lagrangian of each one is given by $L_i = (1/2)m_i\dot{q}_i^2$, and the total Lagrangian is

$$L = \sum_i \frac{1}{2}m_i\dot{q}_i^2 \,. \tag{8.99}$$

In the limit in which the discrete index i becomes a continuous variable **x**, the sum over i becomes an integral over d^3x. These considerations suggest to write the Lagrangian of a scalar field in the form

$$L = \int d^3x\, c\mathcal{L}[\phi, \partial_\mu\phi] \,. \tag{8.100}$$

The function $\mathcal{L}[\phi, \partial_\mu\phi]$ is called the Lagrangian density (although, with a common abuse of notation, it is often called simply the Lagrangian), and we have defined it extracting a factor of c for later convenience. The action is then obtained as

$$\begin{aligned}
S &= \int dt\, L \\
&= c \int dt\, d^3x\, \mathcal{L}[\phi, \partial_\mu\phi] \\
&= \int d^4x\, \mathcal{L}[\phi, \partial_\mu\phi] \,. \tag{8.101}
\end{aligned}$$

Note that, in this way, we reconstructed d^4x as integration measure and, as we have seen in eq. (8.5), this is Lorentz invariant. Therefore, if we take $\mathcal{L}[\phi, \partial_\mu\phi]$ to be Lorentz invariant, the action will also be Lorentz invariant. We will extend the integration over all of space-time, with the boundary conditions on ϕ that it vanishes sufficiently fast both as $t \to \pm\infty$ and as $|\mathbf{x}| \to \infty$, so that we can neglect all boundary terms that will emerge from integration by parts.

We now consider a variation of the field $\phi \to \phi + \delta\phi$. Then $\partial_\mu\phi \to \partial_\mu\phi + \delta(\partial_\mu\phi)$ with $\delta(\partial_\mu\phi) = \partial_\mu(\delta\phi)$, and the variation of the action is

$$\begin{aligned}
\delta S &= \int d^4x \left[\frac{\delta\mathcal{L}}{\delta\phi}\delta\phi + \frac{\delta\mathcal{L}}{\delta(\partial_\mu\phi)}\delta(\partial_\mu\phi)\right] \\
&= \int d^4x \left[\frac{\delta\mathcal{L}}{\delta\phi}\delta\phi + \frac{\delta\mathcal{L}}{\delta(\partial_\mu\phi)}\partial_\mu(\delta\phi)\right] \\
&= \int d^4x \left[\frac{\delta\mathcal{L}}{\delta\phi} - \partial_\mu\frac{\delta\mathcal{L}}{\delta(\partial_\mu\phi)}\right]\delta\phi \,, \tag{8.102}
\end{aligned}$$

where, in the last line, we integrated ∂_μ by parts discarding boundary terms, given our boundary conditions at infinity. The classical evolution of the field is defined by the condition that the classical solution of the equations of motion is an extremum of S, i.e., is such that $\delta S = 0$ for a generic variation $\delta\phi$. This means that the quantity in bracket must vanish. We then obtain the Euler–Lagrange equations for a relativistic scalar field,

$$\boxed{\frac{\delta\mathcal{L}}{\delta\phi} - \partial_\mu \frac{\delta\mathcal{L}}{\delta(\partial_\mu\phi)} = 0\,.}$$

(8.103)

This is the generalization of eq. (8.98) to field theory. As an example, the simplest Lagrangian density of a real scalar field is given by[17]

$$\mathcal{L} = -\frac{1}{2}\left(\partial_\mu\phi\partial^\mu\phi + \mu^2\phi^2\right),$$

(8.104)

with μ a parameter with dimensions of inverse of length, as ∂_μ. Then,

$$\frac{\delta\mathcal{L}}{\delta\phi} = -\mu^2\phi\,,$$

(8.105)

while

$$\frac{\delta\mathcal{L}}{\delta(\partial_\mu\phi)} = -\partial^\mu\phi\,,$$

(8.106)

so eq. (8.103) becomes

$$(-\Box + \mu^2)\phi = 0\,.$$

(8.107)

This is called the *Klein–Gordon equation*. Searching for a solution of the form

$$\phi(x) = \mathcal{A}_k e^{ikx}\,,$$

(8.108)

and using $\partial_\mu e^{ikx} = ik_\mu e^{ikx}$ and $\Box e^{ikx} = -k_\mu k^\mu e^{ikx}$, we get the condition

$$k^2 + \mu^2 = 0\,,$$

(8.109)

where $k^2 = k_\mu k^\mu = -(k^0)^2 + \mathbf{k}^2$. More explicitly, eq. (8.109) then reads

$$(k^0)^2 = \mathbf{k}^2 + \mu^2\,.$$

(8.110)

In Chapter 9 we will study in detail how similar equations for the gauge potential (but without the μ^2 term) give rise to electromagnetic wave solutions.[18]

If we have several scalar fields $\phi_i(x)$, the Lagrangian density depends on ϕ_i and $\partial_\mu\phi_i$ for all values of the index i, and the variation with respect to each of these fields must vanish, so eq. (8.103) simply becomes

$$\frac{\delta\mathcal{L}}{\delta\phi_i} - \partial_\mu \frac{\delta\mathcal{L}}{\delta(\partial_\mu\phi_i)} = 0\,,$$

(8.111)

for each value of the index i. Consider now a four-vector field A_ν. The Lagrangian density will be a function of A_ν and its derivatives $\partial_\mu A_\nu$, $\mathcal{L} = \mathcal{L}[A_\nu, \partial_\mu A_\nu]$, and

$$S = \int d^4x\, \mathcal{L}[A_\nu, \partial_\mu A_\nu]\,.$$

(8.112)

The equations of motion are obtained requiring that the variation of the action with respect to each of the four components of A_ν vanish, and therefore are the same as eq. (8.111), with ϕ_i replaced by A_ν,

$$\boxed{\frac{\delta\mathcal{L}}{\delta A_\nu} - \partial_\mu \frac{\delta\mathcal{L}}{\delta(\partial_\mu A_\nu)} = 0\,.}$$

(8.113)

[17]The overall factor $-1/2$ is irrelevant for the equations of motion, but can be fixed requiring that the corresponding Hamiltonian, or energy density, is correctly normalized, see e.g., Section 3.3.1 of Maggiore (2005).

[18]Comparing with eq. (7.146) we see that eq. (8.110) has a form analogous to the dispersion relation of a relativistic massive particle. This, however, only becomes true in the context of quantum theory. In fact, k^0 and $\mathbf{k}$ have dimensions of inverse length, and can be identified with the energy and momentum of a particle only through the quantum-mechanical relations $\mathcal{E}/c = \hbar k^0$ and $\mathbf{p} = \hbar\mathbf{k}$. Then, eq. (8.110) takes the form (7.147) with the identification $m = \hbar\mu/c$.

8.7.2   Lagrangian of the electromagnetic field

We now show that Maxwell's equations can be derived from an action principle, and we identify the Lagrangian of the electromagnetic field. We work with a covariant formalism, using $A_\mu(x)$ as our fundamental dynamical field. Then, we only need to show that eq. (8.23) is the equation of motion derived from a Lagrangian since, as we have seen, the other two Maxwell's equations contained in eq. (8.33) are an automatic consequence of the introduction of the gauge potentials. We consider the Lagrangian density

$$\mathcal{L} = \mathcal{L}_0 + \mathcal{L}_{\text{int}}, \tag{8.114}$$

where

$$\mathcal{L}_0 = -\frac{\epsilon_0 c}{4} F_{\mu\nu} F^{\mu\nu} \tag{8.115}$$

is the Lagrangian of the free electromagnetic field, while

$$\mathcal{L}_{\text{int}} = \frac{1}{c} A_\mu j^\mu \tag{8.116}$$

describes the interaction of the gauge field with a given external current j^μ. Factorizing a term $\epsilon_0 c$, the corresponding action is therefore[19]

$$S = \epsilon_0 c \int d^4 x \left(-\frac{1}{4} F_{\mu\nu} F^{\mu\nu} + \mu_0 A_\mu j^\mu \right). \tag{8.118}$$

Note that the interaction action was already obtained in eq. (8.73). To derive the equations of motion of this action we must compute the (functional) derivatives that appear in eq. (8.113). We perform the computation in detail. We first compute the derivative of the Lagrangian with respect to $\partial_\mu A_\nu$. In this case, only $\mathcal{L}_0$ contribute, since $\mathcal{L}_{\text{int}}$ depends on A_μ but not on its derivatives.[20] First of all, it is necessary to change the names of the dummy indices μ, ν in eq. (8.115), in order not to mix them up with the indices μ, ν that appear in eq. (8.113). We then write $\mathcal{L}_0 = -(\epsilon_0 c/4) F_{\alpha\beta} F^{\alpha\beta}$. Then,

$$\frac{1}{\epsilon_0 c} \frac{\delta \mathcal{L}_0}{\delta(\partial_\mu A_\nu)} = -\frac{1}{4} \frac{\delta \left(F_{\alpha\beta} F^{\alpha\beta} \right)}{\delta(\partial_\mu A_\nu)}$$
$$= -\frac{1}{4} \frac{\delta F_{\alpha\beta}}{\delta(\partial_\mu A_\nu)} F^{\alpha\beta} - \frac{1}{4} F_{\alpha\beta} \frac{\delta F^{\alpha\beta}}{\delta(\partial_\mu A_\nu)}. \tag{8.119}$$

We now observe that the two terms in the last line are equal, since the indices α, β can be raised and lowered with the Minkowski metric that commutes with the derivatives with respect to any field.[21] Then

$$\frac{1}{\epsilon_0 c} \frac{\delta \mathcal{L}_0}{\delta(\partial_\mu A_\nu)} = -\frac{1}{2} F^{\alpha\beta} \frac{\delta F_{\alpha\beta}}{\delta(\partial_\mu A_\nu)}$$
$$= -\frac{1}{2} F^{\alpha\beta} \frac{\delta \left(\partial_\alpha A_\beta - \partial_\beta A_\alpha \right)}{\delta(\partial_\mu A_\nu)}$$
$$= -F^{\alpha\beta} \frac{\delta \left(\partial_\alpha A_\beta \right)}{\delta(\partial_\mu A_\nu)}, \tag{8.121}$$

---

[19] Recall that our metric is $\eta^{\mu\nu} = (-1, 1, 1, 1)$. If one uses instead the opposite signature, $\eta^{\mu\nu} = (1, -1, -1, -1)$, the term $F_{\mu\nu} F^{\mu\nu} = \eta_{\mu\rho} \eta_{\nu\sigma} F^{\rho\sigma} F^{\mu\nu}$ is unchanged because it involves two factors of the metric, while $A_\mu j^\mu = \eta_{\mu\nu} A^\nu j^\mu$ changes sign, so the action becomes

$$S = -\epsilon_0 c \tag{8.117}$$
$$\times \int d^4 x \left(\frac{1}{4} F_{\mu\nu} F^{\mu\nu} + \mu_0 A_\mu j^\mu \right).$$

[20] Note that, in eq. (8.98), $\delta L / \delta q_i$ is the derivative with respect to q_i at fixed $\dot{q}_i$, while $\delta L / \delta \dot{q}_i$ is the derivative with respect to $\dot{q}_i$ at fixed q_i. Similarly, in eq. (8.111) [or in eq. (8.113)] $\delta \mathcal{L} / \delta \phi_i$ is taken at fixed $\partial_\mu \phi_i$ and $\delta \mathcal{L} / \delta(\partial_\mu \phi_i)$ is taken at fixed ϕ_i.

[21] Explicitly,

$$F_{\alpha\beta} \frac{\delta F^{\alpha\beta}}{\delta(\partial_\mu A_\nu)}$$
$$= \eta_{\alpha\alpha'} \eta_{\beta\beta'} F^{\alpha'\beta'} \frac{\delta F^{\alpha\beta}}{\delta(\partial_\mu A_\nu)}$$
$$= F^{\alpha'\beta'} \frac{\delta(\eta_{\alpha\alpha'} \eta_{\beta\beta'} F^{\alpha\beta})}{\delta(\partial_\mu A_\nu)}$$
$$= F^{\alpha'\beta'} \frac{\delta F_{\alpha'\beta'}}{\delta(\partial_\mu A_\nu)}$$
$$= F^{\alpha\beta} \frac{\delta F_{\alpha\beta}}{\delta(\partial_\mu A_\nu)}. \tag{8.120}$$

where, in the last line, we used the fact that $F^{\alpha\beta}$ is antisymmetric under $\alpha \leftrightarrow \beta$, so,

$$
\begin{aligned}
F^{\alpha\beta}\partial_\beta A_\alpha &= F^{\beta\alpha}\partial_\alpha A_\beta \\
&= -F^{\alpha\beta}\partial_\alpha A_\beta \,,
\end{aligned}
\tag{8.122}
$$

where we first renamed the dummy indices as $\alpha \to \beta$ and $\beta \to \alpha$ and we then used $F^{\beta\alpha} = -F^{\alpha\beta}$. We next use[22]

$$
\frac{\delta\left(\partial_\alpha A_\beta\right)}{\delta(\partial_\mu A_\nu)} = \delta^\mu_\alpha \delta^\nu_\beta \,,
\tag{8.123}
$$

and we get

$$
\frac{1}{\epsilon_0 c}\frac{\delta\mathcal{L}_0}{\delta(\partial_\mu A_\nu)} = -F^{\mu\nu} \,.
\tag{8.124}
$$

The derivative of $\mathcal{L}$ with respect to A_ν at fixed $\partial_\mu A_\nu$ is easily evaluated: now $\mathcal{L}_0$ does not contribute, since it depends only on the derivatives of the gauge field, and the contribution only comes from $\mathcal{L}_{\text{int}}$, which gives

$$
\begin{aligned}
\frac{1}{\epsilon_0 c}\frac{\delta\mathcal{L}_{\text{int}}}{\delta A_\nu} &= \mu_0 \left(\frac{\delta A_\mu}{\delta A_\nu}\right) j^\mu \\
&= \mu_0 \delta^\nu_\mu j^\mu \\
&= \mu_0 j^\nu \,.
\end{aligned}
\tag{8.125}
$$

Inserting these results into eq. (8.113) we finally get

$$
\partial_\mu F^{\mu\nu} = -\mu_0 j^\nu \,,
\tag{8.126}
$$

and we have therefore recovered eq. (8.23). This shows that the Lagrangian given by eqs. (8.114)–(8.116) is indeed the Lagrangian of the electromagnetic field interacting with an external current.[23] In particular, we have found that the action of the free electromagnetic field is

$$
\begin{aligned}
S_0 &= -\frac{\epsilon_0 c}{4}\int d^4 x\, F_{\mu\nu}F^{\mu\nu} \\
&= \frac{\epsilon_0}{2}\int dt d^3 x\,\left(\mathbf{E}^2 - c^2\mathbf{B}^2\right) \,,
\end{aligned}
\tag{8.127}
$$

where we used eqs. (8.17) and (8.20), as well as $dx^0 = cdt$. Observe that the action of the free electromagnetic field is gauge invariant since, under the gauge transformation (8.14), $F_{\mu\nu}$ is invariant. The interaction term in eq. (8.118) is also gauge invariant. Indeed, applying ∂_ν to eq. (8.126), we get

$$
\partial_\nu\partial_\mu F^{\mu\nu} = -\mu_0\partial_\nu j^\nu \,.
\tag{8.128}
$$

However, $F^{\mu\nu}$ is antisymmetric in μ, ν, while the operator $\partial_\mu\partial_\nu$ is symmetric (as usual, we assume that it acts on differentiable functions on which the derivatives commute). Then $\partial_\nu\partial_\mu F^{\mu\nu}$ vanishes automatically, and eq. (8.128) implies $\partial_\nu j^\nu = 0$. This is in fact nothing but the derivation of current conservation from the equations of motion, that we have

[22] Recall, from the discussion around eqs. (7.104)–(7.108), that the derivative with respect to a quantity with lower index produces an object with upper index, so μ, ν on the right-hand side of eq. (8.123) must be in the upper position. Since $\delta\left(\partial_\alpha A_\beta\right)/\delta(\partial_\mu A_\nu)$ is equal to one if $\alpha = \mu$ and $\beta = \nu$ and is zero otherwise, the result is a product of Kronecker deltas.

[23] More precisely, this shows that this is a possible choice for such a Lagrangian. In general, the Lagrangian that reproduces a given equation of motion is not unique. In the classical mechanics of non-relativistic systems, if we add a total time derivative to the Lagrangian in eq. (8.96), the variational principle is not affected, since a total time derivative in the Lagrangian gives a boundary term in the action, and the variation of the action is computed keeping $q_i(t)$ fixed on the boundary, i.e., at $t = t_i$ and at $t = t_f$. Similarly, the addition to the Lagrangian in eq. (8.101) or in eq. (8.112) of a term of the form $\partial_\mu K^\mu$, with K^μ an arbitrary function of the fields, does not affect the equations of motion, because it is a (three-dimensional) boundary term. Also note that, at the level of eq. (8.118), we could multiply the action by an arbitrary multiplicative factor without affecting the equations of motion. A simple way to fix the correct normalization, which is indeed given by the terms $\epsilon_0 c$ in eq. (8.118), is obtained by also including the dynamics of the point particles. This just amounts to requiring that the interaction term is normalized as in eq. (8.73), since we already saw in Section 8.6.2 that this gives the Lorentz force equation with the correct numerical factors. Otherwise, we will see in the following how the Lagrangian determines the energy density (or, more generally, the energy-momentum tensor), and then one can fix the normalization of the Lagrangian requiring the correct normalization for the energy density of the electromagnetic field, which is given by eq. (3.41). The two procedures are of course equivalent, since the normalization of the energy density of the electromagnetic field was also obtained by comparison with the mechanical energy of point particles, see eqs. (3.39) and (3.40).

already found in Section 3.2.1 and, in the covariant formalism, in Section 8.1. As we have already seen in eq. (8.75), current conservation (together with suitable boundary conditions at infinity, either on θ or on the current), implies that the interaction action is gauge invariant.

In eq. (8.118), or in eq. (8.126), j^μ is a given current; we have not specified the dynamics of the charges that give rise to it, and in this sense we referred to it generically as an "external" current. We can develop the action principle further, by including the dynamics of these charges. For instance, if we consider N classical relativistic point charges, each one described by its world-line $x_a^\mu(\tau)$ (with $a = 1, \ldots, N$), then, for each charge, the current j_a^μ will be given by eq. (8.3) (with q, $x^\mu(\tau)$ and u^μ replaced by q_a, $x_a^\mu(\tau)$ and u_a^μ, respectively), and the free action of these particles will be given by eq. (7.134). Then, the total action describing the free dynamics of N charged point particles, the free dynamics of the electromagnetic field, and the interaction among the electromagnetic field and the particles, is

$$
S = -\sum_{a=1}^{N} m_a c^2 \int dt \sqrt{1 - \frac{v_a^2}{c^2}} - \frac{\epsilon_0 c}{4} \int d^4x \, F_{\mu\nu} F^{\mu\nu}
$$
$$
+ \frac{1}{c} \sum_{a=1}^{N} \int d^4x \, A_\mu(x) j_a^\mu(x) , \tag{8.129}
$$

where

$$
j_a^\mu(x) = q_a \int c d\tau \, u_a^\mu(\tau) \delta^{(4)}[x - x_a(\tau)] , \tag{8.130}
$$

and $u_a^\mu(\tau) = dx_a^\mu/d\tau$. Carrying out the integral over d^4x in the interaction term using the Dirac delta in $j_a^\mu(x)$, we can rewrite this as

$$
S = -\sum_{a=1}^{N} m_a c^2 \int dt \sqrt{1 - \frac{v_a^2}{c^2}} - \frac{\epsilon_0 c}{4} \int d^4x \, F_{\mu\nu} F^{\mu\nu}
$$
$$
+ \sum_{a=1}^{N} q_a \int d\tau \, u_a^\mu(\tau) A_\mu[x_a(\tau)] . \tag{8.131}
$$

This is the same action that we considered in Section 8.6.2, see in particular eqs. (8.69) and (8.70), except that, there, we considered a particle in a given external electromagnetic field, while here we have also included the dynamics of the electromagnetic field.

Actually, the action for the free electromagnetic field that we have found in this section is in a sense more fundamental than the point particle action, and of the interaction term between A_μ and a point particle. In fact, at the quantum level, the point-like approximation for the charged particles is no longer fundamental, and one rather describes also the particles in terms of fields. The fundamental Lagrangian then consists of the Lagrangian (8.115) for the free electromagnetic field, the appropriate Lagrangians for the matter fields (which depend on the spin of the charged particles considered), and of suitable interaction terms between these matter fields and the electromagnetic field, all constructed so as to preserve gauge invariance (see e.g., Chapter 3 of Maggiore (2005) for details). At distance scales sufficiently large, where the particles can be considered point-like, these more fundamental descriptions reduce to eq. (8.129). This has the conceptually important implication that Maxwell's equations, in the form (3.1)–(3.4) or in the covariant form (8.23, 8.33), are not truly fundamental at the scale of elementary particles. There, a more fundamental description of the coupling of the electromagnetic field to the matter field is required.[24] In any case, since at these scales quantum mechanics also enters the game, the subject is more appropriately treated directly in the context of quantum field theory.

8.7.3   Noether's theorem

Noether's theorem expresses the relation between symmetries and conservation laws in classical field theory. We discuss it here briefly, following closely Section 3.2 of Maggiore (2005), to which we refer the reader for more detailed discussions and derivations.[25] We consider a field theory with fields $\phi_i(x)$ and action S. The notation $\phi_i(x)$ here is completely generic and could refer, for instance, to a set of Lorentz scalar fields or, as will be more relevant in our case, to the four components of the four-vector field $A_\mu(x)$. We consider an infinitesimal transformation of the coordinates and of the fields, parametrized by a set of infinitesimal parameters ϵ^a, with $a = 1, \ldots, N$, of the general form

$$x^\mu \quad \rightarrow \quad x'^\mu = x^\mu + \epsilon^a \mathcal{A}_a^\mu(x) \,, \tag{8.132}$$

$$\phi_i(x) \quad \rightarrow \quad \phi_i'(x') = \phi_i(x) + \epsilon^a \mathcal{F}_{i,a}(\phi, \partial\phi) \,, \tag{8.133}$$

with $\mathcal{A}_a^\mu(x)$ (not to be confused with the gauge field) a given function of the coordinates, and $\mathcal{F}_{i,a}(\phi, \partial\phi)$ a given functional of the fields and of their derivatives. An important distinction is between "global" and "local" transformations. A transformation is called global if its parameters ϵ^a are constants; it is called local if they are taken to be arbitrary functions of space-time, $\epsilon^a(x)$.

As an example, space-time translations are transformation in which (both for infinitesimal and finite transformations) the coordinates change as

$$x^\mu \rightarrow x'^\mu = x^\mu + \epsilon^\mu \,, \tag{8.134}$$

with ϵ^μ a constant (so, they are global transformations), while all fields, independently of their properties under Lorentz transformations, transform as "scalars under translations," i.e., as

$$\phi_i(x) \rightarrow \phi_i'(x') = \phi_i(x) \,, \tag{8.135}$$

as we have already seen in eq. (6.22) for the case of spatial translations, see in particular the discussion in Note 3 on page 137. Equation (8.134) can be rewritten as

$$x^\mu \rightarrow x^\mu + \epsilon^\nu \delta_\nu^\mu \,. \tag{8.136}$$

[24]A note for the advanced reader. For spin 1/2 particles, such as electrons and protons, at the level of what is called the Dirac Lagrangian, the coupling to the electromagnetic field indeed turns out to be of the form (8.116), with the current j^μ given in terms of the fields describing these particles. In contrast, for a spin-0 charged particle, an extra interaction term, quadratic both in the gauge fields and in the field describing the particle, is present; see Maggiore (2005), eqs. (3.170) and (3.174). This extra term gives rise to a corresponding extra term in the equation of motion, which then is no longer of the form (8.23).

[25]When comparing the results, note that Maggiore (2005) uses units $\epsilon_0 = \mu_0 = c = 1$, and the opposite metric signature $\eta_{\mu\nu} = (+,-,-,-)$. These conventions are the most common in quantum field theory.

Therefore, in this case the index a in eqs. (8.132) and (8.133) is just a Lorentz index, and we have

$$\mathcal{A}_\nu^\mu = \delta_\nu^\mu , \qquad\qquad \mathcal{F}_{i,\nu} = 0 . \qquad (8.137)$$

Another example is given by spatial rotations. On the spatial coordinates x^i, the infinitesimal form of a spatial rotation can be written as in eq. (1.153), while $x^0 \to x^0$. We can rewrite this in the form (8.132), where the role of the index a is played by the pair of indices (i,j) that identify the plane in which the spatial rotation is performed, and ϵ^a is identified with the parameters ω^{ij} defined in eq. (1.149) (recall that, for spatial rotations, we can write the spatial indices equivalently as upper or lower indices). Then, eq. (8.132) can be written as

$$x^0 \quad\to\quad x^0 + \sum_{i<j} \omega^{ij} \mathcal{A}_{ij}^0(x) , \qquad (8.138)$$

$$x^k \quad\to\quad x^k + \sum_{i<j} \omega^{ij} \mathcal{A}_{ij}^k(x) , \qquad (8.139)$$

(where the restriction to $i < j$ avoids a double counting, since the parameters ω^{ij} with $i > j$ are fixed in terms of those with $i < j$ by $\omega^{ij} = -\omega^{ji}$), with[26]

$$\mathcal{A}_{ij}^0(x) = 0 , \qquad \mathcal{A}_{ij}^k(x) = \delta^{ik} x^j - \delta^{jk} x^i . \qquad (8.140)$$

The corresponding expression for $\mathcal{F}$ in eq. (8.133) depends on the type of field considered. For a field scalar under rotations, such as the scalar gauge potential ϕ, writing eq. (8.133) as

$$\phi(x) \to \phi'(x') = \phi(x) + \sum_{i<j} \omega^{ij} \mathcal{F}_{0,ij} , \qquad (8.141)$$

we have $\mathcal{F}_{0,ij} = 0$. For a vector field, the finite transformation was given in eq. (7.94), so the infinitesimal transformation, written for instance for the vector gauge potential $\mathbf{A}$, is

$$A^k(x) \to A'^k(x') = A^k(x) + \omega^{kj} A_j(x) . \qquad (8.142)$$

Similarly to eqs. (8.139) and (8.140), this can be written as

$$A^k(x) \to A'^k(x') = A^k(x) + \sum_{i<j} \omega^{ij} \mathcal{F}_{ij}^k[A(x)] , \qquad (8.143)$$

with

$$\mathcal{F}_{ij}^k[A(x)] = \delta^{ik} A^j(x) - \delta^{jk} A^i(x) . \qquad (8.144)$$

One could proceed similarly for the full set of Lorentz transformations, but we will limit ourselves to space-time translations and spatial rotations from which, as we will see, Noether's theorem will allow us to compute the energy, momentum, and angular momentum of a given field theory so, in particular, of the electromagnetic field.

Equations (8.132, 8.133) define a symmetry transformation of the theory if they leave the action $S(\phi)$ invariant, for any configuration of the

[26]The fact that this expression for $\mathcal{A}_{ij}^k(x)$, once inserted into eq. (8.139), correctly gives eq. (1.153), can be shown writing

$$\sum_{i<j} \omega^{ij} (\delta^{ik} x^j - \delta^{jk} x^i)$$

$$= \sum_{i<j} \omega^{ij} \delta^{ik} x^j - \sum_{j<i} \omega^{ji} \delta^{ik} x^j$$

$$= \sum_{i<j} \omega^{ij} \delta^{ik} x^j + \sum_{i>j} \omega^{ij} \delta^{ik} x^j$$

$$= \sum_{i,j} \omega^{ij} \delta^{ik} x^j$$

$$= \sum_j \omega^{kj} x^j ,$$

where, for the second term, in the second line we renamed the dummy indices $i \leftrightarrow j$, and we then used $\omega^{ij} = -\omega^{ji}$.

fields ϕ_i. Note that we are not assuming that the fields ϕ_i satisfy the classical equations of motion. A symmetry by definition leaves the action invariant for every field configuration, solution or not of the equations of motion.

Now, suppose that eqs. (8.132) and (8.133) are a global, but not a local, symmetry of our theory, i.e., they leave the action invariant if ϵ is constant, but not if we allow ϵ to depend on x. Then, Noether's theorem states that, for each value of the index $a = 1, \ldots, N$ (i.e., for each independent parameter of the transformation), there is a conserved current j_a^μ, i.e., a current that satisfies

$$\partial_\mu j_a^\mu = 0. \tag{8.145}$$

This implies the existence of a corresponding set of conserved charges

$$Q_a = \frac{1}{c} \int d^3x \, j_a^0. \tag{8.146}$$

In fact, taking the time derivative and using eq. (8.145),

$$\begin{aligned} \frac{dQ_a}{dt} &= \frac{1}{c} \int d^3x \, \partial_0 j_a^0(x) \\ &= -\frac{1}{c} \int d^3x \, \partial_i j_a^i(x), \end{aligned} \tag{8.147}$$

and this is a boundary term, representing the flux entering and leaving the volume of integration. In particular, if we integrate over all space with the boundary condition that j_a^i vanishes sufficiently fast at infinity (or if we integrate over a finite volume with $j_a^i(x) = 0$ on the boundary, so that there is no incoming or outgoing flux), the charge Q_a is conserved. All this is identical to the conservation of the electric charge that we first found, in a non-covariant formalism, in Section 3.2.1, and that we found again, with the covariant formalism, in eqs. (8.10) and (8.11). The notations "charge" and "current" in the context of the Noether theorem are indeed borrowed from this example. However, as we have seen, the index a here can also be a Lorentz index, as in the case of space-time translations, or can represent a pair of antisymmetric spatial indices, as in the case of rotations, so the Lorentz transformation properties of the "currents" $j_a^i(x)$ and "charges" Q_a also depend on the nature of this index.[27]

Noether's theorem also provides an explicit expression of the currents $j_a^\mu(x)$, in terms of the Lagrangian density of the theory and of the functions $\mathcal{A}_a^\mu(x)$ and $\mathcal{F}_{i,a}(\phi, \partial\phi)$ that enter in eqs. (8.132) and (8.133),[28]

$$\boxed{j_a^\mu = \frac{\delta \mathcal{L}}{\delta(\partial_\mu \phi_i)} \left[\mathcal{A}_a^\rho(x) \partial_\rho \phi_i - \mathcal{F}_{i,a}(\phi, \partial\phi) \right] - \mathcal{L}\, \mathcal{A}_a^\mu(x).} \tag{8.148}$$

[27] Also observe that the multiplicative factor in front of the charge is, at this stage, arbitrary, since, if a quantity is conserved, multiplying it by a constant it remains conserved. In eq. (8.146) we have chosen a multiplicative factor $1/c$, as in eq. (8.11).

[28] See Section 3.2 of Maggiore (2005) for a proof of Noether's theorem and a derivation of eq. (8.148).

Energy-momentum tensor of the electromagnetic field from Noether's theorem

We now specialize the above machinery to translations, which are symmetries of Maxwell's theory (and of all standard field theories). In this case, we have seen that the index a is actually a Lorentz index, so the corresponding four conserved currents $j^{\mu}_{(\nu)}$ form a Lorentz tensor. We define

$$\theta^{\mu}{}_{\nu} = -cj^{\mu}_{(\nu)} \, . \tag{8.149}$$

This is the field-theory definition of the energy-momentum tensor.[29] Then, inserting eq. (8.137) into eq. (8.148) and raising the ν index as $\theta^{\mu\nu} = \eta^{\nu\rho}\theta^{\mu}{}_{\rho}$, we get

$$\theta^{\mu\nu} = -c\left[\frac{\delta\mathcal{L}}{\delta(\partial_{\mu}\phi_i)}\partial^{\nu}\phi_i - \eta^{\mu\nu}\mathcal{L}\right] \, . \tag{8.150}$$

Equation (8.145) then states that, when $\theta^{\mu\nu}$ is evaluated on a solution of the equations of motion, it satisfies[30]

$$\partial_{\mu}\theta^{\mu\nu} = 0 \, . \tag{8.151}$$

According to eq. (8.146), the conserved charge associated with the energy-momentum tensor is

$$P^{\nu} \equiv \frac{1}{c}\int d^3x \, \theta^{0\nu} \, , \tag{8.152}$$

and this is the definition of four-momentum in classical field theory. A field configuration, solution of the equations of motion, carries an energy $E = P^0$ and a spatial momentum P^i which can be calculated using eqs. (8.150) and (8.152).

We can now apply this to the free electromagnetic field. In this case, eq. (8.150) becomes

$$\theta^{\mu\nu} = -c\left[\frac{\delta\mathcal{L}}{\delta(\partial_{\mu}A_{\rho})}\partial^{\nu}A_{\rho} - \eta^{\mu\nu}\mathcal{L}\right] \, , \tag{8.153}$$

and $\mathcal{L}$ is given by eq. (8.115). The derivative $\delta\mathcal{L}/\delta(\partial_{\mu}A_{\rho})$, which appears in eq. (8.153), was already computed in eq. (8.124), so we get

$$\theta^{\mu\nu} = -\epsilon_0 c^2\left[-F^{\mu\rho}\partial^{\nu}A_{\rho} + \frac{1}{4}\eta^{\mu\nu}F_{\alpha\beta}F^{\alpha\beta}\right] \, . \tag{8.154}$$

We next write $\partial^{\nu}A_{\rho} = (\partial^{\nu}A_{\rho} - \partial_{\rho}A^{\nu}) + \partial_{\rho}A^{\nu} = F^{\nu}{}_{\rho} + \partial_{\rho}A^{\nu}$. Then

$$\theta^{\mu\nu} = -\epsilon_0 c^2\left[F^{\mu\rho}F_{\rho}{}^{\nu} + \frac{1}{4}\eta^{\mu\nu}F_{\alpha\beta}F^{\alpha\beta}\right] + \epsilon_0 c^2 F^{\mu\rho}\partial_{\rho}A^{\nu} \, . \tag{8.155}$$

The first term in this expression is just the energy-momentum $T^{\mu\nu}$ that was already written in eq. (8.34); as we showed there, this tensor contains, in a covariant form, the energy density of the electromagnetic field,

[29] As we will see below, the factor $-c$ is chosen so as to eventually obtain the same normalization as in eq. (8.34). In particular, the minus sign is related to our signature $\eta_{\mu\nu} = (-,+,+,+)$. As already mentioned, by itself Noether's theorem does not fix the normalization of the conserved current; if a current is conserved, i.e., $\partial_{\mu}j^{\mu} = 0$, any multiple of it is also conserved.

[30] Observe that, at this stage, $\theta^{\mu\nu}$ is not necessarily symmetric in μ,ν, and the contraction of the index of the partial derivative must be done with the first index of $\theta^{\mu\nu}$, while ν was the equivalent of the index a in eq. (8.145). We will see below how to "improve" the energy-momentum tensor so that it becomes symmetric, when the expression obtained from eq. (8.150) is not symmetric.

given by T^{00}, the Poynting vector, given by $S^i = cT^{0i}$, and the Maxwell stress tensor T^{ij}, and satisfies the conservation equations (8.39) that summarizes, in a covariant form, energy conservation and momentum conservation. In particular, in the absence of sources, which is the case that we are considering here, on the solutions of the equations of motion it satisfies $\partial_\mu T^{\mu\nu} = 0$ (or, since it is symmetric in μ, ν, $\partial_\mu T^{\nu\mu} = 0$ and, after renaming the indices $\mu \leftrightarrow \nu$, $\partial_\nu T^{\mu\nu} = 0$).

The extra term in eq. (8.155) is, at first sight, quite puzzling, since it is not even gauge-invariant (also note that it is not symmetric in μ, ν). However, using the equation of motion $\partial_\rho F^{\mu\rho} = 0$ (appropriate to the fact that we have derived $\theta^{\mu\nu}$ from Noether's theorem using the Lagrangian of the free electromagnetic field) we see that, under a gauge transformation (8.14), it induces a change in $\theta^{\mu\nu}$ given by

$$
\begin{aligned}
\theta^{\mu\nu} &\rightarrow \theta^{\mu\nu} - \epsilon_0 c^2 F^{\mu\rho} \partial_\rho \partial^\nu \theta \\
&= \theta^{\mu\nu} - \epsilon_0 c^2 \partial_\rho \left(F^{\mu\rho} \partial^\nu \theta \right),
\end{aligned} \tag{8.156}
$$

so the conserved charge [i.e., the four-momentum P^ν given by eq. (8.152)] changes as

$$
\begin{aligned}
P^\nu &\rightarrow P^\nu - \epsilon_0 c \int d^3x \, \partial_\rho \left(F^{0\rho} \partial^\nu \theta \right) \\
&= P^\nu - \epsilon_0 c \int d^3x \, \partial_i \left(F^{0i} \partial^\nu \theta \right).
\end{aligned} \tag{8.157}
$$

The additional term is a total spatial derivative which integrates to zero (assuming, as always, that the field decreases sufficiently fast at infinity or, in a finite volume, that it vanishes at the boundary of the integration volume), so the extra term in eq. (8.155) does not contribute to the four-momentum (which then, in particular, is gauge invariant), and $\theta^{\mu\nu}$ gives the same charges as the explicitly gauge-invariant tensor $T^{\mu\nu}$ so, from this point of view, they are physically equivalent.

In general, Noether's theorem provides an explicit expression for the conserved current, but this need not be unique. For instance, once we have found an energy momentum tensor $\theta^{\mu\nu}$ such that $\partial_\mu \theta^{\mu\nu} = 0$, we can consider the "improved" energy-momentum tensor

$$
T^{\mu\nu} = \theta^{\mu\nu} + \partial_\rho \mathcal{A}^{\rho\mu\nu}, \tag{8.158}
$$

where $\mathcal{A}^{\rho\mu\nu}$ is an arbitrary tensor antisymmetric in the indices ρ, μ. This new tensor is still conserved: $\partial_\mu \partial_\rho \mathcal{A}^{\rho\mu\nu} = 0$ because of the antisymmetry in ρ, μ. Furthermore, for $\mu = 0$, $\partial_\rho \mathcal{A}^{\rho 0\nu} = \partial_i \mathcal{A}^{i0\nu}$ is a spatial divergence, and therefore this term does not contribute to the four-momentum (8.152) if the fields vanish sufficiently fast at infinity. This is precisely what happened here, with $\mathcal{A}^{\rho\mu\nu} = \epsilon_0 c^2 F^{\mu\rho} A^\nu$.[31] We then need some physical input to choose the "correct" form, if we want to define a local energy density (also see the discussion in Note 11 on page 47); in this case, the requirement of gauge invariance selects $T^{\mu\nu}$. In general, $\mathcal{A}^{\rho\mu\nu}$ can be chosen so that $T^{\mu\nu}$ is symmetric in cases when $\theta^{\mu\nu}$ is not.[32]

[31] Indeed, using the equation of motion $\partial_\rho F^{\mu\rho} = 0$ appropriate to the fact that we are working in vacuum, the term $F^{\mu\rho} \partial_\rho A^\nu$ in eq. (8.155) can be rewritten as $\partial_\rho (F^{\mu\rho} A^\nu)$.

[32] A note for the advanced reader. In General Relativity, the energy-momentum tensor is defined in terms of a functional derivative of the Lagrangian with respect to a generic metric $g_{\mu\nu}$. This automatically gives a symmetric energy-momentum tensor. In the case of electromagnetism in curved space, after taking the derivative with respect to $g_{\mu\nu}$ and specializing $g_{\mu\nu}$ to the flat-space Minkowski metric $\eta_{\mu\nu}$, this procedure gives precisely the tensor $T^{\mu\nu}$ given in eq. (8.34). Eventually, this can be taken as the best way of uniquely resolving the ambiguity in the definition of the energy-momentum tensor.

Angular momentum of the electromagnetic field from Noether's theorem

We can repeat the same procedure for the invariance under rotations, using eqs. (8.138)–(8.144). We write the "charge" associated with rotations in the (i, j) plane as J_{ij}, so[33]

$$J_{ij} = \int d^3x \, j_{ij}^0 \,. \tag{8.159}$$

The corresponding angular momentum is then given by

$$J_i = \frac{1}{2}\epsilon_{ijk}J_{jk} \,. \tag{8.160}$$

We now specialize to the electromagnetic field, adding a subscript "em" to the various quantities. Then eq. (8.148) gives

$$(j_{\rm em})_{ij}^0 = \frac{\delta\mathcal{L}}{\delta(\partial_0 A_\nu)}\left[\mathcal{A}_{ij}^\rho(x)\partial_\rho A_\nu - \mathcal{F}_{\nu,ij}(A)\right] - \mathcal{L}\,\mathcal{A}_{ij}^0(x)\,. \tag{8.161}$$

From eq. (8.124), $\delta\mathcal{L}/\delta(\partial_0 A_\nu) = -\epsilon_0 cF^{0\nu}$. It therefore vanishes for $\nu = 0$, and the only contribution comes when ν is a spatial index, that we denote by l.[34] Using eq. (8.17), $\delta\mathcal{L}/\delta(\partial_0 A_l) = -\epsilon_0 E^l$. Furthermore, from eq. (8.140), $\mathcal{A}_{ij}^0(x) = 0$. Then,

$$(j_{\rm em})_{ij}^0 = -\epsilon_0 E_l\left[\mathcal{A}_{ij}^k(x)\partial_k A_l - \mathcal{F}_{l,ij}(A)\right]\,. \tag{8.162}$$

Inserting the explicit expressions of $\mathcal{A}_{ij}^k$ and $\mathcal{F}_{l,ij}$ from eqs. (8.140) and (8.144), we get[35]

$$(J_{\rm em})_{ij} = \epsilon_0 \int d^3x \left[E_l(x_i\partial_j - x_j\partial_i)A_l + (E_iA_j - E_jA_i)\right]\,. \tag{8.163}$$

We can further manipulate the second term in this expression writing

$$\int d^3x \,(E_iA_j - E_jA_i) = \int d^3x \, E_l[A_j(\partial_l x_i) - A_i(\partial_l x_j)]\,, \tag{8.164}$$

since $\partial_l x_i = \delta_{li}$. We next integrate ∂_l by parts and use the fact that, in vacuum $\partial_l E_l = \boldsymbol{\nabla}\cdot\mathbf{E} = 0$. Then, neglecting the boundary terms,

$$\int d^3x \,(E_iA_j - E_jA_i) = \int d^3x \, E_l(-x_i\partial_l A_j + x_j\partial_l A_i)\,. \tag{8.165}$$

Inserting this into eq. (8.163) and using eq. (8.160), we get

$$\begin{aligned}
(J_{\rm em})_k &= \epsilon_0 \int d^3x \,\epsilon_{ijk}x_i(\partial_j A_l - \partial_l A_j)E_l \\
&= \epsilon_0 \int d^3x \,\epsilon_{ijk}x_i\epsilon_{jlm}B_m E_l \\
&= \epsilon_0 \int d^3x \,\epsilon_{ijk}x_i(\mathbf{E}\times\mathbf{B})_j \\
&= \epsilon_0 \int d^3x \,[\mathbf{x}\times(\mathbf{E}\times\mathbf{B})]_k\,.
\end{aligned} \tag{8.166}$$

[33] Once again, the overall normalization of the charge and the current cannot be fixed from the Noether theorem since, if a quantity is conserved, any multiple of it is still conserved, and we fix it so that the result will eventually agree with the one found in eq. (3.76). In particular, we do not insert in eq. (8.159) a factor $1/c$ as in eq. (8.146). The powers of c are actually fixed from dimensional analysis, in this case by the requirement of obtaining a quantity with the dimensions of angular momentum.

[34] Incidentally, this shows that A_0 is not a real dynamical variable, since its time derivative does not appear in the Lagrangian. This has important implications for the Hamiltonian formalism and for the quantization of the theory, see Maggiore (2005), Section 4.3.2.

[35] A note for the advanced reader. The first term in eq. (8.163) depends only on the transformation of the coordinates, through the term $\mathcal{A}_{ij}^k(x)$. A similar term will appear for any field theory, independently of the transformation properties of the field. Upon quantization, it corresponds to the orbital angular momentum of the field, as can already be realized from the appearance of the operator $(x_i\partial_j - x_j\partial_i)$ which, in quantum mechanics, is related to the angular momentum operator,

$$\begin{aligned}
\hat{L}_i &= -i\hbar\epsilon_{ijk}x_j\partial_k \\
&= -(i/2)\hbar\epsilon_{ijk}(x_j\partial_k - x_k\partial_j)\,.
\end{aligned}$$

The term $(E_iA_j - E_jA_i)$, instead, comes from the term proportional to $\mathcal{F}_{l,ij}(A)$ in eq. (8.162), and is therefore specific to the vector nature of the field $\mathbf{A}$. Upon quantization, it corresponds to the spin part of the total angular momentum. See Maggiore (2005), eqs. (4.81)–(4.90).

This agrees with eq. (3.76). We have therefore recovered, using Noether's theorem, the expression for the angular momentum of the electromagnetic field that we found in eq. (3.76) by extracting a conservation law directly from Maxwell's equation. The derivation from Noether's theorem emphasizes that angular momentum is the conserved quantity associated with the invariance under spatial rotations.

8.8   Solved problems

Problem 8.1.   Lorentz invariance of the charge associated with a current distribution

In eq. (8.11) we found that the electric charge associated with a generic current distribution can be written as

$$Q = \frac{1}{c} \int d^3x \, j^0(t, \mathbf{x}) \,, \tag{8.167}$$

and we anticipated that this expression is Lorentz invariant. This is not evident a priori, since the integral is only over the spatial variables, and the integrand is the $\mu = 0$ component of a four-vector. However, the Lorentz invariance of this expression can be proven as follows.[36] Equation (8.167) is written with respect to a specific reference frame, let's call it K, which uses coordinates $(t, \mathbf{x})$. We then introduce a four-vector n^μ that, in this reference frame, is given by $n^\mu = (-1, 0, 0, 0)$, or, equivalently, $n_\mu = (1, 0, 0, 0)$. We next observe that eq. (8.167) can be rewritten as

$$Q = \frac{1}{c} \int d^4x \, j^\mu(x) \, \partial_\mu \left[\theta(n_\nu x^\nu) \right] \,, \tag{8.168}$$

where θ is the Heaviside theta function (1.67). In fact, in the K reference frame, $\theta(n_\nu x^\nu) = \theta(x^0)$, so

$$\partial_\mu \left[\theta(n_\nu x^\nu) \right] = \partial_\mu \theta(x^0)$$
$$= \delta^0_\mu \, \delta(x^0) \,, \tag{8.169}$$

where $\delta(x^0)$ is the Dirac delta, and we used eq. (1.70). Inserted into eq. (8.168), eq. (8.169) gives back eq. (8.167). However, the form (8.168) is more convenient to study the behavior of Q under Lorentz transformations. Using $\partial_\mu j^\mu = 0$, we can rewrite this as

$$Q = \frac{1}{c} \int d^4x \, \partial_\mu \left[j^\mu(x) \, \theta(n_\nu x^\nu) \right] \,. \tag{8.170}$$

Note that, even if the current $j^\mu(x) = j^\mu(t, \mathbf{x})$ is localized in space, it is not localized in time (i.e., we are not assuming that the four-current vanishes at $t \to \pm\infty$), so the integral in eq. (8.170) can be reduced to a boundary term, but this boundary term would be non-vanishing. We will then still keep it, for the moment, in the form of a four-dimensional integral.

We now perform a transformation to a new reference frame K'. Then, $x^\mu \to x'^\mu = \Lambda^\mu{}_\nu x^\nu$, see eq. (7.43), and similarly $n_\mu \to n'_\mu = \Lambda_\mu{}^\nu n_\nu$. Denoting the value of the transformed charge by Q', we therefore have

$$Q' = \frac{1}{c} \int d^4x' \, \frac{\partial}{\partial x'^\mu} \left[j^\mu(x') \, \theta(n'_\nu x''^\nu) \right] \,. \tag{8.171}$$

[36] We follow the derivation given in Section 6 of Weinberg (1972).

However, in eq. (8.171), x'^μ is a dummy integration variable, that we can just rename x^μ (note that here we also make use of the fact that we are integrating over all of space-time, so the integration domain is unchanged under the transformation $x^\mu \to \Lambda^\mu{}_\nu x^\nu$). Therefore, in the frame K', the value of the charge is

$$Q' = \frac{1}{c} \int d^4x \, \partial_\mu \left[j^\mu(x) \, \theta(n'_\nu x^\nu) \right] , \qquad (8.172)$$

and

$$Q' - Q = \frac{1}{c} \int d^4x \, \partial_\mu \left\{ j^\mu(x) \left[\theta(n'_\nu x^\nu) - \theta(n_\nu x^\nu) \right] \right\} . \qquad (8.173)$$

The crucial point is that the difference of theta functions in the square brackets vanishes at $t \to \pm\infty$ at fixed $\mathbf{x}$; actually, for fixed $|\mathbf{x}|$, it is even a function with compact support in t, since it becomes exactly zero when x^0 is sufficiently large, so that both $n'_\nu x^\nu$ and $n_\nu x^\nu$ are positive, or when x_0 is negative and sufficiently large in absolute value, so that $n'_\nu x^\nu$ and $n_\nu x^\nu$ are both negative. Since $j^\mu(x)$ vanishes at $|\mathbf{x}| \to \infty$ at fixed t (because we assume that the current is localized in space, or at least decreases sufficiently fast as $|\mathbf{x}| \to \infty$), the integrand in eq. (8.173) now vanishes on the whole boundary of the four-dimensional integration region. Therefore, now Gauss's theorem ensures that $Q' - Q = 0$, so Q is Lorentz invariant.

Having established the Lorentz invariance of Q, we can search for an explicitly Lorentz-invariant expression that reduces to eq. (8.11) in the K frame which uses the variables $(t, \mathbf{x})$. This will then give an expression for Q valid in any frame. To this purpose, we define a covariant four-vector d^3S_μ from the condition that, in the frame K where eq. (8.167) holds,

$$d^3S_\mu = d^3x(1, 0, 0, 0) . \qquad (8.174)$$

In a generic frame, d^3S_μ is then obtained transforming it as a covariant four-vector. Then, eq. (8.167) can be written in an explicitly Lorentz-invariant form as

$$Q = \frac{1}{c} \int d^3S_\mu \, j^\mu . \qquad (8.175)$$

An explicitly covariant expression from d^3S_μ can be obtained as follows.[37] First, as a simpler example, consider the integration along a one-dimensional curve embedded in four-dimensional Minkowski space. The curve can be parametrized in terms of a single variable ξ, so that the position in space-time of a point along the curve is given by assigning a function $x^\mu(\xi)$. For instance, if $x^\mu(\xi)$ describes the trajectory of a massive particle, a natural choice for ξ is the proper time τ, as we discussed in Section 7.4.1. Then, for instance, the line integral of a four-vector field $V^\mu(x)$, along the curve $\mathcal{C}$ defined by the function $x^\mu(\xi)$, can be written as

$$\int_{\mathcal{C}} dx^\mu V_\mu(x) = \int_{\mathcal{C}} d\xi \, \frac{\partial x^\mu(\xi)}{\partial \xi} V_\mu[x(\xi)] . \qquad (8.176)$$

This can be generalized to two-dimensional surfaces, or three-dimensional volumes, embedded in four-dimensional Minkowski space. A two-dimensional surface is parametrized by two parameters (ξ_1, ξ_2), so that the position in four-dimensional space of the point of the surface identified by (ξ_1, ξ_2) is given assigning the functions $x^\mu(\xi_1, \xi_2)$. The two-dimensional surface element $d^2s_{\mu\nu}$ can then be written as

$$d^2s_{\mu\nu} = \frac{1}{2!} \epsilon_{\mu\nu\rho\sigma} \frac{\partial(x^\rho, x^\sigma)}{\partial(\xi_1, \xi_2)} d\xi_1 d\xi_2 , \qquad (8.177)$$

[37]This part is more mathematical and can be skipped at first reading. Proofs and extended discussions can be found in most textbook dealing with Riemannian geometry.

where $\partial(x^\rho, x^\sigma)/\partial(\xi_1, \xi_2)$ is the Jacobian of the transformation,

$$\frac{\partial(x^\rho, x^\sigma)}{\partial(\xi_1, \xi_2)} = \begin{vmatrix} \frac{\partial x^\rho}{\partial \xi_1} & \frac{\partial x^\rho}{\partial \xi_2} \\ \frac{\partial x^\sigma}{\partial \xi_1} & \frac{\partial x^\sigma}{\partial \xi_2} \end{vmatrix} . \tag{8.178}$$

Similarly, a point in a three-dimensional volume embedded in four-dimensional Minkowski space can be parametrized by three parameters $\xi = (\xi_1, \xi_2, \xi_3)$, so that the position in four-dimensional space of the point of the surface identified by (ξ_1, ξ_2, ξ_3) is given by the functions $x^\mu(\xi_1, \xi_2, \xi_3)$. The volume element can then be written, in an explicitly covariant form, as

$$d^3 S_\mu = \frac{1}{3!} \epsilon_{\mu\nu\rho\sigma} \frac{\partial(x^\nu, x^\rho, x^\sigma)}{\partial(\xi_1, \xi_2, \xi_3)} d\xi_1 d\xi_2 d\xi_3 , \tag{8.179}$$

where, again $\partial(x^\nu, x^\rho, x^\sigma)/\partial(\xi_1, \xi_2, \xi_3)$ is the Jacobian of the transformation. In the frame where the volume is parametrized by the choices $x^1(\xi) = \xi_1$, $x^2(\xi) = \xi_2$ and $x^3(\xi) = \xi_3$, we get $d^3 S_\mu = d\xi_1 d\xi_2 d\xi_3 (1, 0, 0, 0)$, i.e., we get back eq. (8.174), with ξ_i identified with x_i.

Problem 8.2. Covariance of $\int d^3x\, T^{0\nu}$

We can proceed similarly to show that, given an energy-momentum $T^{\mu\nu}$ that satisfies $\partial_\mu T^{\mu\nu} = 0$, the quantity defined by

$$P^\nu = \frac{1}{c} \int d^3x\, T^{0\nu} , \tag{8.180}$$

is indeed a four-vector, as the notation P^ν suggests. To this purpose, proceeding as in eqs. (8.168)–(8.170), we rewrite it as

$$P^\nu = \frac{1}{c} \int d^4x\, \partial_\mu \left[T^{\mu\nu}(x)\, \theta(nx) \right] , \tag{8.181}$$

where $nx = n_\rho x^\rho$. Then, under a Lorentz-transformation, $P^\nu \to P'^\nu$, where

$$P'^\nu = \frac{1}{c} \int d^4x\, \Lambda^\nu{}_\sigma \partial_\mu \left[T^{\mu\sigma}(x)\, \theta(n'x) \right] , \tag{8.182}$$

and we used the fact that $\partial_\mu \left[T^{\mu\nu}(x)\, \theta(nx) \right]$ has a single free index ν and therefore transforms with a single matrix $\Lambda^\nu{}_\sigma$. Then

$$P'^\nu - P^\nu = \frac{1}{c} \int d^4x\, \partial_\mu \left[\Lambda^\nu{}_\sigma T^{\mu\sigma}(x)\, \theta(n'x) - \delta^\nu_\sigma T^{\mu\sigma}(x)\, \theta(nx) \right] . \tag{8.183}$$

We now consider an infinitesimal Lorentz transformation, $\Lambda^\nu{}_\sigma = \delta^\nu_\sigma + \omega^\nu{}_\sigma$, so

$$\begin{aligned} P'^\nu - P^\nu =\ & \frac{1}{c} \delta^\nu_\sigma \int d^4x\, \partial_\mu \left\{ T^{\mu\sigma}(x) \left[\theta(n'x) - \theta(nx) \right] \right\} \\ & + \omega^\nu{}_\sigma \frac{1}{c} \int d^4x\, \partial_\mu \left[T^{\mu\sigma}(x)\, \theta(n'x) \right] . \end{aligned} \tag{8.184}$$

The term on the right-hand side in the first line vanishes by the same argument that we used for the electric charge: $T^{\mu\nu}$ is localized in space, while the difference of theta functions vanishes at large $|\mathbf{x}_0|$ and fixed $|\mathbf{x}|$, so Gauss's theorem implies that this integral is a vanishing boundary term. In the second line, to first order in ω we can replace n' by n, since we have already a factor $\omega^\nu{}_\sigma$ in front. Then,

$$
\begin{aligned}
P'^\nu - P^\nu &= \omega^\nu{}_\sigma \frac{1}{c} \int d^4x\, \partial_\mu \left[T^{\mu\sigma}(x)\, \theta(nx) \right] \\
&= \omega^\nu{}_\sigma P^\sigma \, .
\end{aligned}
\tag{8.185}
$$

This is precisely the transformation of a four-vector under infinitesimal Lorentz transformation. Constructing a finite Lorentz transformation as a sequence of infinitesimal transformations, this implies that P^ν transforms as a four-vector also under finite Lorentz transformation. Therefore P^ν, defined by eq. (8.180), is indeed a four-vector.

Similarly to eq. (8.175), we can then write it in an explicitly Lorentz-covariant form as

$$
P_{\text{em}}^\nu = \frac{1}{c} \int d^3 S_\mu \, T^{\mu\nu} \, ,
\tag{8.186}
$$

where, as in eq. (8.174), $d^3 S_\mu$ is defined from the condition that, in the frame K where eq. (8.48) holds, $d^3 S_\mu = d^3 x (1, 0, 0, 0)$, and, in a generic frame, it is obtained requiring that it transforms as a four-vector. Its covariant expression is given in eq. (8.179).

Problem 8.3. Relativistic motion in a constant electric field

We compute here the relativistic motion of a charged particle in a constant electric field. Setting $\mathbf{E} = E\hat{\mathbf{x}}$, the equation of motion (3.6), that, as we have seen in this chapter, when supplemented by $\mathbf{p} = \gamma(v)m\mathbf{v}$ is fully relativistic, becomes

$$
\frac{d\mathbf{p}}{dt} = qE\hat{\mathbf{x}} \, .
\tag{8.187}
$$

Setting the initial condition $\mathbf{p}(t = 0) = 0$, we then have

$$
p_x(t) = qEt \, ,
\tag{8.188}
$$

while $p_y(t) = p_z(t) = 0$. Writing $p_i = \gamma m v_i$, we therefore have $v_y(t) = v_z(t) = 0$ and, denoting $v_x(t)$ simply as $v(t)$, eq. (8.188) becomes

$$
\frac{mv(t)}{\sqrt{1 - v^2(t)/c^2}} = qEt \, .
\tag{8.189}
$$

This can be solved for $v(t)$, obtaining

$$
\frac{v(t)}{c} = \frac{qEt}{\sqrt{m^2 c^2 + (qEt)^2}} \, .
\tag{8.190}
$$

In the limit $qEt \ll mc$ this reduces to

$$
v(t) \simeq \frac{qEt}{m} \, ,
\tag{8.191}
$$

which is the non-relativistic result for a particle subject to the constant force $F = qE$. However, the right-hand side of eq. (8.190) is always smaller than one, so $v(t)$ is always smaller than c and, as $t \to \infty$, $v(t) \to c$. Writing $v(t) = dx/dt$, and integrating eq. (8.190) with the initial condition $x(0) = 0$, gives

$$
x(t) = \frac{mc^2}{qE} \left(\sqrt{1 + \frac{q^2 E^2 t^2}{m^2 c^2}} - 1 \right) \, ,
\tag{8.192}
$$

which interpolates between the non-relativistic behavior

$$x(t) \simeq \frac{1}{2}at^2 \,, \tag{8.193}$$

(with $a = qE/m$) at small t, and the ultra-relativistic behavior $x(t) \simeq ct$ at large t.

As we will see in Chapter 10, an accelerated particle actually radiates electromagnetic waves, and therefore loses energy. The above computation is therefore valid only as long as the associated energy losses can be neglected.

Problem 8.4. Relativistic motion in a constant magnetic field

We next consider the motion of a massive particle in a constant magnetic field $\mathbf{B}$. The relativistic equation of motion (3.6), together with $\mathbf{p} = m\gamma\mathbf{v}$, now gives

$$m\frac{d(\gamma\mathbf{v})}{dt} = q\mathbf{v}\times\mathbf{B} \,. \tag{8.194}$$

Taking the scalar product with $\mathbf{v}$ and using the fact that $\mathbf{v}\cdot(\mathbf{v}\times\mathbf{B}) = 0$ we get

$$m\mathbf{v}\cdot\frac{d(\gamma\mathbf{v})}{dt} = 0 \,. \tag{8.195}$$

This implies that

$$
\begin{aligned}
0 &= \mathbf{v}\cdot\left(\frac{d\gamma}{dt}\mathbf{v} + \gamma\frac{d\mathbf{v}}{dt}\right) \\
&= v^2\frac{d\gamma}{dt} + \frac{1}{2}\gamma\frac{dv^2}{dt} \,.
\end{aligned}
\tag{8.196}
$$

Using

$$
\begin{aligned}
\frac{d\gamma}{dt} &= \frac{d}{dt}\left(1 - \frac{v^2}{c^2}\right)^{-1/2} \\
&= \frac{\gamma^3}{2c^2}\frac{dv^2}{dt} \,,
\end{aligned}
\tag{8.197}
$$

eq. (8.196) becomes

$$\gamma^3\frac{dv^2}{dt} = 0 \,, \tag{8.198}$$

and therefore

$$\frac{dv^2}{dt} = 0 \,. \tag{8.199}$$

The modulus of the velocity of a particle in a magnetic field therefore stays constant, even in the full relativistic setting. Since the energy $\mathcal{E}$ is just a function of v^2 through $\mathcal{E} = \gamma mc^2$, the energy is also constant. This result could have also been obtained from the $\mu = 0$ component of eq. (8.62), which is given explicitly by eq. (8.66), and becomes $d\mathcal{E}/dt = 0$ when $\mathbf{E} = 0$.

Since v^2 is constant, also γ is constant, and we can extract it from the time derivative in eq. (8.194). Setting $\mathbf{B} = B\hat{\mathbf{z}}$, we then obtain

$$\frac{d\mathbf{v}}{dt} = \omega\mathbf{v}\times\hat{\mathbf{z}} \,, \tag{8.200}$$

where

$$\omega = \frac{qB}{\gamma m} \,, \tag{8.201}$$

or, in terms of the energy $\mathcal{E} = \gamma mc^2$ of the particle,

$$\omega = \frac{qBc^2}{\mathcal{E}} \, . \tag{8.202}$$

In the non-relativistic limit, eq. (8.201) becomes

$$\omega = \frac{qB}{m} \, , \qquad \text{(non-relativistic)}, \tag{8.203}$$

which is also called the *cyclotron frequency*. In components, eq. (8.200) reads

$$\frac{dv_x}{dt} = \omega v_y \, , \tag{8.204}$$

$$\frac{dv_y}{dt} = -\omega v_x \, , \tag{8.205}$$

while $dv_z/dt = 0$. The solution is v_z constant, and (choosing for definiteness the origin of time so that $v_x = 0$ when $t = 0$)

$$v_x(t) = v_\perp \sin \omega t \, , \tag{8.206}$$

$$v_y(t) = v_\perp \cos \omega t \, , \tag{8.207}$$

where $v_\perp$ is a constant, which represents the modulus of the velocity in the (x, y) plane, so that $v^2 = v_\perp^2 + v_z^2$ (as we have seen, both v^2 and v_z^2 are constant). A further integration gives

$$x(t) = x_0 - \frac{v_\perp}{\omega} \cos \omega t \, , \tag{8.208}$$

$$y(t) = y_0 + \frac{v_\perp}{\omega} \sin \omega t \, , \tag{8.209}$$

so the particle moves with angular velocity ω, given by eq. (8.201), on a circle in the (x, y) plane, centered in a generic point (x_0, y_0) and of radius $r = v_\perp/|\omega|$. From eq. (8.201), we get

$$r = \frac{m\gamma v_\perp}{|q|B} \, . \tag{8.210}$$

For a given velocity $v_\perp$ of the particle in the (x, y) plane, the larger is the values of B, the smaller is the radius of the circle to which the particle's motion is confined. With a non-vanishing value of v_z, we have also $z(t) = z_0 + v_z t$, and the motion in three-dimensional space is actually helicoidal.

Just as for the motion in an electric field, we have here neglected the fact that an accelerated particle radiates electromagnetic waves so, again, the previous computation is valid only as long as the associated energy losses can be neglected. For a non-relativistic particle in circular motion we will compute their effect in Problem 10.2, while the radiation emitted in the full relativistic setting will be discussed in Section 10.6.3.

Electromagnetic waves in vacuum

In this chapter, we study Maxwell's equations in the absence of sources. We will then discover that, even in the absence of sources, there are non-trivial solutions that describe electromagnetic waves propagating across empty space. We will study them in the Lorenz gauge and in the Coulomb gauge. The former treatment allows us to maintain Lorentz covariance explicitly at each stage. However, it will be more subtle to understand how to eliminate spurious polarizations, and remain with the two physical polarizations that characterize electromagnetic waves. Working in the Coulomb gauge, in contrast, we will deal from the beginning with only the two physical polarizations, at the price of losing explicit Lorentz covariance in the intermediate steps. The two approaches are complementary, technically and conceptually, and are both important to understanding electromagnetic waves.[1] We begin, however, with a discussion of wave equations in a simpler setting, involving only a Lorentz scalar field, rather than the full electromagnetic field.

[1]The interplay between working with the physical degrees of freedom at the price of losing explicit Lorentz covariance, or using a Lorentz-covariant formalism at the price of having to deal with spurious degrees of freedoms, is also a recurrent theme in the quantization of electrodynamics.

9.1 Wave equations

Let us begin by studying an equation of the form

$$\Box f \equiv \left[-\frac{1}{c^2} \frac{\partial^2}{\partial t^2} + \boldsymbol{\nabla}^2 \right] f = 0 \,, \tag{9.1}$$

for some scalar function $f(x) = f(t, \mathbf{x})$. It is interesting to compare this equation to a Laplace equation $\boldsymbol{\nabla}^2 f(\mathbf{x}) = 0$. If we set the boundary condition that $f(\mathbf{x})$ vanishes as $|\mathbf{x}| \to \infty$, it can be proved that the only solution of $\boldsymbol{\nabla}^2 f(\mathbf{x}) = 0$ is $f = 0$. In order to have a non-vanishing solution, a source term is needed, as in eq. (3.93), and in this case we saw that (for a localized source) the solution is given by eq. (4.16). In contrast, because of the opposite sign of the time and space derivatives, eq. (9.10) has non-trivial solutions even in the absence of sources.

In a space-time with just one spatial dimension, the most general wave solution can be found quite easily: consider the equation

$$\left[-\frac{1}{c^2} \frac{\partial^2}{\partial t^2} + \frac{\partial^2}{\partial x^2} \right] f(t, x) = 0 \,, \tag{9.2}$$

for a function of two variables t and x. Defining $x_{\pm} = x \pm ct$ and

$\partial_\pm = \partial/\partial x_\pm$, this equation can be rewritten as

$$\partial_+\partial_- f(x_+,x_-)=0\,. \tag{9.3}$$

The most general solution is

$$f(x_+,x_-)=f_1(x_-)+f_2(x_+)\,, \tag{9.4}$$

for arbitrary functions f_1 and f_2. A function $f_1(x_-)=f_1(x-ct)$ describes a wave moving rightward with speed c: if this function at $t=0$ has a given value in $x=0$, at a subsequent time $t=t_0$ it will have the same value in $x=ct_0$, so the whole function is simply translated, and advances with speed c toward the positive direction of the x axis. Similarly, $f_2(x_+)$ describes a left-moving wave, which again travels with speed c.

If we have a function of all three spatial variables plus time, $f(t,\mathbf{x})$, that obeys $\Box f=0$, there are therefore particular solutions of the form $f(t,x,y,z)=f_1(x\pm ct)$, independent of the (y,z) coordinates, that describe a plane wavefront that moves in the x direction (leftward or rightward) at the speed of light. To study the most general wave-like solution, in more than one spatial dimension, it is convenient to use Fourier analysis, that we already introduced in Section 1.5. In particular, any function $f(\mathbf{x})$ (subject to conditions such as belonging to $L^1(\mathbb{R}^3)$, the space of functions whose absolute value is integrable over $\mathbb{R}^3$) can be decomposed in a superposition of modes $e^{i\mathbf{k}\cdot\mathbf{x}}$ with coefficients $\tilde{f}(\mathbf{k})$, as in eq. (1.100), and similarly any function of time [again, belonging e.g., to $L^1(\mathbb{R})$] can be decomposed as in eq. (1.104). For a function of the four-dimensional coordinates $x^\mu=(x^0,\mathbf{x})$, the Fourier decomposition can be written as in eq. (1.105) that, in a four-vector notation, reads

$$f(x)=\int\frac{d^4k}{(2\pi)^4}\,\tilde{f}(k)e^{ikx}\,, \tag{9.5}$$

where $k^\mu=(k^0,\mathbf{k})$ is a four-vector and $kx=k_\mu x^\mu$. If $f(x)$ is real, then $\tilde{f}^*(k)=\tilde{f}(-k)$. For a scalar function $f(x)$, let us search a solution of $\Box f=0$ by making the ansatz

$$f(x)=f_k\,e^{ikx}\,, \tag{9.6}$$

where f_k is the amplitude (which cannot be determined by the equation, since it is just an overall constant). We then insert eq. (9.6) into $\Box f=0$ to see if, and under what conditions on k, this is indeed a solution. We observe that[2]

$$\partial_\mu e^{ikx}=ik_\mu e^{ikx}\,, \tag{9.7}$$

and

$$\Box e^{ikx}=-k^2 e^{ikx}\,, \tag{9.8}$$

where $k^2=k^\mu k_\mu$. We see that e^{ikx} is a solution of $\Box e^{ikx}=0$ if, and only if, $k^2=0$, i.e., $-(k^0)^2+\mathbf{k}^2=0$,[3] so we get

$$(k^0)^2=\mathbf{k}^2\,. \tag{9.9}$$

[2]Explicitly,
$$\begin{aligned}\partial_\mu e^{ikx}&=\frac{\partial}{\partial x^\mu}e^{ik_\nu x^\nu}\\&=ik_\nu\frac{\partial x^\nu}{\partial x^\mu}e^{ikx}\\&=ik_\nu\delta^\nu_\mu e^{ikx}\\&=ik_\mu e^{ikx}\,,\end{aligned}$$
and
$$\begin{aligned}\Box e^{ikx}&=\partial^\mu\partial_\mu e^{ikx}\\&=\partial^\mu(ik_\mu e^{ikx})\\&=-k^\mu k_\mu e^{ikx}\\&=-k^2 e^{ikx}\,.\end{aligned}$$

[3]Observe that the Minkowskian signature of the wave equation was essential to have a non-trivial solution. The same procedure applied to the equation $\nabla^2 f(\mathbf{x})=0$ would give $\mathbf{k}^2=0$ which, for $\mathbf{k}^2=k_1^2+k_2^2+k_3^2$, has only the solution $\mathbf{k}=0$. The Fourier mode with $\mathbf{k}=0$ corresponds to a solution constant in space, which is eliminated (in the sense that its coefficient $f_\mathbf{k}$ is set to zero) by the boundary condition that $f(\mathbf{x})$ vanishes at $|\mathbf{x}|\to\infty$.

Comparing with eq. (7.146) and identifying k^0 with E/c and $\mathbf{k}$ with $\mathbf{p}$ (apart from a common overall constant which is also needed for dimensional reasons since, from eq. (9.5), k^μ has dimensions of inverse length), we see that this is formally the same as the relation between energy and momentum of a massless particle. To develop the connection between fields and particles the formalism of quantum field theory is really needed so, within our classical context, we cannot push this analogy too much.[4] At the quantum field theory level, however, one indeed finds that the quanta of a field that obeys an equation such as $\Box f = 0$ correspond to massless particles.

[4] As already discussed in Note 18 on page 201, the correct quantum-mechanical relations are in fact $E/c = \hbar k^0$ and $\mathbf{p} = \hbar \mathbf{k}$, and require the reduced Planck constant $\hbar$ that enters in quantum mechanics.

9.2   Electromagnetic waves in the Lorenz gauge

As we have seen, the introduction of the gauge potentials and the use of the covariant formalism has the advantage of explicitly unveiling the Lorentz symmetry of electromagnetism. It also has the technical advantage of considerably simplifying the equations. We will therefore adopt it for our treatment of electromagnetic waves.[5] We have seen that, before making any choice of gauge, the first couple of Maxwell's equations is given by eq. (8.28). In this form, each of the four equations (corresponding to $\nu = 0, 1, 2, 3$) involves all four components of A^ν. Similarly, the original Maxwell's equations in the form (3.8)–(3.11) mix the components of $\mathbf{E}$ and $\mathbf{B}$. However, once formulated in terms of gauge potentials, electromagnetism is invariant under the gauge transformation (8.14), and we can fix this freedom to impose the Lorenz gauge $\partial_\mu A^\mu = 0$, see eq. (8.29). In this gauge, eq. (8.28) becomes simply eq. (8.30), where the four components of A^μ are decoupled. We now set $j^\mu = 0$, so we study this equation in the absence of sources,

[5] The gauge potential A^μ also turns out to be the fundamental quantity for describing the electromagnetic field at the quantum level, a subject in which, however, we cannot enter in this course.

$$\Box A^\mu \equiv \left[-\frac{1}{c^2}\frac{\partial^2}{\partial t^2} + \boldsymbol{\nabla}^2 \right] A^\mu = 0 \,. \tag{9.10}$$

For the electromagnetic potential $A^\mu(x)$ the solution is slightly more complex compared to eq. (9.6), because we must take into account that, besides the x dependence, A^μ also carries a four-vector index. Therefore, we rather look for elementary solutions of the form

$$\begin{aligned} A^\mu(x) &= \mathcal{A}_k\,\epsilon^\mu(k)e^{ikx} \\ &= \mathcal{A}_k\,\epsilon^\mu(k)e^{-i\omega t + i\mathbf{k}\cdot\mathbf{x}} \,, \end{aligned} \tag{9.11}$$

where $\mathcal{A}_k$ is the amplitude and, in the last line, we defined ω from $k^0 = \omega/c$. The four-vector ϵ^μ, called the *polarization* four-vector, carries the Lorentz index and is normalized as $|\eta_{\mu\nu}\epsilon^\mu\epsilon^\nu| = 1$ (except in the special case in which it is a null vector, i.e., $\eta_{\mu\nu}\epsilon^\mu\epsilon^\nu = 0$, which will be analyzed separately below). The dependence on $x = (ct, \mathbf{x})$ is only carried by the exponential. A wave with the simple temporal dependence given in eq. (9.11) is called monochromatic, since it has just a single frequency. Its spatial dependence corresponds to a *plane wave*, i.e., a wave that is constant in the direction transverse to the propagation direction $\hat{\mathbf{k}}$. More general functions are obtained by superpositions of the form (9.5), with arbitrary coefficients, as we will see in more details below. Since plane waves extend indefinitely in the transverse direction, they are a mathematical idealization, and realistic wave solutions are obtained from a superposition of plane waves which results in a finite extent in the transverse direction. Note that, since the exponential depends on k, a priori, to obtain a solution, we must admit the possibility of a dependence of ϵ^μ on k, too. Once again, the reality condition will be taken care of by the superposition with the complex conjugate solution $A_k^*[\epsilon^\mu(k)]^* e^{-ikx}$.

Inserting the ansatz (9.11) into eq. (9.10), ϵ^μ, which is independent of x, simply goes through the $\Box$ operator and, using eq. (9.8), we get again the condition

$$k^2 = 0 \,, \tag{9.12}$$

just as in the scalar case discussed in Section 9.1. However, we are not done yet, since eq. (9.10) was obtained imposing that A^μ satisfies the Lorenz gauge condition $\partial_\mu A^\mu = 0$, so our ansatz must also satisfy it. Using eq. (9.7)

$$\partial_\mu[\epsilon^\mu(k)e^{ikx}] = ik_\mu\epsilon^\mu(k)e^{ikx} \,, \tag{9.13}$$

and therefore we must require

$$k_\mu\epsilon^\mu(k) \equiv \eta_{\mu\nu}k^\mu\epsilon^\nu(k) = 0 \,. \tag{9.14}$$

We see that ϵ^μ depends indeed on k: it must be orthogonal (with respect to the scalar product defined by the metric $\eta_{\mu\nu}$) to k^μ.

Apparently, we have found that, for a given k^μ, there are three independent solutions, corresponding to the three independent solutions of eq. (9.14). For instance, with a rotation we can always set $\mathbf{k}$ along the positive z axis; then, since $k^2 = 0$, we have $k_z = +k_0$ (choosing for definiteness $k^0 > 0$; the case $k^0 < 0$ can be treated in the same way) and

$$k^\mu = k^0(1, 0, 0, 1) \,. \tag{9.15}$$

Given this form of k^μ, two solutions of eq. (9.14) are immediately found, and are given by

$$\epsilon^\mu_{(1)}(k) = (0, 1, 0, 0) \,, \qquad \epsilon^\mu_{(2)}(k) = (0, 0, 1, 0) \,, \tag{9.16}$$

where the normalization factors have been chosen so that $\eta_{\mu\nu}\epsilon^\mu\epsilon^\nu = 1$. These are called *transverse* polarizations, since the corresponding spatial vectors $\boldsymbol{\epsilon}_{(1)} = (1, 0, 0)$ and $\boldsymbol{\epsilon}_{(2)} = (0, 1, 0)$ are orthogonal, in a three-dimensional sense, to the vector $\hat{\mathbf{k}} = (0, 0, 1)$.

The third solution, recalling the expression $\eta_{\mu\nu} = (-1, 1, 1, 1)$ for the metric, is

$$\epsilon^\mu_{(3)}(k) = (1, 0, 0, 1) \,. \tag{9.17}$$

Notice that $\epsilon^\mu_{(3)}(k) \propto k^\mu$, and the fact that $\epsilon^\mu_{(3)}(k)k_\mu = 0$ can also be seen as a consequence of $k^\mu k_\mu = 0$. In particular, for the spatial components, we have $\boldsymbol{\epsilon}_{(3)} \propto \mathbf{k}$. This is called a *longitudinal* polarization. The four-vector (9.17) has zero norm, $\eta_{\mu\nu}\epsilon^\mu\epsilon^\nu = 0$, so it cannot be normalized imposing $|\eta_{\mu\nu}\epsilon^\mu\epsilon^\nu| = 1$. We then simply choose eq. (9.17) as our third solution [any rescaling of it then corresponds to a redefinition of the amplitude $\mathcal{A}_k$ associated with this solution in eq. (9.11)]. However, the longitudinal polarization is non-physical and can be eliminated with a further gauge transformation. In fact, after having reached the Lorenz gauge $\partial_\mu A^\mu = 0$, we can perform a further gauge transformation

$$A^\mu \rightarrow A'^\mu = A^\mu - \partial^\mu\theta, \tag{9.18}$$

and, if we choose θ such that $\Box\theta = 0$, we still have $\partial_\mu A'^\mu = 0$. In other words, the Lorenz gauge condition does not completely fix the freedom of performing gauge transformations, and we can still make a further gauge transformation with θ such that $\Box\theta = 0$. Working in Fourier space, we can therefore consider a function $\theta(x) = ae^{ikx}$, with a an arbitrary complex constant (once again, the reality condition is ensured by the complex conjugate term in the superposition of modes). Requiring $\Box\theta = 0$ gives again $k^2 = 0$. Under this gauge transformation,

$$A^\mu \rightarrow A^\mu - \partial^\mu(ae^{ikx}) = A^\mu - iak^\mu e^{ikx}. \tag{9.19}$$

Therefore, a solution proportional to $k^\mu e^{ikx}$, with $k^2 = 0$, can always be removed with a residual gauge transformation. We can therefore set it to zero without loss of generality.[6]

In conclusion, in vacuum, electromagnetic waves are described by a superposition of terms of the form (9.11) and of its complex conjugate, with $k^2 = 0$ and $k_\mu\epsilon^\mu(k) = 0$. The latter condition admits as solutions the two transverse polarizations and a longitudinal polarization $\epsilon^\mu(k) \propto k^\mu$; however, the latter can be set to zero with a residual gauge transformation, so the electromagnetic field has only two degrees of freedom, corresponding to the two transverse polarizations. In a frame where k^μ is given by eq. (9.15), a basis for these two transverse polarization is given by eq. (9.16). This is called the basis of *linear polarizations*. Another useful basis is given by *circular polarizations*, defined as

$$\epsilon^\mu_{(+)}(k) = \frac{1}{\sqrt{2}}(0, 1, i, 0), \qquad \epsilon^\mu_{(-)}(k) = \frac{1}{\sqrt{2}}(0, 1, -i, 0). \tag{9.20}$$

Given that the temporal components of these four-vector vanish, we can more simply say that, in the frame where $\mathbf{k} = (0, 0, k)$, the linear polarizations are

$$\hat{\mathbf{e}}_{(1)} = (1, 0, 0), \qquad \hat{\mathbf{e}}_{(2)} = (0, 1, 0), \tag{9.21}$$

while the circular polarizations are

$$\hat{\mathbf{e}}_{(+)} = \frac{1}{\sqrt{2}}(1, i, 0), \qquad \hat{\mathbf{e}}_{(-)} = \frac{1}{\sqrt{2}}(1, -i, 0). \tag{9.22}$$

[6]If one had not noticed the existence of this residual gauge transformation and would have kept the solution proportional to $\epsilon^\mu_{(3)}(k)$ in the computation of gauge invariant quantities such as the electric and magnetic field, one would have found that the contributions proportional to $\epsilon^\mu_{(3)}$ "miracolously" cancel. This cancelation is simply due to the fact that the $\epsilon^\mu_{(3)}(k)$ term can be set to zero with a gauge transformation, and therefore cannot affect gauge-invariant quantities.

The physical polarization vectors are therefore orthogonal (in a three-dimensional sense) to the propagation direction $\hat{\mathbf{k}}$. In Section 9.5 we will discuss linear, circular (and elliptic) polarizations further, and we will understand the reason for these names.

9.3   Electromagnetic waves in the Coulomb gauge

It is interesting to compare these results with an analysis in the Coulomb gauge, $\boldsymbol{\nabla}\cdot\mathbf{A} = 0$. In this case, the relevant equations are (3.93) and (3.94) that, in vacuum, become

$$\nabla^2\phi \;=\; 0\,, \tag{9.23}$$

$$\Box\mathbf{A} \;=\; \frac{1}{c^2}\boldsymbol{\nabla}\frac{\partial\phi}{\partial t}\,. \tag{9.24}$$

With the boundary condition that ϕ vanishes at spatial infinity, the only solution of eq. (9.23) is $\phi = 0$. Thus, *in vacuum*, we can set

$$\phi = 0\,, \qquad \boldsymbol{\nabla}\cdot\mathbf{A} = 0\,. \tag{9.25}$$

In fact, we could have also reached this conclusion with a more complete choice of gauge, as follows.[7] First of all, starting from a generic field configuration A_μ, we can find a gauge transformation $A_\mu \to A'_\mu$ such that $A'_0 = 0$. It is given simply by

$$A_\mu(t,\mathbf{x}) \to A'_\mu(t,\mathbf{x}) = A_\mu(t,\mathbf{x}) - \partial_\mu\int^t cdt'\, A_0(t',\mathbf{x})\,, \tag{9.26}$$

since

$$\begin{aligned} A'_0(t,\mathbf{x}) &= A_0(t,\mathbf{x}) - \partial_t\int^t dt'\, A_0(t',\mathbf{x}) \\ &= 0\,. \end{aligned} \tag{9.27}$$

After that, we still have the freedom of performing a gauge transformation with θ independent of t, because this does not modify the condition $A'_0 = 0$. We then perform a further gauge transformation which sends A'_μ into a new field $A''_\mu = A'_\mu - \partial_\mu\theta$, choosing

$$\theta(\mathbf{x}) = -\int\frac{d^3y}{4\pi|\mathbf{x}-\mathbf{y}|}\frac{\partial A'^i(t,\mathbf{y})}{\partial y^i}\,. \tag{9.28}$$

Despite the dependence on time of $A'^i(t,\mathbf{y})$, the integral on the right-hand side is actually independent of t. In fact, in this gauge $E^i = -\partial_t A'^i$, since $A'_0 = 0$. Then the vacuum Maxwell equation $\partial_i E^i = 0$ implies $\partial_t\partial_i A'^i = 0$. It then follows from eq. (9.28) that $\partial_t\theta = 0$.

Furthermore, from $A''_i = A'_i - \partial_i\theta$ it follows that

$$\boldsymbol{\nabla}\cdot\mathbf{A}'' = \boldsymbol{\nabla}\cdot\mathbf{A}' - \nabla^2\theta\,. \tag{9.29}$$

Using in eq. (9.28) the identity

$$\nabla_{\mathbf{x}}^2 \left(\frac{-1}{4\pi|\mathbf{x}-\mathbf{y}|} \right) = \delta^3(\mathbf{x}-\mathbf{y}), \qquad (9.30)$$

that we derived in eq. (1.90), we get

$$\nabla^2\theta = \boldsymbol{\nabla}\cdot\mathbf{A}', \qquad (9.31)$$

and therefore $\boldsymbol{\nabla}\cdot\mathbf{A}'' = 0$. We have therefore used the gauge freedom to set

$$\boxed{A_0 = 0 \qquad \boldsymbol{\nabla}\cdot\mathbf{A} = 0,} \qquad (9.32)$$

(where we have eventually removed the double primes on $\mathbf{A}$). This gauge is called the *radiation gauge*. Note that it implies the Lorenz gauge $\partial_\mu A^\mu = 0$, as well as the Coulomb gauge $\boldsymbol{\nabla}\cdot\mathbf{A} = 0$. Thus, both the Lorenz and the Coulomb gauge do not fix the gauge freedom completely. In contrast, in the radiation gauge there is no residual gauge freedom.

In any case, whether we directly fix the radiation gauge, or else we only fix the Coulomb gauge and then discover that $\phi = 0$ from the solution of eq. (9.23), we end up with eqs. (9.24) and (9.25). Since $\phi = 0$, the former simply becomes $\Box\mathbf{A} = 0$, so, in the end, $\mathbf{A}(t,\mathbf{x})$ must solve the two equations

$$\Box\mathbf{A} = 0, \qquad (9.33)$$
$$\boldsymbol{\nabla}\cdot\mathbf{A} = 0. \qquad (9.34)$$

We can then proceed in a way completely analogous to the discussion in Section 9.2. Working in Fourier space, the solutions of eq. (9.33) are superpositions of

$$\boxed{\mathbf{A}(t,\mathbf{x}) = \mathcal{A}_{\mathbf{k}}\,\hat{\mathbf{e}}(\mathbf{k})e^{ikx},} \qquad (9.35)$$

and its complex conjugate, with $k^2 = 0$, i.e., $(k^0)^2 = |\mathbf{k}|^2$. The quantity $\mathcal{A}_{\mathbf{k}}$ is the amplitude, while the polarization vector is normalized as

$$\hat{\mathbf{e}}(\mathbf{k})\cdot\hat{\mathbf{e}}^*(\mathbf{k}) = 1, \qquad (9.36)$$

which becomes simply $\hat{\mathbf{e}}(\mathbf{k})\cdot\hat{\mathbf{e}}(\mathbf{k}) = 1$ if we take a real polarization vector. Equation (9.34) requires that

$$\boxed{\hat{\mathbf{e}}(\mathbf{k})\cdot\mathbf{k} = 0.} \qquad (9.37)$$

Therefore, we again find the two transverse polarizations that we have already found in the Lorentz gauge. The most general solution is given by a superposition of the elementary solutions, labeled by $\mathbf{k}$ and by an index $\lambda = 1, 2$ that identifies the two independent solutions of eq. (9.37), with arbitrary amplitudes $\mathcal{A}_{\mathbf{k},\lambda}$,

$$\mathbf{A}(t,\mathbf{x}) = \int \frac{d^3 k}{(2\pi)^3} \sum_{\lambda=1,2} \left[\hat{\mathbf{e}}(\mathbf{k},\lambda) \mathcal{A}_{\mathbf{k},\lambda} e^{ikx} + \hat{\mathbf{e}}^*(\mathbf{k},\lambda) \mathcal{A}^*_{\mathbf{k},\lambda} e^{-ikx} \right]_{k^0 = +|\mathbf{k}|} . \tag{9.38}$$

Note that the equation $k^2 = 0$ has two solutions, $k^0 = \pm|\mathbf{k}|$. However, once we consider the general superposition of plane waves e^{ikx} and e^{-ikx}, we can limit ourselves to $k^0 = +|\mathbf{k}|$. For example, when $k^0 = -|\mathbf{k}|$ we have

$$
\begin{aligned}
\left(e^{ikx}\right)_{k^0=-|\mathbf{k}|} &= \left(e^{-ik^0 x^0 + i\mathbf{k}\cdot\mathbf{x}}\right)_{k^0=-|\mathbf{k}|} \\
&= \exp\{i|\mathbf{k}|x^0 + i\mathbf{k}\cdot\mathbf{x}\} .
\end{aligned} \tag{9.39}
$$

This is the same as a term e^{-ikx} with $k^0 = +|\mathbf{k}|$, i.e.,

$$\exp\{i|\mathbf{k}|x^0 - i\mathbf{k}\cdot\mathbf{x}\} , \tag{9.40}$$

after renaming the integration variable $\mathbf{k}$ into $-\mathbf{k}$. As in eq. (9.11), we define ω from $\omega/c = k^0$. Since we can restrict to $k^0 = +|\mathbf{k}|$, we have

$$\frac{\omega}{c} = k^0 = +|\mathbf{k}| . \tag{9.41}$$

We can then rewrite eq. (9.38) more simply as

$$\mathbf{A}(t,\mathbf{x}) = \int \frac{d^3 k}{(2\pi)^3} \sum_{\lambda=1,2} \left[\hat{\mathbf{e}}(\mathbf{k},\lambda) \mathcal{A}_{\mathbf{k},\lambda} e^{-i\omega t + i\mathbf{k}\cdot\mathbf{x}} + \hat{\mathbf{e}}^*(\mathbf{k},\lambda) \mathcal{A}^*_{\mathbf{k},\lambda} e^{i\omega t - i\mathbf{k}\cdot\mathbf{x}} \right] . \tag{9.42}$$

The vectors $\hat{\mathbf{e}}(\mathbf{k},\lambda)$, with $\lambda = 1, 2$, in eq. (9.42) are the two independent solutions of $\mathbf{k}\cdot\hat{\mathbf{e}}(\mathbf{k}) = 0$. As a basis, we can use for instance the linear polarizations, or the circular polarizations; for $\mathbf{k}$ pointing along the $\hat{\mathbf{z}}$ direction, they are given by eq. (9.21) and eq. (9.22), respectively.

9.4   Solutions for E and B

We can now immediately read the solutions for $\mathbf{E}$ and $\mathbf{B}$ from the solutions for the gauge potential. It is simpler to work in the Coulomb gauge, where $\phi = 0$, and $\mathbf{A}$ is given by eq. (9.35) together with the condition (9.37). From eq. (3.80), and eq. (3.83) with $\phi = 0$, we have

$$\mathbf{E} = -\frac{\partial \mathbf{A}}{\partial t} , \qquad \mathbf{B} = \nabla \times \mathbf{A} . \tag{9.43}$$

We take for $\mathbf{A}$ a solution of the form

$$\mathbf{A}(t,\mathbf{x}) = \mathcal{A}_{\mathbf{k}} \, \hat{\mathbf{e}}(\mathbf{k}) \, e^{-i\omega t + i\mathbf{k}\cdot\mathbf{x}} , \tag{9.44}$$

with $\mathbf{k}\cdot\hat{\mathbf{e}}(\mathbf{k}) = 0$. We also use $\mathbf{k} = (\omega/c)\hat{\mathbf{k}}$, that follows from eq. (9.41). Then, we get

$$
\begin{aligned}
\mathbf{E}(t,\mathbf{x}) &= \hat{\mathbf{e}}(\mathbf{k}) \, i\omega \mathcal{A}_{\mathbf{k}} \, e^{-i\omega t + i\mathbf{k}\cdot\mathbf{x}} , \tag{9.45} \\
\mathbf{B}(t,\mathbf{x}) &= [\hat{\mathbf{k}} \times \hat{\mathbf{e}}(\mathbf{k})] \frac{i\omega}{c} \mathcal{A}_{\mathbf{k}} \, e^{-i\omega t + i\mathbf{k}\cdot\mathbf{x}} . \tag{9.46}
\end{aligned}
$$

Defining $E_\mathbf{k} = i\omega \mathcal{A}_\mathbf{k}$, we can rewrite this more compactly as

$$\mathbf{E}(t, \mathbf{x}) = \hat{\mathbf{e}}(\mathbf{k}) \, E_\mathbf{k} \, e^{-i\omega t + i\mathbf{k}\cdot\mathbf{x}} \,, \tag{9.47}$$

$$c\mathbf{B}(t, \mathbf{x}) = [\hat{\mathbf{k}} \times \hat{\mathbf{e}}(\mathbf{k})] \, E_\mathbf{k} \, e^{-i\omega t + i\mathbf{k}\cdot\mathbf{x}} \,. \tag{9.48}$$

As always, the actual field will be given by the real part of these complex expressions

$$\mathbf{E}(t, \mathbf{x}) = \mathrm{Re}\left[\hat{\mathbf{e}}(\mathbf{k}) \, E_\mathbf{k} \, e^{-i\omega t + i\mathbf{k}\cdot\mathbf{x}}\right] \,, \tag{9.49}$$

$$c\mathbf{B}(t, \mathbf{x}) = \mathrm{Re}\left[[\hat{\mathbf{k}} \times \hat{\mathbf{e}}(\mathbf{k})] \, E_\mathbf{k} \, e^{-i\omega t + i\mathbf{k}\cdot\mathbf{x}}\right] \,. \tag{9.50}$$

Equations (9.47) and (9.48) show the basic features of electromagnetic waves. Both the electric and magnetic field are transverse to the propagation direction, and satisfy

$$c\mathbf{B}(t, \mathbf{x}) = \hat{\mathbf{k}} \times \mathbf{E}(t, \mathbf{x}) \,, \tag{9.51}$$

so they are orthogonal to each other, and their moduli are related by $|\mathbf{E}| = c|\mathbf{B}|$. The dependence on space and time is through the phase factor $e^{-i\varphi(t, \mathbf{x})}$, where

$$\varphi(t, \mathbf{x}) = \omega t - \mathbf{k}\cdot\mathbf{x} \,. \tag{9.52}$$

Using eq. (9.41), we can write

$$\varphi(t, \mathbf{x}) = -\frac{\omega}{c}(\hat{\mathbf{k}}\cdot\mathbf{x} - ct) \,, \tag{9.53}$$

and we see that the surfaces of constant phase, i.e., the surfaces where $\varphi(t, \mathbf{x})$ is constant, travel at the speed of light.[8] The quantity ω is the frequency of the wave, since the wave is unchanged under a time translation $t \to t + T$ with $T = 2\pi/\omega$, while its wavelength λ is given by

$$\lambda = \frac{2\pi}{|\mathbf{k}|} \,, \tag{9.54}$$

since the wave is periodic if we translate it by λ along the propagation direction, i.e., under $\mathbf{x} \to \mathbf{x} + \lambda\hat{\mathbf{k}}$. The vector $\mathbf{k}$, with dimensions of inverse length, is called the *wavenumber*. From eqs. (9.41) and (9.54), we have the relation

$$\frac{\omega}{c} = \frac{2\pi}{\lambda} \,. \tag{9.55}$$

It is also useful to introduce the reduced wavelength $\lambdabar = \lambda/(2\pi)$, so that

$$\lambdabar = \frac{c}{\omega} \,. \tag{9.56}$$

From eq. (3.44), and the fact that in any plane wave $B \equiv |\mathbf{B}|$ and $E \equiv |\mathbf{E}|$ are related by $cB = E$, it follows that the energy density $u(t, \mathbf{x})$ is

$$u(t, \mathbf{x}) = \epsilon_0 |\mathbf{E}(t, \mathbf{x})|^2 \,. \tag{9.57}$$

[8] We will expand on this in Section 15.2, where, beside this notion of "phase velocity," we will introduce the group velocity of a wave-packet, for waves propagating in a generic medium.

Similarly, from eq. (3.56), the momentum density of an electromagnetic wave is

$$
\begin{aligned}
\mathbf{g}(t, \mathbf{x}) &= \epsilon_0 \, \mathbf{E}(t, \mathbf{x}) \times \mathbf{B}(t, \mathbf{x}) \\
&= \frac{\epsilon_0}{c} \, |\mathbf{E}(t, \mathbf{x})|^2 \, \hat{\mathbf{k}} ,
\end{aligned}
\tag{9.58}
$$

and the Poynting vector is given by

$$
\mathbf{S}(t, \mathbf{x}) = \epsilon_0 c |\mathbf{E}(t, \mathbf{x})|^2 \, \hat{\mathbf{k}} .
\tag{9.59}
$$

Note that

$$
\mathbf{g}(t, \mathbf{x}) = \frac{1}{c} \, u(t, \mathbf{x}) \, \hat{\mathbf{k}} ,
\tag{9.60}
$$

so the relation between the energy density and the momentum density of a plane electromagnetic wave is the same as the relation between energy and momentum for a massless particle (see eq. (3.73) with $v = c$). This fact, that will have a full explanation in a quantum theory of electromagnetism, can already suggest us that, at the quantum level, an electromagnetic wave will be a collection of massless particles, the photons.

For a monochromatic electromagnetic wave, we are typically interested in the energy density and momentum density averaged in time over a period of the wave, at a fixed point in space. We denote this average by a bracket $\langle \ldots \rangle$, so

$$
\begin{aligned}
\langle u(t, \mathbf{x}) \rangle &= \epsilon_0 \left\langle |\mathbf{E}(t, \mathbf{x})|^2 \right\rangle ,
\end{aligned}
\tag{9.61}
$$

$$
\begin{aligned}
\langle \mathbf{g}(t, \mathbf{x}) \rangle &= \frac{\epsilon_0}{c} \left\langle |\mathbf{E}(t, \mathbf{x})|^2 \right\rangle \hat{\mathbf{k}} .
\end{aligned}
\tag{9.62}
$$

Using eq. (9.49),

$$
\mathbf{E}(t, \mathbf{x}) = \frac{1}{2} \left[\hat{\mathbf{e}}(\mathbf{k}) \, E_{\mathbf{k}} \, e^{-i\omega t + i\mathbf{k}\cdot\mathbf{x}} + \hat{\mathbf{e}}^*(\mathbf{k}) \, E_{\mathbf{k}}^* \, e^{i\omega t - i\mathbf{k}\cdot\mathbf{x}} \right] ,
\tag{9.63}
$$

so

$$
\begin{aligned}
\left\langle |\mathbf{E}(t, \mathbf{x})|^2 \right\rangle &= \frac{1}{4} \Big\langle \left[\hat{\mathbf{e}}(\mathbf{k}) \, E_{\mathbf{k}} \, e^{-i\omega t + i\mathbf{k}\cdot\mathbf{x}} + \hat{\mathbf{e}}^*(\mathbf{k}) \, E_{\mathbf{k}}^* \, e^{i\omega t - i\mathbf{k}\cdot\mathbf{x}} \right] \\
&\qquad \cdot \left[\hat{\mathbf{e}}(\mathbf{k}) \, E_{\mathbf{k}} \, e^{-i\omega t + i\mathbf{k}\cdot\mathbf{x}} + \hat{\mathbf{e}}^*(\mathbf{k}) \, E_{\mathbf{k}}^* \, e^{i\omega t - i\mathbf{k}\cdot\mathbf{x}} \right] \Big\rangle .
\end{aligned}
\tag{9.64}
$$

After taking the scalar product, the terms proportional to $e^{-2i\omega t}$ and to $e^{+2i\omega t}$ average to zero, since $\langle \sin(2\omega t) \rangle = 0$ and $\langle \cos(2\omega t) \rangle = 0$, so only the cross terms, which are independent of time, remain. Then, for a monochromatic electromagnetic wave,

$$
\left\langle |\mathbf{E}(t, \mathbf{x})|^2 \right\rangle = \frac{1}{2} |E_{\mathbf{k}}|^2 .
\tag{9.65}
$$

As an alternative route, rather than working with the gauge potentials, we could have derived wave equations for $\mathbf{E}$ and $\mathbf{B}$ directly from Maxwell's equations in vacuum,

$$
\boldsymbol{\nabla}\cdot\mathbf{E} = 0 ,
\tag{9.66}
$$

$$
\boldsymbol{\nabla}\times\mathbf{B} - \frac{1}{c^2} \frac{\partial \mathbf{E}}{\partial t} = 0 ,
\tag{9.67}
$$

$$
\boldsymbol{\nabla}\cdot\mathbf{B} = 0 ,
\tag{9.68}
$$

$$
\boldsymbol{\nabla}\times\mathbf{E} + \frac{\partial \mathbf{B}}{\partial t} = 0 .
\tag{9.69}
$$

We take the curl of eq. (9.69), and we use

$$
\begin{aligned}
[\boldsymbol{\nabla} \times (\boldsymbol{\nabla} \times \mathbf{E})]_i &= \epsilon_{ijk}\partial_j(\epsilon_{klm}\partial_l E_m) \\
&= (\delta_{il}\delta_{jm} - \delta_{im}\delta_{jl})\partial_j\partial_l E_m \\
&= \partial_i(\boldsymbol{\nabla}\cdot\mathbf{E}) - \boldsymbol{\nabla}^2 E_i \,.
\end{aligned}
\tag{9.70}
$$

or, in vector form,

$$
\boldsymbol{\nabla} \times (\boldsymbol{\nabla} \times \mathbf{E}) = \boldsymbol{\nabla}(\boldsymbol{\nabla}\cdot\mathbf{E}) - \boldsymbol{\nabla}^2\mathbf{E} \,.
\tag{9.71}
$$

Therefore we get

$$
\boldsymbol{\nabla}(\boldsymbol{\nabla}\cdot\mathbf{E}) - \boldsymbol{\nabla}^2\mathbf{E} + \frac{\partial}{\partial t}(\boldsymbol{\nabla} \times \mathbf{B}) = 0 \,.
\tag{9.72}
$$

Using eqs. (9.66) and (9.67) we then obtain

$$
\left(-\frac{1}{c^2}\frac{\partial^2}{\partial t^2} + \boldsymbol{\nabla}^2\right)\mathbf{E} = 0 \,,
\tag{9.73}
$$

i.e., $\Box\mathbf{E} = 0$. Similarly, taking the curl of eq. (9.67) and using eqs. (9.68) and (9.69), we get

$$
\left(-\frac{1}{c^2}\frac{\partial^2}{\partial t^2} + \boldsymbol{\nabla}^2\right)\mathbf{B} = 0 \,.
\tag{9.74}
$$

Note, however, that eqs. (9.73) and (9.74) are a consequence of the full Maxwell's equations (9.66)–(9.69), but are not equivalent to them. The general solution of eqs. (9.73) and (9.74) is a superposition of plane waves of the form

$$
\begin{aligned}
\mathbf{E}(t, \mathbf{x}) &= \mathbf{E_k}\, e^{-i\omega t + i\mathbf{k}\cdot\mathbf{x}} \,, \\
\mathbf{B}(t, \mathbf{x}) &= \mathbf{B_k}\, e^{-i\omega t + i\mathbf{k}\cdot\mathbf{x}} \,,
\end{aligned}
\tag{9.75}
\tag{9.76}
$$

with $\omega/c = \mathbf{k}$ but $\mathbf{E_k}$ and $\mathbf{B_k}$ independent, and arbitrary. Once we put these solutions back into the full set of Maxwell's equations, eq. (9.66) further imposes

$$
\mathbf{k}\cdot\mathbf{E_k} = 0 \,,
\tag{9.77}
$$

while eq. (9.68) gives

$$
\mathbf{k}\cdot\mathbf{B_k} = 0 \,.
\tag{9.78}
$$

Finally, eq. (9.67) gives

$$
\mathbf{k}\times\mathbf{B_k} + \frac{\omega}{c^2}\mathbf{E_k} = 0 \,,
\tag{9.79}
$$

which, using $\omega/c = \mathbf{k}$, becomes

$$
\mathbf{E_k} = -c\hat{\mathbf{k}} \times \mathbf{B_k} \,,
\tag{9.80}
$$

while eq. (9.69) gives the equivalent equation

$$
c\mathbf{B_k} = \hat{\mathbf{k}} \times \mathbf{E_k} \,.
\tag{9.81}
$$

We have therefore recovered the solution (9.47), (9.48).[9]

[9]In Note 19 on page 395 we will show that, actually, this result can have been obtained using just eqs. (9.67) and (9.69), rather than the full set of vacuum Maxwell's equation.

9.5   Polarization of light

Consider a monochromatic wave propagating along the z direction, so $\hat{\mathbf{k}} = \hat{\mathbf{z}}$, with $E_{\mathbf{k}} = E$ real, and consider the basis of the linear polarizations (9.21), $\hat{\mathbf{e}}_{(1)} = \hat{\mathbf{x}}$, $\hat{\mathbf{e}}_{(2)} = \hat{\mathbf{y}}$. Since these are real, for a wave linearly polarized along the $\hat{\mathbf{x}}$ axis eq. (9.49) gives

$$\mathbf{E}(t, \mathbf{x}) = E\cos(\omega t - kz)\,\hat{\mathbf{x}}\,, \tag{9.82}$$

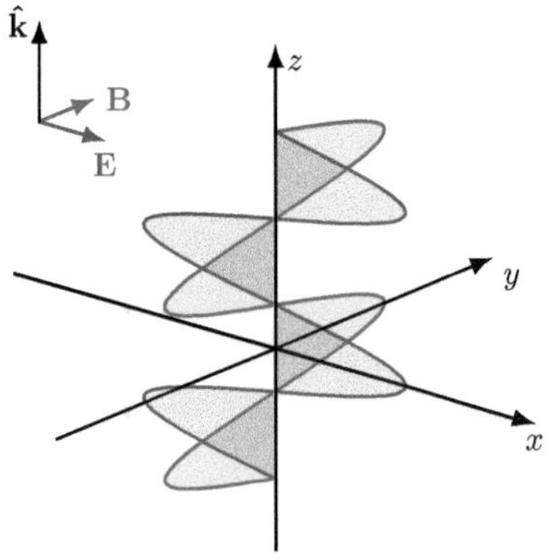

and eq. (9.50) gives $c\mathbf{B}(t, \mathbf{x}) = E\cos(\omega t - kz)\,\hat{\mathbf{y}}$. Thus, as a function of time (for given z), the electric field oscillates along a fixed direction in the (x, y) plane, which in this case is the $\hat{\mathbf{x}}$ axis, and the magnetic field along the orthogonal direction in the (x, y) plane, in this case the $\hat{\mathbf{y}}$ axis, and similarly as a function of z at fixed t, as illustrated in Fig. 9.1. In the same way, for a wave linearly polarized along the $\hat{\mathbf{y}}$ axis,

$$\mathbf{E}(t, \mathbf{x}) = E\cos(\omega t - kz)\,\hat{\mathbf{y}}\,, \tag{9.83}$$

and $c\mathbf{B}(t, \mathbf{x}) = -E\cos(\omega t - kz)\,\hat{\mathbf{x}}$. So, in these cases, the electric field (as well as the magnetic field) oscillates along a fixed direction. This is the origin of the name "linear polarization." A combination with real coefficients of the solutions (9.82) and (9.83) gives a solution that oscillates along a generic direction in the (x, y) plane.

Consider now the polarizations vectors (9.22). In this case, $\hat{\mathbf{e}}_{(+)} = (\hat{\mathbf{x}} + i\hat{\mathbf{y}})/\sqrt{2}$. Then eq. (9.49) gives

$$\mathbf{E}(t, \mathbf{x}) = E\frac{1}{\sqrt{2}}\left[\cos(\omega t - kz)\,\hat{\mathbf{x}} + \sin(\omega t - kz)\,\hat{\mathbf{y}}\right]\,. \tag{9.84}$$

This represents a vector that, as t increases for fixed z, rotates counterclockwise in the (x, y) plane, describing a circle; hence, the name circular polarization. Equivalently, as z increases for fixed t, it rotates clockwise in the (x, y) plane, as illustrated in Fig. 9.2. With respect to the wave propagating along the $+\hat{\mathbf{z}}$ direction, this is called a *right* circular polarization. Similarly, with $\hat{\mathbf{e}}_{(-)}$ the polarization vector rotates clockwise in the (x, y) plane as t increases for fixed z, and describes *left* circular polarization. The corresponding magnetic field is given by eq. (9.48), and rotates so as to remain orthogonal to $\mathbf{E}$, in the (x, y) plane transverse to the propagation direction $\hat{\mathbf{k}}$.

The most general case, for a monochromatic wave propagating along the $+\hat{\mathbf{z}}$ axis, is given by

$$\mathbf{E}(t, \mathbf{x}) = \mathrm{Re}\left[(E_1\hat{\mathbf{x}} + E_2\hat{\mathbf{y}})e^{-i(\omega t - kz)}\right]\,, \tag{9.85}$$

where E_1 and E_2 are arbitrary complex quantities. Writing

$$E_1 = A_1 e^{-i\delta_1}\,, \qquad E_2 = A_2 e^{-i\delta_2}\,, \tag{9.86}$$

with A_1, A_2, δ_1 and δ_2 real, and using the notation $\varphi = \omega t - kz$, we get

$$\mathbf{E}(t, \mathbf{x}) = E_x(t, z)\hat{\mathbf{x}} + E_y(t, z)\hat{\mathbf{y}}\,, \tag{9.87}$$

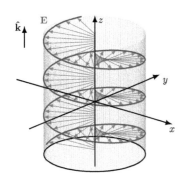

Fig. 9.1 A linearly polarized electromagnetic wave.

Fig. 9.2 A circularly polarized electromagnetic wave.

where

$$E_x(t, z) = A_1 \cos(\varphi + \delta_1), \qquad (9.88)$$
$$E_y(t, z) = A_2 \cos(\varphi + \delta_2), \qquad (9.89)$$

and the dependence on (t, z) enters through φ. With simple trigonometry, we then find

$$\frac{E_x}{A_1} \sin \delta_2 - \frac{E_y}{A_2} \sin \delta_1 = \cos \varphi \sin \delta, \qquad (9.90)$$

$$\frac{E_x}{A_1} \cos \delta_2 - \frac{E_y}{A_2} \cos \delta_1 = \sin \varphi \sin \delta, \qquad (9.91)$$

where $\delta = \delta_2 - \delta_1$. Squaring the two terms and summing them, we get

$$\left(\frac{E_x}{A_1}\right)^2 + \left(\frac{E_y}{A_2}\right)^2 - 2 \left(\frac{E_x}{A_1}\right) \left(\frac{E_y}{A_2}\right) \cos \delta = \sin^2 \delta. \qquad (9.92)$$

This is the equation of an ellipse in the (E_x, E_y) plane, with semi-axes A_1 and A_2. Correspondingly, light is said to be *elliptically polarized*, see Fig. 9.3.

For $\delta = \pi/2$ (or, more generally, for $\delta = m\pi/2$ with $m = \pm 1, \pm 3, \ldots$), we get

$$\left(\frac{E_x}{A_1}\right)^2 + \left(\frac{E_y}{A_2}\right)^2 = 1, \qquad (9.93)$$

so the semi-axes are aligned with the E_x and E_y axes. If, furthermore, $A_1 = A_2$, the ellipse becomes a circle, and we get back circular polarization. If, instead, $\delta = m\pi$ with $m = 0, \pm 1, \pm 2, \ldots$ we get

$$\left(\frac{E_x}{A_1}\right)^2 + \left(\frac{E_y}{A_2}\right)^2 \pm 2 \left(\frac{E_x}{A_1}\right) \left(\frac{E_y}{A_2}\right) = 0, \qquad (9.94)$$

(with the plus sign for m odd and the minus for m even) so that

$$\left(\frac{E_x}{A_1} \pm \frac{E_y}{A_2}\right)^2 = 0, \qquad (9.95)$$

and therefore

$$\frac{E_y}{E_x} = \mp \frac{A_2}{A_1}, \qquad (9.96)$$

is fixed. Therefore, the electric field does not rotate in the (E_x, E_y) plane, and we get back linear polarization.

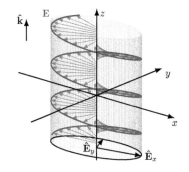

Fig. 9.3 An elliptically polarized electromagnetic wave.

9.6 Doppler effect and light aberration

We have seen that the space and time dependence of a monochromatic electromagnetic wave is given by the factor e^{ikx} (and its complex conjugate), where

$$k^\mu = (k^0, \mathbf{k}) = \frac{\omega}{c}(1, \hat{\mathbf{k}}). \qquad (9.97)$$

While the speed of light c is the same for all observers, the frequency ω and the propagation direction $\hat{\mathbf{k}}$ depend on the observer, in a way determined by the fact that k^μ is a four-vector. As we will see in this section, this gives rise to the relativistic Doppler effect and to the aberration of light.

To fix the geometry, consider a source S, such as a star, and let K_s denote an inertial frame comoving with the star. In this frame the star therefore has zero velocity. We denote by $(t_s, \mathbf{x}_s)$ the coordinates of the source in this frame, and by $\hat{\mathbf{n}}_s$ the unit vector from the origin of this reference frame, toward the star. Consider now a second reference frame $K_{\rm obs}$, moving with uniform velocity $\mathbf{v}$ with respect to K_s. In this frame, the star has a velocity $\mathbf{v}_s = -\mathbf{v}$. We fix the origins of the two reference frames so that, at a given time t_0, they coincide, and we call $O_{\rm obs}$ an observer that sits at the origin of the frame $K_{\rm obs}$. We denote by $(t_{\rm obs}, \mathbf{x}_{\rm obs})$ the coordinates of this frame, and by $\hat{\mathbf{n}}_{\rm obs}$ the unit vector from the observer to the source, in this frame, at the given time considered. From eqs. (7.22) and (7.23), written for a boost in a generic direction, the coordinates of the source in the two frames are related by

$$x^0_{\rm obs} = \gamma_s(x^0_s + \boldsymbol{\beta}_s \cdot \mathbf{x}_s), \tag{9.98}$$

$$x_{\parallel,{\rm obs}} = \gamma_s(x_{\parallel,s} + \beta_s x^0_s), \tag{9.99}$$

$$\mathbf{x}_{\perp,{\rm obs}} = \mathbf{x}_{\perp,s}, \tag{9.100}$$

where $\gamma_s = \gamma(v_s)$, and we have spilt the vector part of the equation into the components parallel and perpendicular to $\boldsymbol{\beta}_s$, $\mathbf{x} = \mathbf{x}_\perp + x_\parallel \hat{\boldsymbol{\beta}}_s$. Note that we have chosen the signs so that, if the source is at rest in the frame K_s, it moves in the direction $+\hat{\boldsymbol{\beta}}_s$ in the frame $K_{\rm obs}$.

In the reference frame K_s, the four-momentum of the light that propagates from the source to the origin of the coordinate system is

$$k^\mu_s = \frac{\omega_s}{c}(1, -\hat{\mathbf{n}}_s). \tag{9.101}$$

Note that $\hat{\mathbf{n}}_s$ is defined to point from the observer toward the source, while light propagates from the source to the observer, so $\hat{\mathbf{k}}_s = -\hat{\mathbf{n}}_s$. In the reference frame $K_{\rm obs}$, the four-momentum of the light emitted from the source and received at time t_0 by the observer $O_{\rm obs}$ is

$$k^\mu_{\rm obs} = \frac{\omega_{\rm obs}}{c}(1, -\hat{\mathbf{n}}_{\rm obs}). \tag{9.102}$$

The four-momenta $k^\mu_{\rm obs}$ and k^μ_s are related by a Lorentz transformation, completely analogous to eqs. (9.98)–(9.100). Splitting again the vector part into the components parallel and perpendicular to $\boldsymbol{\beta}_s = \mathbf{v}_s/c$, we have

$$k^0_{\rm obs} = \gamma_s\left(k^0_s + \boldsymbol{\beta}_s \cdot \mathbf{k}_s\right), \tag{9.103}$$

$$k_{\parallel,{\rm obs}} = \gamma_s\left(k_{\parallel,s} + \beta_s k^0_s\right), \tag{9.104}$$

$$\mathbf{k}_{\perp,{\rm obs}} = \mathbf{k}_{\perp,s}. \tag{9.105}$$

The inversion of these equations gives

$$k_s^0 = \gamma_s \left(k_{\text{obs}}^0 - \boldsymbol{\beta}_s \cdot \mathbf{k}_{\text{obs}} \right) , \tag{9.106}$$

$$k_{\parallel,s} = \gamma_s \left(k_{\parallel,\text{obs}} - \beta_s k_{\text{obs}}^0 \right) . \tag{9.107}$$

This could be easily checked analytically, but in fact we can more simply observe that all relations between the quantities in the frame K_s and the corresponding quantities in the frame K_{obs} can be inverted by replacing $\boldsymbol{\beta}_s$ with $-\boldsymbol{\beta}_s$ since, if K_s moves with velocity $\mathbf{v}_s$ with respect to K_{obs}, then K_{obs} moves with velocity $-\mathbf{v}_s$ with respect to K_s. Inserting eqs. (9.101) and (9.102) into eq. (9.106) gives

$$\omega_s = \gamma_s \omega_{\text{obs}} \left(1 + \boldsymbol{\beta}_s \cdot \hat{\mathbf{n}}_{\text{obs}} \right) . \tag{9.108}$$

Therefore

$$\boxed{\omega_{\text{obs}} = \frac{\omega_s}{\gamma_s \left(1 + \boldsymbol{\beta}_s \cdot \hat{\mathbf{n}}_{\text{obs}} \right)} .} \tag{9.109}$$

This equation gives the frequency ω_{obs}, as measured by the inertial observer O_{obs} for which the source has a velocity $\mathbf{v}_s$, as a function of the intrinsic frequency of the source ω_s (i.e., the frequency measured by an inertial observer for which the source is at rest), of the velocity $\mathbf{v}_s$ of the source in the frame of O_{obs}, and of the direction of the source, $\hat{\mathbf{n}}_{\text{obs}}$, again as measured by the observer O_{obs}. Using the explicitly expression for $\gamma(v_s)$ and writing $\boldsymbol{\beta}_s \cdot \hat{\mathbf{n}}_{\text{obs}} = \beta_s \cos\theta_{\text{obs}}$ (where $\beta_s = |\boldsymbol{\beta}_s| > 0$), we can rewrite it as

$$\boxed{\omega_{\text{obs}} = \omega_s \frac{(1 - \beta_s^2)^{1/2}}{1 + \beta_s \cos\theta_{\text{obs}}} .} \tag{9.110}$$

This change of frequency between a frame comoving with the source and a frame where the source has non-zero velocity is called the *Doppler effect*.

In the limit $\beta_s \ll 1$ we can expand eq. (9.110) in powers of β_s. The terms of order β_s and β_s^2 are called the first-order and the second-order Doppler effect, respectively. The first-order Doppler effect is given by

$$\omega_{\text{obs}} \simeq \omega_s \left(1 - \beta_s \cos\theta_{\text{obs}} \right) . \tag{9.111}$$

If the source is moving away from the observer, $\cos\theta_{\text{obs}} > 0$ and then $\omega_{\text{obs}} < \omega_s$. One conventionally says that the frequency of light is red-shifted (this nomenclature, of course, had its origin in the shift of the frequency of visible light, but is now universally used just to mean that the frequency decreases). Conversely, if the source comes toward the observer, $\cos\theta_{\text{obs}} < 0$ and $\omega_{\text{obs}} > \omega_s$, i.e., light is "blueshifted."

Note, however, that for $\theta_{\text{obs}} = \pi/2$, i.e., for a source moving in a direction orthogonal to the line of sight of the observer, there is no first-order effect, and

$$\omega_{\text{obs}} = \omega_s (1 - \beta_s^2)^{1/2} \qquad (\cos\theta_{\text{obs}} = 0) . \tag{9.112}$$

This is called the *transverse* Doppler effect, and always corresponds to a redshift.

Next consider the relation between $\hat{\mathbf{n}}_s$ and $\hat{\mathbf{n}}_{\rm obs}$. The relation between the components of $\hat{\mathbf{n}}_s$ and $\hat{\mathbf{n}}_{\rm obs}$ parallel to β is obtained by inserting eqs. (9.101) and (9.102) into eq. (9.107). Writing

$$\hat{n}_s^{\parallel} = \cos\theta_s \,, \tag{9.113}$$

$$\hat{n}_{\rm obs}^{\parallel} = \cos\theta_{\rm obs} \,, \tag{9.114}$$

and using eq. (9.109), we get

$$\cos\theta_s = \frac{\cos\theta_{\rm obs} + \beta_s}{1 + \beta_s \cos\theta_{\rm obs}} \,. \tag{9.115}$$

Again, this can be inverted by exchanging the labels "obs" and "s" and replacing $\beta_s \to -\beta_s$,

$$\boxed{\cos\theta_{\rm obs} = \frac{\cos\theta_s - \beta_s}{1 - \beta_s \cos\theta_s} \,.} \tag{9.116}$$

Equation (9.116) also implies that

$$\sin\theta_{\rm obs} = \frac{\sin\theta_s}{\gamma_s(1 - \beta_s \cos\theta_s)} \,, \tag{9.117}$$

and

$$\tan\theta_{\rm obs} = \frac{\sin\theta_s}{\gamma_s(\cos\theta_s - \beta_s)} \,. \tag{9.118}$$

Equation (9.116) shows that the direction in which the observers K and K' see the source are not the same. This phenomenon is called the *aberration* of light.

The Doppler and aberration effects already exist in Galilean Relativity, i.e., when the transformation between coordinates of inertial frames is given by eq. (7.1). For sound waves, the Doppler effect is the familiar change of pitch of the siren of an ambulance from when it approaches to when it recedes from us.[10] A familiar example of aberration can be given by the tracks left by the rain on the window of a moving train, which have an inclination with respect to the vertical due to the velocity of the train. In both cases, these effects can be derived as a consequence of the non-relativistic composition law for velocities, as follows.

For the Doppler effect consider a source that, in its rest frame, emits signals with a period T, and therefore a frequency $\omega_s = 2\pi/T$. As illustrated in Fig. 9.4, in the observer frame the source moves at velocity $\mathbf{v}_s$ and, at time t_1, is in the position $\mathbf{x}_1$. At time $t_2 = t_1 + T$ it will be in the position

$$\begin{aligned} \mathbf{x}_2 &= \mathbf{x}_1 + \mathbf{v}_s(t_2 - t_1) \\ &= \mathbf{x}_1 + \mathbf{v}_s T \,, \end{aligned} \tag{9.119}$$

[10]In this case, of course, the role of the speed of light is played by the speed of sound (in the rest frame of the medium where the wave propagates).

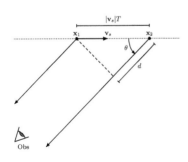

Fig. 9.4 The geometric setting of the Doppler effect discussed in the text.

and, as we see from Fig. 9.4, to reach an observer (located at a distance much larger than $\mathbf{v}_s T$), light must travel an extra distance

$$d = v_s T \cos\theta\,, \tag{9.120}$$

or, in vector form,

$$d = \mathbf{v}_s \cdot \hat{\mathbf{n}}_{\text{obs}} T\,. \tag{9.121}$$

Therefore, the difference in the time of arrival of two signals emitted at times t and $t + T$ is

$$
\begin{aligned}
T_{\text{obs}} &= T + \frac{d}{c} \\
&= T\left(1 + \boldsymbol{\beta}_s \cdot \hat{\mathbf{n}}_{\text{obs}}\right),
\end{aligned}
\tag{9.122}
$$

and therefore $\omega_{\text{obs}} \equiv 2\pi/T_{\text{obs}}$ is related to $\omega_s \equiv 2\pi/T$ by

$$\omega_{\text{obs}} = \frac{\omega_s}{1 + \boldsymbol{\beta}_s \cdot \hat{\mathbf{n}}_{\text{obs}}}\,. \tag{9.123}$$

This agrees with eq. (9.109), except for the factor $1/\gamma_s$ which is a purely relativistic effect and is simply due to the time dilatation effect that we studied in Section 7.2.3: for the observer, the clock on the star goes slower, and therefore it emits the second signal only after a time $\Delta t = 2\pi\gamma_s/\omega_s$, rather than $2\pi/\omega_s$. This reproduces the correct factor $1/\gamma_s$ in eq. (9.109). Note in particular that, in the non-relativistic computation, there is no transverse Doppler effect.

The non-relativistic expression for the aberration can be computed similarly. Consider a frame at rest with respect to the source, in which light has speed c (of course, if we assume Galilean Relativity, the speed of light becomes frame dependent) and let

$$\hat{\mathbf{n}}_s = (\cos\theta_s, \sin\theta_s, 0)\,, \tag{9.124}$$

be the unit vector toward the source in this frame. In this frame, the velocity of a light signal emitted at the source and reaching the origin of the reference is given by the vector

$$\mathbf{c}_s = (-c\cos\theta_s, -c\sin\theta_s, 0)\,. \tag{9.125}$$

In the observer frame, where the source moves away from the observer with velocity v_s along the positive direction of the x axis, using the Galilean composition of velocities we would have

$$\mathbf{c}_{\text{obs}} = (-c\cos\theta_s + v_s, -c\sin\theta_s, 0)\,. \tag{9.126}$$

Therefore, the prediction of Galilean Relativity is that light will be seen to arrive from an angle θ_{obs} such that

$$
\begin{aligned}
\tan\theta_{\text{obs}} &= \frac{c\sin\theta_s}{c\cos\theta_s - v_s} \\
&= \frac{\sin\theta_s}{\cos\theta_s - \beta_s}\,.
\end{aligned}
\tag{9.127}
$$

This agrees with eq. (9.118), except again for the factor $1/\gamma_s$, which is therefore a purely relativistic effect.

Electromagnetic field of moving charges

In this chapter we study the electromagnetic fields generated by moving charges. At the mathematical level, a fundamental tool is provided by the Green's functions of the d'Alembertian operator, in particular the retarded Green's function, that we introduce in Section 10.1. We will then be able to study the field generated by charges with arbitrary motion, and we will discover that, when a charge is accelerated, it produces electromagnetic waves. We will then study in detail the radiation emitted in different situations.

10.1 Advanced and retarded Green's function

To solve radiation problems we use the Green's function method, that we have already introduced in Section 4.1.2 for the case of the Laplace operator. Here, however, the relevant operator is the d'Alembertian operator (3.88) and, as we will see, the situation is richer because it admits different Green's functions. We define the Green's functions $G(x, x')$ of the d'Alembertian as the solutions of the equation[1]

$$\Box_x G(x; x') = \delta^{(4)}(x - x') \,, \tag{10.1}$$

or, more explicitly

$$\left[-\frac{\partial^2}{\partial (x^0)^2} + \nabla_{\mathbf{x}}^2\right] G(x^0, \mathbf{x}; x'^0, \mathbf{x}') = \delta(x^0 - x'^0)\delta^{(3)}(\mathbf{x} - \mathbf{x}') \,. \tag{10.2}$$

Once found a Green's function, a particular solution of an equation such as

$$\Box f(x) = j(x) \tag{10.3}$$

is given by

$$f(x) = \int d^4x' \, G(x; x')j(x') \,, \tag{10.4}$$

as can be checked by applying $\Box_x$ to both sides.[2] The most general solution of eq. (10.3) is then obtained adding the most general solution $f_{\text{hom}}(x)$ of the homogeneous equation $\Box f = 0$, so that

$$f(x) = f_{\text{hom}}(x) + \int d^4x' \, G(x; x')j(x') \,. \tag{10.5}$$

---

[1] Different conventions exist in the literature for the normalization and overall sign of the Green's function. Sometimes the Green's functions of the d'Alembertian are rather defined as the solutions of

$$\Box_x G(x; x') = -4\pi\delta^{(4)}(x - x') \,,$$

to reabsorbe a factor $-1/(4\pi)$ that we will find in eq. (10.24) below. Also notice that we are using the signature $(-, +, +, +)$, so that $\Box = -(1/c^2)\partial_t^2 + \nabla^2$. Sometimes the Greens function is defined by $\Box_x G(x; x') = \delta^{(4)}(x - x')$, using, however, the opposite signature $(+, -, -, -)$, so $\Box = (1/c^2)\partial_t^2 - \nabla^2$; this definition then differs from ours by an overall sign.

[2] This is valid as long as the integral converges. After having found the explicit form of the Green's function, we will discuss the corresponding boundary conditions that need to be imposed on $j(x)$.

Notice that the use of different Green's functions can be reabsorbed into a different choice of solution of the homogeneous equation. Indeed, let $G_1(x; x')$ and $G_2(x; x')$ be two different Green's functions, and define

$$f_1(x) = \int d^4x' \, G_1(x; x') j(x'), \qquad (10.6)$$

$$f_2(x) = \int d^4x' \, G_2(x; x') j(x'). \qquad (10.7)$$

Then $f_2(x) = f_{\text{hom}}(x) + f_1(x)$, where

$$f_{\text{hom}}(x) = \int d^4x' \, [G_2(x; x') - G_1(x; x')] j(x'). \qquad (10.8)$$

Since

$$\begin{aligned}
\Box_x f_{\text{hom}}(x) &= \int d^4x' \, [\delta^{(4)}(x - x') - \delta^{(4)}(x - x')] j(x') \\
&= 0, \qquad\qquad\qquad\qquad\qquad\qquad\qquad\qquad (10.9)
\end{aligned}$$

the solutions $f_2(x)$ and $f_1(x)$ indeed differ by a solution of the homogeneous equation. Observe that, in the case of the Laplacian studied in Section 4.1.2, the homogeneous equation $\nabla^2 \phi = 0$ (with the boundary condition that ϕ vanishes at infinity, that was the physically relevant one in the setting of electrostatics) only has the solution $\phi = 0$, and therefore the Green's function was unique. In contrast, as we saw in Chapter 9, an equation such as $\Box f(x) = 0$ has non-vanishing solutions, corresponding to plane waves, that are the physically relevant solutions in a radiation problem. The physically correct boundary conditions must therefore be such to allow for the possibility of these solutions, and the choice of the homogeneous solution, or, equivalently, of the appropriate Green's function, reflects these boundary conditions.

We use the Green's function technique to compute the electromagnetic field generated by a generic current j^μ. It is convenient to work in the Lorenz gauge $\partial_\mu A^\mu = 0$, so the equation to be solved is eq. (8.30). Given a Green's function $G(x; x')$ of the d'Alembertian operator, a particular solution of the inhomogeneous equation is then

$$A^\mu(x) = -\mu_0 \int d^4x' \, G(x, x') j^\mu(x'), \qquad (10.10)$$

subject, again, to suitable boundary conditions on $j^\mu(x)$, such that the integral converges.

The problem, therefore, amounts to computing the Green's functions of the d'Alembertian operator. Without loss of generality, we can set $x' = 0$ and solve the equation

$$\Box_x G(x) = \delta^{(4)}(x). \qquad (10.11)$$

A convenient way to solve this equation is to perform a Fourier transform only with respect to x^0, writing

$$G(x^0, \mathbf{x}) = \int_{-\infty}^{+\infty} \frac{dk_0}{2\pi} e^{-ik_0 x^0} \tilde{G}(k_0, \mathbf{x}). \qquad (10.12)$$

Then, eq. (10.2) becomes

$$\int_{-\infty}^{+\infty} \frac{dk_0}{2\pi} e^{-ik_0 x^0} (\nabla^2 + k_0^2)\tilde{G}(k_0, \mathbf{x}) = \int_{-\infty}^{+\infty} \frac{dk_0}{2\pi} e^{-ik_0 x^0} \delta^{(3)}(\mathbf{x}),$$

(10.13)

where, on the left-hand side, we inserted eq. (10.12), and, on the right-hand side, we used the integral representation (1.76) of $\delta(x_0)$. Therefore, inverting the Fourier transform with respect to k_0, we get

$$(\nabla^2 + k_0^2)\tilde{G}(k_0, \mathbf{x}) = \delta^{(3)}(\mathbf{x}).$$

(10.14)

The problem is now reduced to computing the Green's function of the three-dimensional operator $(\nabla^2 + k_0^2)$, which is called the Helmholtz operator. To compute this Green's function we observe that eq. (10.14) is invariant under rotations around the origin, where the Dirac delta sits, and therefore $\tilde{G}(k_0, \mathbf{x})$ depends on $\mathbf{x}$ only through $r = |\mathbf{x}|$. We define $f(r)$ from

$$\tilde{G}(k_0, r) = -\frac{1}{4\pi r} f(k_0, r).$$

(10.15)

Extracting explicitly a factor $1/r$ is convenient because, from eq. (1.90), the Laplacian of $1/r$ produces the Dirac delta. Then [suppressing, for notational simplicity, the argument k_0 from $f(k_0, r)$],

$$\nabla^2 \left[\frac{1}{r} f(r) \right] = -4\pi \delta^{(3)}(\mathbf{x}) f(0) + \frac{1}{r} f''(r),$$

(10.16)

where $f' = df/dr$.[3] From this we get

$$(\nabla^2 + k_0^2) G(\mathbf{x}) = \delta^{(3)}(\mathbf{x}) f(0) - \frac{1}{4\pi r}(f'' + k_0^2 f),$$

(10.18)

and therefore, to solve eq. (10.14), we must require $f(0) = 1$ and

$$f'' + k_0^2 f = 0.$$

(10.19)

The most general solution of this equation is

$$f(r) = A e^{ik_0 r} + B e^{-ik_0 r},$$

(10.20)

and the condition $f(0) = 1$ fixes $A + B = 1$, so the most general solution of eq. (10.14) is

$$\tilde{G}(k_0, \mathbf{x}) = -\frac{1}{4\pi r} \left[A e^{ik_0 r} + (1 - A) e^{-ik_0 r} \right].$$

(10.21)

There are therefore two independent solutions, that can be taken to be

$$\tilde{G}^{\pm}(k_0, \mathbf{x}) = -\frac{1}{4\pi r} e^{\pm ik_0 r}.$$

(10.22)

We have therefore found the Green's functions of the Helmholtz operator (10.14), and we see that, for $k_0 \neq 0$, there are two independent Green's functions. For $k_0 = 0$, these two solutions become identical, and reduce

[3] The explicit computation goes as follows:

$$\nabla^2 \left[\frac{1}{r} f(r) \right] = \partial_i \partial_i \left[\frac{1}{r} f(r) \right]$$

$$= \left(\nabla^2 \frac{1}{r} \right) f(r) + 2 \left(\partial_i \frac{1}{r} \right) \partial_i f$$

$$+ \frac{1}{r} \nabla^2 f.$$

(10.17)

To compute $\nabla^2 f$ we use the expression for the Laplacian in spherical coordinates, eq. (1.26), while, to compute the term $\partial_i (1/r) \partial_i f$, we use $\partial_i r = n_i$, see eq. (6.12), so that

$$\left(\partial_i \frac{1}{r} \right) \partial_i f = -\frac{1}{r^2} n_i f'(r) n_i$$

$$= -\frac{1}{r^2} f'(r),$$

since $n_i n_i = 1$. Then, from eq. (1.90)

$$\nabla^2 \left[\frac{1}{r} f(r) \right] = -4\pi \delta^{(3)}(\mathbf{x}) f(r)$$

$$-\frac{2}{r^2} f'(r) + \frac{1}{r^3}(2r f' + r^2 f'')$$

$$= -4\pi \delta^{(3)}(\mathbf{x}) f(0) + \frac{1}{r} f''(r).$$

to the unique Green's function of the Laplace operator, eq. (4.15). This is as expected, since the Helmholtz operator reduces to the Laplace operator when $k_0 = 0$. Inserting eq. (10.22) into eq. (10.12) we get

$$
\begin{aligned}
G^{\pm}(x^0, \mathbf{x}) &= -\frac{1}{4\pi r} \int_{-\infty}^{+\infty} \frac{dk_0}{2\pi} e^{-ik_0(x^0 \mp r)} \\
&= -\frac{1}{4\pi|\mathbf{x}|} \delta(x^0 \mp |\mathbf{x}|).
\end{aligned} \tag{10.23}
$$

The result for a generic second argument x', that we had set to zero, can be obtained from the fact that the Green's function $G(x; x')$ is actually a function only of $x - x'$, because of invariance under space-time translation. Using furthermore ct instead of x^0, we get

$$
\boxed{G^{\pm}(t, \mathbf{x}; t', \mathbf{x}') = -\frac{1}{4\pi|\mathbf{x} - \mathbf{x}'|} \delta\left[c(t - t') \mp |\mathbf{x} - \mathbf{x}'|\right].}
$$
$$\tag{10.24}$$

The corresponding inhomogeneous solutions for A_μ, from eq. (10.10) (using $dx^0 = cdt$ and $\delta(ct) = (1/c)\delta(t)$ to write the final result in terms of t rather than x^0), are

$$
[A^\mu(t, \mathbf{x})]^{\pm} = \frac{\mu_0}{4\pi} \int dt' d^3x' \frac{j^\mu(t', \mathbf{x}')}{|\mathbf{x} - \mathbf{x}'|} \delta\left[t' - (t \mp |\mathbf{x} - \mathbf{x}'|/c)\right]. \tag{10.25}
$$

Consider first the solution $[A^\mu(t, \mathbf{x})]^+$. Performing the integral over dt' with the help of the Dirac delta we see that

$$
[A^\mu(t, \mathbf{x})]^+ = \frac{\mu_0}{4\pi} \int d^3x' \frac{j^\mu(t - |\mathbf{x} - \mathbf{x}'|/c, \mathbf{x}')}{|\mathbf{x} - \mathbf{x}'|}. \tag{10.26}
$$

Therefore, $[A^\mu(t, \mathbf{x})]^+$ depends on the value of the current $j^\mu(t', \mathbf{x}')$ only on the *past light cone* of the space-time point $(t, \mathbf{x})$, i.e., on the points $(t', \mathbf{x}')$ from which a signal, traveling at the speed of light, could arrive at $(t, \mathbf{x})$. The Green's function $G^+(t, \mathbf{x}; t', \mathbf{x}')$ is called the *retarded Green's function*. The retardation effect expresses the fact that a change in the source at a point $\mathbf{x}'$ at time t' cannot affect instantaneously the value of the field at a different point $\mathbf{x}$. Rather, its effect will be felt only at a subsequent time $t = t' + |\mathbf{x} - \mathbf{x}'|/c$, so that $t - t'$ is equal to the time taken by light to travel from $\mathbf{x}'$ to $\mathbf{x}$. This is consistent with Special Relativity, and follows from the relativistic structure of the d'Alembertian.

The other solution is

$$
[A^\mu(t, \mathbf{x})]^- = \frac{\mu_0}{4\pi} \int d^3x' \frac{j^\mu(t + |\mathbf{x} - \mathbf{x}'|/c, \mathbf{x}')}{|\mathbf{x} - \mathbf{x}'|}, \tag{10.27}
$$

and depends on the value of the current $j^\mu(t', \mathbf{x}')$ only on the *future* light cone. The corresponding Green's function $G^-(t, \mathbf{x}; t', \mathbf{x}')$ is called the *advanced* Green's function. Retarded and advanced Green's functions will be also denoted as G_{ret} and G_{adv}, respectively.[4]

First of all, we can check the static limit of these solutions. In the static limit, $[A^\mu(t, \mathbf{x})]^+$ and $[A^\mu(t, \mathbf{x})]^-$ become identical, since $j^\mu(t', \mathbf{x}')$

[4]Having found the explicit form of the Green's functions, we can also check that the integrals in eqs. (10.26) and (10.27) both converge if, for all times t', the source is localized in space, i.e., for all t', $j^\mu(t', \mathbf{x}')$ has compact support in $\mathbf{x}'$. Less stringent conditions could also be used, depending on the problem. For instance, to have a well-defined retarded solution (10.26) at a given time t, it is sufficient that $j^\mu(t', \mathbf{x}')$ is localized in space for all times $t' \le t$. In practice, in most physical situations, $j^\mu(t', \mathbf{x}')$ has compact support in $\mathbf{x}'$ for all values of t', and also switches off if $t' \to -\infty$.

loses the dependence on the first argument, so $j^\mu(t \mp |\mathbf{x} - \mathbf{x}'|/c, \mathbf{x}') = j^\mu(\mathbf{x}')$. Then, using $j^0(\mathbf{x}') = c\rho(\mathbf{x}')$, together with eqs. (8.12) and (8.27), we see that the $\mu = 0$ component of eq. (10.26) reduces to the static solution (4.16) for the scalar potential, while the vector part of eq. (10.26) reduces to the static solution (4.92) for the vector potential.

However, whenever there is an actual time dependence, the solutions are different and, in this case, the advanced solution $[A^\mu(t, \mathbf{x})]^-$ is physically unacceptable. At first, one might think that this is due to the fact that, since its value at time t depends on what the source will do in the future, this solution violates causality.[5] Actually, the reason why this solution is physically unacceptable is somewhat more subtle and is rather related to the possibility of imposing natural boundary conditions, as we now discuss. Using the retarded Green's function, the most general solution of eq. (8.30) can be written as

$$
\begin{aligned}
A^\mu(t, \mathbf{x}) &= A_{\text{in}}^\mu(t, \mathbf{x}) - \mu_0 \int d^4x' \, G_{\text{ret}}(x, x') j^\mu(x') \qquad (10.28) \\
&= A_{\text{in}}^\mu(t, \mathbf{x}) + \frac{\mu_0}{4\pi} \int d^3x' \, \frac{j^\mu(t - |\mathbf{x} - \mathbf{x}'|/c, \mathbf{x}')}{|\mathbf{x} - \mathbf{x}'|},
\end{aligned}
$$

where $A_{\text{in}}^\mu(t, \mathbf{x})$ is a general solution of the homogeneous equation. The physical meaning of $A_{\text{in}}^\mu(t, \mathbf{x})$ can be seen by taking the limit $t \to -\infty$. In this case, also the argument $t - |\mathbf{x} - \mathbf{x}'|/c$ of $j^\mu(t - |\mathbf{x} - \mathbf{x}'|/c, \mathbf{x}')$ goes to $-\infty$ and, if we assume that the source is localized in time, in this limit $j^\mu(t - |\mathbf{x} - \mathbf{x}'|/c, \mathbf{x}')$ goes to zero, for all $\mathbf{x}$ and $\mathbf{x}'$, and the integral vanishes. Therefore, $A_{\text{in}}^\mu(t, \mathbf{x})$ represents the initial value of $A^\mu(t, \mathbf{x})$, at $t \to -\infty$ and $\mathbf{x}$ arbitrary. Notice that the same argument cannot be made for the limit $t \to +\infty$, for all values of the arguments $\mathbf{x}$ of $A^\mu(t, \mathbf{x})$. In particular, we might wish to study the behavior of $A^\mu(t, \mathbf{x})$ as $t \to \infty$ while $r = |\mathbf{x}|$ also goes to infinity, in such a way that $t - r/c$ stays fixed at a given value, that we denote by t_u, smaller than the time at which the source eventually switches off. Then, in this limit $j^\mu(t - |\mathbf{x} - \mathbf{x}'|/c, \mathbf{x}')$ goes to a non-vanishing value $j^\mu(t_u, \mathbf{x}')$, so it can contribute to $A^\mu(t, \mathbf{x})$.

If instead we use the advanced Green's function, we can write the solution as

$$
\begin{aligned}
A^\mu(t, \mathbf{x}) &= A_{\text{out}}^\mu(t, \mathbf{x}) - \mu_0 \int d^4x' \, G_{\text{adv}}(x, x') j^\mu(x') \qquad (10.29) \\
&= A_{\text{out}}^\mu(t, \mathbf{x}) + \frac{\mu_0}{4\pi} \int d^3x' \, \frac{j^\mu(t + |\mathbf{x} - \mathbf{x}'|/c, \mathbf{x}')}{|\mathbf{x} - \mathbf{x}'|},
\end{aligned}
$$

where, again, $A_{\text{out}}^\mu(t, \mathbf{x})$ is a general solution of the homogeneous equation. The same argument now shows that $A_{\text{out}}^\mu(t, \mathbf{x})$ is the value of $A^\mu(t, \mathbf{x})$ at $t \to +\infty$, for all $\mathbf{x}$. We can rewrite this solution as

$$
\begin{aligned}
A^\mu(t, \mathbf{x}) = {}& A_{\text{out}}^\mu(t, \mathbf{x}) - \mu_0 \int d^4x' \, [G_{\text{adv}}(x, x') - G_{\text{ret}}(x, x')] j^\mu(x') \\
& - \mu_0 \int d^4x' \, G_{\text{ret}}(x, x') j^\mu(x'). \qquad (10.30)
\end{aligned}
$$

[5]To be more precise, the gauge potential is not directly observable, so it could a priori be acausal, as long as the corresponding electric and magnetic fields turned out to be causal. An example of this behavior will be discussed in Section 11.1.2. However, for the gauge potential (10.27), the corresponding electric and magnetic fields would also depend on the behavior of the sources on the future light cone.

We now define

$$A_{\text{rad}}^\mu(t,\mathbf{x}) = -\mu_0 \int d^4x' \, [G_{\text{ret}}(x,x') - G_{\text{adv}}(x,x')] j^\mu(x') . \quad (10.31)$$

This is a solution of the homogeneous equation $\Box A^\mu = 0$, by the same argument used in eqs. (10.6)–(10.9). Then, eq. (10.30) can be rewritten as

$$A^\mu(t,\mathbf{x}) = A_{\text{out}}^\mu(t,\mathbf{x}) - A_{\text{rad}}^\mu(t,\mathbf{x}) - \mu_0 \int d^4x' \, G_{\text{ret}}(x,x') j^\mu(x') . \quad (10.32)$$

Since both $A_{\text{out}}^\mu(t,\mathbf{x})$ and $A_{\text{rad}}^\mu(t,\mathbf{x})$ are solutions of the homogeneous equation, also their difference is a solution of the homogeneous equation, so eq. (10.32) is of the form (10.28), with

$$A_{\text{in}}^\mu(t,\mathbf{x}) = A_{\text{out}}^\mu(t,\mathbf{x}) - A_{\text{rad}}^\mu(t,\mathbf{x}) . \quad (10.33)$$

So, the apparently acausal solution (10.29) has been rewritten in terms of the retarded Green's function. This shows that the problem, with the advanced solution, is not the apparent acausality. The advanced solution can be rewritten as an integral that depends only on the behavior of the source on the past light cone. The real problem is in the meaning of the associated homogeneous solution. As we have seen, when we use the retarded solution (10.28), the associated homogeneous solution is the value of the field at $t \to -\infty$. It is easy to specify a physically meaningful expression for $A_{\text{in}}^\mu(t,\mathbf{x})$. For instance, setting $A_{\text{in}}^\mu(t,\mathbf{x}) = 0$ describes a situation where, at $t \to -\infty$, there was no incoming radiation. A system of charges, accelerated by their mutual interactions, will then produce an outgoing radiation that can be computed by setting $A_{\text{in}}^\mu(t,\mathbf{x}) = 0$ in eq. (10.28). In contrast, if we write the solution in the form (10.29), we must specify the function $A_{\text{out}}^\mu(t,\mathbf{x})$, which has the meaning of the limit of the solution $A^\mu(t,\mathbf{x})$ for $t \to +\infty$. First of all, this is not what we typically want to do. In general, we want to specify initial conditions and see how a system evolves, rather than specifying the desired final outcome of the evolution. Furthermore, there is no way of specifying meaningful final conditions. For instance, a mathematically simple choice such as $A_{\text{out}}^\mu(t,\mathbf{x}) = 0$ corresponds, physically, to a situation in which, at $t \to -\infty$, there was radiation coming from spatial infinity and impinging on a system of charges, perfectly tuned so that the charges of the system, accelerated both by their mutual interactions and by the incoming electromagnetic wave, emit outgoing electromagnetic waves that perfectly cancel among each other, leaving a total vanishing outgoing radiation field. Such an initial condition is acceptable mathematically, but not physically.[6] A physically meaningful solution can only be specified writing the solution in the form (10.28), using the retarded Green's function, and specifying the initial field $A_{\text{in}}^\mu(t,\mathbf{x})$ in a way that corresponds to realistic settings, such as no incoming radiation, i.e., $A_{\text{in}}^\mu(t,\mathbf{x}) = 0$, or any other physically realistic choice, such as a given laser pulse arriving on a system of charges.

[6]One can make a parallel with the situation in which a glass falls from a table and breaks into pieces on the floor, with its initial mechanical energy dissipated into heat, which is a form of radiation. The time-reversed solution, where radiation is focused on the pieces of glass scattered on the floor, in such a precise way that they jump back on the table and reassemble into a glass, is a mathematically legitimate solution but, physically, it is meaningless.

We therefore write the physically relevant solution for A^μ as

$$A^\mu(t, \mathbf{x}) = A^\mu_{\text{in}}(t, \mathbf{x}) + \frac{\mu_0}{4\pi} \int d^3 x' \, \frac{j^\mu(t - |\mathbf{x} - \mathbf{x}'|/c, \mathbf{x}')}{|\mathbf{x} - \mathbf{x}'|}, \qquad (10.34)$$

where $A^\mu_{\text{in}}(t, \mathbf{x})$ is a solution of the homogeneous equation describing the incoming field. If we are only interested in the field generated by the source, we can simply set $A^\mu_{\text{in}}(t, \mathbf{x}) = 0$.[7]

Observe that eq. (10.11) is explicitly Lorentz invariant, because both the $\Box$ operator and $\delta^{(4)}(x)$ are Lorentz invariant.[8] In the form (10.24) it is not evident how G^+ and G^- behave under Lorentz transformations, but in fact they are invariant. This can be seen using the property (8.7) of the delta function, that implies that

$$\begin{aligned}\delta(x^2) &= \delta\left[(x^0)^2 - |\mathbf{x}|^2\right] \\ &= \frac{1}{2|\mathbf{x}|}\left[\delta(x^0 - |\mathbf{x}|) + \delta(x^0 + |\mathbf{x}|)\right]. \end{aligned} \qquad (10.36)$$

The expression in the last line is not yet the combination that appears in the retarded or advanced Green's functions. However, we can multiply this by a theta function, defined in eq. (1.67), to obtain[9]

$$\theta(x^0)\delta(x^2) = \frac{1}{2|\mathbf{x}|}\delta(x^0 - |\mathbf{x}|), \qquad (10.38)$$

$$\theta(-x^0)\delta(x^2) = \frac{1}{2|\mathbf{x}|}\delta(x^0 + |\mathbf{x}|). \qquad (10.39)$$

In general, the sign of x^0 is not invariant under Lorentz transformations. However, if $x^0 > 0$ and $x^2 = 0$, the event $(x^0, \mathbf{x})$ is on the light-cone of the event $(x^0 = 0, \mathbf{x} = 0)$. Then, x^0 will remain positive under any Lorentz boost (since the velocity v_0 of the boost is always restricted to be strictly smaller than c). This can be seen from the Lorentz transformation (7.22). Setting for definiteness $\mathbf{x} = (x, 0, 0)$, the condition $(x^0)^2 - \mathbf{x}^2 = 0$ gives $x = \pm x^0$ and, under a Lorentz boost, $x^0 \to x'^0 = \gamma(x^0 + \beta x)$. Since $|x| = x^0$ and $|\beta| < 1$, x'^0 remains positive. We can also see it graphically from Fig. 7.1 on page 161, where the events in the future of the boosted observer are given by the part of the (x^0, x) plane that lies above the line $t' = 0$, and contains all points with $x^0 > 0$ which lie on the light cone of the observer at the origin.

Therefore, the combinations $\theta(x^0)\delta(x^2)$ and, similarly, $\theta(-x^0)\delta(x^2)$, are explicitly Lorentz invariant. In terms of them, the advanced and retarded Green's functions (10.23) can be written as

$$G^\pm(x) = -\frac{1}{2\pi}\theta(\pm x^0)\delta(x^2), \qquad (10.40)$$

or, reinstating the second argument,

$$G^\pm(x; x') = -\frac{1}{2\pi}\theta[\pm(x^0 - x'^0)]\,\delta[(x - x')^2]. \qquad (10.41)$$

[7]As a byproduct of this analysis observe, from eq. (10.33), that

$$A^\mu_{\text{rad}}(t, \mathbf{x}) = A^\mu_{\text{out}}(t, \mathbf{x}) - A^\mu_{\text{in}}(t, \mathbf{x}), \tag{10.35}$$

is the difference between the outgoing and the incoming field, and therefore can be interpreted as the radiation generated by the system. We see, from eq. (10.31), that it is obtained from the combination $G_A = G_{\text{ret}} - G_{\text{adv}}$. We will find this combination again in Section 12.3.5, when we will discuss radiation reaction.

[8]As we have already seen in eqs. (8.4) and (8.5), the invariance of $\delta^{(4)}(x)$ follows from the fact that $\int d^4x\, \delta^{(4)}(x) = 1$, together with the fact that, under a Lorentz transformation $x^\mu \to \Lambda^\mu{}_\nu x^\nu$, $d^4x \to (\det \Lambda)\, d^4x$ and $\det \Lambda = 1$, so d^4x is Lorentz invariant.

[9]We are using the fact that, for $|\mathbf{x}| \neq 0$, $\delta(x^0 - |\mathbf{x}|)$ has its support at x^0 strictly positive, where $\theta(x^0) = 1$, so

$$\theta(x^0)\delta(x^0 - |\mathbf{x}|) = \delta(x^0 - |\mathbf{x}|), \tag{10.37}$$

and similarly $\theta(x^0)\delta(x^0 + |\mathbf{x}|) = 0$. All these relations, however, become ill-defined at $|\mathbf{x}| = 0$ or at $x^0 = 0$, even in the sense of distributions, and can give ambiguous results. For instance, when $\mathbf{x} = 0$, multiplying both sides of eq. (10.37) by a regular function $f(x^0)$ and integrating, the left-hand side gives

$$\int_{-\infty}^{\infty} dx^0\, \theta(x^0)\delta(x^0)f(x^0)$$
$$= \int_0^{\infty} dx^0\, \delta(x^0)f(x^0),$$

and this is an ill-defined integral, since $x^0 = 0$ is at border of the integration domain, and the result depends on the sequence of functions chosen to define the Dirac delta. In contrast, on the right-hand side we get $\int_{-\infty}^{\infty} dx^0\, \delta(x^0)f(x^0)$, which is always well-defined and equal to $f(0)$. Therefore, care must be taken in some situation when using the explicitly covariant form of the Green's functions. In Section 12.3.5 we will provide a careful treatment of these covariant expressions and of their derivatives.

10.2   The Liénard–Wiechert potentials

We now compute the fields generated by a point charge q moving with arbitrary velocity. The corresponding charge and current densities are given by eqs. (8.1) and (8.2) or, in covariant notation, by eq. (8.3). We perform first the computation for the potential $A^0 = \phi/c$. Plugging eq. (8.1) into eq. (10.25) and using the sign corresponding to retarded Green's function, we get

$$\phi(t, \mathbf{x}) = \frac{q}{4\pi\epsilon_0} \int dt' d^3 x' \frac{\delta^{(3)}[\mathbf{x}' - \mathbf{x}_0(t')]}{|\mathbf{x} - \mathbf{x}'|} \delta\left(t' - t + \frac{|\mathbf{x} - \mathbf{x}'|}{c}\right).$$

$$(10.42)$$

We denote the trajectory of the particle by $\mathbf{x}_0(t)$, and its velocity by

$$\mathbf{v}(t) = \frac{d\mathbf{x}_0(t)}{dt}.$$

$$(10.43)$$

We are interested in the field generated by the charge itself, so we have set to zero the solution of the homogeneous equation. Rather than performing first the integral over dt' to reach the form (10.26), for a point charge it is convenient to carry out first the integral over $d^3 x'$ with the help of $\delta^{(3)}[\mathbf{x}' - \mathbf{x}_0(t')]$. This gives

$$\phi(t, \mathbf{x}) = \frac{q}{4\pi\epsilon_0} \int dt' \frac{1}{|\mathbf{x} - \mathbf{x}_0(t')|} \delta\left(t' - t + \frac{|\mathbf{x} - \mathbf{x}_0(t')|}{c}\right).$$

$$(10.44)$$

We now define *retarded time* t_{ret} as the solution of the equation

$$t_{\text{ret}} + \frac{|\mathbf{x} - \mathbf{x}_0(t_{\text{ret}})|}{c} = t,$$

$$(10.45)$$

so the Dirac delta in eq. (10.44) is satisfied for $t' = t_{\text{ret}}$. Note that eq. (10.45) is an implicit definition of t_{ret} as a function of t and $\mathbf{x}$,

$$t_{\text{ret}} = t_{\text{ret}}(t, \mathbf{x}).$$

$$(10.46)$$

Retarded time has a clear physical meaning. If we imagine that the charge continuously emits light signals toward an observer in $\mathbf{x}$, then the signal that reaches the observer at time t was emitted by the charge at an earlier time t_{ret}, when the charge was at the position $\mathbf{x}_0(t_{\text{ret}})$, such that the observation time t is equal to the emission time t_{ret} plus the time $|\mathbf{x} - \mathbf{x}_0(t_{\text{ret}})|/c$ taken by the light signal to reach $\mathbf{x}$ from the position $\mathbf{x}_0(t_{\text{ret}})$. Also note that, for the observer in $\mathbf{x}$, $\mathbf{x}_0(t_{\text{ret}})$ is the apparent position of the charge at time t. It is useful to define

$$\mathbf{R}(t, \mathbf{x}) = \mathbf{x} - \mathbf{x}_0(t),$$

$$(10.47)$$

and $R = |\mathbf{R}|$, so that eq. (10.45) reads

$$t_{\text{ret}} + \frac{1}{c} R(t_{\text{ret}}, \mathbf{x}) = t.$$

$$(10.48)$$

To compute the integral over t' in eq. (10.44) we define

$$
\begin{aligned}
f(t') &\equiv t' - t + \frac{|\mathbf{x} - \mathbf{x}_0(t')|}{c} \\
&= t' - t + \frac{R(t')}{c},
\end{aligned}
\tag{10.49}
$$

[where, for notational simplicity, in the intermediate steps we omit the argument $\mathbf{x}$ in $R(t', \mathbf{x})$], and we observe that the equation $f(t') = 0$ has just a single solution at $t' = t_{\rm ret}$.[10] Therefore

$$
\delta[f(t')] = \frac{1}{|df/dt'|}\delta(t' - t_{\rm ret}).
\tag{10.50}
$$

From eq. (10.49),

$$
\frac{df}{dt'} = 1 + \frac{1}{c}\frac{dR(t')}{dt'}.
\tag{10.51}
$$

We now use[11]

$$
\frac{dR(t')}{dt'} = -\frac{1}{R(t')}\mathbf{v}(t')\cdot\mathbf{R}(t').
\tag{10.52}
$$

Then eq. (10.44) gives (reinstating the argument $\mathbf{x}$ in R)

$$
\begin{aligned}
\phi(t, \mathbf{x}) &= \frac{q}{4\pi\epsilon_0}\int dt' \frac{1}{R(t', \mathbf{x})}\frac{1}{1 - \frac{1}{R(t',\mathbf{x})}\frac{\mathbf{v}(t')}{c}\cdot\mathbf{R}(t',\mathbf{x})}\delta(t' - t_{\rm ret}) \\
&= \frac{q}{4\pi\epsilon_0}\frac{1}{\left[R(t', \mathbf{x}) - \frac{\mathbf{v}(t')}{c}\cdot\mathbf{R}(t',\mathbf{x})\right]_{|t'=t_{\rm ret}(t,\mathbf{x})}}.
\end{aligned}
\tag{10.53}
$$

From eqs. (8.2) and (10.25), exactly the same computation gives $\mathbf{A}$, with $q\mathbf{v}(t_{\rm ret})$ instead of q at the numerator, and μ_0 instead of $1/\epsilon_0$ in front.

In conclusion, the potentials generated by a charge on an arbitrary trajectory $\mathbf{x}_0(t)$ and velocity $\mathbf{v}(t) = d\mathbf{x}_0/dt$, are given by

$$
\boxed{\phi(t, \mathbf{x}) = \frac{1}{4\pi\epsilon_0}\left(\frac{q}{R - \frac{\mathbf{v}}{c}\cdot\mathbf{R}}\right)_{\rm ret}},
\tag{10.54}
$$

and

$$
\boxed{\mathbf{A}(t, \mathbf{x}) = \frac{\mu_0}{4\pi}\left(\frac{q\mathbf{v}}{R - \frac{\mathbf{v}}{c}\cdot\mathbf{R}}\right)_{\rm ret}},
\tag{10.55}
$$

where the subscript "ret" in these expressions indicates that, to get the potentials at time t and position $\mathbf{x}$, the right-hand sides must be evaluated at the retarded time $t_{\rm ret}(t, \mathbf{x})$, defined by eq. (10.45), i.e., in eqs. (10.54) and (10.55), $R = R(t_{\rm ret}(t, \mathbf{x}), \mathbf{x})$. These are the *Liénard–Wiechert potentials*.[12] Also note that the velocity that enters in eq. (10.53) is

$$
\begin{aligned}
\mathbf{v}(t')_{|t'=t_{\rm ret}(t,\mathbf{x})} &= \frac{d\mathbf{x}_0(t')}{dt'}\Big|_{t'=t_{\rm ret}(t,\mathbf{x})} \\
&= \frac{d\mathbf{x}_0(t_{\rm ret})}{dt_{\rm ret}},
\end{aligned}
\tag{10.56}
$$

[10] The fact that $t' = t_{\rm ret}$ is a solution of $f(t') = 0$ follows from the definition (10.45) of $t_{\rm ret}$. This solution is unique because, for a physically acceptable trajectory $\mathbf{x}_0(t)$, such that $|d\mathbf{x}_0/dt| < c$, at fixed $\mathbf{x}$ retarded time $t_{\rm ret}(t, \mathbf{x})$ is a monotonically increasing function of t.

[11] The explicit computation goes as follows:

$$
\begin{aligned}
\frac{dR(t')}{dt'} &= \frac{1}{2R(t')}\frac{d}{dt'}R^2(t') \\
&= \frac{1}{2R(t')}\frac{d}{dt'}\left[\mathbf{x}^2 + \mathbf{x}_0^2(t') - 2\mathbf{x}\cdot\mathbf{x}_0(t')\right] \\
&= \frac{1}{2R(t')}\left[2\mathbf{x}_0(t')\cdot\mathbf{v}(t') - 2\mathbf{x}\cdot\mathbf{v}(t')\right] \\
&= -\frac{1}{R(t')}\mathbf{v}(t')\cdot\mathbf{R}(t').
\end{aligned}
$$

[12] Found by Alfred-Marie Liénard in 1898 and, with an independent method, by Emil Wiechert in 1900.

i.e., is the velocity computed with respect to the natural time variable, $t_{\rm ret}$, of an observer located at the position of the charge, rather than with respect to the time t of the distant observer that measures the field produced.

It can be useful to define also the quantity $\mathbf{R}_a(t, \mathbf{x})$ from

$$\mathbf{R}_a(t, \mathbf{x}) = \mathbf{x} - \mathbf{x}_0[t_{\rm ret}(t, \mathbf{x})], \tag{10.57}$$

so that, comparing with eq. (10.47),

$$\mathbf{R}_a(t, \mathbf{x}) = \mathbf{R}(t_{\rm ret}(t, \mathbf{x}), \mathbf{x}). \tag{10.58}$$

Note that $-\mathbf{R}_a = \mathbf{x}_0(t_{\rm ret}) - \mathbf{x}$ is the *apparent* position of the source, with respect to an observer in $\mathbf{x}$ (the subscript "a" in $\mathbf{R}_a$ indeed stands for "apparent"). It depends on $\mathbf{x}$ both explicitly, and through the dependence of $t_{\rm ret}$ on $\mathbf{x}$. We also define the retarded velocity $\mathbf{v}_r(t, \mathbf{x})$ as

$$\begin{aligned}
\mathbf{v}_r(t, \mathbf{x}) &= \left. \frac{d\mathbf{x}_0(t')}{dt'} \right|_{t' = t_{\rm ret}(t, \mathbf{x})} \\
&= \frac{d\mathbf{x}_0(t_{\rm ret})}{dt_{\rm ret}}.
\end{aligned} \tag{10.59}$$

Therefore, eqs. (10.54) and (10.55) can be written as

$$\boxed{\phi(t, \mathbf{x}) = \frac{1}{4\pi\epsilon_0} \frac{q}{R_a(t, \mathbf{x}) - \mathbf{v}_r(t, \mathbf{x}) \cdot \mathbf{R}_a(t, \mathbf{x})/c},} \tag{10.60}$$

and

$$\boxed{\mathbf{A}(t, \mathbf{x}) = \frac{\mu_0}{4\pi} \frac{q\mathbf{v}_r(t, \mathbf{x})}{R_a(t, \mathbf{x}) - \mathbf{v}_r(t, \mathbf{x}) \cdot \mathbf{R}_a(t, \mathbf{x})/c}.} \tag{10.61}$$

Observe that, in the Liénard–Wiechert potentials, the gauge potentials are related by

$$\mathbf{A}(t, \mathbf{x}) = \frac{\mathbf{v}_r(t, \mathbf{x})}{c^2} \phi(t, \mathbf{x}). \tag{10.62}$$

It can be useful to write $\mathbf{R}_a$ in terms of its modulus R_a and the unit vector $\hat{\mathbf{R}}_a \equiv \mathbf{R}_a/R_a$. Then, in particular, eq. (10.60) reads

$$\phi(t, \mathbf{x}) = \frac{q}{4\pi\epsilon_0} \frac{1}{R_a(t, \mathbf{x})} \left[\frac{1}{1 - \mathbf{v}_r(t, \mathbf{x}) \cdot \hat{\mathbf{R}}_a(t, \mathbf{x})/c} \right]. \tag{10.63}$$

If the terms in brackets were not there, this would just be a "retarded Coulomb potential," in which the instantaneous position of the source is replaced by its retarded position. The emergence of the extra velocity-dependent term in brackets, that came out from the computation, can also be understood considering the simpler case of a particle moving with constant speed, and performing a Lorentz boost of the Coulomb potential, as we discuss in Section 10.3.

10.3 Fields of charge in uniform motion

In the simple case of a charge moving with constant speed, the general formalism based on the Liénard–Wiechert potentials is not really necessary. To compute the field generated by this source it is much simpler to observe that the result can be obtained by performing a Lorentz boost from the inertial frame where the charge is at rest to the inertial frame where it has speed $\mathbf{v} = v\hat{\mathbf{x}}$. Consider first a frame K_s (the "source" frame) where the charge is at rest at the origin, and denote the coordinates in this frame by $(t_s, \mathbf{x}_s)$. The potentials ϕ_s and $\mathbf{A}_s$ in this frame are given by

$$\phi_s(t_s, \mathbf{x}_s) = \frac{1}{4\pi\epsilon_0} \frac{q}{\sqrt{x_s^2 + y_s^2 + z_s^2}}, \qquad (10.64)$$

$$\mathbf{A}_s(t_s, \mathbf{x}_s) = 0. \qquad (10.65)$$

Let K be a frame, with coordinates $(t, \mathbf{x})$, such that, at $t = 0$, the charge is at the origin with velocity $\mathbf{v} = v\hat{\mathbf{x}}$. The relation between the coordinates of the two frames are given by eqs. (7.24) and (7.25), that we rewrite here in the notation

$$t = \gamma\left(t_s + \frac{v}{c^2}x_s\right), \qquad (10.66)$$

$$x = \gamma(x_s + vt_s). \qquad (10.67)$$

These relations can be inverted to give

$$t_s = \gamma\left(t - \frac{v}{c^2}x\right), \qquad (10.68)$$

$$x_s = \gamma(x - vt), \qquad (10.69)$$

while $y = y_s$ and $z = z_s$. So, in particular,

$$x_s^2 + y_s^2 + z_s^2 = \gamma^2(x - vt)^2 + y^2 + z^2. \qquad (10.70)$$

Similarly, since $A^0 = \phi/c$ and $\mathbf{A}$ are the components of a four-vector, the potentials transform as

$$\phi = \gamma\left[\phi_s + v(A_x)_s\right], \qquad (10.71)$$

$$A_x = \gamma\left[(A_x)_s + \frac{v}{c^2}\phi_s\right], \qquad (10.72)$$

while $A_y = (A_y)_s$ and $A_z = (A_z)_s$. We use the label "s" for the source frame, and we reserve ϕ and $\mathbf{A}$ for the fields in the observer frame. Note that these are the transformations of the potentials at the same space-time point P, whose coordinates will have different numerical values in different frames. Since $\mathbf{A}_s = 0$, for $\phi(t, \mathbf{x})$ we get

$$\phi(t, \mathbf{x}) = \gamma\phi_s(t_s, \mathbf{x}_s)$$

$$= \frac{1}{4\pi\epsilon_0} \frac{\gamma q}{\sqrt{x_s^2 + y_s^2 + z_s^2}}. \qquad (10.73)$$

We now make use of the fact that $(t, \mathbf{x})$ and $(t_s, \mathbf{x}_s)$ refer to the same space-time point seen in two different Lorentz frame, so are related by eqs. (10.68) and (10.69), and therefore by eq. (10.70). We then obtain

$$\phi(t, \mathbf{x}) = \frac{1}{4\pi\epsilon_0} \frac{\gamma q}{\sqrt{\gamma^2(x - vt)^2 + y^2 + z^2}}. \qquad (10.74)$$

Similarly,

$$
\begin{aligned}
A_x(t, \mathbf{x}) &= \gamma \frac{v}{c^2} \phi_s \\
&= \frac{1}{4\pi\epsilon_0 c^2} \frac{q\gamma v}{\sqrt{x_s^2 + y_s^2 + z_s^2}} \\
&= \frac{\mu_0}{4\pi} \frac{q\gamma v}{\sqrt{x_s^2 + y_s^2 + z_s^2}}, \qquad (10.75)
\end{aligned}
$$

while $A_y = A_z = 0$. Since $\mathbf{v} = v\hat{\mathbf{x}}$, in vector form we have

$$\mathbf{A}(t, \mathbf{x}) = \frac{\mu_0}{4\pi} \frac{q\gamma \mathbf{v}}{\sqrt{\gamma^2(x - vt)^2 + y^2 + z^2}}. \qquad (10.76)$$

It is instructive to rederive these results using the Liénard–Wiechert potentials. We therefore set $\mathbf{x}_0(t) = \mathbf{v}t$, with $\mathbf{v} = v\hat{\mathbf{x}}$, in eqs. (10.54) and (10.55). For a generic trajectory, the main technical difficulty, when applying the general formalism based on the Liénard–Wiechert potentials, is to solve eq. (10.45) to get t_{ret} as a function of t and $\mathbf{x}$. However, for a uniform motion, this can be performed analytically. Equation (10.45) in this case gives

$$
\begin{aligned}
c^2(t - t_{\mathrm{ret}})^2 &= |\mathbf{x} - vt_{\mathrm{ret}}\hat{\mathbf{x}}|^2 \\
&= (x - vt_{\mathrm{ret}})^2 + y^2 + z^2. \qquad (10.77)
\end{aligned}
$$

This is a second-order equation in t_{ret}, or equivalently in $(t - t_{\mathrm{ret}})$, whose solution can be written as[13]

$$c(t - t_{\mathrm{ret}}) = \gamma^2 \frac{v}{c}(x - vt) + \gamma\sqrt{\gamma^2(x - vt)^2 + y^2 + z^2}. \qquad (10.78)$$

Note that, from eq. (10.48), $R(t_{\mathrm{ret}}) = c(t - t_{\mathrm{ret}})$ and therefore, for uniform motion,

$$R(t_{\mathrm{ret}}) = \gamma^2 \frac{v}{c}(x - vt) + \gamma\sqrt{\gamma^2(x - vt)^2 + y^2 + z^2}, \qquad (10.79)$$

while[14]

$$\frac{\mathbf{v}}{c} \cdot \mathbf{R}(t_{\mathrm{ret}}) = \frac{v}{c}(x - vt) + \frac{v^2}{c^2} R(t_{\mathrm{ret}}). \qquad (10.80)$$

Therefore

$$
\begin{aligned}
R(t_{\mathrm{ret}}) - \frac{\mathbf{v}}{c} \cdot \mathbf{R}(t_{\mathrm{ret}}) &= \left(1 - \frac{v^2}{c^2}\right) R(t_{\mathrm{ret}}) - \frac{v}{c}(x - vt) \\
&= \frac{1}{\gamma^2} R(t_{\mathrm{ret}}) - \frac{v}{c}(x - vt) \\
&= \frac{1}{\gamma}\sqrt{\gamma^2(x - vt)^2 + y^2 + z^2}, \qquad (10.81)
\end{aligned}
$$

---

[13]The plus sign in front of the square root is fixed by the fact that, for $v = 0$, this expression reduces to $t - t_{\mathrm{ret}} = +r/c$.

[14]Explicitly,

$$
\begin{aligned}
\frac{\mathbf{v}}{c} \cdot \mathbf{R}(t_{\mathrm{ret}}) &= \frac{v}{c}\hat{\mathbf{x}} \cdot \mathbf{R}(t_{\mathrm{ret}}) \\
&= \frac{v}{c}[x - x_0(t_{\mathrm{ret}})] \\
&= \frac{v}{c}(x - vt_{\mathrm{ret}}) \\
&= \frac{v}{c}[(x - vt) + v(t - t_{\mathrm{ret}})] \\
&= \frac{v}{c}(x - vt) + \frac{v^2}{c^2} R(t_{\mathrm{ret}}).
\end{aligned}
$$

where, in the second line, we used eq. (10.79). Then, eq. (10.54) gives

$$\phi(t, \mathbf{x}) = \frac{1}{4\pi\epsilon_0} \frac{\gamma q}{\sqrt{\gamma^2(x - vt)^2 + y^2 + z^2}}, \tag{10.82}$$

and eq. (10.55) gives

$$\mathbf{A}(t, \mathbf{x}) = \frac{\mu_0}{4\pi} \frac{q\gamma \mathbf{v}}{\sqrt{\gamma^2(x - vt)^2 + y^2 + z^2}}. \tag{10.83}$$

We have therefore recovered eqs. (10.74) and (10.76).

We can now compute the corresponding expressions for the electric and magnetic fields. Using eq. (3.83), we get

$$\mathbf{E}(t, \mathbf{x}) = \frac{q}{4\pi\epsilon_0} \gamma \frac{\mathbf{x} - \mathbf{x}_0(t)}{\{\gamma^2[x - x_0(t)]^2 + y^2 + z^2\}^{3/2}}, \tag{10.84}$$

where $\mathbf{x}_0(t) = (vt, 0, 0)$ is the position of the charge at time t. We can rewrite the result using $\mathbf{R}(t, \mathbf{x}) = \mathbf{x} - \mathbf{x}_0(t)$, see eq. (10.47), which, for the case of uniform motion, becomes $\mathbf{R}(t, \mathbf{x}) = \mathbf{x} - \mathbf{v}t$. We also define θ as the angle between $\hat{\mathbf{R}}$ and $\hat{\mathbf{v}}$, so that

$$x - x_0(t) = R(t, \mathbf{x}) \cos\theta, \tag{10.85}$$

and

$$y^2 + z^2 = R^2(t, \mathbf{x}) \sin^2\theta. \tag{10.86}$$

Then eq. (10.84) can be rewritten as

$$\mathbf{E}(t, \mathbf{x}) = \frac{q}{4\pi\epsilon_0} \frac{\gamma}{[1 + (\gamma^2 - 1)\cos^2\theta]^{3/2}} \frac{\hat{\mathbf{R}}(t, \mathbf{x})}{R^2(t, \mathbf{x})}, \tag{10.87}$$

or, equivalently, writing $\cos^2\theta = 1 - \sin^2\theta$ and using the explicit expression of γ in terms of v, as

$$\boxed{\mathbf{E}(t, \mathbf{x}) = \frac{q}{4\pi\epsilon_0} \frac{1 - v^2/c^2}{[1 - (v^2/c^2)\sin^2\theta]^{3/2}} \frac{\hat{\mathbf{R}}(t, \mathbf{x})}{R^2(t, \mathbf{x})}.} \tag{10.88}$$

Similarly, using eq. (3.80), we get

$$\boxed{\mathbf{B}(t, \mathbf{x}) = \frac{1}{c^2} \mathbf{v} \times \mathbf{E}(t, \mathbf{x}).} \tag{10.89}$$

Several aspects of this result are noteworthy:

(1) The electric field is directed radially with respect to the instantaneous position of the charge, i.e., is in the direction of $\mathbf{R} = \mathbf{x} - \mathbf{x}_0(t)$. The fact that the field at time t depends on the *instantaneous* position $\mathbf{x}_0(t)$ of the charge at the same time t, rather than on the position at retarded time, is not a sign of action at a distance. Simply, having specified that the motion is at constant speed for all times, the future position of the charge can be perfectly predicted or, in other words, the position of the charge at retarded time $t_{\rm ret}$ determines the position of the charge at the subsequent time t.

(2) The modulus of the electric field is not spherically symmetric: in particular, if, in eq. (10.87) we set $\theta = 0$, i.e., $y = z = 0$ and $x - x_0(t) = R$, the modulus of the electric field becomes

$$E = \frac{q}{4\pi\epsilon_0} \frac{1}{\gamma^2 R^2}, \qquad (\theta = 0), \qquad (10.90)$$

so the field in the direction of motion is reduced by a factor $1/\gamma^2$ compared to the Coulomb field of a charge at rest. On the other hand, if we set $\theta = \pi/2$, i.e., $y^2 + z^2 = R^2$ and $x - x_0(t) = 0$, we get

$$E = \frac{q}{4\pi\epsilon_0} \frac{\gamma}{R^2}, \qquad \left(\theta = \frac{\pi}{2}\right), \qquad (10.91)$$

so the field at a right angle with respect to the direction of motion is enhanced by a factor γ compared to the Coulomb field.

(3) The lines of $\mathbf{B}$ circulate around the direction of motion, as for a steady current. The modulus of $\mathbf{B}$ is stronger at a right angle to the motion and is reduced as we approach the direction of motion, both because of the behavior of E, and because of the $\sin\theta$ factor coming from the vector product $\mathbf{v} \times \mathbf{E}$,

$$\begin{aligned} B &= \frac{1}{c^2} vE\sin\theta \\ &= \frac{\mu_0}{4\pi} \frac{\gamma qv}{R^2} \frac{\sin\theta}{[1 + (\gamma^2 - 1)\cos^2\theta]^{3/2}}, \end{aligned} \qquad (10.92)$$

or, equivalently,

$$B = \frac{\mu_0}{4\pi} \frac{qv}{R^2} \frac{(1 - v^2/c^2)\sin\theta}{[1 - (v^2/c^2)\sin^2\theta]^{3/2}}. \qquad (10.93)$$

In particular, B vanishes at $\theta = 0$, i.e., on the direction of motion.

(4) In the limit of an ultra-relativistic charge, $\gamma \gg 1$, the electric and magnetic fields are concentrated within a small angle $\delta\theta$ around $\theta = \pi/2$. Indeed, the limit $\gamma \gg 1$ at fixed θ has two regimes. If θ is fixed and different from $\pm\pi/2$, so that $\cos\theta \neq 0$, eventually in the limit $\gamma \to \infty$ also $\gamma^2\cos^2\theta \gg 1$, and

$$\frac{\gamma}{[1 + (\gamma^2 - 1)\cos^2\theta]^{3/2}} = \mathcal{O}\left(\frac{1}{\gamma^2}\right). \qquad (10.94)$$

Correspondingly, for the modulus of the electric field we have

$$E = \frac{1}{4\pi\epsilon_0} \frac{q}{R^2} \times \mathcal{O}\left(\frac{1}{\gamma^2}\right), \qquad (10.95)$$

which is smaller than the Coulomb potential of a non-relativistic particle by a factor γ^2. Consider, however, the situation in which we send at the same time $\gamma \to \infty$ and $\theta \to \pi/2$. Writing $\theta = (\pi/2) + \delta\theta$, to first order we have $\cos\theta \simeq -\delta\theta$, and, for $\gamma \to \infty$ and $\delta\theta \to 0$,

$$\frac{\gamma}{[1 + (\gamma^2 - 1)\cos^2\theta]^{3/2}} \simeq \frac{\gamma}{[1 + (\gamma\delta\theta)^2]^{3/2}}. \qquad (10.96)$$

If the limit $\gamma \to \infty$ and $\delta\theta \to 0$ is taken so that $\gamma\delta\theta$ stays finite, say $\gamma\delta\theta \lesssim \mathcal{O}(1)$, then, rather than reproducing the behavior (10.94), this expression grows as γ, and

$$E = \frac{1}{4\pi\epsilon_0} \frac{q}{R^2} \times \mathcal{O}(\gamma) \,. \tag{10.97}$$

This is larger by a factor $\mathcal{O}(\gamma)$ compared to the Coulomb potential of a non-relativistic particle. Therefore, there are two regimes, separated by the region $\gamma\delta\theta = \mathcal{O}(1)$. For $\gamma\delta\theta \gg 1$ we are in the regime (10.95) and E is very small (compared to the Coulomb potential of a static charge), while, for $\gamma\delta\theta \lesssim \mathcal{O}(1)$, E is very large. The electric field of an ultra-relativistic particle is therefore focused in a small region, with opening angle $\delta\theta \sim 1/\gamma$, in the plane transverse to the velocity of the charge.

10.4 Radiation field from accelerated charges

The fields generated by a charge in arbitrary motion can now in principle be obtained inserting eqs. (10.60) and (10.61) into the expression of $\mathbf{E}$ and $\mathbf{B}$ in terms of the gauge potentials, eqs. (3.83) and (3.80). The computation, however, still involves some subtleties; the main point, when taking the derivatives with respect to t and $\mathbf{x}$ of the gauge potentials, is to correctly account for the dependence on t and on $\mathbf{x}$ that enters implicitly through $t_{\rm ret}(t, \mathbf{x})$. To this purpose, we first compute $\partial t_{\rm ret}/\partial t$ (at constant $\mathbf{x}$) by differentiating eq. (10.48). We use $R_a(t, \mathbf{x})$ defined in eq. (10.57), so that $\mathbf{R}(t_{\rm ret}, \mathbf{x}) = \mathbf{R}_a(t, \mathbf{x})$. Then

$$\frac{\partial t_{\rm ret}}{\partial t} + \frac{1}{c}\frac{\partial}{\partial t} R_a(t, \mathbf{x}) = 1 \,. \tag{10.98}$$

We next observe that[15]

$$\frac{\partial}{\partial t} R_a(t, \mathbf{x}) = -\hat{\mathbf{R}}_a(t, \mathbf{x})\cdot\mathbf{v}_r(t) \frac{\partial t_{\rm ret}}{\partial t} \,. \tag{10.99}$$

Inserting this into eq. (10.98) and solving for $\partial t_{\rm ret}/\partial t$ we get

$$\frac{\partial t_{\rm ret}}{\partial t} = \frac{1}{1 - \hat{\mathbf{R}}_a(t, \mathbf{x})\cdot\mathbf{v}_r(t)/c} \,. \tag{10.100}$$

We proceed in the same way to compute $\partial_i t_{\rm ret}(t, \mathbf{x})$. Differentiating eq. (10.48) with respect to ∂_i, at constant t, we get[16]

$$-c\partial_i t_{\rm ret} = (\hat{\mathbf{R}}_a)_i - \hat{\mathbf{R}}_a\cdot\mathbf{v}_r(t)\partial_i t_{\rm ret} \,, \tag{10.101}$$

and therefore

$$\boldsymbol{\nabla} t_{\rm ret} = -\frac{1}{c}\frac{\hat{\mathbf{R}}_a}{1 - \hat{\mathbf{R}}_a\cdot\mathbf{v}_r(t)/c} \,. \tag{10.102}$$

[15] Explicitly,

$$\frac{\partial}{\partial t} R_a(t, \mathbf{x}) = \frac{1}{2R_a(t, \mathbf{x})}\frac{\partial}{\partial t} R_a^2(t, \mathbf{x})$$

$$= \frac{1}{2R_a(t, \mathbf{x})}\frac{\partial}{\partial t}|\mathbf{x} - \mathbf{x}_0(t_{\rm ret})|^2$$

$$= \frac{1}{2R_a(t, \mathbf{x})}\frac{\partial}{\partial t}$$
$$\times [\mathbf{x}^2 + \mathbf{x}_0^2(t_{\rm ret}) - 2\mathbf{x}\cdot\mathbf{x}_0(t_{\rm ret})]$$

$$= -\frac{\mathbf{x} - \mathbf{x}_0(t_{\rm ret})}{R_a(t, \mathbf{x})}\cdot\frac{d\mathbf{x}_0(t_{\rm ret})}{dt}$$

$$= -\hat{\mathbf{R}}_a(t, \mathbf{x})\cdot\frac{d\mathbf{x}_0(t_{\rm ret})}{dt_{\rm ret}}\frac{\partial t_{\rm ret}}{\partial t}$$

$$= -\hat{\mathbf{R}}_a(t, \mathbf{x})\cdot\left(\frac{d\mathbf{x}_0(t')}{dt'}\right)_{t'=t_{\rm ret}}\frac{\partial t_{\rm ret}}{\partial t}$$

$$= -\hat{\mathbf{R}}_a(t, \mathbf{x})\cdot\mathbf{v}_r(t)\frac{\partial t_{\rm ret}}{\partial t} \,,$$

where, in the last line, we used the definition (10.59) of $\mathbf{v}_r(t)$.

[16] This is obtained similarly, writing

$$-c\partial_i t_{\rm ret} = \partial_i R_a$$

$$= \frac{1}{2R_a(t, \mathbf{x})}\partial_i |\mathbf{x} - \mathbf{x}_0[t_{\rm ret}(t, \mathbf{x})]|^2$$

$$= \frac{1}{2R_a(t, \mathbf{x})}\partial_i\{\mathbf{x}^2 + \mathbf{x}_0^2[t_{\rm ret}(t, \mathbf{x})]$$
$$-2\mathbf{x}\cdot\mathbf{x}_0[t_{\rm ret}(t, \mathbf{x})]\}$$

$$= \frac{1}{2R_a(t, \mathbf{x})}\left\{2x_i + \frac{d\mathbf{x}_0^2(t_{\rm ret})}{dt_{\rm ret}}\partial_i t_{\rm ret}\right.$$
$$\left.-2[\mathbf{x}_0(t_{\rm ret})]_i - 2\mathbf{x}\cdot\frac{d\mathbf{x}_0(t_{\rm ret})}{dt_{\rm ret}}\partial_i t_{\rm ret}\right\}$$

$$= (\hat{\mathbf{R}}_a)_i + \frac{[\mathbf{x}_0(t_{\rm ret}) - \mathbf{x}]\cdot\mathbf{v}_r(t)}{R_a(t, \mathbf{x})}\partial_i t_{\rm ret}$$

$$= (\hat{\mathbf{R}}_a)_i - \hat{\mathbf{R}}_a\cdot\mathbf{v}_r(t)\partial_i t_{\rm ret} \,.$$

Using eqs. (10.100) and (10.102), one can explicitly compute all spatial and temporal derivatives of the quantities $\mathbf{R}_a(t, \mathbf{x})$ and $\mathbf{v}_r(t, \mathbf{x})$ that appear in eqs. (10.60) and (10.61) and obtain $\mathbf{E}$ and $\mathbf{B}$ from eqs. (3.80) and (3.83). The rest of the computation is long, but in principle straightforward. The result for the electric field can be written as

$$\mathbf{E}(t, \mathbf{x}) = \mathbf{E}_v(t, \mathbf{x}) + \mathbf{E}_{\mathrm{rad}}(t, \mathbf{x}) \,. \qquad (10.103)$$

The term $\mathbf{E}_v$ depends on the retarded position and velocity of the charge (the subscript v indeed stands for "velocity"), and is given by

$$\mathbf{E}_v(t, \mathbf{x}) = \frac{1}{4\pi\epsilon_0} \frac{q}{R_a^2} \frac{\hat{\mathbf{R}}_a - \mathbf{v}_r/c}{\gamma^2(1 - \hat{\mathbf{R}}_a \cdot \mathbf{v}_r/c)^3} \,, \qquad (10.104)$$

where, for notational simplicity, here and in the following equations, we do not explicitly write that R_a and $\hat{\mathbf{R}}_a$ are actually functions of $(t, \mathbf{x})$, and that $\mathbf{v}_r$ is a function of t.

The term $\mathbf{E}_{\mathrm{rad}}$, in contrast, depends on the retarded position, velocity, *and acceleration* of the charge, and is given by

$$\mathbf{E}_{\mathrm{rad}}(t, \mathbf{x}) = \frac{1}{4\pi\epsilon_0} \frac{q}{R_a} \frac{[\dot{\mathbf{v}}_r \times (\hat{\mathbf{R}}_a - \mathbf{v}_r/c)] \times \hat{\mathbf{R}}_a}{c^2(1 - \hat{\mathbf{R}}_a \cdot \mathbf{v}_r/c)^3} \,, \qquad (10.105)$$

where

$$\dot{\mathbf{v}}_r = \frac{d\mathbf{v}_r(t')}{dt'}\Big|_{t'=t_{\mathrm{ret}}} \qquad (10.106)$$

$$= \frac{d^2\mathbf{x}_0(t_{\mathrm{ret}})}{d^2 t_{\mathrm{ret}}} \,, \qquad (10.107)$$

and, in the second line, we used eq. (10.59). The subscript "rad" in $\mathbf{E}_{\mathrm{rad}}$ stands for "radiation," for reasons that we will discuss below. The result for the magnetic field can be written as

$$\mathbf{B}(t, \mathbf{x}) = \frac{1}{c}\hat{\mathbf{R}}_a \times \mathbf{E}(t, \mathbf{x}) \,. \qquad (10.108)$$

Therefore, it can also be split into two terms,

$$\mathbf{B}(t, \mathbf{x}) = \mathbf{B}_v(t, \mathbf{x}) + \mathbf{B}_{\mathrm{rad}}(t, \mathbf{x}) \,, \qquad (10.109)$$

where

$$\mathbf{B}_v(t, \mathbf{x}) = \frac{1}{c}\hat{\mathbf{R}}_a \times \mathbf{E}_v(t, \mathbf{x}) \qquad (10.110)$$

depends only on retarded position and velocity, while

$$\mathbf{B}_{\mathrm{rad}}(t, \mathbf{x}) = \frac{1}{c}\hat{\mathbf{R}}_a \times \mathbf{E}_{\mathrm{rad}}(t, \mathbf{x}) \qquad (10.111)$$

depends also on the retarded acceleration. Performing explicitly the vector products, and using μ_0 instead of ϵ_0, we get

$$\mathbf{B}_v(t,\mathbf{x}) = \frac{\mu_0}{4\pi} \frac{q}{R_a^2} \frac{\mathbf{v}_r \times \hat{\mathbf{R}}_a}{\gamma^2(1 - \hat{\mathbf{R}}_a \cdot \mathbf{v}_r/c)^3}, \qquad (10.112)$$

and

$$\mathbf{B}_{\mathrm{rad}}(t,\mathbf{x}) = \frac{\mu_0}{4\pi} \frac{q}{R_a} \frac{(\mathbf{v}_r \times \hat{\mathbf{R}}_a)(\dot{\mathbf{v}}_r \cdot \hat{\mathbf{R}}_a) + c(1 - \hat{\mathbf{R}}_a \cdot \mathbf{v}_r/c)\dot{\mathbf{v}}_r \times \hat{\mathbf{R}}_a}{c^2(1 - \hat{\mathbf{R}}_a \cdot \mathbf{v}_r/c)^3}.$$
$$(10.113)$$

Note that the first term in the numerator of eq. (10.113) depends on the component of the acceleration parallel to $\hat{\mathbf{R}}_a$, while the second on the component transverse to $\hat{\mathbf{R}}_a$. For $\mathbf{v}(t) = \mathbf{v}$, constant, $\mathbf{E}_{\mathrm{rad}}$ vanishes, while $\mathbf{E}_v$ reduces to the result given in eq. (10.84).[17]

A crucial difference between the two terms in eq. (10.103) is their behavior at large distances from the source. If $R_a \to \infty$, $\mathbf{E}_v$ decays as $1/R_a^2$, just as the Coulomb field, to which it reduces in the non-relativistic limit. From eq. (10.110), the same $1/R_a^2$ behavior at large R_a then holds for $\mathbf{B}_v$. Therefore, in the absence of acceleration, the Poynting vector (3.34) decays as $1/R_a^4$. The total flux radiated at large distances is given by the right-hand side of eq. (3.35); since $ds = R_a^2 d\Omega$ and $|\mathbf{S}| \sim 1/R_a^4$, the flux at infinity vanishes. This part of the field is therefore known as the *non-radiative* part (or the *induction* part). In contrast, $\mathbf{E}_{\mathrm{rad}}$ and $\mathbf{B}_{\mathrm{rad}}$ decay only as $1/R_a$ at large distances. Their combined contribution to the Poynting vector then goes as $1/R_a^2$, and the flux at infinity is non-vanishing. This means that energy is radiated away at infinity, and this part of the field is called *radiative*. We will compute the radiated power, in a full relativistic setting, in Section 10.6, after having first studied the non-relativistic limit in Section 10.5, and we will then discuss the radiation field in greater detail in Chapter 11.

From eq. (10.111) we see that $\mathbf{B}_{\mathrm{rad}}$ is orthogonal to both $\mathbf{E}_{\mathrm{rad}}$ and $\hat{\mathbf{R}}_a$. Similarly, because of eq. (10.110), $\mathbf{B}_v$ is orthogonal to both $\mathbf{E}_v$ and $\hat{\mathbf{R}}_a$. However, for the radiative field, we see from eq. (10.105) that even $\mathbf{E}_{\mathrm{rad}}$ is orthogonal to $\hat{\mathbf{R}}_a$. Therefore, in the radiative part of the electromagnetic field, $\mathbf{E}$, $\mathbf{B}$, and $\hat{\mathbf{R}}_a$ form an orthogonal system of vectors, and eq. (10.111) then implies

$$c|\mathbf{B}_{\mathrm{rad}}| = |\mathbf{E}_{\mathrm{rad}}|. \qquad (10.115)$$

In contrast, we see from eq. (10.104) that $\mathbf{E}_v$ is not orthogonal to $\hat{\mathbf{R}}_a$ (actually, in the non-relativistic limit, it becomes exactly parallel to $\hat{\mathbf{R}}_a$, and reduces to the radial Coulomb field), and then eq. (10.110) implies that

$$c|\mathbf{B}_v| < |\mathbf{E}_v|, \qquad (10.116)$$

with $\mathbf{B}_v$ vanishing in the limit $\mathbf{v}_r \to 0$, as we see from eq. (10.112).

---

[17]This can be shown using eq. (10.81) to write the denominator in eq. (10.104) in the form given in eq. (10.84) while, for the numerator, we observe that, for constant $\mathbf{v}$,

$$\mathbf{R}_a - R_a \mathbf{v}_r/c$$
$$= \mathbf{x} - \mathbf{x}_0(t_{\mathrm{ret}}) - c(t - t_{\mathrm{ret}})\mathbf{v}/c$$
$$= \mathbf{x} - [\mathbf{x}_0(t_{\mathrm{ret}}) + \mathbf{v}(t - t_{\mathrm{ret}})]$$
$$= \mathbf{x} - \mathbf{x}_0(t), \qquad (10.114)$$

where, for constant $\mathbf{v}$, we have used $R_a(t) = R(t_{\mathrm{ret}}) = c(t - t_{\mathrm{ret}})$ from eq. (10.79).

10.5   Radiation from non-relativistic charges. Larmor formula

Now consider the field generated by a non-relativistic particle that moves in a bounded region of space. In this case, in eq. (10.57), $|\mathbf{x}_0(t_{\text{ret}})| < d$ for some length-scale d so, in the limit $|\mathbf{x}| \to \infty$, also $|\mathbf{R}_a| \to \infty$. Then, from eqs. (10.103)–(10.105), to lowest order in v/c and lowest order in $1/R_a$ the electric field becomes

$$\begin{aligned} \mathbf{E}(t, \mathbf{x}) &\simeq \mathbf{E}_{\text{rad}}(t, \mathbf{x}) \\ &\simeq \frac{1}{4\pi\epsilon_0 c^2} \frac{q}{R_a} [\mathbf{a}(t_{\text{ret}}) \times \hat{\mathbf{R}}_a] \times \hat{\mathbf{R}}_a \,, \end{aligned} \qquad (10.117)$$

where we have written the retarded acceleration $\dot{\mathbf{v}}_r$ as $\mathbf{a}(t_{\text{ret}})$.

We now observe that, given any vector $\mathbf{V}$ and any unit vector $\hat{\mathbf{n}}$, we can always write

$$\begin{aligned} \mathbf{V} &= \hat{\mathbf{n}}(\hat{\mathbf{n}}\cdot\mathbf{V}) + [\mathbf{V} - \hat{\mathbf{n}}(\hat{\mathbf{n}}\cdot\mathbf{V})] \\ &= \mathbf{V}_{\|}(\hat{\mathbf{n}}) + \mathbf{V}_{\perp}(\hat{\mathbf{n}}) \,, \end{aligned} \qquad (10.118)$$

where

$$\begin{aligned} \mathbf{V}_{\|}(\hat{\mathbf{n}}) &\equiv \hat{\mathbf{n}}(\hat{\mathbf{n}}\cdot\mathbf{V}) \,, & (10.119) \\ \mathbf{V}_{\perp}(\hat{\mathbf{n}}) &\equiv \mathbf{V} - \hat{\mathbf{n}}(\hat{\mathbf{n}}\cdot\mathbf{V}) \,. & (10.120) \end{aligned}$$

The vector $\mathbf{V}_{\|}(\hat{\mathbf{n}})$ is the projection of $\mathbf{V}$ in the direction of $\hat{\mathbf{n}}$, while $\mathbf{V}_{\perp}(\hat{\mathbf{n}})$ is transverse to $\hat{\mathbf{n}}$, since

$$\begin{aligned} \hat{\mathbf{n}}\cdot[\mathbf{V} - \hat{\mathbf{n}}(\hat{\mathbf{n}}\cdot\mathbf{V})] &= (\hat{\mathbf{n}}\cdot\mathbf{V}) - (\hat{\mathbf{n}}\cdot\mathbf{V}) \\ &= 0 \,, \end{aligned} \qquad (10.121)$$

Using eq. (1.9) we see that we can also rewrite $\mathbf{V}_{\perp}(\hat{\mathbf{n}})$ in the form

$$\mathbf{V}_{\perp}(\hat{\mathbf{n}}) = -\hat{\mathbf{n}}\times(\hat{\mathbf{n}}\times\mathbf{V}) \,, \qquad (10.122)$$

or, equivalently,

$$\mathbf{V}_{\perp}(\hat{\mathbf{n}}) = -(\mathbf{V}\times\hat{\mathbf{n}})\times\hat{\mathbf{n}} \,. \qquad (10.123)$$

We then decompose $\mathbf{a}(t_{\text{ret}})$ into its parts orthogonal and parallel with respect to $\hat{\mathbf{R}}_a$,

$$\mathbf{a}(t_{\text{ret}}) = \mathbf{a}_{\perp}(t_{\text{ret}}) + \mathbf{a}_{\|}(t_{\text{ret}}) \,. \qquad (10.124)$$

In eq. (10.117) the term $\mathbf{a}_{\|}(t_{\text{ret}})$ gives a vanishing contribution when taking the vector product with $\hat{\mathbf{R}}_a$, while, from eq. (10.123),

$$\left[\mathbf{a}_{\perp}(t_{\text{ret}}) \times \hat{\mathbf{R}}_a\right] \times \hat{\mathbf{R}}_a = -\mathbf{a}_{\perp}(t_{\text{ret}}) \,. \qquad (10.125)$$

Therefore, to lowest order in v/c and lowest order in $1/R_a$,

$$\mathbf{E}(t, \mathbf{x}) \simeq -\frac{1}{4\pi\epsilon_0 c^2} \frac{q}{R_a(t)} \mathbf{a}_{\perp}[t - R_a(t)/c] \,. \qquad (10.126)$$

Note that $q\mathbf{x}_0(t)$ is the dipole moment $\mathbf{d}(t)$ of the charge, so eq. (10.126) can be rewritten as

$$\mathbf{E}(t, \mathbf{x}) \simeq -\frac{1}{4\pi\epsilon_0 c^2} \frac{1}{R_a(t)} \ddot{\mathbf{d}}_\perp[t - R_a(t)/c]. \tag{10.127}$$

We see that, to lowest order in v/c, the radiation field is generated by the dipole moment of the charge.[18]
  Two main features of this result are:

(1) As we already remarked, at large distances the field generated by an accelerated charge decays as $1/R_a$. It is therefore a radiative field, contrary to the Coulomb field of a static charge, or of a charge in uniform motion, that decays as $1/R_a^2$.

(2) For a non-relativistic charge moving in a bounded region, $|\mathbf{x}_0(t)| < d$, the electric field at a point $\mathbf{x}$, with $|\mathbf{x}| \gg d$, is proportional (and opposite) to the component of the acceleration (computed at retarded time) transverse to the line of sight from the observer at $\mathbf{x}$ to the apparent position of the charge. In particular, a charge accelerating in straight line does not emit radiation in that direction. More generally, a non-relativistic charge does not radiate in the direction of its apparent acceleration.

We also observe that, in the limit $|\mathbf{x}| \gg d$,

$$\begin{aligned}
\mathbf{R}_a(t, \mathbf{x}) &= \mathbf{x} - \mathbf{x}_0[t_{\rm ret}(t, \mathbf{x})] \\
&\simeq \mathbf{x} - \mathbf{x}_0(t) \\
&= \mathbf{R}(t, \mathbf{x}),
\end{aligned} \tag{10.128}$$

since the distance $|\mathbf{x} - \mathbf{x}_0(t)|$ is very large compared to d, while $|\mathbf{x}_0(t)|$ and $|\mathbf{x}_0[t_{\rm ret}(t, \mathbf{x})]|$ are both smaller than d. Therefore, in the limit of large distances from the source, to leading order eq. (10.126) is equivalent to

$$\mathbf{E}(t, \mathbf{x}) \simeq -\frac{1}{4\pi\epsilon_0 c^2} \frac{q}{R(t)} \mathbf{a}_\perp[t - R(t)/c], \tag{10.129}$$

or, more explicitly,

$$\mathbf{E}(t, \mathbf{x}) \simeq -\frac{1}{4\pi\epsilon_0 c^2} \frac{q}{|\mathbf{x} - \mathbf{x}_0(t)|} \ddot{\mathbf{x}}_{0\perp} \left(t - \frac{|\mathbf{x} - \mathbf{x}_0(t)|}{c} \right). \tag{10.130}$$

Note that $R(t, \mathbf{x}) = |\mathbf{x} - \mathbf{x}_0(t)|$ is the distance between the point $\mathbf{x}$ where we compute the fields and the instantaneous position of the source.
  Actually, in the limit in which $r \equiv |\mathbf{x}| \gg d$, to lowest order we can simply neglect altogether the term $\mathbf{x}_0(t)$ in $\mathbf{R}(t)$ and replace $\mathbf{R}(t)$ with r, which is the distance of the point $\mathbf{x}$ to a fixed origin, that it is convenient to choose inside the region where the motion of the source is localized. Then, we can write simply

$$\mathbf{E}(t, r) \simeq -\frac{1}{4\pi\epsilon_0 c^2} \frac{q}{r} \mathbf{a}_\perp(t - r/c), \tag{10.131}$$

[18] As we discussed below eq. (6.21), the value of the dipole moment of a system with non-vanishing total charge changes if we shift the origin of the coordinate system: if $\mathbf{x}_0(t) \to \mathbf{x}_0(t) + \mathbf{s}$, with $\mathbf{s}$ a constant vector, then $\mathbf{d} \to \mathbf{d} + q\mathbf{s}$. However, the extra term is constant in time, and does not affect $\ddot{\mathbf{d}}$.

or

$$\mathbf{E}(t, r) \simeq -\frac{1}{4\pi\epsilon_0 c^2}\frac{1}{r}\ddot{\mathbf{d}}_\perp(t - r/c)\,. \qquad (10.132)$$

In terms of μ_0, we can also write this as

$$\mathbf{E}(t, \mathbf{x}) = -\frac{\mu_0}{4\pi}\frac{1}{r}\ddot{\mathbf{d}}_\perp(t - r/c)\,. \qquad (10.133)$$

In Section 11.2 we will see in detail how eq. (10.132) emerges as the lowest-order term in a systematic expansion in v/c, with higher-order terms parametrized by time derivatives of higher and higher multipole moments.

The notation $\mathbf{d}_\perp$ is useful to stress that, physically, only the component of $\mathbf{d}$ transverse to the line of sight contributes. For explicit computations, however, it can be more useful to leave it in the original form (10.117), in terms of a triple vector product. In the approximation in which we replace $\mathbf{R}_a$ by $\mathbf{x} = r\hat{\mathbf{n}}$, from eq. (10.122) we have

$$\mathbf{d}_\perp = -\hat{\mathbf{n}} \times (\hat{\mathbf{n}} \times \mathbf{d})\,. \qquad (10.134)$$

Then, we can rewrite eq. (10.127) as

$$\mathbf{E}(t, \mathbf{x}) \simeq \frac{1}{4\pi\epsilon_0 c^2}\frac{1}{r}\,\hat{\mathbf{n}} \times [\hat{\mathbf{n}} \times \ddot{\mathbf{d}}(t_{\mathrm{ret}})]\,, \qquad (10.135)$$

where, in the same approximation, $t_{\mathrm{ret}} = t - (r/c)$. Another useful form of this result is obtained using the identity (1.9),

$$\mathbf{E}(t, \mathbf{x}) \simeq -\frac{1}{4\pi\epsilon_0 c^2}\frac{1}{r}\left\{\ddot{\mathbf{d}}(t_{\mathrm{ret}}) - [\hat{\mathbf{n}}\cdot\ddot{\mathbf{d}}(t_{\mathrm{ret}})]\hat{\mathbf{n}}\right\}\,. \qquad (10.136)$$

From eq. (10.135) we also see that the modulus of the electric field can be written as

$$|\mathbf{E}| \simeq \frac{1}{4\pi\epsilon_0 c^2}\frac{1}{r}\,|\hat{\mathbf{n}} \times \ddot{\mathbf{d}}(t_{\mathrm{ret}})|\,. \qquad (10.137)$$

For the magnetic field, to leading order we get

$$\mathbf{B}(t, \mathbf{x}) \simeq -\frac{1}{4\pi\epsilon_0 c^3}\frac{1}{r}\,\hat{\mathbf{n}} \times \ddot{\mathbf{d}}(t_{\mathrm{ret}})\,, \qquad (10.138)$$

where we used eq. (10.108) (with $\hat{\mathbf{R}}_a$ replaced by $\hat{\mathbf{n}}$) and eq. (10.135), and we expanded the triple vector product using eq. (1.9). Using μ_0 instead of ϵ_0 in the expression for the magnetic field,

$$\mathbf{B}(t, \mathbf{x}) \simeq -\frac{\mu_0}{4\pi c}\frac{1}{r}\,\hat{\mathbf{n}} \times \ddot{\mathbf{d}}(t_{\mathrm{ret}})\,. \qquad (10.139)$$

We next compute the power radiated. Using eq. (10.108), with $\hat{\mathbf{R}}_a$ replaced by $\hat{\mathbf{n}}$, for the Poynting vector (3.34) we get

$$\mathbf{S} = \frac{1}{\mu_0 c} |\mathbf{E}|^2 \, \hat{\mathbf{n}} \,. \qquad (10.140)$$

In eq. (3.35) we saw that the energy flowing per unit time out of a volume V is given by $\int_{\partial V} d\mathbf{s} \cdot \mathbf{S}$ (note that the energy flowing *out* of the volume is given by this expression with the plus sign so that, when $\int_{\partial V} d\mathbf{s} \cdot \mathbf{S}$ is positive, according to eq. (3.35) the energy inside the volume V decreases). For the volume V we take a sphere of large radius r, so $d\mathbf{s} = r^2 d\Omega \, \hat{\mathbf{n}}$. Since the energy flowing out per unit time is the power P radiated, we see that the power dP radiated at time t through a surface at distance r from the source and within an infinitesimal solid angle $d\Omega$ is

$$
\begin{aligned}
dP(t; \theta, \phi) &= \frac{1}{\mu_0 c} \frac{\mu_0^2}{(4\pi)^2} \frac{1}{r^2} |\hat{\mathbf{n}} \times \ddot{\mathbf{d}}(t_{\text{ret}})|^2 \, r^2 d\Omega \\
&= \frac{\mu_0}{(4\pi)^2 c} |\hat{\mathbf{n}} \times \ddot{\mathbf{d}}(t_{\text{ret}})|^2 d\Omega \,.
\end{aligned}
\qquad (10.141)
$$

Therefore, the angular distribution of the radiated power is

$$\frac{dP(t; \theta, \phi)}{d\Omega} = \frac{\mu_0}{(4\pi)^2 c} |\hat{\mathbf{n}} \times \ddot{\mathbf{d}}(t_{\text{ret}})|^2 \,. \qquad (10.142)$$

At a given time t, we choose polar coordinates with the polar axis in the direction of $\ddot{\mathbf{d}}(t_{\text{ret}})$, so $|\hat{\mathbf{n}} \times \ddot{\mathbf{d}}(t_{\text{ret}})| = |\ddot{\mathbf{d}}(t_{\text{ret}})| \sin \theta$. Then,

$$\frac{dP(t; \theta)}{d\Omega} = \frac{\mu_0}{(4\pi)^2 c} |\ddot{\mathbf{d}}(t_{\text{ret}})|^2 \sin^2 \theta \,. \qquad (10.143)$$

Note that the angular distribution is independent of ϕ, because the setting is invariant under rotations around the direction of $\ddot{\mathbf{d}}(t_{\text{ret}})$. We can rewrite this in terms of ϵ_0, as[19]

$$\boxed{\frac{dP(t; \theta)}{d\Omega} = \frac{1}{4\pi\epsilon_0} \frac{1}{4\pi c^3} |\ddot{\mathbf{d}}(t_{\text{ret}})|^2 \sin^2 \theta \,.} \qquad (10.144)$$

The integration over the solid angle is performed using

$$\int_0^{2\pi} d\phi \int_{-1}^{1} d\cos\theta \, \sin^2 \theta = 2\pi \times \frac{4}{3} \,. \qquad (10.145)$$

We then obtain *Larmor's formula*, either in the form

$$P(t) = \frac{\mu_0}{6\pi c} |\ddot{\mathbf{d}}(t_{\text{ret}})|^2 \,, \qquad (10.146)$$

or, in terms of ϵ_0, as[20,21]

$$\boxed{P(t) = \frac{1}{4\pi\epsilon_0} \frac{2}{3c^3} |\ddot{\mathbf{d}}(t_{\text{ret}})|^2 \,.} \qquad (10.148)$$

[19] We give the most important results both in terms of μ_0 and in terms of ϵ_0. In the latter case, factorizing a factor $4\pi\epsilon_0$ allows us to quickly pass from SI units to Gaussian units, just setting $4\pi\epsilon_0 = 1$, see Appendix A for details.

[20] Observe that this is the power radiated instantaneously at time t, through a sphere of fixed radius r, and has been obtained neglecting altogether $|\mathbf{x}_0(t)|$ with respect to r. Correspondingly, we have approximated $t_{\text{ret}} = t - (r/c)$. If one wants to integrate the power (10.148) over a given period of time, one must make sure that the condition $|\mathbf{x}_0(t)| \ll r$ is valid all along the time period considered; if not, one must go back to eq. (10.127) and take into account the actual time dependence of $R_a(t)$. However, if we are interested in the formal limit $r \to \infty$, the issue does not arise.

[21] An alternative derivation, useful also in the generalization to higher multipoles that we will study in Section 11.2.2, is obtained starting from eq. (10.132) and using the expression (10.120) for the transverse part of a vector. This gives, for the power,

$$
\begin{aligned}
P &= \frac{1}{4\pi\epsilon_0} \frac{1}{4\pi c^3} \int d\Omega \left| \ddot{\mathbf{d}} - \hat{\mathbf{n}}(\ddot{\mathbf{d}} \cdot \hat{\mathbf{n}}) \right|^2 \bigg|_{t_{\text{ret}}} \\
&= \frac{1}{4\pi\epsilon_0 c^3} \int \frac{d\Omega}{4\pi} \left[|\ddot{\mathbf{d}}|^2 - (\ddot{\mathbf{d}} \cdot \hat{\mathbf{n}})^2 \right]_{t_{\text{ret}}} \\
&= \frac{1}{4\pi\epsilon_0 c^3} \left[|\ddot{\mathbf{d}}|^2 - \ddot{d}_i \ddot{d}_j \int \frac{d\Omega}{4\pi} \hat{n}_i \hat{n}_j \right]_{t_{\text{ret}}} \,.
\end{aligned}
\qquad (10.147)
$$

The remaining angular integral can be performed using eq. (6.49), and we get eq. (10.148).

10.6   Power radiated by relativistic sources

10.6.1   Relativistic generalization of Larmor's formula

After having discussed the power radiated in the non-relativistic limit, we now go back to the general results of Section 10.4, and we compute the radiated power in the full relativistic setting. The most straightforward approach consists in taking the electric and magnetic fields in the radiation zone from eqs. (10.105) and (10.108). For the Poynting vector (3.34), at time t and position $\mathbf{x}$, we then find

$$
\begin{aligned}
\mathbf{S}(t, \mathbf{x}) \;&=\; \frac{1}{\mu_0 c} |\mathbf{E}(t, \mathbf{x})|^2\, \hat{\mathbf{R}}_a \\[2mm]
&=\; \epsilon_0 c \left(\frac{q}{4\pi\epsilon_0} \right)^2 \frac{\hat{\mathbf{R}}_a}{R_a^2} \frac{\left| [\dot{\mathbf{v}}_r \times (\hat{\mathbf{R}}_a - \mathbf{v}_r/c)] \times \hat{\mathbf{R}}_a \right|^2}{c^4 (1 - \hat{\mathbf{R}}_a \cdot \mathbf{v}_r/c)^6}\,, \quad (10.149)
\end{aligned}
$$

where, to keep the notation simple, we have omitted the arguments $(t, \mathbf{x})$ in $R_a(t, \mathbf{x})$ and $\hat{\mathbf{R}}_a(t, \mathbf{x})$, and the argument t in $\mathbf{v}_r(t)$ and $\dot{\mathbf{v}}_r(t)$. Given the time t and position $\mathbf{x}$ at which we observe the radiation, and given the trajectory $\mathbf{x}_0(t)$ of the particle, eq. (10.45) determines the time t_{ret} at which the radiation was emitted, and therefore the position $\mathbf{x}_0(t_{\text{ret}})$ of the particle at the time of emission. By definition of $\mathbf{R}_a(t, \mathbf{x})$, the distance between this position and the observation point, $|\mathbf{x} - \mathbf{x}_0(t_{\text{ret}})|$, is equal to $R_a(t, \mathbf{x})$, see eq. (10.57). Now consider a sphere of radius $R_a(t, \mathbf{x})$ centered on the position $\mathbf{x}_0(t_{\text{ret}})$ of the particle at time of emission. The power radiated per unit solid angle through this sphere is

$$
\begin{aligned}
\frac{d\mathcal{E}}{dt\, d\Omega} \;&=\; \mathbf{S} \cdot R_a^2 \hat{\mathbf{R}}_a \\[2mm]
&=\; \frac{1}{4\pi\epsilon_0} \frac{1}{c^3} \frac{q^2}{4\pi} \frac{\left| [\dot{\mathbf{v}}_r \times (\hat{\mathbf{R}}_a - \mathbf{v}_r/c)] \times \hat{\mathbf{R}}_a \right|^2}{(1 - \hat{\mathbf{R}}_a \cdot \mathbf{v}_r/c)^6}\,, \quad (10.150)
\end{aligned}
$$

where we denote by $\mathcal{E}$ the energy in the electromagnetic field. On the left-hand side, we have explicitly written the radiated power in the form $d\mathcal{E}/dt$, to stress that this is the energy per unit time interval dt. This is the time relevant for the distant observer, located at a distance R_a from the charge. This is the quantity that we have denoted simply by P in the previous chapters,

$$
P = \frac{d\mathcal{E}}{dt}\,. \quad (10.151)
$$

When it is useful to stress that this is the power measured by the distant observer, we will call it the "received" power, and we will use the notation P_r.

However, from the point of view of the charge that emits the radiation, $d\mathcal{E}/dt_{\text{ret}}$ is more relevant, since this determines the rate at which the charge loses energy, with respects to the time measured by an observer located at the instantaneous position of the charge (and at rest with respect to the distant observer); in particular, t_{ret} is related to the proper time τ of the charge by $d\tau = dt_{\text{ret}}/\gamma(v)$, where v is the instantaneous velocity of the charge, with respect to this observer. We call this the "emitted" power, and we denote it by P_e,

$$P_e = \frac{d\mathcal{E}}{dt_{\text{ret}}}. \tag{10.152}$$

The relation between d/dt and d/dt_{ret}, for fixed value of the point $\mathbf{x}$ where the radiation field is observed, was already computed in eq. (10.100). Then,

$$P_e = (1 - \hat{\mathbf{R}}{\cdot}\mathbf{v}_r/c)P_r. \tag{10.153}$$

Which quantity is most appropriate, among P_e and P_r, depends on the physical situation considered. In the rest of this section we will focus on P_e (using the more explicit notation $d\mathcal{E}/dt_{\text{ret}}$, for clarity) which, as we will see, is Lorentz invariant and has an elegant expression that makes its Lorentz invariance explicit.

Using eq. (10.153), and expressing the result in terms of t_{ret} by using $\mathbf{R}(t_{\text{ret}}, \mathbf{x})$ instead of $\mathbf{R}_a(t, \mathbf{x})$ [see eq. (10.58)],

$$\frac{d\mathcal{E}}{dt_{\text{ret}}d\Omega} = \frac{1}{4\pi\epsilon_0}\frac{1}{c^3}\frac{q^2}{4\pi}\frac{\left|[\dot{\mathbf{v}}_r \times (\hat{\mathbf{R}} - \mathbf{v}_r/c)] \times \hat{\mathbf{R}}\right|^2}{(1 - \hat{\mathbf{R}}{\cdot}\mathbf{v}_r/c)^5}, \tag{10.154}$$

where $\hat{\mathbf{R}} = \hat{\mathbf{R}}(t_{\text{ret}}, \mathbf{x})$. Note that everything here is expressed in terms of t_{ret} rather than t, since, as we saw in eqs. (10.59) and (10.107), $\mathbf{v}_r = d\mathbf{x}_0(t_{\text{ret}})/dt_{\text{ret}}$ and $\dot{\mathbf{v}}_r = d^2\mathbf{x}_0(t_{\text{ret}})/d^2 t_{\text{ret}}$.

The integration over the angles can now be performed, writing $\hat{\mathbf{R}} = (\sin\theta\cos\phi, \sin\theta\sin\phi, \cos\theta)$, and gives

$$\frac{d\mathcal{E}}{dt_{\text{ret}}} = \frac{1}{4\pi\epsilon_0}\frac{2q^2}{3c^3}\gamma^6\left[|\dot{\mathbf{v}}_r|^2 - \frac{|\mathbf{v}_r \times \dot{\mathbf{v}}_r|^2}{c^2}\right], \tag{10.155}$$

which we can also write as

$$\boxed{\frac{d\mathcal{E}}{dt_{\text{ret}}} = \frac{1}{4\pi\epsilon_0}\frac{2q^2}{3c^3}\gamma^6\left[a^2 - \frac{|\mathbf{v}\times\mathbf{a}|^2}{c^2}\right]_{t=t_{\text{ret}}},} \tag{10.156}$$

where $\mathbf{a} = \dot{\mathbf{v}}$ and $a = |\mathbf{a}|$. In the limit $v/c \to 0$ we have $\gamma \to 1$, while the second term in the bracket is suppressed by a factor $(v/c)^2$ with respect to the first. Then, writing $q^2a^2 = q^2|\ddot{\mathbf{x}}_0|^2 = |\ddot{\mathbf{d}}^2|$, we recover the Larmor formula (10.148) [note that, in the limit $v/c \to 0$, P_e is the same as $P_r \equiv P$, as we see from eq. (10.153)]. In this sense, eq. (10.156) is also referred to as the *relativistic Larmor formula*. Observe that the factor γ^6 means that the power radiated by a relativistic charge is highly enhanced, compared to the non-relativistic case.

We can rewrite eq. (10.156) expanding[22]

$$|\mathbf{v}\times\mathbf{a}|^2 = v^2a^2 - (\mathbf{v}{\cdot}\mathbf{a})^2. \tag{10.157}$$

[22]This is obtained by writing

$$|\mathbf{v}\times\mathbf{a}|^2 = (\epsilon_{ijk}v_ja_k)(\epsilon_{ilm}v_la_m),$$

and using

$$\epsilon_{ijk}\epsilon_{ilm} = \delta_{jl}\delta_{km} - \delta_{jm}\delta_{kl}.$$

Then

$$\frac{d\mathcal{E}}{dt_{\mathrm{ret}}} = \frac{1}{4\pi\epsilon_0} \frac{2q^2}{3c^3}\gamma^6 \left[\frac{a^2}{\gamma^2} + \frac{(\mathbf{v}\cdot\mathbf{a})^2}{c^2}\right]_{t=t_{\mathrm{ret}}} . \qquad (10.158)$$

We can also use $\mathbf{v}\cdot\mathbf{a} = v\,dv/dt$,[23] to rewrite this as

$$\frac{d\mathcal{E}}{dt_{\mathrm{ret}}} = \frac{1}{4\pi\epsilon_0} \frac{2q^2}{3c^3}\gamma^6 \left[\frac{a^2}{\gamma^2} + \frac{v^2}{c^2}\left(\frac{dv}{dt}\right)^2\right]_{t=t_{\mathrm{ret}}} . \qquad (10.160)$$

Another useful form is obtained by decomposing $\mathbf{a}$ into its component parallel and transverse to $\mathbf{v}$, as in eq. (10.118), $\mathbf{a} = \mathbf{a}_\parallel + \mathbf{a}_\perp$. Then $a^2 = a_\parallel^2 + a_\perp^2$, where $a_\parallel = |\mathbf{a}_\parallel|$ and $a_\perp = |\mathbf{a}_\perp|$, while $\mathbf{v}\cdot\mathbf{a} = va_\parallel$, so eq. (10.158) becomes

$$\boxed{\frac{d\mathcal{E}}{dt_{\mathrm{ret}}} = \frac{1}{4\pi\epsilon_0} \frac{2q^2}{3c^3}\gamma^4 \left(a_\perp^2 + \gamma^2 a_\parallel^2\right)_{t=t_{\mathrm{ret}}} .} \qquad (10.161)$$

An alternative and elegant derivation of the relativistic Larmor formula is obtained if one realizes that $d\mathcal{E}/dt_{\mathrm{ret}}$ is a Lorentz-invariant quantity. To show this, consider an inertial frame K where the charge is instantaneously at rest at a given value t_{ret} of retarded time (defined with respect to a far observer at rest with respect to K), so $\mathbf{v}(t_{\mathrm{ret}}) = 0$, although $\dot{\mathbf{v}}(t_{\mathrm{ret}}) \neq 0$ since the charge is accelerating. In the frame K, we denote by $d\mathcal{E}$ the energy emitted by the charge in the interval between t_{ret} and $t_{\mathrm{ret}} + dt_{\mathrm{ret}}$ (and which will therefore be seen by the distant observer in the interval between t and $t + dt$), and by $d\mathbf{P}_{\mathrm{em}}$ the radiated momentum. However, in the limit $v/c \to 0$, and therefore for a particle instantaneously at rest, the radiation is given by the electric dipole term. The symmetry of the dipole radiation implies that the momenta carried away by the radiation in opposite directions are equal in magnitude and opposite in direction, so the momentum radiated in the interval between t_{ret} and $t_{\mathrm{ret}} + dt_{\mathrm{ret}}$ vanishes, $d\mathbf{P}_{\mathrm{em}} = 0$.

Let K' be a boosted frame, where the instantaneous velocity of the charge is $\mathbf{v}_0$, and let t'_{ret} be the retarded time measured in this frame. As we have shown in Solved Problem 8.2, the energy and momentum of the electromagnetic field form a four-vector so, from eq. (7.49), in the frame K' the radiated energy is

$$d\mathcal{E}' = \gamma(v_0)\left(d\mathcal{E} + \boldsymbol{\beta}\cdot d\mathbf{P}_{\mathrm{em}}\right) . \qquad (10.162)$$

However, since $d\mathbf{P}_{\mathrm{em}} = 0$, we simply have $d\mathcal{E}' = \gamma(v_0)d\mathcal{E}$. At the same time, at the position $\mathbf{x} = 0$ where the particle instantaneously sits in the K frame, the transformation of the "local" time variables t_{ret} is $dt'_{\mathrm{ret}} = \gamma(v_0)dt_{\mathrm{ret}}$. Therefore, $d\mathcal{E}'/dt'_{\mathrm{ret}} = d\mathcal{E}/dt_{\mathrm{ret}}$, so $d\mathcal{E}/dt_{\mathrm{ret}}$ is Lorentz invariant.

We now use the fact that, in the K frame where the charge is instantaneously at rest, the radiated power is given by the non-relativistic Larmor formula (10.148): indeed, given that in this frame $v = 0$, the higher-order corrections in v/c are identically zero, and the non-relativistic

---

[23]This follows from

$$\begin{aligned}\mathbf{v}\cdot\frac{d\mathbf{v}}{dt} &= v^i\frac{dv^i}{dt} = \frac{1}{2}\frac{d}{dt}(v^iv^i) \\ &= \frac{1}{2}\frac{d}{dt}(v^2) = v\frac{dv}{dt} .\end{aligned} \quad (10.159)$$

Note that $dv/dt = d|\mathbf{v}|/dt$ and, of course, this in general different from $a \equiv |\mathbf{a}| = |d\mathbf{v}/dt|$. As we see from eq. (10.159), the equality $dv/dt = a$ implies $\mathbf{v}\cdot\mathbf{a} = va$ and therefore only holds when $\mathbf{v}$ and $\mathbf{a}$ are parallel.

Larmor formula becomes exact. Then, to find the radiated power in a generic boosted frame, it is sufficient to find a Lorentz-invariant expression that reduces to the Larmor formula (10.148) in the instantaneous rest frame of the particle.[24] To this purpose, we consider the Lorentz-invariant quantity

$$\frac{dp_\mu}{d\tau}\frac{dp^\mu}{d\tau} = -\left(\frac{dp^0}{d\tau}\right)^2 + \left(\frac{d\mathbf{p}}{d\tau}\right)^2, \qquad (10.163)$$

where τ is the proper time of the radiating charge and p^μ is its relativistic four-momentum, so $p^0 = \gamma(v)mc$ and $\mathbf{p} = \gamma(v)m\mathbf{v}$. It is very natural to consider the quantity in eq. (10.163), when looking for a Lorentz-invariant expression that reproduces eq. (10.160). Indeed, a point particle, without internal structure, is only characterized by its four-momentum p^μ and by its proper time τ. The invariant $p_\mu p^\mu$ is equal to $-m^2c^2$, so it is just a fixed number, independent of the velocity and acceleration of the particle, while $p_\mu dp^\mu/d\tau = (1/2)d(p_\mu p^\mu)/d\tau = 0$. Therefore, the only quantity that can be formed, which is quadratic in the velocities and accelerations, is $(dp_\mu/d\tau)(dp^\mu/d\tau)$. Writing $dp^0/d\tau$ and $dp^i/d\tau$ in terms of $dv/d\tau$ and $dv^i/d\tau$, where as usual $v = |\mathbf{v}|$, we get[25]

$$\frac{1}{m^2}\frac{dp_\nu}{d\tau}\frac{dp^\nu}{d\tau} = \gamma^6\left[\frac{1}{\gamma^2}\left(\frac{dv^i(t)}{dt}\right)^2 + \frac{v^2}{c^2}\left(\frac{dv(t)}{dt}\right)^2\right]_{t=t_{\rm ret}}. \qquad (10.166)$$

Comparing with eq. (10.160), we see that

$$\boxed{\frac{d\mathcal{E}}{dt_{\rm ret}} = \frac{1}{4\pi\epsilon_0}\frac{2q^2}{3m^2c^3}\left(\frac{dp_\nu}{d\tau}\frac{dp^\nu}{d\tau}\right).} \qquad (10.167)$$

Equation (10.167) therefore provides an expression for the power radiated by a point charge in an arbitrary relativistic motion, equivalent to eq. (10.156), but written in an explicitly Lorentz-invariant form. Using $d\tau = dt_{\rm ret}/\gamma$, and recalling that $u^0 = \gamma c$, we can rewrite eq. (10.167) as

$$\frac{d\mathcal{E}}{d\tau} = \frac{1}{4\pi\epsilon_0}\frac{2q^2}{3m^2c^4}\left(\frac{dp_\nu}{d\tau}\frac{dp^\nu}{d\tau}\right)u^0. \qquad (10.168)$$

We then recognize that this is the $\mu = 0$ component of a covariant equation,

$$\boxed{\frac{dP_{\rm em}^\mu}{d\tau} = \frac{1}{4\pi\epsilon_0}\frac{2q^2}{3m^3c^5}\left(\frac{dp_\nu}{d\tau}\frac{dp^\nu}{d\tau}\right)p^\mu,} \qquad (10.169)$$

where $P_{\rm em}^\mu = (\mathcal{E}/c, \mathbf{P}_{\rm em})$ is the four-vector describing the electromagnetic energy and momentum, radiated by a charge with four-momentum p^μ and four-velocity $u^\mu = p^\mu/m$. The spatial components of this equation give the radiated momentum.[26]

[24]The uniqueness of this covariantization procedure is ensured by the fact that, given the value of a quantity in a frame, in this case the power in the rest frame, and its transformation properties under Lorentz transformation, the value in any other Lorentz frame is uniquely determined.

[25]Explicitly, $d\gamma/d\tau = (\gamma^3/c^2)v\,dv/d\tau$, so

$$\frac{dp^0}{d\tau} = \frac{m\gamma^3}{c}v\frac{dv}{d\tau}, \qquad (10.164)$$

while

$$\frac{dp^i}{d\tau} = \frac{m\gamma^3}{c^2}v\frac{dv}{d\tau}v^i + m\gamma\frac{dv^i}{d\tau}. \qquad (10.165)$$

Therefore

$$\frac{1}{m^2}\frac{dp_\nu}{d\tau}\frac{dp^\nu}{d\tau} = -\frac{\gamma^6}{c^2}v^2\left(\frac{dv}{d\tau}\right)^2$$

$$+\frac{\gamma^6}{c^4}v^4\left(\frac{dv}{d\tau}\right)^2 + \gamma^2\left(\frac{dv^i}{d\tau}\right)^2$$

$$+2\frac{\gamma^4}{c^2}v^2\left(\frac{dv}{d\tau}\right)^2$$

$$=\frac{\gamma^4 v^2}{c^2}\left(\frac{dv}{d\tau}\right)^2 + \gamma^2\left(\frac{dv^i}{d\tau}\right)^2,$$

where, in the second line, we used $v^i dv^i/d\tau = v\,dv/d\tau$, see eq. (10.159). Using $d\tau = dt_{\rm ret}/\gamma$, see the discussion following eq. (10.150), we get eq. (10.166).

[26]It should be stressed that these results have been obtained assuming that the charge that radiates is exactly point-like, since we have used the Liénard–Wiechert potentials, which make use of eqs. (8.1) and (8.2). So, while eqs. (10.167) and (10.169) might look as "exact" results valid for arbitrary relativistic motion, one should bear in mind that they are exact only in this, highly idealized, approximation. As we will see in Section 11.2, for an extended charge distribution the radiation emitted has a further dependence on all its higher-order charge and current multipoles, which give contributions suppressed by higher powers of v/c. So, while in the non-relativistic limit the leading term is indeed given by the non-relativistic Larmor formula (10.148), the exact result at all orders in v/c is given by eq. (10.167) only in the approximation when one models the charge distribution as point-like.

10.6.2   Acceleration parallel to the velocity

We now go back to eq. (10.154), to discuss in more detail the angular dependence in some simple cases. In this subsection we consider the situation in which the acceleration is parallel (or antiparallel) to the velocity. We then write

$$\mathbf{v} = v\hat{\mathbf{v}}, \qquad \mathbf{a} = a\hat{\mathbf{v}}, \tag{10.170}$$

and we define θ from $\hat{\mathbf{R}}\cdot\hat{\mathbf{v}} = \cos\theta$, so θ is the angle between the velocity and the direction of observation. The case where $\mathbf{v}$ and $\mathbf{a}$ are anti-parallel can be included choosing $v > 0$ and $a < 0$.[27] Then, in the numerator of eq. (10.154),[28]

$$[\dot{\mathbf{v}} \times (\hat{\mathbf{R}} - \mathbf{v}/c)] \times \hat{\mathbf{R}} = a(\cos\theta\hat{\mathbf{R}} - \hat{\mathbf{v}}). \tag{10.171}$$

Therefore

$$\left| [\dot{\mathbf{v}} \times (\hat{\mathbf{R}} - \mathbf{v}/c)] \times \hat{\mathbf{R}} \right|^2 = a^2(\cos^2\theta + 1 - 2\cos^2\theta)$$
$$= a^2 \sin^2\theta. \tag{10.172}$$

The angular distribution of the radiated power is then given by

$$\boxed{\frac{dP_{e,\text{parallel}}(t_{\text{ret}})}{d\Omega} = \frac{1}{4\pi\epsilon_0}\frac{q^2 a^2}{4\pi c^3}\frac{\sin^2\theta}{(1 - \beta\cos\theta)^5},} \tag{10.173}$$

where $P_e(t_{\text{ret}}) = d\mathcal{E}/dt_{\text{ret}}$ is the "emitted" power, see eq. (10.152), and the subscript "parallel" stresses that this result is valid when $\mathbf{a}$ is parallel (or antiparallel) to $\mathbf{v}$. This distribution is very peculiar, for $\beta \to 1$. Indeed, in the forward direction, i.e., at $\theta = 0$, the radiation emitted vanishes exactly because of the $\sin^2\theta$ in the numerator. However, for β close to one, the denominator strongly enhances the radiation at small angles. Therefore, the distribution has a peak at a value of θ non-vanishing but very small and (taking into account that the distribution is invariant under rotations around the direction of the acceleration) the radiation is focused into a narrow cone close to the forward direction. Fig. 10.1 shows the function

$$f_\|(\theta) = \frac{\sin^2\theta}{(1 - \beta\cos\theta)^5}, \tag{10.174}$$

for $\beta = 0$, $\beta = 0.6$ and $\beta = 0.9$. The maximum of the right-hand side of eq. (10.173), as a function of θ, is given by

$$\cos\theta_{\max} = \frac{-1 + \sqrt{1 + 15\beta^2}}{3\beta}. \tag{10.175}$$

Inverting the relation $\gamma^2 = 1/(1 - \beta^2)$ we have $\beta^2 = 1 - 1/\gamma^2$, so, for large γ,

$$\beta = 1 - \frac{1}{2\gamma^2} + \mathcal{O}\left(\frac{1}{\gamma^4}\right). \tag{10.176}$$

[27]Note that, in contrast, we always keep $v > 0$, otherwise we have to change correspondingly the definition of θ, if we want θ to remain the angle between acceleration and velocity. When $v = 0$, we define θ from $\hat{\mathbf{R}}\cdot\hat{\mathbf{a}} = \cos\theta$, so the case $v = 0$ is obtained as the limit $v \to 0$ with $a > 0$.

[28]Explicitly,

$$[\dot{\mathbf{v}} \times (\hat{\mathbf{R}} - \mathbf{v}/c)] \times \hat{\mathbf{R}}$$
$$= a[\hat{\mathbf{v}} \times (\hat{\mathbf{R}} - \beta\hat{\mathbf{v}})] \times \hat{\mathbf{R}}$$
$$= a(\hat{\mathbf{v}} \times \hat{\mathbf{R}}) \times \hat{\mathbf{R}}$$
$$= a\left[(\hat{\mathbf{v}}\cdot\hat{\mathbf{R}})\hat{\mathbf{R}} - (\hat{\mathbf{R}}\cdot\hat{\mathbf{R}})\hat{\mathbf{v}}\right]$$
$$= a(\cos\theta\hat{\mathbf{R}} - \hat{\mathbf{v}}),$$

where $\beta = v/c$, and we used $\hat{\mathbf{v}} \times \hat{\mathbf{v}} = 0$ and eq. (1.10).

Inserting this into eq. (10.175) and expanding the numerator and the denominator to first order in $1/\gamma^2$, we get

$$\cos\theta_{\max} \simeq 1 - \frac{1}{8\gamma^2}, \qquad (\gamma \gg 1). \qquad (10.177)$$

This confirms that, for large γ, the peak of the distribution is at an angle $\theta_{\max}$ such that $\cos\theta_{\max}$ is very close to one, i.e., $\theta_{\max}$ is very close to zero. Writing $\cos\theta_{\max} \simeq 1 - (1/2)\theta_{\max}^2$, we get

$$\theta_{\max} \simeq \pm\frac{1}{2\gamma} + \mathcal{O}\left(\frac{1}{\gamma^2}\right). \qquad (10.178)$$

The angular width of the peak is also of order $1/\gamma$. Indeed, in the limit $\theta \ll 1$ and $\gamma \gg 1$ we can write

$$\frac{\sin^2\theta}{(1 - \beta\cos\theta)^5} \simeq \frac{\theta^2}{(1 - \beta + \beta\theta^2/2)^5}$$

$$\simeq 32\gamma^8 \frac{\theta^2\gamma^2}{(1 + \theta^2\gamma^2)^5}, \qquad (10.179)$$

where, in the second line, we used eq. (10.176) (note that we made no assumption on the product $\theta\gamma$). Therefore

$$\frac{dP_{e,\text{parallel}}(t_{\text{ret}})}{d\Omega} \simeq \frac{1}{4\pi\epsilon_0}\frac{8q^2a^2}{\pi c^3}\gamma^8\frac{\theta^2\gamma^2}{(1 + \theta^2\gamma^2)^5}, \qquad (\gamma \gg 1,\ \theta \ll 1). \qquad (10.180)$$

From this expression, we can verify again that the maximum is at $\theta\gamma = 1/2$, and we see that the width $\Delta\theta$ of the distribution is of order $1/\gamma$. Therefore, when the acceleration is parallel to the velocity, and the particle is highly relativistic, the radiation is focused into a very narrow cone in the direction of motion, peaked at an opening angle $\theta_{\max} \simeq 1/(2\gamma)$, and with a width of order $1/\gamma$.

Observe that the case of acceleration antiparallel to the velocity can be simply obtained replacing $a \to -a$ in eq. (10.170). However, since eq. (10.173) is unchanged under $a \to -a$, the result is the same when the particle is accelerated or decelerated in the direction of its velocity. The latter situation typically takes place when a relativistic electron hits a target, that rapidly decelerates it. The corresponding radiation, that classically is described by eq. (10.173), is called *bremsstrahlung*, or "braking radiation." More generally, bremsstrahlung takes place because of the acceleration of a charge in the Coulomb field of another charge and is also called *free-free emission*.

The integration of eq. (10.173) over the solid angle is given by an elementary integral,

$$\int_{-1}^{1} d\cos\theta \int_{0}^{2\pi} d\phi\, \frac{\sin^2\theta}{(1 - \beta\cos\theta)^5} = 2\pi \int_{-1}^{1} dx\, \frac{1 - x^2}{(1 - \beta x)^5}$$

$$= \frac{8\pi}{3}\left(\frac{1}{1 - \beta^2}\right)^3$$

$$= \frac{8\pi}{3}\gamma^6. \qquad (10.181)$$

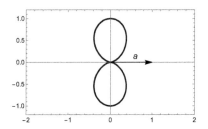

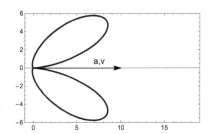

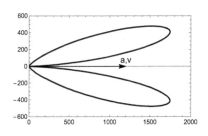

Fig. 10.1 A polar plot of the function $f_{\parallel}(\theta)$ for $\beta = 0$ (upper panel), $\beta = 0.6$ (middle panel), and $\beta = 0.9$ (lower panel). The direction of **a** (and, when non-zero, of **v**) is shown by the arrows. All plots are unchanged if we invert the direction of the acceleration. The full distributions in three-dimensional space are rotationally symmetric around the horizontal axis, corresponding to the fact that the distributions are independent of the polar angle ϕ. Note the difference in scales between the three panels.

Therefore, the emitted power radiated instantaneously when the acceleration is parallel (or anti-parallel) to the velocity is

$$P_{e,\text{parallel}} = \frac{1}{4\pi\epsilon_0} \frac{2q^2 a^2}{3c^3} \gamma^6 \,. \tag{10.182}$$

For $\gamma \to 1$ this reduces to the Larmor's formula (10.148), as it should. We could have also obtained this result from eq. (10.160), using the fact that, when the acceleration is parallel to the velocity, $dv/dt \equiv d|\mathbf{v}|/dt$ becomes the same as $a \equiv |d\mathbf{v}/dt|$.

Instead of expressing the power in terms of the acceleration, it can be useful to express it in terms of $|d\mathbf{p}/dt|$, i.e., of the force applied to the particle in order to accelerate it (equivalently, we could use $d\mathbf{p}/d\tau = \gamma d\mathbf{p}/dt$). Proceeding as in eq. (10.164) and using the fact that, when the acceleration is parallel to the velocity, the modulus of the acceleration is related to the modulus of the velocity by $a = dv/dt$, we have[29]

$$\frac{dp}{dt} = m\gamma^3 a \,, \tag{10.183}$$

and therefore

$$\left|\frac{d\mathbf{p}}{dt}\right| = \gamma^3 ma \,, \qquad (\mathbf{a} \parallel \mathbf{v}) \,. \tag{10.184}$$

Using this to eliminate a in favor of $|d\mathbf{p}/dt|$ from eq. (10.182), we get

$$P_{e,\text{parallel}} = \frac{1}{4\pi\epsilon_0} \frac{2q^2}{3m^2 c^3} \left|\frac{d\mathbf{p}}{dt}\right|^2 \,. \tag{10.185}$$

10.6.3   Acceleration perpendicular to the velocity

We next consider the case in which the acceleration is perpendicular to the velocity, as for a particle accelerated in a circular ring, which is a common situation in particle accelerators. In this case we set the instantaneous velocity $\mathbf{v}$ along the z axis, and the acceleration $\mathbf{a}$ along the x axis, so

$$\mathbf{v} = v\hat{\mathbf{z}} \,, \qquad \mathbf{a} = a\hat{\mathbf{x}} \,. \tag{10.186}$$

We define the polar angles θ, ϕ with respect to the $\hat{\mathbf{z}}$ axis, so the generic direction of observation is given by

$$\hat{\mathbf{R}} = \sin\theta\cos\phi\hat{\mathbf{x}} + \sin\theta\sin\phi\hat{\mathbf{y}} + \cos\theta\hat{\mathbf{z}} \,. \tag{10.187}$$

Note that we still have

$$\hat{\mathbf{R}}\cdot\hat{\mathbf{v}} = \cos\theta \,, \tag{10.188}$$

as in the setting of the previous subsection. However, now the numerator of eq. (10.154) depends also on the ϕ angle: carrying out the triple vector product,

$$[\hat{\mathbf{x}} \times (\hat{\mathbf{R}} - \beta\hat{\mathbf{z}})] \times \hat{\mathbf{R}} = [\sin^2\theta\cos^2\phi - (1 - \beta\cos\theta)]\hat{\mathbf{x}}$$
$$+ \sin^2\theta\sin\phi\cos\phi\,\hat{\mathbf{y}} + \sin\theta\cos\phi(\cos\theta - \beta)\hat{\mathbf{z}} \,, \tag{10.189}$$

[29] Explicitly,

$$\begin{aligned}
\frac{1}{m}\frac{dp}{dt} &= \frac{d(\gamma v)}{dt} \\
&= \frac{d\gamma}{dt}v + \gamma\frac{dv}{dt} \\
&= \frac{\gamma^3 v^2}{c^2}a + \gamma a \\
&= \gamma^3 a \,,
\end{aligned}$$

and, taking the modulus squared and combining the various terms,

$$\left|[\hat{\mathbf{x}} \times (\hat{\mathbf{R}} - \beta\hat{\mathbf{z}})] \times \hat{\mathbf{R}}\right|^2 = (1 - \beta\cos\theta)^2 - (1 - \beta^2)\sin^2\theta\cos^2\phi . \quad (10.190)$$

Inserting this into eq. (10.154) and writing $1 - \beta^2 = 1/\gamma^2$, we get

$$\frac{dP_{e,\text{circ}}(t_{\text{ret}})}{d\Omega} = \frac{1}{4\pi\epsilon_0}\frac{q^2 a^2}{4\pi c^3}\frac{1}{(1 - \beta\cos\theta)^3}\left[1 - \frac{\sin^2\theta\cos^2\phi}{\gamma^2(1 - \beta\cos\theta)^2}\right] ,$$

$$(10.191)$$

where we added the subscript "circ" to the emitted power P_e to stress that this is the result when, instantaneously, $\mathbf{a} \perp \mathbf{v}$, so, in particular, for a circular motion.

In Fig. 10.2 we show, for $\beta = 0, 0.6$, and 0.9, the function

$$f_\perp(\theta) = \frac{1}{(1 - \beta\cos\theta)^3}\left[1 - \frac{(1 - \beta^2)\sin^2\theta}{(1 - \beta\cos\theta)^2}\right] , \quad (10.192)$$

that, according to eq. (10.191), determines the distribution $dP_{\text{circ}}/d\Omega$ in θ, in the plane $\phi = 0$.

Integrating eq. (10.191) over the solid angle, we get

$$P_{e,\text{circ}} = \frac{1}{4\pi\epsilon_0}\frac{2q^2 a^2}{3c^3}\gamma^4 . \quad (10.193)$$

Comparing with eq. (10.182) we see that, for fixed acceleration a, the power radiated when the acceleration is parallel to the velocity is larger than that radiated in a circular motion, by a factor γ^2. However, it is usually more significant to compare the power radiated for a fixed external force. To this purpose, similarly to eq. (10.184), we can rewrite P_{circ} expressing a in terms of $d\mathbf{p}/dt$. For circular motion, the modulus v of the velocity does not change (so γ does not change) and $d\mathbf{p}/dt = \gamma m a$, so

$$\left|\frac{d\mathbf{p}}{dt}\right| = \gamma m a , \quad (\mathbf{a} \perp \mathbf{v}) . \quad (10.194)$$

Using this to eliminate a in favor of $|d\mathbf{p}/dt|$ from eq. (10.193), we get

$$P_{e,\text{circ}} = \frac{1}{4\pi\epsilon_0}\frac{2q^2}{3m^2 c^3}\gamma^2\left|\frac{d\mathbf{p}}{dt}\right|^2 . \quad (10.195)$$

Comparing with eq. (10.185) we see that, at fixed $|d\mathbf{p}/dt|$, the situation is opposite and now $P_{e,\text{circ}}$ is larger than $P_{e,\text{parallel}}$ by a factor γ^2. Therefore, much more external power is needed to overcome radiation losses in circular accelerators, compared to linear accelerators.

In the non-relativistic limit $\beta \to 0$, eq. (10.191) becomes

$$\frac{dP_{e,\text{circ}}(t_{\text{ret}})}{d\Omega} \simeq \frac{1}{4\pi\epsilon_0}\frac{q^2 a^2}{4\pi c^3}\left(1 - \sin^2\theta\cos^2\phi\right) . \quad (10.196)$$

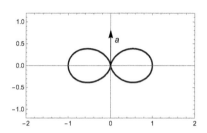

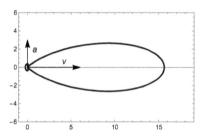

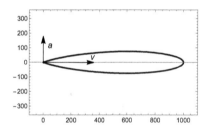

Fig. 10.2 A polar plot of the function $f_\perp(\theta)$ for $\beta = 0$ (upper panel, the same as the upper panel in Fig. 10.1, with the acceleration now set on the vertical axis), $\beta = 0.6$ (middle panel), and $\beta = 0.9$ (lower panel). The direction of $\mathbf{a}$ (and, when non-zero, of $\mathbf{v}$) is shown by the arrows. All plots are unchanged if we invert the direction of the acceleration. Note the difference in scales between the three panels.

[30]The cyclotron is a particle accelerator, invented by E. Lawrence in 1929–1930, where the charged particles are kept in an outward-bound spiral orbit by a magnetic field, and are accelerated by a time-varying electric field. The fast temporal variation of the electric field is synchronized with the outward inspiral motion of the particle, so that the particles undergo several cycles of acceleration. This is possible only as long as the particle is non-relativistic, and the frequency for the motion in a magnetic field, which in general is given by $\omega = qB/m\gamma$ [see eq. (8.201)], reduces to $\omega = qB/m$ and becomes independent of the velocity (and is indeed called the cyclotron frequency). To keep accelerating the particles when they become relativistic, the most successful solution turned out to be to keep the particle in a circular orbit of fixed radius by increasing the magnetic field during the acceleration phase; this led to the *synchrotron*. Cyclotrons were the most powerful particle accelerators until the 1950s, when they were superseded by synchrotrons (and other variants of the idea, such as synchrocyclotrons), but are still used in medical applications. Currently, the largest accelerator in the world is the Large Hadron Collider (LHC) at CERN, which is a synchrotron-type accelerator.

The radiation emitted in the non-relativistic limit by a charged particle in a circular or quasi-circular orbit (which, in practice, is obtained when the particle is moving in the plane perpendicular to an external magnetic field) is called *cyclotron radiation*. The limit $\beta \to 1$ (i.e., $\gamma \to \infty$) of eq. (10.191) defines instead *synchrotron radiation*.[30] Once again, as $\beta \to 1$, the radiation is focused into a narrow forward cone, because of the focusing effect of the denominator. Note, however, that now eq. (10.191) is non-vanishing for $\theta = 0$. Just as we have done for eq. (10.180), we can expand eq. (10.191) for $\gamma \to \infty$ and $\theta \to 0$, without the need of assuming anything for the product $\theta\gamma$, writing

$$1 - \beta \cos\theta \simeq 1 - \beta\left(1 - \frac{\theta^2}{2}\right)$$
$$\simeq \frac{1}{2\gamma^2}(1 + \theta^2\gamma^2), \qquad (10.197)$$

where we used eq. (10.176). Then, for $\gamma \gg 1$ and $\theta \ll 1$, eq. (10.191) becomes

$$\frac{dP_{e,\text{circ}}(t_{\text{ret}})}{d\Omega} \simeq \frac{1}{4\pi\epsilon_0}\frac{2q^2a^2}{\pi c^3}\frac{\gamma^6}{(1+\theta^2\gamma^2)^3}\left[1 - \frac{4\theta^2\gamma^2\cos^2\phi}{(1+\theta^2\gamma^2)^2}\right], \quad (10.198)$$

[to be compared with eq. (10.180)], so now the maximum is at $\theta = 0$, while the width of the distribution is again $\Delta\theta \sim 1/\gamma$, as is also seen from Fig. 10.2. Observe that, for $\theta\gamma > 0$, the term in brackets vanishes when

$$1 + (\theta\gamma)^2 = 2(\theta\gamma)|\cos\phi|. \qquad (10.199)$$

Solving the second-degree equation (10.199) with respect to $\theta\gamma$, gives

$$\theta\gamma = |\cos\phi| \pm \sqrt{\cos^2\phi - 1}. \qquad (10.200)$$

This has real solutions only if $\cos^2\phi = 1$, and then $\theta\gamma = 1$, so the distribution vanishes for $\theta = 1/\gamma$ and $\phi = 0$ or $\phi = \pi$.

In Fig. 10.3 we show a zoom of the the middle panel of Fig. 10.2, where $\beta = 0.6$ and $\cos\phi = 1$, that shows that the distribution indeed vanishes at a critical value of θ, and also has a smaller lobe, mostly in the backward direction, but partially tilted forward.

10.7 Solved problems

Problem 10.1. Dipole oscillating along the z axis

As a first simple application, we compute the radiation emitted by an electric dipole with charge q, oscillating with amplitude a along the z axis,

$$\mathbf{d}(t) = qa\hat{\mathbf{z}}\cos\omega_s t. \qquad (10.201)$$

We assume $a\omega_s \ll c$, so the non-relativistic formulas of Section 10.5 apply. We compute the radiation in the direction $\hat{\mathbf{n}}$, given in terms of θ and ϕ by

$$\hat{\mathbf{n}} = (\sin\theta\cos\phi, \sin\theta\sin\phi, \cos\theta). \qquad (10.202)$$

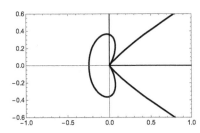

Fig. 10.3 A zoom of the middle panel of Fig. 10.2, with $\beta = 0.6$.

Then, using eq. (10.136),

$$\mathbf{E}(t, \mathbf{x}) = \frac{qa\omega_s^2}{4\pi\epsilon_0 c^2} \frac{1}{r} \cos[\omega_s(t - r/c)]$$
$$\times [-\hat{\mathbf{x}} \sin\theta \cos\theta \cos\phi - \hat{\mathbf{y}} \sin\theta \cos\theta \sin\phi + \hat{\mathbf{z}} \sin^2\theta]. \quad (10.203)$$

First of all observe that, in the dipole approximation, $\mathbf{E}(t, \mathbf{x})$ oscillates in time at a frequency ω equal to the frequency ω_s of the source. We also observe that the electric field vanishes on the z axis, i.e., for $\theta = 0$. The same holds for the magnetic field, since $c\mathbf{B} = \hat{\mathbf{n}} \times \mathbf{E}$. Clearly, there is no radiation in this direction since there is no acceleration of the source transverse to the z axis. In the (x, y) plane we have $\theta = \pi/2$, and we see from eq. (10.203) that $\mathbf{E}$ is linearly polarized along the $\hat{\mathbf{z}}$ axis. For the angular distribution of the power radiated, eq. (10.144) gives

$$\frac{dP}{d\Omega}(t) = \frac{1}{4\pi\epsilon_0} \frac{q^2 a^2 \omega^4}{4\pi c^3} \cos^2[\omega(t - r/c)] \sin^2\theta, \quad (10.204)$$

where we used the fact that the frequency of the source, ω_s, is the same as the frequency ω at which the electric field oscillates. If we perform a time average over one period, using

$$\frac{1}{2\pi} \int_0^{2\pi} d\alpha \cos^2\alpha = \frac{1}{2}, \quad (10.205)$$

the factor $\cos^2[\omega(t - r/c)]$ is replaced by $1/2$, and

$$\left\langle \frac{dP}{d\Omega} \right\rangle = \frac{1}{4\pi\epsilon_0} \frac{q^2 a^2 \omega^4}{8\pi c^3} \sin^2\theta. \quad (10.206)$$

Performing the angular integral, we get

$$\langle P \rangle = \frac{1}{4\pi\epsilon_0} \frac{q^2 a^2 \omega^4}{3c^3}. \quad (10.207)$$

Observe that, in the dipole radiation from a periodic source with frequency ω_s, the radiated power grows as ω_s^4 (which, for dipole radiation, is the same as ω^4, with ω the frequency of the radiation). This is due to the fact that each derivative brings a factor of ω_s, so $\ddot{d} \propto \omega_s^2$ and $P \propto \ddot{d}^2 \propto \omega_s^4$.

Problem 10.2. Radiation emitted by a non-relativistic charge in circular orbit

As the next application, we consider a charged particle moving counterclockwise on a circular orbit of radius a in the (x, y) plane, with frequency ω_s (for instance, the particle could be kept in a circular orbit by the action of an external magnetic field). We assume again that the velocity $v = \omega_s a \ll c$, so we can use the non-relativistic approximation to the particle motion, and we can use Larmor's formula, which is valid to lowest order in v/c. We write

$$\mathbf{x}_0(t) = a(\cos\omega_s t, \sin\omega_s t, 0), \quad (10.208)$$

so

$$\ddot{\mathbf{d}}(t) = -qa\omega_s^2(\cos\omega_s t, \sin\omega_s t, 0). \quad (10.209)$$

The radiative part of the electric field is obtained again from eq. (10.136), and we compute the radiation emitted in the direction of the unit vector $\hat{\mathbf{n}}$, given in terms of the polar angles θ, ϕ by eq. (10.202). Then, eq. (10.136) gives

$$E_x(t, r, \theta, \phi) = \frac{qa\omega_s^2}{(4\pi\epsilon_0)c^2 r} \left[\cos\omega_s t_{\text{ret}} - \sin^2\theta \cos\phi \cos(\omega_s t_{\text{ret}} - \phi) \right], \quad (10.210)$$

$$E_y(t, r, \theta, \phi) = \frac{qa\omega_s^2}{(4\pi\epsilon_0)c^2 r} \left[\sin\omega_s t_{\text{ret}} - \sin^2\theta \sin\phi \cos(\omega_s t_{\text{ret}} - \phi) \right], \quad (10.211)$$

$$E_z(t, r, \theta, \phi) = -\frac{qa\omega_s^2}{(4\pi\epsilon_0)c^2 r} \sin\theta \cos\theta \cos(\omega_s t_{\text{ret}} - \phi), \quad (10.212)$$

where as usual, to leading order in a/r retarded time becomes $t_{\text{ret}} = t - (r/c)$. First of all, we observe that, just as in the case of a dipole oscillating along the z axis, the frequency ω at which the electric field oscillates is the same as the frequency ω_s at which the source rotates. As we will see in more generality in Section 11.2, for dipole radiation the frequency of the electromagnetic waves generated by a monochromatic source is indeed always equal to the frequency of the source (as we will see, in general this is no longer true for the radiation generated by higher-order multipoles).

Along the positive $\hat{\mathbf{z}}$ axis, where $\theta = 0$, we get (writing henceforth ω instead of ω_s)

$$\mathbf{E}(t, r, \theta = 0) = \frac{qa\omega^2}{(4\pi\epsilon_0)c^2 r} (\cos\omega t_{\text{ret}}, \sin\omega t_{\text{ret}}, 0). \quad (10.213)$$

Therefore, in this direction, light is circularly polarized, with the electric field rotating counterclockwise with respect to the $+\hat{\mathbf{z}}$ direction. This corresponds to right circularly polarized light, see eq. (9.84). At $\theta = \pi$, the result for $\mathbf{E}$ is still given by eq. (10.213). However, now the propagation direction is $-\hat{\mathbf{z}}$ and, with respect to this propagation direction, eq. (10.213) is left circularly polarized light. If we rather set $\theta = \pi/2$, so that we look at the radiation emitted in the (x, y) plane, from eqs. (10.210)–(10.212) we get

$$\mathbf{E}\left(t, r, \theta = \frac{\pi}{2}, \phi\right) = \frac{qa\omega^2}{(4\pi\epsilon_0)c^2 r} \sin(\omega t_{\text{ret}} - \phi)(-\sin\phi, \cos\phi, 0). \quad (10.214)$$

Therefore, in this case the electromagnetic radiation is linearly polarized along the direction of the tangent vector to a circle in the (x, y) plane (oriented toward the counterclockwise direction).

To compute the radiated power we first observe, from eq. (10.209) that

$$|\ddot{\mathbf{d}}(t)|^2 = q^2 a^2 \omega_s^4 \quad (10.215)$$

is actually independent of time. Then, according to eq. (10.144), the radiated power is also time-independent, and given by

$$\frac{dP(\theta)}{d\Omega} = \frac{1}{4\pi\epsilon_0} \frac{q^2 a^2 \omega^4}{4\pi c^3} \sin^2\theta, \quad (10.216)$$

where we used $\omega_s = \omega$ to write the power in terms of the frequency of the radiation. The $\sin^2\theta$ dependence of the power is typical of dipole radiation, as we see from the general result (10.144). Performing the angular integral,

$$P = \frac{1}{4\pi\epsilon_0} \frac{2q^2 a^2 \omega^4}{3c^3}. \quad (10.217)$$

Note that this is twice as large as the result in eq. (10.207), corresponding to the fact that a circular motion in the (x, y) plane can be seen as a superposition of two oscillators, one oscillating along the x axis and one along the y axis.

Problem 10.3. Radiative collapse of a classical model of the hydrogen atom

Historically, the problem of the radiation emitted by a charge in a circular orbit was also important to show the inadequacy of a classical model of the atom. Consider a classical model of the hydrogen atom, with the electron moving in a circular orbit around the proton. The system radiates electromagnetic energy to infinity according to eq. (10.217). This energy must be taken from the mechanical energy $\mathcal{E}_{mech}$ of the system, which therefore must decrease, according to[31]

$$P = -\frac{d\mathcal{E}_{mech}}{dt}.$$ (10.218)

The mechanical energy is given by

$$\mathcal{E}_{mech} = \frac{1}{2}mv^2 - \frac{1}{4\pi\epsilon_0}\frac{e^2}{r},$$ (10.219)

where r is the relative distance between the electron and the proton and m the reduced mass. Actually, since $m_p \simeq 2000 m_e$, to good approximation we can take the proton at rest at the origin, and $m \simeq m_e$.

For an electron in a circular orbit of radius r we have $v = \omega r$ and the modulus of the acceleration $\mathbf{a}$ is $|\mathbf{a}| = \omega^2 r$, so $\mathbf{F} = m\mathbf{a}$ gives Kepler's law

$$\omega^2 = \frac{e^2}{4\pi\epsilon_0}\frac{1}{mr^3}.$$ (10.220)

Then

$$\mathcal{E}_{mech} = \frac{1}{2}m\omega^2 r^2 - \frac{1}{4\pi\epsilon_0}\frac{e^2}{r}$$

$$= -\frac{1}{4\pi\epsilon_0}\frac{e^2}{2r},$$ (10.221)

which is the result, familiar from elementary classical mechanics, that in the Coulomb potential the kinetic energy is $1/2$ of the absolute value of the potential energy. Note that the total energy is negative, as it should for a bound state. In order to balance the energy radiated, $\mathcal{E}_{mech}$ must decrease, i.e., must become more negative, so, according to eq. (10.221), r gets smaller until, eventually, the electron collapses on the nucleus.

The radiated power (10.217) has been computed under the assumption that the electron is kept by an external force on an exactly circular orbit. In the case of the (classical!) hydrogen atom, we have just seen that r will actually decrease to compensate for the emitted radiation, so $r = r(t)$. Still, eq. (10.217) remains a good approximation to the actual radiated power as long as the induced radial velocity $\dot{r}$ is much smaller, in absolute value, than the tangential component of the velocity, i.e., as long as $|\dot{r}| \ll \omega r$; furthermore, the use of eq. (10.217) is justified as long as the motion is non-relativistic, so $\omega r \ll c$. We will check the validity of these conditions a posteriori. Let us compute the evolution of $r(t)$ in this regime. Combining eq. (10.217) (with $q = -e$) with eqs. (10.218) and (10.221) we get

$$\frac{2e^2\omega^4 r^2}{3c^3} = -\frac{d}{dt}\left(\frac{-e^2}{2r}\right).$$ (10.222)

Using eq. (10.220) we can rewrite it as

$$\dot{r} = -\frac{4}{3}\frac{1}{m^2c^3}\left(\frac{e^2}{4\pi\epsilon_0}\right)^2\frac{1}{r^2}.$$ (10.223)

This integrates to

$$r^3 = r^3(0) - \frac{4}{m^2 c^3} \left(\frac{e^2}{4\pi\epsilon_0} \right)^2 t , \qquad (10.224)$$

where $r(0)$ is the value of the radius at $t = 0$, that we take equal to the Bohr radius of the hydrogen atom, $r_B \simeq 0.53 \times 10^{-8}$ cm. Therefore, in this approximation, the electron collapses on the nucleus in a time

$$\tau = \frac{1}{4} \left(mc^2 r_B \right)^2 \left(\frac{4\pi\epsilon_0}{e^2} \right)^2 \frac{r_B}{c} . \qquad (10.225)$$

We could directly plug in the numbers, and find $\tau \simeq 1.5 \times 10^{-11}$ s. Actually, it is more instructive to rewrite eq. (10.225) introducing some combinations that enter at the quantum level, even if our computation is purely classical. An important combination is the fine structure constant, already introduced in eq. (5.34),

$$\alpha = \frac{1}{4\pi\epsilon_0} \frac{e^2}{\hbar c} . \qquad (10.226)$$

This quantity is dimensionless, and, numerically, $\alpha \simeq 1/137$. We borrow from quantum mechanics the information that the Bohr radius is given by

$$r_B = \frac{1}{\alpha} \frac{\hbar}{mc} . \qquad (10.227)$$

Therefore, we have the identity

$$mc^2 r_B \left(\frac{4\pi\epsilon_0}{e^2} \right) = \frac{1}{\alpha^2} , \qquad (10.228)$$

and eq. (10.225) can be rewritten as

$$\tau = \frac{1}{4\alpha^4} \frac{r_B}{c} . \qquad (10.229)$$

Note that r_B/c is the time that light takes to cross the size of the atom. It is an extremely small number, about 0.53×10^{-10} m$/(3 \times 10^8$ m/s$) \simeq 1.8 \times 10^{-19}$ s. Even if τ is enhanced, with respect to this, by a factor $1/\alpha^4 \simeq (137)^4$, we still get the very small value $\sim 10^{-11}$ s given previously. This means that, in classical electromagnetism, atoms are unstable to emission of electromagnetic radiation, and would collapse in about 10^{-11} s! One can check, from the previous expressions, that the conditions $|\dot{r}| \ll \omega r$ and $\omega r \ll 1$ both break down only at $r \simeq \alpha^2 r_B$, so, when the size of the atom has become about $5 \times 10^{-5} r_B$, which is already of order of the size of the nucleus. Therefore, the conclusion on the collapse of the atom is not an artifact of these approximations (furthermore, a relativistic particle radiates away energy even faster).

This instability to emission of electromagnetic radiation is one of several difficulties of a classical model of the atom and shows that, at these scales, classical physics is inadequate. As we now know, at these scales the correct description is provided by quantum mechanics.

Radiation from localized sources

In the previous chapter we discussed the radiation generated by a point-like charge, with arbitrary motion. We now expand on the previous discussion, studying the radiation generated by a localized, but otherwise generic, distribution of charges and currents. This will also provide the basis for the multipole expansion of the radiation field, that we will present in Section 11.2, while in Section 11.3 we will give a first simple discussion of relevant scales defining the near and far zone. The dynamics of relativistic particles in the near zone will be discussed in more detail in Chapter 12, where we will tackle the rather technical issue of how relativistic effects, and the back-reaction due to the emitted radiation, affect the dynamics of the sources in the near zone.

11.1 Far zone fields for generic velocities

11.1.1 Computation in the Lorenz gauge

We begin by computing the radiation emitted by a generic charge distribution. In this subsection we perform the computation in the Lorenz gauge (we will compare with the computation in the Coulomb gauge in Section 11.1.2). We therefore start from eq. (10.34),

$$A^{\mu}(t, \mathbf{x}) = \frac{\mu_0}{4\pi} \int d^3 x' \, \frac{j^{\mu}(t - |\mathbf{x} - \mathbf{x}'|/c, \mathbf{x}')}{|\mathbf{x} - \mathbf{x}'|} \,, \qquad (11.1)$$

where we have set to zero the solution of the homogeneous equation, representing incoming radiation, since we want to compute the radiation produced by the current j^{μ}. Note that eq. (11.1) is valid for sources with arbitrary velocities, since we have used the exact (retarded) Green's function of the d'Alembertian. As in Chapter 6, we denote by d the size of the spatial region in which the source is localized. Therefore, in eq. (11.1), $j^{\mu}(t_{\rm ret}, \mathbf{x}')$ vanishes for $|\mathbf{x}'| > d$, so the integration variable $\mathbf{x}'$ is effectively limited to $|\mathbf{x}'| < d$. We write, as usual, $|\mathbf{x}| = r$ and $\mathbf{x} = r\hat{\mathbf{n}}$, and we use the large r expansion (6.3). We are interested only in the radiation field, that decays as $1/r$. Then, in eq. (11.1) we can replace $1/|\mathbf{x} - \mathbf{x}'|$ simply by $1/r$, since the subsequent terms of the expansion will give contributions of order $1/r^2$ and higher. In contrast, in the argument of j^{μ} we must also keep the next term, which does not vanish for $r \to \infty$. Thus,

$$A^\mu(t, \mathbf{x}) = \frac{\mu_0}{4\pi} \frac{1}{r} \int d^3x'\, j^\mu \left(t - \frac{r}{c} + \frac{\hat{\mathbf{n}} \cdot \mathbf{x}'}{c}, \mathbf{x}' \right) + \mathcal{O}\left(\frac{1}{r^2} \right). \qquad (11.2)$$

We now perform a Fourier transform with respect to time, writing

$$\rho(t, \mathbf{x}) = \int \frac{d\omega}{2\pi} e^{-i\omega t}\, \tilde{\rho}(\omega, \mathbf{x}), \qquad (11.3)$$

$$\mathbf{j}(t, \mathbf{x}) = \int \frac{d\omega}{2\pi} e^{-i\omega t}\, \tilde{\mathbf{j}}(\omega, \mathbf{x}). \qquad (11.4)$$

Then, restricting to the $1/r$ term, and recalling eqs. (8.9) and (8.12),

$$\phi(t, \mathbf{x}) = \frac{1}{4\pi\epsilon_0} \frac{1}{r} \int d^3x' \int \frac{d\omega}{2\pi}\, \tilde{\rho}(\omega, \mathbf{x}')\, e^{-i\omega[t - (r/c) + \hat{\mathbf{n}} \cdot \mathbf{x}'/c]}, \,(11.5)$$

$$\mathbf{A}(t, \mathbf{x}) = \frac{\mu_0}{4\pi} \frac{1}{r} \int d^3x' \int \frac{d\omega}{2\pi}\, \tilde{\mathbf{j}}(\omega, \mathbf{x}')\, e^{-i\omega[t - (r/c) + \hat{\mathbf{n}} \cdot \mathbf{x}'/c]}. \quad (11.6)$$

From these expressions it is clear why, in the exponential, we could not neglect the term $\hat{\mathbf{n}} \cdot \mathbf{x}'$, that comes from eq. (6.3). Even if $r \gg |\hat{\mathbf{n}} \cdot \mathbf{x}'|$, still the fact that, inside a phase, $\omega r/c$ is much larger than $\omega \hat{\mathbf{n}} \cdot \mathbf{x}'/c$ is a priori irrelevant, since phases are defined only modulo 2π.[1] In contrast, the term $\mathcal{O}(d^2/r)$ in eq. (6.3) goes to zero as $r \to \infty$, and is therefore negligible even in the phase.

We can now compute the electric field in the far zone, using eq. (3.83). Let us compute first $\nabla\phi$. Recalling that $\mathbf{x} = r\hat{\mathbf{n}}$, in eq. (11.5) the ∇ operator acts on the factor $1/r$ in front, as well as on the factors r and $\hat{\mathbf{n}}$ in the exponential. Again, we are only interested in the part of $\mathbf{E}$ that decreases as $1/r$. The derivative of the overall $1/r$ factor gives a term proportional to $1/r^2$, and we can neglect it. Similarly,

$$\partial_i\, e^{-i\omega\hat{\mathbf{n}} \cdot \mathbf{x}'/c} = -\frac{i\omega}{c}\, (\partial_i n_j) x_j'\, e^{-i\omega\hat{\mathbf{n}} \cdot \mathbf{x}'/c}, \qquad (11.7)$$

and[2]

$$\partial_i n_j = \frac{1}{r}\, (\delta_{ij} - n_i n_j). \qquad (11.8)$$

Therefore, when combined with the overall $1/r$ factor in front of the exponential, the term obtained from $\partial_i n_j$ gives again an overall contribution of order $1/r^2$ as $r \to \infty$. The only contribution to the radiation field therefore comes from the derivative of the factor r in the exponential. Notice that this factor is a consequence of the retardation effect. If it were not for this retardation term, there would be no $1/r$ term in the electric field, and therefore no radiation at infinity. Using

$$\partial_i e^{i\omega r/c} = (i\omega/c)(\partial_i r)\, e^{i\omega r/c}$$
$$= (i\omega/c) n_i\, e^{i\omega r/c}, \qquad (11.9)$$

where we used again eq. (6.12), we see that, to order $1/r$,

$$\nabla\phi(t, \mathbf{x}) = \frac{1}{4\pi\epsilon_0} \frac{1}{r} \int d^3x' \int \frac{d\omega}{2\pi} \frac{i\omega}{c} \hat{\mathbf{n}}\, \tilde{\rho}(\omega, \mathbf{x}')\, e^{-i\omega[t - (r/c) + \hat{\mathbf{n}} \cdot \mathbf{x}'/c]}. $$
$$(11.10)$$

[1]We will see in Section 11.2 that, for non-relativistic sources, the integral gets a significant contribution only for values of the integration variable ω such that the term $\omega \hat{\mathbf{n}} \cdot \mathbf{x}'/c$ is indeed small, and we will be able to expand the exponential in powers of it. For generic source velocities, however, this term must be kept in its full form.

[2]Explicitly,

$$\begin{aligned} \partial_i n_j &= \partial_i \left(\frac{x_j}{r} \right) \\ &= \frac{\delta_{ij}}{r} - \frac{x_j}{r^2} \partial_i r \\ &= \frac{1}{r} \left(\delta_{ij} - \frac{x_i x_j}{r^2} \right) \\ &= \frac{1}{r} \left(\delta_{ij} - n_i n_j \right), \end{aligned}$$

where we used eq. (6.12).

The computation of $\partial \mathbf{A}/\partial t$ is straightforward and, keeping again only the $1/r$ terms, we get

$$
\mathbf{E}(t,\mathbf{x}) = \frac{1}{4\pi\epsilon_0}\frac{1}{r}\int d^3x' \int \frac{d\omega}{2\pi}\frac{i\omega}{c^2} e^{-i\omega[t-(r/c)+\hat{\mathbf{n}}\cdot\mathbf{x}'/c]}
$$
$$
\times \left[\tilde{\mathbf{j}}(\omega,\mathbf{x}') - c\hat{\mathbf{n}}\tilde{\rho}(\omega,\mathbf{x}') \right]
$$
$$
= \frac{1}{4\pi\epsilon_0 c^2}\frac{1}{r}\int \frac{d\omega}{2\pi} e^{-i\omega[t-(r/c)]} i\omega \int d^3x' e^{-i\omega\hat{\mathbf{n}}\cdot\mathbf{x}'/c} \left[\tilde{\mathbf{j}}(\omega,\mathbf{x}') - c\hat{\mathbf{n}}\tilde{\rho}(\omega,\mathbf{x}') \right]
$$
$$
= \frac{1}{4\pi\epsilon_0 c^2}\frac{1}{r}\int \frac{d\omega}{2\pi} e^{-i\omega[t-(r/c)]} i\omega \left[\tilde{\mathbf{j}}(\omega,\omega\hat{\mathbf{n}}/c) - c\hat{\mathbf{n}}\tilde{\rho}(\omega,\omega\hat{\mathbf{n}}/c) \right], \quad (11.11)
$$

where, in the last equality, we expressed the result in terms of the full Fourier transform with respect to both time and space,

$$
\rho(t,\mathbf{x}) = \int \frac{d\omega}{2\pi}\frac{d^3k}{(2\pi)^3} e^{-i\omega t+i\mathbf{k}\cdot\mathbf{x}} \tilde{\rho}(\omega,\mathbf{k}), \qquad (11.12)
$$

$$
\mathbf{j}(t,\mathbf{x}) = \int \frac{d\omega}{2\pi}\frac{d^3k}{(2\pi)^3} e^{-i\omega t+i\mathbf{k}\cdot\mathbf{x}} \tilde{\mathbf{j}}(\omega,\mathbf{k}), \qquad (11.13)
$$

see eq. (1.105). The inversion of these Fourier transforms gives

$$
\tilde{\rho}(\omega,\mathbf{k}) = \int dt d^3x\, e^{i\omega t-i\mathbf{k}\cdot\mathbf{x}} \rho(t,\mathbf{x}), \qquad (11.14)
$$

$$
\tilde{\mathbf{j}}(\omega,\mathbf{k}) = \int dt d^3x\, e^{i\omega t-i\mathbf{k}\cdot\mathbf{x}} \mathbf{j}(t,\mathbf{x}), \qquad (11.15)
$$

see eq. (1.106). We now observe that the continuity equation (3.22), written in Fourier space (for ω and $\mathbf{k}$ generic), becomes

$$
\omega\tilde{\rho}(\omega,\mathbf{k}) = \mathbf{k}\cdot\tilde{\mathbf{j}}(\omega,\mathbf{k}). \qquad (11.16)
$$

In eq. (11.11), however, $\mathbf{k}$ and ω are related by $\mathbf{k} = \omega\hat{\mathbf{n}}/c$, and in this case the continuity equation gives

$$
c\tilde{\rho}(\omega,\omega\hat{\mathbf{n}}/c) = \hat{\mathbf{n}}\cdot\tilde{\mathbf{j}}(\omega,\omega\hat{\mathbf{n}}/c). \qquad (11.17)
$$

Then, we get

$$
\mathbf{E}(t,\mathbf{x}) = \frac{1}{4\pi\epsilon_0 c^2}\frac{1}{r}\int \frac{d\omega}{2\pi} e^{-i\omega[t-(r/c)]} i\omega
$$
$$
\times \left\{ \tilde{\mathbf{j}}\left(\omega,\frac{\omega\hat{\mathbf{n}}}{c}\right) - \hat{\mathbf{n}}\left[\hat{\mathbf{n}}\cdot\tilde{\mathbf{j}}\left(\omega,\frac{\omega\hat{\mathbf{n}}}{c}\right)\right] \right\}. \qquad (11.18)
$$

According to eq. (10.120), the expression in braces is just the component of the vector $\tilde{\mathbf{j}}(\omega,\omega\hat{\mathbf{n}}/c)$ transverse to the momentum $\mathbf{k} = \omega\hat{\mathbf{n}}/c$ or, equivalently, to the unit vector $\hat{\mathbf{n}}$,

$$
\tilde{\mathbf{j}}_\perp\left(\omega,\frac{\omega\hat{\mathbf{n}}}{c}\right) \equiv \tilde{\mathbf{j}}\left(\omega,\frac{\omega\hat{\mathbf{n}}}{c}\right) - \hat{\mathbf{n}}\left[\hat{\mathbf{n}}\cdot\tilde{\mathbf{j}}\left(\omega,\frac{\omega\hat{\mathbf{n}}}{c}\right)\right]. \qquad (11.19)
$$

Using eq. (10.122), we can also rewrite it in the equivalent form

$$
\tilde{\mathbf{j}}_\perp\left(\omega,\frac{\omega\hat{\mathbf{n}}}{c}\right) = -\hat{\mathbf{n}}\times\left[\hat{\mathbf{n}}\times\tilde{\mathbf{j}}\left(\omega,\frac{\omega\hat{\mathbf{n}}}{c}\right)\right]. \qquad (11.20)
$$

Then, the $1/r$ part of the electric field generated by a localized source can be written as

$$\mathbf{E}(t,\mathbf{x}) = \frac{1}{4\pi\epsilon_0 c^2} \frac{1}{r} \int \frac{d\omega}{2\pi} e^{-i\omega[t-(r/c)]} \, i\omega \, \tilde{\mathbf{j}}_\perp(\omega, \omega\hat{\mathbf{n}}/c). \qquad (11.21)$$

We finally observe that $i\omega e^{-i\omega t} = -\partial_t e^{-i\omega t}$. Then, transforming back to $\mathbf{j}(t,\mathbf{x})$,

$$
\begin{aligned}
\mathbf{E}(t,\mathbf{x}) &= -\frac{1}{4\pi\epsilon_0 c^2} \frac{1}{r} \partial_t \int \frac{d\omega}{2\pi} e^{-i\omega[t-(r/c)]} \\
&\quad \times \int d^3x' dt' \, e^{i\omega t' - i\omega \hat{\mathbf{n}} \cdot \mathbf{x}'/c} \mathbf{j}_\perp(t',\mathbf{x}'). \quad (11.22)
\end{aligned}
$$

We can carry out the integrals over $d\omega$ and dt', writing

$$
\begin{aligned}
\mathbf{E}(t,\mathbf{x}) &= -\frac{1}{4\pi\epsilon_0} \frac{1}{c^2 r} \partial_t \int d^3x' dt' \, \mathbf{j}_\perp(t',\mathbf{x}') \int \frac{d\omega}{2\pi} e^{-i\omega(t-r/c-t'+\hat{\mathbf{n}}\cdot\mathbf{x}'/c)} \\
&= -\frac{1}{4\pi\epsilon_0} \frac{1}{c^2 r} \partial_t \int d^3x' dt' \, \mathbf{j}_\perp(t',\mathbf{x}') \, \delta[t' - (t - r/c + \hat{\mathbf{n}}\cdot\mathbf{x}'/c)].
\end{aligned}
$$
$$(11.23)$$

We therefore arrive at our final result,

$$\boxed{\mathbf{E}(t,\mathbf{x}) = -\frac{1}{4\pi\epsilon_0 c^2} \frac{1}{r} \partial_t \int d^3x' \, \mathbf{j}_\perp(t - r/c + \hat{\mathbf{n}}\cdot\mathbf{x}'/c, \mathbf{x}'),} \qquad (11.24)$$

or, in terms of μ_0,

$$\mathbf{E}(t,\mathbf{x}) = -\frac{\mu_0}{4\pi} \frac{1}{r} \partial_t \int d^3x' \, \mathbf{j}_\perp(t - r/c + \hat{\mathbf{n}}\cdot\mathbf{x}'/c, \mathbf{x}'). \qquad (11.25)$$

We see that the radiation field is generated by the time-varying component of the current transverse to the line of sight. The only approximation made to arrive at this result is that we have kept just the term that at large distances decreases as $1/r$, while we have made no assumption on the motion of the charges, that can be fully relativistic. This generalizes the result found in Section 10.5, e.g., eq. (10.131), where we found that, to lowest order in v/c, and for a single point charge, the radiative component of the electric field is generated by the component of the acceleration of the charge transverse to the line of sight. Note also that $t - r/c + \hat{\mathbf{n}}\cdot\mathbf{x}'/c$ is just $t_{\rm ret}$, in the large r limit that we have used.

The magnetic field $\mathbf{B} = \boldsymbol{\nabla} \times \mathbf{A}$ can be computed similarly, from eq. (11.6). Writing $B_i = \epsilon_{ijk}\partial_j A_k$, again the only term proportional to $1/r$ is obtained when ∂_j acts on the factor $e^{i\omega r/c}$ in eq. (11.6), while the action on $1/r$ and on the factor $\hat{\mathbf{n}}$ that appears in $e^{-i\omega\hat{\mathbf{n}}\cdot\mathbf{x}'/c}$ gives terms of order $1/r^2$. Using eq. (11.9), we get

$$
\begin{aligned}
\mathbf{B}(t,\mathbf{x}) &= \frac{\mu_0}{4\pi} \frac{1}{r} \int d^3x' \int \frac{d\omega}{2\pi} \frac{i\omega}{c} \hat{\mathbf{n}} \times \tilde{\mathbf{j}}(\omega, \mathbf{x}') \, e^{-i\omega[t-(r/c)+\hat{\mathbf{n}}\cdot\mathbf{x}'/c]} \\
&= \frac{\mu_0}{4\pi} \hat{\mathbf{n}} \times \left[\frac{1}{r} \int d^3x' \int \frac{d\omega}{2\pi} \frac{i\omega}{c} \tilde{\mathbf{j}}(\omega, \mathbf{x}') \, e^{-i\omega[t-(r/c)+\hat{\mathbf{n}}\cdot\mathbf{x}'/c]} \right]. \quad (11.26)
\end{aligned}
$$

Comparing with the first line in eq. (11.11) (and using $\hat{\mathbf{n}} \times \tilde{\mathbf{j}} = \hat{\mathbf{n}} \times \tilde{\mathbf{j}}_\perp$) we see that, in the far region,

$$c\mathbf{B} = \hat{\mathbf{n}} \times \mathbf{E} . \tag{11.27}$$

Once again, observe that this is valid for sources with arbitrary velocities, i.e., is an exact result for the $1/r$ part of the electric and magnetic fields.

It can also be useful to observe, from eq. (11.21), that the Fourier transform with respect to time of $\mathbf{E}(t, \mathbf{x})$ is given by

$$\tilde{\mathbf{E}}(\omega, \mathbf{x}) = \frac{1}{4\pi\epsilon_0} \frac{i\omega\, e^{i\omega r/c}}{c^2 r} \tilde{\mathbf{j}}_\perp(\omega, \omega\hat{\mathbf{n}}/c) , \tag{11.28}$$

and, correspondingly,

$$\tilde{\mathbf{B}}(\omega, \mathbf{x}) = \frac{\mu_0}{4\pi} \frac{i\omega\, e^{i\omega r/c}}{cr} \hat{\mathbf{n}} \times \tilde{\mathbf{j}}_\perp(\omega, \omega\hat{\mathbf{n}}/c) , \tag{11.29}$$

where we recall that $\mathbf{x} = r\hat{\mathbf{n}}$.

11.1.2 Computation in the Coulomb gauge

It is instructive to repeat the computation in the Coulomb gauge, where the potentials obey eqs. (3.93) and (3.94). Let us start from ϕ. For a time-independent $\rho(\mathbf{x})$, we found the solution in eq. (4.16), using the Green's function of the Laplacian. This is immediately generalized to a source $\rho(t, \mathbf{x})$, since the Laplacian only acts on the spatial coordinates, so the solution of eq. (3.93) with a time-dependent source $\rho(t, \mathbf{x})$ is

$$\phi(t, \mathbf{x}) = \frac{1}{4\pi\epsilon_0} \int d^3x' \frac{\rho(t, \mathbf{x}')}{|\mathbf{x} - \mathbf{x}'|} . \tag{11.30}$$

In the large r limit,

$$\phi(t, \mathbf{x}) = \frac{1}{4\pi\epsilon_0} \frac{Q}{r} + \mathcal{O}\left(\frac{1}{r^2}\right) , \tag{11.31}$$

where Q is the total charge of the system. Therefore, the leading term in $\boldsymbol{\nabla}\phi$ goes as $1/r^2$ and does not contribute to the radiation field.[3] Indeed, we already remarked above that the $1/r$ term in the electric field comes from the retardation effect that is responsible for the term $e^{i\omega r/c}$ in eqs. (11.5) and (11.6). In the Coulomb gauge, where ϕ satisfies a Laplace equation, $\phi(t, \mathbf{x})$ is determined by $\rho(t, \mathbf{x}')$ at the same time t, i.e., is an instantaneous, rather than a retarded, solution. Correspondingly, there is no contribution from ϕ to the radiation at infinity.[4]

Let us turn now to eq. (3.94). Using eq. (11.30) together with the conservation equation (3.22), we get[5]

$$\frac{\partial}{\partial t}\phi(t, \mathbf{x}) = -\frac{1}{4\pi\epsilon_0} \boldsymbol{\nabla} \cdot \int d^3x' \frac{\mathbf{j}(t, \mathbf{x}')}{|\mathbf{x} - \mathbf{x}'|} . \tag{11.32}$$

Plugging eq. (11.32) into eq. (3.94) we get

$$\Box\mathbf{A} = -\mu_0 \left\{ \mathbf{j}(t, \mathbf{x}) + \boldsymbol{\nabla} \left[\boldsymbol{\nabla} \cdot \int d^3x' \frac{1}{4\pi|\mathbf{x} - \mathbf{x}'|}\mathbf{j}(t, \mathbf{x}') \right] \right\} . \tag{11.33}$$

[3] Furthermore, this contribution is time independent, since, for a system of charges moving in a bounded region, without charges flowing through the boundary coming from (or escaping to) infinity, the total charge is conserved. The first time-dependent term in the expansion of eq. (11.30) for large r comes from the dipole $4\pi\epsilon_0\phi(t, \mathbf{x}) = Q/r + \mathbf{d}(t)\cdot\hat{\mathbf{r}}/r^2 + \dots$. The first time-dependent term in ϕ is therefore proportional to $1/r^2$, and contributes to a time-dependent term $\mathcal{O}(1/r^3)$ in $\mathbf{E}$.

[4] One might be puzzled by the fact that $\phi(t, \mathbf{x})$ is determined by the instantaneous, rather than retarded, value of the charge density, since this seems to violate the postulate of Special Relativity that information cannot be transmitted faster than the speed of light. However, one should not forget that the gauge potentials are not directly observable quantities. The observable quantities are the electric and magnetic fields. Since these are gauge invariant we already know, before performing it explicitly, that the computation of $\mathbf{E}$ and $\mathbf{B}$ in the Coulomb gauge will give the same result that we found in the previous section in the Lorenz gauge. In particular, the electric field at large distances will be given by eq. (11.24), where indeed the electric field at time t depends on the behavior of the source at time $t_{\rm ret}$ (that, in the large r limit, we have approximated by $t - r/c + \hat{\mathbf{n}}\cdot\mathbf{x}'/c$).

[5] Explicitly,

$$\begin{aligned}
(4\pi\epsilon_0) &\frac{\partial}{\partial t}\phi(t, \mathbf{x}) \\
&= \int d^3x' \frac{1}{|\mathbf{x} - \mathbf{x}'|} \partial_t\rho(t, \mathbf{x}') \\
&= -\int d^3x' \frac{1}{|\mathbf{x} - \mathbf{x}'|} \boldsymbol{\nabla}_{\mathbf{x}'}\cdot\mathbf{j}(t, \mathbf{x}') \\
&= \int d^3x' \left(\boldsymbol{\nabla}_{\mathbf{x}'}\frac{1}{|\mathbf{x} - \mathbf{x}'|}\right) \cdot\mathbf{j}(t, \mathbf{x}') \\
&= -\int d^3x' \left(\boldsymbol{\nabla}_{\mathbf{x}}\frac{1}{|\mathbf{x} - \mathbf{x}'|}\right) \cdot\mathbf{j}(t, \mathbf{x}') \\
&= -\boldsymbol{\nabla}\cdot\int d^3x' \frac{1}{|\mathbf{x} - \mathbf{x}'|}\mathbf{j}(t, \mathbf{x}') ,
\end{aligned}$$

where we integrated $\boldsymbol{\nabla}_{\mathbf{x}'}$ by parts using the fact that we are considering a source localized in space, so $\mathbf{j}(t, \mathbf{x})$ has compact spatial support.

We now use the decomposition of a vector field into its longitudinal and transverse parts. As discussed in detail in Solved Problem 11.1, a general vector field $\mathbf{V}(x)$ can be decomposed as

$$\mathbf{V}(\mathbf{x}) = \mathbf{V}_\perp(\mathbf{x}) + \mathbf{V}_\parallel(\mathbf{x}) \,, \tag{11.34}$$

where

$$\mathbf{V}_\perp(\mathbf{x}) = \boldsymbol{\nabla} \times \left[\boldsymbol{\nabla} \times \int d^3x' \, \frac{\mathbf{V}(\mathbf{x}')}{4\pi|\mathbf{x}-\mathbf{x}'|} \right] \,, \tag{11.35}$$

$$\mathbf{V}_\parallel(\mathbf{x}) = -\boldsymbol{\nabla} \left[\boldsymbol{\nabla} \cdot \int d^3x' \, \frac{\mathbf{V}(\mathbf{x}')}{4\pi|\mathbf{x}-\mathbf{x}'|} \right] \,. \tag{11.36}$$

The decomposition is such that $\boldsymbol{\nabla} \cdot \mathbf{V}_\perp = 0$ and $\boldsymbol{\nabla} \times \mathbf{V}_\parallel = 0$, and defines the transverse and longitudinal components of the vector field. In Fourier space, eqs. (11.35) and (11.36) become

$$\tilde{\mathbf{V}}_\perp(\mathbf{k}) = \tilde{\mathbf{V}}(\mathbf{k}) - \hat{\mathbf{k}}\left[\hat{\mathbf{k}}\cdot\tilde{\mathbf{V}}(\mathbf{k})\right] \,, \qquad \tilde{\mathbf{V}}_\parallel(\mathbf{k}) = \hat{\mathbf{k}}\left[\hat{\mathbf{k}}\cdot\tilde{\mathbf{V}}(\mathbf{k})\right] \,, \tag{11.37}$$

and these quantities satisfy $\mathbf{k}\cdot\tilde{\mathbf{V}}_\perp(\mathbf{k}) = 0$, $\mathbf{k} \times \tilde{\mathbf{V}}_\parallel(\mathbf{k}) = 0$ (see Solved Problem 11.1 for details). So, eq. (11.33) can be written as

$$\Box\mathbf{A} = -\mu_0 \left[\mathbf{j}(t,\mathbf{x}) - \mathbf{j}_\parallel(t,\mathbf{x})\right] \,, \tag{11.38}$$

and, since $\mathbf{j}(t,\mathbf{x}) = \mathbf{j}_\perp(t,\mathbf{x}) + \mathbf{j}_\parallel(t,\mathbf{x})$,

$$\boxed{\Box\mathbf{A} = -\mu_0\mathbf{j}_\perp(t,\mathbf{x}) \,.} \tag{11.39}$$

Notice that this equation is consistent with the fact that we have derived it in the Coulomb gauge $\boldsymbol{\nabla}\cdot\mathbf{A} = 0$, precisely because, on the right-hand side, the source satisfies $\boldsymbol{\nabla}\cdot\mathbf{j}_\perp = 0$.[6] Therefore, the quantity denoted by $\mathbf{j}_\perp(t,\mathbf{x})$ in the previous section is precisely the divergence-less part of the vector field $\mathbf{j}(t,\mathbf{x})$, defined as in eq. (11.35).

Equation (11.39) can be solved using the Green's function method. As in eq. (11.1), we use the retarded Green's function (10.24) of the d'Alembertian, so

$$\mathbf{A}(t,\mathbf{x}) = \frac{\mu_0}{4\pi}\int d^3x' \, \frac{\mathbf{j}_\perp(t-|\mathbf{x}-\mathbf{x}'|/c,\mathbf{x}')}{|\mathbf{x}-\mathbf{x}'|} \,. \tag{11.43}$$

Performing the Fourier transform of $\mathbf{j}_\perp$ with respect to time, we can rewrite this as

$$\mathbf{A}(t,\mathbf{x}) = \frac{\mu_0}{4\pi}\int d^3x' \int \frac{d\omega}{2\pi} \, e^{-i\omega(t-|\mathbf{x}-\mathbf{x}'|/c)} \, \frac{\tilde{\mathbf{j}}_\perp(\omega,\mathbf{x}')}{|\mathbf{x}-\mathbf{x}'|} \,. \tag{11.44}$$

We next take the large r limit, keeping only the terms that will contribute to the $1/r$ part, so

$$\begin{aligned} \mathbf{A}(t,\mathbf{x}) &= \frac{\mu_0}{4\pi}\frac{1}{r}\int d^3x' \int \frac{d\omega}{2\pi} \, e^{-i\omega(t-r/c+\hat{\mathbf{n}}\cdot\mathbf{x}'/c)} \, \tilde{\mathbf{j}}_\perp(\omega,\mathbf{x}') \\ &= \frac{\mu_0}{4\pi}\frac{1}{r}\int \frac{d\omega}{2\pi} \, e^{-i\omega(t-r/c)} \int d^3x' \, e^{-i\omega\hat{\mathbf{n}}\cdot\mathbf{x}'/c}\tilde{\mathbf{j}}_\perp(\omega,\mathbf{x}') \\ &= \frac{\mu_0}{4\pi}\frac{1}{r}\int \frac{d\omega}{2\pi} \, e^{-i\omega(t-r/c)}\tilde{\mathbf{j}}_\perp(\omega,\omega\hat{\mathbf{n}}/c) \,. \end{aligned} \tag{11.45}$$

---

[6]Observe that the quantity $\tilde{\mathbf{j}}_\perp(\omega,\omega\hat{\mathbf{n}}/c)$ that appeared in the Lorenz gauge computation, and that was defined in eq. (11.19), is the transverse part of $\tilde{\mathbf{j}}(\omega,\omega\hat{\mathbf{n}}/c)$ with respect to the unit vector $\hat{\mathbf{n}}$, i.e., it satisfies

$$\hat{\mathbf{n}}\cdot\tilde{\mathbf{j}}_\perp(\omega,\omega\hat{\mathbf{n}}/c) = 0 \,. \tag{11.40}$$

This is just the Fourier transform, with respect to both time and space, of the condition

$$\boldsymbol{\nabla}\cdot\mathbf{j}_\perp(t,\mathbf{x}) = 0 \,, \tag{11.41}$$

that defines the quantity $\mathbf{j}_\perp(t,\mathbf{x})$ that appears in eq. (11.35). In fact, from eq. (11.40) we get

$$\begin{aligned} 0 &= \hat{\mathbf{n}}\cdot\tilde{\mathbf{j}}_\perp(\omega,\omega\hat{\mathbf{n}}/c) \\ &= \int d^3x'dt' \, e^{i\omega t'-i\omega\hat{\mathbf{n}}\cdot\mathbf{x}'/c} \\ &\quad \times \hat{\mathbf{n}}\cdot\mathbf{j}_\perp(t',\mathbf{x}') \\ &= \frac{ic}{\omega}\int d^3x'dt' \, e^{i\omega t'} \\ &\quad \times \left[\boldsymbol{\nabla}_{\mathbf{x}'}e^{-i\omega\hat{\mathbf{n}}\cdot\mathbf{x}'/c}\right]\cdot\mathbf{j}_\perp(t',\mathbf{x}') \\ &= -\frac{ic}{\omega}\int d^3x'dt' \, e^{i\omega t'-i\omega\hat{\mathbf{n}}\cdot\mathbf{x}'/c} \\ &\quad \times \boldsymbol{\nabla}_{\mathbf{x}'}\cdot\mathbf{j}_\perp(t',\mathbf{x}') \,, \end{aligned} \tag{11.42}$$

where in the last line we have integrated by parts, assuming that $\mathbf{j}_\perp(t',\mathbf{x}')$ is localized. Therefore, $\hat{\mathbf{n}}\cdot\tilde{\mathbf{j}}_\perp(\omega,\omega\hat{\mathbf{n}}/c) = 0$ is equivalent to $\boldsymbol{\nabla}_{\mathbf{x}}\cdot\mathbf{j}_\perp(t,\mathbf{x}) = 0$. There is actually a subtlety for $\omega = 0$, since, from eq. (11.42), the vanishing of $\boldsymbol{\nabla}_{\mathbf{x}'}\cdot\mathbf{j}_\perp(t',\mathbf{x}')$ only implies the vanishing of $\omega\hat{\mathbf{n}}\cdot\mathbf{j}_\perp(\omega,\omega\hat{\mathbf{n}}/c)$, and this is automatically satisfied if $\omega = 0$, without the need of imposing $\hat{\mathbf{n}}\cdot\tilde{\mathbf{j}}_\perp(\omega,\omega\hat{\mathbf{n}}/c)|_{\omega=0} = 0$. However, a Fourier mode with $\mathbf{k} = \omega\hat{\mathbf{n}}/c = 0$ corresponds to a spatially constant term, which is eliminated by the boundary condition that $\mathbf{j}(t,\mathbf{x})$ is localized in space.

Finally, we have seen that, in the Coulomb gauge, the term $-\boldsymbol{\nabla}\phi$ does not contribute to order $1/r$, so keeping only the term $O(1/r)$ we get

$$
\begin{aligned}
\mathbf{E}(t,\mathbf{x}) &= -\frac{\partial \mathbf{A}}{\partial t} \\
&= \frac{\mu_0}{4\pi}\frac{1}{r}\int\frac{d\omega}{2\pi}\,e^{-i\omega[t-(r/c)]}\,i\omega\,\tilde{\mathbf{j}}_\perp(\omega,\omega\hat{\mathbf{n}}/c) \quad (11.46) \\
&= \frac{1}{4\pi\epsilon_0 c^2}\frac{1}{r}\int\frac{d\omega}{2\pi}\,e^{-i\omega[t-(r/c)]}\,i\omega\,\tilde{\mathbf{j}}_\perp(\omega,\omega\hat{\mathbf{n}}/c)\,. \quad (11.47)
\end{aligned}
$$

This agrees with eq. (11.21) [and therefore also with eq. (11.24)], as it should for a gauge invariant quantity computed in two different gauges.

11.1.3 Radiated power and spectral distribution

As we saw in eq. (3.35), the energy radiated per unit time through an infinitesimal surface $d\mathbf{s}$ is given by $d\mathbf{s}\cdot\mathbf{S}$, where $\mathbf{S}$ is the Poynting vector, given by eq. (3.34). We consider a sphere at large distance from the region where the source is localized, with its origin inside the localization region, so $d\mathbf{s} = r^2 d\Omega\,\hat{\mathbf{n}}$, where $\hat{\mathbf{n}}$ is the unit normal in the radial direction, and we take the large r limit. For the electric field we then use eq. (11.25), which is exact as far as the term $1/r$ is concerned while, for the magnetic field, we have seen the $1/r$ contribution is given by eq. (11.27). Then, to order $1/r^2$, the exact result for the Poynting vector is

$$
\begin{aligned}
\mathbf{S} &= \frac{1}{\mu_0 c}E^2\hat{\mathbf{n}} \\
&= \frac{\mu_0}{16\pi^2 c}\frac{1}{r^2}\left|\partial_t\int d^3x'\,\mathbf{j}_\perp(t_{\text{ret}},\mathbf{x}')\right|^2\hat{\mathbf{n}}\,, \quad (11.48)
\end{aligned}
$$

or, in terms of ϵ_0,

$$
\mathbf{S} = \frac{1}{4\pi\epsilon_0}\frac{1}{4\pi c^3 r^2}\left|\partial_t\int d^3x'\,\mathbf{j}_\perp(t_{\text{ret}},\mathbf{x}')\right|^2\hat{\mathbf{n}}\,. \quad (11.49)
$$

The power $d\mathcal{E}/dt$ radiated in the solid angle $d\Omega$ is obtained performing the scalar product with $d\mathbf{s} = r^2 d\Omega\,\hat{\mathbf{n}}$,[7] so

$$
\frac{d\mathcal{E}}{dt d\Omega} = \frac{\mu_0}{16\pi^2 c}\left|\partial_t\int d^3x'\,\mathbf{j}_\perp(t_{\text{ret}},\mathbf{x}')\right|^2\,, \quad (11.50)
$$

or, in terms of ϵ_0,

$$
\boxed{\frac{d\mathcal{E}}{dt d\Omega} = \frac{1}{4\pi\epsilon_0}\frac{1}{4\pi c^3}\left|\partial_t\int d^3x'\,\mathbf{j}_\perp(t_{\text{ret}},\mathbf{x}')\right|^2\,.} \quad (11.51)
$$

[7]Observe that here we are considering the power per unit time t, rather than per unit retarded time t_{ret}, i.e., the "received" power P, see eq. (10.151) and the discussion below it. Observe also that we denote the energy by $\mathcal{E}$, reserving the symbol E for the modulus of the electric field.

[8]Explicitly,

$$
\int_{-\infty}^{+\infty} dt \, |f(t)|^2
$$

$$
= \int_{-\infty}^{+\infty} dt \int_{-\infty}^{+\infty} \frac{d\omega}{2\pi} \, \tilde{f}(\omega) e^{-i\omega t}
$$

$$
\times \int_{-\infty}^{+\infty} \frac{d\omega'}{2\pi} \, \tilde{f}^*(\omega') e^{i\omega' t}
$$

$$
= \int_{-\infty}^{+\infty} \frac{d\omega}{2\pi} \frac{d\omega'}{2\pi} \, \tilde{f}(\omega) \tilde{f}^*(\omega')
$$

$$
\times \int_{-\infty}^{+\infty} dt \, e^{i(\omega' - \omega)t}
$$

$$
= \int_{-\infty}^{+\infty} \frac{d\omega}{2\pi} \frac{d\omega'}{2\pi} \, \tilde{f}(\omega) \tilde{f}^*(\omega')
$$

$$
\times 2\pi \delta(\omega' - \omega)
$$

$$
= \int_{-\infty}^{+\infty} \frac{d\omega}{2\pi} \, |\tilde{f}(\omega)|^2 ,
$$

[9]In the physics literature, it is also often called the Parseval theorem. More accurately, the Parseval theorem is a discrete version of this result, based on Fourier series rather than on Fourier integrals.

[10]Explicitly,

$$
\tilde{\mathbf{f}}(\omega) = \int_{-\infty}^{+\infty} dt \, e^{i\omega t}
$$

$$
\times \partial_t \int d^3x' \, \mathbf{j}_\perp(t_{\mathrm{ret}}, \mathbf{x}')
$$

$$
= -\int_{-\infty}^{+\infty} dt \, (\partial_t e^{i\omega t})
$$

$$
\times \int d^3x' \, \mathbf{j}_\perp(t_{\mathrm{ret}}, \mathbf{x}')
$$

$$
= -i\omega \int_{-\infty}^{+\infty} dt \, e^{i\omega t}
$$

$$
\times \int d^3x' \, \mathbf{j}_\perp(t_{\mathrm{ret}}, \mathbf{x}')
$$

$$
= -i\omega \int d^3x' \int_{-\infty}^{+\infty} dt
$$

$$
\times e^{i\omega(t_{\mathrm{ret}} + r/c - \hat{\mathbf{n}} \cdot \mathbf{x}'/c)} \, \mathbf{j}_\perp(t_{\mathrm{ret}}, \mathbf{x}')
$$

$$
= -i\omega e^{i\omega r/c} \int d^3x' \, e^{-i\omega \hat{\mathbf{n}} \cdot \mathbf{x}'/c}
$$

$$
\times \int_{-\infty}^{+\infty} dt_{\mathrm{ret}} \, e^{i\omega t_{\mathrm{ret}}} \, \mathbf{j}_\perp(t_{\mathrm{ret}}, \mathbf{x}')
$$

$$
= -i\omega e^{i\omega r/c} \tilde{\mathbf{j}}_\perp(\omega, \omega \hat{\mathbf{n}}/c) ,
$$

where we have used the expression of t_{ret} valid at large r, $t_{\mathrm{ret}} = t - r/c + \hat{\mathbf{n}} \cdot \mathbf{x}'/c$, to express t in terms of t_{ret}, and then the fact that, at fixed $\mathbf{x}'$, $\int_{-\infty}^{+\infty} dt = \int_{-\infty}^{+\infty} dt_{\mathrm{ret}}$.

Consider now a source that acts only for a finite amount of time. In this case, the total radiated energy per unit angle is finite, and is given by

$$
\frac{d\mathcal{E}}{d\Omega} = \int_{-\infty}^{+\infty} dt \, \frac{d\mathcal{E}}{dt \, d\Omega}
$$

$$
= \frac{1}{4\pi\epsilon_0} \frac{1}{4\pi c^3} \int_{-\infty}^{+\infty} dt \left| \partial_t \int d^3x' \, \mathbf{j}_\perp(t_{\mathrm{ret}}, \mathbf{x}') \right|^2 . \quad (11.52)
$$

We now use the fact that, given a square-integrable function $f(t)$, we have the identity[8]

$$
\int_{-\infty}^{+\infty} dt \, |f(t)|^2 = \int_{-\infty}^{+\infty} \frac{d\omega}{2\pi} \, |\tilde{f}(\omega)|^2 , \quad (11.53)
$$

which is known as the *Plancherel theorem*.[9] If, furthermore, $f(t)$ is real, then $\tilde{f}(\omega) = \tilde{f}^*(-\omega)$, so $|\tilde{f}(\omega)|^2 = |\tilde{f}(-\omega)|^2$, and eq. (11.53) can be written as

$$
\int_{-\infty}^{+\infty} dt \, [f(t)]^2 = 2 \int_0^{\infty} \frac{d\omega}{2\pi} \, |\tilde{f}(\omega)|^2 . \quad (11.54)
$$

We can apply this to the real (vector) function

$$
\mathbf{f}(t) \equiv \partial_t \int d^3x' \, \mathbf{j}_\perp(t_{\mathrm{ret}}, \mathbf{x}') , \quad (11.55)
$$

that appears in eq. (11.51). (Note that this function is real; the modulus that appears in eq. (11.51) refers, of course, to the modulus in the sense of vectors, $|\mathbf{f}|^2 = f_i f_i$). Its Fourier transform is given by[10]

$$
\tilde{\mathbf{f}}(\omega) = -i\omega e^{i\omega r/c} \tilde{\mathbf{j}}_\perp(\omega, \omega \hat{\mathbf{n}}/c) . \quad (11.56)
$$

Then, from eqs. (11.52), (11.54), and (11.56),

$$
\frac{d\mathcal{E}}{d\Omega} = \frac{1}{4\pi\epsilon_0} \frac{1}{2\pi c^3} \int_0^{\infty} \frac{d\omega}{2\pi} \, \omega^2 |\tilde{\mathbf{j}}_\perp(\omega, \omega \hat{\mathbf{n}}/c)|^2 . \quad (11.57)
$$

Note that the Fourier modes $\tilde{\mathbf{j}}_\perp$ are complex, and $|\tilde{\mathbf{j}}_\perp|^2$ is now a notation for $|(\tilde{\mathbf{j}}_\perp)_i (\tilde{\mathbf{j}}_\perp)_i|$, i.e., we take the modulus square of the vector and the modulus of the complex number. We can rewrite eq. (11.57) in the form

$$
\frac{d\mathcal{E}}{d\Omega \, d\omega} = \frac{1}{4\pi\epsilon_0} \frac{1}{4\pi^2 c^3} \, \omega^2 |\tilde{\mathbf{j}}_\perp(\omega, \omega \hat{\mathbf{n}}/c)|^2 , \quad (11.58)
$$

or, in terms of μ_0,

$$
\frac{d\mathcal{E}}{d\Omega \, d\omega} = \frac{\mu_0}{16\pi^3 c} \, \omega^2 |\tilde{\mathbf{j}}_\perp(\omega, \omega \hat{\mathbf{n}}/c)|^2 . \quad (11.59)
$$

This gives the total energy radiated by the source, per unit solid angle and unit frequency, i.e., the energy spectrum per unit solid angle. The total energy spectrum is obtained integrating over the solid angle,

$$
\frac{d\mathcal{E}}{d\omega} = \frac{\mu_0}{16\pi^3 c} \, \omega^2 \int d\Omega \, |\tilde{\mathbf{j}}_\perp(\omega, \omega \hat{\mathbf{n}}/c)|^2 . \quad (11.60)
$$

Observe that the angular dependence, over which we integrate, enters through the unit vector in the radial direction, $\hat{\mathbf{n}}$. In polar coordinates, $d\Omega = d\cos\theta d\phi$, and

$$\hat{\mathbf{n}} = (\sin\theta\cos\phi, \sin\theta\sin\phi, \cos\theta)\,. \tag{11.61}$$

Writing the result in terms of $\tilde{\mathbf{j}}_\perp$ gives a nicely compact expression. Furthermore, we see from eq. (11.28) that the direction of $\tilde{\mathbf{E}}(\omega,\mathbf{x})$ is the same as that of $\tilde{\mathbf{j}}_\perp$, so this expression also allows us to easily read the polarization of the radiation. Two alternative expressions, however, can also be useful. First, using eq. (11.20) and observing that, for any vector $\mathbf{v}$,

$$|\hat{\mathbf{n}}\times(\hat{\mathbf{n}}\times\mathbf{v})| = |\hat{\mathbf{n}}\times\mathbf{v}|\,, \tag{11.62}$$

we can rewrite eq. (11.58) as[11]

$$\boxed{\frac{d\mathcal{E}}{d\Omega d\omega} = \frac{1}{4\pi\epsilon_0}\frac{1}{4\pi^2 c^3}\omega^2\left|\hat{\mathbf{n}}\times\tilde{\mathbf{j}}\left(\omega,\frac{\omega\hat{\mathbf{n}}}{c}\right)\right|^2\,.} \tag{11.65}$$

Alternatively, we can rewrite the result in terms of $\tilde{\mathbf{j}}$ and $\tilde{\rho}$, using eq. (11.19) and observing that the continuity equation in momentum space, eq. (11.16), gives

$$\hat{\mathbf{n}}\cdot\tilde{\mathbf{j}}(\omega,\omega\hat{\mathbf{n}}/c) = c\tilde{\rho}(\omega,\omega\hat{\mathbf{n}}/c)\,, \tag{11.66}$$

and therefore

$$\tilde{\mathbf{j}}_\perp(\omega,\omega\hat{\mathbf{n}}/c) = \tilde{\mathbf{j}}(\omega,\omega\hat{\mathbf{n}}/c) - \hat{\mathbf{n}}c\tilde{\rho}(\omega,\omega\hat{\mathbf{n}}/c)\,. \tag{11.67}$$

[We have indeed simply undone the passages that led from eq. (11.11) to eq. (11.18)]. Then, suppressing temporarily for notational simplicity the argument $(\omega,\omega\hat{\mathbf{n}}/c)$,

$$\begin{aligned}|\tilde{\mathbf{j}}_\perp|^2 &= (\tilde{\mathbf{j}} - \hat{\mathbf{n}}c\tilde{\rho})\cdot(\tilde{\mathbf{j}} - \hat{\mathbf{n}}c\tilde{\rho})^* \\ &= |\tilde{\mathbf{j}}|^2 + c^2|\tilde{\rho}|^2 - 2\mathrm{Re}\left[c\tilde{\rho}^*\hat{\mathbf{n}}\cdot\tilde{\mathbf{j}}\right]\,.\end{aligned} \tag{11.68}$$

From eq. (11.67), together with $\hat{\mathbf{n}}\cdot\tilde{\mathbf{j}}_\perp = 0$, it follows that

$$\hat{\mathbf{n}}\cdot\tilde{\mathbf{j}} = c\tilde{\rho}\,, \tag{11.69}$$

and therefore eq. (11.68) becomes

$$|\tilde{\mathbf{j}}_\perp|^2 = |\tilde{\mathbf{j}}|^2 - c^2|\tilde{\rho}|^2\,. \tag{11.70}$$

Therefore, we can also rewrite eq. (11.59) as

$$\frac{d\mathcal{E}}{d\Omega d\omega} = \frac{\mu_0}{16\pi^3 c}\omega^2\left(|\tilde{\mathbf{j}}(\omega,\omega\hat{\mathbf{n}}/c)|^2 - c^2|\tilde{\rho}(\omega,\omega\hat{\mathbf{n}}/c)|^2\right)\,, \tag{11.71}$$

or, in terms of ϵ_0,

$$\boxed{\frac{d\mathcal{E}}{d\Omega d\omega} = \frac{1}{4\pi\epsilon_0}\frac{1}{4\pi^2 c^3}\omega^2\left(|\tilde{\mathbf{j}}(\omega,\omega\hat{\mathbf{n}}/c)|^2 - c^2|\tilde{\rho}(\omega,\omega\hat{\mathbf{n}}/c)|^2\right)\,.} \tag{11.72}$$

[11]Note that we have defined the spectral density $d\mathcal{E}/d\omega d\Omega$ as the quantity that gives the total radiated energy per unit solid angle, when integrated over $d\omega$ from $\omega = 0$ to $\omega = \infty$ (rather than from $-\infty$ to $+\infty$), see eq. (11.57). This is also called a "one-sided spectral density." If one rather uses a "two-sided spectral density," which is the quantity that gives the the total radiated energy per unit angle when integrated over $d\omega$ from $\omega = -\infty$ to $\omega = \infty$, then eq. (11.58) is replaced by

$$\frac{d\mathcal{E}}{d\Omega d\omega} = \frac{1}{4\pi\epsilon_0}\frac{1}{8\pi^2 c^3}\omega^2|\tilde{\mathbf{j}}_\perp(\omega,\omega\hat{\mathbf{n}}/c)|^2\,, \tag{11.63}$$

and eq. (11.65) is replaced by

$$\begin{aligned}\frac{d\mathcal{E}}{d\Omega d\omega} &= \frac{1}{4\pi\epsilon_0}\frac{1}{8\pi^2 c^3} \\ &\times\omega^2\left|\hat{\mathbf{n}}\times\tilde{\mathbf{j}}\left(\omega,\frac{\omega\hat{\mathbf{n}}}{c}\right)\right|^2\,.\end{aligned} \tag{11.64}$$

11.2 Low-velocity limit and multipole expansion of the radiation field

We now go back to the expression for the large-r limit of the gauge potentials. We work in the Lorenz gauge, so we use eqs. (11.5) and (11.6), that we rewrite here, inverting the order of the integrals, as,

$$\phi(t,\mathbf{x}) = \frac{1}{4\pi\epsilon_0}\frac{1}{r}\int \frac{d\omega'}{2\pi}\,e^{-i\omega'(t-r/c)}\int d^3x'\,\tilde{\rho}(\omega',\mathbf{x}')\,e^{-i\omega'\hat{\mathbf{n}}\cdot\mathbf{x}'/c},\quad (11.73)$$

$$\mathbf{A}(t,\mathbf{x}) = \frac{1}{4\pi\epsilon_0 c^2}\frac{1}{r}\int \frac{d\omega'}{2\pi}\,e^{-i\omega'(t-r/c)}\int d^3x'\,\tilde{\mathbf{j}}(\omega',\mathbf{x}')\,e^{-i\omega'\hat{\mathbf{n}}\cdot\mathbf{x}'/c}.\quad(11.74)$$

We have put a prime also over ω, to stress that it is an integration variable. To obtain these expressions we had assumed that the source, given in general by a distribution of moving charges, is localized in a region of space $|\mathbf{x}'| < d$, and we kept only the leading term in the limit $r \gg d$, which is $\mathcal{O}(1/r)$. On the other hand, the computation of this $1/r$ term was exact. In particular, we made no assumption on the velocity of the particles that creates this distribution of charge and current. We now consider the case in which the velocities of the charges are non-relativistic, $v \ll c$.

Consider first the case in which the source performs a simple harmonic motion, with angular frequency ω_s confined to a region of size d; for instance, a point charge in a circular orbit of radius d and angular frequency ω_s. The typical velocity v of such a charge is of order $\omega_s d$ and the condition $v/c \ll 1$ becomes

$$\frac{\omega_s d}{c} \ll 1.\quad (11.75)$$

More generally, a source with more complex internal motions will be characterized by a superposition of Fourier modes, and will generate a distribution of charge density and current density which is not monochromatic, but rather described by $\tilde{\rho}(\omega',\mathbf{x}')$ and $\tilde{\mathbf{j}}(\omega',\mathbf{x}')$. The non-relativistic limit is applicable when the Fourier modes $\tilde{\rho}(\omega',\mathbf{x}')$ and $\tilde{\mathbf{j}}(\omega',\mathbf{x}')$ are sizable only for values of ω' such that $\omega'd/c \ll 1$, and go quickly to zero for larger values of ω'. This means that only values of ω' such that $\omega'd/c \ll 1$ contribute significantly to the integrals over $d\omega'$ in eqs. (11.73) and (11.74).[12] At the same time, the integration over d^3x' in eqs. (11.73) and (11.74) is restricted to the region $|\mathbf{x}'| < d$, since the source is localized. Therefore, the term $\omega'\hat{\mathbf{n}}\cdot\mathbf{x}'/c$, that appears in the exponentials in eqs. (11.73) and (11.74), is always much smaller than one (in absolute value) whenever the rest of the integrand is sizable, and we can expand the exponentials in powers of it. As we will see, this gives rise to an expansion in time-dependent (or "radiative") multipoles.

Before entering into the technicalities of this expansion, let us better understand its physical meaning. As we will confirm below with the explicit computation, the frequency ω of the radiation emitted by a source whose motion has a typical frequency ω_s, will also be of order of

[12]Note that here we are only concerned with the "internal motions" of the source, i.e., a superposition of periodic motions, with non-zero frequencies ω', localized within a bounded region of space (for which, therefore, the Fourier transform is well defined). Any bulk motion of the center-of-mass of the source can then be included performing first the computation of the electromagnetic field in a reference frame where the center-of-mass is at rest, and then performing a boost, similarly to what we did in Section 10.3 for a point charge.

ω_s (with a numerical coefficient of order one that depends on the order of the multipole expansion, see below; we already saw in Problems 10.1 and 10.2 that, for dipole radiation, $\omega = \omega_s$) and therefore $\omega \sim \omega_s \sim v/d$. In terms of the reduced wavelength $\lambdabar = c/\omega$ defined in eq. (9.56), this means

$$\lambdabar \sim \frac{c}{v} d \,. \tag{11.76}$$

In a non-relativistic system we have $v \ll c$, and therefore the reduced wavelength of the radiation emitted is much larger than the size of the system that generates it:

$$\boxed{\text{non-relativistic sources} \quad \Longrightarrow \quad \lambdabar \gg d \,.} \tag{11.77}$$

When the reduced wavelength is much larger than the size of the system, we expect that we do not need to know the internal motions of the source in all its details, and only the coarse features of the distribution of the charge and current densities, as encoded in their lowest multipole moments, should be relevant. This physical intuition will be confirmed by the computation that we perform below, where we will verify that the expansion in powers of $\omega' \hat{\mathbf{n}} \cdot \mathbf{x}' / c$ gives rise to an expansion in time-dependent multipole moments. We will then compare with the expansion in static multipole moments that we performed in Chapter 6.

We now perform explicitly the expansion of the scalar and vector potentials.[13] From eq. (11.73),

$$\phi(t, \mathbf{x}) = \frac{1}{4\pi\epsilon_0} \frac{1}{r} \int \frac{d\omega'}{2\pi} e^{-i\omega'(t - r/c)} \int d^3x'\, \tilde{\rho}(\omega', \mathbf{x}')$$

$$\times \left[1 - \frac{i\omega'}{c} \hat{n}_i x'_i + \frac{1}{2} \left(\frac{-i\omega'}{c} \right)^2 \hat{n}_i \hat{n}_j x'_i x'_j + \dots \right]. \tag{11.78}$$

The first term gives simply the total charge q of the source,

$$\int \frac{d\omega'}{2\pi} e^{-i\omega'(t - r/c)} \int d^3x'\, \tilde{\rho}(\omega', \mathbf{x}')$$

$$= \int d^3x' \int \frac{d\omega'}{2\pi} e^{-i\omega'(t - r/c)} \tilde{\rho}(\omega', \mathbf{x}')$$

$$= \int d^3x'\, \rho(t - r/c, \mathbf{x}')$$

$$= q(t - r/c) \,. \tag{11.79}$$

However, we assumed that the motion of the source is localized inside a finite volume, so no charges are escaping to infinity (or coming from infinity). Therefore the charge is actually time independent,

$$q(t - r/c) = q \,. \tag{11.80}$$

The second term in the expansion is related to the time derivative of the

[13] Note that, if we are interested only in the electric and magnetic fields, we could directly perform the expansion in powers of $\omega' \hat{\mathbf{n}} \cdot \mathbf{x}' / c$ on the electric field, starting from eq. (11.22), and recover the magnetic field in the far region from eq. (11.27). However, the structure of the expansion is somewhat clearer, conceptually, starting from the gauge potentials. Furthermore, especially at the quantum level, one can be interested in the multipole expansion of the gauge potentials themselves.

electric dipole of the charge distribution. Indeed,

$$
\int \frac{d\omega'}{2\pi} e^{-i\omega'(t-r/c)} \int d^3x' \, \tilde{\rho}(\omega', \mathbf{x}') \left(\frac{-i\omega'}{c} \right) \hat{n}_i x_i'
$$

$$
= \frac{\hat{n}_i}{c} \frac{\partial}{\partial t} \int \frac{d\omega'}{2\pi} e^{-i\omega'(t-r/c)} \int d^3x' \, \tilde{\rho}(\omega', \mathbf{x}') x_i'
$$

$$
= \frac{\hat{n}_i}{c} \frac{\partial}{\partial t} \int d^3x' \, \rho(t - r/c, \mathbf{x}') x_i'
$$

$$
= \frac{\hat{n}_i}{c} \frac{\partial}{\partial t} d_i(t - r/c), \tag{11.81}
$$

where the time-dependent electric dipole moment of a charge distribution is defined by

$$
\mathbf{d}(t) = \int d^3x' \, \rho(t, \mathbf{x}') \mathbf{x}', \tag{11.82}
$$

which generalizes the definition (6.8) that we introduced for the static case.[14] The third term of the expansion can be transformed as follows:

$$
\int \frac{d\omega'}{2\pi} e^{-i\omega'(t-r/c)} \int d^3x' \, \tilde{\rho}(\omega', \mathbf{x}') \left(\frac{-i\omega'}{c} \right)^2 \hat{n}_i \hat{n}_j x_i' x_j'
$$

$$
= \frac{\hat{n}_i \hat{n}_j}{c^2} \frac{\partial^2}{\partial t^2} \int d^3x' \, \rho(t - r/c, \mathbf{x}') x_i' x_j' . \tag{11.83}
$$

The result therefore depends on the second moment of the charge distribution,

$$
D_{ij}(t) \equiv \int d^3x' \, \rho(t, \mathbf{x}') x_i' x_j' . \tag{11.84}
$$

Actually, just as in the static case, it is also convenient to introduce the quadrupole moment of the charge distribution, which is defined by

$$
Q_{ij}(t) = \int d^3x' \, \rho(t, \mathbf{x}') \left(3x_i' x_j' - \delta_{ij} |\mathbf{x}'|^2 \right) . \tag{11.85}
$$

Equation (11.85) is the generalization of eq. (6.18) to a time-dependent charge density. Comparing eqs. (11.84) and (11.85),

$$
D_{ij}(t) = \frac{1}{3} Q_{ij}(t) + \frac{1}{3} \delta_{ij} \int d^3x' \, |\mathbf{x}'|^2 \rho(t, \mathbf{x}') . \tag{11.86}
$$

For the moment, we will however write the result in terms of D_{ij}, which gives a more compact expression. Putting together the various terms, we get

$$
\phi(t, \mathbf{x}) = \frac{1}{4\pi\epsilon_0} \frac{1}{r} \left[q + \frac{1}{c} \hat{n}_i \dot{d}_i(t - r/c) + \frac{1}{2} \frac{\hat{n}_i \hat{n}_j}{c^2} \ddot{D}_{ij}(t - r/c) + \dots \right] .
$$

$$
\tag{11.87}
$$

The analogous expansion for $\mathbf{A}$ is obtained from eq. (11.74).

$$
\begin{aligned}
A_i(t, \mathbf{x}) \;=\;& \frac{1}{4\pi\epsilon_0 c^2}\frac{1}{r}\left[\int d^3x'\, j_i(t - r/c, \mathbf{x}')\right.\\
&\left. + \frac{\hat{n}_j}{c}\frac{\partial}{\partial t}\int d^3x'\, j_i(t - r/c, \mathbf{x}')x'_j + \ldots\right].
\end{aligned}
$$

$$\tag{11.88}$$

The electric field in the radiation zone is obtained inserting these expressions for the gauge potentials in eq. (3.83), and the magnetic field in the radiation zone is then obtained from eq. (11.27). We are only interested in the terms $\mathcal{O}(1/r)$ in $\mathbf{E}$, and in the corresponding terms in $\mathbf{B}$, since these are the only ones that give a contribution $\mathcal{O}(1/r^2)$ to the Poynting vector and therefore to the radiation flux at infinity. Let us begin by considering the contribution of $\boldsymbol{\nabla}\phi$ to $\mathbf{E}$. When the gradient acts on the overall $1/r$ factor in eq. (11.87) it produces a term $\mathcal{O}(1/r^2)$, which does not contribute to the radiation field. Then, in particular, the Coulomb term proportional to q/r in eq. (11.87) does not contribute to the radiation field, and

$$
\boldsymbol{\nabla}\phi = \frac{1}{4\pi\epsilon_0}\frac{1}{r}\boldsymbol{\nabla}\left[\frac{1}{c}\hat{n}_i \dot{d}_i(t - r/c) + \frac{1}{2}\frac{\hat{n}_i\hat{n}_j}{c^2}\ddot{D}_{ij}(t - r/c) + \ldots\right] + \mathcal{O}\left(\frac{1}{r^2}\right).
$$

$$\tag{11.89}$$

Similarly, from eq. (11.8) we see that when the gradient acts on the factors $\hat{n}_i$, $\hat{n}_i\hat{n}_j$, and so on, that appear inside the brackets, it generates an extra $1/r$ term, so that, overall, the corresponding contribution to $\mathbf{E}$ is again $\mathcal{O}(1/r^2)$. Therefore, we only need to apply the gradient to the multipole moments,

$$
\begin{aligned}
\boldsymbol{\nabla}\phi \;=\;& \frac{1}{4\pi\epsilon_0}\frac{1}{r}\left[\frac{1}{c}\hat{n}_i\boldsymbol{\nabla}\dot{d}_i(t - r/c) + \frac{1}{2}\frac{\hat{n}_i\hat{n}_j}{c^2}\boldsymbol{\nabla}\ddot{D}_{ij}(t - r/c) + \ldots\right]\\
& + \mathcal{O}\left(\frac{1}{r^2}\right).
\end{aligned}
$$

$$\tag{11.90}$$

The multipole moments depend on the spatial variables only through r and, for a generic function $f(t - r/c)$, we have

$$
\begin{aligned}
\boldsymbol{\nabla}f(t - r/c) \;&=\; \hat{n}\,\partial_r f(t - r/c)\\
&=\; -\frac{\hat{n}}{c}\,\partial_t f(t - r/c),
\end{aligned}
$$

$$\tag{11.91}$$

so in this case the gradient does not produces extra factors of $1/r$. Then

$$
\begin{aligned}
-\boldsymbol{\nabla}\phi \;=\;& \frac{\hat{n}}{4\pi\epsilon_0}\frac{1}{r}\left[\frac{1}{c^2}\hat{n}_i\ddot{d}_i(t - r/c) + \frac{1}{2}\frac{\hat{n}_i\hat{n}_j}{c^3}\dddot{D}_{ij}(t - r/c) + \ldots\right]\\
& + \mathcal{O}\left(\frac{1}{r^2}\right).
\end{aligned}
$$

$$\tag{11.92}$$

There is, therefore, an infinite series of terms contributing to the electric field at order $1/r$, generated by the higher and higher multipoles. These contributions correspond to an expansion in powers of v/c. In fact, for a source with a generic structure, for dimensional reasons each higher-order multipole carries one more power of the only length-scale in the problem, which is the typical size d of the system. For instance, the monopole is just the total electric charge q; apart from dimensionless numbers, the electric dipole d_i is of order qd, the quadrupole Q_{ij} is of order qd^2, etc. Furthermore, each time derivative produces one power of the typical frequency of the system, so, for instance, $\dot{d}_i(t-r/c) \sim \omega_s qd \sim qv$, where $v \sim \omega_s d$ is the typical velocity of the internal motions of the source. Similarly, $\ddot{d}_i(t-r/c) \sim \omega_s^2 qd \sim qv^2/d$, and $\dddot{D}_{ij}(t-r/c) \sim \omega_s^3 qd^2 \sim qv^3/d$. Then, we see that, in eq. (11.92),

$$\frac{1}{c^2}\hat{n}_i \ddot{d}_i(t-r/c) \sim \frac{q}{d}\frac{v^2}{c^2}\,, \qquad (11.93)$$

$$\frac{1}{c^3}\hat{n}_i \hat{n}_j \dddot{D}_{ij}(t-r/c) \sim \frac{q}{d}\frac{v^3}{c^3}\,, \qquad (11.94)$$

and so on: each further term in the bracket has one more time derivative, giving an extra ω_s factor; an extra factor d coming simply because each subsequent multipole moment has an extra power of x^i in its definition; and an extra factor $1/c$ from the expansion of $e^{-i\omega' \hat{\mathbf{n}} \cdot \mathbf{x}'/c}$ leading, overall, to an extra factor of order $\omega_s d/c \sim v/c$ with respect to the previous term.

Similar estimates can be applied to the contribution of $\mathbf{A}$ to the electric field. From eq. (11.88),

$$-\frac{\partial \mathbf{A}}{\partial t} = -\frac{1}{4\pi\epsilon_0}\frac{1}{r}\left[\frac{1}{c^2}\partial_t \int d^3x'\, \mathbf{j}(t-r/c, \mathbf{x}')\right.$$
$$\left. +\frac{\hat{n}_j}{c^3}\partial_t^2 \int d^3x'\, \mathbf{j}(t-r/c, \mathbf{x}')x'_j + \dots\right]\,. \qquad (11.95)$$

We now observe that $\mathbf{j}$ is of order qv, so, in order of magnitude,

$$\frac{1}{c^2}\partial_t \mathbf{j}(t-r/c, \mathbf{x}') \sim \frac{\omega_s qv}{c^2} \sim \frac{q}{d}\frac{v^2}{c^2}\,, \qquad (11.96)$$

and

$$\frac{\hat{n}_j}{c^3}\partial_t^2 \mathbf{j}(t-r/c, \mathbf{x}')x'_j \sim \frac{\omega_s^2 qvd}{c^3} \sim \frac{q}{d}\frac{v^3}{c^3}\,, \qquad (11.97)$$

and so on for the higher-order terms. In summary:

- The structure of the multipole expansion of the radiation field is quite different from the multipole expansion of the static fields studied in Chapter 6. In the static case, we were interested in the fields at distances r large compared to the size d of the system, and we wanted to compute the corrections, subleading in d/r, with respect to the leading Coulomb term. These corrections are parametrized by the multipoles of the source, which in this case are time-independent and are therefore often called the static multipoles. For instance, the electric dipole produces a contribution to the scalar potential ϕ of order $1/r^2$, see eq. (6.9), compared to the $1/r$ behavior of the Coulomb potential. In terms of the electric field, the leading term is the Coulomb term proportional to $1/r^2$ and the expansion in static multipoles has the general structure

$$E \sim \frac{q}{r^2} \left[1 + \mathcal{O}\left(\frac{d}{r}\right) + \mathcal{O}\left(\frac{d^2}{r^2}\right) + \ldots \right]. \qquad (11.98)$$

In contrast, in the multipole expansion of the radiation field, the gauge potentials and the electric and magnetic field receive an infinite series of contributions from all multipoles, already at order $1/r$, associated with terms of higher and higher order in v/c. As we have seen in eqs. (11.93) and (11.96), this expansion starts from order v^2/c^2, and vanishes in the static limit $v = 0$. Schematically, for the electric field

$$E \sim \frac{1}{r} \frac{q}{d} \left[\mathcal{O}\left(\frac{v^2}{c^2}\right) + \mathcal{O}\left(\frac{v^3}{c^3}\right) + \ldots \right]. \qquad (11.99)$$

For static fields $v/c \to 0$, and therefore the $1/r$ term vanishes, leaving us with the expansion (11.98) that starts from order $1/r^2$.

- The multipoles that appear in the expansion of the radiation field are functions of retarded time $t - r/c$, and are sometimes called "radiative multipoles," to distinguish them from the static multipoles that enter in the expansion of the static fields. It is precisely the dependence on retarded time (and therefore, eventually, the fact that light travels at a finite speed) that is responsible for the appearance of the $1/r$ terms in the radiation field. Electromagnetic radiation is a property of a relativistic theory.

- The would-be leading term in the expansion of the scalar potential (11.87) is the Coulomb term $[1/(4\pi\epsilon_0)] (q/r)$. However, this term does not depend on time, and gives no contribution to the $1/r$ part of the electric field at infinity. The leading term to the radiation field at infinity coming from the scalar potential is therefore the dipole term, while the leading contribution from the vector potential is given by the term proportional to the current in eq. (11.88). Comparing their contributions to the electric field, we see that these two terms are of the same order in v/c, compare eqs. (11.93) and (11.96). Indeed, we will see in the next section that both terms can be expressed in terms of the electric dipole. The next-to-leading terms are given by the electric quadrupole term in eq. (11.87) and the term involving $j_i(t - r/c, \mathbf{x}')x'_j$ in eq. (11.88), which are both suppressed by an extra power of v/c, compared to the leading terms.

Having understood the general structure of the multipole expansion of the radiation field, in the next subsections we will study in detail the leading and next-to-leading terms.

11.2.1   Electric dipole radiation

As we have seen, since the term proportional to q/r in eq. (11.87) does not depend on retarded time, when taking its gradient to compute $\mathbf{E}$ we get only a contribution of order $1/r^2$, which therefore does not contribute to the radiation field. The terms giving the leading contributions to the radiation field are obtained setting

$$\phi(t, \mathbf{x}) = \frac{1}{4\pi\epsilon_0} \frac{1}{cr} \hat{n}_i \dot{d}_i(t - r/c) \,, \qquad (11.100)$$

and, from eq. (11.88),

$$A_i(t, \mathbf{x}) = \frac{1}{4\pi\epsilon_0} \frac{1}{c^2 r} \int d^3 x' j_i(t - r/c, \mathbf{x}') \,. \qquad (11.101)$$

This expression for A_i can be rewritten in terms of $\dot{d}_i$ by making use of the obvious identity $\partial_k x_i = \delta_{ki}$ and using the conservation equation (3.22),

$$
\begin{aligned}
\int d^3 x \, j_i(t, \mathbf{x}) &= \int d^3 x \, j_k(t, \mathbf{x}) \partial_k x_i \\
&= -\int d^3 x \, x_i \partial_k j_k(t, \mathbf{x}) \\
&= +\partial_t \int d^3 x \, \rho(t, \mathbf{x}) x_i \\
&= \dot{d}_i(t) \,. \qquad (11.102)
\end{aligned}
$$

In the second line we integrated by parts, discarding the boundary term because the source is localized. Note that these are the same manipulations that we used in eq. (6.31), except that there we were considering the static case, so the time derivative vanished. Therefore, to leading order in v/c,

$$A_i(t, \mathbf{x}) = \frac{1}{4\pi\epsilon_0} \frac{1}{c^2 r} \dot{d}_i(t - r/c) \,. \qquad (11.103)$$

We see that both ϕ and $\mathbf{A}$, to this order in v/c, are determined by the time derivative of the electric dipole. The corresponding radiation is therefore called *dipole radiation*.

The corresponding contributions to the electric field, keeping only the terms $\mathcal{O}(1/r)$, are given by

$$-\boldsymbol{\nabla}\phi = \frac{\hat{\mathbf{n}}}{4\pi\epsilon_0} \frac{1}{c^2 r} \hat{n}_i \ddot{d}_i(t - r/c) \,, \qquad (11.104)$$

as we have already computed in eq. (11.92), and

$$-\frac{\partial \mathbf{A}}{\partial t} = -\frac{1}{4\pi\epsilon_0} \frac{1}{c^2 r} \ddot{\mathbf{d}}(t - r/c) \,. \qquad (11.105)$$

Therefore, for dipole radiation

$$
\begin{aligned}
\mathbf{E}(t, \mathbf{x}) &= -\frac{\partial \mathbf{A}}{\partial t} - \boldsymbol{\nabla}\phi \\
&= -\frac{1}{4\pi\epsilon_0} \frac{1}{c^2 r} \left\{ \ddot{\mathbf{d}}(t_{\text{ret}}) - \hat{\mathbf{n}} \left[\hat{\mathbf{n}} \cdot \ddot{\mathbf{d}}(t_{\text{ret}}) \right] \right\} \,. \qquad (11.106)
\end{aligned}
$$

We have therefore recovered eq. (10.136), as the lowest-order term in an expansion in v/c. We can then rewrite this result in the equivalent forms (10.132) or (10.135). Observe also that the retarded time that appears in these formulas is $t_{\rm ret} = t - r/c$, where r is the distance to the fixed origin of the coordinate system, which has been set inside the region where the source is localized. The variation of the delay across different points of the sources, which is expressed by the factors $e^{-i\omega' \hat{\mathbf{n}}\cdot\mathbf{x}'/c}$ in eqs. (11.73) and (11.74), has been already taken into account, to this order in v/c, by the expansion of the exponential.

The radiative part of the magnetic field is given by $c\mathbf{B} = \hat{\mathbf{n}} \times \mathbf{E}$, since, for the $1/r$ part of the electromagnetic field, eq. (11.27) is an exact relation valid to all orders in v/c. The situation is therefore completely analogous to that discussed in Section 10.5, with the only distinction that, in Section 10.5, which was only concerned with the lowest-order term for large distances and small velocities, one could have used different equivalent forms for the distance to the source and for retarded time, compare for instance eqs. (10.127), (10.130), and (10.131). In contrast, in the context of the systematic multipole expansion discussed here, the only natural choice is to use r for the distance to the source and, correspondingly, $t - (r/c)$ for retarded time. In particular, for the Poynting vector we write

$$
\begin{aligned}
\mathbf{S} &= \frac{1}{\mu_0 c} E^2 \hat{\mathbf{n}} \\
&= \frac{1}{4\pi\epsilon_0} \frac{1}{4\pi c^3 r^2} \left| \ddot{\mathbf{d}}_\perp (t - r/c) \right|^2 \hat{\mathbf{n}}.
\end{aligned} \tag{11.107}
$$

Then, the power per solid angle $dP/d\Omega$ radiated in the dipole approximation is written as

$$
\frac{dP}{d\Omega} = \frac{1}{4\pi\epsilon_0} \frac{1}{4\pi c^3} \left| \ddot{\mathbf{d}}_\perp (t - r/c) \right|^2, \tag{11.108}
$$

and the total radiated power is

$$
\boxed{ P(t) = \frac{1}{4\pi\epsilon_0} \frac{2}{3c^3} |\dddot{\mathbf{d}}(t - r/c)|^2, } \tag{11.109}
$$

compare with eqs. (10.144) and (10.148). We have therefore rederived the Larmor formula (10.148), as the lowest-order term in a systematic expansion in v/c of the $1/r$ contribution to the electric and magnetic fields.

11.2.2   Radiation from charge quadrupole and magnetic dipole

We now look at the next-to-leading term. For ϕ, this is given by the second moment of the charge distribution, denoted by D_{ij}, see eq. (11.87), while for A_i it is given by the term proportional to $j_i(t - r/c, \mathbf{x}')x'_j$ in eq. (11.88). We can rewrite the latter in a physically more transparent form with a straightforward generalization to the time-dependent case of the manipulations performed in eqs. (6.33) and (6.34),

$$
\begin{aligned}
\int d^3x\, j_i(t,\mathbf{x})x_j &= \int d^3x\, x_j j_k(t,\mathbf{x})\partial_k x_i \\
&= -\int d^3x\, x_i \partial_k [x_j j_k(t,\mathbf{x})] \quad\quad (11.110) \\
&= -\int d^3x\, x_i j_j(t,\mathbf{x}) + \partial_t \int d^3x\, \rho(t,\mathbf{x}) x_i x_j \,,
\end{aligned}
$$

where, again, we have used $\boldsymbol{\nabla}\!\cdot\!\mathbf{j} = -\partial_t \rho$. Therefore

$$
\int d^3x\, [j_i(t,\mathbf{x})x_j + j_j(t,\mathbf{x})x_i] = \dot{D}_{ij}(t)\,. \quad\quad (11.111)
$$

We then split $j_i(t,\mathbf{x})x_j$ into its symmetric and antisymmetric parts,

$$
\begin{aligned}
\int d^3x\, j_i(t,\mathbf{x})x_j &= \frac{1}{2}\int d^3x\, [j_i(t,\mathbf{x})x_j + j_j(t,\mathbf{x})x_i] \\
&\quad + \frac{1}{2}\int d^3x\, [j_i(t,\mathbf{x})x_j - j_j(t,\mathbf{x})x_i] \,, \quad (11.112)
\end{aligned}
$$

and we define the antisymmetric tensor

$$
m_{ij}(t) = \frac{1}{2}\int d^3x\, [x_i j_j(t,\mathbf{x}) - x_j j_i(t,\mathbf{x})]\,. \quad\quad (11.113)
$$

The magnetic dipole $m_i(t)$ is then defined as

$$
m_i = \frac{1}{2}\epsilon_{ijk} m_{jk}(t)\,, \quad\quad (11.114)
$$

and therefore

$$
\mathbf{m}(t) = \frac{1}{2}\int d^3x\, \mathbf{x}\times\mathbf{j}(t,\mathbf{x})\,. \qu\quad (11.115)
$$

Equation (11.114) can be inverted, as usual, as $m_{ij} = \epsilon_{ijk} m_k$. Therefore

$$
\int d^3x\, j_i(t,\mathbf{x})x_j = \frac{1}{2}\dot{D}_{ij}(t) - \epsilon_{ijk} m_k(t)\,. \quad\quad (11.116)
$$

Equations (11.115) and (11.116) generalize eqs. (6.36) and (6.37) to the time-dependent case.

Therefore, including only the leading and next-to-leading order in v/c, ϕ is given by eq. (11.87), while, $\mathbf{A}$ is obtained from eq. (11.88),

$$
\begin{aligned}
A_i(t,\mathbf{x}) &= \frac{1}{4\pi\epsilon_0 c^2}\frac{1}{r} \quad\quad\quad\quad\quad\quad\quad\quad\quad\quad\quad\quad (11.117)\\
&\quad \times \left[\dot{d}_i(t-r/c) + \frac{1}{2c}\ddot{D}_{ij}(t-r/c)\hat{n}_j + \frac{1}{c}\epsilon_{ijk}\dot{m}_j(t-r/c)\hat{n}_k\right],
\end{aligned}
$$

where we used eq. (11.102) for the leading term. We now write

$$
D_{ij}(t) = \frac{1}{3}Q_{ij}(t) + \delta_{ij} f(t)\,, \qu\quad (11.118)
$$

where $f(t)$ can be read from eq. (11.86). Then, up to next-to-leading order,

$$
\begin{aligned}
(4\pi\epsilon_0)\phi(t,\mathbf{x}) \;=\;& \frac{q}{r} + \frac{1}{rc}\hat{n}_i\dot{d}_i(t-r/c) + \frac{\hat{n}_i\hat{n}_j}{6rc^2}\ddot{Q}_{ij}(t-r/c) \\
& + \frac{\ddot{f}(t-r/c)}{2rc^2}\,,
\end{aligned} \tag{11.119}
$$

$$
\begin{aligned}
(4\pi\epsilon_0 c^2)A_i(t,\mathbf{x}) \;=\;& \frac{1}{r}\dot{d}_i(t-r/c) + \frac{1}{6rc}\dddot{Q}_{ij}(t-r/c)\hat{n}_j \\
& + \frac{1}{rc}\epsilon_{ijk}\dot{m}_j(t-r/c)\hat{n}_k + \frac{\ddot{f}(t-r/c)}{2rc}\hat{n}_i\,.
\end{aligned} \tag{11.120}
$$

We next observe that

$$
\begin{aligned}
\partial_i\dot{f}(t-r/c) \;&=\; [\partial_t\dot{f}(t-r/c)]\partial_i(t-r/c) \\
\;&=\; -\frac{\hat{n}_i}{c}\ddot{f}(t-r/c)\,,
\end{aligned} \tag{11.121}
$$

so,

$$
\partial_i\frac{\dot{f}(t-r/c)}{2r} = -\frac{\hat{n}_i}{2rc}\ddot{f}(t-r/c) + \mathcal{O}\!\left(\frac{1}{r^2}\right). \tag{11.122}
$$

Therefore, to the order $1/r$ at which we are working, we can rewrite

$$
\begin{aligned}
\phi(t,\mathbf{x}) \;=\;& \frac{1}{4\pi\epsilon_0}\left[\frac{q}{r} + \frac{1}{rc}\hat{n}_i\dot{d}_i(t-r/c) + \frac{\hat{n}_i\hat{n}_j}{6rc^2}\ddot{Q}_{ij}(t-r/c)\right] \\
& + \frac{\partial}{\partial t}\left[\frac{\dot{f}(t-r/c)}{(4\pi\epsilon_0)2rc^2}\right]
\end{aligned} \tag{11.123}
$$

$$
\begin{aligned}
A_i(t,\mathbf{x}) \;=\;& \frac{1}{4\pi\epsilon_0 c^2}\left[\frac{1}{r}\dot{d}_i(t-r/c) + \frac{1}{6rc}\dddot{Q}_{ij}(t-r/c)\hat{n}_j\right. \\
& \left. + \frac{1}{cr}\epsilon_{ijk}\dot{m}_j(t-r/c)\hat{n}_k\right] \\
& - \partial_i\left[\frac{\dot{f}(t-r/c)}{(4\pi\epsilon_0)2rc^2}\right].
\end{aligned} \tag{11.124}
$$

Comparing with eq. (3.86), we see that the extra terms proportional to $\dot{f}$ corresponds to a gauge transformation with

$$
\theta(t,r) = \frac{\dot{f}(t-r/c)}{(4\pi\epsilon_0)2rc^2}\,. \tag{11.125}
$$

Since gauge-equivalent potentials give the same electric and magnetic fields, we can simply drop these extra terms, and write

$$
\phi(t,\mathbf{x}) = \frac{1}{4\pi\epsilon_0}\frac{1}{r}\left[q + \frac{1}{c}\hat{n}_i\dot{d}_i(t-r/c) + \frac{\hat{n}_i\hat{n}_j}{6c^2}\ddot{Q}_{ij}(t-r/c)\right], \tag{11.126}
$$

$$
\begin{aligned}
A_i(t,\mathbf{x}) \;=\;& \frac{1}{4\pi\epsilon_0 c^2}\frac{1}{r}\left[\dot{d}_i(t-r/c) + \frac{1}{6c}\dddot{Q}_{ij}(t-r/c)\hat{n}_j\right. \\
& \left. + \frac{1}{c}\epsilon_{ijk}\dot{m}_j(t-r/c)\hat{n}_k\right].
\end{aligned} \tag{11.127}
$$

Note also that, again to leading order in $1/r$,

$$\Box\theta \;=\; \frac{\Box f(t-r/c)}{2rc} + \mathcal{O}\left(\frac{1}{r^2}\right)$$

$$=\; \mathcal{O}\left(\frac{1}{r^2}\right), \tag{11.128}$$

since $\Box f(t-r/c) = 0$. Therefore, under this gauge transformation, as far as the $1/r$ terms are concerned, the gauge potentials A_μ remain in the Lorenz gauge (that we have used in the computations of this section), see the discussion below eq. (9.18).

As we have already discussed, the term q/r in ϕ does not contribute to the radiation field, so the leading term is given by the contribution of a time-varying dipole, both in ϕ and $\mathbf{A}$. We see that, at next-to-leading order, the radiation is generated by the time variation of the quadrupole moment of the charge distribution, and of the magnetic dipole moment.

The expansion could in principle be carried out to generic order. We already understand, from the computation at this order, that there will be two families of multipole moments that contribute: the multipole moments of the charge distribution, such as the charge dipole, charge quadrupole, and so on; and the multipole moments of the current distribution, of which the magnetic moment is the lowest-order term.

From the gauge potentials we can now compute the electric field up to next-to-leading order. Introducing the vector

$$Q_i \equiv Q_{ij}\hat{n}_j, \tag{11.129}$$

the result is

$$\mathbf{E}(t,\mathbf{x}) = \frac{1}{4\pi\epsilon_0 c^2}\frac{1}{r}\left[\hat{\mathbf{n}}\times(\hat{\mathbf{n}}\times\ddot{\mathbf{d}}) + \frac{1}{6c}\hat{\mathbf{n}}\times(\hat{\mathbf{n}}\times\dddot{\mathbf{Q}}) + \frac{1}{c}\hat{\mathbf{n}}\times\ddot{\mathbf{m}}\right]_{t-r/c},$$
$$\tag{11.130}$$

where the subscript indicates that all quantities on the right-hand side must be evaluated at $t - r/c$. To lowest order, we recover the dipole electric field, in the form given in eq. (10.135). The magnetic field in the radiation zone is then obtained, as usual, from $c\mathbf{B} = \hat{\mathbf{n}}\times\mathbf{E}$. The power radiated, integrated over the solid angle, can be obtained with a computation analogous to that in eq. (10.147). It can be more convenient to use eq. (1.9) to rewrite $\mathbf{E}$ as

$$\mathbf{E}(t,\mathbf{x}) = -\frac{1}{4\pi\epsilon_0 c^2}\frac{1}{r}\left\{[\ddot{\mathbf{d}} - \hat{\mathbf{n}}(\ddot{\mathbf{d}}\cdot\hat{\mathbf{n}})] + \frac{1}{6c}[\dddot{\mathbf{Q}} - \hat{\mathbf{n}}(\dddot{\mathbf{Q}}\cdot\hat{\mathbf{n}})] + \frac{1}{c}\ddot{\mathbf{m}}\times\hat{\mathbf{n}}\right\}_{t-r/c}.$$
$$\tag{11.131}$$

We can then compute $|\mathbf{E}|^2$ and integrate it over the solid angle. Recalling that $Q_i = Q_{ij}\hat{n}_j$ contains a factor of $\hat{\mathbf{n}}$, we see that, when we integrate over $d\Omega$ (and make use of $\hat{\mathbf{n}}\cdot\hat{\mathbf{n}} = 1$), most of the integrals can be performed with the identity (6.49), except for the square of the quadrupole, that has up to four $\hat{\mathbf{n}}$ factors. These can be performed with the identity

$$\int\frac{d\Omega}{4\pi}\hat{n}_i\hat{n}_j\hat{n}_k\hat{n}_l = \frac{1}{15}\left(\delta_{ij}\delta_{kl} + \delta_{ik}\delta_{jl} + \delta_{il}\delta_{jk}\right). \tag{11.132}$$

Similarly to eq. (6.49), this identity can be obtained most efficiently by first fixing the tensor structure from symmetry arguments: the right-hand side must be a tensor, symmetric in all its pairs of indices, constructed with combinations of two occurrences of δ_{ij}. The overall coefficient is then obtained by contracting among them the indices (i, j) and (k, l). Note that the integral over $d\Omega$ of an odd number of $\hat{n}$ factors, such as $\int d\Omega\, n_i$ or $\int d\Omega\, n_i n_j n_k$ vanishes because the integrand is odd under $\hat{n} \to -\hat{n}$ (and the integration over the sphere, for each $\hat{n}$, contains also $-\hat{n}$). In this way we see immediately that the mixed term between the electric dipole and the magnetic dipole, and between the electric dipole and the quadrupole vanish. The mixed term between the quadrupole and the magnetic dipole, after integration over the angles, becomes proportional to $\epsilon_{ijk} Q_{ij} m_k$, which vanishes because Q_{ij} is symmetric. Therefore, the radiated power separates into an electric dipole term, an electric quadrupole term, and a magnetic dipole term, without mixed terms. The remaining computation is straightforward, and gives

$$ P(t) = \frac{1}{4\pi\epsilon_0} \left[\frac{2}{3c^3} |\ddot{\mathbf{d}}|^2 + \frac{1}{180c^5} \dddot{Q}_{ij} \dddot{Q}_{ij} + \frac{2}{3c^5} |\ddot{\mathbf{m}}|^2 \right]_{t-r/c} . \qquad (11.133) $$

The first term was already obtained in eq. (11.109). As we already saw in the estimates above eqs. (11.93) and (11.94), $\dddot{Q}_{ij}$ is smaller than $\ddot{d}_i$ by a factor $\mathcal{O}(v)$. Also, the magnetic dipole is smaller than the electric dipole by a factor $\mathcal{O}(v)$, as we can see comparing their respective definitions, eqs. (11.82) and (11.115) and recalling eq. (3.26) (which is valid for a point charge but extends to a charge distribution, with v the typical velocities of internal motions).[15] Then, we see that both the electric quadrupole and the magnetic dipole contributions to the radiated power are smaller by a factor $\mathcal{O}(v^2/c^2)$, with respect to the power radiated by the electric dipole. This is, of course, as expected, since we have seen that the expansion in radiative multipoles is an expansion in powers of v/c, and the terms in eq. (11.133) are quadratic in the multipole moments.

For a point charge evolving on a prescribed trajectory $\mathbf{x}_0(t)$, inserting eq. (8.1) into eq. (11.85), we get

$$ Q_{ij}(t) = q \left[3x_{0i}(t) x_{0j}(t) - |\mathbf{x}_0(t)|^2 \delta_{ij} \right] . \qquad (11.134) $$

Consider, for example, a charge performing a simple harmonic motion along the z axis, with amplitude A,

$$ \mathbf{x}(t) = \hat{\mathbf{z}} A \cos(\omega_s t) . \qquad (11.135) $$

Then, the only non-vanishing components of Q_{ij} are

$$ \begin{aligned} Q_{11} &= -qA^2 \cos^2 \omega_s t , & (11.136) \\ Q_{22} &= -qA^2 \cos^2 \omega_s t , & (11.137) \\ Q_{33} &= 2qA^2 \cos^2 \omega_s t . & (11.138) \end{aligned} $$

Since $\cos^2 \omega_s t = [1 + \cos(2\omega_s t)]/2$, the time-dependent part, that contributes to the derivatives $\dddot{Q}$ in eq. (11.131) oscillates as $\cos(2\omega_s t)$, and therefore the quadrupole radiation is at a frequency $\omega = 2\omega_s$.

[15] We will see this in more details in eqs. (12.92) and (12.93).

11.3 Near zone, far zone and induction zone

In the previous section we have studied the large r limit, assuming that r is sufficiently large, so that only the terms order of $1/r$ must be retained, and the $1/r^2$ terms can be neglected. In this region, we have found that the radiation field is given by plane waves, with the electric and magnetic field transverse to the propagation direction, sourced by the time-varying multipole moment.

Now we ask how large must r be so that this expansion is valid. Consider for instance the scalar potential ϕ. From the form (11.126), that was already obtained keeping only the term $\mathcal{O}(1/r)$, we see that the various terms in the expansion have the form $F(t - r/c)/r$, such as $\dot{d}_i(t - r/c)/r$, $\ddot{Q}_{ij}(t - r/c)/r$, and so on, multiplied by factors $\hat{n}_i$. The electric field is obtained from $\nabla\phi$. In particular, this produces terms of the form

$$\frac{\partial}{\partial r}\frac{F(t - r/c)}{r} = -\frac{F(t - r/c)}{r^2} + \frac{1}{r}\frac{\partial}{\partial r}F(t - r/c). \tag{11.139}$$

Writing $u = t - r/c$,

$$\begin{aligned} \frac{\partial}{\partial r}F(t - r/c) &= \frac{dF(u)}{du}\frac{\partial u}{\partial r} \\ &= -\frac{1}{c}\frac{dF(u)}{du}, \end{aligned} \tag{11.140}$$

while

$$\begin{aligned} \frac{\partial}{\partial t}F(t - r/c) &= \frac{dF(u)}{du}\frac{\partial u}{\partial t} \\ &= \frac{dF(u)}{du}, \end{aligned} \tag{11.141}$$

and therefore,

$$\frac{\partial}{\partial r}F(t - r/c) = -\frac{1}{c}\frac{\partial}{\partial t}F(t - r/c). \tag{11.142}$$

If the typical frequency of the system is ω_s, $\partial_t F$ is of order $\omega_s F$. Therefore, in order of magnitude,

$$\frac{1}{r}\frac{\partial}{\partial r}F(t - r/c) \sim \frac{\omega_s}{cr}F(t - r/c). \tag{11.143}$$

Therefore, the term $F(t - r/c)/r^2$ in the right-hand side of eq. (11.139) can be neglected with respect to the second term if

$$\frac{1}{r^2} \ll \frac{\omega_s}{cr}, \tag{11.144}$$

and therefore $r \gg c/\omega_s$. Since the frequency ω of the radiation emitted is also of order ω_s, we can write this as

$$\boxed{r \gg \frac{c}{\omega} = \lambdabar.} \tag{11.145}$$

This defines the *far zone*, or radiation zone. Under this condition, all other contributions of order $1/r^2$ to the electric field can be neglected. In particular, we can neglect also the terms $\mathcal{O}(1/r^2)$ in eq. (11.2) coming from the expansion of $1/|\mathbf{x} - \mathbf{x}'|$ to higher order,

$$\frac{1}{|\mathbf{x} - \mathbf{x}'|} = \frac{1}{r} + \mathcal{O}\left(\frac{|\mathbf{x}'|}{r^2}\right). \tag{11.146}$$

Upon integration over the source, the last term becomes $\mathcal{O}(d/r^2)$, where d (not to be confused with the dipole moment) is the size of the source. This term is negligible, with respect to the $1/r$ terms that we have keep, as long as $d/r \ll 1$. As we saw in eq. (11.76), the reduced wavelength $\lambdabar$ is of order $(c/v)\, d$, so, $\lambdabar \geq d$ (and, for non-relativistic source, $\lambdabar \gg d$). Therefore, in the far zone $r \gg \lambdabar$, the condition $r \gg d$ is also automatically satisfied, so also the $1/r^2$ from eq. (11.146) are negligible.

From these estimates we see that the region outside a source of size d and typical frequency ω_s, which emits radiation with a reduced wavelength $\lambdabar = c/\omega \sim c/\omega_s$, can be separated into a *far region*,

$$r \gg \lambdabar, \qquad \text{(far region)}, \tag{11.147}$$

and a *near region*,

$$r \ll \lambdabar, \qquad \text{(near region)}. \tag{11.148}$$

For non-relativistic sources we have $d \ll \lambdabar$, and one can further introduce the *near outer region*, defined by

$$d \leq r \ll \lambdabar, \qquad \text{(near outer region)}, \tag{11.149}$$

i.e., that part of the near region which is outside the source. The intermediate region $r \sim \lambdabar$ is called the *induction region*. In the far region, as we have seen, the multipole expansion has the form (11.126), (11.127), and consists of terms of the form $1/r$ times functions of retarded time $t - r/c$. In contrast, in the near region the situation is reversed, and retardation effects are negligible. This means that, to lowest order, we can approximate eq. (11.1) as

$$A^\mu(t, \mathbf{x}) \simeq \frac{\mu_0}{4\pi} \int d^3x'\, \frac{j^\mu(t, \mathbf{x}')}{|\mathbf{x} - \mathbf{x}'|}, \tag{11.150}$$

where we have replaced $t - |\mathbf{x} - \mathbf{x}'|/c$ with t in the first argument of j^μ.[16] In the near outer region, where retardation effects are negligible and furthermore $r \gg d$, we can then expand $1/|\mathbf{x} - \mathbf{x}'|$ as in eq. (6.4), or, more generally, as in eq. (6.26). For a generic time-dependent source, we then recover the multipole expansion for static fields, described in Chapter 6, except that the source term $j^\mu(\mathbf{x}')$ is replaced by $j^\mu(t, \mathbf{x}')$, with the gauge potentials and electromagnetic fields following the instantaneous time behavior of the sources. For instance,

$$\begin{aligned} \phi(t, \mathbf{x}) &= \frac{1}{4\pi\epsilon_0} \int d^3x'\, \frac{\rho(t - |\mathbf{x} - \mathbf{x}'|/c, \mathbf{x}')}{|\mathbf{x} - \mathbf{x}'|} \\ &\simeq \frac{1}{4\pi\epsilon_0} \int d^3x'\, \frac{\rho(t, \mathbf{x}')}{|\mathbf{x} - \mathbf{x}'|}. \end{aligned} \tag{11.151}$$

[16] In Chapter 12 we will see how to improve systematically over this approximation.

Performing the same steps as in the multipole expansion of static fields, in the near outer region we then get

$$\phi(t, \mathbf{x}) = \frac{1}{4\pi\epsilon_0} \left[\frac{q}{r} + \frac{n_i d_i(t)}{r^2} + \frac{n_i n_j Q_{ij}(t)}{2r^3} + \dots \right],$$ (11.152)

which is the same as eq. (6.20), except that d_i and Q_{ij} are replaced by $d_i(t)$ and $Q_{ij}(t)$, and similarly for the higher-order multipoles. In the same way, for the vector potential in the near region we get

$$\begin{aligned} \mathbf{A}(t, \mathbf{x}) &= \frac{\mu_0}{4\pi} \int d^3 x' \frac{\mathbf{j}(t - |\mathbf{x} - \mathbf{x}'|/c, \mathbf{x}')}{|\mathbf{x} - \mathbf{x}'|} \\ &\simeq \frac{\mu_0}{4\pi} \int d^3 x' \frac{\mathbf{j}(t, \mathbf{x}')}{|\mathbf{x} - \mathbf{x}'|}, \end{aligned}$$ (11.153)

whose expansion, in the near outer region, gives the magnetic dipole term plus terms that depend on higher-order magnetic multipoles,

$$\mathbf{A}(t, \mathbf{x}) = \frac{\mu_0}{4\pi} \frac{\mathbf{m}(t) \times \mathbf{x}}{r^3} + \dots,$$ (11.154)

which is the same as eq. (6.38) with $\mathbf{m}$ replaced by $\mathbf{m}(t)$.

Equations (11.152) and (11.154) are the generalization of eqs. (6.20) and (6.38), respectively, from static sources to time-dependent sources in their near outer region. In the induction region, the spatial and time dependence of the fields gradually evolve from that of the near outer region to that of the far region, and no comparatively simple approximation to eq. (11.1) is possible.

11.4   Solved problems

Problem 11.1. Decomposition of a vector field into transverse and longitudinal parts

In this Solved Problem we discuss the decomposition of a vector field into its transverse and longitudinal parts, also known as the *Helmholtz decomposition* (or Helmholtz theorem). We used this decomposition in Section 11.1.2 in the computation of the radiation field performed in the Coulomb gauge, but this is, more generally, a useful mathematical tool.

In eq. (10.118) we saw how to decompose a vector into its longitudinal and transverse parts. This decomposition can be generalized to a vector field $\mathbf{V}(\mathbf{x})$, writing

$$\mathbf{V}(\mathbf{x}) = \mathbf{V}_{\parallel}(\mathbf{x}) + \mathbf{V}_{\perp}(\mathbf{x}),$$ (11.155)

where

$$\begin{aligned} \mathbf{V}_{\parallel}(\mathbf{x}) &= -\boldsymbol{\nabla} \left[\boldsymbol{\nabla} \cdot \int d^3 x' \frac{\mathbf{V}(\mathbf{x}')}{4\pi|\mathbf{x} - \mathbf{x}'|} \right], \end{aligned}$$ (11.156)

$$\begin{aligned} \mathbf{V}_{\perp}(\mathbf{x}) &= \boldsymbol{\nabla} \times \left[\boldsymbol{\nabla} \times \int d^3 x' \frac{\mathbf{V}(\mathbf{x}')}{4\pi|\mathbf{x} - \mathbf{x}'|} \right], \end{aligned}$$ (11.157)

as long as the integral over d^3x', in eqs. (11.156) and (11.157) converges (writing $d^3x' = r'^2 dr' d\Omega$ we see that this is ensured if $\mathbf{V}(\mathbf{x})$ vanishes faster than $1/|\mathbf{x}|^2$ as $|\mathbf{x}| \to \infty$). Indeed, using eq. (1.7), we see that, in components

$$V_{\perp,i}(\mathbf{x}) = \partial_i \partial_j \int d^3x' \frac{V_j(\mathbf{x}')}{4\pi|\mathbf{x} - \mathbf{x}'|} - \boldsymbol{\nabla}^2 \int d^3x' \frac{V_i(\mathbf{x}')}{4\pi|\mathbf{x} - \mathbf{x}'|}. \qquad (11.158)$$

Then, using eq. (1.91),

$$V_{\perp,i}(\mathbf{x}) = V_i(\mathbf{x}) + \partial_i \partial_j \int d^3x' \frac{V_j(\mathbf{x}')}{4\pi|\mathbf{x} - \mathbf{x}'|}. \qquad (11.159)$$

On the other hand, in components eq. (11.156) reads

$$V_{\|,i}(\mathbf{x}) = -\partial_i \partial_j \int d^3x' \frac{V_j(\mathbf{x}')}{4\pi|\mathbf{x} - \mathbf{x}'|}, \qquad (11.160)$$

so, eq. (11.155) indeed holds. The usefulness of this decomposition comes from the fact that, since $\mathbf{V}_\perp$ is the curl of a vector field and $\mathbf{V}_\|$ is the gradient of a scalar field, they satisfy

$$\boldsymbol{\nabla} \times \mathbf{V}_\| = 0, \qquad (11.161)$$

$$\boldsymbol{\nabla}\cdot\mathbf{V}_\perp = 0. \qquad (11.162)$$

Eqs. (11.155)–(11.157) therefore provide the decomposition of the vector field $\mathbf{V}(\mathbf{x})$ into its curl-free and divergence-less parts. This decomposition is used in many contexts in physics, so it can be useful to elaborate on it a bit further. First of all, note that $\mathbf{V}_\|(\mathbf{x})$ and $\mathbf{V}_\perp(\mathbf{x})$ are non-local functionals of $\mathbf{V}(\mathbf{x})$, in the sense that their value in $\mathbf{x}$ depends on the values of $\mathbf{V}(\mathbf{x}')$ for all $\mathbf{x}'$, and not only on the value of $\mathbf{V}(\mathbf{x}')$ and of a finite number of its derivatives at $\mathbf{x}' = \mathbf{x}$. They are indeed given by a convolution with the Green's function (4.15) of the Laplacian. These relations become, however, local in Fourier space. In fact, using eq. (1.91), we see that eq. (11.160) implies

$$\boldsymbol{\nabla}^2 V_{\|,i}(\mathbf{x}) = \partial_i \partial_j V_j. \qquad (11.163)$$

Writing

$$\mathbf{V}(\mathbf{x}) = \int \frac{d^3k}{(2\pi)^3} \tilde{\mathbf{V}}(\mathbf{k}) e^{i\mathbf{k}\cdot\mathbf{x}}, \qquad (11.164)$$

and similarly for $\mathbf{V}_\|(\mathbf{x})$ and $\mathbf{V}_\perp(\mathbf{x})$, eq. (11.163) becomes

$$-k^2 \tilde{V}_{\|,i}(\mathbf{k}) = -k_i k_j \tilde{V}_j(\mathbf{k}), \qquad (11.165)$$

where $k^2 \equiv |\mathbf{k}|^2$. Therefore (for $\mathbf{k} \neq 0$),

$$\boxed{\tilde{V}_{\|,i}(\mathbf{k}) = \frac{k_i k_j}{k^2} \tilde{V}_j(\mathbf{k}).} \qquad (11.166)$$

Writing $\tilde{V}_{\perp,i}(\mathbf{k}) = \tilde{V}_i(\mathbf{k}) - \tilde{V}_{\|,i}(\mathbf{k})$, it also follows that

$$\boxed{\tilde{V}_{\perp,i}(\mathbf{k}) = \left(\delta_{ij} - \frac{k_i k_j}{k^2} \right) \tilde{V}_j(\mathbf{k}).} \qquad (11.167)$$

In terms of the unit vector $\hat{\mathbf{k}} = \mathbf{k}/k$, eqs. (11.166) and (11.167) read

$$\tilde{\mathbf{V}}_\|(\mathbf{k}) = \hat{\mathbf{k}} \left[\hat{\mathbf{k}}\cdot\tilde{\mathbf{V}}(\mathbf{k}) \right], \qquad (11.168)$$

Equation (11.172) can be shown as follows:

$$\left[\nabla_{\mathbf{x}} \times \int d^3x' \frac{\mathbf{V}(\mathbf{x}')}{4\pi|\mathbf{x}-\mathbf{x}'|}\right]_i$$

$$= \epsilon_{ijk}\partial_j \int d^3x' \frac{V_k(\mathbf{x}')}{4\pi|\mathbf{x}-\mathbf{x}'|}$$

$$= \epsilon_{ijk}\int d^3x'\, V_k(\mathbf{x}') \frac{\partial}{\partial x^j}\frac{1}{4\pi|\mathbf{x}-\mathbf{x}'|}$$

$$= -\epsilon_{ijk}\int d^3x'\, V_k(\mathbf{x}') \frac{\partial}{\partial x'^j}\frac{1}{4\pi|\mathbf{x}-\mathbf{x}'|}$$

$$= \epsilon_{ijk}\int d^3x' \frac{\partial V_k(\mathbf{x}')}{\partial x'^j}\frac{1}{4\pi|\mathbf{x}-\mathbf{x}'|},$$

where, in the last line, we integrated by part and we observed that, by assumption, $\mathbf{V}(\mathbf{x})$ vanishes faster than $1/|\mathbf{x}|^2$ as $|\mathbf{x}| \to \infty$ (so that the integral in eqs. (11.157) and (11.156) converges), and therefore the boundary term vanishes. Similar manipulations can be used to prove eq. (11.171).

[18]Observe that we could have derived directly the decomposition (11.174) without passing through eqs. (11.156) and (11.157). The procedure is almost equivalent, except that the boundary conditions to be imposed are slightly different: the integrals in eq. (11.174) converge if both $(\nabla\cdot\mathbf{V})(\mathbf{x})$ and $(\nabla\times\mathbf{V})(\mathbf{x})$ go to zero faster than $1/|\mathbf{x}|^2$ as $|\mathbf{x}| \to \infty$. Furthermore, the uniqueness theorem discussed in Solved Problem (4.1.5), eqs. (4.45)–(4.47), ensure that this decomposition is unique as long as $\mathbf{V}(\mathbf{x})$ goes to zero as $|\mathbf{x}| \to \infty$. Therefore, eq. (11.174) is valid if $\mathbf{V}(\mathbf{x})$ goes to zero (with any speed) as $|\mathbf{x}| \to \infty$, and $(\nabla\cdot\mathbf{V})(\mathbf{x})$ and $(\nabla\times\mathbf{V})(\mathbf{x})$ go to zero faster than $1/|\mathbf{x}|^2$. In contrast, the decomposition (11.155)–(11.156) assumes that $\mathbf{V}(\mathbf{x})$ goes to zero faster than $1/|\mathbf{x}|^2$, which is more restrictive.

[19]Note that all the results of this decomposition are also valid for time-dependent fields, simply replacing $\mathbf{A}(\mathbf{x})$ by $\mathbf{A}(t,\mathbf{x})$, since the time variable never enter any of these equations.

and

$$\tilde{\mathbf{V}}_{\perp}(\mathbf{k}) = \tilde{\mathbf{V}}(\mathbf{k}) - \hat{\mathbf{k}}\left[\hat{\mathbf{k}}\cdot\tilde{\mathbf{V}}(\mathbf{k})\right]. \tag{11.169}$$

Comparing with eqs. (10.119) and (10.120) we see that $\tilde{\mathbf{V}}_{\parallel}(\mathbf{k})$ is the component of $\tilde{\mathbf{V}}(\mathbf{k})$ parallel to $\mathbf{k}$, while $\tilde{\mathbf{V}}_{\perp}(\mathbf{k})$ is the component of $\tilde{\mathbf{V}}(\mathbf{k})$ transverse to $\mathbf{k}$. Then, $\tilde{\mathbf{V}}_{\perp}(\mathbf{k})$ is called the transverse part of $\tilde{\mathbf{V}}(\mathbf{k})$ (since it is transverse to $\mathbf{k}$) and $\tilde{\mathbf{V}}_{\parallel}(\mathbf{k})$ is also called the longitudinal part of $\tilde{\mathbf{V}}(\mathbf{k})$. Note that $\tilde{\mathbf{V}}_{\parallel}(\mathbf{k})$ and $\tilde{\mathbf{V}}_{\perp}(\mathbf{k})$ are determined by $\tilde{\mathbf{V}}(\mathbf{k})$ with the same value of $\mathbf{k}$, i.e., in wavenumber space these relations are local. In wavenumber space, eqs. (11.162) and (11.161) become

$$\mathbf{k} \times \tilde{\mathbf{V}}_{\parallel}(\mathbf{k}) = 0, \qquad \mathbf{k}\cdot\tilde{\mathbf{V}}_{\perp}(\mathbf{k}) = 0, \tag{11.170}$$

and the validity of these relations can be immediately checked from the explicit form (11.168,11.169). Another useful form of the decomposition is obtained observing that[17]

$$\nabla\cdot\int d^3x' \frac{\mathbf{V}(\mathbf{x}')}{4\pi|\mathbf{x}-\mathbf{x}'|} = \int d^3x' \frac{\nabla_{\mathbf{x}'}\cdot\mathbf{V}(\mathbf{x}')}{4\pi|\mathbf{x}-\mathbf{x}'|}, \tag{11.171}$$

$$\nabla_{\mathbf{x}} \times\int d^3x' \frac{\mathbf{V}(\mathbf{x}')}{4\pi|\mathbf{x}-\mathbf{x}'|} = \int d^3x' \frac{\nabla_{\mathbf{x}'} \times \mathbf{V}(\mathbf{x}')}{4\pi|\mathbf{x}-\mathbf{x}'|}. \tag{11.172}$$

Therefore, introducing the notation

$$f(\mathbf{x}) = \nabla\cdot\mathbf{V}(\mathbf{x}), \qquad \mathbf{w}(\mathbf{x}) = \nabla\times\mathbf{V}(\mathbf{x}), \tag{11.173}$$

the decomposition (11.155)–(11.156) can be rewritten as[18]

$$\boxed{\mathbf{V}(\mathbf{x}) = -\nabla\left(\int d^3x' \frac{f(\mathbf{x}')}{4\pi|\mathbf{x}-\mathbf{x}'|}\right) + \nabla\times\left(\int d^3x' \frac{\mathbf{w}(\mathbf{x}')}{4\pi|\mathbf{x}-\mathbf{x}'|}\right).} \tag{11.174}$$

Recall that, in Solved Problem 4.1.5, we proved that, in $\mathbb{R}^3$, a vector field $\mathbf{V}(\mathbf{x})$ is determined uniquely by its curl and its divergence (under the assumption that it vanishes sufficiently fast at infinity). Equation (11.174) shows explicitly how to compute it, in terms of its divergence $f(\mathbf{x})$ and its curl $\mathbf{w}(\mathbf{x})$.

Applying this decomposition to the vector gauge potential $\mathbf{A}$, we can write

$$\begin{aligned}\mathbf{A}(\mathbf{x}) &= \mathbf{A}_{\perp}(\mathbf{x}) + \mathbf{A}_{\parallel}(\mathbf{x})\\ &= \mathbf{A}_{\perp}(\mathbf{x}) - \nabla\alpha,\end{aligned} \tag{11.175}$$

where, using $\mathbf{w}(\mathbf{x}) \equiv \nabla\times\mathbf{A}(\mathbf{x}) = \mathbf{B}(\mathbf{x})$,

$$\mathbf{A}_{\perp}(\mathbf{x}) = \nabla\times\left(\int d^3x' \frac{\mathbf{B}(\mathbf{x}')}{4\pi|\mathbf{x}-\mathbf{x}'|}\right), \tag{11.176}$$

and

$$\alpha = \int d^3x' \frac{(\nabla\cdot\mathbf{A})(\mathbf{x}')}{4\pi|\mathbf{x}-\mathbf{x}'|}. \tag{11.177}$$

Under a gauge transformation (3.86), $\mathbf{A}$ transforms as $\mathbf{A} \to \mathbf{A} - \nabla\theta$. Since the additional term is a gradient, it is reabsorbed into a transformation of $\alpha(\mathbf{x})$,

$$\mathbf{A}_{\perp}(\mathbf{x}) \to \mathbf{A}_{\perp}(\mathbf{x}), \tag{11.178}$$

$$\alpha(\mathbf{x}) \to \alpha(\mathbf{x}) + \theta(\mathbf{x}). \tag{11.179}$$

This shows that $\alpha(x)$, and therefore $\mathbf{A}_{\parallel}(\mathbf{x})$, is a pure gauge degree of freedom, that can be set to zero with a gauge transformation. In contrast, $\mathbf{A}_{\perp}(\mathbf{x})$ is gauge invariant. This was clear already from eq. (11.176), where $\mathbf{A}_{\perp}(\mathbf{x})$ is written in terms of the magnetic field $\mathbf{B}$, which is gauge invariant.[19]

Post-Newtonian expansion and radiation reaction

In this chapter we focus more closely on the dynamics in the near zone, $r \ll \lambda$. As we have discussed in Section 11.3, in the near zone, in the lowest-order approximation, retardation effects can be neglected, and the electromagnetic field depends on the instantaneous value of the sources. In this limit, we recover the Newtonian dynamics. We now want to understand how to systematically improve over this zero-th order approximation, performing a systematic expansion for small retardation effects. This will give rise to the so-called post-Newtonian (PN) expansion.[1] We will then study the back-reaction problem: a system of charges that emits electromagnetic radiation at infinity loses energy, and this affects the dynamics of the particles in the near zone. This will manifest itself through "radiation-reaction" forces in the near region. In this context, we will also have to deal with divergences that appear for point charges. We will see that a conceptually clean treatment of the problem only emerges by using renormalization techniques, typical of quantum field theory, despite our purely classical context (we already found a similar situation in Section 5.2.2). This chapter treats advanced subjects and some parts are very technical, and is meant for advanced readers.

12.1 Expansion for small retardation

We first discuss how to set up, in general, the expansion for small retardation effects, while in the subsequent sections we will explicitly perform the computation of the first leading terms. To this purpose, we start from eq. (11.1), that we rewrite here separately for the scalar and vector potentials, in the form

$$\phi(t, \mathbf{x}) = \frac{1}{4\pi\epsilon_0} \int d^3x' \frac{\rho(t - |\mathbf{x} - \mathbf{x}'|/c, \mathbf{x}')}{|\mathbf{x} - \mathbf{x}'|}, \qquad (12.1)$$

$$\mathbf{A}(t, \mathbf{x}) = \frac{1}{4\pi\epsilon_0} \frac{1}{c^2} \int d^3x' \frac{\mathbf{j}(t - |\mathbf{x} - \mathbf{x}'|/c, \mathbf{x}')}{|\mathbf{x} - \mathbf{x}'|}. \qquad (12.2)$$

As we have seen, these equations are exact, in the sense that they are valid for arbitrary velocities of the internal motions of the source, and only assume that the source is localized, i.e., that $\rho(t, \mathbf{x})$ and $\mathbf{j}(t, \mathbf{x})$ have compact support in $\mathbf{x}$ (or, more generally, that they decrease sufficiently fast as $|\mathbf{x}| \to \infty$, so that the integrals converge). We also recall that these expressions have been derived fixing the Lorenz gauge, $\partial_\mu A^\mu = 0$.

[1] The name "PN expansion" is borrowed from the corresponding expansion that is performed in General Relativity, where it corresponds to an expansion beyond the Newtonian gravitational potential (see e.g., Chapter 5 of Maggiore (2007) or Poisson and Will (2014) for an introduction). We will use this terminology also in the context of electrodynamics, to denote, more generally, the expansion beyond Newtonian dynamics. Both in electrodynamics and in gravity, the Newtonian dynamics is obtained in the formal limit $c \to \infty$ and, as we will see, the PN expansion corresponds to an expansion in powers of (v/c), with v the typical velocity of the inner motions of the source. Since in electrodynamics the Newtonian dynamics is governed by the Coulomb potential, in this case, instead of "post-Newtonian expansion," the name "post-Coulombian expansion" is sometimes used, see e.g., Chapter 12 of Poisson and Will (2014).

The PN expansion is an expansion for small retardation effects, that improves systematically on the approximation used in eqs. (11.151) and (11.153), where retardation effects were simply neglected. This is obtained by expanding the charge density and current distribution in eqs. (12.1) and (12.2) as

$$
\rho(t - |\mathbf{x} - \mathbf{x}'|/c, \mathbf{x}') = \rho(t, \mathbf{x}') - \frac{|\mathbf{x} - \mathbf{x}'|}{c} \partial_t \rho(t, \mathbf{x}')
$$
$$
+ \frac{1}{2} \left(\frac{|\mathbf{x} - \mathbf{x}'|}{c} \right)^2 \partial_t^2 \rho(t, \mathbf{x}') + \dots , \quad (12.3)
$$

and [2]

$$
\mathbf{j}(t - |\mathbf{x} - \mathbf{x}'|/c, \mathbf{x}') = \mathbf{j}(t, \mathbf{x}') - \frac{|\mathbf{x} - \mathbf{x}'|}{c} \partial_t \mathbf{j}(t, \mathbf{x}') + \dots . \quad (12.4)
$$

If we denote by ω_s the typical frequency of the source, each time derivative brings a factor of order ω_s, so eqs. (12.3) and (12.4) are actually an expansion in the parameter $\omega_s |\mathbf{x} - \mathbf{x}'|/c$. As we have seen in Chapter 11, ω_s is of order of the typical frequency ω of the radiation emitted, and $\omega/c = 1/\lambdabar$. Therefore, the expansion is valid as long as $|\mathbf{x} - \mathbf{x}'| \ll \lambdabar$. On the other hand, $\mathbf{x}'$ is an integration variable that, for a localized source, is restricted to values $|\mathbf{x}'| < d$; since $\lambdabar = (c/v)d$, we have $d \ll \lambdabar$ for the non-relativistic sources that we consider in the PN expansion. Therefore, the condition $|\mathbf{x} - \mathbf{x}'| \ll \lambdabar$ is equivalent to the condition $|\mathbf{x}| \equiv r \ll \lambdabar$. In conclusion, *the PN expansion is valid in the near zone, $r \ll \lambdabar$, and breaks down in the far zone.* In particular, the PN approximation cannot be used to compute directly the radiation field at infinity (although, as we will see, it can be used to compute how the emission of radiation at infinity affects the dynamics of the charges in the near zone).

From the above discussion we see that the PN approximation is valid in the near region $r \ll \lambdabar$, without any assumption on the relative values of r and d. In the outer near region, $d \ll r \ll \lambdabar$, we could combine the PN expansion with a multipole expansion, analogous to the expansion in static multipoles of Chapter 6. However, a main application of the PN expansion is that it allows us to compute the electromagnetic fields in the inner near region, i.e., in the region where the charges are localized, $r < d \ll \lambdabar$. In turn, this allows us to compute how a set of charges evolves under their mutual interaction, beyond the Newtonian approximation. Therefore, in this case no multipole expansion can be performed. Below we will examine the general form of the PN expansion, valid also in the inner near region.

In the inner near region, the expansion in eqs. (12.3) and (12.4) actually becomes an expansion in powers of v/c. In fact, in the inner near region both $|\mathbf{x}|$ and $|\mathbf{x}'|$ are of order d, so the expansion parameter $\omega_s |\mathbf{x} - \mathbf{x}'|/c$ becomes of order $\omega_s d/c \sim v/c$. In general, a term suppressed by a factor $(v/c)^{2n}$, with respect to the Newtonian dynamics, is called a term of order nPN, with half-integer PN orders representing odd powers of v/c; e.g., the 1PN order gives corrections of order $(v/c)^2$

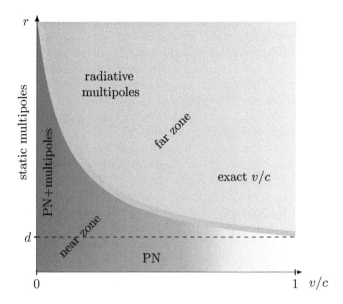

Fig. 12.1 The different expansion regimes discussed in the text.

to the Newtonian result, while the 1.5PN term is suppressed by $(v/c)^3$. This notation, which is standard in the context of the PN expansion, is also useful because, as we will see below, "half-integer" PN orders are associated with radiation reaction, and this notation allows us to single them out more clearly.

In the previous chapters we have examined several different approximations to eqs. (12.1) and (12.2) and, to put the PN expansion in the correct perspective, it is useful to recall and compare them. It is convenient to display the different regimes in the plane $(v/c, r)$, as in Fig. 12.1, where v is the typical velocity of the internal motions of the source, and r is the distance at which we compute the electromagnetic field. In this plane, an important region is determined by the condition $r \simeq \lambdabar$. As we saw in eq. (11.76), $\lambdabar$ is of order $(c/v)d$, so this corresponds to $r \simeq (c/v)d$. This region is shown in Fig. 12.1 as a shaded thick curve, which diverges for $v/c \to 0$ and decreases as v/c increases, until it reaches the value $r \simeq d$ when $v/c \to 1$. The part of the plot well above this curve is the far region (or far zone) $r \gg \lambdabar$, while that well below it is the near region (or near zone). The shaded part $r \simeq \lambdabar$ is the induction region, where the transition between the near and the far regions takes place. We have further marked, on the vertical axis, the value $r = d$, that separates the inner near region from the outer near region.

In Chapter 6 we considered static sources, and we performed an expansion for $r \gg d$. Within the plot in Fig. 12.1, a perfectly static source corresponds to $v/c = 0$ and therefore to the vertical axis. In this case, the "near" zone extends all the way up to spatial infinity and, for $r \gg d$, the appropriate tool is the expansion in static multipoles, as in eqs. (6.20) and (6.30). Clearly, the assumption of exactly static sources is an idealization. A system of interacting charges will undergo accelerations because of their mutual interactions and will have non-zero velocities. If we denote by ω_s the typical order of magnitude of the frequencies of such motions, this system will generate electromagnetic waves with typical frequency $\omega \simeq \omega_s$ (as we have seen, with numerical coefficients of order one that depend on the multipole involved), and the corresponding value of $\lambdabar$ is equal to $c/\omega \simeq c/\omega_s$, and is finite, so the expansion in static multipoles is actually only valid for $d \ll r \ll c/\omega_s$.

For generic, non-zero values of v/c, at $r \gg \lambdabar$, i.e., above the shaded curve in Fig. 12.1, we are in the far zone, and retardation effects are crucial: as we have seen in Section 11.1, they are responsible for the presence of terms in the electric and magnetic fields that decrease only as $1/r$, i.e., of the radiation field. The appropriate treatment here is the one developed in Section 11.1 for sources with arbitrary velocity. If, furthermore, $v/c \ll 1$, we can perform an expansion in radiative multipoles, as we discussed in Section 11.2.

In the part of the plot below the shaded curve we are in the near zone, $r \ll \lambdabar$. On the vertical axis, at $v/c = 0$, we have the Newtonian dynamics. As we have seen, if furthermore $r \gg d$ we can expand in static multipoles, while for $r \leq d$ we must deal with the exact Newtonian dynamics. As we move away from the vertical axis, while still staying in the near zone, as long as $v/c \ll 1$ retardation effects can be included perturbatively, using the PN expansion that we will discuss in this chapter. If, furthermore, $r \gg d$, we can combine the PN expansion with a multipole expansion.

12.2 Dynamics to order $(v/c)^2$

After these general considerations, we are ready to perform the explicit computation of the first non-trivial corrections to the Newtonian dynamics of a system of point charges, which, as we will see, correspond to the 1PN order. For the scalar potential, we insert eq. (12.3) into eq. (12.1). This gives

$$\phi(t, \mathbf{x}) = \phi_N(t, \mathbf{x}) + \phi_{0.5\text{PN}}(t, \mathbf{x}) + \phi_{1\text{PN}}(t, \mathbf{x}) + \dots , \quad (12.5)$$

where[3]

[3]Once extracted from the integral, the time derivative in eq. (12.7) becomes a total time derivative d/dt given that, in this case, the integral is independent of $\mathbf{x}$, while for the 1PN and higher-order terms it remains a partial time derivative ∂_t.

$$\phi_N(t, \mathbf{x}) = \frac{1}{4\pi\epsilon_0} \int d^3x' \, \frac{\rho(t, \mathbf{x}')}{|\mathbf{x} - \mathbf{x}'|} , \quad (12.6)$$

$$\phi_{0.5\text{PN}}(t, \mathbf{x}) = -\frac{1}{4\pi\epsilon_0} \frac{1}{c} \frac{d}{dt} \int d^3x' \, \rho(t, \mathbf{x}') , \quad (12.7)$$

$$\phi_{1\text{PN}}(t, \mathbf{x}) = \frac{1}{4\pi\epsilon_0} \frac{1}{2c^2} \partial_t^2 \int d^3x' \, \rho(t, \mathbf{x}')|\mathbf{x} - \mathbf{x}'| . \quad (12.8)$$

The term $\phi_N(t, \mathbf{x})$ is just the instantaneous Coulomb potential that describes the Newtonian dynamics, with the potential ϕ_N at time t determined by the charge density at the same value of t, while $\phi_{0.5\text{PN}}(t, \mathbf{x})$

and $\phi_{1PN}(t, \mathbf{x})$ are 0.5PN correction and the 1PN correction, respectively. Observe that, in practice, the order at which a given term enters the PN expansion can be read simply from the powers of $1/c$ in front of it. So, for, instance, $\phi_{1PN}(t, \mathbf{x})$ in eq. (12.8) has an explicit factor $1/c^2$. In any computation, such as in the equations of motion a system of point particles that we will do below, the required powers of v will appear automatically for dimensional reasons, so this term will give a correction of order v^2/c^2 to the Newtonian result.

We next observe that $\phi_{0.5PN}(t, \mathbf{x})$ actually vanishes because, for a localized system of charges, the total charge

$$Q = \int d^3x'\, \rho(t, \mathbf{x}'), \tag{12.9}$$

is conserved, so $dQ/dt = 0$. Therefore, there is no 0.5PN correction to the scalar potential.

For the vector potential we can perform the same expansion, plugging eq. (12.4) into eq. (12.2). However, because of the explicit $1/c^2$ term in eq. (12.2), the expansion at 1PN order is obtained using simply the lowest-order term in eq. (12.4),

$$\mathbf{A}_{1PN}(t, \mathbf{x}) = \frac{1}{4\pi\epsilon_0} \frac{1}{c^2} \int d^3x' \frac{\mathbf{j}(t, \mathbf{x}')}{|\mathbf{x} - \mathbf{x}'|}. \tag{12.10}$$

We now specialize to a system of point charges. In this case the charge density and current distributions are given by

$$\rho(t, \mathbf{x}) = \sum_{a=1}^{N} q_a \delta^{(3)}[\mathbf{x} - \mathbf{x}_a(t)], \tag{12.11}$$

$$\mathbf{j}(t, \mathbf{x}) = \sum_{a=1}^{N} q_a \mathbf{v}_a(t) \delta^{(3)}[\mathbf{x} - \mathbf{x}_a(t)]. \tag{12.12}$$

The action of a single charged point particle interacting with an external electromagnetic field was already given in eq. (8.70). We find it useful to write the interaction term in the form (8.74), as an integral over dt and d^3x; we will indeed see below that some aspects of the computation are clearer if we deal with the fields $\phi(t, \mathbf{x})$ and $\mathbf{A}(t, \mathbf{x})$ at a generic space-time point, rather than with the fields $\phi[t, \mathbf{x}_a(t)]$ and $\mathbf{A}[t, \mathbf{x}_a(t)]$ evaluated on the particle trajectory. Furthermore, we recall that eq. (8.74) gives the interaction of a given charge and current density with an *external* electromagnetic field. To stress this, we rewrite eq. (8.74) in the notation

$$S_{\text{int}} = \int dt d^3x \left[-\rho(t, \mathbf{x})\phi_{\text{ext}}(t, \mathbf{x}) + \mathbf{j}(t, \mathbf{x}) \cdot \mathbf{A}_{\text{ext}}(t, \mathbf{x}) \right]. \tag{12.13}$$

When we are interested in the interaction between a set of point charges, it is natural to assume that the a-th charge is subject to the potentials generated by all other charges, except the a-th charge itself. Therefore, we will exclude self-interaction terms, analogous to the self-energy terms discussed and subtracted in Section 5.2.2. It should be acknowledged that this is a somewhat "ad hoc" prescription. In Section 12.3.2 we will come back to these self-force terms, and we will give a deeper discussion of why, in the present computation, in which we work to 1PN order, they can indeed be discarded. Then, for a system of point charges, the action (8.70) becomes

$$S = S_{\text{free}} + S_{\text{int}}, \qquad (12.14)$$

where

$$S_{\text{free}} = -\sum_{a=1}^{N} m_a c^2 \int dt \sqrt{1 - \frac{v_a^2(t)}{c^2}}, \qquad (12.15)$$

and

$$S_{\text{int}} = \frac{1}{2} \sum_{a=1}^{N} \int dt d^3x \left[-\rho_a(t, \mathbf{x}) \phi_{a,\text{ext}}(t, \mathbf{x}) + \mathbf{j}_a(t, \mathbf{x}) \cdot \mathbf{A}_{a,\text{ext}}(t, \mathbf{x}) \right], \qquad (12.16)$$

where ρ_a, $\mathbf{j}_a$ are the charge and current density of the a-th particle,

$$\rho_a(t, \mathbf{x}) = q_a \delta^{(3)}[\mathbf{x} - \mathbf{x}_a(t)], \qquad (12.17)$$
$$\mathbf{j}_a(t, \mathbf{x}) = q_a \mathbf{v}_a(t) \delta^{(3)}[\mathbf{x} - \mathbf{x}_a(t)], \qquad (12.18)$$

while $\phi_{a,\text{ext}}$, $\mathbf{A}_{a,\text{ext}}$ are the gauge potentials generated by all other particles, except the a-th particle itself, which therefore are seen as "external" gauge potential by the a-th charge. So, for instance, the function $\phi_{1,\text{ext}}(t, \mathbf{x})$ is the scalar potential generated, at the point $\mathbf{x}$ and at time t, by the particles $2, \ldots, N$, which have positions $\mathbf{x}_2(t), \ldots, \mathbf{x}_N(t)$; the function $\phi_{2,\text{ext}}(t, \mathbf{x})$ is the potential generated by the particles $1, 3, \ldots, N$, with positions $\mathbf{x}_1(t), \mathbf{x}_3(t), \ldots, \mathbf{x}_N(t)$, and so on. Note that the time dependence of $\phi_{a,\text{ext}}(t, \mathbf{x})$ and $\mathbf{A}_{a,\text{ext}}(t, \mathbf{x})$ enters through the time dependence of the position of the other particles $\mathbf{x}_b(t)$, with $b \neq a$, that generate the potential felt by the particle a. Similarly to eq. (5.9), the overall factor $1/2$ in front of eq. (12.16) compensates for the double counting of the particles pairs.

It is useful at this point to recall that the interaction action (12.13) is invariant under the gauge transformation (3.86), that, in the present notation, reads

$$\mathbf{A}_{\text{ext}}(t, \mathbf{x}) \rightarrow \mathbf{A}_{\text{ext}}(t, \mathbf{x}) - \boldsymbol{\nabla}\theta, \qquad (12.19)$$

$$\phi_{\text{ext}}(t, \mathbf{x}) \rightarrow \phi_{\text{ext}}(t, \mathbf{x}) + \frac{\partial \theta}{\partial t}. \qquad (12.20)$$

As we saw in eq. (8.75), this is a consequence of the conservation equation $\partial_\mu j^\mu = 0$, or equivalently, of eq. (3.22). Indeed, under the transformation (12.19), (12.20), the interaction action (12.13) changes as

$$\begin{aligned} S_{\text{int}} &\rightarrow S_{\text{int}} + \int dt d^3x \left[-\rho(t, \mathbf{x}) \frac{\partial \theta}{\partial t} - \mathbf{j}(t, \mathbf{x}) \cdot \boldsymbol{\nabla}\theta \right] \\ &= S_{\text{int}} + \int dt d^3x \, \theta(t, \mathbf{x}) \left[\frac{\partial}{\partial t} \rho(t, \mathbf{x}) + \boldsymbol{\nabla} \cdot \mathbf{j}(t, \mathbf{x}) \right], \end{aligned} \qquad (12.21)$$

where we integrated by parts discarding the boundary term, using the fact that the system of charges is localized in space. The term in bracket then vanishes, because of eq. (3.22). We now observe that, in eq. (12.16), we can perform a gauge transformation independently on each term of the sum, i.e., for each $(\phi_{a,\text{ext}}, \mathbf{A}_{a,\text{ext}})$. Indeed, if we transform

$$\mathbf{A}_{a,\text{ext}} \rightarrow \mathbf{A}_{a,\text{ext}} - \boldsymbol{\nabla}\theta_a \,, \tag{12.22}$$

$$\phi_{a,\text{ext}}(t, \mathbf{x}) \rightarrow \phi_{a,\text{ext}}(t, \mathbf{x}) + \frac{\partial \theta_a}{\partial t} \,, \tag{12.23}$$

with an independent gauge function θ_a for each a, the a-th term in the sum in eq. (12.16) changes as

$$\int dt d^3x \, [-\rho_a(t, \mathbf{x})\phi_{a,\text{ext}}(t, \mathbf{x}) + \mathbf{j}_a(t, \mathbf{x}){\cdot}\mathbf{A}_{a,\text{ext}}(t, \mathbf{x})]$$

$$\rightarrow \int dt d^3x \, [-\rho_a(t, \mathbf{x})\phi_{a,\text{ext}}(t, \mathbf{x}) + \mathbf{j}_a(t, \mathbf{x}){\cdot}\mathbf{A}_{a,\text{ext}}(t, \mathbf{x})]$$

$$+ \int dt d^3x \, \theta_a(t, \mathbf{x}) \left[\frac{\partial}{\partial t}\rho_a(t, \mathbf{x}) + \boldsymbol{\nabla}{\cdot}\mathbf{j}_a(t, \mathbf{x}) \right] \,, \tag{12.24}$$

and for each particle separately the term in bracket vanishes, as we saw in eq. (3.30).[4] We will make use of this extended gauge freedom below.

12.2.1   The gauge potentials to 1PN order

We can now compute explicitly the expression for the gauge potential to 1PN order. We consider first the scalar potential $\phi_{a,\text{ext}}(t, \mathbf{x})$. For the Newtonian term, using eq. (12.6), and eq. (12.11) with the self-term excluded, we get

$$\phi_{a,\text{ext}}(t, \mathbf{x})|_N = \frac{1}{4\pi\epsilon_0} \int d^3x' \sum_{b \neq a} q_b \delta^{(3)} \left[\mathbf{x}' - \mathbf{x}_b(t) \right] \frac{1}{|\mathbf{x} - \mathbf{x}'|}$$

$$= \frac{1}{4\pi\epsilon_0} \sum_{b \neq a} \frac{q_b}{|\mathbf{x} - \mathbf{x}_b(t)|} \,, \tag{12.25}$$

which, of course, is just the Newtonian potential already computed in eq. (5.8), except that now the particles that generate this potential have coordinates $\mathbf{x}_b(t)$, functions of time.

For the 1PN term, from eq. (12.8),[5]

$$\phi_{a,\text{ext}}(t, \mathbf{x})|_{1\text{PN}} = \frac{1}{4\pi\epsilon_0} \frac{1}{2c^2} \partial_t^2 \int d^3x' \, \rho(t, \mathbf{x}')|\mathbf{x} - \mathbf{x}'|$$

$$= \frac{1}{4\pi\epsilon_0} \frac{1}{2c^2} \sum_{b \neq a} q_b \partial_t^2 \int d^3x' \, \delta^{(3)} \left[\mathbf{x}' - \mathbf{x}_b(t) \right] |\mathbf{x} - \mathbf{x}'|$$

$$= \frac{1}{4\pi\epsilon_0} \frac{1}{2c^2} \sum_{b \neq a} q_b \partial_t^2 |\mathbf{x} - \mathbf{x}_b(t)| \,. \tag{12.26}$$

Therefore, to 1PN order, in the Lorenz gauge in which we are working,

$$\phi_{a,\text{ext}}(t, \mathbf{x}) = \frac{1}{4\pi\epsilon_0} \sum_{b \neq a} q_b \left[\frac{1}{|\mathbf{x} - \mathbf{x}_b(t)|} + \frac{1}{2c^2}\partial_t^2 |\mathbf{x} - \mathbf{x}_b(t)| \right] \,. \tag{12.27}$$

[4]The implicit assumption is that we have a set of charges that interact among them electromagnetically, but retain their individuality; for instance, they do not merge together, and they do not decay into other charged particles, because of electromagnetic or other interactions.

[5]Observe that, even if, eventually, in eq. (12.16) $\phi_{a,\text{ext}}(t, \mathbf{x})$ is multiplied by $\rho_a(t, \mathbf{x})$, which is given by eq. (12.17) and therefore is proportional to $\delta^{(3)}[\mathbf{x}-\mathbf{x}_a(t)]$, still, in eq. (12.26) we cannot yet replace $\mathbf{x}$ by $\mathbf{x}_a(t)$. In eq. (12.26), the ∂_t^2 operator acts only on $\mathbf{x}_b(t)$ and not on $\mathbf{x}$, since $\phi_{a,\text{ext}}(t, \mathbf{x})$ is the potential at a generic point $\mathbf{x}$ in space, with a time-dependence due to the fact that it is generated by a set of charges q_b, with $b \neq a$, on time-dependent trajectories $\mathbf{x}_b(t)$. If one would replace $\mathbf{x} = \mathbf{x}_a(t)$ in eq. (12.26) before taking the time derivatives, one would make a mistake, introducing spurious time derivatives of the function $\mathbf{x}_a(t)$. Similar considerations hold for the gauge transformations that we will perform below on $\phi_{a,\text{ext}}(t, \mathbf{x})$ and $\mathbf{A}_{a,\text{ext}}(t, \mathbf{x})$, with a gauge function $\theta_a(t, \mathbf{x})$ that will also depend on t and $\mathbf{x}$ through $|\mathbf{x} - \mathbf{x}_b(t)|$, see eq. (12.29). When we take the time derivative and spatial gradient of such a function, we must make clear the distinction between the $\mathbf{x}$ dependence and the t dependence, which is not possible if we replace from the start $\mathbf{x}$ by $\mathbf{x}_a(t)$. For these reasons, it is necessary to perform the computation starting from the expression (12.16) of the interaction action, without yet performing the integral over d^3x with the use of the delta functions $\delta^{(3)}[\mathbf{x} - \mathbf{x}_a(t)]$ which is present in $\rho_a(t, \mathbf{x})$ and $\mathbf{j}_a(t, \mathbf{x})$.

Similarly, plugging eq. (12.12) into eq. (12.10), and discarding again the term with $b = a$, we find, again to 1PN order,

$$
\begin{aligned}
\mathbf{A}_{a,\mathrm{ext}}(t, \mathbf{x}) &= \frac{1}{4\pi\epsilon_0} \frac{1}{c^2} \sum_{b\neq a} q_b \mathbf{v}_b(t) \int d^3 x'\, \delta^{(3)} \left[\mathbf{x}' - \mathbf{x}_b(t)\right] \frac{1}{|\mathbf{x} - \mathbf{x}'|} \\
&= \frac{1}{4\pi\epsilon_0} \frac{1}{c^2} \sum_{b\neq a} \frac{q_b \mathbf{v}_b(t)}{|\mathbf{x} - \mathbf{x}_b(t)|} \,.
\end{aligned}
\tag{12.28}
$$

If now one just proceeded by computing explicitly the second time derivative of $|\mathbf{x} - \mathbf{x}_b(t)|$ in eq. (12.27), one would find terms that depend on the accelerations $d^2\mathbf{x}_b/dt^2$, and therefore the corresponding interaction action (12.16) would also appear to depend on the accelerations of the particles, leading to equations of motion involving time derivatives of order higher than the second. As we will discuss in Section 12.2.3, such a dependence on higher-order derivatives is in principle unavoidable at higher orders in the PN expansion; however, it can be eliminated, order by order in the PN expansion, expressing these higher derivatives in terms of the positions and velocities, by using the equations of motion to lower orders. While this procedure becomes necessary to higher orders of the PN expansion, at the 1PN order at which we are working it is much simpler to observe that these higher-derivative terms can be eliminated with a gauge transformation of the form (12.22, 12.23), choosing as gauge function

$$
\theta_a(t, \mathbf{x}) = -\frac{1}{4\pi\epsilon_0} \frac{1}{2c^2} \sum_{b\neq a} q_b \partial_t |\mathbf{x} - \mathbf{x}_b(t)| \,.
\tag{12.29}
$$

This function is chosen so as to eliminate the term involving ∂_t^2 in eq. (12.27), since, under this gauge transformation,

$$
\begin{aligned}
\phi_a(t, \mathbf{x}) &\rightarrow \phi_a(t, \mathbf{x}) + \partial_t \theta_a \\
&= \phi_{a,\mathrm{ext}}(t, \mathbf{x}) - \frac{1}{4\pi\epsilon_0} \frac{1}{2c^2} \sum_{b\neq a} q_b \partial_t^2 |\mathbf{x} - \mathbf{x}_b(t)| \,.
\end{aligned}
\tag{12.30}
$$

Therefore, in the new gauge,

$$
\phi_{a,\mathrm{ext}}(t, \mathbf{x}) = \frac{1}{4\pi\epsilon_0} \sum_{b\neq a} \frac{q_b}{|\mathbf{x} - \mathbf{x}_b(t)|} \,,
\tag{12.31}
$$

which is just the Newtonian potential. In other words, this gauge is chosen so that $\phi_{a,\mathrm{ext}}(t, \mathbf{x})|_{\mathrm{1PN}} = 0$, which has been possible because the term $\phi_{a,\mathrm{ext}}(t, \mathbf{x})|_{\mathrm{1PN}}$ in eq. (12.26) is a total time derivative. Under this gauge transformation $\mathbf{A}_{a,\mathrm{ext}}(t, \mathbf{x})$ picks an extra term,

$$
\begin{aligned}
\mathbf{A}_{a,\mathrm{ext}}(t, \mathbf{x}) &\rightarrow \mathbf{A}_{a,\mathrm{ext}}(t, \mathbf{x}) - \boldsymbol{\nabla}\theta_a(t, \mathbf{x}) \\
&= \frac{1}{4\pi\epsilon_0} \frac{1}{c^2} \sum_{b\neq a} q_b \left[\frac{\mathbf{v}_b(t)}{|\mathbf{x} - \mathbf{x}_b(t)|} + \frac{1}{2}\partial_t \boldsymbol{\nabla} |\mathbf{x} - \mathbf{x}_b(t)| \right] \,.
\end{aligned}
\tag{12.32}
$$

We next use

$$
\boldsymbol{\nabla} |\mathbf{x} - \mathbf{x}_b(t)| = \frac{\mathbf{x} - \mathbf{x}_b(t)}{|\mathbf{x} - \mathbf{x}_b(t)|} \,,
\tag{12.33}
$$

and[6]

$$\partial_t \left[\frac{\mathbf{x} - \mathbf{x}_b(t)}{|\mathbf{x} - \mathbf{x}_b(t)|} \right] = -\frac{\mathbf{v}_b(t)}{|\mathbf{x} - \mathbf{x}_b(t)|} + \frac{\mathbf{x} - \mathbf{x}_b(t)}{|\mathbf{x} - \mathbf{x}_b(t)|^3} \left\{ [\mathbf{x} - \mathbf{x}_b(t)] \cdot \mathbf{v}_b(t) \right\} .$$

(12.36)

So, in the new gauge,

$$\mathbf{A}_{a,\mathrm{ext}}(t, \mathbf{x}) = \frac{1}{4\pi\epsilon_0} \frac{1}{2c^2} \sum_{b \neq a} q_b \frac{\mathbf{v}_b(t)}{|\mathbf{x} - \mathbf{x}_b(t)|}$$

(12.37)

$$+ \frac{1}{4\pi\epsilon_0} \frac{1}{2c^2} \sum_{b \neq a} q_b \frac{\mathbf{x} - \mathbf{x}_b(t)}{|\mathbf{x} - \mathbf{x}_b(t)|^3} \left\{ [\mathbf{x} - \mathbf{x}_b(t)] \cdot \mathbf{v}_b(t) \right\} .$$

From this expression, we can verify explicitly that

$$\boldsymbol{\nabla} \cdot \mathbf{A}_{a,\mathrm{ext}}(t, \mathbf{x}) = 0 ,$$

(12.38)

and therefore we have simply reached the Coulomb gauge (3.92) for each of the $\mathbf{A}_{a,\mathrm{ext}}$, at 1PN order. So, we started from the general expression for the gauge potentials in the Lorenz gauge, where eqs. (12.1) and (12.2) hold; we have computed them explicitly to 1PN order, and we have then performed a gauge transformation that puts these expressions for the gauge potentials in the Coulomb gauge. Note that, since $\boldsymbol{\nabla} \cdot \mathbf{A}_{a,\mathrm{ext}} = 0$ but $\partial_t \phi_{a,\mathrm{ext}} \neq 0$, these potentials no longer satisfy the Lorenz gauge condition $\partial_\mu A_a^\mu = 0$. Equivalently, one could have performed the PN expansion working from the start in the Coulomb gauge (which can be done to all orders in the PN expansion), using eq. (3.93) for $\phi_{a,\mathrm{ext}}$, sourced by $\rho_{a,\mathrm{ext}} \equiv \sum_{b \neq a} \rho_b$. This immediately gives eq. (12.31), in fact to all PN orders. We can then insert this expression for $\phi_{a,\mathrm{ext}}$ into eq. (3.94) and solve it in an expansion for small retardation effect. To 1PN, only the lowest-order term of this expansion is needed, and gives eq. (12.37).

12.2.2 Effective dynamics of a system of point charges

Equations (12.31) and (12.37) allow us to eliminate the gauge fields in terms of the variables $\mathbf{x}_a(t)$ and $\mathbf{v}_a(t)$ which describe the charged particles, up to 1PN order. This process can in principle be carried out to all orders in the PN expansion. In this way, in the near region the coupled dynamics of the electromagnetic field and the charged particles can be expressed entirely in terms of the degrees of freedom $\mathbf{x}_a(t)$, $\mathbf{v}_a(t)$ describing the charged particles, with all these variables evaluated at the same value t of time, as in Newtonian mechanics.

The elimination of $\phi(t, \mathbf{x})$ and $\mathbf{A}(t, \mathbf{x})$ in favor of $\mathbf{x}_a(t)$, $\mathbf{v}_a(t)$ could in principle be performed either at the level of the action or at the level of the equations of motion and, naively, one might think that the two procedures should be equivalent. Further reflection, however, shows that, at least at a generic order of the PN expansion, this cannot be the case. If we perform the elimination at the level of the action, starting from

[6]The explicit computations go as follows. Equation (12.33) is obtained writing

$$\boldsymbol{\nabla}|\mathbf{x} - \mathbf{x}_b(t)| = \frac{\boldsymbol{\nabla}|\mathbf{x} - \mathbf{x}_b(t)|^2}{2|\mathbf{x} - \mathbf{x}_b(t)|}$$

$$= \frac{\boldsymbol{\nabla} [\mathbf{x} \cdot \mathbf{x} - 2\mathbf{x} \cdot \mathbf{x}_b(t) + \mathbf{x}_b(t) \cdot \mathbf{x}_b(t)]}{2|\mathbf{x} - \mathbf{x}_b(t)|}$$

$$= \frac{\mathbf{x} - \mathbf{x}_b(t)}{|\mathbf{x} - \mathbf{x}_b(t)|} ,$$

(12.34)

while eq. (12.36) follows from

$$\partial_t \left[\frac{\mathbf{x} - \mathbf{x}_b(t)}{|\mathbf{x} - \mathbf{x}_b(t)|} \right] = -\frac{\mathbf{v}_b(t)}{|\mathbf{x} - \mathbf{x}_b(t)|}$$

$$- \frac{\mathbf{x} - \mathbf{x}_b(t)}{|\mathbf{x} - \mathbf{x}_b(t)|^2} \partial_t |\mathbf{x} - \mathbf{x}_b(t)| ,$$

and last terms is computed writing

$$\partial_t |\mathbf{x} - \mathbf{x}_b(t)| = \frac{\partial_t |\mathbf{x} - \mathbf{x}_b(t)|^2}{2|\mathbf{x} - \mathbf{x}_b(t)|}$$

$$= \frac{\partial_t [\mathbf{x} \cdot \mathbf{x} - 2\mathbf{x} \cdot \mathbf{x}_b(t) + \mathbf{x}_b(t) \cdot \mathbf{x}_b(t)]}{2|\mathbf{x} - \mathbf{x}_b(t)|}$$

$$= -\frac{[\mathbf{x} - \mathbf{x}_b(t)] \cdot \mathbf{v}_b(t)}{|\mathbf{x} - \mathbf{x}_b(t)|} .$$

(12.35)

[7]The proof is a standard result from classical mechanics: given a system with generalized coordinates $q_i(t)$ and Lagrangian $L(q_i, \dot{q}_i)$ (with no explicit time dependence), the conjugate momentum is $p_i = \delta L/\delta \dot{q}_i$ and the Hamiltonian is $H = p_i \dot{q}_i - L$, while the equation of motion is

$$\frac{d}{dt}\frac{\delta L}{\delta \dot{q}_i} - \frac{\delta L}{\delta q_i} = 0.$$

In terms of p_i, this means that $\dot{p}_i = \delta L/\delta q_i$. Then

$$\begin{aligned}\frac{dH}{dt} &= \dot{p}_i \dot{q}_i + p_i \ddot{q}_i - \frac{dL}{dt} \\ &= \frac{\delta L}{\delta q_i}\dot{q}_i + p_i \ddot{q}_i - \frac{dL}{dt}.\end{aligned}$$

However

$$\begin{aligned}\frac{d}{dt} L(q_i, \dot{q}_i) &= \frac{\delta L}{\delta q_i}\frac{dq_i}{dt} + \frac{\delta L}{\delta \dot{q}_i}\frac{d\dot{q}_i}{dt} \\ &= \frac{\delta L}{\delta q_i}\dot{q}_i + p_i \ddot{q}_i,\end{aligned}$$

and therefore $dH/dt = 0$.

eq. (12.16), we eventually end up with a dynamics described by an action, and therefore a Lagrangian, that depends only on the position and velocities of the particles [we will see in Section 12.2.3 how higher-order derivatives, that in principle can also appear at a generic order of the PN expansion, can be re-expressed order by order in the expansion in terms of the positions and velocities $\mathbf{x}_a(t)$ and $\mathbf{v}_a(t)$]. Such a Lagrangian has no explicit time dependence and therefore the corresponding Hamiltonian is automatically conserved on the solutions of the equations of motion.[7] Therefore, a Lagrangian description cannot catch dissipative terms, i.e., terms that describe the decrease of the energy of the system. However, we know that, beyond some PN order, dissipative terms must be present, to account for the energy lost by the system of charges to electromagnetic radiation. To obtain them, we cannot work at the level of the Lagrangian, but we must rather work at the level of the equations of motion, starting from the Lorentz force equation (8.83), which is the equation of motion derived from the full Lagrangian (8.71), and eliminate the electric and magnetic fields from it, using their expressions obtained from the solution of the gauge potentials in terms of $\mathbf{x}_a(t)$, $\mathbf{v}_a(t)$, order by order in the PN expansion. This procedure produces the full answer, including both conservative and dissipative terms.

In this section we explicitly carry out the elimination of the gauge potentials at 1PN order, both at the level of the equations of motion and at the level of the Lagrangian, and we will find that, to this order, the two procedures are equivalent, and that the dynamics is conservative. As we will see in Section 12.3.3, dissipative terms start from 1.5PN order. Beyond 1PN order, one must therefore work at the level of equations of motion, in order to obtain the full result including the conservative and the dissipative terms.

Effective dynamics from the Lorentz force equation

We begin by performing the explicit computation using the equations of motion (8.83). We observe that

$$\begin{aligned}(\mathbf{v} \times \mathbf{B})_i &= [\mathbf{v} \times (\boldsymbol{\nabla}\times\mathbf{A})]_i \\ &= \epsilon_{ijk}\epsilon_{klm}v_j \partial_l A_m \\ &= v_j \partial_i A_j - v_j \partial_j A_i,\end{aligned}\tag{12.39}$$

where we used the identity (1.7). Then, the Lorentz force equation (8.83) can be written as

$$\frac{d}{dt}(\gamma_a m_a \mathbf{v}_a) = \tag{12.40}$$
$$\left(-q_a \boldsymbol{\nabla}_{\mathbf{x}}\phi_{a,\text{ext}} - q_a \frac{\partial \mathbf{A}_{a,\text{ext}}}{\partial t} + q_a v_{a,j}\boldsymbol{\nabla}_{\mathbf{x}}A^j_{a,\text{ext}} - q_a v^j_a \partial_j \mathbf{A}_{a,\text{ext}}\right)_{\mathbf{x}=\mathbf{x}_a(t)},$$

where $\gamma_a = \gamma(v_a)$ and we have stressed that, in the force acting on the a-th particle, the electric and magnetic fields are evaluated on the position of the particle, i.e., for $\mathbf{x} = \mathbf{x}_a(t)$ (and we recall that spatial indices can be equivalently written as upper or lower indices). Using

$$
\begin{aligned}
\frac{d}{dt}\mathbf{A}_{a,\text{ext}}[t, \mathbf{x}_a(t)] &= \frac{\partial \mathbf{A}_{a,\text{ext}}[t, \mathbf{x}_a(t)]}{\partial t} + \frac{dx_a^j(t)}{dt}\left[\frac{\partial \mathbf{A}_{a,\text{ext}}(t, \mathbf{x})}{\partial x^j}\right]_{\mathbf{x}=\mathbf{x}_a(t)} \\
&= \left[\frac{\partial \mathbf{A}_{a,\text{ext}}(t, \mathbf{x})}{\partial t} + v_a^j\partial_j \mathbf{A}_{a,\text{ext}}(t, \mathbf{x})\right]_{\mathbf{x}=\mathbf{x}_a(t)}, \quad (12.41)
\end{aligned}
$$

we see that eq. (12.40) can be rewritten as

$$
\begin{aligned}
\frac{d}{dt}&\{\gamma_a m_a \mathbf{v}_a + q_a \mathbf{A}_{a,\text{ext}}[t, \mathbf{x}_a(t)]\} \\
&= \left[-q_a\boldsymbol{\nabla}_{\mathbf{x}}\phi_{a,\text{ext}}(t, \mathbf{x}) + q_a v_{a,j}\boldsymbol{\nabla}_{\mathbf{x}}A_{a,\text{ext}}^j(t, \mathbf{x})\right]_{\mathbf{x}=\mathbf{x}_a(t)}. \quad (12.42)
\end{aligned}
$$

For consistency with the fact that we are working to 1PN order, we also expand the left-hand side as in eq. (12.67) keeping only the correction v_a^2/c^2, so we write

$$
\begin{aligned}
\frac{d}{dt}&\left\{\left(1 + \frac{v_a^2}{2c^2}\right) m_a \mathbf{v}_a + q_a \mathbf{A}_{a,\text{ext}}[t, \mathbf{x}_a(t)]\right\} \\
&= \left[-q_a\boldsymbol{\nabla}_{\mathbf{x}}\phi_{a,\text{ext}}(t, \mathbf{x}) + q_a v_{a,j}\boldsymbol{\nabla}_{\mathbf{x}}A_{a,\text{ext}}^j(t, \mathbf{x})\right]_{\mathbf{x}=\mathbf{x}_a(t)}. \quad (12.43)
\end{aligned}
$$

On the right-hand side, $\phi_{a,\text{ext}}$ and $A_{a,\text{ext}}^j$ are given by eqs. (12.31) and (12.37), respectively. After having computed their gradients with respect to the variable $\mathbf{x}$ [using eq. (12.33)], we finally set $\mathbf{x} = \mathbf{x}_a(t)$.

It is now convenient to use the notation

$$
\phi_a(\mathbf{x}_1, \ldots, \mathbf{x}_N) = \frac{1}{4\pi\epsilon_0}\sum_{b\neq a}\frac{q_b}{|\mathbf{x}_a(t) - \mathbf{x}_b(t)|}, \quad (12.44)
$$

and

$$
\begin{aligned}
\mathbf{A}_a(\mathbf{x}_1, \ldots, \mathbf{x}_N; \mathbf{v}_1, \ldots \mathbf{v}_N) &= \frac{1}{4\pi\epsilon_0}\frac{1}{2c^2}\sum_{b\neq a}q_b\frac{\mathbf{v}_b(t)}{|\mathbf{x}_a(t) - \mathbf{x}_b(t)|} \quad (12.45) \\
&+ \frac{1}{4\pi\epsilon_0}\frac{1}{2c^2}\sum_{b\neq a}q_b\frac{\mathbf{x}_a(t) - \mathbf{x}_b(t)}{|\mathbf{x}_a(t) - \mathbf{x}_b(t)|^3}\{[\mathbf{x}_a(t) - \mathbf{x}_b(t)]\cdot\mathbf{v}_b(t)\},
\end{aligned}
$$

which automatically takes into account that the gauge potentials are evaluated at $\mathbf{x} = \mathbf{x}_a(t)$ and treats them symmetrically with respect to the positions (and velocities) of all particles. We also introduce the notation

$$
\mathbf{r}_{ab}(t) = \mathbf{x}_a(t) - \mathbf{x}_b(t), \quad (12.46)
$$

and $r_{ab} = |\mathbf{r}_{ab}|$, $\hat{\mathbf{r}}_{ab}/r_{ab}$, so that

$$
\phi_a(\mathbf{x}_1, \ldots, \mathbf{x}_N) = \frac{1}{4\pi\epsilon_0}\sum_{b\neq a}\frac{q_b}{r_{ab}}, \quad (12.47)
$$

and

$$
\mathbf{A}_a(\mathbf{x}_1, \ldots, \mathbf{x}_N; \mathbf{v}_1, \ldots \mathbf{v}_N) = \frac{1}{4\pi\epsilon_0}\frac{1}{2c^2}\sum_{b\neq a}\frac{q_b}{r_{ab}}[\mathbf{v}_b + \hat{\mathbf{r}}_{ab}(\hat{\mathbf{r}}_{ab}\cdot\mathbf{v}_b)].
$$
$$
(12.48)
$$

Then, eq. (12.43) reads

$$\frac{d}{dt}\left[\left(1+\frac{v_a^2}{2c^2}\right)m_a\mathbf{v}_a + q_a\mathbf{A}_a(\mathbf{x}_1,\ldots,\mathbf{x}_N;\mathbf{v}_1,\ldots\mathbf{v}_N)\right] = -\boldsymbol{\nabla}_a\phi_a$$

$$+\sum_{b\neq a}^N \frac{q_a q_b}{r_{ab}^2}\frac{\mathbf{v}_a(\hat{\mathbf{r}}_{ab}\cdot\mathbf{v}_b)+\mathbf{v}_b(\hat{\mathbf{r}}_{ab}\cdot\mathbf{v}_a)-\hat{\mathbf{r}}_{ab}\left[(\mathbf{v}_a\cdot\mathbf{v}_b)+3(\hat{\mathbf{r}}_{ab}\cdot\mathbf{v}_a)(\hat{\mathbf{r}}_{ab}\cdot\mathbf{v}_b)\right]}{8\pi\epsilon_0 c^2}.$$

$$(12.49)$$

Observe that the quantity inside d/dt on the left-hand side is just the conjugate momentum (expanded to 1PN order), compare with eq. (8.78). More explicitly, eq. (12.49) reads

$$\frac{d}{dt}\left[\left(1+\frac{v_a^2}{2c^2}\right)m_a\mathbf{v}_a + \frac{1}{4\pi\epsilon_0}\frac{1}{2c^2}\sum_{b\neq a}\frac{q_a q_b}{r_{ab}}\left[\mathbf{v}_b+\hat{\mathbf{r}}_{ab}(\hat{\mathbf{r}}_{ab}\cdot\mathbf{v}_b)\right]\right]$$

$$=\frac{1}{4\pi\epsilon_0}\sum_{b\neq a}^N\frac{q_a q_b}{r_{ab}^2} \tag{12.50}$$

$$\times\left[\hat{\mathbf{r}}_{ab}+\frac{\mathbf{v}_a(\hat{\mathbf{r}}_{ab}\cdot\mathbf{v}_b)+\mathbf{v}_b(\hat{\mathbf{r}}_{ab}\cdot\mathbf{v}_a)-\hat{\mathbf{r}}_{ab}\left[(\mathbf{v}_a\cdot\mathbf{v}_b)+3(\hat{\mathbf{r}}_{ab}\cdot\mathbf{v}_a)(\hat{\mathbf{r}}_{ab}\cdot\mathbf{v}_b)\right]}{2c^2}\right].$$

These are the 1PN equations of motion for a set of charged particles.

The Darwin Lagrangian

We now perform the elimination of the gauge fields in terms of $\mathbf{x}_a(t)$, $\mathbf{v}_a(t)$ at the level of the action, again at 1PN level. This is done inserting eqs. (12.31) and (12.37) into eq. (12.16), and carrying out the integrals over d^3x with the help of the Dirac deltas present in ρ_a and $\mathbf{j}_a$. The result is

$$S_{\rm int} = \frac{1}{8\pi\epsilon_0}\sum_{a=1}^N\sum_{b\neq a}^N\int dt\,\frac{q_a q_b}{r_{ab}}\left[-1+\frac{\mathbf{v}_a\cdot\mathbf{v}_b+(\hat{\mathbf{r}}_{ab}\cdot\mathbf{v}_a)(\hat{\mathbf{r}}_{ab}\cdot\mathbf{v}_b)}{2c^2}\right].$$

$$(12.51)$$

Since we are working to 1PN order, i.e., corrections up to order $(v/c)^2$ to the Newtonian result, we also expand the square-root in eq. (12.15) keeping only the $(v/c)^2$ correction to the Newtonian kinetic energy,

$$-m_a c^2\sqrt{1-\frac{v_a^2}{c^2}} = -m_a c^2 + \frac{1}{2}m_a v_a^2 + \frac{1}{8}m_a\frac{v_a^4}{c^2} + \ldots \tag{12.52}$$

As far as we are interested in using the Lagrangian to derive the equations of motion, the term $-m_a c^2$ is a constant, and can be dropped (of course, if we rather want to compute the Hamiltonian and therefore the energy, this gives the rest energy of the particle). Then, summing up, we get the Lagrangian of a system of point charges, up to 1PN order,

$$L = L_N + L_{\rm 1PN}, \tag{12.53}$$

where the Newtonian term is

$$L_N = \sum_{a=1}^{N} \frac{1}{2} m_a v_a^2 - \frac{1}{8\pi\epsilon_0} \sum_{a=1}^{N} \sum_{b\neq a}^{N} \frac{q_a q_b}{r_{ab}}, \qquad (12.54)$$

and the 1PN correction is

$$L_{1PN} = \sum_{a=1}^{N} \frac{m_a v_a^4}{8c^2} + \frac{1}{4\pi\epsilon_0} \frac{1}{4c^2} \sum_{a=1}^{N} \sum_{b\neq a}^{N} \frac{q_a q_b}{r_{ab}} \left[\mathbf{v}_a \cdot \mathbf{v}_b + (\hat{\mathbf{r}}_{ab} \cdot \mathbf{v}_a)(\hat{\mathbf{r}}_{ab} \cdot \mathbf{v}_b) \right].$$

$$(12.55)$$

The total Lagrangian up to 1PN order is called the *Darwin Lagrangian*.[8] From eqs. (12.44) and (12.48), we see that we can also rewrite the result as

$$L_N = \sum_{a=1}^{N} \left[\frac{1}{2} m_a v_a^2 - \frac{1}{2} q_a \phi_a(\mathbf{x}_1, \dots, \mathbf{x}_N) \right], \qquad (12.56)$$

and

$$L_{1PN} = \sum_{a=1}^{N} \frac{m_a v_a^4}{8c^2} + \frac{1}{2} \sum_{a=1}^{N} q_a \mathbf{v}_a \cdot \mathbf{A}_a(\mathbf{x}_1, \dots, \mathbf{x}_N; \mathbf{v}_1, \dots \mathbf{v}_N). \qquad (12.57)$$

Note that, in the Coulomb gauge that we are using, ϕ_a is a purely Newtonian term, with no 1PN correction, while $\mathbf{A}_a$ is a purely 1PN term.

We can now check that eq. (12.49) is indeed the equation of motion derived from the Darwin Lagrangian. To keep the notation simple, we perform the computation explicitly limiting ourselves to the case $N = 2$. Then,

$$L_N = \frac{1}{2} m_1 v_1^2 + \frac{1}{2} m_2 v_2^2 - \frac{1}{4\pi\epsilon_0} \frac{q_1 q_2}{r_{12}}, \qquad (12.58)$$

$$L_{1PN} = \frac{1}{8c^2}(m_1 v_1^4 + m_2 v_2^4)$$

$$+ \frac{1}{4\pi\epsilon_0} \frac{1}{2c^2} \frac{q_1 q_2}{r_{12}} \left[\mathbf{v}_1 \cdot \mathbf{v}_2 + (\hat{\mathbf{r}}_{12} \cdot \mathbf{v}_1)(\hat{\mathbf{r}}_{12} \cdot \mathbf{v}_2) \right]. \qquad (12.59)$$

We now compute the equation of motion, for instance taking the variation with respect to the variables of the first particle, separating the Newtonian and the 1PN terms,

$$\left[\frac{d}{dt} \left(\frac{\delta L_N}{\delta \mathbf{v}_1} \right) - \frac{\delta L_N}{\delta \mathbf{x}_1} \right] + \left[\frac{d}{dt} \left(\frac{\delta L_{1PN}}{\delta \mathbf{v}_1} \right) - \frac{\delta L_{1PN}}{\delta \mathbf{x}_1} \right] = 0. \qquad (12.60)$$

In the case of two particles we use the simpler notation $\mathbf{r}_{12} = r\hat{\mathbf{n}}$. The variations with respect to $\mathbf{x}_1$ are computed using

$$\frac{\partial}{\partial x_1^i} \frac{1}{r} = -\frac{n_i}{r^2}, \qquad (12.61)$$

[8] After Charles Galton Darwin (grandson of the great Charles Darwin), who first wrote it in 1920.

and (as in eq. (11.8) with $\mathbf{x}$ replaced by $\mathbf{x}_1 - \mathbf{x}_2$)

$$\frac{\partial n_i}{\partial x_1^j} = \frac{1}{r}(\delta_{ij} - n_i n_j). \tag{12.62}$$

The Newtonian term gives the standard result,

$$\frac{d}{dt}\left(\frac{\delta L_N}{\delta \mathbf{v}_1}\right) - \frac{\delta L_N}{\delta \mathbf{x}_1} = \frac{d}{dt}(m_1 \mathbf{v}_1) - \frac{q_1 q_2}{4\pi\epsilon_0}\frac{\hat{\mathbf{n}}}{r^2}, \tag{12.63}$$

while

$$\frac{d}{dt}\left(\frac{\delta L_{1PN}}{\delta \mathbf{v}_1}\right) = \frac{d}{dt}\left\{\frac{m_1 v_1^2}{2c^2}\mathbf{v}_1 + \frac{1}{4\pi\epsilon_0}\frac{1}{2c^2}\frac{q_1 q_2}{r}\left[\mathbf{v}_2 + \hat{\mathbf{n}}(\hat{\mathbf{n}}\cdot\mathbf{v}_2)\right]\right\}, \tag{12.64}$$

and

$$\frac{\delta L_{1PN}}{\delta \mathbf{x}_1} = \frac{1}{4\pi\epsilon_0}\frac{q_1 q_2}{r^2}\frac{1}{2c^2} \\ \times \left\{\mathbf{v}_1(\hat{\mathbf{n}}\cdot\mathbf{v}_2) + \mathbf{v}_2(\hat{\mathbf{n}}\cdot\mathbf{v}_1) - \hat{\mathbf{n}}\left[(\mathbf{v}_1\cdot\mathbf{v}_2) + 3(\hat{\mathbf{n}}\cdot\mathbf{v}_1)(\hat{\mathbf{n}}\cdot\mathbf{v}_2)\right]\right\}. \tag{12.65}$$

Plugging these expressions into eq. (12.60), we recover eq. (12.50) for the case $a = 1$ and $N = 2$. The computation for generic N is analogous.

This shows that the equations of motion at 1PN order, given in eq. (12.50) and obtained by eliminating from the Lorentz force equation the electric and magnetic field in terms of the variables $\mathbf{x}_a(t)$, $\mathbf{v}_a(t)$ that describe the charges, can also be derived from a Lagrangian (obtained by eliminating the gauge fields at the level of the action describing the coupled dynamics of the charges and the electromagnetic field). Therefore, the corresponding 1PN Hamiltonian, whose explicit form we will compute below, is conserved on the solution of the equations of motion, see Note 7 on page 304, and the dynamics up to 1PN order is conservative.

The Hamiltonian to 1PN order

To complete the discussion of the dynamics at 1PN order, we compute the corresponding Hamiltonian. To this purpose, we first compute the conjugate momentum $\mathbf{P}_a$, that we already introduced in eq. (8.78) for a generic vector potential. In the present case, from eqs. (12.53)–(12.55),

$$\mathbf{P}_a \equiv \frac{\delta L}{\partial \mathbf{v}_a} \tag{12.66}$$

$$= \left(1 + \frac{v_a^2}{2c^2}\right)m_a\mathbf{v}_a + \frac{1}{4\pi\epsilon_0}\frac{1}{2c^2}\sum_{b\neq a}^{N}\frac{q_a q_b}{r_{ab}}\left[\mathbf{v}_b + \hat{\mathbf{r}}_{ab}(\hat{\mathbf{r}}_{ab}\cdot\mathbf{v}_b)\right].$$

Comparing with eq. (12.48), we see that this is just the same as

$$\mathbf{P}_a = \left(1 + \frac{v_a^2}{2c^2}\right)m_a\mathbf{v}_a + q_a\mathbf{A}_{a,\mathrm{ext}}[t, \mathbf{x}_a(t)], \tag{12.67}$$

in agreement with eq. (8.78) (with $\mathbf{p} = \gamma m\mathbf{v}$ expanded to second order in v/c). The Hamiltonian is then computed from

$$H = \sum_{a=1}^{N}\mathbf{P}_a\cdot\mathbf{v}_a - L, \tag{12.68}$$

where $\mathbf{v}_a$ is written in terms of $\mathbf{P}_a$ inverting eq. (12.66). To the 1PN order at which we are working, this can be done writing eq. (12.66) as

$$\mathbf{v}_a = \frac{\mathbf{P}_a}{m_a} - \frac{v_a^2}{2c^2}\mathbf{v}_a - \frac{1}{4\pi\epsilon_0}\frac{1}{2c^2}\sum_{b\neq a}^{N}\frac{q_aq_b}{m_ar_{ab}}\left[\mathbf{v}_b + \hat{\mathbf{r}}_{ab}(\hat{\mathbf{r}}_{ab}\cdot\mathbf{v}_b)\right] , \quad (12.69)$$

and substituting, in terms proportional to $1/c^2$, the lowest-order relation $\mathbf{v}_a = \mathbf{P}_a/m_a$, since the terms neglected are of higher order. This gives

$$\mathbf{v}_a = \frac{\mathbf{P}_a}{m_a} - \frac{P_a^2}{2m_a^3c^2}\mathbf{P}_a - \frac{1}{4\pi\epsilon_0}\frac{1}{2c^2}\sum_{b\neq a}^{N}\frac{q_aq_b}{m_am_br_{ab}}\left[\mathbf{P}_b + \hat{\mathbf{r}}_{ab}(\hat{\mathbf{r}}_{ab}\cdot\mathbf{P}_b)\right] .$$
$$(12.70)$$

Plugging this into eqs. (12.68) and (12.55) we get

$$H = \sum_{a=1}^{N}\frac{P_a^2}{2m_a}\left(1 - \frac{P_a^2}{4m_a^2c^2}\right) + \frac{1}{8\pi\epsilon_0}\sum_{a=1}^{N}\sum_{b\neq a}^{N}\frac{q_aq_b}{r_{ab}} \qquad (12.71)$$

$$-\frac{1}{8\pi\epsilon_0}\frac{1}{2c^2}\sum_{a=1}^{N}\sum_{b\neq a}^{N}\frac{q_aq_b}{m_am_br_{ab}}\left[\mathbf{P}_a\cdot\mathbf{P}_b + (\hat{\mathbf{r}}_{ab}\cdot\mathbf{P}_a)(\hat{\mathbf{r}}_{ab}\cdot\mathbf{P}_b)\right] .$$

In particular, for a two-body system,

$$H = \frac{P_1^2}{2m_1}\left(1 - \frac{P_1^2}{4m_1^2c^2}\right) + \frac{P_2^2}{2m_2}\left(1 - \frac{P_2^2}{4m_2^2c^2}\right)$$

$$+ \frac{1}{4\pi\epsilon_0}\frac{q_1q_2}{r}\left[1 - \frac{\mathbf{P}_1\cdot\mathbf{P}_2 + (\hat{\mathbf{r}}\cdot\mathbf{P}_1)(\hat{\mathbf{r}}\cdot\mathbf{P}_2)}{2m_1m_2c^2}\right] , \qquad (12.72)$$

where $\mathbf{r} = \mathbf{r}_{12}$. An alternative derivation is obtained using eq. (8.87), that, for a single particle in an external potential, we rewrite in the form

$$H = mc^2\sqrt{1 + \left(\frac{\mathbf{P} - q\mathbf{A}}{m_ac}\right)^2} + q\phi , \qquad (12.73)$$

where the potentials ϕ and $\mathbf{A}$ must be computed on the position $\mathbf{x}(t)$ of the particle. Expanding the square root to second order[9]

$$H = mc^2 + \frac{(\mathbf{P} - q\mathbf{A})^2}{2m} - \frac{(\mathbf{P} - q\mathbf{A})^4}{8m^3c^2} + \ldots + q\phi . \qquad (12.74)$$

The first term is the energy associated with the rest-mass, which, in a non-relativistic context, we subtract. To 1PN order, the Hamiltonian of the a-th particle, in the fields $\phi_{a,\text{ext}}$, $\mathbf{A}_{a,\text{ext}}$ generated by the other particles, is therefore

$$H_a = \frac{P_a^2}{2m_a} - \frac{P_a^4}{8m_a^3c^2} + q_a\phi_{a,\text{ext}} - \frac{q_a}{m_a}\mathbf{P}_a\cdot\mathbf{A}_{a,\text{ext}} , \qquad (12.75)$$

where we subtracted the rest mass contribution, and we took into account the fact that the vector potential $\mathbf{A}_{a,\text{ext}}$ generated by the non-relativistic charges q_b (with $b \neq a$) is proportional $1/c^2$, see eq. (12.37)

[9]With the obvious notation $(\mathbf{P} - q\mathbf{A})^4 = [(\mathbf{P} - q\mathbf{A})^2]^2$.

and therefore, to 1PN order, it contributes only through the cross product with $\mathbf{P}_a$ in the expansion of the quadratic term. To this order, the total Hamiltonian is therefore

$$
H = \sum_{a=1}^{N} \left[\frac{P_a^2}{2m_a} - \frac{P_a^4}{8m_a^3 c^2} + \frac{1}{2} q_a \phi_{a,\text{ext}} - \frac{q_a}{2m_a} \mathbf{P}_a \cdot \mathbf{A}_{a,\text{ext}} \right] ,
$$

(12.76)

where, as before, the factors of $1/2$ compensate the double counting of the interaction term: otherwise, e.g., the term in the Hamiltonian proportional to $q_1 q_2 / r_{12}$ would be counted once when we compute the energy of particle 1 in the potential generated by particle 2, and once when we compute the energy of particle 2 in the potential generated by particle 1. In this expression, the potentials $\phi_{a,\text{ext}}$ and $\mathbf{A}_{a,\text{ext}}$ must be evaluated on the position of the a-th particle, i.e., they are given, more precisely, by $\phi_a[t, \mathbf{x}_a(t)]$ and $\mathbf{A}_{a,\text{ext}}[t, \mathbf{x}_a(t)]$. Reading them from eqs. (12.31) and (12.37), with $\mathbf{x}$ replaced by $\mathbf{x}_a(t)$, and using, at this order, $\mathbf{v}_a = \mathbf{P}_a / m_a$, $\mathbf{v}_b = \mathbf{P}_b / m_b$, we get back eq. (12.71).

12.2.3   Reduction of order of the equations of motion

As we discussed after eq. (12.28), in general, in the PN expansion, we find terms with derivatives higher than the second in the Lagrangian. In the Lorenz gauge, we see from eqs. (12.1)–(12.4) that, at nPN order, ϕ contains times derivatives of the positions of the particles up to ∂_t^{2n} and $\mathbf{A}$ starts to contribute from 1PN and, at nPN order, contains times derivatives up to ∂_t^{2n-2}. At the 1PN level, we were able to eliminate the higher-order derivatives with a gauge transformation to the Coulomb gauge. However, this is no longer possible at higher orders. To eliminate a term proportional to ∂_t^{2n} in ϕ with a gauge transformation $\phi \to \phi + (1/c)\partial_t \theta$, we need terms proportional to ∂_t^{2n-1} in θ. Then, the same gauge transformation applied to the vector potential, $\mathbf{A} \to \mathbf{A} - \boldsymbol{\nabla}\theta$, induces terms proportional to ∂_t^{2n-1} in $\mathbf{A}$ (that adds up to the derivatives up to ∂_t^{2n-2} that were already present in $\mathbf{A}$ before the gauge transformation), so, even if we can eliminate altogether higher-order derivatives from ϕ choosing the Coulomb gauge, beyond 1PN order, second- and higher-order derivatives of $\mathbf{x}_a(t)$ remain in $\mathbf{A}$, and therefore in the Lagrangian. Correspondingly, the equations of motion will be of order higher than second, involving third time derivatives of the position, as well as higher and higher time derivatives as we increase the PN order. At first one might think that this is very problematic, since the evolution would no longer be determined by giving, as initial conditions, the initial positions and velocities of the particles; even worse, at each new order of the PN expansion we would need more and more initial conditions, involving higher and higher time derivatives of the position. Furthermore, higher-order differential equations are typically plagued by all kind of instabilities. Actually, this catastrophe is only apparent, and the equations of motion can be systematically reduced to second-order equations, at each order of the PN expansion. As a schematic example of this "reduction-of-order" procedure, consider an equation of the form

$$\ddot{q} = f_0(q, \dot{q}) + \frac{1}{c^2} f_1(q, \dot{q}) \, \dddot{q} + \mathcal{O}\left(\frac{1}{c^4}\right), \tag{12.77}$$

for some dynamical variable $q(t)$. Taking a time derivative of this equation gives

$$\dddot{q} = \frac{d}{dt} f_0(q, \dot{q}) + \frac{1}{c^2} \frac{d}{dt} [f_1(q, \dot{q}) \, \dddot{q}] + \mathcal{O}\left(\frac{1}{c^4}\right). \tag{12.78}$$

We can then use this to replace $\dddot{q}$ on the right-hand side of eq. (12.77), which gives

$$
\begin{aligned}
\ddot{q} = {} & f_0(q, \dot{q}) \\
& + \frac{1}{c^2} f_1(q, \dot{q}) \left\{ \frac{d}{dt} f_0(q, \dot{q}) + \frac{1}{c^2} \frac{d}{dt} [f_1(q, \dot{q}) \, \dddot{q}] + \mathcal{O}\left(\frac{1}{c^4}\right) \right\} \\
& + \mathcal{O}\left(\frac{1}{c^4}\right).
\end{aligned}
\tag{12.79}
$$

However, to order $1/c^2$, this is the same as

$$\ddot{q} = f_0(q, \dot{q}) + \frac{1}{c^2} f_1(q, \dot{q}) \frac{d}{dt} f_0(q, \dot{q}) + \mathcal{O}\left(\frac{1}{c^4}\right), \tag{12.80}$$

in which, on the right-hand side, only derivatives up to $\ddot{q}$ enter. So, in practice, to order $1/c^2$, we can simply use the zeroth-order equation of motion $\ddot{q} = f_0(q, \dot{q})$ to compute the term $\dddot{q}$ in eq. (12.77), since the latter is already multiplied by $1/c^2$. Observe that $\ddot{q}$ also appears on the right-hand side of eq. (12.80), since

$$\frac{d}{dt} f_0(q, \dot{q}) = \frac{\partial f_0}{\partial q} \dot{q} + \frac{\partial f_0}{\partial \dot{q}} \ddot{q}. \tag{12.81}$$

Then, solving for $\ddot{q}$, the equation can then be re-arranged in the form

$$\ddot{q} = f_0(q, \dot{q}) + \frac{1}{c^2} g_1(q, \dot{q}) + \mathcal{O}\left(\frac{1}{c^4}\right), \tag{12.82}$$

with a different functions $g_1(q, \dot{q})$, given explicitly by

$$g_1(q, \dot{q}) = f_1(q, \dot{q}) \left[f_0(q, \dot{q}) \frac{\partial f_0}{\partial \dot{q}} + \frac{\partial f_0}{\partial q} \dot{q} \right]. \tag{12.83}$$

This kind of procedure can be carried out order by order, keeping the equations of motion always of second order.

As a related application of this reduction-of-order procedure, observe, from eq. (12.48), that $\mathbf{A}_{a,\text{ext}}[t, \mathbf{x}_a(t)]$ depends on $\mathbf{v}_b(t)$, for all $b \neq a$. Then, on the left-hand side of eq. (12.49) appear also the accelerations

$d\mathbf{v}_b/dt$ of all particles, so it might look that, even if the equations are just of second order, still the equation of motion of the a-th particle involves the acceleration of all other particles, giving rise to a very complicated coupled set of equations. However, since, in eq. (12.49), $d\mathbf{v}_b/dt$ (with $b \neq a$) only appears in a term of order $1/c^2$, it can be replaced by its expression computed using the equations of motion for the b-th particle at Newtonian order, so as a function of positions only. So, in the end, the acceleration of the a-th particle depends on the position and velocities of all other particles, but not on their accelerations.

12.3 Self-force and radiation reaction

This section is very advanced and should definitely be skipped at first reading.

In the previous section, when computing the conservative dynamics of a system of point particles up to 1PN order, we have excluded "self-force" terms, claiming that, for a system of particles, eq. (12.13) must be replaced by eq. (12.16), in which the charge q_a feels the effect of the fields generated by the charges q_b with $b \neq a$, but is not subject to the force produced by its own field. While this might sound eminently reasonable, it must be acknowledged that this is an "ad hoc" prescription, that we have super-imposed on the formalism. In reality, Maxwell's equations instruct us to compute the electric and magnetic fields generated by the *total* charge and current densities, and then these electric and magnetic fields act on the charges according to the Lorentz force [in its relativistic form, (8.62), or (8.65)]. In Section 12.3.2 we will perform the computation of the 1PN dynamics including these self-force terms and we will show that, up to 1PN, the various self-force contributions either vanish or can be reabsorbed into a redefinition (more technically, a "renormalization") of the mass of the particle. On the one hand, this will confirm the results of the previous section, putting them on a firmer conceptual ground. On the other hand, these results will pave the way for the study of radiation reaction at 1.5PN order in Section 12.3.3, where we will see that the inclusion of self-force terms is necessary to get the correct result. If one would simply discard them by hand, one would not get the energy loss to radiation corresponding to the Larmor formula. These self-energy terms are therefore absolutely real and, beyond the 1PN level (where, a posteriori, the naive approach of throwing them away turns out to give the correct answer) they must be included. In Section 12.3.4 we will see how to obtain radiation reaction to all orders from a covariantization of the 1.5PN result and finally, in Section 12.3.5, we will show how mass renormalization and radiation reaction can be obtained from a single, fully covariant, computation, that gets rid completely of any notion of extended electron.

First, however, in Section 12.3.1 we compare two different general frameworks that can be used to address these problems, either based on an extended classical model of the electron, or using regularization and renormalization techniques borrowed from quantum field theory.

12.3.1 Classical extended electron models vs. regularization schemes

The basic problem that we will have to face, when dealing with self-forces, is related to the assumption of exactly point-like charges. As we already saw in Section 5.2.2, this leads to a divergent expression for the self-energy of the particle. Historically, the problem has been first tackled trying to build a classical model of an extended electron. However, no consistent and convincing classical electron model has ever emerged, despite significant effort, starting already from works of Abraham, Lorentz, and Poincaré at the beginning of the 20th century. In the Abraham–Lorentz–Poincaré model, the electron was modeled as a uniformly charged spherical shell. However, hypothetical mechanical forces must be introduced to stabilize the electron against the electrostatic repulsion among its parts. Even when this is done, and the forces are chosen so as to render stable a configuration with spherical symmetry, the model still turns out to be unstable under non-spherical perturbations.[10] Furthermore, and most importantly, these hypothetical stabilizing mechanical forces do not correspond to anything in Nature. We now understand that a consistent theory of the structure of elementary particles can only be obtained in the framework of quantum field theory.

Here we will then take a different approach to the problem, inspired indeed by quantum field theory, in which these divergences are dealt with by using renormalization theory.[11] In this approach, one starts by *regularizing* the theory, which means that we introduce a length-scale ℓ that smooths out the divergences so that, to begin with, we deal with well-defined mathematical expressions. For instance, in Section 5.2.2, we regularized the self-energy of a charge distribution by removing the Fourier modes with wavenumber $|\mathbf{k}| > \pi/\ell$, see eqs. (5.27)–(5.31). Regularization must be understood just as a mathematical step. It is not meant to correspond, in any way, to a physical model of an extended classical electron, and physical results are only obtained in the limit $\ell \to 0$. This eliminates any concern about the the need for hypothetical stabilizing mechanical forces, or on the stability of the extended electron model under perturbations. The issue, now, rather becomes how to take the limit $\ell \to 0$ so as to recover finite results. This is obtained realizing that, once we need to introduce a cutoff ℓ in the theory, the various parameters that we have introduced at the level of the action (or, in our classical context, of the equations of motion), such as the mass and charge of the particle, are not yet the observable quantities, but rather just parameters that enter in the intermediate steps of the computation ("bare parameters", in the field theory jargon), that can a priori also depend on this cutoff. For instance, the action (7.133) must be replaced by

$$S_{\text{free}} = -m_0(\ell)c^2 \int d\tau + \dots. \tag{12.84}$$

where we admit that the parameter previously denoted by m and in-

[10]See Pearle (1982) for a review of these approaches, and Damour (2017) for a historical discussion of the contribution of Poincaré to the extended electron model.

[11]It is interesting to observe that this approach was already advocated by Dirac (1938), when the understanding of the divergences in quantum field theory was still quite limited. From Dirac's 1938 paper: "We shall retain Maxwell's theory to describe the field right up to the point-singularity which represents our electron and shall try to get over the difficulties associated with the infinite energy by a process of direct omission or subtraction of unwanted terms, somewhat similar to what has been used in the theory of the positron. Our aim will be not so much to get a model of the electron as to get a simple scheme of equations that can be used to calculate all the results that can be obtained from experiment."

terpreted as the mass, could a priori also depend on ℓ, and is not yet the physical, observed, mass. To stress this change of perspective, we denote it by m_0 instead of m, and we call it the "bare" mass. The dots in eq. (12.84) indicate that we also admit the possibility of adding extra terms with a different structure, that we will discuss below. The crucial point is that $m_0(\ell)$ is not observable. A charged particle, with charge q_a, always comes with its own electric field and, as we saw in eq. (5.31), this produces a self-energy term $[1/(4\pi\epsilon_0)](q_a^2/\ell)$. Therefore, the total rest-energy of a particle, labeled by an index a, is

$$E_{a,\text{rest}} = m_{0,a}(\ell)c^2 + \frac{1}{4\pi\epsilon_0}\frac{q_a^2}{\ell}. \tag{12.85}$$

This means that the actual observable mass of the particle, that we call the "renormalized" mass, is given by

$$m_a = m_{0,a}(\ell) + \frac{1}{4\pi\epsilon_0 c^2}\frac{q_a^2}{\ell}. \tag{12.86}$$

The logic of renormalization is that the bare mass $m_{0,a}(\ell)$ must be seen as a quantity which is completely in our hands when we define the theory, and which is chosen so that it cancels the divergence of the self-energy term for $\ell \to 0$, leaving us with a finite result for the renormalized mass, equal to the observed value. By themselves, the two separate terms on the right-hand side of eq. (12.86) have no physical meaning. They both separately diverge in the limit $\ell \to 0$, just in a way that their divergences cancel in the sum, and their value before removing the cutoff depends on the regularization scheme that we have chosen. Only their sum is physical, and finite.[12]

It should be stressed that this point of view, based on renormalization, is not just an optional possibility, when dealing with the divergences that appear in the point particle limit. At the scale of elementary particles, eventually quantum mechanics and quantum field theory enter the game. At that level, divergences such as that discussed above are unavoidable, and can only be cured through renormalization. The mass renormalization discussed above will then automatically take place, together with the renormalization of other parameters (such as the electron charge). From this perspective, the attempts at constructing a classical extended model of the electron, where the self-energy term is interpreted as an actual finite physical contribution to the total mass, looks futile. At the quantum level, the self-energy term will anyhow be divergent, and this divergence can only be cured through renormalization.

This quantum field theory approach also allows us to clarify another aspect that has plagued the attempts at constructing classical extended models of the electron, which is related to Lorentz invariance. A charged shell which is spherical in its rest frame would be deformed, by the Lorentz contraction, from the point of view of a boosted observer. Therefore a rigid extended model, such as that based on a rigid charged shell initially devised by Abraham in 1903–1904, is not Lorentz invariant. As a result, the self-energy contribution to the energy and the corresponding self-contribution to the momentum (obtained from the momentum

[12]Notice, in particular, that $m_{0,a}(\ell) = m_a - [1/(4\pi\epsilon_0 c^2)]q_a^2/\ell$ goes to minus infinity as $\ell \to 0^+$ so, for ℓ sufficiently small, it is negative. Most of the confusion, in some literature, comes from interpreting $m_{0,a}(\ell)$ as a "mechanical mass", and the term proportional to q_a^2/ℓ as an "electromagnetic mass." This nomenclature is already misleading, since it implicitly suggests that these two quantities have, separately, an intrinsic physical meaning (typically leading these texts to statements such as that a negative "mechanical" mass is unacceptable). In the logic of renormalization, which is the standard tool of quantum field theory, neither of them has any separate physical meaning. Their value, and even their sign, depends on the regularization scheme used, and only their sum is physically meaningful.

associated with the electromagnetic field of the particle, in a frame where it has a velocity $\mathbf{v}$) do not form a four-vector. For a rigid charged shell of radius b, whose center-of-mass moves with four-velocity $u^\mu = (\gamma c, \gamma \mathbf{v})$, the self-energy and self-momentum due to the electromagnetic field of the particle are given by[13]

$$p_{\text{em}}^\mu = \frac{1}{4\pi\epsilon_0} \frac{e^2}{2bc^2} u^\mu + \frac{1}{4\pi\epsilon_0} \frac{1}{3} \frac{e^2}{2bc^2} \left(\frac{v^2/c^2}{\sqrt{1+(v/c)^2}}, \mathbf{v} \right). \qquad (12.87)$$

[13]See eqs. (7.16) and (11.29) of Pearle (1982).

Note that the four-vector notation p_{em}^μ here is an abuse of notation, since, on the right-hand side, the term proportional to u^μ is a four-vector, but the second term, written giving explicitly its temporal and spatial components, is not. Poincaré, in 1905–1906, added suitable mechanical stresses, which produce a "mechanical energy and momentum." Also these, separately, do not form a four-vector, but can be chosen so as to cancel exactly the second term in eq. (12.87). So, adding this to the mechanical four-momentum $m_0 u^\mu$ of the center of mass, in the Poincaré model the extended electron has a total mechanical four-momentum

$$p_{\text{mech}}^\mu = m_0 u^\mu - \frac{1}{4\pi\epsilon_0} \frac{1}{3} \frac{e^2}{2bc^2} \left(\frac{v^2/c^2}{\sqrt{1+(v/c)^2}}, \mathbf{v} \right). \qquad (12.88)$$

The sum of the two terms, $p^\mu = p_{\text{em}}^\mu + p_{\text{mech}}^\mu$, is a four-vector, that is interpreted as the total four-momentum of the extended electron,

$$p^\mu = \left(m_0 + \frac{1}{4\pi\epsilon_0} \frac{e^2}{2bc^2} \right) u^\mu, \qquad (12.89)$$

so that Lorentz invariance is recovered. Apart from the fact that this model turns out to be unstable under non-spherical perturbations, and therefore eventually is not viable even classically, within our modern perspective it is clear that the whole construction is very artificial and has nothing to do with the actual description of an elementary particle in quantum field theory.

However, what Poincaré did can be reinterpreted, in the framework of renormalization in quantum field theory, as follows. When one regularizes a theory, the symmetries of the original action may or may not be respected by the regularization process. For instance, if we regularize the Dirac deltas in eqs. (12.17) and (12.18) using an extended rigid model of the electron, our regularization breaks Lorentz invariance since, as we have discussed, a rigid spherical electron is not consistent with Lorentz invariance. The same happens if we regularize the action of a point particle by imposing a cutoff on the Fourier modes, restricting to modes with $|\mathbf{k}| < \pi/\ell$. Again this is not a Lorentz-invariant condition, since the value of the wavenumber $\mathbf{k}$, and of its modulus, changes under a Lorentz boost, so the above condition can only be valid in a specific frame. In general, there is nothing wrong with using a regularization that breaks one of the symmetries of the theory, and this is commonly done in quantum field theory computations. Simply, in this case one must admit that the bare quantities (that, we should remember, are not physical observables, but just mathematical entities that we choose at our will) do not have to respect that symmetry either, and will rather be adjusted so as to recover that symmetry at the level of renormalized quantities. So, for instance, if we start from the point-particle action (12.84), that in this context is called the "bare" action, and we use a regularization that breaks Lorentz invariance, we must admit the presence of other terms ["counterterms," in the quantum field theory jargon, indicated generically by the dots in eq. (12.84)], that do not need to respect Lorentz invariance, and that will be adjusted so as to recover Lorentz invariance at the level of the renormalized theory. So, what Poincaré actually did, from this perspective, is equivalent to starting from the bare action that corresponds to the Abraham model of the electron, interpreted now just as a form of regularization of a point charge (that breaks Lorentz invariance), and adding to it a counterterm which is also not Lorentz invariant, and is adjusted so as to obtain Lorentz invariance for the renormalized quantities.

In Section 12.3.2 we will perform a similar but conceptually more transparent computation, as follows. We will start from the charge and current density of a point-particle, and we will regularize them by imposing a cutoff $|\mathbf{k}| < \pi/\ell$ over the wavenumbers of their Fourier modes. This will be the equivalent of an extended classical electron model, since it amounts to smoothing out, over a distance of order ℓ, the Dirac deltas in eqs. (12.17) and (12.18), except that we make it clear that this is just a regularization, and the limit $\ell \to 0$ must be taken in the end, so it should not be interpreted as an actual model of a classical extended electron. As we already mentioned, this regularization breaks Lorentz invariance. We will then compute explicitly the corresponding divergences in the self-energy and in the self-momentum (which would provide the result analogous to eqs. (12.88) and (12.89) with our regularization of the point particle, rather than with the Abraham–Lorentz–Poincaré model) and we will then show how to renormalize the theory with a simple non-Lorentz-invariant counterterm, so as to recover Lorentz invariance for the renormalized quantities.

The use of a renormalization scheme that involves counterterms that are not Lorentz invariant might be unfamiliar even to many advanced readers.[14] However, there is no need to break Lorentz invariance with the regularization. After working out mass renormalization with the above regularization, that breaks Lorentz invariance (and that corresponds to the naive idea of an extended classical electron model), in Section 12.3.5 we will show how to regularize and renormalize the divergences associated with point particles in a fully Lorentz-invariant manner, recovering mass renormalization in a way that maintains Lorentz symmetry manifest at each stage. The latter procedure will be completely in line with standard quantum field theory computations, and is in fact the best starting point for including also quantum effects.

[14]In quantum field theory, a cutoff over wavenumbers, $|\mathbf{k}| < \Lambda$ (or over momenta, where, at the quantum level, the momentum $\mathbf{p}$ is related to the wavenumber $\mathbf{k}$ by $\mathbf{p} = \hbar\mathbf{k}$), is typically used only for qualitative discussions, precisely because it breaks Lorentz invariance, and for actual computations in general one prefers to use other schemes, such as Pauli–Villars or dimensional regularization, to avoid the need of dealing with counterterms that do not respect Lorentz invariance. However, there are situations where it is necessary to use a regularization that breaks Lorentz invariance. In particular, for non-perturbative computations in quantum chromodynamics (the fundamental theory of strong interactions), the best regularization consists in putting the theory on a space-time lattice (furthermore, rotating from Minkoswki to Euclidean space). In this case the full (Euclidean) Lorentz invariance is broken to a subgroup of discrete rotations and, to recover Lorentz invariance in the continuum limit, one must introduce counterterms that are not Lorentz invariant, and only respect this smaller symmetry group. There are also more complex situations, where a symmetry broken by the regularization is not recovered when the cutoff is removed. This gives rise to quantum field theory *anomalies*, but these will not concern us here.

12.3.2 Self-energy and mass renormalization

In this subsection we extend the discussion of Section 12.2, including now also the self-forces, for a collection of charges modeled as extended charge distributions. As discussed above, in the case of elementary particles this must be considered merely as a regularization, and we are only interested in taking eventually the point-particle limit, eliminating the divergences through the renormalization procedure. Some aspects of the formalism, however, can also be useful for actual macroscopic bodies, where the extended charge distribution is really the physical distribution of the body, rather than a mathematical trick for regularizing a point particle.[15] So, in the following each particle will be described by a generic charge density $\rho_a(t, \mathbf{x})$, localized in a volume $V_a(t)$ (whose position changes in time because the position of the particle changes), that we take small compared to the overall volume in which is localized the system of charges. We assume that the volumes V_a corresponding to the different charges are non-overlapping during the whole time span for which we follow the time evolution, i.e., that the particles do not merge (nor disintegrate). This assumption was already implicit in the computation of Section 12.2, see Note 4 on page 301. The position of the a-th charged body is then defined by its "center-of-charge" coordinate

$$\mathbf{x}_a(t) = \frac{1}{q_a} \int d^3x \, \rho_a(t, \mathbf{x}) \mathbf{x} \,. \tag{12.90}$$

Note that the integration is actually only over the volume $V_a(t)$ where the particle is localized, but we can extend it to all of space, because anyhow $\rho_a(t, \mathbf{x})$ vanishes outside $V_a(t)$. This has the advantage that the time dependence of $\mathbf{x}_a(t)$ enters only through $\rho_a(t, \mathbf{x})$. The velocity and acceleration of the particles are given by $\mathbf{v}_a(t) = d\mathbf{x}_a/dt$ and $\mathbf{a}(t) = d^2\mathbf{x}_a/dt^2$. Using the continuity equation (3.30),

$$\begin{aligned}
\frac{dx_a^i}{dt} &= \frac{1}{q_a} \int d^3x \, \partial_t \rho_a(t, \mathbf{x}) x^i \\
&= -\frac{1}{q_a} \int d^3x \left[\partial_k j_a^k(t, \mathbf{x}) \right] x^i \\
&= +\frac{1}{q_a} \int d^3x \, j_a^k(t, \mathbf{x}) \partial_k x^i \\
&= \frac{1}{q_a} \int d^3x \, j_a^i(t, \mathbf{x}) \,, \tag{12.91}
\end{aligned}$$

where we integrated by parts (neglecting the boundary term since ρ_a is localized) and we used $\partial_k x^i = \delta_k^i$. Therefore

$$\mathbf{v}_a(t) = \frac{1}{q_a} \int d^3x \, \mathbf{j}_a(t, \mathbf{x}) \,. \tag{12.92}$$

In the point-like limit, using eqs. (12.17) and (12.18), eqs. (12.90) and (12.92) correctly reduce to the position and velocity of a point charge. The simplest model of a current distribution consistent with eq. (12.92) is given by

$$\mathbf{j}_a(t, \mathbf{x}) = \rho_a(t, \mathbf{x}) \mathbf{v}_a(t) \,. \tag{12.93}$$

[15]A very similar formalism is useful in Newtonian gravity (with the charge density replaced by the mass density) to describe self-gravitating objects, see Damour (1987) for pioneering work and Poisson and Will (2014) for recent textbook discussion.

This corresponds to a charge distribution that moves with a velocity $\mathbf{v}_a(t)$, without superimposed internal motions. Taking a different modelization would simply complicate the explicit computation, adding terms that vanish when, at the end of the computation, we eventually take the point-particle limit. We will therefore assume the form (12.93) of the current density, for our extended model of the current and charge distribution.

We regularize the theory by putting a cutoff $|\mathbf{k}| < \pi/\ell$, i.e., setting to zero all Fourier modes $\tilde{\rho}(t, \mathbf{k})$ of the charge distribution with wavenumber higher that π/ℓ, as we already did in Section 5.2.2. In coordinate space this corresponds to smoothing the Dirac delta, with most of its support being concentrated up to a distance of order ℓ from its center. The point-like limit is recovered as $\ell \to 0$. As we already mentioned, this is not a Lorentz-invariant regularization, since the value of $|\mathbf{k}|$ changes under a Lorentz boost, so the condition $|\mathbf{k}| < \pi/\ell$ is not Lorentz invariant.

With this regularization, the contribution of the self-field of the particle to its rest energy was already computed in Section 5.2.2 and is given by the second term on the right-hand side of eq. (12.85). This self-energy contribution is reabsorbed into a mass renormalization, given in eq. (12.86). We now compute the contribution of the self-field of the particle to its spatial momentum, to understand how the full four-momentum renormalizes. To this purpose, we consider the equation of motion of a charge and current distribution written in the form of the Lorentz "force" equation (3.68). For a single extended object, we rewrite it as

$$\frac{d\mathbf{p}_a}{dt} = \int_V d^3x \left(\rho\mathbf{E} + \mathbf{j}\times\mathbf{B} \right). \tag{12.94}$$

The crucial point is that, in the relation $\mathbf{p}_a = \gamma(v_a)m_a\mathbf{v}_a$, the mass must again be taken as a bare parameter, that depends on the cutoff and that will be adjusted so as to obtain the desired value for the renormalized mass. Furthermore, since our regularization breaks Lorentz invariance, this bare parameter is a priori different from the one that enters in the energy, and that we denoted by $m_{0,a}(\ell)$ in eq. (12.86). We will then denote it by $\tilde{m}_{0,a}(\ell)$.[16] Then, including also the self-force term that we have omitted in our treatment in Section 12.2, the equation of motion of the a-th particle becomes

$$\frac{d}{dt}\left[\gamma(v_a)\tilde{m}_{0,a}(\ell)\mathbf{v}_a\right] = \mathbf{F}_{a,\text{ext}}$$
$$+ \int d^3x \left[\rho_a(t, \mathbf{x})\mathbf{E}_a(t, \mathbf{x}) + \mathbf{j}_a(t, \mathbf{x})\times\mathbf{B}_a(t, \mathbf{x})\right], \tag{12.95}$$

where $\mathbf{F}_{a,\text{ext}}$ is the contribution to the equation of motion of the a-th particle from the other particles, while $\mathbf{E}_a$ and $\mathbf{B}_a$ are the electric and magnetic fields generated by the a-th particle itself. In Section 12.2 we studied this equation with only $\mathbf{F}_{a,\text{ext}}$ on the right-hand side, and we threw away by hand the self-force, i.e., the effect of $\mathbf{E}_a$ and $\mathbf{B}_a$ on the a-th particle itself. The new aspect of this computation is that we now take it into account explicitly, and we will see how, eventually, the result of Section 12.2 can be justified.

[16]We will see below how to obtain these different bare masses for the energy and momentum from the addition of a non-Lorentz-invariant counterterm in the action.

These self-fields are given by

$$\mathbf{E}_a = -\boldsymbol{\nabla}\phi_a - \partial_t \mathbf{A}_a , \qquad (12.96)$$

$$\mathbf{B}_a = \boldsymbol{\nabla}\times\mathbf{A}_a , \qquad (12.97)$$

where ϕ_a and $\mathbf{A}_a$ are, respectively, the scalar and vector potentials generated by the particle a. From eqs. (12.1) and (12.2), in the Lorenz gauge they are given by

$$\phi_a(t,\mathbf{x}) = \frac{1}{4\pi\epsilon_0} \int d^3x' \frac{\rho_a(t-|\mathbf{x}-\mathbf{x}'|/c,\mathbf{x}')}{|\mathbf{x}-\mathbf{x}'|} , \qquad (12.98)$$

$$\mathbf{A}_a(t,\mathbf{x}) = \frac{1}{4\pi\epsilon_0} \frac{1}{c^2} \int d^3x' \frac{\mathbf{j}_a(t-|\mathbf{x}-\mathbf{x}'|/c,\mathbf{x}')}{|\mathbf{x}-\mathbf{x}'|} . \qquad (12.99)$$

We consider first the contribution to the self-force in eq. (12.95) coming from the electric field, up to 1PN order. Expanding eq. (12.98) up to 1PN, and working in components, the contribution of the scalar potential to the right-hand side of eq. (12.95) is

$$\int d^3x\, \rho_a(t,\mathbf{x})\left[-\partial_i\phi_a(t,\mathbf{x})\right] = -\frac{1}{4\pi\epsilon_0} \int d^3x\, d^3x'\, \rho_a(t,\mathbf{x})$$
$$\times \partial_i \left[\frac{\rho_a(t-|\mathbf{x}-\mathbf{x}'|/c,\mathbf{x}')}{|\mathbf{x}-\mathbf{x}'|}\right]$$
$$= -\frac{1}{4\pi\epsilon_0} \int d^3x\, d^3x'\, \rho_a(t,\mathbf{x})$$
$$\times \partial_i \left[\frac{\rho_a(t,\mathbf{x}')}{|\mathbf{x}-\mathbf{x}'|} - \frac{1}{c}\partial_t\rho_a(t,\mathbf{x}') + \frac{|\mathbf{x}-\mathbf{x}'|}{2c^2}\partial_t^2\rho_a(t,\mathbf{x}') + \dots\right]. \qquad (12.100)$$

The first term of this expansion is the Newtonian self-force, since it is the only term that survives in the limit $c \to \infty$ (even when we will include the contribution due to $\mathbf{A}$, since $\mathbf{A}$ starts from order $1/c^2$). However, computing explicitly the derivative $\partial_i = \partial/\partial x^i$,

$$\int d^3x\, d^3x'\, \rho_a(t,\mathbf{x})\partial_i\frac{\rho_a(t,\mathbf{x}')}{|\mathbf{x}-\mathbf{x}'|} = -\int d^3x\, d^3x'\, \rho_a(t,\mathbf{x})\frac{x_i - x_i'}{|\mathbf{x}-\mathbf{x}'|^3}\rho_a(t,\mathbf{x}') ,$$
$$(12.101)$$

and this expression vanishes, independently of the functional form of the charge density $\rho_a(t,\mathbf{x})$, since the integrand is odd under the exchange $\mathbf{x} \leftrightarrow \mathbf{x}'$ (while, since the integrals in d^3x and d^3x' can be extended to all $\mathbb{R}^3$, the integration domain is invariant under such exchange). Therefore, *the Newtonian self-force vanishes, for any extended charge distribution.*

Consider now the second term in the expansion in eq. (12.100). This is formally a term of order 0.5PN, since it is proportional to $1/c$, but again it vanishes, simply because ∂_i is the derivative with respect to $\mathbf{x}$, and it acts on $\partial_t\rho_a(t,\mathbf{x}')$, which depends on $\mathbf{x}'$ but not on $\mathbf{x}$. The third term, which is a 1PN correction, requires a more involved computation.

We manipulate it as follows:

$$
\int d^3x \, \rho_a(t, \mathbf{x}) \left[-\partial_i \phi_a(t, \mathbf{x})\right]_{1PN}
$$

$$
= -\frac{1}{4\pi\epsilon_0} \frac{1}{2c^2} \int d^3x \, d^3x' \, \rho_a(t, \mathbf{x}) \left(\partial_i |\mathbf{x} - \mathbf{x}'|\right) \partial_t^2 \rho_a(t, \mathbf{x}')
$$

$$
= -\frac{1}{4\pi\epsilon_0} \frac{1}{2c^2} \int d^3x \, d^3x' \, \rho_a(t, \mathbf{x}) \frac{x_i - x_i'}{|\mathbf{x} - \mathbf{x}'|} \partial_t \left(-\frac{\partial j_a^k(t, \mathbf{x}')}{\partial x_k'}\right)
$$

$$
= -\frac{1}{4\pi\epsilon_0} \frac{1}{2c^2} \int d^3x \, d^3x' \, \rho_a(t, \mathbf{x}) \left(\frac{\partial}{\partial x_k'} \frac{x_i - x_i'}{|\mathbf{x} - \mathbf{x}'|}\right) \partial_t j_a^k(t, \mathbf{x}')
$$

$$
= \frac{1}{4\pi\epsilon_0} \frac{1}{2c^2} \int d^3x \, d^3x' \, \rho_a(t, \mathbf{x}) \partial_t j_a^k(t, \mathbf{x}')
$$

$$
\times \frac{1}{|\mathbf{x} - \mathbf{x}'|} \left[\delta_{ik} - \frac{(x_i - x_i')(x_k - x_k')}{|\mathbf{x} - \mathbf{x}'|^2}\right] , \qquad (12.102)
$$

where, to go from the second to the third line, we used the continuity equation (3.30), and in the next line we integrated $\partial/\partial x_k'$ by parts.

The 1PN contribution to the electric field coming from $\mathbf{A}_a$ is obtained neglecting retardation in $\mathbf{j}_a(t - |\mathbf{x} - \mathbf{x}'|/c, \mathbf{x}')$ in eq. (12.99), since $\mathbf{A}_a$ is already proportional to $1/c^2$, so is given by

$$
-\partial_t A_a^k(t, \mathbf{x}) = -\frac{1}{4\pi\epsilon_0} \frac{1}{c^2} \int d^3x' \frac{\partial_t j_a^k(t, \mathbf{x}')}{|\mathbf{x} - \mathbf{x}'|} . \qquad (12.103)
$$

Putting together the contribution from $-\boldsymbol{\nabla}\phi_a$ and the contribution of $-\partial_t \mathbf{A}_a$ to the electric field $\mathbf{E}_a$, we find that, at 1PN order,

$$
\left[\int d^3x \, \rho_a(t, \mathbf{x}) E_a^i(t, \mathbf{x})\right]_{1PN} =
$$

$$
= -\frac{1}{4\pi\epsilon_0} \frac{1}{2c^2} \int d^3x \, d^3x' \, \rho_a(t, \mathbf{x}) \partial_t j_a^k(t, \mathbf{x}')
$$

$$
\times \frac{1}{|\mathbf{x} - \mathbf{x}'|} \left[\delta_{ik} + \frac{(x_i - x_i')(x_k - x_k')}{|\mathbf{x} - \mathbf{x}'|^2}\right] . \qquad (12.104)
$$

The above steps are valid for a generic function $\mathbf{j}(t, \mathbf{x})$. We now insert the modelization (12.93). Then, in eq. (12.104),

$$
\partial_t j_a(t, \mathbf{x}') = \rho_a(t, \mathbf{x}') \dot{\mathbf{v}}_a(t) + [\partial_t \rho_a(t, \mathbf{x}')] \, \mathbf{v}_a(t)
$$

$$
= \rho_a(t, \mathbf{x}') \dot{\mathbf{v}}_a(t) - [\boldsymbol{\nabla}_{\mathbf{x}'} \cdot \mathbf{j}_a(t, \mathbf{x}')] \, \mathbf{v}_a . \qquad (12.105)
$$

The second term on the right-hand side is of order $\mathbf{v}_a^2$, since $\mathbf{j}_a$ is proportional to $\mathbf{v}_a$, so

$$
\partial_t \mathbf{j}_a(t, \mathbf{x}') = \rho_a(t, \mathbf{x}') \dot{\mathbf{v}}_a(t) + \mathcal{O}\left(\mathbf{v}_a^2\right) . \qquad (12.106)
$$

We insert this into eq. (12.104) and we limit ourselves to the contributions linear in $\mathbf{v}_a$. This will be sufficient to understand how renormalization works for the spatial momentum, and can in principle be extended order by order to include higher powers of $\mathbf{v}_a$ (in the explicitly covariant computation that we will perform later, all these higher-order terms will be automatically included). Then, we get

$$\left[\int d^3x \, \rho_a(t,\mathbf{x}) E_a^i(t,\mathbf{x}) \right]_{1PN} =$$

$$= -\frac{1}{4\pi\epsilon_0} \frac{1}{2c^2} \dot{v}_a^j(t) \int d^3x d^3x' \, \rho_a(t,\mathbf{x}) \rho_a(t,\mathbf{x}')$$

$$\times \frac{1}{|\mathbf{x}-\mathbf{x}'|} \left[\delta_{ij} + \frac{(x_i - x_i')(x_j - x_j')}{|\mathbf{x}-\mathbf{x}'|^2} \right]. \quad (12.107)$$

It is convenient to choose $\rho(t,\mathbf{x})$ so that it is spherically symmetric, i.e., invariant under rotations around the center of the distribution, defined by eq. (12.90).[17] Then, by symmetry, the integral in eq. (12.107) can only be proportional to δ_{ij}, since, if $\rho_a(t,\mathbf{x})$ and $\rho_a(t,\mathbf{x}')$ are spherically symmetric, there are no privileged directions inside the integral.[18] Therefore, in the integrand, we can replace

$$(x_i - x_i')(x_j - x_j') \rightarrow \frac{1}{3}|\mathbf{x}-\mathbf{x}'|^2 \, \delta_{ij} \,, \quad (12.108)$$

and we get

$$\left[\int d^3x \, \rho_a(t,\mathbf{x}) \mathbf{E}_a(t,\mathbf{x}) \right]_{1PN} = -\frac{1}{4\pi\epsilon_0} \frac{2}{3} \frac{\dot{\mathbf{v}}_a(t)}{c^2} \int d^3x d^3x' \, \frac{\rho_a(t,\mathbf{x})\rho_a(t,\mathbf{x}')}{|\mathbf{x}-\mathbf{x}'|}. \quad (12.109)$$

If we replace here $\rho_a(t,\mathbf{x})$ and $\rho_a(t,\mathbf{x}')$ by Dirac deltas, the integral diverges. However, this is the same integral that we met in Section 5.2.2, and with our regularization in which we set to zero the Fourier modes with $|\mathbf{k}| < \pi/\ell$, it is given by [compare eqs. (5.25) and (5.31)]

$$\frac{1}{2} \int d^3x d^3x' \, \frac{\rho_a(t,\mathbf{x})\rho_a(t,\mathbf{x}')}{|\mathbf{x}-\mathbf{x}'|} = \frac{q_a^2}{\ell}. \quad (12.110)$$

Inserting eq. (12.110) into eq. (12.109), we get

$$\left[\int d^3x \, \rho_a(t,\mathbf{x}) \mathbf{E}_a(t,\mathbf{x}) \right]_{1PN} = -\frac{4}{3} \left(\frac{1}{4\pi\epsilon_0} \frac{q_a^2}{\ell} \right) \frac{\dot{\mathbf{v}}_a(t)}{c^2}. \quad (12.111)$$

Since we are limiting ourselves to terms linear in $\mathbf{v}_a$, the contribution of the term $\mathbf{j}_a \times \mathbf{B}_a$ in eq. (12.95) can be neglected, since $\mathbf{B}_a$, as $\mathbf{A}_a$, is proportional to $\mathbf{j}_a$ and therefore $\mathbf{j}_a \times \mathbf{B}_a$ is quadratic in $\mathbf{j}_a$ and therefore in $\mathbf{v}_a$. Then, to linear order in $\mathbf{v}_a$, the equation of motion (12.95) reads

$$\tilde{m}_{0,a}(\ell)\dot{\mathbf{v}}_a = -\frac{4}{3} \left(\frac{1}{4\pi\epsilon_0} \frac{q_a^2}{\ell} \right) \frac{\dot{\mathbf{v}}_a}{c^2} + \mathbf{F}_{a,\text{ext}}. \quad (12.112)$$

We can rewrite this as

$$\left[\tilde{m}_{0,a}(\ell) + \frac{4}{3} \left(\frac{1}{4\pi\epsilon_0 c^2} \right) \frac{q_a^2}{\ell} \right] \dot{\mathbf{v}}_a = \mathbf{F}_{a,\text{ext}}, \quad (12.113)$$

and we see that the self-force term can be reabsorbed into a renormalization of the mass, choosing the bare parameter $\tilde{m}_{0,a}(\ell)$ so that

$$m_a = \tilde{m}_{0,a}(\ell) + \frac{4}{3} \left(\frac{1}{4\pi\epsilon_0 c^2} \right) \frac{q_a^2}{\ell}, \quad (12.114)$$

[17]Recall that $\rho(t,\mathbf{x})$ is just a regularization of the Dirac delta. Our regularization has been defined by the condition that the Fourier modes with $|\mathbf{k}| > \pi/\ell$ vanish, and this condition is invariant under spatial rotations. We are now further requiring that the non-vanishing Fourier modes $\tilde{\rho}(t,\mathbf{k})$ actually depend only on $|\mathbf{k}|$, rather than of the full vector $\mathbf{k}$, so as to give a rotationally invariant distribution $\rho(t,\mathbf{x})$. This assumption is just useful to simplify the computation. In any case, any deviation from spherical symmetry would give vanishing contribution when removing the cutoff and approaching the Dirac delta distribution, which is spherically symmetric.

[18]We can check this explicitly, choosing a reference frame so that, at a given time t, $\mathbf{x}_a(t) = 0$, so rotations around $\mathbf{x}_a$ are the same as rotations around the origin of the reference frame. Then, a spherically symmetric function $\rho_a(t,\mathbf{x})$ is invariant under any transformation that leaves $|\mathbf{x}|$ invariant; one such transformation is $\mathbf{x} = (x,y,z) \rightarrow (-x,y,z)$. If, in the integral in eq. (12.107), we transform simultaneously $\mathbf{x} = (x,y,z) \rightarrow (-x,y,z)$ and $\mathbf{x}' = (x',y',z') \rightarrow (-x',y',z')$, the factors $\rho_a(t,\mathbf{x})$, $\rho_a(t,\mathbf{x}')$ are invariant. Also $|\mathbf{x}-\mathbf{x}'|$ is invariant, since $|\mathbf{x}-\mathbf{x}'|^2 = (x-x')^2 + (y-y')^2 + (z-z')^2$ is unchanged if, simultaneously, $x \rightarrow -x$ and $x' \rightarrow -x'$. In contrast, the term $(x_i - x_i')(x_j - x_j')$ with $i = 1$ and $j \neq 1$ changes sign; therefore, the part of the integrand proportional to $(x_i - x_i')(x_j - x_j')$ is odd under this transformation, and its contribution to the integral vanishes. Similarly, all other terms with $i \neq j$ vanish. When $i = j$, in contrast, the result is independent of the value of i, again because of rotational symmetry: the integral with $i = j = 1$, which involves $(x - x')^2$, is the same as that with $i = j = 2$, which involves $(y-y')^2$ or that with $i = j = z$, which involves $(z - z')^2$.

where m_a is the observed value of the mass of the charged particle, say of the electron.

Note that the self-field contribution to the momentum differ by a factor of 4/3 from the self-field contribution to the rest energy, resulting in the different 4/3 factor between the second term on the right-hand sides of eqs. (12.86) and (12.114). As we already anticipated, within the renormalization logic, there is nothing surprising about it. Simply, we have broken Lorentz invariance with the regularization, and we must then use two different bare parameters $m_{0,a}(\ell)$ and $\tilde{m}_{0,a}(\ell)$, associated with energy and to momentum, respectively, in order to recover the same value of the renormalized mass m_a.

The conclusion of this section is that the "naive" treatment of Section 12.2, in which the self-energy term where simply discarded when studying the 1PN dynamics, eventually gives the correct result because, when the self-energy terms are correctly taken into account, to 1PN order they can just be reabsorbed into a renormalization of the mass. This requires first a regularization of the theory. If, as we have done in this section, we use a regularization that breaks Lorentz invariance, then we must use two different bare mass terms for energy and for spatial momentum, in order to reabsorbe the divergences. However (despite the fact that the extra 4/3 factor between eqs. (12.86) and (12.114) has created often confusion, to the extent of being called "the infamous 4/3 factor") within a proper approach based on regularization and renormalization this is just a minor technical point, of no special consequence. In Section 12.3.5 we will show how to regularize and renormalize the theory in a fully Lorentz-covariant manner, and then a single bare mass term will be sufficient to renormalize the four-momentum.

To conclude this section we observe that, in the classical theory that we are considering, we can equivalently discuss renormalization at the level of the equations of motion, as we have done, or at the level of the action. To make contact with the quantum field theory treatment, where one rather works at the level of the action, it is useful to show explicitly the form of a bare action that corresponds to the introduction of two different bare mass terms for the rest energy and for spatial momentum. Consider the bare action (suppressing for simplicity the label a of the particle)

$$S_{\text{bare}} = -m_0(\ell)c^2 \int d\tau - \frac{1}{2}[m_0(\ell) - \tilde{m}_0(\ell)] \int dt\, v^2$$
$$+q \int d\tau\, u_\mu(\tau) A^\mu[x(\tau)]. \tag{12.115}$$

The first and third terms were already given in eq. (8.69). The first term describes the action of a free particle, with m now replaced by a bare parameter $m_0(\ell)$, and the third describes its interaction with the electromagnetic field.[19] In particular, the third term is responsible for the form (8.1, 8.2) of the charge and current densities (or, equivalently, of the covariant expression (8.3) for j^μ), as we saw in eq. (8.73). Supplemented with the regularization $|\mathbf{k}| < \pi/\ell$ on the Fourier modes of the

[19]In a full quantum setting, also the charge q should be replaced by a bare parameter $q_0(\ell)$ and renormalized, although this will not concern us at the classical level.

Dirac delta, this action was the starting point of our computation. The second term in eq. (12.115) is proportional to the non-relativistic action of a free particle and is therefore a counterterm which is not Lorentz-invariant. The corresponding Lagrangian, limiting ourselves to the free part, is

$$L = -m_0(\ell)c^2\sqrt{1 - \frac{v^2}{c^2}} - \frac{1}{2}[m_0(\ell) - \tilde{m}_0(\ell)]v^2 . \qquad (12.116)$$

Keeping the terms up to order v^2 (which are enough to compute the rest energy and the term in the momentum linear in the velocity), we have

$$L = -m_0(\ell)c^2 + \frac{1}{2}\tilde{m}_0(\ell)v^2 + \mathcal{O}(v^4) . \qquad (12.117)$$

Therefore the bare rest energy is still $E_{\rm rest} = m_0(\ell)c^2$, while, to linear order in $\mathbf{v}$, the bare momentum of a free particle is $\mathbf{p} = \delta L/\delta\mathbf{v} = \tilde{m}_0(\ell)\mathbf{v}$.

12.3.3 Radiation reaction at 1.5PN order

As we have discussed in Section 12.1, the radiation field itself cannot be computed within the PN expansion: radiation appears in the far zone, while the PN expansion is only valid in the near zone. However, the PN expansion allows us to study the dynamics of the system of charges in the near zone so, from the PN expansion, we must be able to see that the mechanical energy of the system of charges decreases, so as to compensate for the energy that is carried away by the electromagnetic waves radiated by the system. In Section 12.2 we found that, up to 1PN order, the dynamics of a system of point particles is conservative. The existence of dissipative effects can be found by pushing the PN expansion up to 1.5PN order, i.e., up to terms proportional to $1/c^3$. This could have already been anticipated from the fact that the power radiated by a non-relativistic charge is proportional $1/c^3$, as we see from Larmor's formula (10.148), while its relativistic generalization (10.156) has a more complicated dependence on c that, again, when expanded in powers of $1/c$, starts with the $\mathcal{O}(1/c^3)$ Larmor's term.

In this section we then compute the 1.5PN contribution to the gauge potentials, and therefore to the equations of motion. In Section 12.3.4 we will show how a covariantization of the result gives the expression for radiation reaction to all orders, leading to the so-called Abraham–Lorentz–Dirac (ALD) equation. In Section 12.3.5 we will finally show how mass renormalization and the ALD equation can be derived in a unified treatment, with mass renormalization obtained from an explicitly Lorentz-invariant computation.

Given that we are looking for dissipative terms, it will be important to compute the effect on the equations of motion starting from the Lorentz force equation, rather than working at the level of the action. We then expand eqs. (12.1) and (12.2) to 1.5PN order, and insert them in the equations of motion. The 1.5PN contributions are

$$\phi_{1.5\text{PN}}(t, \mathbf{x}) = -\frac{1}{4\pi\epsilon_0}\frac{1}{6c^3}\partial_t^3\int d^3x'\,\rho(t, \mathbf{x}')|\mathbf{x} - \mathbf{x}'|^2\,, \quad (12.118)$$

$$\mathbf{A}_{1.5\text{PN}}(t, \mathbf{x}) = -\frac{1}{4\pi\epsilon_0}\frac{1}{c^3}\frac{d}{dt}\int d^3x'\,\mathbf{j}(t, \mathbf{x}')\,. \quad (12.119)$$

Using eq. (11.102), the vector potential can also be rewritten in terms of the electric dipole moment $\mathbf{d}(t)$ as

$$\mathbf{A}_{1.5\text{PN}}(t, \mathbf{x}) = -\frac{1}{4\pi\epsilon_0}\frac{1}{c^3}\ddot{\mathbf{d}}(t)\,. \quad (12.120)$$

Note that $\mathbf{A}_{1.5\text{PN}}(t, \mathbf{x})$ actually depends only on t, and is independent of $\mathbf{x}$. To compute the corresponding electric field we observe that

$$\begin{aligned}\boldsymbol{\nabla}\phi_{1.5\text{PN}} &= -\frac{1}{4\pi\epsilon_0}\frac{1}{6c^3}\partial_t^3\int d^3x'\,\rho(t, \mathbf{x}')2(\mathbf{x} - \mathbf{x}') \\ &= -\frac{1}{4\pi\epsilon_0}\frac{1}{3c^3}\partial_t^3\left[Q\mathbf{x} - \mathbf{d}(t)\right] \\ &= \frac{1}{4\pi\epsilon_0}\frac{1}{3c^3}\dddot{\mathbf{d}}(t)\,, \end{aligned} \quad (12.121)$$

where we observed that the term $Q\mathbf{x}$ is independent of time and gives zero when we apply ∂_t to it. Then

$$\begin{aligned}\mathbf{E}_{1.5\text{PN}}(t, \mathbf{x}) &= -\boldsymbol{\nabla}\phi_{1.5\text{PN}} - \partial_t\mathbf{A}_{1.5\text{PN}} \\ &= \frac{1}{4\pi\epsilon_0}\frac{2}{3c^3}\dddot{\mathbf{d}}(t)\,. \end{aligned} \quad (12.122)$$

For the magnetic field we have $\mathbf{B}_{1.5\text{PN}}(t, \mathbf{x}) = 0$ since $\mathbf{A}_{1.5\text{PN}}$ is independent of $\mathbf{x}$, and therefore $\boldsymbol{\nabla}\times\mathbf{A}_{1.5\text{PN}} = 0$. Putting these expressions in the Lorentz force equation, we get the 1.5PN contribution to the equation of motion on the a-th particle,

$$\boxed{\left(\frac{d\mathbf{p}_a}{dt}\right)_{1.5\text{PN}} = \frac{1}{4\pi\epsilon_0}\frac{2q_a}{3c^3}\dddot{\mathbf{d}}(t)\,,} \quad (12.123)$$

where, as usual, $\mathbf{p}_a = \gamma_a m_a\mathbf{v}_a$. Combining this with eq. (12.50), we can write the equation of motion, up to 1.5PN order, in the form

$$m_a\frac{d\mathbf{v}_a}{dt} = (\mathbf{F}_N + \mathbf{F}_{1\text{PN}} + \mathbf{F}_{1.5\text{PN}})_a\,, \quad (12.124)$$

where

$$(\mathbf{F}_N)_a = -\frac{1}{4\pi\epsilon_0}\sum_{b\neq a}^{N}\frac{q_a q_b}{r_{ab}^2}\hat{\mathbf{r}}_{ab}\,, \quad (12.125)$$

is the Newtonian force on the a-th particle, $(\mathbf{F}_{1\text{PN}})_a$ is obtained collecting all the terms proportional to $1/c^2$ in eq. (12.50), and

$$(\mathbf{F}_{1.5\text{PN}})_a = \frac{1}{4\pi\epsilon_0}\frac{2q_a}{3c^3}\dddot{\mathbf{d}}(t)\,. \quad (12.126)$$

The notation in eq. (12.124) is suggestive of the Newtonian equation of motion $\mathbf{F} = m\mathbf{a}$. The analogy, however, is purely formal. For instance, in $\mathbf{F}_{1\text{PN}}$ we have collected all terms proportional to $1/c^2$ in eq. (12.50), including the term that comes from the expansion of γ_a in $\mathbf{p} = \gamma_a m_a \mathbf{v}_a$, which has nothing to do with the interaction of the a-th particle with the other particles. This notation, however, will be useful in the steps that we perform below.

We now compute how the energy changes in time. First of all, we must define energy in this context. This is done by looking at the conservative part of the dynamics (as obtained eliminating the gauge fields in favor of $\mathbf{x}_a(t)$ and $\mathbf{v}_a(t)$ at the level of the action), so that we can define a Lagrangian and the corresponding Hamiltonian. In our case, since we are working up to 1.5PN order, the conservative dynamics include the terms up to 1PN order, and the Hamiltonian is given by eq. (12.71). We can then write the Hamiltonian as

$$H = H_N + H_{1\text{PN}}, \tag{12.127}$$

where

$$H_N = \sum_{a=1}^{N} \frac{P_a^2}{2m_a} + \frac{1}{8\pi\epsilon_0} \sum_{a=1}^{N} \sum_{b\neq a}^{N} \frac{q_a q_b}{r_{ab}}, \tag{12.128}$$

while $H_{1\text{PN}}$ can be obtained writing $\mathbf{P}_a$ as in eq. (12.66), and then collecting the terms $1/c^2$ terms in eq. (12.71). So, the corresponding energy is

$$E = E_N + E_{1\text{PN}}, \tag{12.129}$$

where the Newtonian part is given by the usual form

$$E_N = \sum_{a=1}^{N} \frac{1}{2} m_a v_a^2 + U, \tag{12.130}$$

with the potential energy U given by

$$U = \frac{1}{8\pi\epsilon_0} \sum_{a=1}^{N} \sum_{b\neq a}^{N} \frac{q_a q_b}{r_{ab}}, \tag{12.131}$$

while $E_{1\text{PN}}$ can be written in terms of positions and velocities (rather than positions and momenta) by using eq. (12.66) in H_N (and using simply $\mathbf{P}_a = m_a v_a$ in $H_{1\text{PN}}$) and collecting the terms proportional to $1/c^2$. We now write

$$
\begin{aligned}
\frac{dE}{dt} &= \frac{d}{dt} \left[\sum_{a=1}^{N} \frac{1}{2} m_a v_a^2 + U \right] + \frac{dE_{1\text{PN}}}{dt} \\
&= \sum_{a=1}^{N} \frac{d(m_a \mathbf{v}_a)}{dt} \cdot \mathbf{v}_a + \frac{dU}{dt} + \frac{dE_{1\text{PN}}}{dt} \\
&= \sum_{a=1}^{N} (\mathbf{F}_N + \mathbf{F}_{1\text{PN}} + \mathbf{F}_{1.5\text{PN}})_a \cdot \mathbf{v}_a + \frac{dU}{dt} + \frac{dE_{1\text{PN}}}{dt}, \tag{12.132}
\end{aligned}
$$

where, in the last line, we used eq. (12.124). The crucial point, now, is that, as we have seen, the dynamics up to 1PN order is conservative, so $E = E_N + E_{1PN}$ is conserved if we use the equations of motion up to 1PN order, i.e., $dE/dt = 0$ when $\mathbf{F}_{1.5PN}$ is not included in the equation. Therefore,

$$\sum_{a=1}^{N} (\mathbf{F}_N + \mathbf{F}_{1PN})_a \cdot \mathbf{v}_a + \frac{dU}{dt} + \frac{dE_{1PN}}{dt} = 0. \tag{12.133}$$

Then, eq. (12.132) gives

$$\frac{dE}{dt} = \sum_{a=1}^{N} (\mathbf{F}_{1.5PN})_a \cdot \mathbf{v}_a. \tag{12.134}$$

Note that, formally, $(\mathbf{F}_{1.5PN})_a \cdot \mathbf{v}_a$ is the same as the work that would be made by a Newtonian force $\mathbf{F}_{1.5PN}$ on the non-relativistic particle a, in the context of non-relativistic mechanics. Once again, the analogy is purely formal, and comes from the fact that we have written the equation of motion, including non-relativistic corrections up to order $(v/c)^3$, in the form (12.124), where $\mathbf{F}_N + \mathbf{F}_{1PN}$ formally plays the role of a conservative Newtonian force while $\mathbf{F}_{1.5PN}$ of a dissipative force.[20,21]

Inserting eq. (12.126) into eq. (12.134) we get

$$\frac{dE}{dt} = \frac{1}{4\pi\epsilon_0} \frac{2}{3c^3} \dddot{\mathbf{d}}(t) \cdot \sum_{a=1}^{N} q_a \mathbf{v}_a(t). \tag{12.135}$$

However, $q_a \mathbf{v}_a = \dot{\mathbf{d}}_a$, and $\sum_{a=1}^{N} \dot{\mathbf{d}}_a = \dot{\mathbf{d}}$, where $\mathbf{d}$ is the total dipole moment of the system. Therefore, we can rewrite eq. (12.135) as

$$\frac{dE}{dt} = \frac{1}{4\pi\epsilon_0} \frac{2}{3c^3} \dddot{\mathbf{d}} \cdot \dot{\mathbf{d}}$$
$$= \frac{1}{4\pi\epsilon_0} \frac{2}{3c^3} \left[\frac{d}{dt}(\dot{\mathbf{d}} \cdot \ddot{\mathbf{d}}) - |\ddot{\mathbf{d}}|^2 \right]. \tag{12.136}$$

The second term in eq. (12.136) corresponds precisely to minus the energy radiated by the Larmor formula, eq. (10.148), and reproduces the fact that the energy of the system decreases because it radiates electromagnetic waves. To deal with the first term, different options are possible. One possibility is to average this equation over a time T. Denoting this average with angular brackets, we have, in particular,

$$\langle \frac{d}{dt}(\dot{\mathbf{d}} \cdot \ddot{\mathbf{d}}) \rangle = \frac{1}{T} \int_{-T/2}^{T/2} dt\, \frac{d}{dt}(\dot{\mathbf{d}} \cdot \ddot{\mathbf{d}}). \tag{12.137}$$

The right-hand side vanishes if $\mathbf{d}(t)$ is a periodic function with period T, or if we send $T \to \infty$ with $\mathbf{d}(t)$ that vanishes for $t \to \pm\infty$. In that cases, we get

$$\langle \frac{dE}{dt} \rangle = -\frac{1}{4\pi\epsilon_0} \frac{2}{3c^3} \langle |\ddot{\mathbf{d}}|^2 \rangle. \tag{12.138}$$

[20] Indeed, in all standard textbook derivations, after having obtained eq. (12.123), it is simply stated, without much explanation, that the term $\mathbf{F}_{1.5PN}$ on the right-hand side is a force, that makes a work $W_a = (\mathbf{F}_{1.5PN})_a \cdot \mathbf{v}_a$ on the particle a. While this eventually leads to the correct answer, without the more explicit steps that we have performed it would be unclear why one should use a non-relativistic formula such as $W = \mathbf{F} \cdot \mathbf{v}$ in our relativistic context. The derivation that we have provided also makes it clear that the energy that appears in dE/dt, when we compute radiation reaction up to 1.5PN order, is the energy obtained from the conservative part of the dynamics up to 1PN order.

[21] One might ask what happens if we rather insert eqs. (12.118) and (12.120) into the action (12.16), and carry out the same steps that we did in eqs. (12.51)–(12.55) when we computed the 1PN Lagrangian. The result is that the resulting 1.5PN contribution to the Lagrangian is a total time derivative, and therefore does not affect the equations of motion. As we already discussed, the dissipative term cannot be obtained from a Lagrangian, and we must rather eliminate the gauge fields at the level of the equations of motion.

In this way, energy conservation is recovered, at least in an averaged form: eq. (12.138) shows that the energy of the system of charges decreases, exactly in such a way to compensate (when averaged over one period of the source motion) the energy radiated in electromagnetic waves, which, to the order $1/c^3$ to which we are working, is given by the Larmor formula (10.148).[22]

The need for a time averaging, however, is not really satisfying; classically, we expect that energy conservation should be valid instantaneously (and should not be restricted to periodic motions, or to situations where the source becomes static as $t \to \pm\infty$ and energy conservation is only recovered as an average over a time $T \to \infty$). However, we can get rid of the averaging procedure if we define

$$E_{1.5\text{PN}} = -\frac{1}{4\pi\epsilon_0}\frac{2}{3c^3}\dot{\mathbf{d}}\cdot\ddot{\mathbf{d}} \tag{12.139}$$

$$= -\frac{1}{4\pi\epsilon_0}\frac{2}{3c^3}\sum_{a,b=1}^{N}q_a q_b \dot{\mathbf{v}}_a\cdot\ddot{\mathbf{v}}_b.$$

Since this quantity is proportional to $1/c^3$, we can interpret it as a 1.5PN contribution to the energy. Recalling that, on the left-hand side of eq. (12.136), $E = E_N + E_{1\text{PN}}$, we can then rewrite eq. (12.136) as

$$\frac{d}{dt}(E_N + E_{1\text{PN}} + E_{1.5\text{PN}}) = -\frac{1}{4\pi\epsilon_0}\frac{2}{3c^3}|\ddot{\mathbf{d}}|^2. \tag{12.140}$$

In this way, the 1.5PN dynamics produces both a conservative term, that contributes to the energy and is localized in the near region, and the dissipative term that describes the loss of energy to radiation escaping at infinity.[23]

It is interesting to observe that the energy balance equation (12.140) has been obtained using eq. (12.126) for the force $(\mathbf{F}_{1.5\text{PN}})_a$ acting on the a-th particle and, in this expression, the dipole moment $\mathbf{d}$ is the one obtained from the *total* charge density ρ, without excluding the contribution from the a-th particle itself, on which the force acts. In this way we correctly recovered the power radiated to order $1/c^3$, as given by Larmor's formula. If we had excluded the self-term by hand, as we did when computing the 1PN dynamics in Section 12.2, in eq. (12.126) $\mathbf{d}(t)$ would have been replaced by $\sum_{b\neq a}\mathbf{d}_b(t)$, i.e., by $\mathbf{d}(t) - \mathbf{d}_a(t)$, and this, when inserted into eq. (12.134), would not have reproduced correctly the Larmor formula. Rather, on the right-hand side of eq. (12.140), instead of $|\ddot{\mathbf{d}}|^2$, we would have found the combination $(|\ddot{\mathbf{d}}|^2 - \sum_a |\ddot{\mathbf{d}}_a|^2)$. We see that the self-energy terms are indeed physical, and are needed to obtain the correct radiation reaction force. In the computation to order 1PN that we performed in Section 12.2 we had arbitrarily thrown them away. However, this "mistake" turned out to be without consequences because, as we have shown in Section 12.3.2, once one includes them, their contribution to the 1PN dynamics can just be reabsorbed into a

[22] Note that, in the Larmor formula (10.148), we used t to denote the time of a distant observer, and t_{ret} was the retarded time, i.e., the time at which the radiation was produced, which corresponds to the variable that we are denoting by t here.

[23] Observe that, for a conservative system, described by a Lagrangian, there is a natural and unique definition of the energy, in terms of the Hamiltonian corresponding to the given Lagrangian. As we have seen (see Note 7 on page 304), this Hamiltonian is automatically conserved on the solutions of the equations of motion. For a dissipative system there is no such unique definition of energy. However, having at our disposal an energy balance condition such as eq. (12.140), the natural definition is to include in the energy all terms that appear inside the time derivative on the left-hand side of eq. (12.140). Also note that $E_{1.5\text{PN}}$ is itself a total time derivative, since eq. (12.139) can be rewritten as

$$E_{1.5\text{PN}} = -\frac{1}{4\pi\epsilon_0}\frac{1}{3c^3}\frac{d}{dt}|\dot{\mathbf{d}}|^2. \tag{12.141}$$

mass renormalization. We see, however, that at 1.5PN order they give a finite contribution, which is essential to recover the correct energy balance equation. Also note that, since this 1.5PN contribution is finite, we could compute it without specifying a regularization procedure.

Another comment concerns the fact that radiation reaction first appears at order $1/c^3$, i.e., is associated with an odd power of $1/c$. This is related to time reversal invariance. As we showed in Section 3.4, Maxwell's equations are invariant under time reversal. However, as we already discussed (see in particular Note 39 on page 86), the symmetries of the equations are not necessarily the same as the symmetries of their solutions. The other option is that we have a family of solutions, that transform into each other under the given symmetry transformation. In the case of a discrete transformation such as time reversal, $t \rightarrow -t$, this means that we could have two solutions, that transform into each other under time reversal. This is indeed what happens when solving an equation such as $\Box A^\mu = 0$, since, as we have seen in Section 10.1, the d'Alembertian operator has two Green's functions, the advanced and the retarded ones, that are exchanged under time reversal, as we see from eq. (10.24). In our computations we have explicitly broken time reversal invariance by selecting, for physical reasons, the retarded Green's function. Consider now a system of charges, evolving under their mutual interaction. As we have seen, they will emit outgoing electromagnetic waves and, correspondingly, they lose energy. The time-reversal of this solution is obtained "running the film backward," and corresponds to the rather strange situation in which incoming electromagnetic radiation impinges on the system of charges and pumps energy into it, furthermore exactly canceling any outgoing radiation generated by these accelerated charges (just as, in the original setting, there was no incoming radiation on the system). In practice, this is a solution that we will never observe in Nature because it requires incredibly fine-tuned initial conditions on the incoming radiation, but still it is a solution of Maxwell's equations coupled to the sources. Exchanging the retarded with the advanced solution must therefore result in a change of sign of the radiation-reaction force, since, in the time-reversed situation, it will have to describe an energy gain rather than an energy loss. However, the exchange of the advanced and retarded Green's functions (10.24) can be formally obtained with the replacement $c \rightarrow -c$. This means that the radiation-reaction force is necessarily associated with odd powers of $1/c$, i.e., with half-integer orders of the PN expansion. As we saw, the scalar potential vanishes at 0.5PN order because of charge conservation, and the vector potential starts from 1PN order, so the first order at which radiation reaction could have appeared is 1.5PN, as indeed we have found.[24]

[24] In gravity, as described by General Relativity, dipole radiation is absent and the leading term in the radiated power is given by the (mass) quadrupole radiation, and is proportional to $1/c^5$. Correspondingly, also the back-reaction force starts at 2.5PN order. See Maggiore (2007), Section 5.1.7.

12.3.4   The Abraham–Lorentz–Dirac equation

In eq. (12.126) we found the radiation reaction force at leading order in v/c, which turned out to be the 1.5PN order. As we saw, this reproduces the energy lost to electromagnetic waves by a system of charges, as predicted by Larmor's formula (10.148). However, we found in Section 10.6.1 that Larmor's formula is just the power radiated to lowest non-trivial order in v/c, and the full result for the power radiated by a point charge, to all orders in v/c, is given by eq. (10.156), or, in covariant form, by eq. (10.169). It must therefore be possible to find an expression for the radiation reaction force, valid to all orders in v/c, which corresponds to the full energy loss (10.156). To obtain it, one could compute explicitly the exact self-field generated by a particle, to all orders in v/c. We will follow this route in Section 12.3.5, where we will see that this is indeed possible but the computation, while instructive, is quite involved. A much simpler procedure, that we follow in this section, consists in looking for a covariant generalization of eqs. (12.124) and (12.126). We focus on a given single particle a, and we rewrite eqs. (12.124) and (12.126) separating the total dipole moment $\mathbf{d}(t) = \sum_{b=1}^{N} \mathbf{d}_b(t)(t)$ as $\mathbf{d}(t) = \mathbf{d}_a(t) + \sum_{b\neq a}^{N} \mathbf{d}_b(t)$, and we reabsorb the term $\sum_{b\neq a}^{N} \mathbf{d}_b$ into the external force exerted on particle a. Then, writing $\mathbf{d}_a(t) = q_a \mathbf{x}_a(t)$, so that $\ddot{\mathbf{d}}_a(t) = q_a \dot{\mathbf{v}}_a(t)$, we rewrite eqs. (12.124) and (12.126) as

$$\frac{d}{dt}\left[\left(1 + \frac{v_a^2}{2c^2}\right)m_a\mathbf{v}_a\right] = \frac{1}{4\pi\epsilon_0}\frac{2q_a^2}{3c^3}\frac{d^2\mathbf{v}_a}{dt^2} + \mathbf{F}_{\text{ext}}, \qquad (12.142)$$

where we also included the expansion of $\gamma(v_a)$ to order v_a^2/c^2, that in eq. (12.124) was formally included in $\mathbf{F}_{1\text{PN}}$, back to where it belongs, as a multiplicative factor for $m_a\mathbf{v}_a$.

We now proceed with the covariantization. In the absence of the backreaction force, the covariantization of eq. (12.142) is just given by eq. (8.62) with $F^{\mu\nu} = F^{\mu\nu}_{\text{ext}}$, i.e.,

$$m_a\dot{u}^\mu = q_a F^{\mu\nu}_{\text{ext}}[x_a(\tau)]u_{a,\nu}, \qquad (12.143)$$

where the dot denotes the derivative with respect to τ, and we have written explicitly that, in the Lorentz force equation, $F^{\mu\nu}_{\text{ext}}(x)$ must be evaluated on the particle world-line. We next observe that, on the right-hand side of eq. (12.142), $d^2v_a^i/dt^2$ is naturally covariantized as the spatial component of $\ddot{u}_a^\mu$, where the dot denotes the derivative with respect to proper time τ_a of the particle. However, this covariantization is not unique, since any expression of the form $\ddot{u}_a^\mu + \alpha u_a^\mu$, with α an arbitrary function of $u_a^\nu, \dot{u}_a^\nu, \ddot{u}_a^\nu$ and possibly of higher-order derivatives, is such that its $\mu = i$ component that reduces to $d^2v_a^i/dt^2$ in a frame where the particle a is instantaneously at rest. This leads us to

$$m_a\dot{u}_a^\mu = \frac{1}{4\pi\epsilon_0}\frac{2q_a^2}{3c^3}\left[\ddot{u}_a^\mu + \alpha(u_a^\nu, \dot{u}_a^\nu, \ddot{u}_a^\nu, \ldots)u_a^\mu\right] + q_a F^{\mu\nu}_{\text{ext}}[x_a(\tau)]u_{a,\nu}. \qquad (12.144)$$

Since u_a^μ is the only four-vector whose spatial components vanish in the rest frame of the charge, no further freedom is possible. The function $\alpha(u^\nu, \dot{u}^\nu, \ddot{u}^\nu, \ldots)$ can then be fixed by observing that, taking the derivative with respect to τ of the relation $u_\mu u^\mu = -c^2$, it follows the identity

$u_\mu \dot{u}^\mu = 0$. Then, multiplying eq. (12.144) by $u_{a,\mu}$, we get the condition

$$
\begin{aligned}
0 &= u_{a,\mu} \left[\ddot{u}_a^\mu + \alpha(u_a^\nu, \dot{u}_a^\nu, \ddot{u}_a^\nu, \ldots) u_a^\mu \right] \\
&= u_{a,\mu} \ddot{u}_a^\mu - c^2 \alpha(u_a^\nu, \dot{u}_a^\nu, \ddot{u}_a^\nu, \ldots),
\end{aligned}
\tag{12.145}
$$

and therefore

$$
\alpha(u_a^\nu, \dot{u}_a^\nu, \ddot{u}_a^\nu, \ldots) = \frac{1}{c^2} u_a^\nu \ddot{u}_{a,\nu}.
\tag{12.146}
$$

Inserting this into eq. (12.144), we finally get

$$
m_a \dot{u}_a^\mu = \frac{1}{4\pi\epsilon_0} \frac{2q_a^2}{3c^3} \left(\ddot{u}_a^\mu + \frac{u_a^\nu \ddot{u}_{a,\nu}}{c^2} u_a^\mu \right) + q_a F_{\text{ext}}^{\mu\nu}[x_a(\tau)] u_{a,\nu}.
\tag{12.147}
$$

We can rewrite eq. (12.147) in an equivalent form by taking one more derivative of the identity $u_\mu \dot{u}^\mu = 0$, to get $\dot{u}_\mu \dot{u}^\mu + u_\mu \ddot{u}^\mu = 0$, and therefore

$$
u_\mu \ddot{u}^\mu = -\dot{u}^2,
\tag{12.148}
$$

so eq. (12.147) can also be rewritten as[25]

$$
\boxed{ m_a \dot{u}_a^\mu = \frac{1}{4\pi\epsilon_0} \frac{2q_a^2}{3c^3} \left(\ddot{u}_a^\mu - \frac{\dot{u}_a^2}{c^2} u_a^\mu \right) + q_a F_{\text{ext}}^{\mu\nu}[x_a(\tau)] u_{a,\nu}. }
$$
$$
\tag{12.149}
$$

Equivalently, eq. (12.147) can also be written as

$$
\boxed{ m_a \dot{u}_a^\mu = \frac{1}{4\pi\epsilon_0} \frac{2q_a^2}{3c^3} \left(\eta^{\mu\nu} + \frac{u_a^\mu u_a^\nu}{c^2} \right) \ddot{u}_{a,\nu} + q_a F_{\text{ext}}^{\mu\nu}[x_a(\tau)] u_{a,\nu}, }
$$
$$
\tag{12.150}
$$

which explicitly displays the tensor structure $(\eta^{\mu\nu} + u_a^\mu u_a^\nu / c^2)$, which is transverse to $u_{a,\mu}$. Equation (12.150), or any of its equivalent forms, is called the *Abraham–Lorentz–Dirac* (ALD) equation. Its non-relativistic limit, including both corrections of order $1/c^2$ and $1/c^3$, is given by eq. (12.142). Actually, in the non-relativistic limit, if one is interested in the energy loss of the particle, one can keep only the leading dissipative effect, which is the term $1/c^3$, neglecting the correction of order $1/c^2$ to the conservative dynamics, and write

$$
m_a \frac{d\mathbf{v}_a}{dt} = \frac{1}{4\pi\epsilon_0} \frac{2q_a^2}{3c^3} \frac{d^2\mathbf{v}_a}{dt^2} + \mathbf{F}_{\text{ext}},
\tag{12.151}
$$

which is called the *Abraham–Lorentz* equation.[26] In terms of the four-momentum p_a^μ, and writing explicitly the dot as $d/d\tau$, eq. (12.149) reads

$$
\begin{aligned}
\frac{dp_a^\mu}{d\tau} &= \frac{1}{4\pi\epsilon_0} \frac{2q_a^2}{3m_a c^3} \frac{d^2 p_a^\mu}{d\tau^2} - \frac{1}{4\pi\epsilon_0} \frac{2q_a^2}{3m_a^3 c^5} \left(\frac{dp_{a,\nu}}{d\tau} \frac{dp_a^\nu}{d\tau} \right) p_a^\mu \\
&\quad + \frac{q_a}{m_a} F_{\text{ext}}^{\mu\nu}[x_a(\tau)] p_{a,\nu}.
\end{aligned}
\tag{12.152}
$$

[25] Observe that the relative sign among the two terms in parenthesis depends on our signature $\eta_{\mu\nu} = (-, +, +, +)$. With the opposite signature, in eq. (12.147) one would have the combination $\left[\ddot{u}_a^\mu - (u_a^\nu \ddot{u}_{a,\nu}/c^2) u_a^\mu \right]$, and in eq. (12.149) would appear the combination $\left[\ddot{u}_a^\mu + (\dot{u}_a^2/c^2) u_a^\mu \right]$.

[26] Equation (12.151) was first found by Lorentz, while the relativistic expression was first found by Abraham in 1904, so the year before Einstein's first paper on Special Relativity. Despite the fact that Special Relativity was yet to be formulated, he could get a relativistic result by using Maxwell's equations (which, as we now know, implicitly contain Special Relativity!), see Rohrlich (2000) for a historical discussion. The covariant form of the equation was first explicitly derived by Dirac in 1938. In Section 12.3.5 we will provide a covariant derivation, conceptually along the lines of Dirac's derivation, although somewhat different technically.

Comparing with eq. (10.169) we see that the second term,

$$F^\mu_{\rm rad} \equiv -\frac{1}{4\pi\epsilon_0}\frac{2q_a^2}{3m_a^3c^5}\left(\frac{dp_{a,\nu}}{d\tau}\frac{dp_a^\nu}{d\tau}\right)p_a^\mu\,, \qquad (12.153)$$

is precisely the negative of the four-momentum radiated by a point particle, to all orders in v/c. This term, therefore, is a dissipative term in the
equation of motion, that accounts for the loss of energy to electromagnetic waves, exactly in v/c. $F^\mu_{\rm rad}$ is therefore called the radiation-reaction
"force" (of course, it is a four-vector, but, following common usage, we
will refer to it as a force, rather than as a "four-force"). The first term,

$$F^\mu_{\rm Schott} \equiv \frac{1}{4\pi\epsilon_0}\frac{2q_a^2}{3m_ac^3}\frac{d^2p_a^\mu}{d\tau^2}\,, \qquad (12.154)$$

is called the *Schott term*. The total self-force, due to the particle self-
field, is therefore the sum of the radiation-reaction term and of the Schott
term,

$$F^\mu_{\rm self} = F^\mu_{\rm rad} + F^\mu_{\rm Schott}\,. \qquad (12.155)$$

We observe that the Schott term is a total time derivative, so eq. (12.152)
can be rewritten as

$$\frac{d}{d\tau}\left(p_a^\mu - \frac{1}{4\pi\epsilon_0}\frac{2q_a^2}{3m_ac^3}\frac{dp_a^\mu}{d\tau}\right) = F^\mu_{\rm rad} + \frac{q_a}{m_a}F^{\mu\nu}_{\rm ext}[x_a(\tau)]p_{a,\nu}\,. \qquad (12.156)$$

To make contact with the discussion of Section 12.3.3, we consider the
first non-vanishing contributions in the non-relativistic limit. Equations (12.153) and (12.154) give

$$F^0_{\rm Schott} = \frac{1}{4\pi\epsilon_0}\frac{2q_a^2}{3c^4}\left(\dot{\mathbf{v}}_a\cdot\mathbf{v}_a + \dot{\mathbf{v}}_a^2\right) + \mathcal{O}\left(\frac{1}{c^6}\right)\,, \qquad (12.157)$$

$$F^0_{\rm rad} = -\frac{1}{4\pi\epsilon_0}\frac{2q_a^2}{3c^4}\dot{\mathbf{v}}_a^2 + \mathcal{O}\left(\frac{1}{c^6}\right)\,, \qquad (12.158)$$

and

$$\mathbf{F}_{\rm Schott} = \frac{1}{4\pi\epsilon_0}\frac{2q_a^2}{3c^3}\ddot{\mathbf{v}}_a + \mathcal{O}\left(\frac{1}{c^5}\right)\,, \qquad (12.159)$$

$$\mathbf{F}_{\rm rad} = \mathcal{O}\left(\frac{1}{c^5}\right)\,. \qquad (12.160)$$

Therefore, to lowest order in v/c, the spatial component of eq. (12.152)
becomes

$$\frac{d\mathbf{p}_a}{dt} = \frac{1}{4\pi\epsilon_0}\frac{2q_a^2}{3c^3}\ddot{\mathbf{v}}_a + \mathbf{F}_{\rm ext}\,, \qquad (12.161)$$

and reproduces the Abraham–Lorentz equation (12.151). Note that, in
this equation, to lowest order, only the Schott term contributes. For the
temporal component, writing $q_a\mathbf{v}_a = \dot{\mathbf{d}}_a$, we get

$$\frac{dE_a}{dt} = \frac{1}{4\pi\epsilon_0}\frac{2}{3c^3}\left[\left(\dddot{\mathbf{d}}_a\cdot\mathbf{d}_a + \ddot{\mathbf{d}}_a^2\right) - \ddot{\mathbf{d}}_a^2\right] + \mathbf{F}_{\rm ext}\cdot\mathbf{v}_a\,. \qquad (12.162)$$

Inside the bracket, the two terms $\ddot{\mathbf{d}}_a^2$ cancel, giving back the first line of eq. (12.136) (with $\dddot{\mathbf{d}} \cdot \mathbf{d}$ replaced by $\ddot{\mathbf{d}}_a \cdot \mathbf{d}_a$, and all the terms involving the other particles reabsorbed in the work $\mathbf{F}_{\rm ext} \cdot \mathbf{v}_a$ made by the external force acting on the a-th particle). However, we see that the two separate contribution from $F^0_{\rm Schott}$ and $F^0_{\rm rad}$ correspond to the separation made in the second line of eq. (12.136). In particular, $F^0_{\rm Schott}$ is a total time derivative, and produces the term that we denoted as $E_{1.5\rm PN}$ in eq. (12.140).

Until this point, everything appears to work very nicely. Pushing the PN expansion up to 1.5PN order, in Section 12.3.3 we have found a radiation-reaction force that describes the loss of energy to electromagnetic waves, as predicted by the Larmor formula. In this section, we have found a covariantization of this result, that has provided an (apparently) exact equation, the ALD equation, that describes the self-force to all orders in v/c, and we have seen that it includes a radiation-reaction force that reproduces the loss of energy given by the exact relativistic formula (10.169), plus conservative terms that generalize to all orders the 1.5PN contribution to the energy, $E_{1.5\rm PN}$, found in eq. (12.140). Further examination of the ALD equation, however, reveals an apparent pathology. The problem is already present in the non-relativistic limit given by the Abraham–Lorentz equation, so let us discuss it first in this simpler setting. Introducing the timescale

$$\tau_a = \frac{1}{4\pi\epsilon_0} \frac{2q_a^2}{3m_a c^3} \,, \tag{12.163}$$

we can rewrite eq. (12.151) as

$$m_a \left(\frac{d\mathbf{v}_a}{dt} - \tau_a \frac{d^2\mathbf{v}_a}{dt^2} \right) = \mathbf{F}_{\rm ext} \,. \tag{12.164}$$

This equation is of second order in $\mathbf{v}_a(t)$, i.e., of third order in $\mathbf{x}_a(t)$ and therefore, to be solved, requires initial conditions on the position, velocity and acceleration. This is already in contrast with the normal situation in classical mechanics. Furthermore, consider this equation when $\mathbf{F}_{\rm ext} = 0$. Then, eq. (12.164) has the obvious solution $d\mathbf{v}_a/dt = 0$, as we aspect for a free particle not subject to any external force. However, it is also satisfied if the acceleration $\mathbf{a}_a = d\mathbf{v}_a/dt$ satisfies

$$\frac{d\mathbf{a}_a}{dt} = \frac{1}{\tau_a} \mathbf{a}_a \,, \tag{12.165}$$

and this, beside $\mathbf{a}_a = 0$, also has the exponentially growing solution

$$\mathbf{a}_a(t) = \mathbf{a}_a(0) e^{t/\tau_a} \,. \tag{12.166}$$

This solution describes a charged particle that accelerates even in the absence of an external force, and this acceleration even grows exponentially in time. Such a "self-accelerating" solution is obviously unphysical. One could simply exclude such solutions by hand (or, e.g., imposing the boundary condition that the acceleration does not diverge as $t \to \infty$).

However, it is possible to modify directly the Abraham–Lorentz equation so that this solution simply does not appear. To this purpose, we recall from eq. (5.32) that, for an electron with charge $q = -e$, the quantity

$$r_0 = \frac{1}{4\pi\epsilon_0} \frac{e^2}{m_e c^2} \qquad (12.167)$$

is called the classical electron radius. Numerically, its value is $r_0 \simeq 2.8 \times 10^{-15}$ m, and therefore it represents a scale typical of the realm of elementary particle physics. The timescale τ_a introduced above, for an electron, becomes $\tau_0 = (2/3)(r_0/c)$. It is therefore of order of the time that light takes to travel across such a small length-scale scale and, numerically, is of order 6×10^{-24} s. The second term in parenthesis in eq. (12.164) is comparable to the first only if the typical time-scale over which the velocity changes in time is of order τ_a, otherwise it is much smaller. Such fast variations, implying relativistic speeds over subatomic distances, necessarily belong to the domain of relativistic quantum mechanics and quantum field theory. In all situations where a classical treatment is justified, the term $\tau_a \ddot{\mathbf{v}}_a$ in eq. (12.164) is much smaller than $\dot{\mathbf{v}}_a$. We can then use a perturbative approach, analogous to the reduction of order discussed in Section 12.2.3: to zero-th order we just neglect the radiation reaction term, writing simply $m_a \dot{\mathbf{v}}_a = \mathbf{F}_{\text{ext}}$. We then use this equation to compute $\ddot{\mathbf{v}}_a$, so that $m_a \ddot{\mathbf{v}}_a = \dot{\mathbf{F}}_{\text{ext}}$, and we plug it into eq. (12.164). This gives

$$m_a \frac{d\mathbf{v}_a}{dt} = \mathbf{F}_{\text{ext}} + \tau_a \frac{d\mathbf{F}_{\text{ext}}}{dt}. \qquad (12.168)$$

Now, if $\mathbf{F}_{\text{ext}} = 0$, the only solution is $d\mathbf{v}_a/dt = 0$. In the range of validity of classical electrodynamics, eq. (12.168) is equivalent to eq. (12.164).

The same problem appears in the full relativistic ALD equation, and the same solution applies. To lowest order eq. (12.150) gives

$$m_a \dot{u}_a^\nu = q_a F_{\text{ext}}^{\nu\sigma}[x_a(\tau)] u_{a,\sigma}. \qquad (12.169)$$

Using this to compute $\ddot{u}_a^\nu$, we get

$$\ddot{u}_a^\nu = \frac{q_a}{m_a} \left(\partial_\rho F_{\text{ext}}^{\nu\sigma}\right)_{x=x_a(\tau)} u_a^\rho u_{a,\sigma} + \frac{q_a^2}{m_a^2} \eta_{\rho\sigma} F_{\text{ext}}^{\nu\sigma}[x_a(\tau)] F_{\text{ext}}^{\rho\alpha}[x_a(\tau)] u_{a,\alpha}. \qquad (12.170)$$

Inserting this into eq. (12.150) and using the definition (12.163), we get

$$m\dot{u}_a^\mu = q_a F_{\text{ext}}^{\mu\nu} u_{a,\nu} + \tau_a \left[q_a (\partial_\rho F_{\text{ext}}^{\mu\sigma}) u_a^\rho u_{a,\sigma} \right. \qquad (12.171)$$

$$\left. + \frac{q_a^2}{m_a} F_{\text{ext}}^{\mu\rho} F_{\rho\sigma}^{\text{ext}} u_a^\sigma + \frac{q_a^2}{m_a c^2} (F_{\rho\sigma}^{\text{ext}} u_a^\sigma)(F_{\text{ext}}^{\rho\nu} u_{a,\nu}) u_a^\mu \right],$$

where it is understood that $F_{\text{ext}}^{\mu\nu}$ and $\partial_\rho F_{\text{ext}}^{\mu\sigma}$ must be computed on the particle world-line. This is called the Landau–Lifshitz form of the ALD equation.[27] It is the covariant generalization of eq. (12.168), to all orders in v/c, and correctly reproduces the fact that $\dot{u}_a^\mu = 0$ when $F_{\text{ext}}^{\mu\nu} = 0$.

[27] See Landau and Lifschits (1975), Section 76.

12.3.5 Covariant derivation of mass renormalization and radiation reaction

In the previous sections we have seen that, at 1PN order, the self-field of a particle produces a divergent contribution, that must be regularized and is eventually reabsorbed into a mass renormalization while, at 1.5PN order, appears a radiation-reaction force, to which the particle self-field gives a finite contribution (that, indeed, we could compute without the need of introducing a regularization). We have then seen how to compute radiation reaction exactly, to all orders in v/c, by covariantizing the lowest-order result. In the computation of mass renormalization, we used a regularization scheme that breaks Lorentz invariance (and that mimics an extended classical electron model), but we recovered Lorentz invariance at the level of renormalized quantities, by introducing a counterterm that is not Lorentz invariant. In this section we show that it is possible to deal with the mass renormalization with a regularization that respects Lorentz invariance and, by the same computation, to obtain the ALD equation from a direct evaluation of the self-field on the particle world-line, without appealing to a covariantization of the lowest-order result.

We start from the Lorentz-covariant equation of motion (8.62) for a point particle of charge q_a, whose world-line is given by $x_a^\mu(\tau)$, and we rewrite it separating explicitly the contribution of the external field and of the self-field of the particle,

$$m_{a,0}(\ell)\frac{du_a^\mu(\tau)}{d\tau} = q_a F_{a,\text{self}}^{\mu\nu}[x_a(\tau)]u_{a,\nu}(\tau) + q_a F_{\text{ext}}^{\mu\nu}[x_a(\tau)]u_{a,\nu}(\tau)\,.$$

(12.172)

We have also anticipated that the mass that appears on the left-hand side of this equation is still a bare mass, with ℓ the corresponding cutoff, and we have stressed that $F_{a,\text{self}}^{\mu\nu}$ and $F_{\text{ext}}^{\mu\nu}$ must be computed on the particle world-line. We then focus on the contribution of the self-field, which is given by

$$F_{a,\text{self}}^{\mu\nu} = \partial^\mu A_{a,\text{self}}^\nu - \partial^\nu A_{a,\text{self}}^\mu\,,$$

(12.173)

where $A_{a,\text{self}}^\mu$ is the gauge field generated by the a-th particle itself. From eq. (10.10),

$$A_{a,\text{self}}^\mu(x) = -\frac{1}{\epsilon_0 c^2}\int d^4x'\, G_{\text{ret}}(x - x')j_a^\mu(x')\,,$$

(12.174)

where $G_{\text{ret}}(x - x')$ is the retarded Green's function, that we write in the explicitly Lorentz-invariant form (10.41),

$$G_{\text{ret}}(x - x') = -\frac{1}{2\pi}\theta(x^0 - x'^0)\,\delta[(x - x')^2]\,.$$

(12.175)

The four-current $j_a^\mu(x)$ associated with a point particle was given in eq. (8.3), that we rewrite here as

$$j_a^\mu(x) = q_a\int_{-\infty}^\infty cd\tau'\, u_a^\mu(\tau')\delta^{(4)}[x - x_a(\tau')]\,.$$

(12.176)

Inserting eq. (12.176) in eq. (12.174), we get

$$A^{\mu}_{a,\text{self}}(x) = -\frac{q_a}{\epsilon_0 c} \int_{-\infty}^{\infty} d\tau'\, u^{\mu}_a(\tau') \int d^4x'\, G_{\text{ret}}(x-x')\delta^{(4)}[x'-x_a(\tau')]\,.$$

$$(12.177)$$

If we now carry out the integral over d^4x' using $\delta^{(4)}[x'-x_a(\tau')]$ we get

$$A^{\mu}_{a,\text{self}}(x) = -\frac{q_a}{\epsilon_0 c} \int_{-\infty}^{\infty} d\tau'\, G_{\text{ret}}[x-x_a(\tau')]u^{\mu}_a(\tau')\,, \qquad (12.178)$$

and (as we will see explicitly in the following computation), if we finally evaluate this expression in $x^{\mu} = x^{\mu}_a(\tau)$, we find a divergence in $A^{\mu}_{a,\text{self}}[x_a(\tau)]$, or in the corresponding result for $F^{\mu\nu}_{a,\text{self}}[x_a(\tau)]$, when the integration variable τ' is equal to τ, due to the Dirac delta on the past light cone that appears in eq. (12.175). From this point of view, the origin of the divergence that we found in Section 12.3.2 is therefore the "collision" between the term $\delta^{(4)}[x'-x_a(\tau')]$ in the current (12.176), and the term $\delta[(x-x')^2]$ in the Green's function. We therefore need to regularize eq. (12.177). The computation that we have described in Section 12.3.2 was based on the idea of regularizing the current $j^{\mu}_a(x)$ given in eq. (12.176), replacing $\delta^{(4)}[x'-x_a(\tau')]$ with a smoothed version of the Dirac delta, which, in practice, we did by expanding it in Fourier modes and setting to zero the modes with $|\mathbf{k}| > \pi/\ell$. This regularization corresponds to the intuitive idea of an extended classical electron model, although we have repeatedly stressed that it must only be considered as a regularization scheme, and the limit $\ell \to 0$ must eventually be taken, canceling the divergence against an ℓ-dependence of the bare parameters. As we have discussed, this regularization is not Lorentz invariant, but Lorentz invariance is recovered for the renormalized quantities by introducing two different bare masses for the temporal and spatial components of p^{μ} or, equivalently, by adding to the bare action a counterterm which is not Lorentz invariant (and, in fact, is simply proportional to the non-relativistic action of a free particle).

An alternative procedure, that we will follow in this section, consists in regularizing the retarded Green's function in eq. (12.175), rather than the current $j^{\mu}(x)$. As we will see, this can be done preserving explicitly Lorentz invariance for the divergent part, so that a single bare mass term will be sufficient to reabsorb the divergences in the spatial and in the temporal components of the equation of motion.

Using eq. (12.178) for $A^{\mu}_{a,\text{self}}(x)$, with the retarded Green's function suitably regularized, we write

$$
\begin{aligned}
F^{\mu\nu}_{a,\text{self}}(x) &= -\frac{q_a}{\epsilon_0 c} \int_{-\infty}^{\infty} d\tau' \left\{ \partial^{\mu} G_{\text{ret}}[x-x_a(\tau')]u^{\nu}_a(\tau') \right. \\
&\qquad\qquad \left. -\partial^{\nu} G_{\text{ret}}[x-x_a(\tau')]u^{\mu}_a(\tau') \right\} \qquad (12.179) \\
&= -\frac{q_a}{\epsilon_0 c} \int_{-\infty}^{\infty} d\tau' \left[\eta^{\mu\beta}u^{\nu}_a(\tau') - \eta^{\nu\beta}u^{\mu}_a(\tau') \right] \partial_{\beta} G_{\text{ret}}[x-x_a(\tau')] \\
&= -\frac{q_a}{\epsilon_0 c} (\eta^{\mu\beta}\eta^{\nu\alpha} - \eta^{\nu\beta}\eta^{\mu\alpha}) \int_{-\infty}^{\infty} d\tau'\, u_{a,\alpha}(\tau')\partial_{\beta} G_{\text{ret}}[x-x_a(\tau')]\,.
\end{aligned}
$$

Then, evaluating $F_{a,\text{self}}^{\mu\nu}(x)$ on the particle world-line,

$$
\begin{aligned}
F_{a,\text{self}}^{\mu\nu}[x_a(\tau)] &= -\frac{q_a}{\epsilon_0 c}\left(\eta^{\mu\beta}\eta^{\nu\alpha} - \eta^{\nu\beta}\eta^{\mu\alpha}\right) \\
&\times \int_{-\infty}^{\infty} d\tau'\, u_{a,\alpha}(\tau')\left(\partial_\beta G_{\text{ret}}[x - x_a(\tau')]\right)_{x=x_a(\tau)}. \quad (12.180)
\end{aligned}
$$

After evaluating this expression explicitly, with a given regularization, we will insert it in eq. (12.172) and, as we will see, we will obtain a term which diverges when we remove the regularization, which is reabsorbed in a mass renormalization, and a finite contribution that gives the self-force term of the ALD equation (12.149).

To regularize G_{ret}, it is convenient to introduce first the symmetric and antisymmetric combinations of the retarded and advanced Green's functions,

$$
G_S(x) = \frac{1}{2}\left[G_{\text{ret}}(x) + G_{\text{adv}}(x)\right], \quad (12.181)
$$

$$
G_A(x) = \frac{1}{2}\left[G_{\text{ret}}(x) - G_{\text{adv}}(x)\right], \quad (12.182)
$$

so

$$
G_{\text{ret}}(x) = G_S(x) + G_A(x). \quad (12.183)
$$

Using the covariant expressions (10.40) for the retarded and advanced Green's functions, we see that

$$
G_S(x) = -\frac{1}{4\pi}\,\delta(x^2). \quad (12.184)
$$

Since this quantity depends only on the Lorentz-invariant variable x^2, it is easy to regularize it in a Lorentz-invariant manner, simply replacing the Dirac delta by any of its regularizations discussed in Section 1.4, so we replace eq. (12.184) by

$$
G_S(x) = -\frac{1}{4\pi}\,\delta_{\text{reg}}(x^2). \quad (12.185)
$$

We will use, for definiteness, the gaussian regularization (1.55), that we rewrite here in the notation

$$
\delta_{\text{reg}}(z) = \frac{1}{\sqrt{2\pi}\,a}\,e^{-z^2/(2a^2)}, \quad (12.186)
$$

with a a regularization parameter to be eventually sent to zero.

For the antisymmetric combination, a proper treatment involves some subtlety. Using eq. (10.40),

$$
G_A(x) = -\frac{1}{4\pi}\,\epsilon(x^0)\delta(x^2), \quad (12.187)
$$

where $\epsilon(x_0) = \theta(x^0) - \theta(-x^0)$. However, the product of a distribution such as $\theta(x^0)$, or $\epsilon(x^0)$, and a distribution such as $\delta(x^2) = \delta[(x^0)^2 - |\mathbf{x}|^2]$, is ill defined on the tip of the light cone, where $|\mathbf{x}| = 0$. As will be clear from the following computation, this is exactly the region from where all the result for the self-field comes, and a naive computation using eq. (12.187), even with $\delta(x^2)$ replaced by $\delta_{\text{reg}}(x^2)$, would eventually lead to ill-defined expressions (as we will see in Note 30 on page 341 and in Note 33 on page 343). In order to deal with this expression in a clean way, we then proceed as follows. We begin by recalling the identity, valid in the sense of distributions,

$$\lim_{\epsilon \to 0^+} \frac{1}{x \pm i\epsilon} = \text{P}\frac{1}{x} \mp i\pi\delta(x), \qquad (12.188)$$

where the symbol P denotes the principal part. We now introduce a time-like four-vector $\eta^\mu = \epsilon(1,0,0,0)$, with $\epsilon \to 0^+$, and we consider the combination $x^\mu - i\eta^\mu = (x^0 - i\epsilon, \mathbf{x})$. Then, to $\mathcal{O}(\epsilon)$,

$$\begin{aligned} (x - i\eta)^2 &= -(x^0 - i\epsilon)^2 + \mathbf{x}^2 \\ &= x^2 + 2i\epsilon x^0. \end{aligned} \qquad (12.189)$$

Since $\epsilon > 0$, $2\epsilon x^0$ is an infinitesimal quantity whose sign is the same as that of x^0. Therefore, from eq. (12.188),

$$\lim_{\eta \to 0^+} \frac{1}{(x - i\eta)^2} = \text{P}\frac{1}{x^2} - i\pi\epsilon(x^0)\delta(x^2), \qquad (12.190)$$

where the limit $\eta \to 0^+$ is defined by sending $\epsilon \to 0^+$ in $\eta^\mu = \epsilon(1,0,0,0)$, and $\epsilon(x^0)$ (not to be confused with the infinitesimal constant ϵ; both notations are standard and we do not change either of them) is the sign of x^0, i.e., $\epsilon(x^0) = \theta(x^0) - \theta(-x^0)$. Similarly, to $\mathcal{O}(\epsilon)$, $(x + i\eta)^2 = x^2 - 2i\epsilon x^0$, and therefore

$$\lim_{\eta \to 0^+} \frac{1}{(x + i\eta)^2} = \text{P}\frac{1}{x^2} + i\pi\epsilon(x^0)\delta(x^2). \qquad (12.191)$$

Taking the difference of eqs. (12.191) and (12.190), the principal part cancels and we get

$$\lim_{\eta \to 0^+} \left[\frac{1}{(x + i\eta)^2} - \frac{1}{(x - i\eta)^2} \right] = 2i\pi\epsilon(x^0)\delta(x^2). \qquad (12.192)$$

Therefore, the combination $\epsilon(x^0)\delta(x^2)$ has the representation

$$\epsilon(x^0)\delta(x^2) = -\frac{i}{2\pi} \lim_{\eta \to 0^+} \left[\frac{1}{(x + i\eta)^2} - \frac{1}{(x - i\eta)^2} \right]. \qquad (12.193)$$

We will then use this expression, with finite η, as a regularization of the whole combination $\epsilon(x^0)\delta(x^2)$, defining

$$\left[\epsilon(x^0)\delta(x^2) \right]_{\text{reg}} = -\frac{i}{2\pi} \left[\frac{1}{(x + i\eta)^2} - \frac{1}{(x - i\eta)^2} \right], \qquad (12.194)$$

and we will then take the limit $\eta \to 0^+$ only at the end of the computation. Note that this regularization formally breaks Lorentz invariance, since, if the four-vector η^μ has the form $\eta^\mu = \epsilon(1,0,0,0)$ in a reference frame, it will have a different form in a boosted frame. Therefore, the previous computation implicitly assumes a given reference frame. However, in the limit $\epsilon \to 0^+$, we expect Lorentz invariance to be recovered. We will check this point in the following.

From eq. (12.187), the regularized Green's function $G_A(x)$ can then be written as

$$
\begin{aligned}
G_A(x) &= -\frac{1}{4\pi} \left[\epsilon(x^0) \delta(x^2) \right]_{\text{reg}} \\
&= \frac{i}{8\pi^2} \left[\frac{1}{(x+i\eta)^2} - \frac{1}{(x-i\eta)^2} \right].
\end{aligned} \quad (12.195)
$$

[28] Observe that we are using two different regulators for G_S and for G_A. This is justified because the computation of the contribution from G_S and from G_A are completely independent, and the former will turn out to be divergent, while the latter will be finite when we send the regulator to zero.

We are now ready to carry out the computation.[28] We consider first the contribution of G_S to $F_{a,\text{self}}^{\mu\nu}[x_a(\tau)]$, defining

$$
\begin{aligned}
F_{a,S}^{\mu\nu}[x_a(\tau)] &\equiv -\frac{q_a}{\epsilon_0 c} \left(\eta^{\mu\beta} \eta^{\nu\alpha} - \eta^{\nu\beta} \eta^{\mu\alpha} \right) \\
&\quad \times \int_{-\infty}^{\infty} d\tau'\, u_{a,\alpha}(\tau') \left(\partial_\beta G_S[x - x_a(\tau')] \right)_{x=x_a(\tau)},
\end{aligned} \quad (12.196)
$$

where G_S is the regularized Green's function (12.185). From eq. (12.185), $G_S(x-x')$ is actually a function only of the combination $w = (x-x')^2$, so

$$
\begin{aligned}
\partial_\beta G_S(x-x') &= \left[\partial_\beta (x-x')^2 \right] \frac{d}{dw} G_S(w) \\
&= -\frac{1}{2\pi} (x-x')_\beta \frac{d\delta_{\text{reg}}(w)}{dw},
\end{aligned} \quad (12.197)
$$

and therefore

$$
\begin{aligned}
\left(\partial_\beta G_S[x - x_a(\tau')] \right)_{x=x_a(\tau)} &= -\frac{1}{2\pi} \left[x_a(\tau) - x_a(\tau') \right]_\beta \\
&\quad \times \left(\frac{d\delta_{\text{reg}}(w)}{dw} \right) \Big|_{w=w_a(\tau,\tau')},
\end{aligned} \quad (12.198)
$$

where $w_a(\tau,\tau') \equiv [x_a(\tau) - x_a(\tau')]^2$. Then,

$$
\begin{aligned}
F_{a,S}^{\mu\nu}[x_a(\tau)] &= -\frac{q_a}{2\pi\epsilon_0 c} \left(\eta^{\mu\beta} \eta^{\nu\alpha} - \eta^{\nu\beta} \eta^{\mu\alpha} \right) \\
&\quad \times \int_{-\infty}^{\infty} d\tau'\, u_{a,\alpha}(\tau') \left[x_a(\tau') - x_a(\tau) \right]_\beta \left(\frac{d\delta_{\text{reg}}(w)}{dw} \right) \Big|_{w=w_a(\tau,\tau')}.
\end{aligned} \quad (12.199)
$$

We now expand the integrand around $\tau' = \tau$. Even if the integration is over $\tau' \in [-\infty, \infty]$, we will see in a moment that, when we remove the regularization and $\delta_{\text{reg}}(w) \to \delta(w)$, only a finite number of terms in the expansion (in fact, only one term) gives a non-vanishing contribution, so the expansion will provide us with the full answer. Writing $\tau' = \tau + \sigma$, we have

$$
u_{a,\alpha}(\tau') = u_{a,\alpha}(\tau) + \sigma \dot{u}_{a,\alpha}(\tau) + \frac{1}{2} \sigma^2 \ddot{u}_{a,\alpha}(\tau) + \mathcal{O}(\sigma^3), \quad (12.200)
$$

and

$$x_{a,\beta}(\tau+\sigma) - x_{a,\beta}(\tau) = \sigma u_{a,\beta}(\tau) + \frac{1}{2}\sigma^2 \dot{u}_{a,\beta}(\tau) + \frac{1}{6}\sigma^3 \ddot{u}_{a,\beta}(\tau) + \mathcal{O}(\sigma^4).$$
(12.201)

Then,

$$(\eta^{\mu\beta}\eta^{\nu\alpha} - \eta^{\nu\beta}\eta^{\mu\alpha})u_{a,\alpha}(\tau')\,[x_a(\tau') - x_a(\tau)]_\beta$$
$$= \frac{1}{2}\sigma^2\,[u_a^\mu(\tau)\dot{u}_a^\nu(\tau) - u_a^\nu(\tau)\dot{u}_a^\mu(\tau)] + \frac{1}{3}\sigma^3\,[u_a^\mu(\tau)\ddot{u}_a^\nu(\tau) - u_a^\nu(\tau)\ddot{u}_a^\mu(\tau)]$$
$$+\mathcal{O}(\sigma^4),$$
(12.202)

where we used $u_\mu u^\mu = -c^2$ and $u_\mu \dot{u}^\mu = 0$ (which follows applying $d/d\tau$ to $u_\mu u^\mu = -c^2$). Note that the term linear in σ canceled because it is proportional to $u_{a,\alpha}(\tau)u_{a,\beta}(\tau)$ which, upon contraction with $(\eta^{\mu\beta}\eta^{\nu\alpha} - \eta^{\nu\beta}\eta^{\mu\alpha})$, gives zero.

To expand the derivative of the Dirac delta in eq. (12.199) we observe, from eq. (12.201), that

$$w_a(\tau,\tau') = -c^2\sigma^2 + \mathcal{O}(\sigma^4).$$
(12.203)

Then, apart from term that will give a vanishing contribution when we remove the regularization,[29]

$$\left(\frac{d\delta_{\mathrm{reg}}(w)}{dw}\right)\Big|_{w=w_a(\tau,\tau')} = \left(\frac{d\delta_{\mathrm{reg}}(w)}{dw}\right)\Big|_{w=-c^2\sigma^2}.$$
(12.204)

We now introduce $z = -w/c^2$ and we use the properties $\delta(-z) = \delta(z)$ and $\delta(c^2 z) = (1/c^2)\delta(z)$ that, apart at most for terms that vanish as we remove the regularization, also hold for the regularized Dirac delta [independently of the specific choice (12.186)]. We then get

$$\left(\frac{d\delta_{\mathrm{reg}}(w)}{dw}\right)\Big|_{w=w_a(\tau,\tau')} = -\frac{1}{c^4}\left(\frac{d\delta_{\mathrm{reg}}(z)}{dz}\right)\Big|_{z=\sigma^2}.$$
(12.205)

Since τ is fixed, in eq. (12.199) the integral in $d\tau'$ is the same as an integral in $d\sigma$. Then, inserting eqs. (12.202) and (12.205) into eq. (12.199), we are left with the evaluation of integrals of the form

$$A_n = \int_{-\infty}^{\infty} d\sigma\,\sigma^n \left(\frac{d\delta_{\mathrm{reg}}(z)}{dz}\right)\Big|_{z=\sigma^2},$$
(12.206)

for $n \geq 2$. For n odd, these integrals vanish by parity since $d\delta_{\mathrm{reg}}(z)/dz$ is evaluated in $z = \sigma^2$, so at a point that does not change under $\sigma \to -\sigma$. For n even, we write

$$A_n = 2\int_0^\infty d\sigma\,\sigma^n \left(\frac{d\delta_{\mathrm{reg}}(z)}{dz}\right)\Big|_{z=\sigma^2},$$
(12.207)

since the integrand is even in σ. It is now necessary to carry out the integral using an explicit regularization of the Dirac delta and remove the regularization only afterwards. If, instead, we removed the regularization inside the integral, we would end up with a derivative of the Dirac

[29]The vanishing of these terms is a consequence of the property (1.60) of the Dirac delta, so that, for instance $\delta(x + x^2) = \delta(x)/|1 + x|$, which is the same as $\delta(x)$. For the regularized Dirac delta, we can write $\delta_{\mathrm{reg}}(x+x^2) = \delta_{\mathrm{reg}}(x)/[1 + \mathcal{O}(x)]$. One can then check the effect of these corrections on the final result, writing

$$\left(\frac{d\delta_{\mathrm{reg}}(w)}{dw}\right)\Big|_{w=-c^2\sigma^2+\mathcal{O}(\sigma^4)}$$
$$= \frac{1}{1+\mathcal{O}(\sigma^2)}\left(\frac{d\delta_{\mathrm{reg}}(w)}{dw}\right)\Big|_{w=-c^2\sigma^2}.$$

Then, repeating the computation that we will carry out below, in eq. (12.208), one would find that σ^{n+2} is replaced by $\sigma^{n+2}/[1 + \mathcal{O}(\sigma^2)]$, and, in the second line of eq. (12.208), we would get

$$-\frac{a^{(n-3)/2}}{\sqrt{2\pi}}\int_0^\infty du\,\frac{u^{(n+1)/2}}{1+c_1 au}e^{-u^2/2},$$

for some constant c_1. We would then confirm that, in the limit $a \to 0$, all terms with $n \geq 4$ are vanishing, and the term $n = 2$ still gives the same divergent contribution proportional to $a^{-1/2}$, and even the same finite part, since $a^{-1/2}[1 + \mathcal{O}(a)]$ goes to $a^{-1/2}$. Analogous considerations hold for the manipulations leading to eq. (12.205).

delta evaluated at the boundary of the integration domain, which is an ill-defined quantity. Using, for definiteness, the regularization (12.186),

$$
\begin{aligned}
A_n &= -\frac{2}{\sqrt{2\pi}\,a^3}\int_0^\infty d\sigma\,\sigma^{n+2}e^{-\sigma^4/(2a^2)} \\
&= -\frac{1}{\sqrt{2\pi}}a^{(n-3)/2}\int_0^\infty du\,u^{(n+1)/2}e^{-u^2/2}\,, \qquad (12.208)
\end{aligned}
$$

where, in the second line, we introduced $u = \sigma^2/a$. We now observe that, among the values of n even and such that $n \geq 2$, in the limit $a \to 0$, A_n diverges for $n = 2$, while it vanishes for all $n \geq 4$. We also note that the regularization parameter a has the same dimensions as $z = -(x - x')^2/c^2$, so is a length squared over c^2. To make contact with our previous notation, we then write it as $a = \ell^2/c^2$, where ℓ is a regularization parameter with dimensions of length, so removing the regularization corresponds to sending $\ell \to 0$. Then, from eq. (12.208), $A_2 = a_2 c/\ell$, where a_2 is a dimensionless constant (whose precise value, in this case, is $a_2 = -2^{1/4}\Gamma(5/4)/\sqrt{2\pi}$, but has no special meaning since it depends on the regularization used).

Therefore, since only $n = 2$ contributes, the result for $F_{a,S}^{\mu\nu}[x_a(\tau)]$, to all orders in the expansion in σ, is

$$
\boxed{F_{a,S}^{\mu\nu}[x_a(\tau)] = \frac{q_a}{4\pi\epsilon_0 c^4}\frac{a_2}{\ell}\left[u_a^\mu(\tau)\dot{u}_a^\nu(\tau) - u_a^\nu(\tau)\dot{u}_a^\mu(\tau)\right]\,,} \qquad (12.209)
$$

apart from terms that vanish as $\ell \to 0$. We now insert this into the equation of motion (12.172), and use the identities $u_a^\nu(\tau)u_{a,\nu}(\tau) = -c^2$ and $\dot{u}_a^\nu(\tau)u_{a,\nu}(\tau) = 0$. Then, we get

$$
\begin{aligned}
m_{a,0}(\ell)\frac{du_a^\mu(\tau)}{d\tau} &= \frac{q_a^2}{4\pi\epsilon_0 c^2}\frac{a_2}{\ell}\frac{du_a^\mu}{d\tau} + q_a F_{a,A}^{\mu\nu}[x_a(\tau)]u_{a,\nu}(\tau) \\
&\quad + q_a F_{\text{ext}}^{\mu\nu}[x_a(\tau)]u_{a,\nu}(\tau)\,. \qquad (12.210)
\end{aligned}
$$

We can rewrite this as

$$
\begin{aligned}
\left[m_{a,0}(\ell) - \frac{q_a^2}{4\pi\epsilon_0 c^2}\frac{a_2}{\ell}\right]\frac{du^\mu}{d\tau} &= q_a F_{a,A}^{\mu\nu}[x_a(\tau)]u_{a,\nu}(\tau) \\
&\quad + q_a F_{\text{ext}}^{\mu\nu}[x_a(\tau)]u_{a,\nu}(\tau)\,, \qquad (12.211)
\end{aligned}
$$

and we see that the divergence is reabsorbed in a renormalization of the mass,

$$
m_a = m_{a,0}(\ell) - \frac{q_a^2}{4\pi\epsilon_0 c^2}\frac{a_2}{\ell}\,, \qquad (12.212)
$$

so that we simply have

$$
\frac{dp^\mu}{d\tau} = q_a F_{a,A}^{\mu\nu}[x_a(\tau)]u_{a,\nu}(\tau) + q_a F_{\text{ext}}^{\mu\nu}[x_a(\tau)]u_{a,\nu}(\tau)\,, \qquad (12.213)
$$

with $p^\mu = m_a u^\mu$. Note that, having preserved Lorentz invariance in the regularization, the spatial and temporal components of p^μ renormalize in the same way, and the divergence is reabsorbed into a single bare mass term.

We next turn our attention to the contribution from $F_{a,A}^{\mu\nu}$, which depends on the anti-symmetric combination (12.187) of the retarded and advanced Green's functions. We define

$$F_{a,A}^{\mu\nu}[x_a(\tau)] \equiv -\frac{q_a}{\epsilon_0 c} (\eta^{\mu\beta}\eta^{\nu\alpha} - \eta^{\nu\beta}\eta^{\mu\alpha}) \tag{12.214}$$

$$\times \int_{-\infty}^{\infty} d\tau' \, u_{a,\alpha}(\tau') \, (\partial_\beta G_A[x - x_a(\tau')])_{x=x_a(\tau)} \,,$$

where G_A is the regularized Green's function (12.195). Then

$$\partial_\beta G_A(x) = -\frac{i}{4\pi^2} \left[\frac{(x+i\eta)_\beta}{[(x+i\eta)^2]^2} - \frac{(x-i\eta)_\beta}{[(x-i\eta)^2]^2} \right]. \tag{12.215}$$

We now observe that, as in eq. (12.189), to first order in ϵ

$$[(x-i\eta)^2]^2 = (x^2 + 2i\epsilon x^0)^2$$
$$= x^2(x^2 + 4i\epsilon x^0)$$
$$= x^2(x - 2i\eta)^2 \,, \tag{12.216}$$

and similarly for $[(x+i\eta)^2]^2$. Then, apart from terms $\mathcal{O}(\eta)$ that will eventually give no contribution when we eventually take the limit $\eta \to 0^+$ (and redefining $\eta \to \eta/2$, given that anyhow it is a parameter used only to take the limit $\eta \to 0^+$),

$$\partial_\beta G_A(x) = -\frac{i}{4\pi^2} \frac{x_\beta}{x^2} \left[\frac{1}{(x+i\eta)^2} - \frac{1}{(x-i\eta)^2} \right]. \tag{12.217}$$

Comparing with eq. (12.194), we see that

$$\partial_\beta G_A(x) = \frac{1}{2\pi} \frac{x_\beta}{x^2} \left[\epsilon(x^0)\delta(x^2) \right]_{\text{reg}}. \tag{12.218}$$

If we remove the regularization at this stage, we therefore get[30]

$$\partial_\beta G_A(x) = \frac{1}{2\pi} \frac{x_\beta}{x^2} \epsilon(x^0)\delta(x^2). \tag{12.221}$$

However, we still need to keep $\partial_\beta G_A$ in the regularized form (12.217), in order to properly carry out the integral over τ' in eq. (12.214). In eq. (12.214) we need $\partial_\beta G_A(x)$ evaluated in $x = x_a(\tau) - x_a(\tau')$, that we prefer to write as $-[x_a(\tau') - x_a(\tau)]$. Equation (12.214) then becomes

$$F_{a,A}^{\mu\nu}[x_a(\tau)] = -\frac{q_a}{\epsilon_0 c} \frac{i}{4\pi^2} \tag{12.222}$$

$$\times \int_{-\infty}^{\infty} d\tau' \, (\eta^{\mu\beta}\eta^{\nu\alpha} - \eta^{\nu\beta}\eta^{\mu\alpha}) u_{a,\alpha}(\tau') \, [x_a(\tau') - x_a(\tau)]_\beta$$

$$\frac{1}{[x_a(\tau') - x_a(\tau)]^2} \left[\frac{1}{(x_a(\tau') - x_a(\tau) - i\eta)^2} - \frac{1}{(x_a(\tau') - x_a(\tau) + i\eta)^2} \right].$$

[30]Observe that, if one removes the regularization already in eq. (12.195), writing formally

$$G_A(x) = -\frac{1}{4\pi} \epsilon(x^0)\delta(x^2) \,,$$

and attempts to compute $\partial_\beta G_A$ from this expression, one is confronted with ill-defined expressions, and it is not possible to get eq. (12.221) in any clean way. A regularization of the full combination $\epsilon(x^0)\delta(x^2)$ is necessary to perform a proper computation. As a check of eq. (12.218) observe, from eq. (12.195), that it can be rewritten as

$$\partial_\beta G_A = -\frac{2x_\beta}{x^2} G_A. \tag{12.219}$$

Then, using $\partial^\beta x_\beta = 4$,

$$\partial^\beta \partial_\beta G_A = -\frac{8G_A}{x^2} + 4x_\beta \frac{x^\beta}{(x^2)^2} G_A$$
$$- 2\frac{x_\beta}{x^2} \left(-2\frac{x^\beta}{x^2} G_A \right)$$
$$= -8\frac{G_A}{x^2} + \frac{4G_A}{x^2} + \frac{4G_A}{x^2} = 0. \tag{12.220}$$

Therefore $\Box G_A = 0$, as it should, since G_A is the difference of two Green's functions. Note that the above manipulations become ill-defined at $x^2 = 0$. However, the validity of $\Box G_A = 0$ for all values of x^μ, including when $x^2 = 0$, can be proven using the regularized expression (12.215), and sending $\eta \to 0^+$ only at the end.

We now write $\tau' = \tau + \sigma$ and we expand in powers of σ. The expansion of the term in the second line is given by eq. (12.202). To expand the term in the third line, we begin by observing that, working up to $\mathcal{O}(\sigma^2)$

$$
\begin{aligned}
[x_a(\tau') - x_a(\tau) - i\eta]^2 &= (\sigma u_a(\tau) - i\eta)^2 \\
&= -c^2(\sigma^2 - 2i\epsilon\sigma u^0/c^2). \quad (12.223)
\end{aligned}
$$

For a physical trajectory $u^0 = dx^0/d\tau > 0$, and therefore, defining $\epsilon' = 2\epsilon u^0/c^2$ (so that $\epsilon \to 0^+$ corresponds to $\epsilon' \to 0^+$), and finally renaming ϵ' as ϵ,

$$
\begin{aligned}
&\frac{1}{(x_a(\tau') - x_a(\tau) - i\eta)^2} - \frac{1}{(x_a(\tau') - x_a(\tau) + i\eta)^2} \\
&= -\frac{1}{c^2}\left(\frac{1}{\sigma^2 - i\epsilon\sigma} - \frac{1}{\sigma^2 + i\epsilon\sigma}\right) \\
&= -\frac{2i\pi}{c^2}\left[\epsilon(\sigma)\delta(\sigma^2)\right]_{\text{reg}}, \quad (12.224)
\end{aligned}
$$

where, with the same steps leading to eq. (12.194),

$$
\left[\epsilon(\sigma)\delta(\sigma^2)\right]_{\text{reg}} = -\frac{i}{2\pi}\left[\frac{1}{\sigma^2 - i\epsilon\sigma} - \frac{1}{\sigma^2 + i\epsilon\sigma}\right]. \quad (12.225)
$$

Finally, $[x_a(\tau') - x_a(\tau)]^2 = -c^2\sigma^2 + \mathcal{O}(\sigma^4)$. Then, eq. (12.222) becomes

$$
\begin{aligned}
F_{a,A}^{\mu\nu}[x_a(\tau)] &= -\frac{q_a}{\epsilon_0 c}\frac{i}{4\pi^2}\int_{-\infty}^{\infty} d\sigma \left\{\frac{1}{2}\sigma^2\left[u_a^\mu(\tau)\dot{u}_a^\nu(\tau) - u_a^\nu(\tau)\dot{u}_a^\mu(\tau)\right]\right. \\
&\quad + \frac{1}{3}\sigma^3\left[u_a^\mu(\tau)\ddot{u}_a^\nu(\tau) - u_a^\nu(\tau)\ddot{u}_a^\mu(\tau)\right] + \mathcal{O}(\sigma^4)\right\} \\
&\quad \times \frac{2i\pi}{c^4}\frac{1}{\sigma^2 + \mathcal{O}(\sigma^4)}\left[\epsilon(\sigma)\delta(\sigma^2)\right]_{\text{reg}} \\
&= \frac{q_a}{4\pi\epsilon_0 c}\frac{1}{c^4}\int_{-\infty}^{\infty} d\sigma \left[\epsilon(\sigma)\delta(\sigma^2)\right]_{\text{reg}}\left\{\left[u_a^\mu(\tau)\dot{u}_a^\nu(\tau) - u_a^\nu(\tau)\dot{u}_a^\mu(\tau)\right]\right. \\
&\quad \left. + \frac{2}{3}\sigma\left[u_a^\mu(\tau)\ddot{u}_a^\nu(\tau) - u_a^\nu(\tau)\ddot{u}_a^\mu(\tau)\right] + \mathcal{O}(\sigma^2)\right\}. \quad (12.226)
\end{aligned}
$$

We are therefore led to evaluate the integrals

$$
B_n = \int_{-\infty}^{\infty} d\sigma\, \sigma^n \left[\epsilon(\sigma)\delta(\sigma^2)\right]_{\text{reg}}, \quad (12.227)
$$

with $n \geq 0$. From eq. (12.225), $\left[\epsilon(\sigma)\delta(\sigma^2)\right]_{\text{reg}}$ is an odd function of σ, and therefore, B_n vanishes for n even. For n odd and $n \geq 3$ we can remove the regularization, replacing $\left[\epsilon(\sigma)\delta(\sigma^2)\right]_{\text{reg}}$ by $\epsilon(\sigma)\delta(\sigma^2)$, and we get $B_n = 0$.[31] Therefore, the only finite contribution comes from $n = 1$,

---

[31] This can be shown writing $\delta(\sigma^2) = \delta(\sigma)/(2|\sigma|)$ and using $\epsilon(\sigma)|\sigma| = \sigma$. Then, for n odd and $n \geq 3$,

$$
B_n = \frac{1}{2}\int_{-\infty}^{\infty} d\sigma\, \sigma^{n-1}\delta(\sigma), \quad (12.228)
$$

and all these integral vanish.

and is given by

$$
\begin{aligned}
B_1 &= \int_{-\infty}^{\infty} d\sigma \, \sigma \, \left[\epsilon(\sigma)\delta(\sigma^2) \right]_{\mathrm{reg}} \\
&= 2 \int_0^{\infty} d\sigma \, \sigma \left(-\frac{i}{2\pi} \right) \left(\frac{1}{\sigma^2 - i\epsilon\sigma} - \frac{1}{\sigma^2 + i\epsilon\sigma} \right) \\
&= -\frac{i}{\pi} \int_0^{\infty} d\sigma \, \sigma \, \frac{2i\epsilon\sigma}{\sigma^4 + \epsilon^2\sigma^2} \\
&= \frac{2}{\pi} \epsilon \int_0^{\infty} d\sigma \, \frac{1}{\sigma^2 + \epsilon^2} .
\end{aligned}
\tag{12.229}
$$

Writing $\sigma = \epsilon t$, ϵ cancels and

$$
\begin{aligned}
B_1 &= \frac{2}{\pi} \int_0^{\infty} dt \, \frac{1}{t^2 + 1} \\
&= 1 .
\end{aligned}
\tag{12.230}
$$

Therefore, eq. (12.226) gives the exact result for $F_{a,A}^{\mu\nu}[x_a(\tau)]$,

$$
\boxed{\; F_{a,A}^{\mu\nu}[x_a(\tau)] = \frac{q_a}{4\pi\epsilon_0 c}\frac{1}{c^4}\frac{2}{3} \left[u_a^{\mu}(\tau)\ddot{u}_a^{\nu}(\tau) - u_a^{\nu}(\tau)\ddot{u}_a^{\mu}(\tau) \right] . \;}
\tag{12.231}
$$

Note that this result is finite, contrary to $F_{a,A}^{\mu\nu}[x_a(\tau)]$ that was divergent, see eq. (12.209). Also observe that the result is fully covariant for the finite part as well, so, even if our regularization (12.194) breaks Lorentz invariance through the introduction of a fixed four-vector $\eta^{\mu} = \epsilon(1,0,0,0)$, Lorentz invariance is recovered as $\epsilon \to 0^+$, i.e., when the regularization is removed.[32],[33] We now plug this into eq. (12.210) and we observe that, from $u^{\nu}u_{\nu} = -c^2$ it follow, by taking a derivative with respect to τ, that $u^{\nu}\dot{u}_{\nu} = 0$ and, taking one more derivative, $u^{\nu}\ddot{u}_{\nu} = -\dot{u}^{\nu}\dot{u}_{\nu} \equiv -\dot{u}^2$. Then

$$
F_{a,A}^{\mu\nu}[x_a(\tau)]u_{a,\nu}(\tau) = \frac{q_a}{4\pi\epsilon_0}\frac{2}{3c^3} \left[\ddot{u}_a^{\mu}(\tau) - \frac{\dot{u}^2(\tau)}{c^2}u^{\mu}(\tau) \right] ,
\tag{12.232}
$$

and eq. (12.213) becomes

$$
\boxed{\; \frac{dp^{\mu}}{d\tau} = \frac{1}{4\pi\epsilon_0}\frac{2q_a^2}{3c^3} \left[\ddot{u}_a^{\mu}(\tau) - \frac{\dot{u}_a^2(\tau)}{c^2}u_a^{\mu}(\tau) \right] + q_a F_{\mathrm{ext}}^{\mu\nu}[x_a(\tau)]u_{a,\nu}(\tau) . \;}
\tag{12.233}
$$

We have therefore recovered the ALD equation (12.149). The derivation of this section, although technically more involved, has the advantage of unifying the results of the previous sections in a single, conceptually clean, framework. Mass renormalization comes from the symmetric combination of the advanced and retarded Green's functions, which gives a divergent contribution, so it requires a regularization (that, for the symmetric combination of Green's functions, can be performed in a fully Lorentz-invariant manner), while the self-force term in the ALD equation is due the anti-symmetric combination of the advanced and retarded Green's functions, and is a finite contribution.

[32]This is an unavoidable consequence of the fact that the result that we have found for $F_{a,A}^{\mu\nu}[x_a(\tau)]$ is finite. In quantum field theory, there are situations, known as *anomalies*, where a symmetry broken by the regularization is not recovered when the regularization is removed, but this only happens when factors of order ϵ describing the correction to the result when the cutoff is removed combine with terms of order $1/\epsilon$ from divergences, to produce a finite result.

[33]It is also interesting to observe that, if we had not been careful to regularize the product $\epsilon(x^0)\delta(x^2)$, instead of eq. (12.229) we would have now found the formal expression $B_1 = \int_{-\infty}^{\infty} d\sigma \, \sigma \, \epsilon(\sigma)\delta(\sigma^2)$. In the literature, this has been manipulated writing it as $B_1 = 2\int_0^{\infty} d\sigma \, \sigma \, \delta(\sigma^2)$, since the integrand is even and, for $\sigma > 0$, $\epsilon(\sigma) = 1$. For $\sigma > 0$ we also have $\delta(\sigma^2) = (1/2\sigma)\delta(\sigma)$, so we end up with $B_1 = \int_0^{\infty} d\sigma \, \delta(\sigma)$. However, this integral is ill-defined, and has been typically dealt with by introducing a prescription on $\delta(\sigma)$ only at this stage, such as replacing it with $\delta(\sigma - \epsilon)$, with $\epsilon > 0$, so that the integral becomes equal to one. At this level, such a prescription is completely arbitrary and is only motivated by the desire to obtain the ALD equation with the correct coefficient $2/3$, that we already know from the derivation performed using the the covariantization (or from the non-relativistic limit (12.126), where it is just the factor $2/3$ that appears in Larmor's formula). Our computation shows how the correct prescription emerges automatically by regularizing the product $\epsilon(x^0)\delta(x^2)$ and always working with regularized quantities.

Electromagnetic fields in material media

Until now we have studied the equations of electromagnetism "in vacuum," in the sense that the source terms, when they were not absent altogether (as when we studied the propagation of electromagnetic waves), were due to localized charge distributions, often even idealized as point charges, and we were interested in the electromagnetic field outside the sources. We now begin our study of electromagnetic fields in materials. At a microscopic level, the electromagnetic field has large variations, for instance near atoms in a solid, and it depends on a myriad of details. However, to understand the behavior of a material on larger scales, we do not need to know all these short-distance details. For instance, in a typical solid, there will be large fluctuations of the electromagnetic field at the atomic scale, i.e., at the scale of the Bohr radius, $r_B \sim 10^{-8}$ cm. However, we are typically interested in the collective behavior of the system on a macroscopic scale, say, for instance, 1 cm. We can then choose a scale L intermediate between the atomic and the macroscopic scale, for instance $L \sim 100 r_B \sim 10^{-6}$ cm, and average the electric and magnetic fields over such a scale. This results in "smoothed-out" fields, that govern the behavior of the material at such larger scales.

This is a standard approach in physics: in general, it would be impossible to describe any physical system if we needed to know all details of what happens at short-distance scales. Furthermore, as we go to shorter and shorter scales, we encounter new phenomena and new physical laws; for instance, at some point, going toward the atomic scale, classical physics must give way to quantum mechanics, and if we look even closer and closer into atoms and nuclei we enter the regime of relativistic quantum field theory, we discover new interactions, and so on. The basic approach is therefore to develop an effective theory, valid at the length-scales at which we are interested, in which all short-distance details are taken into account in an "average" manner. Another important aspect is that our ignorance of the physical laws at short scales should be encoded into an effective description, involving a few phenomenological parameters that can be fixed by comparison with experiments; it is true that, for describing material media, we cannot really dispense with quantum mechanics (we have seen for instance in Problem 10.2 that a classical model of the atom miserably fails, collapsing in about 10^{-11} s because of emission of electromagnetic radiation); still, at least at a first level of analysis, we do not want to enter into all the details of the quantum-mechanical interactions at short distances, and we want to have an effective classical description of the property of a material.

13.1   Maxwell's equations for macroscopic fields

As a first step let us see how, from the fundamental, "microscopic", Maxwell's equations, we can derive equations governing the behavior of smoothed fields. We denote by $\mathbf{E}_{\text{micro}}(t, \mathbf{x})$ and $\mathbf{B}_{\text{micro}}(t, \mathbf{x})$ the microscopic electric and magnetic field, respectively, and by $\rho_{\text{micro}}(t, \mathbf{x})$ and $\mathbf{j}_{\text{micro}}(t, \mathbf{x})$ the microscopic charge and current densities. In this notation, Maxwell's equations (3.8)–(3.11) read

$$\nabla \cdot \mathbf{E}_{\text{micro}} = \frac{1}{\epsilon_0} \rho_{\text{micro}}, \tag{13.1}$$

$$\nabla \times \mathbf{B}_{\text{micro}} - \frac{1}{c^2} \frac{\partial \mathbf{E}_{\text{micro}}}{\partial t} = \mu_0 \mathbf{j}_{\text{micro}}, \tag{13.2}$$

$$\nabla \cdot \mathbf{B}_{\text{micro}} = 0, \tag{13.3}$$

$$\nabla \times \mathbf{E}_{\text{micro}} + \frac{\partial \mathbf{B}_{\text{micro}}}{\partial t} = 0. \tag{13.4}$$

We now smooth out these fields, performing a spatial average with the help of a smoothing function $s(\mathbf{x})$. The smoothed electric and magnetic fields, that we denote by $\mathbf{E}$ and $\mathbf{B}$, are defined by

$$\mathbf{E}(t, \mathbf{x}) = \int d^3x' \, s(\mathbf{x} - \mathbf{x}') \mathbf{E}_{\text{micro}}(t, \mathbf{x}'), \tag{13.5}$$

$$\mathbf{B}(t, \mathbf{x}) = \int d^3x' \, s(\mathbf{x} - \mathbf{x}') \mathbf{B}_{\text{micro}}(t, \mathbf{x}'). \tag{13.6}$$

The exact form of the function $s(\mathbf{x} - \mathbf{x}')$ is not important; the important point is that it is approximately constant for $|\mathbf{x} - \mathbf{x}'|$ smaller than the smoothing scale L, vanishes quickly for $|\mathbf{x} - \mathbf{x}'| \gg L$, and is a smooth function on the scale L. We normalize it by

$$\int d^3x' \, s(\mathbf{x}) = 1, \tag{13.7}$$

so that eqs. (13.5) and (13.6) represent an average of the microscopic fields over a region of linear size of order L, centered around the point $\mathbf{x}$. Note that, in terms of the spatial Fourier transform, eq. (13.5) reads

$$\tilde{\mathbf{E}}(t, \mathbf{k}) = \tilde{s}(\mathbf{k}) \tilde{\mathbf{E}}_{\text{micro}}(t, \mathbf{k}). \tag{13.8}$$

The fact that $s(\mathbf{x})$ is a smooth function of the spatial scale L means that its Fourier modes with $|\mathbf{k}| \gg 1/L$ are very small. Therefore, another way of thinking to this smoothing procedure is to say that the role of $\tilde{s}(\mathbf{k})$ is to suppress the Fourier modes with large $|\mathbf{k}|$ in $\tilde{\mathbf{E}}(t, \mathbf{k})$.

We similarly define the smoothed out, or macroscopic, charge and current density,

$$\rho(t, \mathbf{x}) = \int d^3x' \, s(\mathbf{x} - \mathbf{x}') \rho_{\text{micro}}(t, \mathbf{x}'), \qquad (13.9)$$

$$\mathbf{j}(t, \mathbf{x}) = \int d^3x' \, s(\mathbf{x} - \mathbf{x}') \mathbf{j}_{\text{micro}}(t, \mathbf{x}'). \qquad (13.10)$$

It is now straightforward to derive the equations satisfied by the macroscopic fields. Consider for instance eq. (13.1) and apply the smoothing to both sides,

$$\int d^3x' \, s(\mathbf{x} - \mathbf{x}') \boldsymbol{\nabla}_{\mathbf{x}'} \cdot \mathbf{E}_{\text{micro}}(t, \mathbf{x}') = \frac{1}{\epsilon_0} \int d^3x' \, s(\mathbf{x} - \mathbf{x}') \rho_{\text{micro}}(t, \mathbf{x}').$$
$$(13.11)$$

The integral on the right-hand side is just the macroscopic charge density $\rho(t, \mathbf{x})$. On the left-hand side, integrating by parts, we write

$$\int d^3x' s(\mathbf{x} - \mathbf{x}') \boldsymbol{\nabla}_{\mathbf{x}'} \cdot \mathbf{E}_{\text{micro}}(t, \mathbf{x}') = -\int d^3x' \left[\boldsymbol{\nabla}_{\mathbf{x}'} s(\mathbf{x} - \mathbf{x}') \right] \cdot \mathbf{E}_{\text{micro}}(t, \mathbf{x}')$$

$$= \int d^3x' \left[\boldsymbol{\nabla}_{\mathbf{x}} s(\mathbf{x} - \mathbf{x}') \right] \cdot \mathbf{E}_{\text{micro}}(t, \mathbf{x}')$$

$$= \boldsymbol{\nabla}_{\mathbf{x}} \cdot \int d^3x' \, s(\mathbf{x} - \mathbf{x}') \mathbf{E}_{\text{micro}}(t, \mathbf{x}')$$

$$= \boldsymbol{\nabla}_{\mathbf{x}} \cdot \mathbf{E}(t, \mathbf{x}). \qquad (13.12)$$

In other words, the averaging procedure commutes with the operation of taking the divergence. We therefore get the macroscopic Maxwell equation $\boldsymbol{\nabla}_{\mathbf{x}} \cdot \mathbf{E}(t, \mathbf{x}) = (1/\epsilon_0)\rho(t, \mathbf{x})$. The same manipulations can be performed on all other equations, so we finally find the full set of macroscopic Maxwell's equations

$$\boldsymbol{\nabla} \cdot \mathbf{E} = \frac{1}{\epsilon_0}\rho, \qquad (13.13)$$

$$\boldsymbol{\nabla} \times \mathbf{B} - \frac{1}{c^2}\frac{\partial \mathbf{E}}{\partial t} = \mu_0 \mathbf{j}, \qquad (13.14)$$

$$\boldsymbol{\nabla} \cdot \mathbf{B} = 0, \qquad (13.15)$$

$$\boldsymbol{\nabla} \times \mathbf{E} + \frac{\partial \mathbf{B}}{\partial t} = 0. \qquad (13.16)$$

The result looks extremely simple: the macroscopic electric and magnetic field satisfy the same Maxwell's equations as the microscopic fields, with the microscopic charge and current densities replaced by their macroscopic counterparts.[1] The simplicity of this result, however, is deceptive. In reality, to use these equations, we must have a model for the macroscopic charge and current densities, and this is where the real difficulty resides. The huge variety and complexity of materials in condensed matter physics emerges from the variety of behaviors of the macroscopic charge and current densities. A detailed study belongs to a course of condensed matter physics. However, a broad classification of the simplest situations is possible, and will be discussed in the following sections.

[1] As a historical note, this averaging procedure is due to Lorentz. However, his path was in the reverse order. He started from Maxwell's equations, as they had been empirically established at the macroscopic level, and realized that the theory could also be applied to the microscopic domain, with the macroscopic equations derived, through this averaging procedure, from the more fundamental microscopic equations.

13.2 The macroscopic charge density: free and bound charges

We begin with the macroscopic charge density. A first useful distinction, for the electric charges in a material, is between those that are free to move and those that are bound to atoms or molecules. In particular, in a metal, some of the electrons are free to move, and this is what gives them their conduction properties. In insulators, free charges can be externally added (by "doping" them), so also for them this distinction is useful.

Let us consider the effect of bound charges in a simple case such an ensemble of molecules, where the interaction between individual molecules is weak. Each molecule is electrically neutral; however, its electric dipole moment will in general be non-vanishing. We denote by $\mathbf{P}$ the electric dipole moment per unit volume, obtained averaging the individual dipoles of the molecules over a given macroscopic volume (i.e., a volume with a linear size L much larger than the atomic scale, but small compared to the scales at which we will measure the smoothed fields; e.g., $L \sim 100 r_B \sim 10^{-6}$ cm as in the discussion above). The vector $\mathbf{P}$ is known as the *polarization* vector. In general, $\mathbf{P}$ is a function of the point $\mathbf{x}$ around which we take a spatial average and, because of the smoothing implied by the spatial average, it only varies significantly on the scale L. From the multipole expansion (11.152), we see that a dipole moment $\mathbf{d}(t)$ located at the point $\mathbf{x}'$ generates in the point $\mathbf{x}$ a potential

$$\phi(t, \mathbf{x}) = \frac{1}{4\pi\epsilon_0} \frac{\mathbf{d}(t) \cdot (\mathbf{x} - \mathbf{x}')}{|\mathbf{x} - \mathbf{x}'|^3} . \tag{13.17}$$

Therefore, a distribution of electric dipoles with electric dipole moment density $\mathbf{P}(t, \mathbf{x})$ within a material body of finite volume V generates a potential[2]

$$\phi(t, \mathbf{x}) = \frac{1}{4\pi\epsilon_0} \int_V d^3 x' \frac{\mathbf{P}(t, \mathbf{x}') \cdot (\mathbf{x} - \mathbf{x}')}{|\mathbf{x} - \mathbf{x}'|^3} . \tag{13.18}$$

We now observe, from eq. (4.19), that

$$\frac{\mathbf{x} - \mathbf{x}'}{|\mathbf{x} - \mathbf{x}'|^3} = \boldsymbol{\nabla}_{\mathbf{x}'} \frac{1}{|\mathbf{x} - \mathbf{x}'|} . \tag{13.19}$$

Then, integrating by parts,

$$\begin{aligned}
\phi(t, \mathbf{x}) &= \frac{1}{4\pi\epsilon_0} \int_V d^3 x' \, \mathbf{P}(t, \mathbf{x}') \cdot \boldsymbol{\nabla}_{\mathbf{x}'} \frac{1}{|\mathbf{x} - \mathbf{x}'|} \tag{13.20} \\
&= -\frac{1}{4\pi\epsilon_0} \int_V d^3 x' \frac{\boldsymbol{\nabla}_{\mathbf{x}'} \cdot \mathbf{P}(t, \mathbf{x}')}{|\mathbf{x} - \mathbf{x}'|} + \frac{1}{4\pi\epsilon_0} \int_{\partial V} d^2 s' \frac{\mathbf{P}(t, \mathbf{x}')}{|\mathbf{x} - \mathbf{x}'|} .
\end{aligned}$$

Comparison with the second line of eq. (11.151) shows that a spatially-varying polarization vector generates a near-region field equal to that produced by a charge density

$$\boxed{\rho_{\text{pol}}(t, \mathbf{x}) = -\boldsymbol{\nabla} \cdot \mathbf{P}(t, \mathbf{x}) ,} \tag{13.21}$$

[2]Observe that eq. (11.152) is valid also in the near outer region, i.e., as long as we are outside the source. Therefore, given a small volume $d^3 x' \ll L^3$, which contains an electric dipole $\mathbf{P}(t, \mathbf{x}') d^3 x'$, it can be used to compute the field outside it and, in the limit in which $d^3 x'$ is actually infinitesimal, it gives the correct result everywhere. Note that, by construction, $\mathbf{P}(t, \mathbf{x}')$ is smooth already on a scale L, much larger than the linear size of the infinitesimal volume $d^3 x'$.

as well as a surface charge density, on the boundary of the material, given by

$$\sigma_{\rm pol} = \hat{\mathbf{n}}_s \cdot \mathbf{P}(t, \mathbf{x}) \,, \tag{13.22}$$

where $\hat{\mathbf{n}}_s$ is the outer normal to the surface element $d^2 s$. Since this charge density arises from electric dipoles, each of which is electrically neutral, if we integrate this total charge over a volume V, we must find that the net charge inside the volume plus that on the surface is zero. Indeed, using Gauss theorem,

$$
\begin{aligned}
\int_V d^3x \, \rho_{\rm pol}(t, \mathbf{x}) &= -\int_V d^3x \, \boldsymbol{\nabla} \cdot \mathbf{P}(t, \mathbf{x}) \\
&= -\int_{\partial V} d^2 s \, \hat{\mathbf{n}}_s \cdot \mathbf{P}(t, \mathbf{x}) \,,
\end{aligned}
\tag{13.23}
$$

and this is precisely minus the surface integral of the surface charge (13.22). These volume and surface charges are present, despite the fact that each dipole is neutral, because of the different way in which the dipoles arrange themselves spatially, as in the schematic example shown in Fig. 13.1.

13.3   The macroscopic current density

A similar analysis can be performed for the current density. In the simplest modelization, one writes

$$\mathbf{j} = \mathbf{j}_{\rm free} + \mathbf{j}_{\rm pol} + \mathbf{j}_{\rm mag} \,. \tag{13.24}$$

The term $\mathbf{j}_{\rm free}$ is the current carried by the electrons and ions that are not bound to each other and move under the action of electromagnetic fields, whose density is given by $\rho_{\rm free}$. It satisfies a separate conservation equation

$$\frac{\partial \rho_{\rm free}}{\partial t} + \boldsymbol{\nabla} \cdot \mathbf{j}_{\rm free} = 0 \,, \tag{13.25}$$

that expresses the fact that the variation in time of the density of free charges at a point $\mathbf{x}$, or, more precisely, in a small volume centered at $\mathbf{x}$, is given by the flux of free charges flowing through it.[3] In solids, $\mathbf{j}_{\rm free}$ is relevant in conductors and semiconductors.

The second term in eq. (13.24) is the current generated by a time-dependent polarization density. According to eq. (13.21),

$$\frac{\partial \rho_{\rm pol}}{\partial t} = -\boldsymbol{\nabla} \cdot \frac{\partial \mathbf{P}}{\partial t} \,, \tag{13.26}$$

Therefore, defining a current density

$$\mathbf{j}_{\rm pol}(t, \mathbf{x}) = \frac{\partial \mathbf{P}(t, \mathbf{x})}{\partial t} \,, \tag{13.27}$$

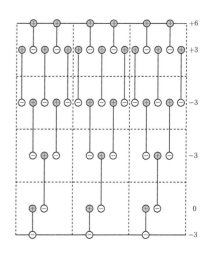

Fig. 13.1 A schematic example of a distribution of dipoles resulting in a volume charge density and a surface charge density. The dotted squares correspond to the volumes of size L used to smooth out the charge distribution. The gradient of the dipole distribution (here represented by the variation of the number of dipoles straddling through adjacent volumes in the vertical direction) induces a net charge in the volumes in the interior of the material (the sum over the net charges in each horizontal block of cell is marked explicitly), which is compensated by the surface charge. If $\boldsymbol{\nabla} \cdot \mathbf{P} = 0$ (which, in the above picture, can be obtained if a constant number of dipoles straddles across two neighboring vertical cells), then the charges inside the volume compensate exactly inside each cell, and we only have a surface charge, that integrates to zero.

[3] At the mathematical level, eq. (13.25) is a local equation, so one could take an infinitesimally small volume. Recall, however, that the whole macroscopic description implies a coarse graining over volumes larger that the atomic size, say over regions of size $L \sim 10^2 r_B$, as discussed at the beginning of this chapter. Therefore, equations such as (13.25) and all similar equations below, must always be understood as smoothed over such small volumes.

we also have the separate conservation equation

$$\frac{\partial \rho_{\text{pol}}}{\partial t} + \boldsymbol{\nabla}{\cdot}\mathbf{j}_{\text{pol}} = 0\,. \tag{13.28}$$

Physically, $\mathbf{j}_{\text{pol}}$ describes the fact that the configuration of dipoles can evolve in time, and then the charge density associated with it, as the one shown in Fig. 13.1, evolves with time and therefore generates a current.

The third term, called the *magnetization current*, is due to the density of magnetic dipoles in the materials, which, at the microscopic level, can be due either to the motion of the electrons in atoms, or to the magnetic moments associated with the spin of the electrons and of the nuclei. Its treatment is analogous to that of the density of electric dipole in the previous subsection. To compute the vector potential generated in the near zone by a magnetic dipole at the origin we use eq. (11.154). Then, the vector potential generated in the near zone by a magnetic dipole density $\mathbf{M}(t, \mathbf{x})$ (also called the *magnetization*) in a material of finite volume V is given by

$$
\begin{aligned}
\mathbf{A}(t, \mathbf{x}) &= \frac{\mu_0}{4\pi} \int_V d^3x'\, \frac{\mathbf{M}(t, \mathbf{x}') \times (\mathbf{x} - \mathbf{x}')}{|\mathbf{x} - \mathbf{x}'|^3} \\
&= \frac{\mu_0}{4\pi} \int_V d^3x'\, \mathbf{M}(t, \mathbf{x}') \times \boldsymbol{\nabla}_{\mathbf{x}'} \frac{1}{|\mathbf{x} - \mathbf{x}'|}\,.
\end{aligned} \tag{13.29}
$$

Integrating by parts, we get[4]

$$\mathbf{A}(t, \mathbf{x}) = \frac{\mu_0}{4\pi} \int_V d^3x'\, \frac{\boldsymbol{\nabla}_{\mathbf{x}'} \times \mathbf{M}(t, \mathbf{x}')}{|\mathbf{x} - \mathbf{x}'|} + \frac{\mu_0}{4\pi} \int_{\partial V} d^2s'\, \frac{\mathbf{M}(t, \mathbf{x}') \times \hat{\mathbf{n}}'_s}{|\mathbf{x} - \mathbf{x}'|}\,. \tag{13.31}$$

Comparing with eq. (11.153) we see that a magnetic dipole density is effectively equivalent to a magnetization current

$$\boxed{\mathbf{j}_{\text{mag}}(t, \mathbf{x}) = \boldsymbol{\nabla} \times \mathbf{M}(t, \mathbf{x})\,,} \tag{13.32}$$

and a surface current density

$$\boxed{\mathbf{K}_{\text{mag}}(t, \mathbf{x}) = \mathbf{M}(t, \mathbf{x}) \times \hat{\mathbf{n}}_s\,,} \tag{13.33}$$

where, again, $\hat{\mathbf{n}}_s$ is the outer normal to the surface element d^2s. Observe that, being a curl,

$$\boldsymbol{\nabla}{\cdot}\mathbf{j}_{\text{mag}} = 0\,, \tag{13.34}$$

so there is no time-varying charge associated with it in a conservation equation. This is a consequence of the fact that Gauss' law (3.3) implies that there are no magnetic charges, so $\mathbf{j}_{\text{mag}}$ can only be generated by magnetic dipoles.

A possible extension of the simple modelization (13.24) takes into account the effect of convection, i.e., the transport of charges in gases and liquids through bulk motions of the fluid. Examples of situations

---

[4]The integration by parts is performed more clearly working in components and observing that, for any differentiable function $f(\mathbf{x})$,

$$
\begin{aligned}
&\int_V d^3x\, \epsilon_{ijk} M_j \partial_k f \\
&= -\int_V d^3x\, \epsilon_{ijk} (\partial_k M_j) f \\
&\quad + \int_{\partial V} d^2s_k \epsilon_{ijk} M_j f \\
&= \int_V d^3x\, \epsilon_{ijk} (\partial_j M_k) f \\
&\quad + \int_{\partial V} d^2s_k \epsilon_{ijk} M_j f\,. \tag{13.30}
\end{aligned}
$$

where this takes places are provided by molten metals and plasmas, by convective motions in the interior of stars, or by clouds carrying free electrons that move through the atmosphere under the action of winds. In general, if $\rho(t, \mathbf{x})$ is the density of a fluid (density of mass, or of charge) and $\mathbf{u}(t, \mathbf{x})$ is its velocity field, the corresponding convective current is $\mathbf{j}_{\text{conv}}(t, \mathbf{x}) = \rho(t, \mathbf{x})\mathbf{u}(t, \mathbf{x})$. Free charges can be transported by convective motions, and this gives rise to the convective current

$$\mathbf{j}_{\text{conv,free}}(t, \mathbf{x}) = \rho_{\text{free}}(t, \mathbf{x})\mathbf{u}(t, \mathbf{x}) \,. \tag{13.35}$$

The total flow of free charges is therefore described by the current

$$\mathbf{j}_{\text{free}}(t, \mathbf{x}) = \mathbf{j}_{\text{cond,free}}(t, \mathbf{x}) + \mathbf{j}_{\text{conv,free}}(t, \mathbf{x}) \,, \tag{13.36}$$

where $\mathbf{j}_{\text{cond,free}}$ is the contribution from electric conduction, for which earlier, in the absence of convection, we used the notation $\mathbf{j}_{\text{free}}(t, \mathbf{x})$. The difference between $\mathbf{j}_{\text{cond,free}}$ and $\mathbf{j}_{\text{conv,free}}$ is that the former is generated by free charges moving under the action of electric fields, while the latter is due to free charges moving under the action of mechanical forces. The continuity equation for the free charges reads[5]

$$\frac{\partial \rho_{\text{free}}}{\partial t} + \mathbf{\nabla} \cdot [\mathbf{j}_{\text{cond,free}}(t, \mathbf{x}) + \mathbf{j}_{\text{conv,free}}(t, \mathbf{x})] = 0 \,, \tag{13.37}$$

and expresses the fact that the variation in time of the density of free charges in a small volume is given by the flux of free charges flowing through it either because of conduction, or because of convection.[6]

13.4 Maxwell's equations in material media

We now have all the elements for writing Maxwell's equations in material media. We start from Gauss's law (13.13), and we write the source term as $\rho = \rho_{\text{free}} + \rho_{\text{pol}}$.[7] Then, using eq. (13.21),

$$\begin{aligned} \epsilon_0 \mathbf{\nabla} \cdot \mathbf{E} &= \rho_{\text{free}} + \rho_{\text{pol}} \\ &= \rho_{\text{free}} - \mathbf{\nabla} \cdot \mathbf{P} \,. \end{aligned} \tag{13.39}$$

It is convenient to define the *electric displacement* vector $\mathbf{D}$ as

$$\boxed{\mathbf{D} = \epsilon_0 \mathbf{E} + \mathbf{P} \,,} \tag{13.40}$$

so that Gauss's law can be rewritten as

$$\boxed{\mathbf{\nabla} \cdot \mathbf{D} = \rho_{\text{free}} \,.} \tag{13.41}$$

We proceed similarly for the Ampère–Maxwell equation (13.14). Using eq. (13.24), together with eqs. (13.27) and (13.32), we get

$$\mathbf{\nabla} \times \mathbf{B} - \frac{1}{c^2} \frac{\partial \mathbf{E}}{\partial t} = \mu_0 \left(\mathbf{j}_{\text{free}} + \frac{\partial \mathbf{P}}{\partial t} + \mathbf{\nabla} \times \mathbf{M} \right) \,, \tag{13.42}$$

[5] As we will see below eq. (13.42), this continuity equation is actually a consequence of Maxwell's equation, just as the conservation equation for free charges discussed in Section 3.2.1.

[6] Polarization charges, being bound to the molecules, can also be transported by convective motions. The corresponding convective current is $\mathbf{j}_{\text{conv,pol}}(t, \mathbf{x}) = \rho_{\text{pol}}(t, \mathbf{x})\mathbf{u}(t, \mathbf{x})$, and can be further added to the right-hand side of the macroscopic Ampère–Maxwell equation. Note that $\mathbf{\nabla} \cdot \mathbf{j}_{\text{conv,pol}} = 0$. This can be shown observing that the continuity equation for the polarization charge now reads

$$\partial_t \rho_{\text{pol}} + \mathbf{\nabla} \cdot [\mathbf{j}_{\text{cond,pol}} + \mathbf{j}_{\text{conv,pol}}] = 0 \,, \tag{13.38}$$

where $\mathbf{j}_{\text{cond,pol}}$ is given by eq. (13.27). However, from eqs. (13.21) and (13.27), we have $\partial_t \rho_{\text{pol}} + \mathbf{\nabla} \cdot \mathbf{j}_{\text{cond,pol}} = 0$ automatically, and therefore

$$\mathbf{\nabla} \cdot \mathbf{j}_{\text{conv,pol}} = 0 \,.$$

[7] The notation ρ_{free} is quite conventional. However, more precisely, this refers to any other source of charge, not related to the neutral dipoles inside the material. For instance, it could refer to an external charge that we have placed inside an otherwise neutral dielectric, as in Solved Problem 13.1, or to charges external to the material, that contribute to the electric field inside it.

where, if we also include convection, $\mathbf{j}_{\text{free}}$ is given by eq. (13.36). Note that, taking the divergence of eq. (13.42) and using eqs. (8.27) and (13.39), one gets the continuity equation (13.37), as expected.

Similarly to what we have done for the electric dipole contribution in Gauss's law, the magnetization current can be reabsorbed into a redefinition of the magnetic field,

$$\mathbf{H} = \frac{1}{\mu_0}\mathbf{B} - \mathbf{M}. \tag{13.43}$$

Then, using also eq. (13.40) to eliminate $\mathbf{E}$ in favor of $\mathbf{D}$, eq. (13.42) becomes

$$\boldsymbol{\nabla} \times \mathbf{H} - \frac{\partial \mathbf{D}}{\partial t} = \mathbf{j}_{\text{free}}. \tag{13.44}$$

Thus, the two Maxwell's equations involving the sources can be rewritten more naturally in terms of $\mathbf{D}$ and $\mathbf{H}$.[8] The full set of macroscopic Maxwell's equations in material media therefore reads

$$
\begin{aligned}
\boldsymbol{\nabla}{\cdot}\mathbf{D} &= \rho_{\text{free}}, & (13.45)\\
\boldsymbol{\nabla} \times \mathbf{H} - \frac{\partial \mathbf{D}}{\partial t} &= \mathbf{j}_{\text{free}}, & (13.46)\\
\boldsymbol{\nabla}{\cdot}\mathbf{B} &= 0, & (13.47)\\
\boldsymbol{\nabla} \times \mathbf{E} + \frac{\partial \mathbf{B}}{\partial t} &= 0. & (13.48)
\end{aligned}
$$

[8]Note that there is no universal convention on the nomenclature of these fields. In some books $\mathbf{H}$ is called the "magnetic field" and $\mathbf{B}$ the magnetic induction. This reflects the historical development, when it was not yet understood that $\mathbf{B}$ is a more fundamental entity than $\mathbf{H}$. We will rather follow the alternative convention of calling $\mathbf{B}$ the magnetic field, while $\mathbf{H}$ will be called simply, the "H" field.
Also note that $\mathbf{H}$ does not have the same dimensions as $\mathbf{B}$, since μ_0 is not dimensionless. Rather, from eq. (13.44) we see that $\boldsymbol{\nabla} \times \mathbf{H}$ has the same dimensions as $\mathbf{j}_{\text{free}}$. Recalling that $\mathbf{j}_{\text{free}}$ is a current per unit surface [see the discussion below eq. (2.21)], we see that, in the SI system, $\mathbf{H}$ has dimensions of a current divided by a length, and is therefore measured in A/m. Similarly, from eq. (13.41), in the SI system $\mathbf{D}$ is measured in C/m^2.

Observe that, in the second pair of Maxwell's equations, still enter $\mathbf{B}$ and $\mathbf{E}$ rather than $\mathbf{H}$ and $\mathbf{D}$. Also note that, for instance, in general $\boldsymbol{\nabla}{\cdot}\mathbf{H} \neq 0$, since $\boldsymbol{\nabla}{\cdot}\mathbf{M}$ needs not be zero. So, in general, there is no vector potential whose curl gives $\mathbf{H}$, and therefore no scalar potential in terms of which we could express the $\mathbf{D}$ field.

As a first simple application, consider electrostatics in materials. In this case, the two equations to be solved are

$$\boldsymbol{\nabla}{\cdot}\mathbf{D} = \rho_{\text{free}}, \qquad \boldsymbol{\nabla} \times \mathbf{E} = 0. \tag{13.49}$$

Using eq. (13.40), we can rewrite them as

$$\boldsymbol{\nabla}{\cdot}\mathbf{D} = \rho_{\text{free}}, \qquad \boldsymbol{\nabla} \times \mathbf{D} = \boldsymbol{\nabla} \times \mathbf{P}. \tag{13.50}$$

Comparing with eqs. (4.1) and (4.2) we see that the equations governing electrostatics in material are not just obtained replacing $\mathbf{E}$ with $\mathbf{D}/\epsilon_0$ (and ρ with ρ_{free}) in eq. (4.1), since eq. (4.2) states that $\boldsymbol{\nabla}{\times}\mathbf{E} = 0$ while, from eq. (13.50), $\boldsymbol{\nabla} \times \mathbf{D} \neq 0$. This also implies that we cannot write $\mathbf{D}$ as the gradient of a scalar function. However, we have already seen in Solved Problem 11.1 that a vector field is determined uniquely by its divergence and its curl, through Helmholtz's theorem in the form given in eq. (11.174). From (13.50) we can therefore immediately write the solution for $\mathbf{D}$,

$$\mathbf{D}(\mathbf{x}) = -\boldsymbol{\nabla}\left(\int d^3x' \frac{\rho_{\text{free}}(\mathbf{x}')}{4\pi|\mathbf{x}-\mathbf{x}'|}\right) + \boldsymbol{\nabla}\times\left(\int d^3x' \frac{(\boldsymbol{\nabla}\times\mathbf{P})(\mathbf{x}')}{4\pi|\mathbf{x}-\mathbf{x}'|}\right), \tag{13.51}$$

to be compared with eq. (4.22) in the vacuum case. Using eq. (13.40) we can rewrite this solution in terms of **E**, as

$$
\begin{aligned}
\mathbf{E}(\mathbf{x}) \;=\; & -\frac{1}{4\pi\epsilon_0}\boldsymbol{\nabla}\left(\int d^3x'\,\frac{\rho_{\text{free}}(\mathbf{x}')}{|\mathbf{x}-\mathbf{x}'|}\right) \\
& +\frac{1}{4\pi\epsilon_0}\boldsymbol{\nabla}\times\left(\int d^3x'\,\frac{(\boldsymbol{\nabla}\times\mathbf{P})(\mathbf{x}')}{|\mathbf{x}-\mathbf{x}'|}\right) - \frac{1}{\epsilon_0}\mathbf{P}(\mathbf{x})\,.
\end{aligned}
\tag{13.52}
$$

In vacuum (in the sense used here, i.e., in the absence of bound charges or any other complexity due to a macroscopic material) we have $\mathbf{P}=0$ and $\rho=\rho_{\text{free}}$, and we recover of course eq. (4.18) or, equivalently, eq. (4.22). In Solved Problem 13.1 we will apply these results to the case of the interface between two simple linear dielectrics.

We can proceed similarly for magnetostatics. In this case the two equations to be solved are

$$
\boldsymbol{\nabla}\cdot\mathbf{B}=0\,,\qquad \boldsymbol{\nabla}\times\mathbf{H}=\mathbf{j}_{\text{free}}\,.
\tag{13.53}
$$

Using eq. (13.43), we can rewrite them as

$$
\boldsymbol{\nabla}\cdot\mathbf{H}=-\boldsymbol{\nabla}\cdot\mathbf{M}\,,\qquad \boldsymbol{\nabla}\times\mathbf{H}=\mathbf{j}_{\text{free}}\,.
\tag{13.54}
$$

Using again eq. (11.174), the general solution for **H** is

$$
\mathbf{H}(\mathbf{x}) = \boldsymbol{\nabla}\times\left(\int d^3x'\,\frac{\mathbf{j}_{\text{free}}(\mathbf{x}')}{4\pi|\mathbf{x}-\mathbf{x}'|}\right)+\boldsymbol{\nabla}\left(\int d^3x'\,\frac{(\boldsymbol{\nabla}\cdot\mathbf{M})(\mathbf{x}')}{4\pi|\mathbf{x}-\mathbf{x}'|}\right),
\tag{13.55}
$$

or, in terms of **B**,

$$
\begin{aligned}
\mathbf{B}(\mathbf{x}) \;=\; & \frac{\mu_0}{4\pi}\boldsymbol{\nabla}\times\left(\int d^3x'\,\frac{\mathbf{j}_{\text{free}}(\mathbf{x}')}{|\mathbf{x}-\mathbf{x}'|}\right) \\
& +\frac{\mu_0}{4\pi}\boldsymbol{\nabla}\left(\int d^3x'\,\frac{(\boldsymbol{\nabla}\cdot\mathbf{M})(\mathbf{x}')}{|\mathbf{x}-\mathbf{x}'|}\right) + \mu_0\mathbf{M}(\mathbf{x})\,.
\end{aligned}
\tag{13.56}
$$

When the magnetization vanishes and $\mathbf{j}_{\text{free}}$ is the same as the total current $\mathbf{j}$, we recover eq. (4.94).

13.5   Boundary conditions on E, B, D, H

When we have boundaries between different materials, or between a material and the vacuum, to complete the specification of a problem we must also assign the boundary conditions. To this purpose, consider a cylinder of infinitesimal height h straddling the boundary surface between two media, and let A be the area of the faces of the cylinder, with the upper face lying in medium 2 and the lower face in medium 1, as in Fig. 13.2. From eq. (13.45),

$$
\int_V d^3x\,\boldsymbol{\nabla}\cdot\mathbf{D}=\int_V d^3x\,\rho_{\text{free}}\,,
\tag{13.57}
$$

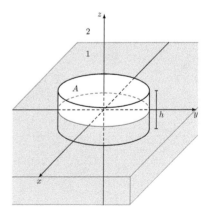

Fig. 13.2 An infinitesimal cylinder across the boundary between two media.

where V is the volume of the cylinder. For definiteness, let us take the z axis along the height of the cylinder, so

$$\int_V d^3x = \int_A dx dy \int_{-h/2}^{+h/2} dz .\tag{13.58}$$

In the limit $h \to 0$, we have

$$\int_{-h/2}^{+h/2} dz \, \rho_{\text{free}} = \sigma_{\text{free}} ,\tag{13.59}$$

where σ_{free} is the surface density of free charges. If, furthermore, we take A sufficiently small, so that σ_{free} can be taken constant over A,[9] then

$$\int_A dx dy \int_{-h/2}^{+h/2} dz \, \rho_{\text{free}} = A \sigma_{\text{free}} .\tag{13.60}$$

On the left-hand side of eq. (13.57), in the limit $h \to 0$ we have

$$
\begin{aligned}
\int_V d^3x \, \boldsymbol{\nabla} \cdot \mathbf{D} &= \int_A dx dy \int_{-h/2}^{+h/2} dz \, \partial_z D_z \\
&\quad + \int_{-h/2}^{+h/2} dz \int_A dx dy \, (\partial_x D_x + \partial_y D_y) \\
&= A[D_z(2) - D_z(1)] + \mathcal{O}(h) ,
\end{aligned}\tag{13.61}
$$

where $D_z(2)$ is the value of D_z as we approach the boundary from the medium 2, and $D_z(1)$ is the value when we approach the boundary from the medium 1. Therefore, sending $h \to 0$ with A sufficiently small so that σ_{free} is constant over it (but still $\sqrt{A}$ much larger than h), we find

$$D_z(2) - D_z(1) = \sigma_{\text{free}} .\tag{13.62}$$

For a boundary with a normal $\hat{\mathbf{n}}$ (pointing from the medium 1 toward the medium 2) in a generic direction, rather than along $\hat{\mathbf{z}}$, we then have

$$\boxed{\hat{\mathbf{n}} \cdot (\mathbf{D}_2 - \mathbf{D}_1) = \sigma_{\text{free}} .}\tag{13.63}$$

Applying the same reasoning to eq. (13.47), we get instead

$$\boxed{\hat{\mathbf{n}} \cdot (\mathbf{B}_2 - \mathbf{B}_1) = 0 .}\tag{13.64}$$

To get the boundary conditions for $\mathbf{E}$ and $\mathbf{D}$ we consider a one-dimensional loop C with the longer side L parallel to the boundary and the shorter side, of length h, straddling across the two materials, as in Fig. 13.3. Integrating eq. (13.48) over a two-dimensional surface S bounded by this loop, we get

$$
\begin{aligned}
0 &= \int_S d\mathbf{s} \cdot \left[\boldsymbol{\nabla} \times \mathbf{E} + \frac{\partial \mathbf{B}}{\partial t} \right] \\
&= \int_S d\mathbf{s} \cdot \boldsymbol{\nabla} \times \mathbf{E} + \frac{\partial}{\partial t} \int_S d\mathbf{s} \cdot \mathbf{B} .
\end{aligned}\tag{13.65}
$$

[9]Note, however, that a formal mathematical limit $A \to 0$ is not physically meaningful. The area A must still be larger than $O(L^2)$, where L is the length-scale introduced above to define the macroscopic equations of motion. The same is implicit for the formal limit $h \to 0$ used below.

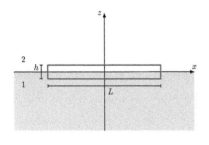

Fig. 13.3 An infinitesimal loop across the boundary between two media.

Let us take again the z axis in the direction of the normal to the interface, which in this case corresponds to the short side of the loop, and the x axis in the direction of the loop, so the long side of the loop goes from $x = -L/2$ to $x = L/2$, at fixed y and z. The normal to the surface $d\mathbf{s}$ is therefore along the y direction, so $d\mathbf{s}{\cdot}\mathbf{B} = dx\,dz\,B_y$. Since B_y is finite at the surface, its surface integral vanishes as $h \to 0$. Then, using Stokes's theorem (1.38), in the limit $h \to 0$ we get

$$
\begin{aligned}
0 &= \int_S d\mathbf{s}{\cdot}\boldsymbol{\nabla} \times \mathbf{E}\\[2mm]
&= \int_{C=\partial S} d\boldsymbol{\ell} \cdot \mathbf{E}\\[2mm]
&= \int_{-L/2}^{L/2} dx\, E_x(x, y, z = -h/2) + \int_{-h/2}^{h/2} dz\, E_z(x = L/2, y, z)\\[2mm]
&\quad + \int_{L/2}^{-L/2} dx\, E_x(x, y, z = h/2) + \int_{h/2}^{-h/2} dz\, E_z(x = -L/2, y, z)\\[2mm]
&= -\int_{-L/2}^{L/2} dx\, [E_x(x, y, z = h/2) - E_x(x, y, z = -h/2)]\\[2mm]
&\quad + \int_{-h/2}^{h/2} dz\, [E_z(x = L/2, y, z) - E_z(x = -L/2, y, z)] \,. \quad (13.66)
\end{aligned}
$$

The last integral vanishes as $h \to 0$, since the electric field is finite on the boundary, while, taking L sufficiently small so that the variation of the field with x can be neglected, we get

$$
0 = L\left[E_x(x, y, z \to 0^+) - E_x(x, y, z \to 0^-)\right], \quad (13.67)
$$

and therefore E_x is continuous across the boundary. The same argument could be made for a loop in the (y, z) plane, leading to the conclusion that also E_y is continuous, so the electric field transverse with respect to the normal to the interface is continuous. For a generic normal $\hat{\mathbf{n}}$, not necessarily oriented along z (again with the convention that it points from the medium 1 toward the medium 2), we can write this as

$$
\boxed{\hat{\mathbf{n}} \times (\mathbf{E}_2 - \mathbf{E}_1) = 0\,.} \quad (13.68)
$$

Finally, we can repeat the same argument for eq. (13.46). Now the integral of **D** over $d\mathbf{s}$ vanishes because **D** is finite on the boundary (just as the integral of **B** in the previous computation). We define the surface current of free charges as

$$
\mathbf{K}_{\text{free}} = \int_{-h/2}^{+h/2} dz\, \mathbf{j}_{\text{free}}, \quad (13.69)
$$

(for an interface whose normal is along the z axis, with dz replaced by $\hat{\mathbf{n}}{\cdot}d\mathbf{x}$ for generic $\hat{\mathbf{n}}$). Then, with the same computations as before, we get

$$
\boxed{\hat{\mathbf{n}} \times (\mathbf{H}_2 - \mathbf{H}_1) = \mathbf{K}_{\text{free}}\,.} \quad (13.70)
$$

13.6   Constitutive relations

Finally, in order to make use of eqs. (13.45)–(13.48), we need to know how $\mathbf{D}$ and $\mathbf{H}$ are related to the fundamental fields $\mathbf{E}$ and $\mathbf{B}$ (i.e., to have a model for $\mathbf{P}$ and $\mathbf{M}$). These are called *constitutive relations*, and it is at this stage that the great variety of materials enters. It should be stressed that such relations are not fundamental, contrary to Maxwell's equations in vacuum. Rather, they are phenomenological relations, assumed to catch the main properties of a material, at least for some range of parameters (such as the strength or frequency of the electric field). We now discuss some common constitutive relations.

13.6.1   Dielectrics

A dielectric is defined as a material with no (or very little) free charges. Since the electric conductivity is given by the free charges, dielectrics are insulators. A *linear dielectric* material is defined by the condition that the polarization $\mathbf{P}$ induced by an applied electric field $\mathbf{E}$ is linear in $\mathbf{E}$. In its simplest form, this means that

$$\mathbf{P}(t,\mathbf{x}) = \epsilon_0 \chi_e \mathbf{E}(t,\mathbf{x}),\tag{13.71}$$

for some constant χ_e, which is called the *electric susceptibility*.[10] From eq. (13.40), this implies that

$$\mathbf{D}(t,\mathbf{x}) = \epsilon \mathbf{E}(t,\mathbf{x}),\tag{13.72}$$

where the *permittivity* of the material, denoted by ϵ, is given by

$$\epsilon = \epsilon_0 (1 + \chi_e).\tag{13.73}$$

The *dielectric constant* (or *relative permittivity*) of the material is defined as $\epsilon_r = \epsilon/\epsilon_0$,[11] so

$$\epsilon_r = 1 + \chi_e.\tag{13.74}$$

This constitutive relation implies a simple relation between the polarization charges and the free charges of the medium: from eqs. (13.71) and (13.72),

$$\mathbf{P}(t,\mathbf{x}) = \frac{\chi_e}{\epsilon_r}\mathbf{D}(t,\mathbf{x}).\tag{13.75}$$

Using eqs. (13.21) and (13.45), we then obtain

$$\begin{aligned}\rho_{\text{pol}}(t,\mathbf{x}) &= -\frac{\chi_e}{\epsilon_r}\boldsymbol{\nabla}{\cdot}\mathbf{D}(t,\mathbf{x})\\ &= -\frac{\chi_e}{1+\chi_e}\rho_{\text{free}}(t,\mathbf{x}).\end{aligned}\tag{13.76}$$

Observe that the density of bound charges has the sign opposite to that of the free charges, and vanishes if $\rho_{\text{free}} = 0$. For a perfect insulator there are no free charges, so also $\rho_{\text{pol}} = 0$. Therefore, in a linear dielectric, all polarization charge resides on the surface of the material.[12] Note

[10]The factor ϵ_0 has been inserted in eq. (13.71) so that χ_e is a dimensionless quantity.

[11]The dielectric constant in the SI system is often also denoted by κ_e.

[12]The situation is the same as illustrated in Fig. 13.1, except that, if $\rho_{\text{free}} = 0$, we have $\boldsymbol{\nabla}{\cdot}\mathbf{D} = 0$ and then, for a linear dielectric, also $\boldsymbol{\nabla}{\cdot}\mathbf{P} = 0$. In the absence of a gradient of the polarization vector, the dipole charges inside the volume cancel each other, and only the surface charge remains.

also that χ_e is always positive because an external static electric field induces an electric dipole moment in the same direction, so $\mathbf{P}$ has the same direction as $\mathbf{E}$. Therefore, in all materials described by eq. (13.72), $\epsilon_r > 1$.

A simple constitutive relation such as eq. (13.72) is, however, only valid for quasi-static fields. More generally, actual linear dielectrics are described by a constitutive relation of the form

$$\tilde{\mathbf{D}}(\omega) = \epsilon(\omega)\tilde{\mathbf{E}}(\omega),\qquad(13.77)$$

where $\tilde{\mathbf{D}}(\omega)$ and $\tilde{\mathbf{E}}(\omega)$ are the Fourier modes of $\mathbf{D}(t)$ and $\mathbf{E}(t)$, respectively and, as it is customary in these equations, we have omitted the tilde over the Fourier transform of $\epsilon(t)$ (the fact that $\epsilon(\omega)$ is the Fourier transform is anyhow clear from its argument ω). In Section 14.2 we will discuss the form of this relation in the time domain and the constraints imposed by causality on the function $\epsilon(\omega)$, while in Section 14.3 we will develop a simple explicit model for $\epsilon(\omega)$.

Note that $\epsilon(\omega)$, being the Fourier transform of a real function, is a complex function. We will refer to $\epsilon(\omega)$ as the permittivity function of the material, and to $\epsilon_r(\omega) \equiv \epsilon(\omega)/\epsilon_0$ as the relative permittivity function (or the dielectric function) of the material. The permittivity ϵ defined in eq. (13.72) is the limit of the permittivity function for $\omega \to 0$, i.e., for static fields and, similarly, the dielectric constant $\epsilon_r = \epsilon/\epsilon_0$ is the limit of the dielectric function $\epsilon_r(\omega)$ for $\omega \to 0$. As an example, Fig. 13.4 shows the real and imaginary parts of the dielectric function of water, at different temperatures. The dielectric constant $\epsilon_r \equiv \epsilon_r(\omega = 0)$ can take values over a broad range.[13]

In anisotropic materials, eq. (13.72) generalizes to $D_i = \epsilon_{ij}E_j$, so the permittivity is promoted to a spatial tensor ϵ_{ij}, and the dielectric constant becomes a tensor $(\epsilon_r)_{ij} \equiv \epsilon_{ij}/\epsilon_0$. Correspondingly, eq. (13.77) has the anisotropic generalization

$$\tilde{D}_i(\omega) = \epsilon_{ij}(\omega)\tilde{E}_j(\omega).\qquad(13.78)$$

As with any phenomenological relation, eqs. (13.72), (13.77), or (13.78) have a finite range of validity. In particular, the assumption of linear response breaks down when the external field $\mathbf{E}$ exceeds a critical threshold, leading to *dielectric breakdown*. Beyond this critical field, current flows in the material and the insulator becomes conductive. For instance, for air, when the electric field exceeds about 10^6 V/m, the air molecules become ionized and air becomes a conductor. Lightning occurs when there is a sufficient build-up of electric charges between the clouds and the ground, so that the threshold value of the electric field is exceeded. Then, air becomes conductor along a path where the charges flow from the cloud to the ground. In Solved Problems 13.1 we will examine some simple examples of electrostatics in dielectrics.

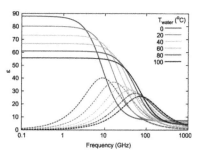

Fig. 13.4 (Relative) permittivity function $\epsilon_r(\omega)$ for water, as a function of frequency $f = \omega/(2\pi)$, for different temperatures from $T = 0\,°C$ to $T = 100\,°C$ [with lower temperatures corresponding to higher values of $\epsilon_r(\omega = 0)$]. Solid lines correspond to the real part, dashed lines to the imaginary part. From Andryieuski *et al.* (2015) (Creative Commons Attribution 4.0 International).

[13]For example, for water, as we see from Fig. 13.4, ϵ_r has a significant dependence on the temperature, and ranges from about $\epsilon_r \simeq 88$ at $0°$ C, to $\epsilon_r \simeq 55$ at $100°$ C, with $\epsilon_r \simeq 80$ at $20°$ C. However, in its gaseous state, at $110°$ C and 1 atm, it drops to $\epsilon_r \simeq 1.012$. For air at 1 atm, $\epsilon_r \simeq 1.00054$, which becomes 1.0548 at 100 atm. For some special materials, ϵ_r can reach values of order 10^3-10^4 (e.g., 2×10^3 for Barium Titanate and larger than 10^4 for Calcium Copper Titanate). We will see in eq. (13.104) that the capacitance of a capacitor filled with a dielectric is enhanced by a factor ϵ_r compared to the vacuum case. Then, materials with such a huge dielectric constant can make super-capacitors.

13.6.2   Metals

In metals, there are both bound and free charges. The effect of bound charges is still described, in a linear approximation, by eq. (13.77), [or by its generalizations (13.78)], that we now write as

$$\tilde{\mathbf{D}}(\omega) = \epsilon_b(\omega)\tilde{\mathbf{E}}(\omega) \,, \tag{13.79}$$

where the subscript "b" in $\epsilon_b(\omega)$ stresses that this is the contribution of the bound electrons. The free electrons, in contrast, generate a current density $\mathbf{j}_{\text{free}}$. The simplest constitutive relation describing the current generated by the free electrons is *Ohm's law*, which states that the steady current of free charges generated by an applied static electric field is given by

$$\mathbf{j}_{\text{free}} = \sigma\mathbf{E} \,, \tag{13.80}$$

where σ is called the *conductivity*. Similarly to eq. (13.72), this relation only holds in the static limit, and can be generalized to a frequency-dependent relation, as

$$\tilde{\mathbf{j}}_{\text{free}}(\omega) = \sigma(\omega)\tilde{\mathbf{E}}(\omega) \,. \tag{13.81}$$

Again, we can generalize to anisotropic media, writing $(\tilde{j}_{\text{free}})_i(\omega) = \sigma_{ij}(\omega)\tilde{E}_j(\omega)$, and such linear relations are valid only up to a maximum value of the electric field. When we use the frequency-dependent conductivity, we will use the notation

$$\sigma(\omega = 0) \equiv \sigma_0 \,, \tag{13.82}$$

for the zero-frequency, or "d.c." conductivity (in contrast to the frequency dependent, or "a.c.", conductivity). In Section 14.4 we will study in detail a simple model for $\sigma(\omega)$.

Let us now consider the pair of Maxwell's equations in matter involving the sources, eqs. (13.45) and (13.46). We perform the Fourier transform with respect to time only, writing for instance

$$\rho_{\text{free}}(t, \mathbf{x}) = \int_{-\infty}^{\infty} \frac{d\omega}{2\pi} e^{-i\omega t} \tilde{\rho}_{\text{free}}(\omega, \mathbf{x}) \,, \tag{13.83}$$

and similarly for all other quantities. Then, eqs. (13.45) and (13.46) become

$$\boldsymbol{\nabla}{\cdot}\tilde{\mathbf{D}}(\omega, \mathbf{x}) = \tilde{\rho}_{\text{free}}(\omega, \mathbf{x}) \,, \tag{13.84}$$

$$\boldsymbol{\nabla} \times \tilde{\mathbf{H}}(\omega, \mathbf{x}) + i\omega\tilde{\mathbf{D}}(\omega, \mathbf{x}) = \tilde{\mathbf{j}}_{\text{free}}(\omega, \mathbf{x}) \,, \tag{13.85}$$

while the continuity equation (13.25) becomes

$$-i\omega\tilde{\rho}_{\text{free}}(\omega, \mathbf{x}) + \boldsymbol{\nabla}{\cdot}\tilde{\mathbf{j}}_{\text{free}}(\omega, \mathbf{x}) = 0 \,. \tag{13.86}$$

Then, eq. (13.84) can be written as

$$\boldsymbol{\nabla}{\cdot}\tilde{\mathbf{D}}(\omega, \mathbf{x}) = \frac{1}{i\omega} \, \boldsymbol{\nabla}{\cdot}\tilde{\mathbf{j}}_{\text{free}}(\omega, \mathbf{x}) \,. \tag{13.87}$$

We now use the constitutive relations (13.79) and (13.81), and eqs. (13.87) and (13.85) become, respectively,

$$\epsilon(\omega)\boldsymbol{\nabla}\cdot\tilde{\mathbf{E}}(\omega,\mathbf{x}) = 0, \tag{13.88}$$

$$\boldsymbol{\nabla}\times\tilde{\mathbf{H}}(\omega,\mathbf{x}) + i\omega\,\epsilon(\omega)\tilde{\mathbf{E}}(\omega,\mathbf{x}) = 0, \tag{13.89}$$

where

$$\epsilon(\omega) \equiv \epsilon_b(\omega) + i\frac{\sigma(\omega)}{\omega}. \tag{13.90}$$

We see that the response function of a metal is not determined separately by the two functions $\epsilon_b(\omega)$ and $\sigma(\omega)$, but only by their combination (13.90) (apart from a function that relates $\mathbf{H}$ to $\mathbf{B}$, that will be introduced in Section 13.6.3, and that can be set to one in many typical situations). The function $\epsilon(\omega)/\epsilon_0$, with $\epsilon(\omega)$ defined by eq. (13.90), is called the *dielectric function of a metal*. The two terms in eq. (13.90) describe the separate contributions of the bound and free electrons. The fact that they combine into a single function is a reflection of the fact that (at least within a classical description) bound electrons oscillate around their equilibrium position with a set of oscillation frequencies ω_i, and a free electron can just be seen as a bound electron with a frequency $\omega_0 = 0$, or anyhow ω_0 much smaller than the other typical frequency scales in the problem, which can be simply added to the set of all other bound electrons. We will see this explicitly in Section 14.5, where we use an explicit model for $\sigma(\omega)$ and $\epsilon_b(\omega)$.

13.6.3   Diamagnetic and paramagnetic materials

Similar to the case of dielectrics, in many materials the magnetization $\mathbf{M}$ is linearly proportional to the external magnetic field $\mathbf{B}$. Because of eq. (13.43), a linear relation between $\mathbf{M}$ and $\mathbf{B}$ implies a linear relation between $\mathbf{M}$ and $\mathbf{H}$. One then defines the *magnetic susceptibility* χ_m from

$$\mathbf{M} = \chi_m\mathbf{H}. \tag{13.91}$$

This is the simplest constitutive relation for magnetic matter. Equation (13.43) then implies a linear relation also between $\mathbf{B}$ and $\mathbf{H}$, which is written as

$$\mathbf{B} = \mu\mathbf{H}, \tag{13.92}$$

where

$$\mu = \mu_0(1 + \chi_m). \tag{13.93}$$

The constant μ is called the (magnetic) *permeability* of the material. Just as in the case of the dielectric constant (or relative electric permittivity) $\epsilon_r = \epsilon/\epsilon_0$, one defines the relative magnetic permeability as μ/μ_0.[14] Note that, for historical reasons, eqs. (13.91) and (13.92) are written as if $\mathbf{H}$ were the fundamental field and $\mathbf{B}$ the derived field, while we now understand that the opposite is true.

[14]When the dielectric constant ϵ_0/ϵ_0 is denoted by κ_e, the relative permeability μ/μ_0 is denoted by κ_m. Note that some texts, such as Garg (2012), reserve the name "magnetic permeability" to μ/μ_0 rather than to μ.

For a medium such that the constitutive relation is given by eq. (13.91), and in static situation, so that $\partial \mathbf{D}/\partial t = 0$, combining eqs. (13.32) and (13.46) we find a relation between the magnetization current and the current of the free electrons,

$$\mathbf{j}_{\mathrm{mag}}(t, \mathbf{x}) = \chi_m \mathbf{j}_{\mathrm{free}} \,, \tag{13.94}$$

to be compared to eq. (13.76) for the dielectric case. Contrary to the electric case, however, χ_m can be either positive or negative. When $\chi_m > 0$ the material is called *paramagnetic*, while when $\chi_m < 0$ is called *diamagnetic*. Paramagnetism arises from the fact that an external magnetic field orients already pre-existing magnetic dipole moments, due e.g. to the spin of the electrons. The pre-existing magnetic moments align with the external magnetic field, so $\chi_m > 0$. Diamagnetism is instead due to the magnetic dipoles induced by the external magnetic field. As we will show in Solved Problem 13.3, the magnetic moment induced in an atom by an external magnetic field is in fact in the direction opposite to the magnetic field.

13.7 Energy conservation

Performing on Maxwell's equations in material bodies, eqs. (13.45)–(13.48), the same manipulations that we performed on the vacuum Maxwell's equations in Section 3.2.2, we find

$$\int_V d^3x \left(\mathbf{E} \cdot \frac{\partial \mathbf{D}}{\partial t} + \mathbf{H} \cdot \frac{\partial \mathbf{B}}{\partial t} \right) + \int_V d^3x\, \mathbf{E} \cdot \mathbf{j}_{\mathrm{free}} = -\int_{\partial V} d\mathbf{s} \cdot (\mathbf{E} \times \mathbf{H}) \,, \tag{13.95}$$

that reduces to eq. (3.35) in vacuum, where $\mathbf{D} = \epsilon_0 \mathbf{E}$ and $\mathbf{H} = \mu_0^{-1} \mathbf{B}$. The term on the right-hand side gives the generalization of the Poynting vector to material media,

$$\mathbf{S} = \mathbf{E} \times \mathbf{H} \,. \tag{13.96}$$

The term $\mathbf{E} \cdot \mathbf{j}_{\mathrm{free}}$ is still the work made on the system of charges by the electric field. Note, however, that the first term on the left-hand side of eq. (13.95) is no longer a total derivative, at least in general. This is not surprising, because energy balance in a medium must now include dissipative processes such as the production of heat. If however the medium is linear, i.e., $\mathbf{D} = \epsilon \mathbf{E}$ and $\mathbf{B} = \mu \mathbf{H}$, and dispersion-less, so that ϵ and μ are independent of time, we can rewrite eq. (13.95) as

$$\frac{d}{dt} \frac{1}{2} \int_V d^3x \left(\epsilon \mathbf{E}^2 + \mu^{-1} \mathbf{B}^2 \right) + \int_V d^3x\, \mathbf{E} \cdot \mathbf{j}_{\mathrm{free}} = -\frac{1}{\mu} \int_{\partial V} d\mathbf{s} \cdot (\mathbf{E} \times \mathbf{B}) \,, \tag{13.97}$$

and we can therefore identify the energy density the electromagnetic field inside the material with

$$\begin{aligned} u &= \frac{1}{2} (\epsilon \mathbf{E}^2 + \mu^{-1} \mathbf{B}^2) \\ &= \frac{1}{2} (\epsilon \mathbf{E}^2 + \mu \mathbf{H}^2) \,, \end{aligned} \tag{13.98}$$

while the Poynting vector (13.96) can also be written as

$$\mathbf{S} = \frac{1}{\mu} \mathbf{E} \times \mathbf{B},\qquad(13.99)$$

to be compared with the expression in vacuum, eq. (3.34).

13.8   Solved problems

Problem 13.1. Electrostatics of dielectrics

We discuss here some aspects of electrostatics of a simple dielectric, with the constitutive relation (13.72), $\mathbf{D}(t,\mathbf{x}) = \epsilon \mathbf{E}(t,\mathbf{x})$, with ϵ constant. In this case, *inside* the dielectric, eq. (13.49) becomes

$$\boldsymbol{\nabla}\cdot\mathbf{E} \;=\; \frac{1}{\epsilon}\rho_{\text{free}},\qquad(13.100)$$

$$\boldsymbol{\nabla}\times\mathbf{E} \;=\; 0.\qquad(13.101)$$

The situation is therefore formally identical to eqs. (4.1) and (4.2), with ϵ_0 replaced by ϵ. So, for instance, if we place a single point charge q inside an infinite dielectric medium, instead of eq. (4.7) we will get

$$\mathbf{E} = \frac{1}{4\pi\epsilon}\frac{q}{r^2}\,\hat{\mathbf{r}}.\qquad(13.102)$$

Formally, this is the same as the field that would be generated, in vacuum, by a charge q/ϵ_r. Since $\epsilon_r \equiv \epsilon/\epsilon_0 > 1$, this means that the charge is screened, compared to the vacuum situation. Physically, what happens is that the presence of the charge q partially orients the dipoles around it so that, on average, any sphere centered around q, beside the charge q itself, also contains an excess of charges of opposite sign, coming from the dipoles, which therefore partially screens it.

Similarly, for a parallel-plate capacitor filled with a dielectric, proceeding as in Solved Problems 4.3, we see that the electric field outside the plates vanishes. The electric field inside the capacitor that, in the vacuum case is given by eq. (4.153), is now given by[15]

$$\mathbf{E} = \frac{\sigma}{\epsilon}\hat{\mathbf{z}},\qquad(13.103)$$

where $\hat{\mathbf{z}}$ is the unit normal to the plates. Therefore, for a parallel-plate capacitor, the capacitance C becomes

$$C = \frac{\epsilon A}{d},\qquad(13.104)$$

so it is larger by a factor ϵ_r compared to the value in eq. (4.158). Filling a capacitor with a dielectric with large dielectric constant is therefore a way of increasing its capacitance.

Inside the dielectric eq. (13.71), together with eq. (13.101) and the assumption that ϵ (and therefore χ_e) is constant, implies

$$\boldsymbol{\nabla}\times\mathbf{P} = 0.\qquad(13.105)$$

Naively, one might then think that, in eq. (13.51), only the integral involving ρ_{free} contributes. Actually, this is no longer true when we consider a realistic

[15]For instance, one can repeat the computation performed in eqs. (4.139)–(4.141), summing over the contribution to the electric field of the individual surface element of each plate. The computation is then formally the same, with ϵ_0 replaced by ϵ.

situation in which the dielectric has a finite extent and therefore a boundary. Consider, for simplicity, the situation in which there is a flat boundary, identified with the plane $z = 0$, separating the region with $z < 0$, filled with a dielectric with permittivity ϵ_1, from that with $z > 0$, which is filled with a dielectric with permittivity ϵ_2. Even if ϵ_1 and ϵ_2 are constant, the permittivity changes abruptly at the interface. We can write

$$\epsilon(z) = \epsilon_1 + (\epsilon_2 - \epsilon_1)\theta(z), \tag{13.106}$$

where, as usual, $\theta(z)$ is the Heaviside theta function. Combining eqs. (13.71), (13.73), and (13.101), we then obtain[16]

$$\boldsymbol{\nabla}\times\mathbf{P} = \delta(z)\,\hat{\mathbf{z}}\times(\mathbf{P}_2 - \mathbf{P}_1), \tag{13.108}$$

and the integral in the second term in eq. (13.51) can be rewritten as

$$\int d^3x' \frac{(\boldsymbol{\nabla}\times\mathbf{P})(\mathbf{x}')}{4\pi|\mathbf{x}-\mathbf{x}'|} = \int_S dx'dy'\frac{\hat{\mathbf{z}}\times(\mathbf{P}_2-\mathbf{P}_1)(\mathbf{x}')}{4\pi|\mathbf{x}-\mathbf{x}'|}, \tag{13.109}$$

where S is the (x', y') plane. For a generic curved boundary surface S, with surface element $d^2\mathbf{s}' = d^2s'\hat{\mathbf{n}}_s(\mathbf{x}')$, this generalizes to

$$\int d^3x' \frac{(\boldsymbol{\nabla}\times\mathbf{P})(\mathbf{x}')}{4\pi|\mathbf{x}-\mathbf{x}'|} = \int_S d^2s'\frac{\hat{\mathbf{n}}_s(\mathbf{x}')\times(\mathbf{P}_2-\mathbf{P}_1)(\mathbf{x}')}{4\pi|\mathbf{x}-\mathbf{x}'|}. \tag{13.110}$$

In particular, if the medium 1 is a dielectric with polarization $\mathbf{P}_1 = \mathbf{P}$ and the medium 2 is the vacuum (so that $\hat{\mathbf{n}}_s$ is the outer normal to the surface of the dielectric), eq. (13.51) becomes

$$\mathbf{D}(\mathbf{x}) = -\boldsymbol{\nabla}\left(\int d^3x' \frac{\rho_{\text{free}}(\mathbf{x}')}{4\pi|\mathbf{x}-\mathbf{x}'|}\right) - \boldsymbol{\nabla}\times\left(\int_S d^2s'\frac{\hat{\mathbf{n}}_s(\mathbf{x}')\times\mathbf{P}(\mathbf{x}')}{4\pi|\mathbf{x}-\mathbf{x}'|}\right). \tag{13.111}$$

If we have an ensemble of dielectrics, labeled by an index i, embedded in free space, we must include the contribution from all their boundaries, so

$$\mathbf{D}(\mathbf{x}) = -\boldsymbol{\nabla}\left(\int d^3x' \frac{\rho_{\text{free}}(\mathbf{x}')}{4\pi|\mathbf{x}-\mathbf{x}'|}\right) - \boldsymbol{\nabla}\times\left(\sum_i\int_{S_i} d^2s'\frac{\hat{\mathbf{n}}_s(\mathbf{x}')\times\mathbf{P}(\mathbf{x}')}{4\pi|\mathbf{x}-\mathbf{x}'|}\right). \tag{13.112}$$

Problem 13.2. Magnetostatics of simple magnetic matter

We can discuss the magnetostatics of simple magnetic matter in a completely analogous manner, with the constitutive relation (13.92), $\mathbf{B} = \mu\mathbf{H}$, where μ is a constant. In this case, *inside* the material, eq. (13.53) becomes

$$\boldsymbol{\nabla}\cdot\mathbf{B} = 0, \tag{13.113}$$
$$\boldsymbol{\nabla}\times\mathbf{B} = \mu\,\mathbf{j}_{\text{free}}. \tag{13.114}$$

The situation is therefore formally the same as eqs. (4.67) and (4.68), with μ_0 replaced by μ, and $\mathbf{j}$ by $\mathbf{j}_{\text{free}}$. Equations (13.91), (13.92), and (13.93) imply that

$$\begin{aligned}
\mathbf{M} &= \frac{\chi_m}{1+\chi_m}\frac{1}{\mu_0}\mathbf{B} \\
&\simeq \chi_m\frac{1}{\mu_0}\mathbf{B}, \tag{13.115}
\end{aligned}$$

---

[16] Explicitly, in components

$$(\boldsymbol{\nabla}\times\mathbf{P})_i = \epsilon_{ijk}(\partial_j\epsilon)E_k, \tag{13.107}$$

where, we used eq. (13.101). Then, using eq. (13.106),

$$\begin{aligned}
\partial_j\epsilon &= \delta_{j3}\frac{d\epsilon(z)}{dz} \\
&= \hat{\mathbf{z}}_j(\epsilon_2-\epsilon_1)\delta(z).
\end{aligned}$$

From eq. (13.73), we have

$$\begin{aligned}
(\epsilon_2-\epsilon_1)\mathbf{E} &= (\epsilon_0\chi_{e,2}-\epsilon_0\chi_{e,1})\mathbf{E} \\
&= \mathbf{P}_2-\mathbf{P}_1,
\end{aligned}$$

where $\mathbf{P}_2$ and $\mathbf{P}_1$ are the polarization vectors of the two dielectrics. Then,

$$\begin{aligned}
\epsilon_{ijk}(\partial_j\epsilon)E_k &= (\epsilon_2-\epsilon_1)\delta(z)\epsilon_{ijk}\hat{\mathbf{z}}_jE_k \\
&= \delta(z)\epsilon_{ijk}\hat{\mathbf{z}}_j(\mathbf{P}_2-\mathbf{P}_1)_k.
\end{aligned}$$

where we used $|\chi_m| \ll 1$. Inside the material, i.e., as far as we do not cross boundaries between different magnetic materials or between a magnetic material and the vacuum, μ (and therefore χ_m) is constant. Then eqs. (13.113) and (13.115) imply that

$$\boldsymbol{\nabla}{\cdot}\mathbf{M} = 0\,. \tag{13.116}$$

This is analogous to eq. (13.105) in the electrostatic case. Just as in the electrostatic case, this is no longer true at the boundaries between different material, where χ_m changes discontinuously. Proceeding as in eq. (13.106), we can consider a planar interface at $z = 0$, where

$$\chi_m(z) = \chi_{m,1} + (\chi_{m,2} - \chi_{m,1})\theta(z)\,. \tag{13.117}$$

Then, using eq. (13.115) together with $\boldsymbol{\nabla}{\cdot}\mathbf{B} = 0$, we get[17]

$$\boldsymbol{\nabla}{\cdot}\mathbf{M} = \delta(z)\,\hat{\mathbf{z}}{\cdot}(\mathbf{M}_2 - \mathbf{M}_1)\,. \tag{13.118}$$

Then, proceeding as in eqs. (13.108)–(13.112), we find that, for an ensemble of magnetic materials embedded in vacuum, eq. (13.55) becomes

$$\mathbf{H}(\mathbf{x}) = \boldsymbol{\nabla}{\times}\left(\int d^3x'\,\frac{\mathbf{j}_{\text{free}}(\mathbf{x}')}{4\pi|\mathbf{x} - \mathbf{x}'|}\right) + \boldsymbol{\nabla}\left(\sum_i \int_{S_i} d^2s\,\frac{(\hat{\mathbf{n}}_s{\cdot}\mathbf{M})(\mathbf{x}')}{4\pi|\mathbf{x} - \mathbf{x}'|}\right)\,. \tag{13.119}$$

Problem 13.3. Diamagnetism

In this Solved Problem we show that an external magnetic field induces, in a classical atom, a magnetic dipole in the opposite direction, leading to diamagnetism. To illustrate the effect, consider an electron with charge $-e$ [recall that $e > 0$ in our conventions, see eq. (2.4)] kept in a circular orbit of radius r by the Coulomb interaction with a nucleus of charge Ze, in a purely classical description.[18] We take the orbit to be in the (x, y) plane, and the electron moving counterclockwise. In the absence of magnetic field, the electron rotates with frequency ω_0 (taken positive by definition) given by

$$m_e\omega_0^2 r = \frac{1}{4\pi\epsilon_0}\frac{Ze^2}{r^2}\,, \tag{13.120}$$

i.e.,

$$\omega_0 = +\left(\frac{1}{4\pi\epsilon_0}\frac{Ze^2}{m_e r^3}\right)^{1/2}\,. \tag{13.121}$$

Since the motion is counterclockwise, the angular momentum of the electron is in the positive z direction, with $L_z = +m_e\omega_0 r^2$. Then, according to eq. (6.43) with $q_a = -e$, the magnetic moment of the electron due to its orbital motion is in the negative z direction,

$$(m_z)_0 = -\frac{1}{2}e\omega_0 r^2\,, \tag{13.122}$$

where the subscript 0 in $(m_z)_0$ stresses that this is the result in the absence of magnetic field. We now add a magnetic field along the positive z direction, $\mathbf{B} = B\hat{\mathbf{z}}$. Using the Lorentz force equation (3.5), the total force acting on the electron becomes

$$\mathbf{F} = -\frac{1}{4\pi\epsilon_0}\frac{Ze^2}{r^2}\hat{\mathbf{r}} - e\mathbf{v}{\times}\mathbf{B}\,. \tag{13.123}$$

For an electron that moves counterclockwise in the (x, y) plane, and a magnetic field $\mathbf{B}$ along the positive z axis, we have $\mathbf{v}{\times}\mathbf{B} = vB\hat{\mathbf{r}}$, so the Lorentz force

[17]Explicitly,

$$\boldsymbol{\nabla}{\cdot}\mathbf{M} = \frac{d\chi_m}{dz}\frac{1}{\mu_0}\hat{\mathbf{z}}{\cdot}\mathbf{B}$$

$$= \delta(z)(\chi_{m,2} - \chi_{m,1})\frac{1}{\mu_0}\hat{\mathbf{z}}{\cdot}\mathbf{B}$$

$$= \delta(z)\,\hat{\mathbf{z}}{\cdot}(\mathbf{M}_2 - \mathbf{M}_1)\,.$$

[18]As we saw in the derivation of eq. (10.225), a purely classical description actually leads to the paradox that the electron would collapse on the nucleus on a very short timescale, because of the emission of radiation. A first-principle computation should therefore be performed within quantum mechanics. However, the classical computation that we present here is already sufficient to understand the sign and the typical size of the perturbation induced by a weak magnetic field, on an equilibrium state determined quantum-mechanically.

has the same inward radial direction as the Coulomb force, and the rotation frequency ω is now determined by the positive solution of

$$m_e \omega^2 r = \frac{1}{4\pi\epsilon_0} \frac{Ze^2}{r^2} + eB\omega r \,. \tag{13.124}$$

This can be rewritten as

$$\omega^2 = \omega_0^2 + 2\omega\omega_L \,, \tag{13.125}$$

where ω_0 is given by eq. (13.121) and

$$\omega_L \equiv \frac{eB}{2m_e} \,, \tag{13.126}$$

is called the *Larmor frequency*. Note that $\omega_L > 0$. In the limit $\omega_L \ll \omega_0$, i.e., for magnetic fields that do not exceed a (rather large) critical value, the solution of this equation for ω is

$$\omega \simeq \omega_0 + \omega_L \,. \tag{13.127}$$

Since both ω_0 and ω_L are positive, the effect of the magnetic field is to increase to rotation frequency, corresponding to the fact that, in our setting, the Lorentz force has the same inward radial direction as the Coulomb force. The resulting magnetic moment is therefore

$$
\begin{aligned}
m_z &= -\frac{1}{2}e\omega r^2 \\
&\simeq (m_z)_0 - \frac{1}{2}e\omega_L r^2 \\
&= (m_z)_0 - \frac{e^2 r^2}{4m_e}B \,, \tag{13.128}
\end{aligned}
$$

or, in vector form,

$$\mathbf{m} = \mathbf{m}_0 - \frac{e^2 r^2}{4m_e}\mathbf{B} \,. \tag{13.129}$$

Therefore, the induced magnetic moment is opposite to the magnetic field. Consider now a collection of atoms with random orientations, in an external magnetic field. The orbital magnetic moment $\mathbf{m}_0$ averages to zero, since the orientations of the atoms are random, while the term induced by the magnetic field has a non-vanishing average value, obtained replacing in eq. (13.129) r^2 by $\langle r_\perp^2 \rangle$, where $r_\perp$ is the projection of the orbital radius on the plane orthogonal to the magnetic field. If the number of atoms per unit volume is n, the magnetization is therefore

$$\mathbf{M} = -\frac{ne^2 \langle r_\perp^2 \rangle}{4m_e}\mathbf{B} \,, \tag{13.130}$$

and therefore, using eqs. (13.91)–(13.93), the magnetic susceptibility is given by

$$\frac{\chi_m}{1 + \chi_m} = -\mu_0 \frac{ne^2 \langle r_\perp^2 \rangle}{4m_e} \,. \tag{13.131}$$

Numerically, $|\chi_m|$ is always a very small number, typically of order 10^{-6}–10^{-5} for paramagnets and 10^{-3} for diamagnetic materials. Then, eq. (13.131) can be written more simply as

$$\chi_m \simeq -\mu_0 \frac{ne^2 \langle r_\perp^2 \rangle}{4m_e} \,. \tag{13.132}$$

Frequency-dependent response of materials

As we have mentioned in Section 13.6, the response of a material, and therefore the corresponding constitutive relation, is in general frequency dependent, as in eqs. (13.77) or (13.81). In this chapter, we will first see how such frequency dependence is significantly constrained by causality and other general principles. We will then discuss explicit models for the frequency dependence of the response functions of dielectrics and of conductors.

14.1 General properties of $\sigma(\omega)$, $\epsilon(\omega)$

To illustrate general properties of the frequency dependence of the response functions, we begin with the relation $\tilde{\mathbf{j}}_{\text{free}}(\omega) = \sigma(\omega)\tilde{\mathbf{E}}(\omega)$. First of all, from the fact that $\mathbf{j}_{\text{free}}(t)$ and $\mathbf{E}(t)$ are real functions, it follows that their Fourier transforms satisfy $\tilde{\mathbf{j}}_{\text{free}}^{*}(\omega) = \tilde{\mathbf{j}}_{\text{free}}(-\omega)$ and $\tilde{\mathbf{E}}^{*}(\omega) = \tilde{\mathbf{E}}(-\omega)$, and therefore

$$\sigma^{*}(\omega) = \sigma(-\omega)\,. \tag{14.1}$$

We separate $\sigma(\omega)$ into its real and imaginary parts,

$$\sigma(\omega) = \sigma_R(\omega) + i\sigma_I(\omega)\,. \tag{14.2}$$

Then eq. (14.1) implies that (for real values of ω)

$$\boxed{\sigma_R(\omega) = \sigma_R(-\omega)\,, \qquad \sigma_I(\omega) = -\sigma_I(-\omega)\,.} \tag{14.3}$$

The same relations hold for $\epsilon(\omega)$. Another property, this one specific to $\sigma(\omega)$, follows from the fact that, according to eq. (13.95), the work made on the free charges of the material by the electric field is

$$\frac{d\mathcal{E}}{dt} = \int_V d^3x\, \mathbf{E}\!\cdot\!\mathbf{j}_{\text{free}}\,. \tag{14.4}$$

This energy is dissipated by the free electrons through collisions in the material, so the energy dissipated per unit volume and unit time is

$$\frac{d\mathcal{E}}{dV\,dt} = \mathbf{E}\!\cdot\!\mathbf{j}_{\text{free}}\,, \tag{14.5}$$

[1]Explicitly:

$$\int_{-\infty}^{+\infty} dt\, \mathbf{E}(t){\cdot}\mathbf{j}_{\text{free}}(t)$$

$$= \int_{-\infty}^{+\infty} dt \int_{-\infty}^{+\infty} \frac{d\omega'}{2\pi} \tilde{\mathbf{E}}(\omega')e^{-i\omega' t}$$

$$\cdot \int_{-\infty}^{+\infty} \frac{d\omega}{2\pi}\sigma(\omega)\tilde{\mathbf{E}}(\omega)e^{-i\omega t}$$

$$= \int_{-\infty}^{+\infty} \frac{d\omega\, d\omega'}{2\pi\,2\pi}\sigma(\omega)\tilde{\mathbf{E}}(\omega){\cdot}\tilde{\mathbf{E}}(\omega')$$

$$\times \int_{-\infty}^{+\infty} dt\, e^{-i(\omega+\omega')t}$$

$$= \int_{-\infty}^{+\infty} \frac{d\omega\, d\omega'}{2\pi\,2\pi}\sigma(\omega)\tilde{\mathbf{E}}(\omega){\cdot}\tilde{\mathbf{E}}(\omega')$$

$$\times 2\pi\delta(\omega+\omega')$$

$$= \int_{-\infty}^{+\infty} \frac{d\omega}{2\pi}\sigma(\omega)\tilde{\mathbf{E}}(\omega){\cdot}\tilde{\mathbf{E}}(-\omega)\,.$$

and the energy per unit volume dissipated from $t=-\infty$ to $t=+\infty$ is[1]

$$\frac{d\mathcal{E}}{dV} = \int_{-\infty}^{+\infty} dt\, \mathbf{E}(t){\cdot}\mathbf{j}_{\text{free}}(t)$$
$$= \int_{-\infty}^{+\infty} \frac{d\omega}{2\pi}\left[\sigma_R(\omega)+i\sigma_I(\omega)\right]\tilde{\mathbf{E}}(\omega){\cdot}\tilde{\mathbf{E}}(-\omega)\,. \quad (14.6)$$

The term $\tilde{\mathbf{E}}^*(\omega)\tilde{\mathbf{E}}(-\omega)$ is an even function of ω, and therefore using eq. (14.3), the integral involving $\sigma_I(\omega)$ vanishes, while that over σ_R is twice the integral from $\omega=0$ to $\omega=\infty$. Using also $\tilde{\mathbf{E}}^*(\omega)=\tilde{\mathbf{E}}(-\omega)$, we finally get

$$\frac{d\mathcal{E}}{dV} = 2\int_0^{+\infty} \frac{d\omega}{2\pi}\sigma_R(\omega)|\tilde{\mathbf{E}}(\omega)|^2\,. \quad (14.7)$$

By the second law of thermodynamics, the dissipated heat must be positive, for an arbitrary function $\tilde{\mathbf{E}}(\omega)$. Therefore, for each Fourier mode, we must have $\sigma_R(\omega)|\tilde{\mathbf{E}}(\omega)|^2>0$, so

$$\boxed{\sigma_R(\omega)>0\,.} \quad (14.8)$$

Another rather general property of $\sigma(\omega)$ is that it goes to zero as $\omega\to\infty$. This is a consequence of the fact that, if an electric field oscillates with a frequency much higher than the natural frequencies of the microscopic medium, the electrons cannot follow its oscillations, and no bulk movement, giving rise to a macroscopic current, is induced. In Section 14.4 we will check this behavior on a simple but explicit microscopic model for $\sigma(\omega)$. In fact, we will see that both $\sigma_R(\omega)$ and $\sigma_I(\omega)$ go to zero as $\omega\to\infty$, but $\sigma_R(\omega)$ goes to zero faster than $\sigma_I(\omega)$.

Analogous considerations can be made for $\epsilon(\omega)$. As in eq. (14.3), the reality of $\epsilon(t)$ implies that

$$\epsilon_R(\omega)=\epsilon_R(-\omega)\,, \qquad \epsilon_I(\omega)=-\epsilon_I(-\omega)\,. \quad (14.9)$$

It is convenient to introduce also $\chi_e(\omega)$ from

$$\frac{\epsilon(\omega)}{\epsilon_0}=1+\chi_e(\omega)\,, \quad (14.10)$$

which generalizes eq. (13.73), so that eq. (13.77) reads

$$\tilde{\mathbf{D}}(\omega)=\epsilon_0\tilde{\mathbf{E}}(\omega)+\epsilon_0\chi_e(\omega)\tilde{\mathbf{E}}(\omega)\,. \quad (14.11)$$

or, equivalently, from the Fourier transform of eq. (13.40),

$$\tilde{\mathbf{P}}(\omega)=\epsilon_0\chi_e(\omega)\tilde{\mathbf{E}}(\omega)\,. \quad (14.12)$$

Writing $\chi_e(\omega)=\chi_{e,R}(\omega)+i\chi_{e,I}(\omega)$, from eq. (14.9), $\chi_{e,R}(\omega)=\chi_{e,R}(-\omega)$ and $\chi_{e,I}(\omega)=-\chi_{e,I}(-\omega)$. If the frequency of the external electric field is too high, it cannot induce a net polarization in the material, so $\chi_e(\omega)$ goes to zero as $\omega\to\infty$, and correspondingly $\epsilon(\omega)/\epsilon_0$ goes to one. We will check this behavior on an explicit microscopic model in Section 14.3.

14.2   Causality constraints and Kramers–Kronig relations

We next consider the constraints imposed by causality. Fourier transforming eq. (13.81) we get[2]

$$\mathbf{j}_{\text{free}}(t) = \int_{-\infty}^{+\infty} dt' \sigma(t - t')\mathbf{E}(t') \,. \tag{14.13}$$

If σ were a generic function of time, this equation would mean that the value of $\mathbf{j}_{\text{free}}(t)$ at a given time t is determined by the applied electric field $\mathbf{E}(t')$ at all possible values of t', both in the past, $t' < t$, and in the future, $t' > t$. This clearly violate causality. We therefore discover that a relation such as eq. (13.81) only makes sense, physically, if $\sigma(t - t') = 0$ for $t' > t$, i.e., if the function $\sigma(t)$ vanishes when its argument t is negative:

$$\sigma(t) = 0 \qquad \text{if } t < 0 \,. \tag{14.14}$$

Then, eq. (14.13) becomes

$$\boxed{\mathbf{j}_{\text{free}}(t) = \int_{-\infty}^{t} dt' \, \sigma(t - t')\mathbf{E}(t') \,,} \tag{14.15}$$

consistent with causality. Equation (14.15) implies that the relation between $\mathbf{j}_{\text{free}}(t)$ and $\mathbf{E}(t)$ is non-local in time: the value of $\mathbf{j}_{\text{free}}$ at time t depends not only on the value of $\mathbf{E}$ (and, possibly, of a finite number of its derivatives) at time t, but rather on the whole past history, i.e., on the whole behavior of $\mathbf{E}(t')$ at $t' < t$. On physical grounds, we expect that $\sigma(t - t')$ will go to zero sufficiently fast as $t' \to -\infty$, i.e., that $\sigma(t)$ goes to zero sufficiently fast when its argument $t \to +\infty$, so that the response of the system only retains a memory of the recent past. In the limit in which $\sigma(\omega)$ becomes independent of the frequency, $\sigma(t)$ becomes proportional to a Dirac delta, $\sigma(t - t') \propto \delta(t - t')$, and eq. (14.13) becomes local in time. Any frequency dependence in $\sigma(\omega)$ will, however, result in a relation between $\mathbf{j}_{\text{free}}(t)$ and $\mathbf{E}(t)$ which is non-local in time.

The condition (14.14) can be translated into a condition on $\sigma(\omega)$, as follows. First of all, we observe that

$$\begin{aligned} \sigma(\omega) &\equiv \int_{-\infty}^{+\infty} dt\, \sigma(t)e^{i\omega t} \\ &= \int_{0}^{+\infty} dt\, \sigma(t)e^{i\omega t} \,, \end{aligned} \tag{14.16}$$

since the integrand vanishes for $t < 0$. This equation can be used to define $\sigma(\omega)$ for *complex* values of ω. Writing $\omega = \omega_R + i\omega_I$,

$$\sigma(\omega) = \int_{0}^{+\infty} dt\, \sigma(t)e^{i\omega_R t}e^{-\omega_I t} \,. \tag{14.17}$$

[2] Explicitly:

$$\begin{aligned} \mathbf{j}_{\text{free}}(t) &= \int_{-\infty}^{+\infty} \frac{d\omega}{2\pi}\, \tilde{\mathbf{j}}_{\text{free}}(\omega)e^{-i\omega t} \\ &= \int_{-\infty}^{+\infty} \frac{d\omega}{2\pi}\, \sigma(\omega)\tilde{\mathbf{E}}(\omega)e^{-i\omega t} \\ &= \int_{-\infty}^{+\infty} \frac{d\omega}{2\pi} \int_{-\infty}^{+\infty} dt''\sigma(t'')e^{i\omega t''} \\ &\quad \times \int_{-\infty}^{+\infty} dt'\, \mathbf{E}(t')e^{i\omega t'}e^{-i\omega t} \\ &= \int_{-\infty}^{+\infty} dt' dt'' \sigma(t'')\mathbf{E}(t') \\ &\quad \times \delta(t' + t'' - t) \\ &= \int_{-\infty}^{+\infty} dt'\sigma(t - t')\mathbf{E}(t') \,. \end{aligned}$$

In the upper half-plane, $\omega_I > 0$, the integral is strongly convergent, since also $t \geq 0$. Therefore the integral is well defined. This reasoning does not hold at $\omega_I = 0$, and we then assume that $\sigma(\omega)$ has no singularity also on the real axis.[3] In that case, we can conclude that *the function $\sigma(\omega)$ is analytic in the upper half-plane.*

This analyticity can be used to prove the *Kramers–Kronig relations,* as follows. We consider the integral

$$\oint_C d\omega' \, \frac{\sigma(\omega')}{\omega' - \omega} \,, \tag{14.18}$$

where ω is real and the contour C is shown in Fig. 14.1. Since $\sigma(\omega')$ is analytic in the upper half-plane, and the contour avoids the singularity in $\omega' = \omega$, the integrand is analytic everywhere inside C and therefore, by the Cauchy theorem, the integral vanishes. As we discussed above, for physical reasons $\sigma(\omega)$ must go to zero as $|\omega| \to \infty$ on the real axis, and this extends to the whole upper half-plane, thanks to the factor $e^{-\omega_I t}$ in eq. (14.17) (note that, in eq. (14.17), the integration domain is over $t \geq 0$). Therefore $\sigma(\omega')/(\omega' - \omega)$ goes to zero faster than $1/|\omega'|$ as $|\omega'| \to \infty$ on the upper half plane, and therefore the contribution from the semi-circle at infinity vanishes. The integral on the real axis, with the small semi-circle excluded, is just the definition of the principal part of the integral, while the integral over the small semicircle, by the Cauchy theorem, is $-i\pi$ (where the minus is due to the fact that the semicircle is followed clockwise) times the residue of the function $\sigma(\omega')/(\omega' - \omega)$ in $\omega' = \omega$, which is just $\sigma(\omega)$. Then,

$$\mathrm{P} \int_{-\infty}^{+\infty} d\omega' \, \frac{\sigma(\omega')}{\omega' - \omega} - i\pi\sigma(\omega) = 0 \,, \tag{14.19}$$

where the symbol P denotes the principal part. Separating into the real and imaginary parts, we get

$$\sigma_R(\omega) = \frac{1}{\pi} \mathrm{P} \int_{-\infty}^{+\infty} d\omega' \, \frac{\sigma_I(\omega')}{\omega' - \omega} \,, \tag{14.20}$$

$$\sigma_I(\omega) = -\frac{1}{\pi} \mathrm{P} \int_{-\infty}^{+\infty} d\omega' \, \frac{\sigma_R(\omega')}{\omega' - \omega} \,. \tag{14.21}$$

These relations (together with similar ones obtained for other functions such as $\epsilon(\omega)$, see later) are examples of *Kramers–Kronig relations,*[4] or *dispersion relations.* Note that the knowledge of $\sigma_R(\omega)$ over the whole real axis fixes $\sigma_I(\omega)$, and vice versa. We can rewrite eq. (14.21) in the form

$$
\begin{aligned}
\sigma_I(\omega) &= -\frac{1}{\pi} \mathrm{P} \int_{-\infty}^{0} d\omega' \, \frac{\sigma_R(\omega')}{\omega' - \omega} - \frac{1}{\pi} \mathrm{P} \int_{0}^{\infty} d\omega' \, \frac{\sigma_R(\omega')}{\omega' - \omega} \\
&= -\frac{1}{\pi} \mathrm{P} \int_{0}^{\infty} d\omega'' \, \frac{\sigma_R(-\omega'')}{-\omega'' - \omega} - \frac{1}{\pi} \mathrm{P} \int_{0}^{\infty} d\omega' \, \frac{\sigma_R(\omega')}{\omega' - \omega} \\
&= -\frac{1}{\pi} \mathrm{P} \int_{0}^{\infty} d\omega' \, \sigma_R(\omega') \left(\frac{1}{\omega' - \omega} - \frac{1}{\omega' + \omega} \right) \\
&= -\frac{2\omega}{\pi} \mathrm{P} \int_{0}^{\infty} d\omega' \, \frac{\sigma_R(\omega')}{\omega'^2 - \omega^2} \,, \tag{14.22}
\end{aligned}
$$

[3]We will check whether this is true on explicit examples, and we will then see how to amend the arguments below, when $\sigma(\omega)$ or $\epsilon(\omega)$ have poles on the real axis, see also Note 6 on page 370.

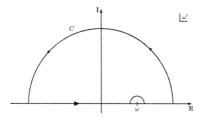

Fig. 14.1 The integration contour in the complex ω' plane discussed in the text.

[4]Found by Ralph Kronig in 1926, and Hans Kramers in 1927.

where, in the second line, we have introduced the variable $\omega'' \equiv -\omega'$ in the first integral, we have then used $\sigma_R(-\omega'') = \sigma_R(\omega'')$, and we finally renamed the integration variable ω'' as ω'. Performing the same manipulations on eq. (14.20) we see that the Kramers–Kronig relations can be rewritten as

$$
\sigma_R(\omega) = \frac{2}{\pi} P \int_0^\infty d\omega' \, \frac{\omega' \sigma_I(\omega')}{\omega'^2 - \omega^2} , \tag{14.23}
$$

$$
\sigma_I(\omega) = -\frac{2\omega}{\pi} P \int_0^\infty d\omega' \, \frac{\sigma_R(\omega')}{\omega'^2 - \omega^2} . \tag{14.24}
$$

From these relations, we can try to extract the high-frequency limit $\omega \to \infty$ of $\sigma_R(\omega)$ and $\sigma_I(\omega)$. Actually, the point is somewhat delicate mathematically, since also the integration region of ω' ranges up to infinity. Consider first the limit $\omega \to \infty$ of eq. (14.24). Let us assume that $\sigma_R(\omega')$ goes to zero sufficiently fast as $\omega' \to \infty$ so that, for large ω, the integral in eq. (14.24) gets most of its contribution from the integration region $\omega' \ll \omega$. In the limit $\omega \to \infty$ we can then try to replace $1/(\omega'^2 - \omega^2)$ by $-1/\omega^2$ inside the integral, which gives

$$
\sigma_I(\omega) \simeq \frac{2}{\pi\omega} \int_0^\infty d\omega' \, \sigma_R(\omega') , \qquad (\omega \to \infty) . \tag{14.25}
$$

This way of extracting the large frequency limit can be correct only if $\sigma_R(\omega)$ goes to zero faster than $1/\omega$ at large ω, so that the integral in eq. (14.25) converges. As we will see, this is the case in the explicit microscopic models that we will study in Section 14.4, where $\sigma_R(\omega) \sim 1/\omega^2$ at large ω. Therefore $\sigma_I(\omega)$ goes to zero as $1/\omega$, with a coefficient determined by the integral of $\sigma_R(\omega')$ over all frequencies. Relations of this form are called *sum rules*.[5]

Another useful sum rule is obtained setting $\omega = 0$ in eq. (14.23). This gives the static conductivity, i.e., $\sigma_R(\omega = 0)$, as an integral over all frequencies of the imaginary part,

$$
\sigma_R(\omega = 0) = \frac{2}{\pi} \int_0^\infty d\omega' \, \frac{\sigma_I(\omega')}{\omega'} . \tag{14.26}
$$

Observe that the property (14.3) implies that, as $\omega \to 0$, $\sigma_I(\omega)$ vanishes at least as ω (or, more generally, as an odd power of ω), and therefore the integral in eq. (14.26) converges.

Similar considerations can be performed for $\epsilon(\omega)$. From eq. (14.11), going back into the time domain, we get

$$
\mathbf{D}(t) = \epsilon_0 \mathbf{E}(t) + \epsilon_0 \int_{-\infty}^{+\infty} dt' \chi_e(t - t') \mathbf{E}(t') . \tag{14.27}
$$

Therefore, causality implies again that $\chi_e(t) = 0$ for $t < 0$, so that eq. (14.27) becomes

$$
\boxed{\mathbf{D}(t) = \epsilon_0 \mathbf{E}(t) + \epsilon_0 \int_{-\infty}^{t} dt' \chi_e(t - t') \mathbf{E}(t') .} \tag{14.28}
$$

[5]Note that we cannot extract the high-frequency limit of $\sigma_R(\omega)$ in the same way as we did in eq. (14.25). If we would naively take the limit $\omega \to \infty$ by replacing $1/(\omega'^2 - \omega^2)$ with $-1/\omega^2$ inside the integral in eq. (14.23), we would get that, at large ω, $\sigma_R(\omega)$ is equal to $-2/(\pi\omega^2)$ times $\int_0^\infty d\omega' \, \omega' \sigma_I(\omega')$. However, having already found from eq. (14.25) that $\sigma_I(\omega')$ goes as $1/\omega'$ at large ω', we see that this integral does not converge, and keeping the term ω'^2 in $1/(\omega'^2 - \omega^2)$ is essential for the convergence. We will see, however, that in the explicit model of Section 14.4 $\sigma_R(\omega)$ still goes as $1/\omega^2$ at high frequencies.

Similarly to what we found for $\sigma(\omega)$, the condition $\chi_e(t) = 0$ for $t > 0$ implies that the Fourier transform $\chi_e(\omega)$ is analytic in the upper-half plane (apart, possibly, from poles on the real axis, that for the moment we assume that are not present).[6] We can then repeat the above derivation and find the corresponding Kramers–Kronig relations,

$$\chi_{e,R}(\omega) = \frac{2}{\pi} \mathrm{P} \int_0^\infty d\omega' \, \frac{\omega' \chi_{e,I}(\omega')}{\omega'^2 - \omega^2} \tag{14.29}$$

$$\chi_{e,I}(\omega) = -\frac{2\omega}{\pi} \mathrm{P} \int_0^\infty d\omega' \, \frac{\chi_{e,R}(\omega')}{\omega'^2 - \omega^2} . \tag{14.30}$$

In terms of the dielectric function $\epsilon_r(\omega) = 1 + \chi_e(\omega)$ these can be rewritten as[7]

$$\epsilon_{r,R}(\omega) = 1 + \frac{2}{\pi} \mathrm{P} \int_0^\infty d\omega' \, \frac{\omega' \epsilon_{r,I}(\omega')}{\omega'^2 - \omega^2} , \tag{14.31}$$

$$\epsilon_{r,I}(\omega) = -\frac{2\omega}{\pi} \mathrm{P} \int_0^\infty d\omega' \, \frac{\epsilon_{r,R}(\omega') - 1}{\omega'^2 - \omega^2} , \tag{14.32}$$

where we wrote $\epsilon_r(\omega) = \epsilon_{r,R}(\omega) + i\epsilon_{r,I}(\omega)$. We will see in Section 14.3 that the extraction of the high-frequency limit from these formulas involves some subtlety.

14.3 The Drude–Lorentz model for $\epsilon(\omega)$

We now consider a simple classical model for the permittivity $\epsilon(\omega)$ of a dielectric, named after Drude and Lorentz. Consider a single electron bound to an atom or a molecule, in the presence of an external electric field $\mathbf{E}(t)$. The simplest classical description corresponds to a damped harmonic oscillator, with frequency ω_0 and damping constant γ_0. Denoting by $\mathbf{x}_0(t)$ its position, the equation of motion is

$$\ddot{\mathbf{x}}_0 + \gamma_0 \dot{\mathbf{x}}_0 + \omega_0^2 \mathbf{x}_0 = -\frac{e\mathbf{E}(t)}{m_e} . \tag{14.33}$$

One should be aware that this classical description is a gross simplification, and the underlying description of the behavior of the electrons is necessarily quantum-mechanical.[8] Still, this simple classical model is useful for a first understanding. Performing the Fourier transform of eq. (14.33) we get

$$\left(-\omega^2 - i\omega\gamma_0 + \omega_0^2\right) \tilde{\mathbf{x}}(\omega) = -\frac{e\tilde{\mathbf{E}}(\omega)}{m_e} . \tag{14.34}$$

The dipole moment of the bound electron is $\mathbf{d}(t) = -e\mathbf{x}(t)$, so $\tilde{\mathbf{d}}(\omega) = -e\tilde{\mathbf{x}}(\omega)$ and, using eq. (14.34),

$$\tilde{\mathbf{d}}(\omega) = \frac{e^2}{m_e} \frac{1}{\omega_0^2 - \omega^2 - i\omega\gamma_0} \tilde{\mathbf{E}}(\omega) . \tag{14.35}$$

[6]This is not always the case for functions that describe the response of a material. For instance, for metals we see from eq. (13.90) that the permittivity function $\epsilon(\omega)$, and therefore also the corresponding expression for $\chi_e(\omega)$, has a pole at $\omega = 0$, since $\sigma(\omega)$ has a finite value at $\omega = 0$ (equal to the static conductivity σ_0). In general, if a function $f(\omega)$ has a simple pole at $\omega = \bar{\omega}$, of the form $f(\omega) \simeq f_0/(\omega - \bar{\omega})$ for $\omega \to \bar{\omega}$, one can write the dispersion relation for the function $g(\omega) = f(\omega) - f_0/(\omega - \bar{\omega})$, from which the pole has been subtracted, and derive the Kramers–Kronig relations for $g(\omega)$; this is equivalent to modifying the contour C in Fig. 14.1 with another semi-arc in the upper half plane, so as to avoid also the pole of the function $f(\omega)$.

[7]Observe that we could not have written the Kramers–Kronig relations directly in terms of $\epsilon_r(\omega)$, because $\epsilon_r(\omega)$ goes to one, rather than to zero, as $|\omega| \to \infty$, and therefore the integral on the semi-circle at infinity, in the contour shown in Fig. 14.1, does not vanish. The dispersion relation must be written in terms of a function that vanishes at infinity, i.e., $\chi_e(\omega) = \epsilon_r(\omega) - 1$. When rewritten in terms of $\epsilon_r(\omega)$ as in eqs. (14.31) and (14.32), this is called a *subtracted dispersion relation*. Another example of subtracted dispersion relation is given by the subtraction of a pole on the real axis, discussed in Note 6.

[8]This is particularly important for semiconductors, where the absence of conduction in the static limit is due to filled bands rather than to a localization of the electrons around individual atoms.

Denoting by n_b the number of bound electrons per unit volume, the polarization $\mathbf{P}(t)$ (i.e., the electric dipole moment per unit volume) is then given by

$$\tilde{\mathbf{P}}(\omega) = \frac{n_b e^2}{m_e} \frac{1}{\omega_0^2 - \omega^2 - i\omega\gamma_0} \tilde{\mathbf{E}}(\omega).$$ (14.36)

Then, from eq. (14.12),

$$\chi_e(\omega) = \frac{\omega_p^2}{\omega_0^2 - \omega^2 - i\omega\gamma_0},$$ (14.37)

where we have defined

$$\boxed{\omega_p^2 = \frac{n_b e^2}{\epsilon_0 m_e},}$$ (14.38)

and the dielectric function is given by

$$\boxed{\epsilon_r(\omega) \equiv \frac{\epsilon(\omega)}{\epsilon_0} = 1 + \frac{\omega_p^2}{\omega_0^2 - \omega^2 - i\omega\gamma_0}.}$$ (14.39)

Observe that ω_p has dimensions of a frequency. For reasons that will become clear in Section 15.3.2, it is called the *plasma frequency* of the material (even when, as in this case, the material is a dielectric rather than a plasma).

A simple improvement of this model, that takes better into account, at least at a heuristic level, the quantum description of the bound electrons, is based on the fact that an electron in the ground state of an atom can be excited to a discrete set of energy levels, corresponding to a set of absorption frequencies ω_i and friction constants γ_i, with probability f_i. We can then improve eq. (14.39), writing

$$\epsilon_r(\omega) = 1 + \omega_p^2 \sum_{i=1}^{N} \frac{f_i}{\omega_i^2 - \omega^2 - i\omega\gamma_i},$$ (14.40)

where, for definiteness, we included N oscillator levels (possibly, with $N \to \infty$). The constants f_i are called the "oscillator strengths" and, as we will see below, the Kramers–Kronig relations enforce the condition

$$\sum_{i=1}^{N} f_i = 1.$$ (14.41)

In a classical description, one could then think to the f_i as the fraction of electrons in the i-th state, in which case one would expect $0 \leq f_i \leq 1$. In a full quantum treatment of the atom, that here we have rather modeled simply as a classical damped oscillator, eqs. (14.40) and (14.41) still hold, but the interpretation of the oscillator strengths f_i is different, and they no longer satisfy $0 \leq f_i \leq 1$.[9]

[9] In a full quantum treatment, the oscillator strengths are related to the transition between quantum states of the atom. If the atom is in a quantum state n, the oscillator strength f_i should actually written as f_{in}, and is determined by transition probability between the given state n and a generic atomic state $i \neq n$, multiplied by a factor that contains the energy difference $E_i - E_n$. Therefore, "absorption oscillator strengths," obtained from transitions to states i with $E_i > E_n$, are positive, while "emission oscillator strengths" are negative. Equation (14.41) still hold, and, in the quantum context, is called the *Thomas–Reiche–Kuhn sum rule*, although the standard normalization of the quantum oscillator strengths is such that it is rather written as $\sum_{i \neq n} f_{in} = Z$, where Z is the total number of electrons in the atoms. In the classical treatment, it is common to reabsorb Z in a rescaling of the f_i, as we have done. Note, however, that now individual oscillator strengths can be negative, and as a consequence the sum rule (14.41) no longer enforces $f_i \leq 1$, either. For a discussion of the Drude–Lorentz model in quantum context, see e.g., Dressel and Grüner (2002), in particular Section 6.1. For a discussion of oscillator strengths in a quantum context and the Thomas–Reiche–Kuhn sum rule, see also Chapter 10 of Rybicki and Lightman (1979).

For bound electrons, all ω_i are strictly positive, and this description is appropriate for a dielectric. We will see later that, in a conductor, the contribution of the free electrons can be described by the same expression, with the addition of electrons in a state with $\omega_i = 0$.

It is instructive to check that these results satisfy the general properties discussed in Sections 14.1 and 14.2. We use for simplicity the expression for $\epsilon(\omega)$ given in eq. (14.39), but this could be repeated for eq. (14.40). Separating the real and imaginary parts, we get (for ω real)

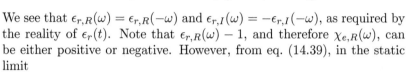

$$\epsilon_{r,R} \equiv \frac{\epsilon_R(\omega)}{\epsilon_0} = 1 + \omega_p^2 \frac{\omega_0^2 - \omega^2}{(\omega_0^2 - \omega^2)^2 + \omega^2\gamma_0^2}, \qquad (14.42)$$

$$\epsilon_{r,I} \equiv \frac{\epsilon_I(\omega)}{\epsilon_0} = \omega_p^2\gamma_0 \frac{\omega}{(\omega_0^2 - \omega^2)^2 + \omega^2\gamma_0^2}. \qquad (14.43)$$

We see that $\epsilon_{r,R}(\omega) = \epsilon_{r,R}(-\omega)$ and $\epsilon_{r,I}(\omega) = -\epsilon_{r,I}(-\omega)$, as required by the reality of $\epsilon_r(t)$. Note that $\epsilon_{r,R}(\omega) - 1$, and therefore $\chi_{e,R}(\omega)$, can be either positive or negative. However, from eq. (14.39), in the static limit

$$\epsilon_r = 1 + \frac{\omega_p^2}{\omega_0^2} > 1, \qquad (14.44)$$

where $\epsilon_r \equiv \epsilon(\omega = 0)/\epsilon_0$. A plot of $\epsilon_{r,R}(\omega)$ and $\epsilon_{r,I}(\omega)$ is shown in Fig. 14.2.[10]

The poles of $\chi_e(\omega)$ in the complex plane are given by the solution of the equation $\omega^2 + i\omega\gamma_0 - \omega_0^2 = 0$, i.e., by

$$\omega_\pm = \pm\sqrt{\omega_0^2 - (\gamma_0/2)^2} - i\frac{\gamma_0}{2}. \qquad (14.45)$$

If $\omega_0 > \gamma_0/2$, the square root root is real. In any case, there are two poles, both in the lower half-plane (even when $\omega_0 \le \gamma_0/2$). Therefore $\chi_e(\omega)$ is analytic in the upper half-plane. As discussed in Section 14.2, these analyticity properties in the complex ω-plane are related to causality: consider the electric susceptibility in the time domain,

$$\chi_e(t) = \int_{-\infty}^{+\infty} \frac{d\omega}{2\pi} \chi_e(\omega)e^{-i\omega t}. \qquad (14.46)$$

Note that the integral converges at $\omega \to \pm\infty$, since at large $|\omega|$ we have $\chi_e(\omega) \propto 1/\omega^2$, see eq. (14.37), and $\chi_e(\omega)$ has no singularity on the real axis, so the integral is well defined. For $t < 0$ we can close the contour in the upper half-plane, since the factor $e^{-i(\omega_R + i\omega_I)t} = e^{-i\omega_R t}e^{\omega_I t}$ ensures the vanishing of the integral on the semi-circle at infinity when $\omega_I \to +\infty$ and $t < 0$. Then, since $\chi_e(\omega)$ is analytic in the upper half-plane, $\chi(t) = 0$ for $t < 0$, as required by causality. For $t > 0$, closing the contour in the lower half plane and picking the residue of the two poles, we find that (for $\omega_0 > \gamma_0/2$) $\chi_e(t)$ is a (real) superposition of terms proportional to $e^{-i\omega_+ t}$ and $e^{-i\omega_- t}$, i.e., is proportional to

$$\chi_e^\pm(t) \propto e^{\pm it\sqrt{\omega_0^2 - (\gamma_0/2)^2}} e^{-(\gamma_0/2)t}. \qquad (14.47)$$

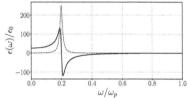

Fig. 14.2 The functions $\epsilon_R(\omega)/\epsilon_0$ (solid line) and $\epsilon_I(\omega)/\epsilon_0$ (dashed line) from eqs. (14.42) and (14.43), as a function of ω/ω_p. We have set for definiteness $\omega_0 = 0.2\omega_p$ and $\gamma_0 = 0.1\omega_0 = 0.02\omega_p$.

[10]Comparing with Fig. 13.4 on page 357 we see that the dielectric permittivity of water is not reproduced by the Drude–Lorentz model. This is due to the fact that the damped harmonic oscillator model that we have considered describes the dipole moment induced by an external electric field, see eq. (14.35). In contrast, water molecules have a permanent electric dipole, i.e., are *polar* molecules. The generation of a polarization vector **P** is then due to the fact that the molecules align themselves with the applied electric field, and is not described by eq. (14.33).

From eq. (14.28), the memory of the past behavior of $\mathbf{E}(t)$ therefore vanishes exponentially, with a decay time $\tau = 2/\gamma_0$. More generally, in the model (14.40), the i-th term in the sum contributes with a decay time $\tau_i = 2/\gamma_i$.

From the explicit result (14.42), (14.43) we can extract the high-frequency limit of the dielectric function,

$$\epsilon_{r,R}(\omega) - 1 = -\frac{\omega_p^2}{\omega^2}, \tag{14.48}$$

$$\epsilon_{r,I}(\omega) = \frac{\omega_p^2 \gamma_0}{\omega^3}, \tag{14.49}$$

so, to lowest non-trivial order, $\epsilon_r(\omega) - 1$ is real, and

$$\epsilon_r(\omega) = 1 - \frac{\omega_p^2}{\omega^2} + \mathcal{O}\left(\frac{1}{\omega^3}\right). \tag{14.50}$$

It is interesting to compare these asymptotic behaviors to those that can be extracted from the Kramers–Kronig relations (14.31, 14.32). In the large ω limit, eq. (14.32) gives

$$\epsilon_{r,I}(\omega) = \frac{2}{\pi\omega} \int_0^\infty d\omega' \, [\epsilon_{r,R}(\omega') - 1] + \mathcal{O}\left(\frac{1}{\omega^3}\right), \qquad (\omega \to \infty), \tag{14.51}$$

where we assumed that $\epsilon_{r,R}(\omega') - 1$ goes to zero sufficiently fast at large ω', so that the integral in eq. (14.51) converges, and we could then replace $1/(\omega'^2 - \omega^2)$ by $-1/\omega^2$ inside the integral. Apparently, this seems to indicate that $\epsilon_{r,I}(\omega)$ goes as $1/\omega$ at large ω while, from the explicit result (14.49), we know that it rather goes as $1/\omega^3$. The solution of this apparent discrepancy is that, for the function (14.42), the integral in eq. (14.51) not only converges, but in fact even vanishes, leaving the $\mathcal{O}(1/\omega^3)$ term as the leading one.[11] So, also from the Kramers–Kronig relations we find that, for large ω, the leading term in $\epsilon_I(\omega)$ is of order $1/\omega^3$. Then, eq. (14.31) shows that

$$\epsilon_{r,R}(\omega) - 1 \simeq -\frac{2}{\pi\omega^2} \int_0^\infty d\omega' \, \omega' \epsilon_{r,I}(\omega'), \qquad (\omega \to \infty), \tag{14.52}$$

where, thanks to the $1/\omega'^3$ asymptotic behavior of $\epsilon_I(\omega')$, the integral converges at large ω', so, in eq. (14.31), in the large ω limit we could replace $1/(\omega'^2 - \omega^2)$ by $-1/\omega^2$. This result is in agreement with the explicit result (14.48). Inserting the asymptotic behavior (14.48) into the left-hand side of eq. (14.52) we see that, independently of the damping details, and therefore of the precise form of the function $\epsilon_{r,I}(\omega)$, we have the sum rule

$$\int_0^\infty d\omega \, \omega \epsilon_{r,I}(\omega) = \frac{\pi}{2}\omega_p^2. \tag{14.53}$$

This is known as the f-sum rule. If we apply it to the more complicated model (14.40), the f-sum rule enforces the condition $\sum_{i=1}^N f_i = 1$.[12] Observe that, for eq. (14.39), the Kramers–Kronig relations are automatically satisfied, as we saw above. This follows from the fact that

---

[11] The proof is non-trivial. Using eq. (14.42), the integral in eq. (14.51) is proportional to the integral

$$I(u_0) \equiv \int_0^\infty du \, \frac{1 - u^2}{(1 - u^2)^2 + u_0^2 u^2},$$

where $u = \omega'/\omega_0$ and $u_0 = \gamma_0/\omega_0$. At first sight, it is not at all obvious that this integral vanishes, furthermore for all values of u_0. Actually, a direct computation of the primitive of the integrand it is quite involved, and it is much simpler to show that the contribution to the integral from the integration region $0 < u < 1$, where the integrand is positive, cancels against the contribution from $1 < u < \infty$, where it is negative. To this purpose, we write $I(u_0) = I_1(u_0) + I_2(u_0)$, where

$$I_1(u_0) = \int_0^1 du \, \frac{1 - u^2}{(1 - u^2)^2 + u_0^2 u^2},$$

$$I_2(u_0) = \int_1^\infty du \, \frac{1 - u^2}{(1 - u^2)^2 + u_0^2 u^2}.$$

In I_1 we make the transformation of variables $u = \tanh\eta$, which gives

$$I_1(u_0) = \int_0^\infty d\eta \, \frac{1}{1 + u_0^2 \sinh^2\eta \cosh^2\eta}.$$

In I_2 we write $u = 1/\tanh\eta$, and we get $I_2 = -I_1$, so, indeed, $I_1 + I_2 = 0$ for all u_0.

[12] Again, the proof is not completely obvious. Inserting eq. (14.43) into eq. (14.53) we get

$$\frac{\pi}{2} = \sum_{i=1}^N f_i \int_0^\infty du \, \frac{u^2}{(u_i^2 - u^2)^2 + u^2},$$

where $u = \omega/\gamma_i$ and $u_i = \omega_i/\gamma_i$. Despite its appearance, the integral is actually independent of u_i, and equal to its value in $u_i = 0$, which is $\pi/2$. This can be shown (with some manipulations) using formula 2.161.3 of Gradshteyn and Ryzhik (1980), and can also be directly checked numerically.

eq. (14.39) has been obtained from an explicit causal model, expressed by eq. (14.33), and therefore the constraints due to causality are automatically satisfied. In contrast, in our classical context, eq. (14.40) is simply an ansatz, not derived by an explicit causal equation of motion, and therefore the Kramers–Kronig relations provide a non-trivial constraint.

Some aspects of these results are more general than the specific Drude–Lorentz model that we have discussed. In particular, the fact that, to lowest order, the asymptotic behavior (14.50) is independent of γ_0 follows from the fact that, at sufficiently large frequencies, the details of the damping mechanism, represented by the term $-i\omega\gamma_0$ in eq. (14.34), become irrelevant, since in any case the inner dynamics of the atom or molecule cannot follow the fast oscillations of the external field.

14.4 The Drude model of conductivity

We next consider a simple microscopic model of the conductivity $\sigma(\omega)$, the *Drude model* (or Drude–Sommerfeld model). Again, the model is purely classical, and treats the free electrons of a conductor as a non-interacting gas of classical electrons. A more realistic treatment would involve quantum mechanics and considerations of the band structure of the solid, and rather belongs to a condensed matter course. However, just as for the Drude–Lorentz model of $\epsilon(\omega)$, the simple classical description that we present here is already useful for a first understanding. The basic physics is quite simple. The free electrons inside a material are accelerated by the external electric field. The collisions with the ions of the material, however, provide a friction term that opposes the acceleration. In an average sense, the equation of motion of an electron is then given by eq. (14.33), where now $\omega_0 = 0$, since there is no restoring force. The friction constant γ_0 is written as $1/\tau$, and τ is identified with the typical time between collisions. Then, the average velocity $\bar{\mathbf{v}}(t)$ of the electrons is governed by the equation

$$\frac{d\bar{\mathbf{v}}(t)}{dt} + \frac{1}{\tau}\bar{\mathbf{v}}(t) = -\frac{e\mathbf{E}(t)}{m_e}\,. \tag{14.54}$$

We write

$$\mathbf{j}_{\text{free}}(t) = -en_f\bar{\mathbf{v}}(t)\,, \tag{14.55}$$

where n_f is the number density of free electrons (with charge $-e$). From eq. (14.54) it then follows that

$$\frac{d\mathbf{j}_{\text{free}}}{dt} + \frac{1}{\tau}\mathbf{j}_{\text{free}}(t) = \frac{n_f e^2}{m_e}\mathbf{E}(t)\,. \tag{14.56}$$

We have also assumed that the electric field is spatially constant over the relevant length-scale, which in this case is the mean free path of the electrons between collisions.[13] Performing the Fourier transform, this gives

$$-i\omega\tilde{\mathbf{j}}_{\text{free}}(\omega) + \frac{1}{\tau}\tilde{\mathbf{j}}_{\text{free}}(\omega) = \frac{n_f e^2}{m_e}\tilde{\mathbf{E}}(\omega)\,, \tag{14.57}$$

[13]More precisely, after performing the Fourier transform as in eq. (14.57), we assume that, at the frequency ω of interest, the Fourier mode $\tilde{\mathbf{E}}(\omega)$ can be taken to be spatially constant. We will come back to this assumption at the end of Section 15.3.1, when we discuss the anomalous skin effect.

and therefore $\tilde{\mathbf{j}}_{\text{free}}(\omega) = \sigma(\omega)\tilde{\mathbf{E}}(\omega)$ with

$$\sigma(\omega) = \frac{n_f e^2 \tau}{m_e} \frac{1}{1 - i\omega\tau}. \tag{14.58}$$

Defining, as in eq. (13.82), $\sigma_0 \equiv \sigma(\omega = 0)$, eq. (14.58) can also be rewritten as

$$\sigma(\omega) = \frac{\sigma_0}{1 - i\omega\tau}, \tag{14.59}$$

where

$$\sigma_0 = \frac{n_f e^2 \tau}{m_e}. \tag{14.60}$$

Separating the real and imaginary parts (for ω real), we get

$$\sigma_R(\omega) = \sigma_0 \frac{1}{1 + \omega^2 \tau^2}, \tag{14.61}$$

$$\sigma_I(\omega) = \sigma_0 \frac{\omega\tau}{1 + \omega^2 \tau^2}. \tag{14.62}$$

We can again check that the general properties discussed in Section 14.1 are satisfied. The reality condition (14.3) is obviously satisfied, and also the condition (14.8), which is a consequence of the second law of thermodynamics. Let us now check the Kramers–Kronig relations. We start from eq. (14.24), that we rewrite in the form [compare with the third line in eq. (14.22)]

$$\begin{aligned}\sigma_I(\omega) &= -\frac{1}{\pi} P \int_0^\infty d\omega'\, \sigma_R(\omega') \left(\frac{1}{\omega' - \omega} - \frac{1}{\omega' + \omega}\right) \\ &= -\frac{\sigma_0}{\pi} P \int_0^\infty d\omega' \frac{1}{1 + \omega'^2\tau^2} \left(\frac{1}{\omega' - \omega} - \frac{1}{\omega' + \omega}\right). \end{aligned} \tag{14.63}$$

We take for definiteness $\omega > 0$. Then, the second term in the parenthesis has no pole, so there the principal value prescription is unnecessary. Introducing $u = \omega'\tau$ and $u_0 = \omega\tau$, we must then compute

$$\begin{aligned}&P \int_0^\infty \frac{d\omega'}{1 + \omega'^2\tau^2} \left(\frac{1}{\omega' - \omega} - \frac{1}{\omega' + \omega}\right) \\ &= P \left(\int_0^\infty \frac{du}{1 + u^2} \frac{1}{u - u_0}\right) - \int_0^\infty \frac{du}{1 + u^2} \frac{1}{u + u_0} \\ &= \lim_{\epsilon \to 0^+} \left(\int_0^{u_0 - \epsilon} \frac{du}{1 + u^2} \frac{1}{u - u_0} + \int_{u_0 + \epsilon}^\infty \frac{du}{1 + u^2} \frac{1}{u - u_0}\right) \\ &\quad - \int_0^\infty \frac{du}{1 + u^2} \frac{1}{u + u_0}, \end{aligned} \tag{14.64}$$

where, in the last equality, we have used the definition of principal value. These integrals can be computed analytically,[14] and the result is

$$\frac{1}{\pi} P \int_0^\infty \frac{d\omega'}{1 + \omega'^2\tau^2} \left(\frac{1}{\omega' - \omega} - \frac{1}{\omega' + \omega}\right) = -\frac{\omega\tau}{1 + \omega^2\tau^2}. \tag{14.65}$$

[14]The explicit computation is as follows. For the first two integrals, taking into account that $u_0 > 0$, $\epsilon > 0$ and $\epsilon < u_0$ (since we want to eventually take the limit $\epsilon \to 0^+$ with u_0 a fixed positive constant), we have

$$\int_0^{u_0-\epsilon} \frac{du}{1 + u^2} \frac{1}{u - u_0} = -\frac{f_1(u_0, \epsilon)}{2(1 + u_0^2)},$$

and

$$\int_{u_0+\epsilon}^\infty \frac{du}{1 + u^2} \frac{1}{u - u_0} = \frac{f_2(u_0, \epsilon)}{2(1 + u_0^2)},$$

where

$$f_1(u_0, \epsilon) = \log\left[1 + (u_0 - \epsilon)^2\right] + 2u_0 \tan^{-1}(u_0 - \epsilon) - 2\log\epsilon + 2\log u_0,$$
$$f_2(u_0, \epsilon) = \log\left[1 + (u_0 + \epsilon)^2\right] + 2u_0 \tan^{-1}(u_0 + \epsilon) - 2\log\epsilon - \pi u_0.$$

As expected, as $\epsilon \to 0^+$, the two integrals are separately divergent, as $\log\epsilon$, corresponding to the fact that, as $\epsilon \to 0^+$, the integration region approaches a pole singularity $1/(u - u_0)$. However, their sum has a finite limit, equal to

$$-\frac{2\log u_0 + \pi u_0}{2(1 + u_0^2)}.$$

The third integral in eq. (14.64) is also elementary and requires no regularization,

$$\int_0^\infty \frac{du}{1 + u^2} \frac{1}{u + u_0} = \frac{-2\log u_0 + \pi u_0}{2(1 + u_0^2)}.$$

Then, putting everything, we get eq. (14.65).

Inserting this into eq. (14.63) we indeed obtain a result for $\sigma_I(\omega)$ that agrees with that given in eq. (14.62), confirming that the Kramers–Kronig relation (14.24) is satisfied, as it should. We can similarly check eq. (14.23).

Several other interesting properties can be explicitly checked from the explicit expression (14.58). In particular we see that, in the complex ω plane, the function $\sigma(\omega)$ has a single pole at $\omega = -i/\tau$. This pole is in the lower half-plane and therefore $\sigma(\omega)$ is analytic in the upper half-plane, as we expected from the general discussion in Section 14.2. Let us then compute the function $\sigma(t)$ that enters in eq. (14.13). We write

$$\sigma(t) = \int_{-\infty}^{+\infty} \frac{d\omega}{2\pi} \sigma(\omega) e^{-i\omega t}$$

$$= \sigma_0 \int_{-\infty}^{+\infty} \frac{d\omega}{2\pi} \frac{1}{1 - i\omega\tau} e^{-i\omega t}. \qquad (14.66)$$

Writing $\omega = \omega_R + i\omega_I$, we have $e^{-i\omega t} = e^{-i\omega_R t} e^{\omega_I t}$. Therefore, for $t < 0$, we can close the contour in the upper half-plane $\omega_I > 0$, where there is no singularity, and the integral vanishes. Therefore $\sigma(t) = 0$ for $t < 0$, in agreement with eq. (14.14). For $t > 0$ we close instead the contour in the lower half-plane. Picking the residue at the pole according to the Cauchy theorem (and taking into account a minus sign because the integration contour runs clockwise, see Fig. 14.3), and writing σ_0 explicitly as in eq. (14.60),

$$\sigma(t) = \frac{n_f e^2 \tau}{m_e} (-2\pi i) \text{Res}_{\omega = -i/\tau} \left[\frac{1}{2\pi} \frac{1}{(-i\tau)(\omega + i/\tau)} e^{-i\omega t} \right]$$

$$= \frac{n_f e^2}{m_e} e^{-t/\tau}. \qquad (14.67)$$

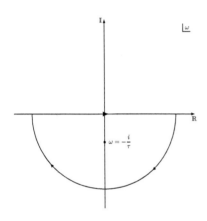

Fig. 14.3 The integration contour in the complex ω plane for $t > 0$. The black dot denotes the position of the pole.

Then, eq. (14.15) becomes

$$\mathbf{j}_{\text{free}}(t) = \frac{n_f e^2}{m_e} \int_{-\infty}^{t} dt' \, e^{-(t-t')/\tau} \, \mathbf{E}(t'). \qquad (14.68)$$

This shows that, in the Drude model, the value of the current at time t is determined by the value of the electric field at all times $t' < t$, weighted with an exponential factor that, basically, tells us that the correlation takes place only over a time interval $t - t'$ of order τ. The value of the electric field at very early times t', such that $t - t' \gg \tau$, has an effect that is exponentially suppressed. In other words, the system has a "memory" of order τ, i.e., of order of the typical collision time. This is completely analogous to the result that we found in eq. (14.47) for the electric susceptibility.

Finally, we can study the high-frequency limit. From eqs. (14.61) and (14.62), in the limit $\omega\tau \gg 1$,

$$\sigma_R(\omega) \simeq \frac{\sigma_0}{\omega^2 \tau^2}, \qquad (14.69)$$

$$\sigma_I(\omega) \simeq \frac{\sigma_0}{\omega\tau}. \qquad (14.70)$$

Therefore, both the real and imaginary parts go to zero, as expected from the general argument discussed at the end of Section 14.1, but σ_R goes to zero faster than σ_I, so that in the high-frequency limit the function $\sigma(\omega)$ becomes purely imaginary. Note also that, using eq. (14.60), in eq. (14.70) τ cancels. This signals that, in the high-frequency limit, the details of the relaxation mechanisms are no longer relevant, to $\mathcal{O}(1/\omega)$. This is due to the fact that, when driven at such a high frequency, the motion of the electrons reverses many times before they have a chance to collide, and this is the dominant mechanism that suppressed the generation of a bulk movement, that would otherwise lead to a macroscopic current.

The sum rule (14.25) is also easily checked,

$$
\begin{aligned}
\sigma_I(\omega) &\simeq \frac{2}{\pi\omega} \int_0^\infty d\omega' \, \sigma_R(\omega') \qquad (\omega \to \infty), \\
&= \frac{2\sigma_0}{\pi\omega} \int_0^\infty d\omega' \, \frac{1}{1 + \omega'^2 \tau^2} \\
&= \frac{\sigma_0}{\omega\tau}, \qquad\qquad\qquad\qquad\qquad (14.71)
\end{aligned}
$$

which correctly reproduces eq. (14.70).

14.5   The dielectric function of metals

As already discussed in Section 13.6.2, in metals we have both free and bound electrons, whose contributions combine into a single response function $\epsilon(\omega)$, the dielectric function of metals, given by eq. (13.90). To understand the physical meaning of this response function, we use the Drude–Lorentz model (14.40) for the contribution of the bound electrons, and the Drude model (14.58) for $\sigma(\omega)$. We now denote by $\epsilon_b(\omega)$ the contribution from the bound electron, that we write as

$$
\frac{\epsilon_b(\omega)}{\epsilon_0} = 1 + \frac{n_b e^2}{\epsilon_0 m_e} \sum_{i=1}^{N} \frac{(f_b)_i}{\omega_i^2 - \omega^2 - i\omega\gamma_i}, \qquad (14.72)
$$

where, for reasons that will become clear soon, we have now denoted by $(f_b)_i$ the quantities that were denoted by f_i in eq. (14.40), and we used the explicit expression $n_b e^2 / \epsilon_0 m_e$ for the quantity that was denoted as ω_p^2 in eq. (14.38). Note that

$$
\sum_{i=1}^{N} (f_b)_i = 1. \qquad (14.73)
$$

This gives, for the full dielectric function $\epsilon_r(\omega) = \epsilon(\omega)/\epsilon_0$ of a metal,

$$
\epsilon_r(\omega) = 1 - \frac{n_f e^2}{\epsilon_0 m_e} \frac{1}{\omega^2 + i\omega\tau^{-1}} + \frac{n_b e^2}{\epsilon_0 m_e} \sum_{i=1}^{N} \frac{(f_b)_i}{\omega_i^2 - \omega^2 - i\omega\gamma_i}, \qquad (14.74)
$$

where n_f and n_b are the number densities of free and bound electrons, respectively. We now introduce the total number density of electrons,

$n = n_f + n_b$, while the plasma frequency, in the presence of both free and bound electrons, is defined as

$$\omega_p^2 = \frac{ne^2}{\epsilon_0 m_e} \, . \tag{14.75}$$

We also define $f_0 = n_f/n$, $f_i = (n_b/n)(f_b)_i$ and $\gamma_0 = 1/\tau$. Then, eq. (14.74) can be rewritten as

$$
\begin{aligned}
\epsilon_r(\omega) &= 1 - \frac{\omega_p^2 f_0}{\omega^2 + i\omega\gamma_0} + \omega_p^2 \sum_{i=1}^{N} \frac{f_i}{\omega_i^2 - \omega^2 - i\omega\gamma_i} \\
&= 1 + \omega_p^2 \sum_{i=0}^{N} \frac{f_i}{\omega_i^2 - \omega^2 - i\omega\gamma_i} \, ,
\end{aligned}
\tag{14.76}
$$

where, in the last line, we included in the sum the index $i = 0$, with $\omega_{i=0} \equiv 0$ and $\gamma_{i=0} \equiv \gamma_0 = 1/\tau$. Note also that

$$
\begin{aligned}
\sum_{i=0}^{N} f_i &= f_0 + \sum_{i=1}^{N} f_i \\
&= \frac{n_f}{n} + \frac{n_b}{n} \sum_{i=1}^{N} (f_b)_i = 1 \, .
\end{aligned}
\tag{14.77}
$$

We see that the contribution of the free electrons to $\epsilon(\omega)$ is formally the same as the contribution of bound electrons with characteristic frequency $\omega_i = 0$, corresponding to the fact that, for free electrons, there is no restoring force.

Observe that, because of the factor

$$\frac{1}{\omega^2 + i\omega\tau^{-1}} = -i\tau \left(\frac{1}{\omega} - \frac{1}{\omega + i\tau^{-1}} \right) , \tag{14.78}$$

for metals the function $\epsilon_r(\omega)$, beside a pole in the lower half-plane at $\omega = -i\tau^{-1}$, also has a pole on the real axis at $\omega = 0$. As we see from eq. (13.90), this holds in general, simply because, for a metal, the d.c. conductivity $\sigma_0 \neq 0$, and is not specific to the model (14.76). It is therefore an example of the situation, discussed in Note 6 on page 370, in which the Kramers–Kronig relations hold for the function from which the pole on the real axis has been subtracted. To subtract the pole, we must consider the function

$$g(\omega) \equiv \epsilon_r(\omega) - \frac{i\sigma_0}{\epsilon_0\omega} \, . \tag{14.79}$$

Writing $g(\omega) = g_R(\omega) + i g_I(\omega)$, we have $g_R(\omega) = \epsilon_{r,R}(\omega)$ and $g_I(\omega) = \epsilon_{r,I}(\omega) - (\sigma_0/\epsilon_0\omega)$. Then, the derivation of the Kramers–Kronig relations (14.31) and (14.32) goes through, with $\epsilon_r(\omega)$ replaced by $g(\omega)$, and we obtain[15]

---

[15] In eq. (14.80), for $\omega \neq 0$ the contribution to the integral proportional to (σ_0/ϵ_0) vanishes, using the definition of principal value of the integral,

$$
\begin{aligned}
P &\int_0^\infty \frac{d\omega'}{\omega'^2 - \omega^2} \\
&\equiv \lim_{\epsilon \to 0^+} \left(\int_0^{\omega-\epsilon} + \int_{\omega+\epsilon}^\infty \right) \frac{d\omega'}{\omega'^2 - \omega^2} \\
&= \frac{1}{2\omega} \lim_{\epsilon \to 0^+} \\
&\quad \left[\log \left| \frac{\omega' - \omega}{\omega' + \omega} \right|_0^{\omega-\epsilon} + \log \left| \frac{\omega' - \omega}{\omega' + \omega} \right|_{\omega+\epsilon}^\infty \right] \\
&= 0 \, .
\end{aligned}
$$

This term must, however, be kept to get the correct limit for $\omega \to 0$ since, according to eq. (14.81), $\omega\epsilon_I(\omega) - (\sigma_0/\epsilon_0)$ vanishes as ω^2 as $\omega \to 0$, ensuring that the integral in eq. (14.80) is well defined also for $\omega \to 0$ [compare with the discussion below eq. (14.26) for $\sigma(\omega)$].

$$\epsilon_{r,R}(\omega) \;=\; 1 + \frac{2}{\pi}\mathrm{P}\int_0^\infty d\omega' \,\frac{\omega'\epsilon_{r,I}(\omega') - (\sigma_0/\epsilon_0)}{\omega'^2 - \omega^2}\,, \qquad (14.80)$$

$$\epsilon_{r,I}(\omega) \;=\; \frac{\sigma_0}{\epsilon_0\omega} - \frac{2\omega}{\pi}\mathrm{P}\int_0^\infty d\omega' \,\frac{\epsilon_{r,R}(\omega') - 1}{\omega'^2 - \omega^2}\,. \qquad (14.81)$$

For metals, it is often a good first approximation to neglect the contribution of the bound electrons, setting $n_b = 0$, i.e., $f_0 = 1$ and $f_i = 0$ for $i \geq 1$. The plasma frequency is then given in terms of the number density of free electrons,

$$\omega_p^2 = \frac{n_f e^2}{\epsilon_0 m_e}\,. \qquad (14.82)$$

In this case, from eq. (14.60), we can also write the plasma frequency in terms of the conductivity at zero frequency $\sigma_0 = \sigma(\omega = 0)$ and of τ, as

$$\omega_p^2 = \frac{\sigma_0}{\epsilon_0 \tau}\,. \qquad (14.83)$$

In this limit, we also find convenient to introduce the notation[16]

$$\gamma_p \equiv \frac{1}{\tau}\,. \qquad (14.84)$$

Then, setting $f_0 = 1$ and $f_i = 0$ for $i \geq 1$, eq. (14.76) can be rewritten as

$$\epsilon_r(\omega) = 1 - \frac{\omega_p^2}{\omega^2 + i\omega\gamma_p}\,. \qquad (14.85)$$

The relation between ω_p and γ_p allows us to classify the materials as good or bad conductors. As we will see in Section 15.3.2, for $\omega_p\tau \gg 1$, i.e., $\omega_p \gg \gamma_p$, a spatial distribution of electrons perturbed from its uniform equilibrium state can oscillates freely for many cycles, with little damping. Therefore, these materials are good conductors. In contrast, for $\omega_p \ll \gamma_p$ we have a poor conductor.[17]

The model (14.76), that assumed a non-interacting gas of classical electrons, can be improved including interactions and a quantum mechanical treatment. Several of these effects, such as the effect of the band structure of the material (i.e., the interaction of the electrons with the lattice of ions), as well as the interaction of the electrons among them and with the coherent vibration of the lattice (described, at the quantum level, by phonons), can be modeled by replacing the electron mass m_e in eq. (14.82) by an effective mass m_*. This leaves the functional form (14.76) unchanged (if we neglect a frequency dependence that actually appears in m_*), although the determination of ω_p from the properties of the material is now more complicated, and ω_p does not depend only on the number density n of the electrons. Furthermore, the net contribution from the positive ion cores is modeled, phenomenologically, with a constant ϵ_∞, so that eq. (14.76) becomes

$$\epsilon_r(\omega) = \epsilon_\infty - \frac{\omega_p^2 f_0}{\omega^2 + i\omega\gamma_0} + \omega_p^2 \sum_{i=1}^{N} \frac{f_i}{\omega_i^2 - \omega^2 - i\omega\gamma_i}\,. \qquad (14.87)$$

[16] The subscript "p" in γ_p (which is more commonly denoted simply as γ), stresses its analogy with ω_p, since they are both properties of the metal, or, as in Section 15.3.2, of a plasma. These quantities are strictly related, and can appear as the real and imaginary parts of a complex frequency, in the combination $\pm\omega_p - i(\gamma_p/2)$, see eq. (15.56).

[17] Writing eq. (14.83) in the form $\omega_p(\omega_p\tau) = \sigma_0/\epsilon_0$ we see that the condition $\omega_p\tau \gg 1$ is equivalent to

$$\frac{\sigma_0}{\epsilon_0} \gg \omega_p\,. \qquad (14.86)$$

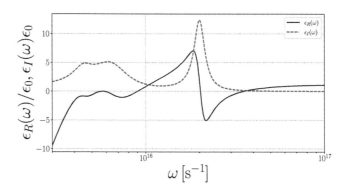

[18]In Figs. 14.4 and 14.5 we used $\omega_p = 13.8 \times 10^{15}\,\mathrm{s}^{-1}$, $\gamma_0 = 0.0058 \times 10^{15}\,\mathrm{s}^{-1}$, $f_0 = 0.760$ and, for $i = 1,\ldots 5$,

$$f_i = \{0.024, 0.010, 0.071, 0.601, 4.384\}$$
$$\omega_i = \{0.630, 1.261, 4.510, 6.538, 20.234\} \times 10^{15}\,\mathrm{s}^{-1},$$
$$\gamma_i = \{0.366, 0.524, 1.322, 3.789, 3.363\} \times 10^{15}\,\mathrm{s}^{-1},$$

and $\epsilon_\infty = 1.2$, from Rakić *et al.* (1998).

Fig. 14.4 The prediction of the Drude–Lorentz model for the real and imaginary parts of $\epsilon(\omega)$ for gold, using a logarithmic scale on the horizontal axis and a linear scale on the vertical axis, and emphasizing a region of frequencies of order ω_p.

For the free electron gas $\epsilon_\infty = 1$ while, for typical metals, $\epsilon_\infty \simeq 1 - 10$. This reflects the fact that the natural oscillation frequencies of the ions are much larger than that of the bound electrons and, at least at optical and UV frequencies, we are not yet in the limit of ω much larger than all natural frequencies ω_i of the system; the constant ϵ_∞ takes into account, phenomenologically, the effect of these high-frequency modes. Separating the real and imaginary parts, we get

$$\epsilon_{r,R}(\omega) = \epsilon_\infty - \frac{\omega_p^2 f_0}{\omega^2 + \gamma_0^2} + \omega_p^2 \sum_{i=1}^{N} \frac{f_i(\omega_i^2 - \omega^2)}{(\omega_i^2 - \omega^2)^2 + \omega^2 \gamma_i^2}, \quad (14.88)$$

$$\epsilon_{r,I}(\omega) = \frac{\omega_p^2 \gamma_0 f_0}{\omega(\omega^2 + \gamma_0^2)} + \omega_p^2 \omega \sum_{i=1}^{N} \frac{f_i \gamma_i}{(\omega_i^2 - \omega^2)^2 + \omega^2 \gamma_i^2}. \quad (14.89)$$

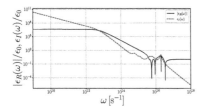

Fig. 14.5 As in Fig. 14.4, on a much larger range of frequencies, and on a log-log scale. Note that now, because of the log-log scale, we plot $|\epsilon_R(\omega)|$; $\epsilon_R(\omega)$ is negative at small ω, and changes sign in correspondence of the downward spikes. The first spike is made of two very close spikes not distinguishable in the figure, so, $\epsilon_R(\omega)$ has five zeros, corresponding to the five frequencies ω_i that we have included; including higher modes would produce more structure and further downward spikes at higher frequencies. At large ω, $\epsilon_R(\omega) \to \epsilon_\infty > 0$. In contrast, $\epsilon_I(\omega)$ is always positive.

Observe that, because of the contribution of the free electrons, the imaginary part $\epsilon_{r,I}(\omega)$ diverges as $1/\omega$ as $\omega \to 0$. The real part, in contrast, saturates to a constant; in particular, if $\gamma_0 \ll \omega_p$, as is typically the case, the contribution from the free electrons to the real part saturates to the value $-f_0(\omega_p/\gamma_0)^2$, which is negative, and large in absolute value. At high frequencies, $\epsilon_{r,I}(\omega)$ vanishes as $1/\omega^3$, and $\epsilon_r(\omega)$ becomes real. In Fig. 14.4 we show the prediction of the Drude–Lorentz model for $\epsilon_{r,R}(\omega)$ and $\epsilon_{r,I}(\omega)$, using the values of ω_p, f_0, γ_0, and the first five values of f_i, ω_i and γ_i in the case of a metal such as gold,[18] on a log-linear scale and a relatively small a range of frequencies of the order of the plasma frequency, to emphasize the structures in this range of frequencies. In Fig. 14.5 we show the result on a much larger frequency range and a log-log scale that emphasizes the asymptotic behaviors.

Electromagnetic waves in material media

15

In Chapter 9 we studied electromagnetic waves propagating in vacuum. In this chapter we study how electromagnetic waves are affected by the properties of the medium in which the propagate, described by frequency-dependent constitutive relations such as those studied in Chapter 14, or by non-trivial boundary conditions, as in the case of electromagnetic waves propagating in waveguides.

15.1 Electromagnetic waves in dielectrics

We first consider the propagation of electromagnetic waves in dielectric materials. Since we want to consider generic dispersive media, we look for monochromatic wave solutions of the form

$$\mathbf{E}(t,\mathbf{x}) = \tilde{\mathbf{E}}(\omega,\mathbf{k})e^{-i\omega t + i\mathbf{k}\cdot\mathbf{x}}, \qquad \mathbf{B}(t,\mathbf{x}) = \tilde{\mathbf{B}}(\omega,\mathbf{k})e^{-i\omega t + i\mathbf{k}\cdot\mathbf{x}}, \quad (15.1)$$

and we write

$$\tilde{\mathbf{D}}(\omega,\mathbf{k}) = \epsilon(\omega)\tilde{\mathbf{E}}(\omega,\mathbf{k}), \qquad \tilde{\mathbf{B}}(\omega,\mathbf{k}) = \mu(\omega)\tilde{\mathbf{H}}(\omega,\mathbf{k}), \quad (15.2)$$

where ϵ and μ are assumed to depend on ω but not on $\mathbf{k}$.[1] We admit a priori that both ω and $\mathbf{k}$ in the ansatz (15.1) could be complex. A frequency $\omega = \omega_R + i\omega_I$, with imaginary part $\omega_I < 0$, corresponds to an excitation that has been set up at an initial time, and then decreases exponentially in time, while $\omega_I > 0$ corresponds to a solution that increases exponentially in time, which can be obtained if the medium pumps energy into the electromagnetic wave, leading to an amplification.[2]

A complex $\mathbf{k}$, in contrast, corresponds to a solution that decreases or increases exponentially in space, from a given surface that will be typically a boundary of the material, where boundary conditions are set. The appropriate solution will then be selected by imposing suitable boundary conditions. For instance, in a semi-infinite medium which extends from $x = 0$ to $x = \infty$, for a wave propagating in the positive x direction we would require that the solution goes to zero as $x \to +\infty$, which selects the decreasing solution. Note that, at this stage, ω and $\mathbf{k}$ are independent variables, that enter the ansatz (15.1). The dispersion relation of the electromagnetic waves in the material, i.e., the relation between ω and $\mathbf{k}$, will be obtained by requiring that the ansatz indeed satisfies Maxwell's equations.

[1] A further generalization of the constitutive relations that we have studied in Chapter 14 can be obtained admitting also a $\mathbf{k}$ dependence in the functions describing the response of the material, e.g., writing $\tilde{\mathbf{D}}(\omega,\mathbf{k}) = \epsilon(\omega,\mathbf{k})\tilde{\mathbf{E}}(\omega,\mathbf{k})$.

[2] As we discussed in Note 9 on page 371, in the context of the quantum Drude–Lorentz model, the oscillator strengths f_i that appear in eq. (14.40) can be positive or negative, and negative values correspond to "emission oscillator strengths", where an excited electron makes a transition to a lower energy state. This leads to pumping energy into the electromagnetic wave; a particularly important example is the mechanism that gives rise to lasers.

For dielectrics, we also set to zero the density of free charges and currents. The computation is then quite similar to that of the propagation of electromagnetic waves in vacuum studied in Chapter 9. Inserting the ansatz (15.1) into Maxwell's equations in material media, eqs. (13.45)–(13.48), we get

$$\epsilon(\omega)\mathbf{k}\cdot\tilde{\mathbf{E}}(\omega,\mathbf{k}) = 0\,,\tag{15.3}$$

$$\mathbf{k}\times\tilde{\mathbf{B}}(\omega,\mathbf{k}) = -\frac{\omega\,n^2(\omega)}{c^2}\tilde{\mathbf{E}}(\omega,\mathbf{k})\,,\tag{15.4}$$

$$\mathbf{k}\cdot\tilde{\mathbf{B}}(\omega,\mathbf{k}) = 0\,,\tag{15.5}$$

$$\mathbf{k}\times\tilde{\mathbf{E}}(\omega,\mathbf{k}) = \omega\tilde{\mathbf{B}}(\omega,\mathbf{k})\,,\tag{15.6}$$

where we have defined the *refraction index* $n(\omega)$ from

$$n(\omega) = c\sqrt{\epsilon(\omega)\mu(\omega)}\,.\tag{15.7}$$

Note that $n(\omega)$ is dimensionless; in vacuum, $\epsilon(\omega)\mu(\omega)$ becomes $\epsilon_0\mu_0$, which is equal to $1/c^2$ [recall eq. (8.27)], and therefore $n(\omega) = 1$. Using $\epsilon_0\mu_0 = 1/c^2$ we can also rewrite eq. (15.7) as

$$\begin{aligned} n^2(\omega) &= \frac{\epsilon(\omega)\mu(\omega)}{\epsilon_0\mu_0}\\ &= \epsilon_r(\omega)\mu_r(\omega)\,,\end{aligned}\tag{15.8}$$

where, as usual, $\epsilon_r(\omega) = \epsilon(\omega)/\epsilon_0$ and $\mu_r(\omega) = \mu(\omega)/\mu_0$.

For the study of electromagnetic waves in dielectrics, we can assume that $\epsilon(\omega)$ never vanishes, as we will discuss in more detail below eq. (15.58). Therefore, eq. (15.3) is equivalent to $\mathbf{k}\cdot\tilde{\mathbf{E}}(\omega,\mathbf{k}) = 0$. Taking the vector product of eq. (15.6) with $\mathbf{k}$ [which, in coordinate space, corresponds to taking its curl, as we did in the derivation of eqs. (9.73) and (9.74)] and using eq. (1.9), together with $\mathbf{k}\cdot\tilde{\mathbf{E}}(\omega,\mathbf{k}) = 0$ and eq. (15.4), we get

$$\begin{aligned} -k^2\tilde{\mathbf{E}}(\omega,\mathbf{k}) &= \omega\mathbf{k}\times\tilde{\mathbf{B}}(\omega,\mathbf{k})\\ &= -\frac{\omega^2}{c^2}n^2(\omega)\tilde{\mathbf{E}}(\omega,\mathbf{k})\,,\end{aligned}\tag{15.9}$$

where $k^2 = \mathbf{k}\cdot\mathbf{k}$.[3] Similarly, from the curl of eq. (15.4), we get

$$k^2\tilde{\mathbf{B}}(\omega,\mathbf{k}) = \frac{\omega^2}{c^2}n^2(\omega)\tilde{\mathbf{B}}(\omega,\mathbf{k})\,.\tag{15.10}$$

We therefore find the dispersion relation

$$\boxed{k^2 = \frac{\omega^2}{c^2}n^2(\omega)\,.}\tag{15.11}$$

In vacuum, where $n(\omega) = 1$, we recover the dispersion relation (9.41). Since ω is now implicitly fixed in terms of $\mathbf{k}$ by the condition (15.11), to label the solutions we can use just $\mathbf{k}$, rather than the pair $(\omega,\mathbf{k})$.[4] We

[3]Note that in this section k will denote the modulus of the three-dimensional vector $\mathbf{k}$, so $k^2 = |\mathbf{k}|^2$ [or, more precisely, for complex $\mathbf{k}$, $k^2 = (\mathbf{k}_R + i\mathbf{k}_I)\cdot(\mathbf{k}_R + i\mathbf{k}_I)$]; in contrast, in a relativistic context, such as in Chapter 9, we used the notation k^2 for $k_\mu k^\mu$.

[4]We assume that, for each $\mathbf{k}$, the solution for ω exists (otherwise there is no plane wave solution) and is unique. If there are several solutions, one can apply the analysis that follows to each of them.

then introduce the simpler notations

$$\mathbf{E_k} = \tilde{\mathbf{E}}(\omega(k), \mathbf{k}), \qquad \mathbf{B_k} = \tilde{\mathbf{B}}(\omega(k), \mathbf{k}). \qquad (15.12)$$

Recalling that we eventually take the real part, we can summarize the ansatz as

$$\mathbf{E}(t, \mathbf{x}) = \mathrm{Re}\left[\mathbf{E_k}e^{-i\omega(k)t+i\mathbf{k}\cdot\mathbf{x}}\right], \qquad (15.13)$$

$$\mathbf{B}(t, \mathbf{x}) = \mathrm{Re}\left[\mathbf{B_k}e^{-i\omega(k)t+i\mathbf{k}\cdot\mathbf{x}}\right], \qquad (15.14)$$

where $\omega(k)$ is given by eq. (15.11). Equation (15.11) is a necessary condition for having a solution, but it is not yet sufficient. To find the full set of conditions on the ansatz, we plug eqs. (15.13) and (15.14) into Maxwell's equations (15.3)–(15.6). From eq. (15.3) (having assumed that $\epsilon(\omega)$ never vanishes) and eq. (15.5) we get

$$\hat{\mathbf{k}}\cdot\mathbf{E_k} = 0, \qquad \hat{\mathbf{k}}\cdot\mathbf{B_k} = 0. \qquad (15.15)$$

This shows that, just as for the electromagnetic waves in vacuum, for electromagnetic waves in dielectrics $\mathbf{E}$ and $\mathbf{B}$ are orthogonal to the propagation direction. Finally, using eq. (15.11), eq. (15.6) can be rewritten as

$$c\mathbf{B_k} = n(\omega)\,\hat{\mathbf{k}} \times \mathbf{E_k}, \qquad (15.16)$$

and the same condition comes from eq. (15.4). This tells us that, just as in vacuum, $\mathbf{E}$ and $c\mathbf{B}$ are orthogonal to each other, but now their modulus is different, compare with eq. (9.51). We have therefore shown that the ansatz (15.13)–(15.14) is indeed a solution of the full set of Maxwell's equations, under the conditions (15.11), (15.15), and (15.16), and therefore dielectric materials sustain the propagation of monochromatic electromagnetic waves.

Observe that, in a dispersive medium, $\epsilon(\omega)$ and $\mu(\omega)$ are necessarily complex. Indeed, from the Kramers–Kronig dispersion relation (14.31) we see that, if $\epsilon_{r,I}(\omega) = 0$ for all ω, then $\epsilon_{r,R} = 1$ independently of the frequency. Therefore, a non-trivial frequency dependence of $\epsilon_R(\omega)$ implies that, at least for some frequencies, $\epsilon_I(\omega) \neq 0$, and similarly for $\mu(\omega)$. Therefore $n(\omega)$ in eq. (15.7) is complex, and the dispersion relation (15.11) is also complex. As a consequence, eq. (15.11) can in general have solutions with both ω and $\mathbf{k}$ complex, $\omega = \omega_R + i\omega_I$ and $\mathbf{k} = \mathbf{k}_R + i\mathbf{k}_I$. The simplest solutions correspond to ω real. In this case the temporal evolution is just an undamped oscillation. Solutions with $\omega_I > 0$ disappear exponentially in time and are therefore less relevant to the free propagation, while solutions with $\omega_I < 0$ are exponentially growing and represent instabilities, where energy is pumped into the electromagnetic wave. In any case, since $n(\omega)$ is complex, even for ω real the corresponding value of $\mathbf{k}$, obtained from eq. (15.11), is in general complex. We write[5]

$$n(\omega) = n_R(\omega) + in_I(\omega), \qquad (15.17)$$

[5]In the literature, another common notation is $n(\omega) = \eta(\omega) + i\kappa(\omega)$. The real part, $n_R(\omega)$ or $\eta(\omega)$, is called the real refractive index, while the imaginary part, $n_I(\omega)$ or $\kappa(\omega)$, is also called the extinction coefficient.

and $\mathbf{k} = \mathbf{k}_R + i\mathbf{k}_I$. Recalling that $k^2 = \mathbf{k}\cdot\mathbf{k} = (\mathbf{k}_R + i\mathbf{k}_I)\cdot(\mathbf{k}_R + i\mathbf{k}_I)$, eq. (15.11) separates into two equations for the real and the imaginary parts,

$$|\mathbf{k}_R|^2 - |\mathbf{k}_I|^2 = \frac{\omega^2}{c^2}\left[n_R^2(\omega) - n_I^2(\omega)\right], \tag{15.18}$$

$$\mathbf{k}_R\cdot\mathbf{k}_I = \frac{\omega^2}{c^2}\,n_R(\omega)n_I(\omega). \tag{15.19}$$

The solution (15.13), for ω real, can then be rewritten as

$$\mathbf{E}(t,\mathbf{x}) = \text{Re}\left[\mathbf{E_k}e^{-i\omega(k)t+i\mathbf{k}_R\cdot\mathbf{x}-\mathbf{k}_I\cdot\mathbf{x}}\right], \tag{15.20}$$

and the same for $\mathbf{B}(t,\mathbf{x})$. The term $e^{-i\omega(k)t+i\mathbf{k}_R\cdot\mathbf{x}}$ gives the phase of the field, while the term $e^{-\mathbf{k}_I\cdot\mathbf{x}}$ affects its amplitude and describes the attenuation of the wave as it propagates in the material. Note that the surfaces of constant phase are perpendicular to $\mathbf{k}_R$ while the surfaces of constant amplitude are perpendicular to $\mathbf{k}_I$. Thus, unless $\mathbf{k}_R$ and $\mathbf{k}_I$ are parallel, these two surfaces are different. Solutions where $\mathbf{k}_R$ and $\mathbf{k}_I$ are not parallel are called *inhomogeneous plane waves*.

In the simpler case of homogeneous plane waves, where $\mathbf{k}_R$ and $\mathbf{k}_I$ are parallel, we can write $\mathbf{k}_R = k_R\hat{\mathbf{k}}$, $\mathbf{k}_I = k_I\hat{\mathbf{k}}$, and eqs. (15.18) and (15.19) give

$$k_R = \frac{\omega}{c}n_R(\omega), \tag{15.21}$$

$$k_I = \frac{\omega}{c}n_I(\omega). \tag{15.22}$$

15.2 Phase velocity and group velocity

We next discuss two distinct notions of velocity related to the propagation of electromagnetic waves. We neglect absorption, so we set $\mathbf{k}_I = 0$, $\mathbf{k}_R = \mathbf{k}$ and $k = |\mathbf{k}|$. Setting $\mathbf{k}_I = 0$ in eq. (15.20), we have

$$\mathbf{E}(t,\mathbf{x}) = \text{Re}\left[\mathbf{E}_k\,e^{-i\varphi(t,\mathbf{x})}\right], \tag{15.23}$$

where the phase φ is given by

$$\varphi(t,\mathbf{x}) = \omega(k)t - \mathbf{k}\cdot\mathbf{x}, \tag{15.24}$$

and the dependence of ω on k is given by the inversion of eq. (15.11). The surfaces of constant phase therefore travel at the velocity $\mathbf{v}_p = v_p\hat{\mathbf{k}}$, where

$$\boxed{v_p(k) = \frac{\omega(k)}{k}.} \tag{15.25}$$

Using eq. (15.21) (and setting $n_R = n$ since we are neglecting all imaginary parts), the phase velocity can be written as

$$v_p(\omega) = \frac{c}{n(\omega)}, \tag{15.26}$$

Usually, $n(\omega) > 1$, so $v_p(\omega) < c$. In some situations, however, $n(\omega)$ can be smaller than one. This is the case, for instance, for a dielectric described by the Drude-Lorenz model, where $\epsilon_r(\omega)$ is given by eq. (14.39). For a dielectric we can set $\mu(\omega) = \mu_0$ so, from eq. (15.8), $n^2(\omega) = \epsilon_r(\omega)$. Therefore, in the approximation in which all imaginary parts are neglected we have $n^2(\omega) = \epsilon_{r,R}(\omega)$, and we see from eq. (14.42) that this becomes smaller than one for $\omega > \omega_0$.[6] This, however, is not in conflict with the postulates of Special Relativity, because monochromatic waves simply do not carry information. In order to transmit information, we must modulate the signal by superposing plane waves into wave packets.[7] The relevant question, therefore, is whether wave packets can transmit information at speed greater than c. According to eq. (15.26), Fourier modes with different values of ω travel at a different phase velocity. To understand the consequences of this, we consider a superposition of plane waves with different wavenumbers $\mathbf{k}$,

$$\mathbf{E}(t, \mathbf{x}) = \int \frac{d^3k}{(2\pi)^3} \, \tilde{\mathbf{E}}(\mathbf{k}) \, e^{-i\omega(\mathbf{k})t + i\mathbf{k}\cdot\mathbf{x}} , \qquad (15.27)$$

where, to be more general, we have assumed that ω depends not only on the modulus $k = |\mathbf{k}|$, but on the full vector $\mathbf{k}$, which could happen, in general, in anisotropic materials, and we take $\mathbf{k}$ and $\omega(k)$ real. Actually, nothing in our considerations will depend on the vector nature of the electric field, and we can more simply study a relation of the form

$$f(t, \mathbf{x}) = \int \frac{d^3k}{(2\pi)^3} \, \tilde{f}(\mathbf{k}) \, e^{-i\omega(\mathbf{k})t + i\mathbf{k}\cdot\mathbf{x}} , \qquad (15.28)$$

for some function f, which could represent a component of the electric field, or an actual scalar function, in which case our analysis would apply also to other kind of waves, such as sound waves. If the function $\tilde{f}(\mathbf{k})$ is completely generic, each Fourier mode will travel at a different velocity, the spatial shape of the signal will be quickly distorted by the propagation, and there is little that can be said in full generality. If, however, $\tilde{f}(\mathbf{k})$ is sharply peaked around a wavenumber $\mathbf{k}_0$, we can expand the frequency as

$$\begin{aligned} \omega(\mathbf{k}) &\simeq \omega(\mathbf{k}_0) + (\mathbf{k} - \mathbf{k}_0)_i \left(\frac{\partial\omega(\mathbf{k})}{\partial k_i}\right)_{\mathbf{k}=\mathbf{k}_0} + \dots \\ &= \omega_0 + (\mathbf{k} - \mathbf{k}_0)\cdot(\boldsymbol{\nabla}_\mathbf{k}\omega)_{\mathbf{k}=\mathbf{k}_0} + \dots , \end{aligned} \qquad (15.29)$$

where $\omega_0 \equiv \omega(\mathbf{k}_0)$. Then

$$f(t, \mathbf{x}) \simeq e^{-i\omega_0 t + i\mathbf{k}_0\cdot\mathbf{x}} \int \frac{d^3k}{(2\pi)^3} \, \tilde{f}(\mathbf{k}) \, e^{i(\mathbf{k}-\mathbf{k}_0)\cdot(\mathbf{x}-\mathbf{v}_g t)} , \qquad (15.30)$$

where

$$\mathbf{v}_g = (\boldsymbol{\nabla}_\mathbf{k}\omega)_{\mathbf{k}=\mathbf{k}_0} \qquad (15.31)$$

is called the *group velocity*. For an isotropic medium, where ω depends only on $k = |\mathbf{k}|$, using eq. (1.23) with $\mathbf{r}$ replaced by $\mathbf{k}$ we see that the

[6]Actually, we see from Fig. 14.2 that for ω just above the resonance frequency ω_0, $\epsilon_R(\omega)/\epsilon_0$ even becomes negative. Note, however, that this happens precisely when $\epsilon_I(\omega)/\epsilon_0$ is large and we cannot neglect the imaginary parts.

[7]An alternative argument is that a purely monochromatic plane wave is just a mathematical idealization. Indeed, from a basic properties of the Fourier transform, if we denote by Δt the duration of a signal and by $\Delta\omega$ the spread in frequencies of its Fourier transform, we have $\Delta t \Delta\omega \gtrsim 1$. Any physical signal that we observe has a finite temporal duration and therefore its Fourier transform is necessarily non-vanishing over an interval of frequencies $\Delta\omega \gtrsim 1/\Delta t$.

group velocity is in the direction $\hat{\mathbf{k}}$ of the propagation, and its modulus is given by

$$v_g(k) = \left(\frac{d\omega(k)}{dk}\right)_{k=k_0}. \qquad (15.32)$$

Equation (15.30) shows that, apart from an overall phase $e^{-i\omega_0 t + i\mathbf{k}_0 \cdot \mathbf{x}}$ (that disappears in quadratic quantities such as $|f|^2$, so in our case in the energy density of the electromagnetic field, which is given by $|\mathbf{E}|^2$), in the approximation in which we stop the expansion in eq. (15.29) to linear order, the shape of the wavepacket remains the same and is just translated in space at a velocity v_g. The group velocity is therefore the quantity relevant to the transport of energy by a packet of electromagnetic waves. Note, however, that this only holds in the approximation where $\tilde{f}(\mathbf{k})$ is sharply peaked around a wavenumber $\mathbf{k}_0$, which justifies the expansion (15.29). Using $k = (\omega/c)n(\omega)$ and writing $d\omega/dk = (dk/d\omega)^{-1}$, we get

$$v_g(\omega) = \frac{c}{n(\omega) + \omega\frac{dn(\omega)}{d\omega}}. \qquad (15.33)$$

For *normal dispersion* we have $n(\omega) > 1$ and $dn/d\omega > 0$. Then $v_g(\omega) < v_p(\omega) < c$. It is possible, however, to have $dn/d\omega$ negative, and large in absolute value. An example is given again by the Drude–Lorentz model (14.39), or by its generalization (14.40): as before, we write $n_R^2(\omega) = \epsilon_{r,R}(\omega)$. Then, from Fig. 14.2 on page 372 we see that, just above the resonance frequency ω_0, the derivative of $\epsilon_{r,R}(\omega)$ becomes negative, and large in absolute value. This behavior is called *anomalous dispersion* and, formally, can give rise to a group velocity larger than c, or even negative. However, again, this simply means that, in such regions, the approximation (15.29) becomes invalid (furthermore, as already mentioned in Note 6, in this regime absorption is large and we cannot neglect it), and the concept of group velocity loses its meaning. The actual evolution of a wavepacket in this regime is more involved, and cannot be captured, even approximately, by a single quantity such as a velocity.

15.3 Electromagnetic waves in metals

We now study electromagnetic waves in conducting materials. We start again from Maxwell's equations in material media, eqs. (13.45)–(13.48). The difference with the treatment for dielectrics of Section 15.1 is that now we must include the density of free charges and currents, ρ_{free} and $\mathbf{j}_{\text{free}}$. We already worked out the corresponding expression for Maxwell's equations that depends on the sources in eqs. (13.88) and (13.89), where $\epsilon(\omega)$ is the dielectric function for metals, given by eq. (13.90). We look for a monochromatic plane wave solution of the form (15.1). The full set of Maxwell's equations (13.88), (13.89), (13.47), and (13.48) then becomes

$$\epsilon(\omega)\mathbf{k}\cdot\tilde{\mathbf{E}}(\omega,\mathbf{k}) = 0, \qquad (15.34)$$

$$\mathbf{k}\times\tilde{\mathbf{B}}(\omega,\mathbf{k}) + \omega\epsilon(\omega)\mu(\omega)\tilde{\mathbf{E}}(\omega,\mathbf{k}) = 0, \qquad (15.35)$$

$$\mathbf{k}\cdot\tilde{\mathbf{B}}(\omega,\mathbf{k}) = 0, \qquad (15.36)$$

$$\mathbf{k}\times\tilde{\mathbf{E}}(\omega,\mathbf{k}) - \omega\tilde{\mathbf{B}}(\omega,\mathbf{k}) = 0. \qquad (15.37)$$

Consider first eq. (15.34). There are two branches of solutions. One is obtained setting $\mathbf{k}\cdot\tilde{\mathbf{E}}(\omega,\mathbf{k}) = 0$. This solution is transverse to the propagation direction, just as we found for electromagnetic waves in vacuum, as well as for electromagnetic waves in dielectrics. We will study this solution in Section 15.3.1. There is, however, another possibility: eq. (15.34) is automatically satisfied, without imposing the transversality condition on $\mathbf{E}$, at a special value $\bar{\omega}$ of the frequency, given by the solution of the equation $\epsilon(\bar{\omega}) = 0$. We will study this solution in Section 15.3.2.[8]

15.3.1 Transverse EM waves

In this section we study the solution of eq. (15.34) obtained requiring that $\mathbf{k}\cdot\tilde{\mathbf{E}}(\omega,\mathbf{k}) = 0$. In this case, the electric field is transverse to the propagation direction. Equation (15.37) tells us that $\tilde{\mathbf{B}}(\omega,\mathbf{k})$ is orthogonal to $\tilde{\mathbf{E}}(\omega,\mathbf{k})$, and eq. (15.36) tells us that it is also orthogonal to the propagation direction. We therefore have the same situation as the electromagnetic waves in vacuum, or in dielectrics, with the electric and magnetic fields orthogonal to the propagation direction and to each other. The dispersion relation $\omega(k)$ can be obtained combining Maxwell's equations, just as we did for dielectrics. Taking the vector product of eq. (15.37) with $\mathbf{k}$ and using eqs. (1.9), (15.34), and (15.35), we get

$$\begin{aligned} -k^2\tilde{\mathbf{E}}(\omega,\mathbf{k}) &= \omega\mathbf{k}\times\tilde{\mathbf{B}}(\omega,\mathbf{k}) \\ &= -\omega^2\epsilon(\omega)\mu(\omega)\tilde{\mathbf{E}}(\omega,\mathbf{k})\,. \end{aligned} \qquad (15.38)$$

Then, the dispersion relation is

$$\boxed{\omega^2\epsilon(\omega)\mu(\omega) = k^2\,,} \qquad (15.39)$$

which, of course, is the same as eq. (15.11) (recalling the definition (15.7) of the refraction index), except that now $\epsilon(\omega)$ is the dielectric function appropriate for a metal, given in eq. (13.90). Let us consider the consequences of this dispersion relation, studying first the low- and high-frequency limits. We limit ourselves to ω real. However, since the function $\epsilon(\omega)$ is complex, the corresponding solution for k will in general be complex, $k = k_R + ik_I$.

In the low-frequency limit $\omega\tau \ll 1$, i.e., $\omega \ll 1/\tau \equiv \gamma_p$, eq. (13.90) gives $\epsilon(\omega) \simeq i\sigma_0/\omega$, where, as usual, $\sigma_0 = \sigma(\omega = 0)$ is the (zero-frequency) conductivity, while $\mu(\omega) \simeq \mu(\omega = 0) \equiv \mu$ becomes the same

[8]We will discuss after eq. (15.58) why, in dielectrics, the corresponding solution obtained setting $\epsilon(\omega) = 0$ does not correspond to a propagating electromagnetic wave.

as the static permeability of the material. Then eq. (15.39) gives (taking the square root with positive imaginary part)

$$k \simeq \frac{1+i}{\sqrt{2}} \sqrt{\sigma_0 \mu \omega}, \tag{15.40}$$

and therefore in this limit the imaginary part of k is

$$k_I = \left(\frac{\sigma_0 \mu \omega}{2}\right)^{1/2}. \tag{15.41}$$

Setting the propagation direction along the positive x axis, and the vacuum-metal interface at $x = 0$, at $x > 0$ the amplitude of the wave therefore decreases as $e^{-k_I x} = e^{-x/\delta_{\text{skin}}}$, where $\delta_{\text{skin}} = 1/k_I$ is called the *skin depth*.[9] From eq. (15.41),

$$\delta_{\text{skin}}(\omega) = \left(\frac{2}{\sigma_0 \mu \omega}\right)^{1/2}, \qquad (\omega \ll \gamma_p). \tag{15.42}$$

Furthermore, for non-magnetic materials, we can approximate $\mu \simeq \mu_0$. For instance, for copper, $\mu/\mu_0 \simeq 0.999994$. Then, writing $\mu \simeq \mu_0 = 1/(\epsilon_0 c^2)$, we get

$$\delta_{\text{skin}}(\omega) = c \left(\frac{2\epsilon_0}{\sigma_0 \omega}\right)^{1/2}, \qquad (\omega \ll \gamma_p). \tag{15.43}$$

To get an idea of typical numbers, in copper at $20\,^{\circ}C$ the collision time is $\tau \simeq 2.4 \times 10^{-14}$ s, so $\gamma_p \simeq 4.2 \times 10^{13}\,\text{s}^{-1}$. The condition $\omega \ll \gamma_p$ is therefore satisfied for $f \ll \gamma_p/(2\pi) \simeq 6\,\text{THz}$. In terms of $\lambda = c/f$, this corresponds to $\lambda \gg 50\,\mu\text{m}$, i.e., wavelengths longer than the mid infrared.[10] In this regime, we can apply eq. (15.43). The static conductivity of Cu at $20\,^{\circ}C$ is $\sigma_0 \simeq 5.96 \times 10^7\,\text{S/m}$.[11] From eqs. (2.12), (2.13), and (4.188) we see that σ_0/ϵ_0 has dimensions of s^{-1}, i.e., has the same dimensions as a frequency, and, for Cu at $20\,^{\circ}C$, we get $\sigma_0/\epsilon_0 \simeq 6.73 \times 10^{18}\,\text{s}^{-1}$. Setting for instance as a reference value $f = 10\,\text{GHz}$, in the microwave region, we get

$$\delta_{\text{skin}}(\omega) \simeq 0.65\,\mu\text{m} \left(\frac{10\,\text{GHz}}{\omega/(2\pi)}\right)^{1/2}, \qquad (\text{Cu}, \frac{\omega}{2\pi} \ll 6\,\text{THz}). \tag{15.44}$$

Also observe that, from eq. (14.83), in copper at room temperature, the plasma frequency is $\omega_p \simeq 1.7 \times 10^{16}\,\text{s}^{-1}$, or $f_p = \omega_p/(2\pi) \simeq 2.7 \times 10^{15}\,\text{Hz}$. The corresponding wavelength is $\lambda_p = 2\pi c/\omega_p \simeq 110$ nm, in the far UV.

We next consider the high frequency limit. There are two frequency scales in the problem, $\gamma_p = 1/\tau$ and ω_p. As mentioned below eq. (14.85), good conductors are characterized by the property $\omega_p \gg \gamma_p$ (as we have seen above, in copper $\omega_p \simeq 1.7 \times 10^{16}\,\text{s}^{-1}$ and $\gamma_p \simeq 4.2 \times 10^{13}\,\text{s}^{-1}$). Therefore, there are two different regimes at $\omega\tau \gg 1$, namely $\gamma_p \ll \omega \ll \omega_p$ and $\omega \gg \omega_p$. We will study the full behavior below. For the moment, we focus on the very high frequency regime $\omega \gg \omega_p$. Then, plugging eq. (14.85) into eq. (15.39) (with $\mu(\omega) \simeq \mu_0$), and keeping the leading

[9]Of course, in this setting, in which we are considering a wave propagating into a metal along the positive x direction, we have chosen the solution for k_I with positive imaginary part, so as to have an exponentially decreasing amplitude. The solution of eq. (15.39) with $k_I = -(\sigma_0\mu\omega/2)^{1/2}$ is eliminated by the boundary condition that the corresponding solution for the electromagnetic wave does not diverges as $x \to \infty$. If we rather consider a setting with the metal at $x < 0$ and a left-moving wave coming from positive x, we would choose the other solution for k_I, $k_I = -(\sigma_0\mu\omega/2)^{1/2}$, to ensure that the electromagnetic wave does not diverge as $x \to -\infty$.

[10]For orientation, in terms of wavelengths, the far UV ranges from 10 to 200 nm, middle UV is 200–300 nm and near UV 300–380 nm. With longer wavelengths we enter the visible range, from violet (380–450 nm) to red (625–750 nm, i.e., 0.625 to 0.75 μm). Then come the near infrared (NIR, about 0.75 to 2.5 μm), middle infrared (MIR, 2.5–10 μm), and far infrared (FIR, 10 μm–1 mm). From 1 mm to about 1 m we are in the domain of microwaves.

[11]The unit of conductivity in the SI system was discussed in Note 49 on page 96, see in particular eq. (4.188).

correction in ω_p^2/ω^2 both in the real and in the imaginary part of $\omega^2\epsilon(\omega)$, we get

$$kc = \omega\left(1 - \frac{\omega_p^2}{2\omega^2}\right) + i\frac{\gamma_p\omega_p^2}{2\omega^2} \, . \tag{15.45}$$

Writing $k = k_R + ik_I$, eq. (15.45) gives

$$\omega^2 \simeq \omega_p^2 + k_R^2 c^2 \, , \tag{15.46}$$

and

$$k_I c = \frac{\gamma_p}{2}\left(\frac{\omega_p}{\omega}\right)^2 \, . \tag{15.47}$$

Equation (15.46) is the dispersion relation of transverse electromagnetic waves in a metal in the limit $\omega \gg \omega_p$, i.e., for $k_R c \gg \omega_p$.[12] Equation (15.47) shows that, in the limit $\omega \gg \omega_p$, $k_I c$ is much smaller than the frequency scale γ_p given by the inverse of the collision time, and goes to zero as $\omega/\omega_p \to \infty$. Therefore, metals become almost transparent to electromagnetic waves for ω much larger than their plasma frequency. For metals, typical values of ω_p are in the UV. For instance, for Cu we have seen that ω_p is in the far UV. For alkali metals the plasma frequency is even smaller; for instance, for thin films of Cesium the transparency already starts in the violet part of the visible spectrum.

Actually, using the simple model (14.85) for the dielectric function of a metal (and setting $\mu(\omega) = \mu_0$), it is not difficult, and instructive, to study the dispersion relation (15.39) for ω generic, rather than only in the limiting cases $\omega \ll \gamma_p$ and $\omega \gg \omega_p$. Inserting eq. (14.85) into eq. (15.39) we get

$$k^2 c^2 = \omega^2 - \frac{\omega^2\omega_p^2}{\omega^2 + i\omega\gamma_p} \, . \tag{15.48}$$

Writing $k = k_R + ik_I$ and separating eq. (15.39) into its real and imaginary parts, the resulting equations can be combined into a second degree equation for the variable k_I^2, whose solution is

$$2k_I^2(\omega)c^2 = \frac{\omega}{\omega^2 + \gamma_p^2}\left\{\left[\omega^2(\omega^2 - \omega_p^2 + \gamma_p^2)^2 + \omega_p^4\gamma_p^2\right]^{1/2} - \omega(\omega^2 - \omega_p^2 + \gamma_p^2)\right\}. \tag{15.49}$$

The corresponding solution for $k_R(\omega)$ is given by[13]

$$k_R(\omega) = \frac{\omega_p^2\gamma_p}{2c^2}\frac{\omega}{(\omega^2 + \gamma_p^2)k_I(\omega)} \, . \tag{15.50}$$

In Fig. 15.1 we plot (on a log-log scale) the corresponding skin depth $\delta_{\text{skin}}(\omega) = 1/k_I(\omega)$, using the values of ω_p and γ_p appropriate for copper at room temperature. We see that there are indeed three distinct regimes: at $\omega \ll \gamma_p$, $\delta_{\text{skin}}(\omega)$ decreases as $1/\sqrt{\omega}$, in agreement with eq. (15.43). From eq. (15.49), in this regime

$$\delta_{\text{skin}}(\omega) \simeq c\frac{(2\gamma_p)^{1/2}}{\omega_p\omega^{1/2}} \, , \qquad (\omega \ll \gamma_p) \, , \tag{15.51}$$

[12]The reader familiar with quantum mechanics can observe that, in quantum mechanics, the energy $\mathcal{E}$ of a particle is related to its frequency by $\mathcal{E} = \hbar\omega$, and the momentum p is related to the (real part of the) wavenumber k by $p = \hbar k$. Then, the above dispersion relation takes the form $\mathcal{E}^2 = m_p^2 c^4 + p^2 c^2$, with $m_p c^2 = \hbar\omega_p$. This is the dispersion relation of a massive particle with mass m_p, see eq. (7.147). In this sense, in a conductor, the photon becomes massive.

[13]In the literature, eq. (15.46) is sometime used to argue that there is no solution for k_R for $\omega < \omega_p$. However, this is incorrect because eq. (15.46) only holds for $\omega \gg \omega_p$ and, as we see from the exact solution (15.49, 15.50), $k_R(\omega)$ is non-vanishing for all values of ω. Note also that the right-hand side of eq. (15.49) is always positive, so $k_I(\omega)$ is well defined for all values of ω.

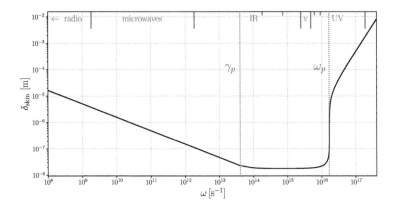

Fig. 15.1 The skin depth $\delta_{\mathrm{skin}}(\omega)$ for the simple model (14.85), using the values of σ_0 and τ for copper at room temperature. The corresponding values of γ_p and ω_p are indicated by the horizontal dotted lines. The bands of the electromagnetic spectrum corresponding to the values of ω are also indicated (the label "v" stands for "visible"). The smaller ticks in the IR and UV regions correspond to the subdivision of the IR into far, middle, and near IR (from left to right), and similarly for the subdivision of the UV into near, middle and far UV (again from left to right).

which, upon use of eq. (14.83), is the same as eq. (15.43). The solution for $\delta_{\mathrm{skin}}(\omega)$ then flattens in the region $\gamma_p \lesssim \omega \lesssim \omega_p$ and, as we approach the plasma frequency, it raises sharply. When $\omega \gg \omega_p$, $\delta_{\mathrm{skin}}(\omega)$ eventually grows as ω^2, in agreement with eq. (15.47) so, for instance, metals can be quite transparent to X-ray radiation. Note, however, that, apart from the fact that have we used the very simplified model (14.85) for the response function, at a more fundamental level the classical analysis used to produce Fig. 15.1 breaks down for X-ray radiation. In this regime a full quantum treatment, based on (coherent and incoherent) Compton scattering, becomes necessary, see also the discussion in Section 16.2.

It should be observed, at this point, that the above computation assumes the validity of Ohm's law in the form (13.81), and of the Drude model of conductivity. As can be seen from the derivation in Section 14.4, this in turn assumes that, at the frequency ω of interest, the electron mean free path ℓ inside the conductor is small compared to the length-scale over which the corresponding Fourier mode of the electric field varies, which is given precisely by the skin depth $\delta_{\mathrm{skin}}(\omega)$. This assumption indeed entered when we assumed that, in eq. (14.57), we could neglect the $\mathbf{x}$ dependence of electric field mode $\tilde{\mathbf{E}}(\omega, \mathbf{x})$. From Fig. 15.1 we see that $\delta_{\mathrm{skin}}(\omega)$ becomes quite low in the regime $\gamma_p \ll \omega \ll \omega$, and in this domain this assumption can fail even at room temperatures (and even more at low temperatures, where the mean free path can increase significantly).[14] In this case, we are in the regime of the *anomalous skin effect*. In this regime the relation between the current and the electric field is more complicated, and is rather given by an integral relation of

---

[14]For instance, for copper at room temperature, the electrons mean free path is $\ell \simeq 4.2 \times 10^{-8}$ m, while at $\lambda = 10\,\mu\mathrm{m}$ (corresponding to $\omega \simeq 1.88 \times 10^{14}\,\mathrm{s}^{-1}$ and $\omega/\gamma_p \simeq 4.5$), eq. (15.49) gives $\delta_{\mathrm{skin}} \simeq 1.8 \times 10^{-8}$ m, so for these frequencies the assumption $\ell \ll \delta_{\mathrm{skin}}(\omega)$ breaks down.

the form

$$\tilde{\mathbf{j}}_{\text{free}}(\omega, \mathbf{x}) = \int d^3x' \sigma(\omega, \mathbf{x} - \mathbf{x}') \tilde{\mathbf{E}}(\omega, \mathbf{x}') \,. \tag{15.52}$$

Upon performing the Fourier transform also with respect to $\mathbf{x}$, this gives

$$\tilde{\mathbf{j}}_{\text{free}}(\omega; \mathbf{k}) = \sigma(\omega, \mathbf{k}) \tilde{\mathbf{E}}(\omega, \mathbf{k}) \,. \tag{15.53}$$

The determination of the kernel $\sigma(\omega, \mathbf{x} - \mathbf{x}')$ is then more complicated, and requires solving a Boltzmann kinetic equation for the non-equilibrium part of the electron distribution function in phase space.

Several other approximations must be improved before comparing Fig. 15.1 to the behavior of an actual metal. In particular, by using eq. (14.85), we have neglected the contribution from bound electrons and, more fundamentally, all our treatment, based on the Drude model, has been purely classical, and has neglected the interaction between the electrons. A full quantum treatment, including the effect of the band structure of the metals, belongs to a solid-state course. However, the simple model that we have discussed in this section already gives a first useful orientation.

15.3.2 Longitudinal EM waves and plasma oscillations

We next investigate the other branch of solutions of eq. (15.34), which exists if there is a value $\bar{\omega}$ of ω such that $\epsilon(\bar{\omega}) = 0$. First of all, we observe that such a solution is physically possible. Using for instance the simple model (14.85) for the response function, the equation $\epsilon(\bar{\omega}) = 0$ becomes

$$\bar{\omega}^2 + i\bar{\omega}\gamma_p - \omega_p^2 = 0 \,, \tag{15.54}$$

whose solutions are

$$\bar{\omega} = \pm\sqrt{\omega_p^2 - \left(\frac{\gamma_p}{2}\right)^2} - i\frac{\gamma_p}{2} \,. \tag{15.55}$$

For metals $\omega_p \gg \gamma_p$, so eq. (15.55) simplifies to

$$\bar{\omega} \simeq \pm\omega_p - \frac{i}{2}\gamma_p \,. \tag{15.56}$$

Therefore, at this special frequency, beside the "usual" transverse electromagnetic waves, there is also a solution of eq. (15.34) where $\mathbf{E}$ is longitudinal,

$$\tilde{\mathbf{E}}(\omega = \bar{\omega}, \mathbf{k}) = \tilde{E}_{\mathbf{k}} \hat{\mathbf{k}} \,. \tag{15.57}$$

If we insert this expression into the other Maxwell's equations we see that, since $\hat{\mathbf{k}} \times \hat{\mathbf{k}} = 0$, eq. (15.37) requires that, on this solution, $\mathbf{B} = 0$. Then, eq. (15.36) is trivially satisfied and, taking into account that $\epsilon(\bar{\omega}) = 0$, also eq. (15.35) is satisfied. We have therefore found a longitudinal electromagnetic wave, in which the electric field has the form

$$\mathbf{E}(t, \mathbf{x}) = e^{-\gamma_p t/2} \text{Re}\left[\tilde{E}_{\mathbf{k}} e^{-i\omega_p t + i\mathbf{k}\cdot\mathbf{x}}\right] \hat{\mathbf{k}} \,, \tag{15.58}$$

i.e., performs oscillations at the plasma frequency, with an amplitude that is exponentially damped in time (since $\gamma_p = 1/\tau > 0$), and is oriented in the propagation direction $\mathbf{k}$, while $\mathbf{B} = 0$. Metals are characterized by the condition $\omega_p \gg \gamma_p$. Therefore, the wave (15.58) performs a large number of oscillations before being significantly damped.

One might ask why we did not consider this solution for dielectrics. Again, one would have a longitudinal solution if there were a frequency $\bar{\omega}$ such that $\epsilon(\bar{\omega}) = 0$ in eq. (15.3). If we use the model (14.39) for the dielectric constant, the equation $\epsilon(\bar{\omega}) = 0$ becomes the same as eq. (15.54), with ω_p^2 replaced by $\omega_0^2 + \omega_p^2$ and γ_p replaced by γ_0. However, while a good conductor is characterized by the condition $\omega_p \gg \gamma_p$, for a dielectric or a poor conductor we are rather in the opposite limit, in which ω_p is at most of the same order as γ_p (and the typical frequency ω_0 for an electron bound in an atom is also in general in the UV, of the same order as ω_p). Any wave oscillating as in eq. (15.58) would therefore get damped on a timescale of at most a few cycles of oscillations, and therefore does not describe an actual wave propagating in the medium.[15]

Note that the Fourier modes $\tilde{\mathbf{E}}(\omega = \bar{\omega}, \mathbf{k})$, in eq. (15.58), all oscillate at the same plasma frequency ω_p (and have the same decay time), independently of $\mathbf{k}$, so the dispersion relation of these modes is $\omega(k) = \omega_p$, independent of k. This is due to our simplified modeling of the response functions $\sigma(\omega)$, which we have taken independent of k.[16]

We now discuss the physics behind the longitudinal solution. From eq. (14.82) we see that, if $n_f = 0$, then $\omega_p = 0$ and there are no oscillations. These oscillations must therefore be related to oscillations of the free charges in the medium, and disappear if there are no free charges. Note that this is different from what happens to the transverse electromagnetic waves that, according to eq. (15.39), in the limit $n_f = 0$, where $\epsilon(\omega) = \epsilon_0$, have the standard dispersion relation of electromagnetic waves in vacuum, $\omega = |\mathbf{k}|c$, see eq. (9.41).[17]

Consider a metal, in which the positively charged ions are, to a first approximation, fixed, while the electrons are free to move. In an equilibrium situation, the macroscopic charge density $\rho_{\text{ions}}(\mathbf{x})$ of the ions and the macroscopic charge density $\rho_f(\mathbf{x})$ of the free electrons are time-independent, and are equal and opposite at each (coarse-grained) point in space, $\rho_{\text{ions}}(\mathbf{x}) = -\rho_f(\mathbf{x})$, so the medium is overall electrically neutral. Suppose now that, either because of statistical fluctuations or because of an external disturbance, some electrons are removed from a region and accumulate into another. Then, the electrons density is lowered in the first region and correspondingly enhanced in the second region. The first region will therefore be overall positively charged, because the positive ion charge density will no longer be fully compensated by the electron charge density, while the second will have an overall negative charge. As a result, there will be a net electric field, pointing from the positively charged region toward the negatively charged region. Under the action of this electric field, all the free electrons of the material will move, and in particular the "cloud" of excess electrons will move back toward the region where there is a net positive charge. If collisions with

---

[15] Actually, for $\omega_p < \gamma_p/2$, there is not even a real part in eq. (15.55), and the solution for $\bar{\omega}$ becomes purely imaginary.

[16] A non-trivial dispersion relation would emerge from a hydrodynamical treatment of the oscillating electron fluid, see Section 10.8 of Jackson (1998), and leads to a dispersion relation $\omega^2(k) = \omega_p^2 + 3\langle v^2 \rangle k^2$, where $\langle v^2 \rangle$ is the average square velocity of the electrons, related to the temperature by $m_e \langle v^2 \rangle = 3kT$.

[17] Recall that, in this computation, we have also set $\mu(\omega) = \mu_0$.

the ions are negligible, it will arrive there with a significant velocity and will swing on the opposite side. The force from the positively charged region will then attract it back, so this cloud of electrons performs a series of oscillations around the fixed position of the positively charged region. Eventually, these oscillations will be damped by the collisions. The electric field generated by these charge inhomogeneities points from the positively charged region toward the electron cloud and is therefore aligned with the direction of motion of the electron cloud (and opposite to it). It is therefore a longitudinal electric field, that points in the same direction as that along which there is a spatial inhomogeneities, i.e., along the direction of the wavenumber $\mathbf{k}$ of the Fourier mode, and oscillates together with the oscillations of the free electron, passing through zero at the moment when the electron cloud passes over the position of the positively charged region, since there the total charge densities of electrons and ions momentarily compensate each other. We therefore have a natural mechanism for creating an oscillating longitudinal electric field. The same can happen in a plasma, where the positive and negative charges can both move freely. In this case, a cloud of excess positive charges and a cloud of excess negative charges would oscillate around their common center of mass. These oscillations are then called *plasma oscillations* (even in the case of metals).

We now compute the frequency of these oscillations, using the Drude model, and we will show that we get precisely the plasma frequency computed above. We consider for definiteness the situation appropriate to a metal, where the ions are fixed and the free charges are provided only be the free electrons, but the argument can be generalized to the case of a plasma, where both the ions and electrons are free to move. We start from eq. (14.56), that we recall here,

$$\left(\frac{\partial}{\partial t} + \frac{1}{\tau}\right)\mathbf{j}_{\text{free}}(t, \mathbf{x}) = \frac{n_f e^2}{m_e}\mathbf{E}(t, \mathbf{x}), \qquad (15.59)$$

[where, compared to eq. (15.59), we have written $\partial\mathbf{j}_{\text{free}}/\partial t$ instead of $d\mathbf{j}_{\text{free}}/dt$ since now we are considering an inhomogeneous situation where $\mathbf{j}_{\text{free}} = \mathbf{j}_{\text{free}}(t, \mathbf{x})$] and we observe that, in this case, $\mathbf{E}$ is not a fixed external electric field, but rather is generated by the positive and negative charges in the metal. It therefore satisfies

$$\boldsymbol{\nabla}\cdot\mathbf{E}(t, \mathbf{x}) = \frac{1}{\epsilon_0}\rho(t, \mathbf{x}), \qquad (15.60)$$

where $\rho(t, \mathbf{x}) = \rho_{\text{ions}}(\mathbf{x}) + \rho_{\text{free}}(t, \mathbf{x})$; note that we have taken a static distribution of ions, as appropriate to a metal, while the electrons provide the freely moving charges and, in this out-of-equilibrium situation, have a time-dependent density. In a homogeneous equilibrium situation, the positive and negative charge densities are independent of time, and compensate each other, so the total density vanishes. However, in the presence of spatial fluctuations, the two densities no longer compensate each other at each point, even if the total charges, i.e., their spatial integrals, are equal and opposite. The charge density and the current density of the free electrons are related by the continuity equation $\nabla \cdot \mathbf{j}_{\text{free}} = -\partial \rho_{\text{free}}/\partial t$. However, since ρ_{ions} is independent of time, we can write it as well as $\nabla \cdot \mathbf{j}_{\text{free}} = -\partial \rho/\partial t$. Taking the divergence of eq. (15.59), we therefore get

$$-\left(\frac{\partial}{\partial t} + \frac{1}{\tau}\right)\frac{\partial \rho}{\partial t} = \frac{n_f e^2}{\epsilon_0 m_e}\rho\,. \qquad (15.61)$$

On the right-hand side we recognize the square of the plasma frequency, eq. (14.82), so we can rewrite eq. (15.61) as

$$\left(\frac{\partial^2}{\partial t^2} + \frac{1}{\tau}\frac{\partial}{\partial t} + \omega_p^2\right)\rho = 0\,. \qquad (15.62)$$

Notice that, in our approximations, this equation is independent of $\mathbf{x}$, i.e., is the same for all spatial Fourier modes. Looking for a solution $\rho(t) \propto e^{-i\bar{\omega}t}$ we get the condition on $\bar{\omega}$,

$$\bar{\omega}^2 + i\frac{\bar{\omega}}{\tau} - \omega_p^2 = 0\,, \qquad (15.63)$$

which is the same as eq. (15.54) (identifying as usual $\gamma_p = 1/\tau$), and therefore has the same solutions, given in eq. (15.55). We have therefore understood that the longitudinal solution of Maxwell's equations in a conductor, found above, is due to the damped oscillations of the charge inhomogeneities.

It could be puzzling the fact that, in plasma oscillations, no magnetic field is generated, since the oscillating electron cloud generates a current. However, from the Ampère–Maxwell law in materials, eq. (13.46), we know that there are two contributions to the magnetic field, one coming from the time derivative of $\mathbf{D}$ and the other from the current of the free charges. In Fourier space, these two contributions are given by the terms proportional to $\tilde{\mathbf{D}}(\omega, \mathbf{x})$ and to $\tilde{\mathbf{j}}_{\text{free}}(\omega, \mathbf{x})$ in eq. (13.85), and they combine to give the term proportional to $\epsilon(\omega)\tilde{\mathbf{E}}(\omega, \mathbf{x})$ in eq. (13.89). We see that the condition $\epsilon(\omega) = 0$ imposes that these two contributions cancel among them, and therefore there is no magnetic field.

15.4   Electromagnetic waves in waveguides

The range of wavelengths of microwaves is about $\lambda \sim 1$ mm to 1 m, corresponding to frequencies $f = c/\lambda$ from 300 MHz (for $\lambda = 1$ m) to 300 GHz (for $\lambda = 1$ mm), and $\omega = 2\pi f$ from about $2 \times 10^9 \, \text{s}^{-1}$ to about $2 \times 10^{12} \, \text{s}^{-1}$.[18] Such wavelengths cannot be transported to large distances with ordinary ac circuits, since, when the dimensions of the circuit become larger than the wavelength, the radiative losses become very large, and an alternative is provided by waveguides, i.e., hollow metallic pipes. Electromagnetic waves can propagate inside such waveguides, and the difference with respect to vacuum propagation comes from the boundary conditions that must be imposed on the fields on the surface of the pipe.

[18] In radio-frequency engineering, microwaves are rather defined, more restrictively, as electromagnetic waves with wavelength between 3 mm to 0.3 m, corresponding to the range 1–100 GHz.

15.4.1 Maxwell's equations in a waveguide

Consider first Maxwell's equations inside the waveguide. Since there are no sources, inside we just have the usual Maxwell's equations in vacuum (we assume that a good vacuum is made inside the waveguide; otherwise one should simply include the corresponding dielectric constant), that we write again here

$$\mathbf{\nabla \cdot E} = 0 \,, \tag{15.64}$$

$$\mathbf{\nabla \times B} - \frac{1}{c^2}\frac{\partial \mathbf{E}}{\partial t} = 0 \,, \tag{15.65}$$

$$\mathbf{\nabla \cdot B} = 0 \,, \tag{15.66}$$

$$\mathbf{\nabla \times E} + \frac{\partial \mathbf{B}}{\partial t} = 0 \,. \tag{15.67}$$

Observe that, for fields that are time-dependent, as electromagnetic waves, eq. (15.64) is implied by eq. (15.65). In fact, taking the divergence of eq. (15.65) and using $\mathbf{\nabla \cdot (\nabla \times B)} = 0$, we obtain $\partial_t(\mathbf{\nabla \cdot E}) = 0$, which, for a field with a time-dependence $e^{-i\omega t}$ with $\omega \neq 0$, implies $\mathbf{\nabla \cdot E} = 0$. Similarly, eq. (15.66) is implied by eq. (15.67). Therefore, for time-dependent fields in vacuum, it is sufficient to consider just eqs. (15.65) and (15.67).[19]

We set the longitudinal direction of the waveguide along the z axis. The boundary conditions break the translation invariance in the (x, y) plane and give a non-trivial structure to the solution in the x and y directions, so we look for a solution in the form of a wave propagating along the z direction, with a generic dependence on the (x, y) variables,

$$\mathbf{E}(t, \mathbf{x}) = \boldsymbol{\mathcal{E}}(x, y)e^{-i\omega t + ikz} \,, \tag{15.72}$$

$$\mathbf{B}(t, \mathbf{x}) = \boldsymbol{\mathcal{B}}(x, y)e^{-i\omega t + ikz} \,. \tag{15.73}$$

On functions of this form, $\partial_t \to -i\omega$ and $\partial_z \to ik$. Then, writing explicitly the equations in components, eq. (15.65) becomes

$$\frac{i\omega}{c^2}\mathcal{E}_x - ik\mathcal{B}_y = -\partial_y \mathcal{B}_z \,, \tag{15.74}$$

$$\frac{i\omega}{c^2}\mathcal{E}_y + ik\mathcal{B}_x = \partial_x \mathcal{B}_z \,, \tag{15.75}$$

$$\partial_x \mathcal{B}_y - \partial_y \mathcal{B}_x = -\frac{i\omega}{c^2}\mathcal{E}_z \,, \tag{15.76}$$

and similarly eq. (15.67) gives

$$i\omega \mathcal{B}_x + ik\mathcal{E}_y = \partial_y \mathcal{E}_z \,, \tag{15.77}$$

$$-i\omega \mathcal{B}_y + ik\mathcal{E}_x = \partial_x \mathcal{E}_z \,, \tag{15.78}$$

$$\partial_x \mathcal{E}_y - \partial_y \mathcal{E}_x = i\omega \mathcal{B}_z \,. \tag{15.79}$$

We now observe that eqs. (15.74) and (15.78) are two algebraic equations for the two variables $\mathcal{E}_x$, and $\mathcal{B}_y$, and similarly eqs. (15.75) and (15.77) are two algebraic equations for the two variables $\mathcal{E}_y$, and $\mathcal{B}_x$. The transverse components of the fields are therefore determined algebraically, in

[19]Indeed, it is instructive to see how the solution for electromagnetic waves in vacuum, that we found in eqs. (9.70)–(9.81) using all four Maxwell's equations, could have been obtained using only eqs. (15.65) and (15.67): inserting the ansatz $\mathbf{E}(t, \mathbf{x}) = \mathbf{E_k}\, e^{-i\omega t + i\mathbf{k \cdot x}}$ and $\mathbf{B}(t, \mathbf{x}) = \mathbf{B_k}\, e^{-i\omega t + i\mathbf{k \cdot x}}$ (where, for the moment, ω and k are independent) into eqs. (15.65) and (15.67) we get

$$\mathbf{k \times B_k} + (\omega/c^2)\mathbf{E_k} = 0 \,, \tag{15.68}$$

$$\mathbf{k \times E_k} - \omega \mathbf{B_k} = 0 \,. \tag{15.69}$$

Solving for $\mathbf{B_k}$ in eq. (15.69), plugging it into eq. (15.68) and expanding the resulting triple vector product, we get

$$(\omega^2 - k^2 c^2)\mathbf{E_k} + k^2 c^2 (\mathbf{E_k \cdot \hat{k}})\mathbf{\hat{k}} = 0 \,. \tag{15.70}$$

Separating $\mathbf{E_k}$ into its transverse and longitudinal parts, $\mathbf{E_k} = \mathbf{E_{k,\perp}} + E_{k,\parallel}\mathbf{\hat{k}}$, we get

$$(\omega^2 - k^2 c^2)\mathbf{E_{k,\perp}} + \omega^2 E_{k,\parallel}\mathbf{\hat{k}} = 0 \,. \tag{15.71}$$

Since $\mathbf{E_{k,\perp}}$ and $\mathbf{\hat{k}}$ are orthogonal, if $\omega \neq 0$ we get the two separate conditions $\omega^2 - k^2 c^2 = 0$ and $E_{k,\parallel} = 0$, i.e., the dispersion relation in vacuum, and the condition $E_{k,\parallel} = 0$ that, before, we had derived using $\mathbf{\nabla \cdot E} = 0$. The same treatment can be made for the magnetic field, solving for $\mathbf{E_k}$ in eq. (15.68) and plugging it into eq. (15.69).

terms of the derivatives of the longitudinal components $\mathcal{E}_z$ and $\mathcal{B}_z$. The solution of these algebraic equations is

$$\gamma^2 \mathcal{E}_x = ik\partial_x \mathcal{E}_z + i\omega\partial_y \mathcal{B}_z\,, \tag{15.80}$$

$$\gamma^2 \mathcal{E}_y = ik\partial_y \mathcal{E}_z - i\omega\partial_x \mathcal{B}_z\,, \tag{15.81}$$

$$\gamma^2 \mathcal{B}_x = ik\partial_x \mathcal{B}_z - \frac{i\omega}{c^2}\partial_y \mathcal{E}_z\,, \tag{15.82}$$

$$\gamma^2 \mathcal{B}_y = ik\partial_y \mathcal{B}_z + \frac{i\omega}{c^2}\partial_x \mathcal{E}_z\,, \tag{15.83}$$

where we have defined

$$\gamma^2 \equiv \left(\frac{\omega}{c}\right)^2 - k^2\,, \tag{15.84}$$

(not to be confused with the Lorentz boost factor, also conventionally denoted by γ). Observing that

$$\hat{\mathbf{z}} \times \boldsymbol{\nabla} = \hat{\mathbf{z}} \times (\hat{\mathbf{x}}\partial_x + \hat{\mathbf{y}}\partial_y + \hat{\mathbf{z}}\partial_z)$$

$$= \hat{\mathbf{y}}\partial_x - \hat{\mathbf{x}}\partial_y\,, \tag{15.85}$$

we can rewrite this more compactly as

$$\gamma^2 \mathcal{E}_i = ik\partial_i \mathcal{E}_z - i\omega(\hat{\mathbf{z}} \times \boldsymbol{\nabla})_i \mathcal{B}_z\,, \tag{15.86}$$

$$\gamma^2 \mathcal{B}_i = ik\partial_i \mathcal{B}_z + \frac{i\omega}{c^2}(\hat{\mathbf{z}} \times \boldsymbol{\nabla})_i \mathcal{E}_z\,, \tag{15.87}$$

where the index i run over the two values $\{x, y\}$. The remaining equations to be satisfied are eqs. (15.76) and (15.79). Inserting eqs. (15.82) and (15.83) into eq. (15.76) we get

$$(\partial_x^2 + \partial_y^2 + \gamma^2)\mathcal{E}_z(x, y) = 0\,, \tag{15.88}$$

which is a Helmholtz equation in two dimensions, of the type already encountered (in three dimensions) in eq. (10.14). Similarly, inserting eqs. (15.80) and (15.81) into eq. (15.79) we get

$$(\partial_x^2 + \partial_y^2 + \gamma^2)\mathcal{B}_z(x, y) = 0\,. \tag{15.89}$$

The problem is therefore reduced to solving eqs. (15.88) and (15.89). The solution for $\mathcal{E}_z(x, y)$ and $\mathcal{B}_z(x, y)$ will then determine all other components through eqs. (15.86) and (15.87) (as long as $\gamma^2 \neq 0$, see below). To solve eqs. (15.88) and (15.89), we must specify the boundary conditions for $\mathcal{E}_z$ and $\mathcal{B}_z$ on the boundary of the vacuum region that we have considered, i.e., on the boundary between the vacuum and the inner surface of the conductor that makes the hollow pipe. We discuss this in the next subsection.

15.4.2 Boundary conditions at the surface of conductors

To study the boundary conditions at the interface between vacuum and a conductor, we begin by observing that eqs. (13.64) and (13.68) are valid even when the media (1) and (2) are the vacuum and a conductor, respectively, since they have been derived from the Maxwell's equations (13.47) and (13.48), that do not depend on the sources. Therefore, the tangential components of $\mathbf{E}$ and the normal component of $\mathbf{B}$ are continuous across the surface separating the vacuum (or a dielectric) from a conductor. However, as we discussed in Section 4.1.6, in the absence of an external applied voltage a conductor quickly reaches an equilibrium situation where any external field is screened, so that inside a conductor there is no current flowing, and $\mathbf{E} = 0$. In the absence of currents (and neglecting magnetic dipoles at the atomic scale, so excluding the case of ferromagnets) there is no magnetic field either, so inside a conductor, in a static situation, $\mathbf{E} = \mathbf{B} = 0$. Then, eqs. (13.64) and (13.68) tell us that, approaching the boundary from the vacuum side, the tangential components of $\mathbf{E}$ and the normal component of $\mathbf{B}$ vanish, so the boundary conditions for a waveguide are

$$\hat{\mathbf{n}} \times \mathbf{E} = 0, \qquad \hat{\mathbf{n}} \cdot \mathbf{B} = 0. \qquad (15.90)$$

To solve eqs. (15.88) and (15.89), we need to extract from this the boundary conditions on E_z and B_z. Consider for instance a point on the boundary where the normal $\hat{\mathbf{n}}$ is along the $\mathbf{x}$ direction (recall that the longitudinal axis of the waveguide has been set along the $\hat{\mathbf{z}}$ direction). Then eq. (15.90) gives, on the boundary, $E_y = E_z = 0$ and $B_x = 0$ and therefore

$$\mathcal{E}_y = \mathcal{E}_z = 0, \qquad \mathcal{B}_x = 0. \qquad (15.91)$$

The boundary condition for $\mathcal{E}_z$ is therefore $\mathcal{E}_z = 0$. For $\mathcal{B}_z$, we use eqs. (15.81) and (15.82). Near the boundary, where $\mathcal{E}_y = 0$ and $\mathcal{B}_x = 0$, they become

$$ik\partial_y\mathcal{E}_z - i\omega\partial_x\mathcal{B}_z = 0, \qquad (15.92)$$

$$ik\partial_x\mathcal{B}_z - \frac{i\omega}{c^2}\partial_y\mathcal{E}_z = 0, \qquad (15.93)$$

which can be combined to give

$$\gamma^2\partial_x\mathcal{B}_z = 0. \qquad (15.94)$$

Therefore (unless $\gamma = 0$), we have the boundary condition $\partial_x\mathcal{B}_z = 0$. More generally, for $\hat{\mathbf{n}}$ generic rather than in the $\hat{\mathbf{x}}$ direction, we get $\hat{\mathbf{n}} \cdot \nabla\mathcal{B}_z = 0$.

15.4.3   TE, TM, and TEM modes

In conclusion, we have to solve the eigenvalue equation

$$-(\partial_x^2 + \partial_y^2)\mathcal{E}_z(x,y) = \gamma^2\mathcal{E}_z(x,y), \qquad (15.95)$$

with the boundary condition

$$(\mathcal{E}_z)|_S = 0, \qquad (15.96)$$

and the eigenvalue equation

$$-(\partial_x^2 + \partial_y^2)\mathcal{B}_z(x,y) = \gamma^2 \mathcal{B}_z(x,y)\,, \tag{15.97}$$

with the boundary condition

$$\hat{\mathbf{n}}\cdot(\boldsymbol{\nabla}\mathcal{B}_z)|_S = 0\,, \tag{15.98}$$

where we have taken the z axis as the propagation direction and $\hat{\mathbf{n}}$ is the normal to the boundary S of the waveguide. These two equations and boundary conditions are independent of each other. The transverse components of the electric and magnetic field are then found from eqs. (15.86) and (15.87). We can distinguish three classes of solutions.

TE modes. These are solutions with $\mathcal{E}_z = 0$, which trivially satisfies eq. (15.95) and the boundary condition (15.96), and $\mathcal{B}_z \neq 0$. Observe, from eq. (15.86), that, even if $\mathcal{E}_z = 0$, as long as $\mathcal{B}_z \neq 0$, $\mathcal{E}_x$ and $\mathcal{E}_y$ are non-zero. In these solutions, therefore, the electric field is transverse, hence the name TE (for "transverse electric"), while the magnetic field has both transverse and longitudinal components. Equation (15.97) with the boundary condition (15.98) is an eigenvalue equation, that has solutions only for a discrete set of values of γ^2. From eq. (15.84), the corresponding dispersion relation is

$$\omega^2 = (k^2 + \gamma^2)c^2\,, \tag{15.99}$$

where $k \equiv k_z$ is a continuous variable, corresponding to the fact that we have assumed a straight infinite waveguide along the z direction, while γ^2 plays the role of $k_x^2 + k_y^2$ and takes discrete values because of the boundary conditions in the x and y directions, exactly as it happens for the vibration modes of a string. Since $\mathcal{E}_z = 0$, eqs. (15.86) and (15.87) simplify to

$$\gamma^2 \mathcal{E}_i = -i\omega(\hat{\mathbf{z}} \times \boldsymbol{\nabla})_i \mathcal{B}_z\,, \tag{15.100}$$

$$\gamma^2 \mathcal{B}_i = ik\partial_i \mathcal{B}_z\,. \tag{15.101}$$

TM modes. These are the solutions with $\mathcal{B}_z = 0$ and $\mathcal{E}_z \neq 0$, so now the magnetic field is transverse, while the electric field has both transverse and longitudinal components. From eqs. (15.86) and (15.87), the transverse component are given by

$$\gamma^2 \mathcal{E}_i = ik\partial_i \mathcal{E}_z\,, \tag{15.102}$$

$$\gamma^2 \mathcal{B}_i = \frac{i\omega}{c^2}(\hat{\mathbf{z}} \times \boldsymbol{\nabla})_i \mathcal{E}_z\,. \tag{15.103}$$

The dispersion relation is again eq. (15.99). However, the eigenvalues γ^2 are, in general, different from those of the TE modes, since γ^2 is determined by eq. (15.97) with the boundary condition (15.98), which is different from the boundary conditions for the TE modes.

TEM modes. Finally, one can search for a solution transverse both in the electric and magnetic fields, $\mathcal{E}_z = \mathcal{B}_z = 0$. From eqs. (15.86) and

(15.87), we see that a non-vanishing solution for the transverse fields is possible only if $\gamma^2 = 0$, so the dispersion relation becomes

$$\omega = kc\,, \tag{15.104}$$

as in vacuum. In this case, eqs. (15.86) and (15.87) become trivial identities $0 = 0$ and do not allow us to determine the transverse component. To determine them, we must rather go back to Maxwell's equations. Consider first eqs. (15.65) and (15.67), in the form (15.74)–(15.79). Setting $\mathcal{E}_z = \mathcal{B}_z = 0$ and $\omega = kc$, eqs. (15.74) and (15.75) give

$$c\mathcal{B}_x = -\mathcal{E}_y\,, \qquad c\mathcal{B}_y = \mathcal{E}_x\,, \tag{15.105}$$

and the same conditions are obtained from eqs. (15.76) and (15.77). Equations (15.78) and (15.79) give

$$\partial_x \mathcal{E}_y - \partial_y \mathcal{E}_x = 0\,, \qquad \partial_x \mathcal{B}_y - \partial_y \mathcal{B}_x = 0\,. \tag{15.106}$$

The remaining equations are the divergence equations (15.64) and (15.66) that, setting $\mathcal{E}_z = \mathcal{B}_z = 0$, become

$$\partial_x \mathcal{E}_x + \partial_y \mathcal{E}_y = 0\,, \qquad \partial_x \mathcal{B}_x + \partial_y \mathcal{B}_y = 0\,. \tag{15.107}$$

We therefore have two two-dimensional fields $\boldsymbol{\mathcal{E}}(x, y)$ and $\boldsymbol{\mathcal{B}}(x, y)$ that satisfy

$$\boldsymbol{\nabla}_T{\cdot}\boldsymbol{\mathcal{E}} = 0\,, \qquad \boldsymbol{\nabla}_T \times \boldsymbol{\mathcal{E}} = 0\,, \tag{15.108}$$

where $\boldsymbol{\nabla}_T = \hat{\mathbf{x}}\partial_x + \hat{\mathbf{y}}\partial_y$ is the two-dimensional gradient in the transverse plane, and similarly $\boldsymbol{\nabla}_T{\cdot}\boldsymbol{\mathcal{B}} = \boldsymbol{\nabla}_T \times \boldsymbol{\mathcal{B}} = 0$. A theorem states that, if a two-dimensional vector field $\boldsymbol{\mathcal{E}}$ satisfies $\boldsymbol{\nabla}_T{\cdot}\boldsymbol{\mathcal{E}} = \boldsymbol{\nabla}_T \times \boldsymbol{\mathcal{E}} = 0$ *in a simply connected domain*, then $\boldsymbol{\mathcal{E}} = 0$. However, this is no longer true if the domain is not simply connected (i.e., if there are closed curves that cannot be deformed continuously to a point). So, for instance, we can take as waveguide a coaxial cable, made of an inner cylindrical conductor whose radius, in the transverse plane, is r_1, and an outer hollow conductor of radius $r_2 > r_1$, so, in the transverse plane, the waveguide consists of the region $r_1^2 < x^2 + y^2 < r_2^2$. In this case, it is possible to have non-zero solutions for $\mathcal{E}_x$, $\mathcal{E}_y$ (and, similarly, for $\mathcal{B}_x$, $\mathcal{B}_y$), with $\mathcal{E}_z = \mathcal{B}_z = 0$, corresponding to modes in which both the electric and the magnetic fields are transverse, called TEM modes.

Example: rectangular waveguide. As an example, consider a waveguide whose transverse section is a rectangle $0 < x < a$, $0 < y < b$. Being simply connected, TEM modes do not exist. The TE modes are the solution of eq. (15.97), with the boundary conditions

$$\frac{\partial \mathcal{B}_z}{\partial x}(x = 0, y) = \frac{\partial \mathcal{B}_z}{\partial x}(x = a, y) = 0\,, \tag{15.109}$$

$$\frac{\partial \mathcal{B}_z}{\partial y}(x, y = 0) = \frac{\partial \mathcal{B}_z}{\partial y}(x, y = b) = 0\,. \tag{15.110}$$

The eigenfunctions are characterized by two integers (m, n), and are

$$(\mathcal{B}_z)_{mn} = \mathcal{B}_{mn} \cos\left(\frac{m\pi}{a}x\right) \cos\left(\frac{n\pi}{b}y\right)\,, \tag{15.111}$$

with $\mathcal{B}_{mn}$ constant, and the corresponding eigenvalues are

$$\gamma_{mn}^2 = \left(\frac{m\pi}{a}\right)^2 + \left(\frac{n\pi}{b}\right)^2 \,. \tag{15.112}$$

The corresponding modes are denoted as TE_{mn}. The TE and TM modes are solutions only if $\gamma^2 \neq 0$, since otherwise eqs. (15.80)–(15.83) degenerate to $0 = 0$ identities and one must rather resort to Maxwell equation in the original form; as we have seen, this gives the condition for the TEM modes, which are absent in a rectangular waveguide. Therefore, the case $m = n = 0$ is excluded, and, taking $a < b$, the lowest TE mode is the mode TE_{01}.

Similarly, the TM modes are the solutions of eq. (15.95) with $\mathcal{E}_z$ vanishing on the boundaries,

$$\mathcal{E}_z(x = 0, y) \;=\; \mathcal{E}_z(x = a, y) = 0\,, \tag{15.113}$$
$$\mathcal{E}_z(x, y = 0) \;=\; \mathcal{E}_z(x, y = b) = 0\,. \tag{15.114}$$

The solutions are

$$(\mathcal{E}_z)_{mn}(x, y) = \mathcal{E}_{mn} \sin\left(\frac{m\pi}{a}x\right) \sin\left(\frac{n\pi}{b}y\right)\,. \tag{15.115}$$

The corresponding eigenvalues are again given by eq. (15.112), but now $m \geq 1$ and $n \geq 1$, since otherwise the solution vanishes. Hence, the lowest TM mode is the mode TM_{11}.

From the dispersion relation

$$\omega_{mn}^2 = k^2 c^2 + \gamma_{mn}^2 c^2 \tag{15.116}$$

it follows that, in the waveguide, the minimum frequency that can propagate is $\omega = c\gamma_{01} = c\pi/b$, and therefore the wavelength $\lambda = 2\pi c/\omega$ must be less than $2b$. Since k is a continuous variable, all frequencies above this limiting value, i.e., all wavelength $\lambda < 2b$ can propagate in the waveguide.

Exercise. Compute the electric and magnetic fields in the transverse plane for the TE_{01} and TM_{11} modes in a rectangular waveguide.

Scattering of electromagnetic radiation

16

In this chapter we discuss the scattering of electromagnetic waves by charged particles. We will examine separately the scattering from free electrons and that from electrons bound in atoms or molecules.

16.1   Scattering cross-section

We begin by defining the scattering cross-section, which provides a quantitative way of describing the outcome of a scattering process. The cross-section is defined, at a fundamental level, with reference to collisions between particles. Consider a beam of particles of type 1, with number density n_1 and velocity v_1, impinging on a target made of particles of type 2 and number density n_2, in the frame where the particles of type 2 are at rest. This could be the case, for instance, of a beam of electron impinging on a block of material. The number of scattering events, dN, that take place in a volume dV of the material in a time interval dt must be proportional to the incoming flux $n_1 v_1$ and to the density of targets n_2. The proportionality constant is, by definition, the cross-section σ,

$$dN = \sigma v_1 n_1 n_2 \, dV dt \,. \qquad (16.1)$$

Dimensional analysis shows immediately that σ has the dimensions of an area. At a more detailed level, we can consider the number $(dN/d\Omega)d\Omega$ of particles scattered within an infinitesimal solid angle $d\Omega = d\cos\theta d\phi$ centered around a given direction, identified by polar angles (θ, ϕ), and define the differential cross-section $d\sigma/d\Omega$ by

$$\frac{dN}{d\Omega} = \frac{d\sigma(\theta, \phi)}{d\Omega} \, v_1 n_1 n_2 \, dV dt \,, \qquad (16.2)$$

so that the total cross-section σ is obtained from

$$\sigma = \int d\Omega \, \frac{d\sigma(\theta, \phi)}{d\Omega} \,. \qquad (16.3)$$

Consider a monochromatic beam of particles, i.e., a beam where all incoming particles have the same energy E. Then $E n_1 v_1$ is the incoming energy flux. Consider first for simplicity elastic scattering, where the incoming particle is scattered from a fixed center, so that its final energy is the same as the initial energy. Then, multiplying by E and integrating over the volume of the target, eq. (16.1) gives

$$EdN = \sigma \, En_1 v_1 dt \int n_2 \, dV \,, \tag{16.4}$$

where, with a slight abuse of notation, we keep the same symbol dN for what is now the number of scattering events differential with respect to dt (rather than with respect to $dtdV$). We now observe that $EdN/dt = d(EN)/dt$ is the scattered energy per unit time, i.e., the scattered power, while $En_1 v_1$ is the incoming energy flux, and $N_2 = \int n_2 \, dV$ is the number of targets. Therefore, the cross-section is equal to the ratio of the scattered power P to the incident energy flux I, per unit target. Therefore, if we consider a wave impinging on a single target,

$$\sigma = \frac{P}{I} \,, \tag{16.5}$$

or, more generally,

$$\boxed{\frac{d\sigma(\theta, \phi)}{d\Omega} = \frac{1}{I} \frac{dP(\theta, \phi)}{d\Omega} \,.} \tag{16.6}$$

We can take this as the definition of the classical cross-section for the elastic scattering of an electromagnetic wave impinging on a single target (the underlying reason being that, at a fundamental level, an electromagnetic wave can be seen as a collection of particles, the photons). Elastic scattering in this case corresponds to the fact that the frequency ω of the scattered electromagnetic wave is the same as the frequency ω_{in} of the incoming radiation (due to the fact that the energy of a photon is related to its frequency by the quantum-mechanical relation $E = \hbar\omega$).

More generally, the incoming energy could be partly absorbed and dissipated into heat in the material, and partly re-radiated, not necessarily at the same frequency as that of the incoming wave. The radiated power will therefore have a frequency spectrum,

$$P = \int d\omega \frac{dP(\omega)}{d\omega} \,, \tag{16.7}$$

and correspondingly we can define the scattering cross-section, differential with respect to frequency, $d\sigma/d\omega$,

$$\frac{d\sigma_{\text{scatt}}(\omega)}{d\omega} = \frac{1}{I} \frac{dP(\omega)}{d\omega} \,, \tag{16.8}$$

as well as the cross-section differential both with respect to frequency and to solid angle,

$$\frac{d\sigma_{\text{scatt}}(\omega; \theta, \phi)}{d\omega d\Omega} = \frac{1}{I} \frac{dP(\omega; \theta, \phi)}{d\omega d\Omega} \,. \tag{16.9}$$

The absorption part, in contrast, will be determined by the term $\mathbf{j} \cdot \mathbf{E}$ in eq. (3.35). If we denote by P_{abs} the energy absorbed per unit time by the material, we can similarly define an absorption cross-section from

$$\sigma_{\text{abs}} = \frac{P_{\text{abs}}}{I} \,. \tag{16.10}$$

For the absorbed energy there is no quantity corresponding to the angles (θ, ϕ) of the emitted radiation, and therefore no sense in which we can consider the differential with respect to the solid angle, as in eq. (16.6). However, we can still consider the power absorbed per unit frequency, and therefore the frequency spectrum

$$\frac{d\sigma_{\mathrm{abs}}(\omega)}{d\omega} = \frac{1}{I}\frac{P_{\mathrm{abs}}(\omega)}{d\omega}. \tag{16.11}$$

For an incoming electromagnetic wave, the incident energy flux is given by the Poynting vector (9.59).

16.2   Scattering on a free electron

We now consider an electromagnetic wave impinging on a free electron initially at rest. The electromagnetic field of the wave accelerates the electron, that therefore emits radiation in all directions. In the classical picture, this is the origin of the scattered wave. The action of the wave on the electron is determined by the Lorentz force (3.6). We assume that the electric field of the wave is not too large, so that the motion induced on the electron remains non-relativistic; we can then neglect the term $\mathbf{v} \times \mathbf{B}$ compared to $\mathbf{E}$, since in an electromagnetic wave $|\mathbf{B}| = |\mathbf{E}|/c$, and $v/c \ll 1$. In this non-relativistic limit, the acceleration of the electron is then given by

$$m_e \ddot{\mathbf{x}}(t) = -e\mathbf{E}[t, \mathbf{x}(t)], \tag{16.12}$$

where $\mathbf{x}(t)$ is the electron position. To understand the limit of validity of this approximation, consider for instance a field $\mathbf{E}(t, \mathbf{x}) = E\hat{\mathbf{x}}\cos(\omega t)$ that does not change appreciably in space in the region over which the electron performs its oscillatory motion. Then the integration of eq. (16.12), with initial condition $x(t = 0) = 0$ and $\dot{x}(t = 0) = 0$, gives

$$\dot{x}(t) = -\frac{eE}{m_e\omega}\sin(\omega t), \qquad x(t) = \frac{eE}{m_e\omega^2}[\cos(\omega t) - 1]. \tag{16.13}$$

The assumption of non-relativistic motion of the electron is therefore satisfied if the electric field is such that

$$\frac{eE}{m_e\omega} \ll c. \tag{16.14}$$

For instance, for visible light with $\lambda = 500$ nm, this limiting field corresponds to 6×10^{12} V/m, which is a quite large field that can only be obtained with picosecond laser pulses. In turn, this implies that

$$x_0 \equiv \frac{eE}{m_e\omega^2} \ll \frac{c}{\omega} = \frac{1}{k}. \tag{16.15}$$

This means that the amplitude x_0 of the oscillations is such that $kx_0 \ll 1$. It is therefore consistent to set $e^{i\mathbf{k}\cdot\mathbf{x}} \simeq 1$ in the expression (9.49) of the electric field of the propagating wave, and write simply

$$\mathbf{E}(t) = \mathrm{Re}\left[\hat{\mathbf{e}}(\mathbf{k})\, E_{\mathbf{k}}\, e^{-i\omega t}\right]. \tag{16.16}$$

The dipole moment of the electron is $\mathbf{d} = -e\mathbf{x}$, so eq. (16.12) gives

$$\ddot{\mathbf{d}}(t) = \frac{e^2 \mathbf{E}(t)}{m_e} .$$ (16.17)

Consider for instance an incoming electromagnetic wave linearly polarized along the $\hat{\mathbf{e}}$ direction, so that $\mathbf{E} = E\hat{\mathbf{e}}$. The radiated power per unit solid angle by a non-relativistic electron is given by Larmor's formula (10.144),

$$\frac{dP(t;\theta)}{d\Omega} = \frac{1}{4\pi\epsilon_0} \frac{1}{4\pi c^3} \frac{e^4 E^2}{m_e^2} \sin^2\theta ,$$ (16.18)

where θ is the polar angle measured from the $\hat{\mathbf{e}}$ axis. The incident energy flux $I = |\mathbf{S}|$ of the incoming wave is given by eq. (10.140), which, using $\mu_0 = 1/(\epsilon_0 c^2)$, can be rewritten as

$$I = c\epsilon_0 E^2 .$$ (16.19)

Therefore,

$$\frac{d\sigma}{d\Omega} = r_0^2 \sin^2\theta ,$$ (16.20)

where we have defined the *classical electron radius*[1]

$$r_0 = \frac{1}{4\pi\epsilon_0} \frac{e^2}{m_e c^2} .$$ (16.21)

Integrating over the angles as in eq. (10.145) we get the *Thomson cross-section* σ_T,

$$\sigma_T = \frac{8\pi}{3} r_0^2 .$$ (16.22)

Notice that the frequency of the radiated electromagnetic wave is the same as the frequency of the incoming wave: an incoming wave oscillating at the frequency ω induces an oscillatory motion of the free electron again at the frequency ω, see eq. (16.13). In turn, as we see from eq. (10.131), this motion generates dipole radiation at the frequency ω. Therefore, the incident and scattered waves have the same frequency.

It should be observed that, in quantum mechanics, this result is modified at sufficiently large frequencies, and the frequency of the scattered wave is lower than that of the incoming wave. This is a simple consequence of energy-momentum conservation, once we borrow from quantum mechanics the information that, at the quantum level, an electromagnetic wave with frequency ω is a collection of particles, the photons, with energy $E = \hbar\omega$ and momentum $\mathbf{k} = \hbar(\omega/c)\hat{\mathbf{k}}$. Consider the scattering process of a photon on an electron initially at rest. Take for instance the (x, y) plane as the scattering plane, with the photon coming along the x axis, and denote by θ the scattering angle of the final photon in

[1] We already met this combination in Note 8 on page 106 in the context of the electron self-energy, as well as in eq. (12.167), in the context of the Abraham–Lorentz equation.

the (x, y) plane, see Fig. 16.1. Then, the initial four momentum of the photon is $k^\mu = (E_\gamma, \mathbf{k})$ with

$$E_\gamma = \hbar\omega, \qquad \mathbf{k} = \frac{\hbar\omega}{c}(1, 0, 0), \qquad (16.23)$$

and the initial four-momentum of the electron is $p^\mu = (E_e, \mathbf{p})$ with

$$E_e = m_e c^2, \qquad \mathbf{p} = 0. \qquad (16.24)$$

The final four-momentum of the photon is $k'^\mu = (E'_\gamma, \mathbf{k}')$, where

$$E'_\gamma = \hbar\omega', \qquad \mathbf{k}' = \frac{\hbar\omega'}{c}(\cos\theta, \sin\theta, 0), \qquad (16.25)$$

corresponding to a scattering angle θ in the (x, y) plane, and a generic frequency ω' that we will determine using energy-momentum conservation. The final four-momentum of the electron is $p'^\mu = (E'_e, \mathbf{p}')$, with

$$E'_e = \sqrt{m_e^2 c^4 + |\mathbf{p}'|^2 c^2}, \qquad \mathbf{p}' = |\mathbf{p}'|(\cos\psi, \sin\psi, 0), \qquad (16.26)$$

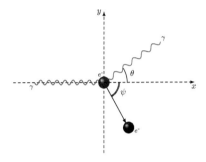

Fig. 16.1 The geometry of the Compton scattering described in the text.

where we denoted by ψ the angle at which the electron recoils with respect to the direction of the incoming photon, see again Fig. 16.1. From energy conservation we have $E_e'^2 = (E_e + E_\gamma - E'_\gamma)^2$, which gives

$$|\mathbf{p}'|^2 c^2 = \hbar^2(\omega - \omega')^2 + 2m_e c^2 \hbar(\omega - \omega'), \qquad (16.27)$$

while momentum conservation along the x and y axes gives, respectively

$$\hbar\omega = \hbar\omega' \cos\theta + c|\mathbf{p}'| \cos\psi, \qquad (16.28)$$
$$0 = \hbar\omega' \sin\theta + c|\mathbf{p}'| \sin\psi, \qquad (16.29)$$

We therefore have three equations for three variables ω', $|\mathbf{p}'|$, and ψ, for a given scattering angle θ. Solving the equations, we get

$$\omega - \omega' = \frac{\hbar}{m_e c^2}\omega\omega'(1 - \cos\theta). \qquad (16.30)$$

This shows that $\omega' \leq \omega$ for all scattering angles, with the equality satisfied only for forward scattering, $\theta = 0$. This is due to the fact that, because of the momentum carried by the photon, the electron recoils in the scattering process (except for $\theta = 0$, where the photon is re-emitted in the same direction), and therefore acquires some kinetic energy. Therefore, part of the energy of the initial photon is transferred to the electron, and the final photon has a smaller energy. Equation (16.30) can be written in a slightly more elegant form in terms of the wavelengths $\lambda = 2\pi c/\omega$ and $\lambda' = 2\pi c/\omega'$, as

$$\lambda' - \lambda = \frac{2\pi\hbar}{m_e c}(1 - \cos\theta). \qquad (16.31)$$

This shift in the wavelength is known as the *Compton effect*, and the length-scale

$$r_C = \frac{\hbar}{m_e c} \tag{16.32}$$

is called the *Compton radius* of the electron. We see that the effect becomes important when the reduced wavelength $\lambdabar = \lambda/(2\pi)$ of the incoming radiation is of order of r_C. Numerically, $r_C \simeq 4 \times 10^{-11}$ cm, and the corresponding wavelengths are in the domain of X rays.

Therefore, the classical Thomson scattering formula is valid for frequencies small (i.e., wavelengths large) compared to those of X rays. When ω becomes of order c/r_C, i.e., when the energy $\hbar\omega$ of the incoming photon becomes of order of the rest energy of the electron $m_e c^2$, the classical formula is no longer valid. Historically, the Compton experiment of scattering of X rays on electrons, that was first carried out between 1919 and 1922, was crucial to show that, at the quantum level, light can behave as particles, leading to the concept of particle-wave duality.

16.3   Scattering on a bound electron

We now consider the scattering of an electromagnetic wave on an electron bound in an atom or in a molecule. A complete description requires quantum mechanics, taking into account the discrete energy level of the bound electron. Furthermore, just as for the scattering on a free electron, if the frequency of the electromagnetic wave is in the X-ray regime or larger, we should also take into account the quantum nature of the electromagnetic field, which eventually requires the use of quantum field theory. Here we discuss the purely classical computation, in which both the electron and the electromagnetic field are treated classically.

Within such a classical approach, a simple description can be obtained by modeling the bound electron as a damped harmonic oscillator with natural frequency ω_0 and damping constant γ_0, forced by the external field due to a monochromatic electromagnetic wave with frequency ω, similarly to what we already did in Section 14.3 when we developed the Drude–Lorentz model for the dielectric constant. As in the case of the free electron, we assume that the amplitude $E = |\mathbf{E}|$ of the electric field satisfies eq. (16.14), so that the motion of the electron under the action of the electromagnetic wave remains non-relativistic. We can then write a Newtonian equation of motion for the electron, using the Lorentz force, and we can neglect the effect of the magnetic field. We have also seen that, for electric fields that satisfy eq. (16.14), the amplitude of the oscillations is such that $kx_0 \ll 1$ (for a free electron, and therefore even more so for a bound electron). We can then approximate $e^{i\mathbf{k}\cdot\mathbf{x}} \simeq 1$ in the expression (9.49) of the electric field of the wave, which is then given simply by eq. (16.16). The equation of motion of the electron, with position $\mathbf{x}_0(t)$, in the presence of the external field $\mathbf{E}(t)$ of the electromagnetic wave, is therefore the one that was already written in eq. (14.33), with $\mathbf{E}(t)$ given by eq. (16.16),

$$\ddot{\mathbf{x}}_0 + \gamma_0 \dot{\mathbf{x}}_0 + \omega_0^2 \mathbf{x}_0 = -\frac{e}{m_e} \operatorname{Re} \left[\hat{\mathbf{e}}(\mathbf{k}) \, E_\mathbf{k} \, e^{-i\omega t} \right] . \qquad (16.33)$$

We take the propagation direction of the electromagnetic wave along the x axis, $\hat{\mathbf{k}} = \hat{\mathbf{x}}$, and we consider a linearly polarized wave with $\hat{\mathbf{e}}(\mathbf{k}) = \hat{\mathbf{z}}$ and $E_\mathbf{k} = E_0$ real. Then, eq. (16.33) becomes

$$\ddot{z}_0(t) + \gamma_0 \dot{z}_0(t) + \omega_0^2 z_0(t) = -\frac{e E_0}{m_e} \operatorname{Re} \left[e^{-i\omega t} \right] . \qquad (16.34)$$

Searching the solution in the form

$$z(t) = \operatorname{Re} \left[\tilde{z}(\omega) e^{-i\omega t} \right] , \qquad (16.35)$$

we get

$$\tilde{z}(\omega) = \frac{e E_0}{m_e} \frac{1}{\omega^2 + i\omega\gamma_0 - \omega_0^2} , \qquad (16.36)$$

so,

$$z(t) = \frac{e E_0}{m_e} \operatorname{Re} \left[\frac{e^{-i\omega t}}{\omega^2 + i\omega\gamma_0 - \omega_0^2} \right] . \qquad (16.37)$$

The z component of the dipole moment is given by $d_z(t) = -e z(t)$, and then

$$
\begin{aligned}
\ddot{d}_z(t) &= \frac{e^2 E_0 \omega^2}{m_e} \operatorname{Re} \left[\frac{e^{-i\omega t}}{\omega^2 + i\omega\gamma_0 - \omega_0^2} \right] \\
&= \left(\frac{e^2 E_0 \omega^2}{m_e} \right) \frac{(\omega^2 - \omega_0^2)\cos\omega t - \omega\gamma_0 \sin\omega t}{(\omega^2 - \omega_0^2)^2 + \omega^2 \gamma_0^2} . \qquad (16.38)
\end{aligned}
$$

Note that the dipole oscillates at the frequency ω of the incoming wave so, again, classically the scattered wave has the same frequency as the incoming wave. Using the Larmor formula (10.144), the radiated power is obtained from $[\ddot{d}_z(t)]^2$. Rather than the power radiated instantaneously, it is convenient to consider the power averaged over one period of oscillation, which better corresponds to what can be actually measured. When we average over one period $T = 2\pi/\omega$ we get

$$
\begin{aligned}
\langle \cos^2(\omega t) \rangle &= \frac{1}{T} \int_0^T dt \, \cos^2(\omega t) \\
&= \frac{1}{2\pi} \int_0^{2\pi} d\alpha \, \cos^2 \alpha \\
&= \frac{1}{2} . \qquad (16.39)
\end{aligned}
$$

Similarly $\langle \sin^2(\omega t) \rangle = 1/2$, while $\langle \sin(\omega t) \cos(\omega t) \rangle = 0$. Therefore,

$$\langle [\ddot{d}_z(t)]^2 \rangle = \frac{1}{2} \left(\frac{e^2 E_0}{m_e} \right)^2 \frac{\omega^4}{(\omega^2 - \omega_0^2)^2 + \omega^2 \gamma_0^2} , \qquad (16.40)$$

and the average radiated power, per unit solid angle, is

$$\frac{dP(\theta)}{d\Omega} = \frac{1}{4\pi\epsilon_0} \frac{1}{8\pi c^3} \left(\frac{e^2 E_0}{m_e} \right)^2 \frac{\omega^4}{(\omega^2 - \omega_0^2)^2 + \omega^2 \gamma_0^2} \sin^2\theta , \qquad (16.41)$$

where θ is the polar angle measured from the z axis. Since we have averaged the power over one cycle, we also average the incident flux over one cycle, and we extend the definition (16.5) to $\sigma = \langle P \rangle / \langle I \rangle$. Equation (9.59) gives

$$\langle I(t) \rangle = \frac{1}{2} \epsilon_0 c E_0^2 . \tag{16.42}$$

The differential scattering cross-section is therefore

$$\frac{d\sigma_{\text{scatt}}(\omega; \theta)}{d\Omega} = r_0^2 \frac{\omega^4}{(\omega^2 - \omega_0^2)^2 + \omega^2 \gamma_0^2} \sin^2 \theta , \tag{16.43}$$

where we have used the definition (16.21) of the classical electron radius, and we wrote explicitly as an argument the frequency of the incident wave. If we set $\omega_0 = \gamma_0 = 0$, we recover the result for a free electron, eq. (16.20), as it should be. Integrating over the angles, we find the total scattering cross-section

$$\sigma_{\text{scatt}}(\omega) = \frac{8\pi}{3} r_0^2 \frac{\omega^4}{(\omega^2 - \omega_0^2)^2 + \omega^2 \gamma_0^2} . \tag{16.44}$$

A plot of the cross-section (16.44) is shown in Fig. 16.2, for $\gamma_0 / \omega_0 = 0.4$.

Several aspects of this result are noteworthy. At low frequencies, $\omega \ll \omega_0$, we have

$$\sigma_{\text{scatt}}(\omega) \simeq \frac{8\pi}{3} r_0^2 \left(\frac{\omega}{\omega_0} \right)^4 , \qquad (\omega \ll \omega_0) . \tag{16.45}$$

At low frequencies, the cross-section is therefore proportional to ω^4. In contrast, when the frequency ω of the electromagnetic matches the natural frequency ω_0 of the oscillator, we are in the resonance condition and

$$\sigma_{\text{scatt}}(\omega = \omega_0) = \frac{8\pi}{3} r_0^2 \left(\frac{\omega_0}{\gamma_0} \right)^2 . \tag{16.46}$$

For $\gamma_0 \ll \omega_0$, the cross-section is therefore greatly enhanced with respect to the Thompson cross-section $\sigma_T = (8\pi/3) r_0^2$. Close to the resonance we can approximate

$$
\begin{aligned}
\sigma_{\text{scatt}}(\omega) &= \frac{8\pi}{3} r_0^2 \frac{\omega^4}{(\omega - \omega_0)^2 (\omega + \omega_0)^2 + \omega^2 \gamma_0^2} \\
&\simeq \frac{8\pi}{3} r_0^2 \frac{\omega_0^4}{(\omega - \omega_0)^2 (2\omega_0)^2 + \omega_0^2 \gamma_0^2} \\
&= \frac{2\pi}{3} r_0^2 \frac{\omega_0^2}{(\omega - \omega_0)^2 + (\gamma_0/2)^2} .
\end{aligned}
\tag{16.47}
$$

From this expression we see that the resonance condition is maintained in a narrow range of frequencies $\Delta \omega \sim \gamma_0 / 2$ centered around ω_0. The peak at $\omega = \omega_0$ is an example or *resonant scattering*, or of a *resonance*.

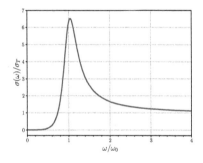

Fig. 16.2 The cross-section $\sigma(\omega)$, normalized to the Thompson cross-section σ_T, as a function of ω/ω_0, for $\gamma_0/\omega_0 = 0.4$.

Finally, at large frequencies, i.e., when $\omega \gg \omega_0$ and $\omega \gg \gamma_0$, the cross-section goes to the Thompson cross-section $\sigma_T = 8\pi r_0^2/3$, corresponding to the limit of scattering on a free particle.

As we have discussed in the free particle case, at sufficiently high frequencies, when $\hbar\omega$ becomes of order $m_e c^2$, the classical computation is no longer valid. A full relativistic quantum computation shows that, beyond that value, $\sigma(\omega)$ decreases as $\log(\omega)/\omega$.

A beautiful consequence of this computation is that it allows us to obtain at least a first understanding (due to Lord Rayleigh, 1871) of why the sky is blue, and why the Sun at sunset looks redder than at noon. Light in the visible spectrum has a frequency that is not high enough to excite electronic transitions in the molecules, so we are in the regime $\omega \ll \omega_0$, where ω_0 is the natural resonance frequency of a molecule. At the same time, ω is much larger than the typical vibrational frequencies of the molecules, which means that the vibrational modes of the nuclei in the molecules cannot follow the fast oscillations of the electromagnetic field, so these modes are not excited, and can be neglected. Another important aspect is that air is sufficiently diluted, with respect to the wavelength of visible light, that the molecules scatter light independently of each other and the computation performed previously, where we considered the scattering from a single charge, applies. The opposite limit, in which a whole ensemble of charges is set into oscillation coherently by the incoming electromagnetic wave, would have required a different computation. So, for scattering of visible light by the atmosphere we are in the situation where eq. (16.45) holds. Compare for instance the scattering cross-section for light at $\lambda = 450$ nm, in the blue region of the visible spectrum, to that of light at $\lambda = 650$ nm (red). We have

$$\left(\frac{\omega_{\text{blue}}}{\omega_{\text{red}}}\right)^4 = \left(\frac{\lambda_{\text{red}}}{\lambda_{\text{blue}}}\right)^4 \simeq 4.35 \,, \tag{16.48}$$

so the cross-section for scattering of blue light is about four times larger than for red light. When we look at the sky, in a direction different from that of the Sun, our eyes receive light that was emitted by the Sun and then scattered toward us by some molecules in the atmosphere. Since blue light is preferentially scattered toward us, we see a blue color. Conversely, at sunset, light must travel through a longer path in the atmosphere, compared to the path when the Sun is at the Zenith. Since, traveling through the atmosphere, blue photons are preferentially removed, we see a redder Sun. Sunset seen from astronauts orbiting the Earth is even redder because the path length through the atmosphere is doubled.[2]

[2]See e.g., https://www.esa.int/ESA_Multimedia/Images/2010/02/Scattered_sunlight.

Electrodynamics in Gaussian units

A

In this appendix we translate the main results and equations of the text into Gaussian units. The conceptual relation between SI and Gaussian units has been discussed in detail in Section 2.2, where we have seen that the crucial difference is that, in the SI system, the unit of current (or the unit of charge) is a fourth independent base unit, with respect to the units of length, time, and mass while, in the Gaussian system, it is a derived unit. As we discussed in Section 2.2, there are actually two variants of the Gaussian system, unrationalized Gaussian units (often referred simply as Gaussian units) and "rationalized Gaussian" (or Heaviside–Lorentz) units. Rationalized Gaussian units are the most common (if not universal) choice in quantum field theory. In this context, one normally sets also $c = 1$ (and even $\hbar = 1$, but $\hbar$ does not appear in our classical equations). As we saw in Section 2.2, the conversion from SI to rationalized Gaussian units is performed with the formal replacements given in eq. (2.46). If we furthermore set $c = 1$, the conversion from SI units to rationalized Gaussian units is therefore trivially performed, simply by making in the SI equations the formal replacements

$$\epsilon_0 \to 1\,, \quad \mu_0 \to 1\,, \quad c \to 1\,. \tag{A.1}$$

We will therefore not consider rationalized Gaussian units further; unrationalized Gaussian units, in contrast, are often used as an alternative to the SI in a classical context, therefore keeping c explicit. Furthermore, there are extra 4π factors, so the conversion is less trivial, and in this appendix we write explicitly most of the main formulas of the main text in (unrationalized) Gaussian units.[1]

As we have seen, the actual relation between charges and fields in Gaussian and SI system is given by eqs. (2.25), (2.32), and (2.33), that we recall here

$$q_{\text{gau}} = \frac{q_{\text{SI}}}{\sqrt{4\pi\epsilon_0}}\,. \tag{A.2}$$

$$\mathbf{E}_{\text{gau}} = \sqrt{4\pi\epsilon_0}\,\mathbf{E}_{\text{SI}}\,, \qquad \mathbf{B}_{\text{gau}} = \sqrt{\frac{4\pi}{\mu_0}}\,\mathbf{B}_{\text{SI}}\,. \tag{A.3}$$

In the Gaussian system, the gauge potentials are introduced from

$$\mathbf{B}_{\text{gau}} = \boldsymbol{\nabla} \times \mathbf{A}_{\text{gau}}\,, \tag{A.4}$$

$$\mathbf{E}_{\text{gau}} = -\boldsymbol{\nabla}\phi_{\text{gau}} - \frac{1}{c}\frac{\partial \mathbf{A}_{\text{gau}}}{\partial t}\,. \tag{A.5}$$

Comparing with the definition of the gauge potentials in the SI system, given in eqs. (3.80) and (3.83), and using eq. (A.3), as well as $\epsilon_0\mu_0 = 1/c^2$, we get the following relation for the gauge potentials,

$$\phi_{\text{gau}} = \sqrt{4\pi\epsilon_0}\,\phi_{\text{SI}}\,, \qquad \mathbf{A}_{\text{gau}} = \sqrt{\frac{4\pi}{\mu_0}}\,\mathbf{A}_{\text{SI}}\,. \tag{A.6}$$

[1] This appendix can also be used as a compact summary of the main equations in the book; they are written here in Gaussian units, but always with a reference to the equation number of the corresponding SI equation in the main text. This allows the reader to quickly find also the desired SI equation.

We have also seen in eq. (2.38) that, independently of the actual relation between quantities in these systems, expressed by the equations given above, a quicker way of passing from the equations in the SI system to those in Gaussian units is to perform the replacements

$$\epsilon_0 \to \frac{1}{4\pi}, \qquad \mu_0 \to \frac{4\pi}{c^2}, \qquad \mathbf{E} \to \mathbf{E} \qquad \mathbf{B} \to \frac{\mathbf{B}}{c}, \tag{A.7}$$

(without changing ρ and $\mathbf{j}$), since this formally transforms Maxwell's equations in the SI system into Maxwell's equations in the Gaussian system.[2] For the gauge potentials, the corresponding formal replacement are[3]

$$\phi \to \phi, \qquad \mathbf{A} \to \frac{\mathbf{A}}{c}. \tag{A.8}$$

In this way, all the equations that we have written in the book can be immediately translated to Gaussian units. In any case, we find it useful to collect together some of the most important results, as a quick reference. We organize this collection of results according to the chapters of the main text where they have been first given in SI units. We henceforth drop the subscript "gau" from all quantities. It is understood that all the formulas in the rest of this Appendix are written in (unrationalized) Gaussian units.

[2]The inverse path, from Gaussian to SI equations, is less straightforward. Once we set ϵ_0 to a dimensionless value, to restore it back requires an input from dimensional analysis. To go from Gaussian to SI it is therefore better to use the actual relations (A.2, A.3).

[3]See, however, eq. (A.99) for the definition of A^μ in terms of ϕ and $\mathbf{A}$ in Gaussian units, and eq. (A.86) for the definition of the magnetic dipole in Gaussian units.

Chapter 3

In Gaussian units, Maxwell's equations (3.1)–(3.4) read

$$\boldsymbol{\nabla}\cdot\mathbf{E} = 4\pi\rho, \tag{A.9}$$

$$\boldsymbol{\nabla}\times\mathbf{B} - \frac{1}{c}\frac{\partial\mathbf{E}}{\partial t} = \frac{4\pi}{c}\mathbf{j}, \tag{A.10}$$

$$\boldsymbol{\nabla}\cdot\mathbf{B} = 0, \tag{A.11}$$

$$\boldsymbol{\nabla}\times\mathbf{E} + \frac{1}{c}\frac{\partial\mathbf{B}}{\partial t} = 0. \tag{A.12}$$

In integrated form, they become

$$\Phi_E(t) = 4\pi Q(t), \tag{A.13}$$

$$\oint_{\partial S} d\boldsymbol{\ell}\cdot\mathbf{B}(t,\mathbf{x}) = \frac{4\pi}{c}I(t) + \frac{1}{c}\frac{d\Phi_E(t)}{dt} \tag{A.14}$$

$$\Phi_B(t) = 0, \tag{A.15}$$

$$\oint_{\partial S} d\boldsymbol{\ell}\cdot\mathbf{E}(t,\mathbf{x}) = -\frac{1}{c}\frac{d\Phi_B(t)}{dt}, \tag{A.16}$$

compare with eqs. (3.12), (3.19), (3.16), and (3.21), where we still have

$$\Phi_E(t) \equiv \int_{\partial V} d\mathbf{s}\cdot\mathbf{E}(t,\mathbf{x}), \tag{A.17}$$

$$\Phi_B(t) \equiv \int_{\partial V} d\mathbf{s}\cdot\mathbf{B}(t,\mathbf{x}) \tag{A.18}$$

and

$$Q_V(t) = \int_V d^3x\,\rho(t,\mathbf{x}), \tag{A.19}$$

$$I(t) = \int_S d\mathbf{s}\cdot\mathbf{j}(t,\mathbf{x}). \tag{A.20}$$

In a non-relativistic setting, where the notion of force makes sense, the Lorentz force equation (3.5) becomes

$$\mathbf{F} = q\left(\mathbf{E} + \frac{\mathbf{v}}{c} \times \mathbf{B}\right),\tag{A.21}$$

or, in a fully relativistic setting,

$$\frac{d\mathbf{p}}{dt} = q\left(\mathbf{E} + \frac{\mathbf{v}}{c} \times \mathbf{B}\right).\tag{A.22}$$

compare with eq. (3.6). The continuity equation for the electric charge still keeps the form (3.22),

$$\frac{\partial \rho}{\partial t} + \mathbf{\nabla}\cdot\mathbf{j} = 0,\tag{A.23}$$

while the Poynting vector and energy conservation, that in SI units are given by eqs. (3.34) and (3.35), become, respectively,

$$\mathbf{S} = \frac{c}{4\pi}\left(\mathbf{E} \times \mathbf{B}\right),\tag{A.24}$$

and

$$\frac{d}{dt}\int_V d^3x\, \frac{\mathbf{E}^2 + \mathbf{B}^2}{8\pi} + \int_V d^3x\, \mathbf{E}\cdot\mathbf{j} = -\int_{\partial V} d\mathbf{s}\cdot\mathbf{S}.\tag{A.25}$$

Therefore, the energy density of the electromagnetic field, eq. (3.43), becomes

$$u(t, \mathbf{x}) = \frac{1}{8\pi}(\mathbf{E}^2 + \mathbf{B}^2).\tag{A.26}$$

The momentum density is

$$\begin{aligned}\mathbf{g} &= \frac{1}{c^2}\mathbf{S}\\ &= \frac{1}{4\pi c}\left(\mathbf{E} \times \mathbf{B}\right),\end{aligned}\tag{A.27}$$

compare with eqs. (3.56) and (3.57), so the electromagnetic field enclosed in a volume V carries a momentum

$$\begin{aligned}\mathbf{P}_{\rm em}(t) &= \int_V d^3x\, \mathbf{g}(t, \mathbf{x})\\ &= \frac{1}{4\pi c}\int_V d^3x\, (\mathbf{E} \times \mathbf{B})(t, \mathbf{x}),\end{aligned}\tag{A.28}$$

compare with eq. (3.67). The Maxwell stress tensor (3.64) becomes

$$T_{ij} = \frac{1}{4\pi}\left[\frac{1}{2}(E^2 + B^2)\delta_{ij} - E_i E_j - B_i B_j\right],\tag{A.29}$$

while the time derivative of the mechanical momentum associated with a charge and current density in an electromagnetic field, eq. (3.68), becomes

$$\frac{d\mathbf{P}_{\rm mech}}{dt} = \int d^3x\left(\rho\mathbf{E} + \frac{1}{c}\mathbf{j}\times\mathbf{B}\right).\tag{A.30}$$

The angular momentum of the electromagnetic field is

$$\mathbf{J}_{\rm em} = \frac{1}{4\pi c}\int d^3x\, \mathbf{x}\times(\mathbf{E} \times \mathbf{B}),\tag{A.31}$$

to be compared with eq. (3.76), while eq. (3.79) becomes

$$\frac{d\mathbf{J}_{\rm mech}}{dt} = \int_V d^3x\, \mathbf{x}\times\left(\rho\mathbf{E} + \frac{1}{c}\mathbf{j}\times\mathbf{B}\right).\tag{A.32}$$

The gauge potentials are introduced from

$$\mathbf{B} = \boldsymbol{\nabla} \times \mathbf{A}, \tag{A.33}$$

$$\mathbf{E} = -\boldsymbol{\nabla}\phi - \frac{1}{c}\frac{\partial \mathbf{A}}{\partial t}, \tag{A.34}$$

compare with eqs. (3.80) and (3.83). In terms of them, the two Maxwell's equations that do not depend on the sources are automatically satisfied, while those that depend on the source, that in the SI system are given by eqs. (3.84) and (3.85), become

$$\nabla^2\phi + \frac{1}{c}\frac{\partial}{\partial t}(\boldsymbol{\nabla}\cdot\mathbf{A}) = -4\pi\rho, \tag{A.35}$$

$$\nabla^2\mathbf{A} - \frac{1}{c^2}\frac{\partial^2\mathbf{A}}{\partial t^2} - \boldsymbol{\nabla}\left(\boldsymbol{\nabla}\cdot\mathbf{A} + \frac{1}{c}\frac{\partial\phi}{\partial t}\right) = -\frac{4\pi}{c}\mathbf{j}. \tag{A.36}$$

The gauge transformation (3.86) reads

$$\mathbf{A} \to \mathbf{A}' = \mathbf{A} - \boldsymbol{\nabla}\theta, \tag{A.37}$$

$$\phi \to \phi' = \phi + \frac{1}{c}\frac{\partial\theta}{\partial t}. \tag{A.38}$$

The Lorenz gauge (3.89) is now defined from

$$\boldsymbol{\nabla}\cdot\mathbf{A} + \frac{1}{c}\frac{\partial\phi}{\partial t} = 0, \tag{A.39}$$

and, in this gauge, eqs. (A.35) and (A.36) become

$$\Box\phi = -4\pi\rho, \tag{A.40}$$

$$\Box\mathbf{A} = -\frac{4\pi}{c}\mathbf{j}, \tag{A.41}$$

compare with eqs. (3.90) and (3.91). The Coulomb gauge still reads $\boldsymbol{\nabla}\cdot\mathbf{A} = 0$ and, in this gauge, eqs. (A.35) and (A.36) become

$$\nabla^2\phi = -4\pi\rho, \tag{A.42}$$

$$\Box\mathbf{A} = -\frac{4\pi}{c}\mathbf{j} + \frac{1}{c}\boldsymbol{\nabla}\frac{\partial\phi}{\partial t}, \tag{A.43}$$

compare with eqs. (3.93) and (3.94).

Chapter 4

Poisson's equation (4.3) becomes

$$\nabla^2\phi = -4\pi\rho. \tag{A.44}$$

The solution can be written in terms of the Green's functions of the Laplacian as

$$\phi(\mathbf{x}) = -4\pi \int d^3x' \, G(\mathbf{x} - \mathbf{x}')\rho(\mathbf{x}'), \tag{A.45}$$

and, given that the Green's function of the Laplacian is still given by eq. (4.15), we get

$$\phi(\mathbf{x}) = \int d^3x' \, \frac{\rho(\mathbf{x}')}{|\mathbf{x} - \mathbf{x}'|}, \tag{A.46}$$

to be compared with eq. (4.16). The corresponding electric field is

$$\mathbf{E}(\mathbf{x}) = \int d^3x' \, \rho(\mathbf{x}')\frac{\mathbf{x} - \mathbf{x}'}{|\mathbf{x} - \mathbf{x}'|^3}, \tag{A.47}$$

to be compared with eq. (4.20) and, for a point charge, the Coulomb force reads

$$\mathbf{F} = \frac{q_1 q_2}{r^2}\hat{\mathbf{r}}\,, \tag{A.48}$$

[as we already wrote in eq. (2.23)], to be compared with eq. (2.6). For the electric field on the surface of a conductor, eq. (4.54) becomes

$$\hat{\mathbf{n}}\cdot\mathbf{E} = 4\pi\sigma\,, \tag{A.49}$$

while eq. (4.60) becomes

$$\frac{d\mathbf{p}_a}{dt} = \frac{1}{4\pi}\int_{\partial V_a} d^2s\left[(\mathbf{E}\cdot\hat{\mathbf{n}})\mathbf{E} - \frac{1}{2}E^2\hat{\mathbf{n}}\right]\,. \tag{A.50}$$

The basic equations of magnetostatics are

$$\boldsymbol{\nabla}\times\mathbf{B} = \frac{4\pi}{c}\mathbf{j}\,, \tag{A.51}$$
$$\boldsymbol{\nabla}\cdot\mathbf{B} = 0\,, \tag{A.52}$$

compare with eqs. (4.67) and (4.68), and the integrated forms are

$$\oint_C d\boldsymbol{\ell}\cdot\mathbf{B}(\mathbf{x}) = \frac{4\pi}{c}I\,, \tag{A.53}$$
$$\oint_S d\mathbf{s}\cdot\mathbf{B}(t,\mathbf{x}) = 0\,, \tag{A.54}$$

compare with eqs. (4.70) and (4.71). The magnetic field generated by an infinite straight wire, eq. (4.79), now reads

$$\mathbf{B}(\rho,\varphi,z) = \frac{2I}{c\rho}\hat{\boldsymbol{\varphi}}\,. \tag{A.55}$$

Equation (4.91) becomes

$$\boldsymbol{\nabla}^2\mathbf{A} = -\frac{4\pi}{c}\mathbf{j}\,, \tag{A.56}$$

whose solution is

$$\mathbf{A}(\mathbf{x}) = \frac{1}{c}\int d^3x'\,\frac{\mathbf{j}(\mathbf{x}')}{|\mathbf{x}-\mathbf{x}'|}\,, \tag{A.57}$$

compare with eq. (4.92). The corresponding solution for $\mathbf{B}$ is

$$\mathbf{B}(\mathbf{x}) = \frac{1}{c}\int d^3x'\,\frac{\mathbf{j}(\mathbf{x}')\times(\mathbf{x}-\mathbf{x}')}{|\mathbf{x}-\mathbf{x}'|^3}\,, \tag{A.58}$$

compare with eq. (4.95). For a thin wire these become, respectively,

$$\mathbf{A}(\mathbf{x}) = \frac{I}{c}\oint_C d\boldsymbol{\ell}\,\frac{1}{|\mathbf{x}-\mathbf{x}(\ell)|}\,, \tag{A.59}$$

and

$$\mathbf{B}(\mathbf{x}) = \frac{I}{c}\oint_C d\boldsymbol{\ell}\times\frac{\mathbf{x}-\mathbf{x}(\ell)}{|\mathbf{x}-\mathbf{x}(\ell)|^3}\,, \tag{A.60}$$

compare with eqs. (4.104) and (4.105). Equations (4.109) and (4.110) become, respectively,

$$\mathbf{F} = \frac{I}{c}\oint_C d\boldsymbol{\ell}\times\mathbf{B}\,, \tag{A.61}$$

and

$$\mathbf{F} = \frac{1}{c}\int d^3x\,\mathbf{j}(\mathbf{x})\times\mathbf{B}(\mathbf{x})\,. \tag{A.62}$$

The force between two parallel wires, eq. (4.112), becomes

$$\frac{d\mathbf{F}_2}{d\ell} = -\frac{2I_1 I_2}{c^2 d}\,\hat{\boldsymbol{\rho}}\,, \tag{A.63}$$

while eqs. (4.118) and (4.119) become, respectively,

$$\mathbf{F}_1 = -\frac{1}{c^2}\int d^3x\, d^3x'\, \mathbf{j}_1(\mathbf{x})\cdot\mathbf{j}_2(\mathbf{x}')\,\frac{\mathbf{x}-\mathbf{x}'}{|\mathbf{x}-\mathbf{x}'|^3}\,, \tag{A.64}$$

and

$$\mathbf{F}_1 = -\frac{I_1 I_2}{c^2}\oint_{C_1}\oint_{C_2} d\boldsymbol{\ell}_1\cdot d\boldsymbol{\ell}_2\,\frac{\mathbf{x}(\ell)-\mathbf{x}(\ell')}{|\mathbf{x}(\ell)-\mathbf{x}(\ell')|^3}\,. \tag{A.65}$$

Equation (4.134) becomes

$$\frac{1}{c}\frac{d\Phi_B}{dt} = -\oint_{C(t)} d\boldsymbol{\ell}\cdot\left(\mathbf{E}+\frac{\mathbf{v}}{c}\times\mathbf{B}\right). \tag{A.66}$$

Chapter 5

The electrostatic energy of a static system of point charges, eq. (5.7), becomes

$$(\mathcal{E}_E)_{\mathrm{p.p.}} = \frac{1}{2}\sum_{a=1}^{N}\sum_{b\neq a}^{N}\frac{q_a q_b}{|\mathbf{x}_a-\mathbf{x}_b|}\,. \tag{A.67}$$

The electrostatic potential felt by the a-th point charge because of the interaction with all other charges is

$$\phi_a(\mathbf{x}_1,\ldots\mathbf{x}_n) = \sum_{b\neq a}^{N}\frac{q_b}{|\mathbf{x}_a-\mathbf{x}_b|}\,, \tag{A.68}$$

compare with eq. (5.8), so eq. (A.67) can also be written as

$$(\mathcal{E}_E)_{\mathrm{p.p.}} = \frac{1}{2}\sum_{a=1}^{N} q_a\phi_a\,, \tag{A.69}$$

so eq. (5.9) is unchanged. Eqs. (5.43)–(5.45), valid for a set of conductors, also stay unchanged. For a continuous charge distribution

$$\mathcal{E}_E = \frac{1}{2}\int d^3x\,\rho(\mathbf{x})\phi(\mathbf{x})\,, \tag{A.70}$$

$$= \frac{1}{2}\int d^3x\, d^3x'\,\frac{\rho(\mathbf{x})\rho(\mathbf{x}')}{|\mathbf{x}-\mathbf{x}'|}\,, \tag{A.71}$$

compare with eqs. (5.15) and (5.16). The corresponding expressions for the magnetic energy are

$$\mathcal{E}_B = \frac{1}{2c}\int d^3x\,\mathbf{j}(\mathbf{x})\cdot\mathbf{A}(\mathbf{x}) \tag{A.72}$$

$$= \frac{1}{2c^2}\int d^3x\, d^3x'\,\frac{\mathbf{j}(\mathbf{x})\cdot\mathbf{j}(\mathbf{x}')}{|\mathbf{x}-\mathbf{x}'|}\,, \tag{A.73}$$

compare with eqs. (5.52) and (5.53). For a set of loops, eq. (5.61) becomes

$$\mathcal{E}_B = \frac{1}{2c}\sum_{a=1}^{N} I_a\Phi_{B,a}\,. \tag{A.74}$$

The mutual inductances and self-inductances are now defined from

$$\frac{1}{c}\Phi_{B,a} = \sum_{b=1}^{N} L_{ab}I_b \,, \tag{A.75}$$

to be compared with eq. (5.63), so that we still have

$$\mathcal{E}_B = \frac{1}{2}\sum_{a=1}^{N} L_{ab}I_aI_b \,, \tag{A.76}$$

as in eq. (5.67). The flux through loop a of the magnetic field $\mathbf{B}_b$ generated by loop b is still given by eq. (5.64),

$$
\begin{aligned}
(\Phi_B)_{a,b} &= \int_{S_a} d\mathbf{s}\cdot\mathbf{B}_b \\
&= \oint_{C_a} d\boldsymbol{\ell}_a\cdot\mathbf{A}_b[\mathbf{x}(\ell_a)] \,.
\end{aligned}
\tag{A.77}
$$

Inserting here the expression for $\mathbf{A}_b$ obtained from eq. (A.59),

$$\mathbf{A}_b[\mathbf{x}(\ell_a)] = \frac{I_b}{c}\oint_{C_b} d\boldsymbol{\ell}_b \frac{1}{|\mathbf{x}(\ell_a) - \mathbf{x}(\ell_b)|} \,, \tag{A.78}$$

we get

$$(\Phi_B)_{a,b} = \frac{I_b}{c}\oint_{C_a}\oint_{C_b} \frac{d\boldsymbol{\ell}_a\cdot d\boldsymbol{\ell}_b}{|\mathbf{x}(\ell_a) - \mathbf{x}(\ell_b)|} \,, \tag{A.79}$$

and therefore eq. (A.75) gives

$$L_{ab} = \frac{1}{c^2}\oint_{C_a}\oint_{C_b} \frac{d\boldsymbol{\ell}_a\cdot d\boldsymbol{\ell}_b}{|\mathbf{x}(\ell_a) - \mathbf{x}(\ell_b)|} \,, \tag{A.80}$$

to be compared with eq. (5.66).

Chapter 6

The expansion in electric multipoles is obtained writing

$$\phi(\mathbf{x}) = \frac{1}{r}\int d^3x'\, \rho(\mathbf{x}')\left[1 + \frac{\hat{\mathbf{n}}\cdot\mathbf{x}'}{r} + \mathcal{O}\left(\frac{d^2}{r^2}\right)\right] \,, \tag{A.81}$$

compare with eq. (6.5). The electric dipole moment is still defined as in eq. (6.8). Then, the electric dipole term of the potential is

$$\phi_{\text{dipole}}(\mathbf{x}) = \frac{\mathbf{d}\cdot\hat{\mathbf{n}}}{r^2} \,, \tag{A.82}$$

and the corresponding electric field is [compare with eq. (6.13)]

$$\mathbf{E}_{\text{dipole}} = \frac{3(\mathbf{d}\cdot\hat{\mathbf{n}})\hat{\mathbf{n}} - \mathbf{d}}{r^3} \,. \tag{A.83}$$

The expansion of the scalar potential up to the electric quadruple term is

$$\phi(\mathbf{x}) = \frac{q}{r} + \frac{n_id_i}{r^2} + \frac{n_in_jQ_{ij}}{2r^3} + \dots \,, \tag{A.84}$$

where the quadrupole Q_{ij} is still defined by eq. (6.18). The expansion in magnetic multipoles is obtained expanding eq. (A.57) as

$$\mathbf{A}(\mathbf{x}) = \frac{1}{cr}\int d^3x'\, \mathbf{j}(\mathbf{x}')\left[1 + \frac{\hat{\mathbf{n}}\cdot\mathbf{x}'}{r} + \mathcal{O}\left(\frac{d^2}{r^2}\right)\right] \,. \tag{A.85}$$

The magnetic dipole moment is now defined as

$$\mathbf{m} = \frac{1}{2c} \int d^3x \; \mathbf{x} \times \mathbf{j}(\mathbf{x}) \,, \tag{A.86}$$

with an extra factor $1/c$ compared to eq. (6.36). Then, for a closed loop of current,

$$\mathbf{m} = \frac{I}{2c} \oint_{\mathcal{C}} \mathbf{x} \times d\boldsymbol{\ell} \,, \tag{A.87}$$

and for a closed planar loop,

$$\mathbf{m} = \frac{IA}{c} \, \hat{\mathbf{n}} \,, \tag{A.88}$$

while, for a charged particle,

$$\mathbf{m} = \frac{q_a}{2m_a c} \, \mathbf{L}_a \,, \tag{A.89}$$

compare with eqs. (6.41)–(6.43). With this definition of magnetic moment, the vector potential and magnetic field generated by a magnetic dipole are, respectively,

$$\mathbf{A}_{\text{dipole}}(\mathbf{x}) = \frac{\mathbf{m} \times \mathbf{x}}{r^3} \,, \tag{A.90}$$

and

$$\mathbf{B}_{\text{dipole}} = \frac{3(\mathbf{m} \cdot \hat{\mathbf{n}})\hat{\mathbf{n}} - \mathbf{m}}{r^3} \,, \tag{A.91}$$

to be compared with eqs. (6.38) and (6.40). For a point-like electric dipole, eq. (6.53) becomes

$$\mathbf{E} = \frac{3(\mathbf{d} \cdot \hat{\mathbf{n}})\hat{\mathbf{n}} - \mathbf{d}}{r^3} - \frac{4\pi}{3} \, \mathbf{d} \, \delta^{(3)}(\mathbf{x}) \,, \tag{A.92}$$

while, for a point-like magnetic dipole, eq. (6.58) becomes

$$\mathbf{B} = \frac{3(\mathbf{m} \cdot \hat{\mathbf{n}})\hat{\mathbf{n}} - \mathbf{m}}{r^3} + \frac{8\pi}{3} \, \mathbf{m} \, \delta^{(3)}(\mathbf{x}) \,. \tag{A.93}$$

Observe that, thanks to the extra factor $1/c$ in eq. (A.86), there is no explicit factor of c in eq. (A.93).

The mechanical potential for the electric dipole still keeps the form (6.61),

$$(U_E)_{\text{dipole}}(\mathbf{x}) = -\mathbf{d} \cdot \mathbf{E}_{\text{ext}}(\mathbf{x}) \,, \tag{A.94}$$

while the electric dipole-dipole interaction (6.92) becomes

$$(U_E)_{\text{dipole-dipole}} = \frac{\mathbf{d}_1 \cdot \mathbf{d}_2 - 3(\mathbf{d}_1 \cdot \hat{\mathbf{r}})(\mathbf{d}_2 \cdot \hat{\mathbf{r}})}{r^3} + \frac{4\pi}{3} \, \mathbf{d}_1 \cdot \mathbf{d}_2 \delta^{(3)}(\mathbf{r}) \,. \tag{A.95}$$

Similarly, we still have[4]

$$(\hat{U}_B)_{\text{dipole}}(\mathbf{x}) = -\mathbf{m} \cdot \mathbf{B}_{\text{ext}}(\mathbf{x}) \,, \tag{A.96}$$

as in eq. (6.98), while eq. (6.105) becomes

$$(\hat{U}_B)_{\text{dipole-dipole}} = \frac{\mathbf{m}_1 \cdot \mathbf{m}_2 - 3(\mathbf{m}_1 \cdot \hat{\mathbf{n}})(\mathbf{m}_2 \cdot \hat{\mathbf{n}})}{r^3} - \frac{8\pi}{3} \, \mathbf{m}_1 \cdot \mathbf{m}_2 \, \delta^{(3)}(\mathbf{r}) \,. \tag{A.97}$$

[4]The extra factor of c in eq. (A.86) is indeed inserted so that there is no explicit factor of c in eq. (A.96).

Chapter 8

In Gaussian units, the four-current j^μ is still written

$$j^\mu = (c\rho, \mathbf{j}) \,, \tag{A.98}$$

as in eq. (8.9). Replacing formally $\mathbf{A} \to \mathbf{A}/c$, see eq. (A.8), eq. (8.12) would become $A^\mu = (\phi/c, \mathbf{A}/c)$. However, it is convenient to get rid of the overall $1/c$ factor, and define A^μ as

$$A^\mu = (\phi, \mathbf{A}) \,. \tag{A.99}$$

$F^{\mu\nu}$ is still given by $F^{\mu\nu} = \partial^\mu A^\nu - \partial^\nu A^\mu$, as in eq. (8.15), where, however, now A^μ is related to the scalar and vector potential as in eq. (A.99). Therefore, instead of eq. (8.16), we have

$$
\begin{aligned}
F^{0i} &= \partial^0 A^i - \partial^i A^0 \\
&= -\frac{1}{c}\partial_t A^i - \partial^i \phi \,,
\end{aligned} \tag{A.100}
$$

and therefore, comparing with the expression of $\mathbf{E}$ in terms of $\mathbf{A}$ and ϕ in the Gaussian system, given in eq. (A.34), we get

$$F^{0i} = E^i \,, \tag{A.101}$$

without the factor $1/c$ that appears in eq. (8.17). For the (ij) components, we still get $F_{ij} = \epsilon_{ijk}B_k$, as in eq. (8.20). Then eq. (8.21) becomes

$$
F^{\mu\nu} = \begin{pmatrix}
0 & E_1 & E_2 & E_3 \\
-E_1 & 0 & B_3 & -B_2 \\
-E_2 & -B_3 & 0 & B_1 \\
-E_3 & B_2 & -B_1 & 0
\end{pmatrix} \,. \tag{A.102}
$$

We see that the Gaussian system treats the electric and magnetic fields on the same footing as components of $F^{\mu\nu}$, without an extra factor of $1/c$ in the electric field, which is more natural from the point of view of Lorentz covariance.

In terms of this $F^{\mu\nu}$, the first pair of Maxwell's equation, that in Gaussian units have the form (A.9, A.10), becomes

$$\partial_\mu F^{\mu\nu} = -\frac{4\pi}{c} j^\nu \,. \tag{A.103}$$

Comparing with eq. (8.23) we see that, in the covariant formalism, the passage from SI to Gaussian equations can be formally performed using, together with the replacements $\epsilon_0 \to 1/(4\pi)$, $\mu_0 \to 4\pi/c^2$, already given in eq. (A.7), also the replacement

$$A^\mu \to \frac{1}{c} A^\mu \,, \tag{A.104}$$

(and therefore also $F^{\mu\nu} \to F^{\mu\nu}/c$). This is a consequence of having rescaled A^μ by an overall factor of c compared to the SI definition, as discussed above eq. (A.99). Then, eq. (8.28) becomes

$$\Box A^\nu - \partial^\nu(\partial_\mu A^\mu) = -\frac{4\pi}{c} j^\nu \,, \tag{A.105}$$

while the Lorenz gauge condition remains $\partial_\mu A^\mu = 0$, so, in the Lorenz gauge, eq. (A.105) becomes

$$\Box A^\mu = -\frac{4\pi}{c} j^\mu \,. \tag{A.106}$$

The definition (8.31) of $\tilde{F}^{\mu\nu}$, as well as the second pair of Maxwell's equation (8.33), are unchanged.

The energy-momentum tensor of the electromagnetic field, eq. (8.34), becomes

$$T^{\mu\nu} = -\frac{1}{4\pi}\left(F^{\mu\rho}F_{\rho}{}^{\nu} + \frac{1}{4}\eta^{\mu\nu}F^{\rho\sigma}F_{\rho\sigma}\right).\tag{A.107}$$

and, in terms of the electric and magnetic fields

$$T^{00} = \frac{1}{8\pi}(E^2 + B^2) = u\,,\tag{A.108}$$

$$T^{0i} = \frac{1}{4\pi}\epsilon^{ijk}E^j B^k = \frac{1}{c}S^i\,,\tag{A.109}$$

[compare with eqs. (8.36) and (8.37)], where the energy density u and the Poynting vector $\mathbf{S}$ in the Gaussian system were already given in eqs. (A.26) and (A.24). Equation (8.39) becomes

$$\partial_\nu T^{\mu\nu} = -\frac{1}{c}F^{\mu\nu}j_\nu\,.\tag{A.110}$$

Its $\mu = 0$ component still gives

$$\partial_t u + \boldsymbol{\nabla}\cdot\mathbf{S} = -\mathbf{E}\cdot\mathbf{j}\,,\tag{A.111}$$

as in eq. (8.41); however, now u, $\mathbf{S}$ and $\mathbf{E}$ are the quantities in the Gaussian system. The $\mu = i$ component, instead, becomes

$$\frac{\partial g_i}{\partial t} + \partial_j T_{ij} = -\left(\rho\mathbf{E} + \frac{1}{c}\mathbf{j}\times\mathbf{B}\right)_i\,,\tag{A.112}$$

where $g^i = T^{0i}/c$. Note the extra factor $1/c$ in front of the current in eq. (A.112), compared to eq. (8.44), completely analogous to the extra $1/c$ factor in eq. (A.21) compared to eq. (3.5).

The transformation of $\mathbf{E}$ and $\mathbf{B}$ under boosts, eqs. (8.57) and (8.58), become

$$\mathbf{E}'_{\parallel} = \mathbf{E}_{\parallel}\,,\qquad \mathbf{E}'_{\perp} = \gamma\left(\mathbf{E}_{\perp} + \frac{\mathbf{v}}{c}\times\mathbf{B}_{\perp}\right),\tag{A.113}$$

and

$$\mathbf{B}'_{\parallel} = \mathbf{B}_{\parallel}\,,\qquad \mathbf{B}'_{\perp} = \gamma\left(\mathbf{B}_{\perp} - \frac{\mathbf{v}}{c}\times\mathbf{E}_{\perp}\right).\tag{A.114}$$

The covariantization of the Lorentz force equation in Gaussian units, eq. (A.22), gives

$$\frac{dp^\mu}{d\tau} = \frac{q}{c}F^{\mu\nu}u_\nu\,,\tag{A.115}$$

to be compared with eq. (8.62). The interaction action (8.68) becomes

$$S_{\text{int}} = \frac{q}{c}\int d\tau\, u_\mu(\tau)A^\mu[x(\tau)]\,,\tag{A.116}$$

and eq. (8.71) becomes

$$L[\mathbf{x}(t), \mathbf{v}(t)] = -mc^2\sqrt{1 - \frac{v^2(t)}{c^2}} - q\phi[t, \mathbf{x}(t)] + \frac{q}{c}\mathbf{v}\cdot\mathbf{A}[t, \mathbf{x}(t)]\,.\tag{A.117}$$

Equations (8.73) and (8.74) become

$$\begin{aligned}S_{\text{int}} &= \frac{1}{c^2}\int d^4x\, j^\mu(x)A_\mu(x)\,,\tag{A.118}\\[6pt] &= \int dt d^3x\left[-\rho(t,\mathbf{x})\phi(t,\mathbf{x}) + \frac{1}{c}\mathbf{j}(t,\mathbf{x})\cdot\mathbf{A}(t,\mathbf{x})\right],\tag{A.119}\end{aligned}$$

and the conjugate momentum $\mathbf{P}$ (8.78) becomes

$$\mathbf{P} = \mathbf{p} + \frac{q}{c}\mathbf{A},\qquad\text{(A.120)}$$

while the Hamiltonian (8.87) becomes

$$H(\mathbf{P},\mathbf{x}) = c\sqrt{(\mathbf{P}-q\mathbf{A}/c)^2 + m^2c^2} + q\phi.\qquad\text{(A.121)}$$

In Gaussian units, the Lagrangian density of the free electromagnetic field, eq. (8.115) [see also eq. (8.127)], becomes

$$\begin{aligned}\mathcal{L}_0 &= -\frac{1}{16\pi c}F_{\mu\nu}F^{\mu\nu} & \text{(A.122)}\\ &= \frac{1}{8\pi c}(\mathbf{E}^2 - \mathbf{B}^2), & \text{(A.123)}\end{aligned}$$

while, as we already saw in eq. (A.118), the interaction Lagrangian density (8.116) becomes[5]

$$\mathcal{L}_{\text{int}} = \frac{1}{c^2}A_\mu j^\mu.\qquad\text{(A.126)}$$

The frequency at which a charged particle rotates in a magnetic field, eqs. (8.201) and (8.202), is given by

$$\omega = \frac{qB}{\gamma mc} = \frac{qBc}{\mathcal{E}},\qquad\text{(A.127)}$$

so the cyclotron frequency (8.203) becomes

$$\omega = \frac{qB}{mc}.\qquad\text{(A.128)}$$

Chapter 9

Equation (9.10) remains $\Box A^\mu = 0$ also in Gaussian units, and therefore the discussion of the solutions goes through without changes. Similarly, also the radiation gauge is still defined by the conditions $A^0 = 0$, $\boldsymbol{\nabla}\cdot\mathbf{A} = 0$, so all the equations of Section 9.3 are unchanged.

In terms of the electric and magnetic field, the wave solution (9.47, 9.48) becomes

$$\begin{aligned}\mathbf{E}(t,\mathbf{x}) &= \hat{\mathbf{e}}(\mathbf{k})\,E_{\mathbf{k}}\,e^{-i\omega t + i\mathbf{k}\cdot\mathbf{x}}, & \text{(A.129)}\\ \mathbf{B}(t,\mathbf{x}) &= [\hat{\mathbf{k}}\times\hat{\mathbf{e}}(\mathbf{k})]\,E_{\mathbf{k}}\,e^{-i\omega t + i\mathbf{k}\cdot\mathbf{x}}. & \text{(A.130)}\end{aligned}$$

and eq. (9.51) becomes

$$\mathbf{B}(t,\mathbf{x}) = \hat{\mathbf{k}}\times\mathbf{E}(t,\mathbf{x}),\qquad\text{(A.131)}$$

so, in particular, now $|\mathbf{B}| = |\mathbf{E}|$. Using eqs. (A.26) and (A.27) we find that the energy density of an electromagnetic wave is

$$u(t,\mathbf{x}) = \frac{|\mathbf{E}|^2}{4\pi},\qquad\text{(A.132)}$$

and the momentum density is

$$\mathbf{g} = \frac{|\mathbf{E}|^2}{4\pi c}\hat{\mathbf{k}},\qquad\text{(A.133)}$$

to be compared with eqs. (9.57) and (9.58).

[5] Observe that we have defined the Lagrangian density $\mathcal{L}$ so that it gives the action upon integration over d^4x, see eq. (8.101). If, instead, one defines it so that

$$S = \int dt d^3x\,\mathcal{L},$$

then eqs. (A.122) and (A.126) become

$$\mathcal{L}_0 = -\frac{1}{16\pi}F_{\mu\nu}F^{\mu\nu},\qquad\text{(A.124)}$$

and

$$\mathcal{L}_{\text{int}} = \frac{1}{c}A_\mu j^\mu.\qquad\text{(A.125)}$$

Also observe [as already discussed before eq. (8.117)] that, if one uses the opposite metric signature to ours, $\mathcal{L}_0$ is still given by eq. (A.124) while eq. (A.125) becomes $\mathcal{L}_{\text{int}} = -(1/c)A_\mu j^\mu$. In this way, we get the Lagrangian density given in eq. (12.85) of Jackson (1998).

Chapter 10

The Green's functions are still defined by eq. (10.1), so the solution of eq. (A.106) is

$$A^\mu(x) = -\frac{4\pi}{c} \int d^4x' \, G(x,x')j^\mu(x'), \tag{A.134}$$

which replaces eq. (10.10). The retarded and advanced Green's functions are still given by eq. (10.24). The retarded solution then reads

$$A^\mu(t,\mathbf{x}) = A^\mu_{\rm in}(t,\mathbf{x}) + \frac{1}{c} \int d^3x' \, \frac{j^\mu(t - |\mathbf{x} - \mathbf{x}'|/c, \mathbf{x}')}{|\mathbf{x} - \mathbf{x}'|}, \tag{A.135}$$

to be compared with eq. (10.34). Using eqs. (A.98) and (A.99), we see that in the static limit eq. (A.135) reduces correctly to eqs. (A.46) and (A.57). The Liénard–Wiechert potentials (10.54) and (10.55) become

$$\phi(t,\mathbf{x}) = \left(\frac{q}{R - \frac{\mathbf{v}}{c}\cdot\mathbf{R}} \right)_{\rm ret}, \tag{A.136}$$

and

$$\mathbf{A}(t,\mathbf{x}) = \left(\frac{q\mathbf{v}/c}{R - \frac{\mathbf{v}}{c}\cdot\mathbf{R}} \right)_{\rm ret}. \tag{A.137}$$

For a charge in uniform motion, eqs. (10.88) and (10.89) become

$$\mathbf{E}(t,\mathbf{x}) = q \frac{1 - v^2/c^2}{[1 - (v^2/c^2)\sin^2\theta]^{3/2}} \frac{\hat{\mathbf{R}}(t,\mathbf{x})}{R^2(t,\mathbf{x})}, \tag{A.138}$$

$$\mathbf{B}(t,\mathbf{x}) = \frac{1}{c}\mathbf{v} \times \mathbf{E}(t,\mathbf{x}). \tag{A.139}$$

The electric field of accelerated charges can be written as in eq. (10.103), where now $\mathbf{E}_v$ and $\mathbf{E}_{\rm rad}$ are given by

$$\mathbf{E}_v(t,\mathbf{x}) = \frac{q}{R_a^2} \frac{\hat{\mathbf{R}}_a - \mathbf{v}_r/c}{\gamma^2(1 - \hat{\mathbf{R}}_a\cdot\mathbf{v}_r/c)^3}, \tag{A.140}$$

$$\mathbf{E}_{\rm rad}(t,\mathbf{x}) = \frac{q}{R_a} \frac{[\dot{\mathbf{v}}_r \times (\hat{\mathbf{R}}_a - \mathbf{v}_r/c)] \times \hat{\mathbf{R}}_a}{c^2(1 - \hat{\mathbf{R}}_a\cdot\mathbf{v}_r/c)^3}, \tag{A.141}$$

instead of eqs. (10.104) and (10.105), while eq. (10.108) becomes

$$\mathbf{B}(t,\mathbf{x}) = \hat{\mathbf{R}}_a \times \mathbf{E}(t,\mathbf{x}). \tag{A.142}$$

Then, writing $\mathbf{B} = \mathbf{B}_v + \mathbf{B}_{\rm rad}$ as in eq. (10.109),

$$\mathbf{B}_v(t,\mathbf{x}) = \frac{1}{c}\frac{q}{R_a^2} \frac{\mathbf{v}_r \times \hat{\mathbf{R}}_a}{\gamma^2(1 - \hat{\mathbf{R}}_a\cdot\mathbf{v}_r/c)^3}, \tag{A.143}$$

and

$$\mathbf{B}_{\rm rad}(t,\mathbf{x}) = \frac{1}{c}\frac{q}{R_a} \frac{(\mathbf{v}_r\times\hat{\mathbf{R}}_a)(\dot{\mathbf{v}}_r\cdot\hat{\mathbf{R}}_a) + c(1 - \hat{\mathbf{R}}_a\cdot\mathbf{v}_r/c)\dot{\mathbf{v}}_r\times\hat{\mathbf{R}}_a}{c^2(1 - \hat{\mathbf{R}}_a\cdot\mathbf{v}_r/c)^3}, \tag{A.144}$$

to be compared with eqs. (10.112) and (10.113). To lowest order in v/c and in $1/r$, the electric field becomes

$$\mathbf{E}(t,r) \simeq -\frac{1}{c^2}\frac{1}{r}\ddot{\mathbf{d}}_\perp(t - r/c) \tag{A.145}$$

$$= \frac{1}{c^2}\frac{1}{r}\hat{\mathbf{n}} \times [\hat{\mathbf{n}} \times \ddot{\mathbf{d}}(t_{\rm ret})], \tag{A.146}$$

to be compared with eqs. (10.132) and (10.135), while eq. (10.139) becomes

$$\mathbf{B}(t,\mathbf{x}) \simeq -\frac{1}{c^2}\frac{1}{r}\,\hat{\mathbf{n}} \times \ddot{\mathbf{d}}(t_{\text{ret}})\,. \tag{A.147}$$

The power radiated per unit solid angle is

$$\frac{dP(t;\theta)}{d\Omega} = \frac{1}{4\pi c^3}\,|\ddot{\mathbf{d}}(t_{\text{ret}})|^2 \sin^2\theta\,, \tag{A.148}$$

and its angular integral gives

$$P(t) = \frac{2}{3c^3}\,|\ddot{\mathbf{d}}(t_{\text{ret}})|^2 \tag{A.149}$$

which is Larmor's formula in Gaussian units, compare with eqs. (10.144) and (10.148). The relativistic Larmor's formula (10.156) or (10.161) becomes

$$\frac{d\mathcal{E}}{dt_{\text{ret}}} = \frac{2q^2}{3c^3}\gamma^6\left[a^2 - \frac{|\mathbf{v}\times\mathbf{a}|^2}{c^2}\right]_{t=t_{\text{ret}}} \tag{A.150}$$

$$= \frac{2q^2}{3c^3}\gamma^4\left(a_\perp^2 + \gamma^2 a_\parallel^2\right)_{t=t_{\text{ret}}}\,, \tag{A.151}$$

and its covariantization (10.169) becomes

$$\frac{dP_{\text{em}}^\mu}{d\tau} = \frac{2q^2}{3m^3c^5}\left(\frac{dp_\nu}{d\tau}\frac{dp^\nu}{d\tau}\right)p^\mu\,. \tag{A.152}$$

The power radiated when the acceleration and the velocity are parallel becomes

$$\frac{dP_{e,\text{parallel}}(t_{\text{ret}})}{d\Omega} = \frac{q^2 a^2}{4\pi c^3}\frac{\sin^2\theta}{(1-\beta\cos\theta)^5}\,, \tag{A.153}$$

compare with eq. (10.173), while, when they are orthogonal, is

$$\frac{dP_{e,\text{circ}}(t_{\text{ret}})}{d\Omega} = \frac{q^2 a^2}{4\pi c^3}\frac{1}{(1-\beta\cos\theta)^3}\left[1 - \frac{\sin^2\theta\cos^2\phi}{\gamma^2(1-\beta\cos\theta)^2}\right]\,, \tag{A.154}$$

compare with eq. (10.191). Finally, in Gaussian units, the fine structure constant is written as

$$\alpha = \frac{e^2}{\hbar c}\,, \tag{A.155}$$

to be compared with eq. (10.226). In the rationalized Gaussian units more commonly used in quantum field theory, $\alpha = e^2/(4\pi\hbar c)$, or simply $\alpha = e^2/(4\pi)$ if one also uses units $\hbar = c = 1$.

Chapter 11

Equation (11.24), which gives the electric field at order $1/r$ in a large distance expansion, but for arbitrary velocities of the source, becomes

$$\mathbf{E}(t,\mathbf{x}) = -\frac{1}{c^2 r}\partial_t\int d^3x'\,\mathbf{j}_\perp(t - r/c + \hat{\mathbf{n}}\cdot\mathbf{x}'/c,\mathbf{x}')\,, \tag{A.156}$$

while eq. (11.27) becomes, as usual, $\mathbf{B} = \hat{\mathbf{n}} \times \mathbf{E}$. Equation (11.39) becomes

$$\Box\mathbf{A} = -\frac{4\pi}{c}\mathbf{j}_\perp(t,\mathbf{x})\,. \tag{A.157}$$

The radiated power, again exact in the velocities of the source, is

$$\frac{d\mathcal{E}}{dtd\Omega} = \frac{1}{4\pi c^3}\left|\partial_t\int d^3x'\,\mathbf{j}_\perp(t_{\text{ret}},\mathbf{x}')\right|^2\,, \tag{A.158}$$

to be compared with eq. (11.51), and the frequency spectrum, eqs. (11.60), (11.65), and (11.72), becomes

$$
\frac{d\mathcal{E}}{d\Omega d\omega} = \frac{1}{4\pi^2 c^3}\,\omega^2 |\tilde{\mathbf{j}}_\perp(\omega, \omega\hat{\mathbf{n}}/c)|^2 \tag{A.159}
$$

$$
= \frac{1}{4\pi^2 c^3}\,\omega^2 \left|\hat{\mathbf{n}}\times\tilde{\mathbf{j}}\,(\omega, \omega\hat{\mathbf{n}}/c)\right|^2 \tag{A.160}
$$

$$
= \frac{1}{4\pi^2 c^3}\,\omega^2 \left(|\tilde{\mathbf{j}}(\omega, \omega\hat{\mathbf{n}}/c)|^2 - c^2|\tilde{\rho}(\omega, \omega\hat{\mathbf{n}}/c)|^2\right). \tag{A.161}
$$

In the low-velocity limit, the multipole expansion of the gauge potentials, eqs. (11.126) and (11.127), becomes[6]

$$
\phi(t, \mathbf{x}) = \frac{1}{r}\left[q + \frac{1}{c}\hat{n}_i \dot{d}_i(t - r/c) + \frac{\hat{n}_i \hat{n}_j}{6c^2}\ddot{Q}_{ij}(t - r/c)\right], \tag{A.162}
$$

$$
A_i(t, \mathbf{x}) = \frac{1}{rc}\left[\dot{d}_i(t - r/c) + \frac{1}{6c}\dddot{Q}_{ij}(t - r/c)\hat{n}_j + \epsilon_{ijk}\dot{m}_j(t - r/c)\hat{n}_k\right].
$$

The corresponding electric field, eq. (11.130), becomes

$$
\mathbf{E}(t, \mathbf{x}) = \frac{1}{c^2 r}\left[\hat{\mathbf{n}} \times (\hat{\mathbf{n}} \times \ddot{\mathbf{d}}) + \frac{1}{6c}\hat{\mathbf{n}} \times (\hat{\mathbf{n}} \times \dddot{\mathbf{Q}}) + \hat{\mathbf{n}} \times \ddot{\mathbf{m}}\right]_{t-r/c}, \tag{A.163}
$$

and the radiated power, eq. (11.133), becomes

$$
P(t) = \left[\frac{2}{3c^3}|\dddot{\mathbf{d}}|^2 + \frac{1}{180c^5}\dddot{Q}_{ij}\dddot{Q}_{ij} + \frac{2}{3c^3}|\ddot{\mathbf{m}}|^2\right]_{t-r/c}. \tag{A.164}
$$

Note that the definition (A.86) of the magnetic dipole in the Gaussian system "hides" a factor of $1/c$ in $\mathbf{m}$, so it is less explicit, but of course still true, that the electric quadrupole and magnetic dipole contributions to the power are both proportional to $1/c^5$, compared to the leading electric dipole term, which is proportional to $1/c^3$.

Chapter 12

To 1PN order, the expression for the gauge potentials in the Coulomb gauge, given in eqs. (12.31) and (12.37), becomes

$$
\phi_{a,\text{ext}}(t, \mathbf{x}) = \sum_{b\neq a}\frac{q_b}{|\mathbf{x} - \mathbf{x}_b(t)|}, \tag{A.165}
$$

$$
\mathbf{A}_{a,\text{ext}}(t, \mathbf{x}) = \frac{1}{2c}\sum_{b\neq a} q_b \frac{\mathbf{v}_b(t)}{|\mathbf{x} - \mathbf{x}_b(t)|} \tag{A.166}
$$

$$
+ \frac{1}{2c}\sum_{b\neq a} q_b \frac{\mathbf{x} - \mathbf{x}_b(t)}{|\mathbf{x} - \mathbf{x}_b(t)|^3}\left\{[\mathbf{x} - \mathbf{x}_b(t)]\cdot\mathbf{v}_b(t)\right\},
$$

so eqs. (12.47) and (12.48) read

$$
\phi_a(\mathbf{x}_1, \ldots, \mathbf{x}_N) = \sum_{b\neq a}\frac{q_b}{r_{ab}}, \tag{A.167}
$$

$$
\mathbf{A}_a(\mathbf{x}_1, \ldots, \mathbf{x}_N; \mathbf{v}_1, \ldots \mathbf{v}_N) = \frac{1}{2c}\sum_{b\neq a}\frac{q_b}{r_{ab}}\left[\mathbf{v}_b + \hat{\mathbf{r}}_{ab}(\hat{\mathbf{r}}_{ab}\cdot\mathbf{v}_b)\right]. \tag{A.168}
$$

[6]Recall that, in the translation from SI to Gaussian units of equation involving the magnetic dipole, one must also take into account the extra factor of c between the definitions (6.36) and (A.86).

The Darwin Lagrangian (12.54, 12.55) reads $L = L_N + L_{1\text{PN}}$ where

$$L_N = \sum_{a=1}^{N} \frac{1}{2}m_a v_a^2 - \frac{1}{2}\sum_{a=1}^{N}\sum_{b\neq a}^{N}\frac{q_a q_b}{r_{ab}}\,, \tag{A.169}$$

$$L_{1\text{PN}} = \sum_{a=1}^{N}\frac{m_a v_a^4}{8c^2} + \frac{1}{4c^2}\sum_{a=1}^{N}\sum_{b\neq a}^{N}\frac{q_a q_b}{r_{ab}}\left[\mathbf{v}_a\cdot\mathbf{v}_b + (\hat{\mathbf{r}}_{ab}\cdot\mathbf{v}_a)(\hat{\mathbf{r}}_{ab}\cdot\mathbf{v}_b)\right]. \tag{A.170}$$

To 1PN order, the conjugate momentum (12.66) is given by

$$\mathbf{P}_a = \left(1 + \frac{v_a^2}{2c^2}\right)m_a\mathbf{v}_a + \frac{1}{2c^2}\sum_{b\neq a}^{N}\frac{q_a q_b}{r_{ab}}\left[\mathbf{v}_b + \hat{\mathbf{r}}_{ab}(\hat{\mathbf{r}}_{ab}\cdot\mathbf{v}_b)\right]\,, \tag{A.171}$$

which can be written as

$$\mathbf{P}_a = \left(1 + \frac{v_a^2}{2c^2}\right)m_a\mathbf{v}_a + \frac{q_a}{c}\mathbf{A}_{a,\text{ext}}[t,\mathbf{x}_a(t)]\,, \tag{A.172}$$

with $\mathbf{A}_{a,\text{ext}}[t,\mathbf{x}_a(t)]$ given by eq. (A.166). The 1PN Hamiltonian (12.71, 12.76) becomes

$$H = \sum_{a=1}^{N}\frac{P_a^2}{2m_a}\left(1 - \frac{P_a^2}{4m_a^2 c^2}\right) + \frac{1}{2}\sum_{a=1}^{N}\sum_{b\neq a}^{N}\frac{q_a q_b}{r_{ab}}$$

$$- \frac{1}{4c^2}\sum_{a=1}^{N}\sum_{b\neq a}^{N}\frac{q_a q_b}{m_a m_b r_{ab}}\left[\mathbf{P}_a\cdot\mathbf{P}_b + (\hat{\mathbf{r}}_{ab}\cdot\mathbf{P}_a)(\hat{\mathbf{r}}_{ab}\cdot\mathbf{P}_b)\right] \tag{A.173}$$

$$= \sum_{a=1}^{N}\left[\frac{P_a^2}{2m_a} - \frac{P_a^4}{8m_a^3 c^2} + \frac{1}{2}q_a\phi_{a,\text{ext}} - \frac{q_a}{2m_a c}\mathbf{P}_a\cdot\mathbf{A}_{a,\text{ext}}\right]. \tag{A.174}$$

The energy balance equation (12.140) now reads

$$\frac{d}{dt}(E_N + E_{1\text{PN}} + E_{1.5\text{PN}}) = -\frac{2}{3c^3}|\ddot{\mathbf{d}}|^2\,, \tag{A.175}$$

where

$$E_N = \sum_{a=1}^{N}\frac{1}{2}m_a v_a^2 + \frac{1}{2}\sum_{a=1}^{N}\sum_{b\neq a}^{N}\frac{q_a q_b}{r_{ab}}\,, \tag{A.176}$$

$$E_{1.5\text{PN}} = -\frac{2}{3c^3}\dot{\mathbf{d}}\cdot\ddot{\mathbf{d}}\,, \tag{A.177}$$

and $E_{1\text{PN}}$ is obtained using eq. (A.171) and then collecting the terms $1/c^2$ in eq. (A.173). The Abraham–Lorentz–Dirac equation (12.150) becomes

$$m_a\dot{u}_a^\mu = \frac{2q_a^2}{3c^3}\left(\eta^{\mu\nu} + \frac{u_a^\mu u_a^\nu}{c^2}\right)\ddot{u}_{a,\nu} + \frac{q_a}{c}F_{\text{ext}}^{\mu\nu}[x_a(\tau)]u_{a,\nu}\,, \tag{A.178}$$

while the Abraham–Lorentz equation (12.151) reads

$$m_a\frac{d\mathbf{v}_a}{dt} = \frac{2q_a^2}{3c^3}\frac{d^2\mathbf{v}_a}{dt^2} + \mathbf{F}_{\text{ext}}\,. \tag{A.179}$$

The radiation-reaction four-force is still written as $F_{\text{self}}^\mu = F_{\text{rad}}^\mu + F_{\text{Schott}}^\mu$, where now

$$F_{\text{rad}}^\mu = -\frac{2q_a^2}{3m_a^3 c^5}\left(\frac{dp_{a,\nu}}{d\tau}\frac{dp_a^\nu}{d\tau}\right)p_a^\mu\,, \tag{A.180}$$

$$F_{\text{Schott}}^\mu = \frac{2q_a^2}{3m_a c^3}\frac{d^2 p_a^\mu}{d\tau^2}\,. \tag{A.181}$$

The Landau–Lifshitz form of the ALD equation is

$$m\dot{u}_a^\mu = \frac{q_a}{c} F_{\text{ext}}^{\mu\nu} u_{a,\nu} + \tau_a \left[\frac{q_a}{c} (\partial_\rho F_{\text{ext}}^{\mu\sigma}) u_a^\rho u_{a,\sigma} \right. \tag{A.182}$$

$$\left. + \frac{q_a^2}{m_a c^2} F_{\text{ext}}^{\mu\rho} F_{\rho\sigma}^{\text{ext}} u_a^\sigma + \frac{q_a^2}{m_a c^4} (F_{\rho\sigma}^{\text{ext}} u_a^\sigma)(F_{\text{ext}}^{\rho\nu} u_{a,\nu}) u_a^\mu \right].$$

where $\tau_a = 2q_a^2/(3m_a c^3)$.

Chapter 13

In Gaussian units eqs. (13.21) and (13.22) remain unchanged because in this case we do not have the factors $1/(4\pi\epsilon_0)$ neither in the Gaussian versions of eqs. (13.18) and (13.20), nor in the Gaussian version of eq. (11.151), so we still have

$$\rho_{\text{pol}}(t, \mathbf{x}) = -\boldsymbol{\nabla}\cdot\mathbf{P}(t, \mathbf{x}), \qquad \sigma_{\text{pol}} = \hat{\mathbf{n}}_s\cdot\mathbf{P}(t, \mathbf{x}). \tag{A.183}$$

Similarly, eq. (13.27) still holds while, reproducing the steps in eqs. (13.29)–(13.32) taking into account the definition (A.86) of the magnetic dipole (that, as we already remarked, has a different factor of c with respect to the SI system), eqs. (13.32) and (13.33) become

$$\mathbf{j}_{\text{mag}}(t, \mathbf{x}) = c\boldsymbol{\nabla} \times \mathbf{M}(t, \mathbf{x}), \qquad \mathbf{K}_{\text{mag}}(t, \mathbf{x}) = c\mathbf{M}(t, \mathbf{x})\times\hat{\mathbf{n}}_s. \tag{A.184}$$

The displacement vector is now defined as

$$\mathbf{D} = \mathbf{E} + 4\pi\mathbf{P}, \tag{A.185}$$

to be compared with eq. (13.40), while the $\mathbf{H}$ field is defined as

$$\mathbf{H} = \mathbf{B} - 4\pi\mathbf{M}, \tag{A.186}$$

to be compared with eq. (13.43). In Gaussian units, the full set of Maxwell's equations in material media then reads

$$\boldsymbol{\nabla}\cdot\mathbf{D} = 4\pi\rho_{\text{free}}, \tag{A.187}$$

$$\boldsymbol{\nabla} \times \mathbf{H} - \frac{1}{c}\frac{\partial\mathbf{D}}{\partial t} = \frac{4\pi}{c}\mathbf{j}_{\text{free}}, \tag{A.188}$$

$$\boldsymbol{\nabla}\cdot\mathbf{B} = 0, \tag{A.189}$$

$$\boldsymbol{\nabla} \times \mathbf{E} + \frac{1}{c}\frac{\partial\mathbf{B}}{\partial t} = 0, \tag{A.190}$$

to be compared with eqs. (13.45)–(13.48). We observe that we can pass from Maxwell's equations in the SI system to Maxwell's equations in the Gaussian system extending the formal replacements (A.7) as follows:

$$\epsilon_0 \to \frac{1}{4\pi}, \qquad \mu_0 \to \frac{4\pi}{c^2}, \qquad \mathbf{E} \to \mathbf{E} \qquad \mathbf{B} \to \frac{\mathbf{B}}{c}, \tag{A.191}$$

$$\mathbf{D} \to \frac{1}{4\pi}\mathbf{D}, \qquad \mathbf{H} \to \frac{c}{4\pi}\mathbf{H}, \tag{A.192}$$

while leaving ρ_{free} and $\mathbf{j}_{\text{free}}$ unchanged. Furthermore, supplementing these replacements with

$$\mathbf{P} \to \mathbf{P}, \qquad \mathbf{M} \to c\mathbf{M}, \tag{A.193}$$

we see that, taking into account eqs. (A.191) and (A.192), eqs. (13.40) and (13.43) go into eqs. (A.185) and (A.186), respectively. We stress, once again, these these are just formal rules to get a quick translation of formulas from

SI to Gaussian units. The actual relations between quantities in the two systems are given by eqs. (2.25), (2.32), and (2.33). Since the electric dipole is proportional to the electric charge, eq. (2.25) also implies that

$$\mathbf{P}_{\text{gau}} = \frac{1}{\sqrt{4\pi\epsilon_0}}\mathbf{P}_{\text{SI}}, \qquad (\text{A.194})$$

and therefore,[7]

$$\mathbf{D}_{\text{gau}} = \sqrt{\frac{4\pi}{\epsilon_0}}\,\mathbf{D}_{\text{SI}}. \qquad (\text{A.195})$$

Similarly, since the magnetic dipole is proportional to the current, and therefore to the charge, using again eq. (2.25), and furthermore including the extra factor of $1/c$ between eqs. (6.36) and (A.86), we get

$$\mathbf{M}_{\text{gau}} = \frac{1}{\sqrt{4\pi\epsilon_0}}\frac{1}{c}\mathbf{M}_{\text{SI}} = \sqrt{\frac{\mu_0}{4\pi}}\,\mathbf{M}_{\text{SI}}, \qquad (\text{A.196})$$

and then[8,9]

$$\mathbf{H}_{\text{gau}} = \sqrt{4\pi\mu_0}\,\mathbf{H}_{\text{SI}}. \qquad (\text{A.197})$$

The equations governing the electrostatics of materials are

$$\boldsymbol{\nabla}\!\cdot\!\mathbf{D} = 4\pi\rho_{\text{free}}, \qquad \boldsymbol{\nabla}\times\mathbf{D} = 4\pi\boldsymbol{\nabla}\times\mathbf{P}, \qquad (\text{A.198})$$

compare with eq. (13.50), and therefore eqs. (13.51) and (13.52) become, respectively,

$$\mathbf{D}(\mathbf{x}) = -\boldsymbol{\nabla}\left(\int d^3x'\,\frac{\rho_{\text{free}}(\mathbf{x}')}{|\mathbf{x}-\mathbf{x}'|}\right) + \boldsymbol{\nabla}\times\left(\int d^3x'\,\frac{(\boldsymbol{\nabla}\times\mathbf{P})(\mathbf{x}')}{|\mathbf{x}-\mathbf{x}'|}\right), \qquad (\text{A.199})$$

$$\mathbf{E}(\mathbf{x}) = -\boldsymbol{\nabla}\left(\int d^3x'\,\frac{\rho_{\text{free}}(\mathbf{x}')}{|\mathbf{x}-\mathbf{x}'|}\right) + \boldsymbol{\nabla}\times\left(\int d^3x'\,\frac{(\boldsymbol{\nabla}\times\mathbf{P})(\mathbf{x}')}{|\mathbf{x}-\mathbf{x}'|}\right) - 4\pi\mathbf{P}(\mathbf{x}). \qquad (\text{A.200})$$

The equations governing the magnetostatics of materials are

$$\boldsymbol{\nabla}\!\cdot\!\mathbf{H} = -4\pi\boldsymbol{\nabla}\!\cdot\!\mathbf{M}, \qquad \boldsymbol{\nabla}\times\mathbf{H} = \frac{4\pi}{c}\mathbf{j}_{\text{free}}, \qquad (\text{A.201})$$

so the general solution for $\mathbf{H}$ and $\mathbf{B}$ are

$$\mathbf{H}(\mathbf{x}) = \frac{1}{c}\boldsymbol{\nabla}\times\left(\int d^3x'\,\frac{\mathbf{j}_{\text{free}}(\mathbf{x}')}{|\mathbf{x}-\mathbf{x}'|}\right) + \boldsymbol{\nabla}\left(\int d^3x'\,\frac{(\boldsymbol{\nabla}\!\cdot\!\mathbf{M})(\mathbf{x}')}{|\mathbf{x}-\mathbf{x}'|}\right), \qquad (\text{A.202})$$

$$\mathbf{B}(\mathbf{x}) = \frac{1}{c}\boldsymbol{\nabla}\times\left(\int d^3x'\,\frac{\mathbf{j}_{\text{free}}(\mathbf{x}')}{|\mathbf{x}-\mathbf{x}'|}\right) + \boldsymbol{\nabla}\left(\int d^3x'\,\frac{(\boldsymbol{\nabla}\!\cdot\!\mathbf{M})(\mathbf{x}')}{|\mathbf{x}-\mathbf{x}'|}\right) + 4\pi\mathbf{M}(\mathbf{x}), \qquad (\text{A.203})$$

compare with eqs. (13.55) and (13.56). The boundary conditions (13.63) and (13.70) become, respectively,

$$\hat{\mathbf{n}}\!\cdot\!(\mathbf{D}_2 - \mathbf{D}_1) = 4\pi\sigma_{\text{free}}, \qquad \hat{\mathbf{n}}\times(\mathbf{H}_2 - \mathbf{H}_1) = \frac{4\pi}{c}\mathbf{K}_{\text{free}}. \qquad (\text{A.204})$$

For a linear dielectric, the electric susceptibility χ_e is defined by

$$\mathbf{P}(t,\mathbf{x}) = \chi_e\mathbf{E}(t,\mathbf{x}), \qquad (\text{A.205})$$

to be compared with eq. (13.71). Then, from eq. (A.185),

$$\mathbf{D}(t,\mathbf{x}) = \epsilon\mathbf{E}(t,\mathbf{x}), \qquad (\text{A.206})$$

[7]Explicitly,

$$\begin{aligned}\mathbf{D}_{\text{gau}} &\equiv \mathbf{E}_{\text{gau}} + 4\pi\mathbf{P}_{\text{gau}}\\ &= \sqrt{4\pi\epsilon_0}\,\mathbf{E}_{\text{SI}} + \frac{4\pi}{\sqrt{4\pi\epsilon_0}}\mathbf{P}_{\text{SI}}\\ &= \sqrt{\frac{4\pi}{\epsilon_0}}\,(\epsilon_0\mathbf{E}_{\text{SI}} + \mathbf{P}_{\text{SI}})\\ &= \sqrt{\frac{4\pi}{\epsilon_0}}\,\mathbf{D}_{\text{SI}}.\end{aligned}$$

[8]Explicitly,

$$\begin{aligned}\mathbf{H}_{\text{gau}} &\equiv \mathbf{B}_{\text{gau}} - 4\pi\mathbf{M}_{\text{gau}}\\ &= \sqrt{\frac{4\pi}{\mu_0}}\,\mathbf{B}_{\text{SI}} - 4\pi\sqrt{\frac{\mu_0}{4\pi}}\,\mathbf{M}_{\text{SI}}\\ &= \sqrt{4\pi\mu_0}\left(\frac{1}{\mu_0}\mathbf{B}_{\text{SI}} - \mathbf{M}_{\text{SI}}\right)\\ &= \sqrt{4\pi\mu_0}\,\mathbf{H}_{\text{SI}}.\end{aligned}$$

[9]From Maxwell's equations eqs. (A.187)–(A.190), and eqs. (A.185) and (A.186), we see that in the Gaussian system, $\mathbf{E}, \mathbf{D}, \mathbf{P}, \mathbf{B}, \mathbf{H}$, and $\mathbf{M}$ all have the same dimensions. In contrast, in the SI system we already saw after eq. (2.33) that, dimensionally, $[\mathbf{B}_{\text{SI}}] = [\mathbf{E}_{\text{SI}}]/[v]$, i.e., there is an extra factor of $1/c$ in $\mathbf{B}_{\text{SI}}$. Similarly, we see from eq. (13.40) that in the SI system $\mathbf{D}$ has the same dimensions as $\mathbf{P}$, but these are different from the dimensions of $\mathbf{E}$ because there is an extra factor of ϵ_0 and, from eq. (13.43), in the SI system $\mathbf{H}$ has the same dimensions as $\mathbf{M}$, but the relation with the dimensions of $\mathbf{B}$ involve a factor $1/\mu_0$.

where the *dielectric constant* ϵ is given by

$$\epsilon = 1 + 4\pi\chi_e\,, \tag{A.207}$$

to be compared with eq. (13.73). Note that, in the Gaussian system, there is no distinction between the permittivity and the dielectric constant (or relative permittivity). Writing eq. (13.72) as $\mathbf{D}_{\mathrm{SI}} = \epsilon_{\mathrm{SI}}\mathbf{E}_{\mathrm{SI}}$ and eq. (A.206) as $\mathbf{D}_{\mathrm{gau}} = \epsilon_{\mathrm{gau}}\mathbf{E}_{\mathrm{gau}}$, and using eqs. (2.32) and (A.195) we see that

$$\epsilon_{\mathrm{SI}}/\epsilon_0 = \epsilon_{\mathrm{gau}}\,, \tag{A.208}$$

so ϵ_{gau} is the same as the relative permittivity, or dielectric constant, of the SI system. Then, comparing eqs. (13.73) and (A.207), written as

$$\epsilon_{\mathrm{SI}} = \epsilon_0[1 + (\chi_e)_{\mathrm{SI}}]\,, \tag{A.209}$$
$$\epsilon_{\mathrm{gau}} = 1 + 4\pi(\chi_e)_{\mathrm{gau}}\,, \tag{A.210}$$

we see that χ_e is dimensionless both in the SI and Gaussian systems but the respective numerical values are different,

$$(\chi_e)_{\mathrm{SI}} = 4\pi(\chi_e)_{\mathrm{gau}}\,. \tag{A.211}$$

In the Gaussian system, eq. (13.76) becomes

$$\begin{aligned}\rho_{\mathrm{pol}}(t,\mathbf{x}) &= -\frac{\chi_e}{\epsilon}\boldsymbol{\nabla}\cdot\mathbf{D}(t,\mathbf{x})\\ &= -\frac{4\pi\chi_e}{1+4\pi\chi_e}\,\rho_{\mathrm{free}}(t,\mathbf{x})\,.\end{aligned} \tag{A.212}$$

We now turn to conductive matter. In the SI system the conductivity is defined by Ohm's law (4.183), $\mathbf{j}_{\mathrm{SI}} = \sigma_{\mathrm{SI}}\mathbf{E}_{\mathrm{SI}}$. Similarly, in the Gaussian system it is defined by $\mathbf{j}_{\mathrm{gau}} = \sigma_{\mathrm{gau}}\mathbf{E}_{\mathrm{gau}}$. Combining eq. (2.32) with

$$\mathbf{j}_{\mathrm{gau}} = \frac{1}{\sqrt{4\pi\epsilon_0}}\,\mathbf{j}_{\mathrm{SI}}\,, \tag{A.213}$$

which follows from eq. (2.25), we obtain[10]

$$\sigma_{\mathrm{gau}} = \frac{\sigma_{\mathrm{SI}}}{4\pi\epsilon_0}\,. \tag{A.214}$$

For metals, eqs. (13.88) and (13.89) become

$$\epsilon(\omega)\boldsymbol{\nabla}\cdot\tilde{\mathbf{E}}(\omega,\mathbf{x}) = 0\,, \tag{A.215}$$
$$\boldsymbol{\nabla}\times\tilde{\mathbf{H}}(\omega,\mathbf{x}) + \frac{i\omega}{c}\,\epsilon(\omega)\tilde{\mathbf{E}}(\omega,\mathbf{x}) = 0\,, \tag{A.216}$$

where

$$\epsilon(\omega) \equiv \epsilon_b(\omega) + i\frac{4\pi\sigma(\omega)}{\omega}\,, \tag{A.217}$$

to be compared with eq. (13.90).[11]

We next consider magnetic matter. In Gaussian units the magnetic susceptibility χ_m and the magnetic permeability μ are still defined from

$$\mathbf{M} = \chi_m\mathbf{H}\,, \qquad \mathbf{B} = \mu\mathbf{H}\,, \tag{A.220}$$

as in eqs. (13.91) and (13.92). Then, from eq. (A.186), in Gaussian units

$$\mu = 1 + 4\pi\chi_m\,, \tag{A.221}$$

---

[10] As we saw in Note 49 on page 96, in the SI system conductivities are measured in siemens per meter (S/m); using eq. (4.188) and comparing with the dimensions of ϵ_0, given in eqs. (2.12) and (2.13), we see that σ_{gau} has dimensions of inverse time, so is measured in s^{-1}.

[11] Observe that, for the bound electrons, in Gaussian system eq. (A.207) holds so, explicitly writing the label "gau," eq. (A.217) reads

$$\begin{aligned}\epsilon_{\mathrm{gau}}(\omega) = 1 &+ 4\pi(\chi_e)_{\mathrm{gau}}(\omega)\\ &+i\frac{4\pi\sigma_{\mathrm{gau}}(\omega)}{\omega}\,,\end{aligned} \tag{A.218}$$

while, from eqs. (13.90) and (13.73)

$$\begin{aligned}\epsilon_{\mathrm{SI}}(\omega)/\epsilon_0 = 1 &+ (\chi_e)_{\mathrm{SI}}(\omega)\\ &+i\frac{\sigma_{\mathrm{SI}}(\omega)/\epsilon_0}{\omega}\,.\end{aligned} \tag{A.219}$$

Using eqs. (A.211) and (A.214) we get back eq. (A.208), confirming the consistency of these relations.

to be compared with eq. (13.93). Proceeding as in the derivation of eq. (A.211), we obtain

$$(\chi_m)_{\mathrm{SI}} = 4\pi(\chi_m)_{\mathrm{gau}}. \tag{A.222}$$

Energy conservation in materials reads

$$\frac{1}{4\pi} \int_V d^3x \left(\mathbf{E} \cdot \frac{\partial \mathbf{D}}{\partial t} + \mathbf{H} \cdot \frac{\partial \mathbf{B}}{\partial t} \right) + \int_V d^3x\, \mathbf{E} \cdot \mathbf{j}_{\mathrm{free}} = -\frac{c}{4\pi} \int_{\partial V} d\mathbf{s} \cdot (\mathbf{E} \times \mathbf{H}), \tag{A.223}$$

to be compared with eq. (13.95). The term on the right-hand side gives the generalization of the Poynting vector to material media,

$$\mathbf{S} = \frac{c}{4\pi} \mathbf{E} \times \mathbf{H}. \tag{A.224}$$

For a simple linear medium with $\mathbf{D} = \epsilon\mathbf{E}$ and $\mathbf{B} = \mu\mathbf{H}$, the energy density of the electromagnetic field is

$$u = \frac{1}{8\pi} \left(\epsilon\mathbf{E}^2 + \mu\mathbf{H}^2 \right), \tag{A.225}$$

to be compared with eq. (13.98). The Larmor frequency (13.126) becomes

$$\omega_L \equiv \frac{eB}{2m_e c}, \tag{A.226}$$

and eq. (13.132) becomes $\chi_m = -ne^2\langle r_\perp^2 \rangle / 4m_e c^2$.

Chapter 14

In Gaussian units, the relation between the dielectric constant and the electric susceptibility is given by eq. (A.207), so its frequency dependent generalization, which in SI units is given by eq. (14.10), becomes

$$\epsilon(\omega) = 1 + 4\pi\chi_e(\omega), \tag{A.227}$$

while eq. (14.12) becomes

$$\tilde{\mathbf{P}}(\omega) = \chi_e(\omega)\tilde{\mathbf{E}}(\omega). \tag{A.228}$$

Equation (14.28) becomes

$$\mathbf{D}(t) = \mathbf{E}(t) + 4\pi \int_{-\infty}^{t} dt'\, \chi_e(t-t')\mathbf{E}(t'). \tag{A.229}$$

The plasma frequency (14.38) is now given by

$$\omega_p^2 = \frac{4\pi n_b e^2}{m_e}, \tag{A.230}$$

and, in the Drude–Lorentz model, we have

$$\chi_e(\omega) = \frac{n_b e^2}{m_e} \frac{1}{\omega_0^2 - \omega^2 - i\omega\gamma_0}, \tag{A.231}$$

and therefore

$$\epsilon(\omega) = 1 + \frac{\omega_p^2}{\omega_0^2 - \omega^2 - i\omega\gamma_0}. \tag{A.232}$$

The conductivity in the Drude model is given by

$$\sigma(\omega) = \frac{n_f e^2 \tau}{m_e} \frac{1}{1 - i\omega\tau}, \tag{A.233}$$

which is formally the same as eq. (14.58). Note that this is consistent with eq. (A.214), given the relation between the charges in SI and Gaussian units, eq. (A.2).

The dielectric function of metals, which in SI units is given in eq. (14.76), is still formally given by

$$\epsilon(\omega) = 1 + \omega_p^2 \sum_{i=0}^{N} \frac{f_i}{\omega_i^2 - \omega^2 - i\omega\gamma_i}, \tag{A.234}$$

where, however, now $\omega_p^2 = 4\pi n e^2/m_e$.

Chapter 15

Searching for a monochromatic wave solutions of the form (15.1), Maxwell's equations in material media, eqs. (A.187)–(A.190), give

$$\epsilon(\omega)\mathbf{k}\cdot\tilde{\mathbf{E}}(\omega, \mathbf{k}) = 0, \tag{A.235}$$

$$\mathbf{k} \times \tilde{\mathbf{B}}(\omega, \mathbf{k}) = -\frac{\omega}{c}n^2(\omega)\tilde{\mathbf{E}}(\omega, \mathbf{k}), \tag{A.236}$$

$$\mathbf{k}\cdot\tilde{\mathbf{B}}(\omega, \mathbf{k}) = 0, \tag{A.237}$$

$$\mathbf{k} \times \tilde{\mathbf{E}}(\omega, \mathbf{k}) = \frac{\omega}{c}\tilde{\mathbf{B}}(\omega, \mathbf{k}), \tag{A.238}$$

where the refraction index $n(\omega)$ is now defined from[12]

$$n(\omega) = \sqrt{\epsilon(\omega)\mu(\omega)}. \tag{A.239}$$

The dispersion relation (15.11) remains unchanged, and therefore also eqs. (15.18) and (15.19), while eq. (15.16) is replaced by

$$\mathbf{B_k} = n(\omega)\,\hat{\mathbf{k}} \times \mathbf{E_k}. \tag{A.240}$$

The dispersion relation (15.39) becomes

$$\omega^2 \epsilon(\omega)\mu(\omega) = k^2 c^2. \tag{A.241}$$

The expression (15.43) for the skin depth becomes

$$\delta_{\text{skin}}(\omega) = \frac{c}{\sqrt{2\pi\sigma_0\omega}}, \qquad (\omega \ll \gamma_p). \tag{A.242}$$

The equations of Section 15.4.1 for the propagation in waveguides are transformed into Gaussian units with the usual replacement $\mathbf{B} \to \mathbf{B}/c$.

[12] Recall that, in Gaussian units, $\epsilon(\omega)$ and $\mu(\omega)$ are dimensionless, see e.g., eqs. (A.210) and (A.221), so $n(\omega)$ remains dimensionless also in Gaussian units.

Chapter 16

Most results of this chapter are independent of whether one uses SI or Gaussian units, except eqs. (16.18) and (16.19) that in Gaussian units read, respectively

$$\frac{dP(t;\theta)}{d\Omega} = \frac{1}{4\pi c^3}\frac{e^4 E^2}{m_e^2} \sin^2\theta, \tag{A.243}$$

and $I = cE^2/4\pi$. Therefore eq. (16.20) still holds, with the classical electron radius defined as

$$r_0 = \frac{e^2}{m_e c^2}. \tag{A.244}$$

Similarly, eqs. (16.41) and (16.42) become

$$\frac{dP(\theta)}{d\Omega} = \frac{1}{8\pi c^3}\left(\frac{e^2 E_0}{m_e}\right)^2 \frac{\omega^4}{(\omega^2 - \omega_0^2)^2 + \omega^2\gamma_0^2} \sin^2\theta, \tag{A.245}$$

and $\langle I(t)\rangle = cE^2/4\pi$. so eq. (16.44) still holds, again with the classical electron radius given by eq. (A.244).

References

Andryieuski, A., Kuznetsova, S., Zhukovsky, S., Kivshar, Y., and Lavri-nenko, A.V. (2015). Water: Promising opportunities for tunable all-dielectric electromagnetic metamaterials. *Scientific Reports*, **5**, 13535.

Damour, T. (1987). The problem of motion in Newtonian and Ein-steinian gravity. In *300 Years of Gravitation*. S. Hawking and W. Israel (eds.), Cambridge University Press, Cambridge.

Damour, T. (2017). Poincaré, the Dynamics of the Electron, and Rel-ativity. *Comptes Rendus Physique*, **18**, 551–562.

Dirac, P. A. M. (1938). Classical theory of radiating electrons. *Proc. R. Soc. Lond. A*, **167**, 148.

Dressel, M. and Grüner, G. (2002). *Electrodynamics of Solids*. Cam-bridge University Press, Cambridge.

Feynman, R. P., Leighton, R. B., and Sands, M. L. (1964). *The Feyn-man Lectures on Physics*, Vol. 2. Addison-Wesley Pub. Co., Reading, Mass.

Garg, A. (2012). *Classical Electromagnetism in a Nutshell*. Princeton University Press, Princeton, USA.

Gradshteyn, I. S. and Ryzhik, I. M. (1980). *Table of Integrals, Series and Products*. Academic Press, London.

Griffiths, D. J. (2004). *Introduction to Quantum Mechanics*, 2nd edn. Pearson, Prentice Hall.

Griffiths, D. J. (2017). *Introduction to Electrodynamics*, 4th edn. Cam-bridge University Press, Cambridge.

Jackson, J. D. (1998). *Classical Electrodynamics*, 3rd edn. Wiley and Sons, USA.

Jackson, J. D. and Okun, L. B. (2001). Historical roots of gauge in-variance. *Rev. Mod. Phys.*, **73**, 663–680.

Landau, L. D. and Lifschits, E. M. (1975). *The Classical Theory of Fields*. Course of Theoretical Physics, Vol. 2. Pergamon Press, Oxford.

Landau, L. D. and Lifschits, E. M. (1984). *Electrodynamics of Contin-uous Media*. Course of Theoretical Physics, Vol. 8, 2nd edn. Pergamon Press, Oxford.

Maggiore, M. (2005). *A Modern Introduction to Quantum Field Theory*. Oxford University Press, Oxford.

Maggiore, M. (2007). *Gravitational Waves. Vol. 1: Theory and Exper-iments*. Oxford University Press, Oxford.

Pearle, P. (1982). Classical electron models. In *Electromagnetism: Paths to Research*. D. Teplitz, (ed.), Springer New York, NY.

Poisson, E. and Will, C. M. (2014). *Gravity. Newtonian, Post-Newtonian, Relativistic*. Cambridge University Press, Cambridge.

Purcell, E. M. and Morin, D. (2013). *Electricity and Magnetism,* 3rd edn. Cambridge University Press, Cambridge.

Rakić, D., Djurišić, D., Elazar, J. M., and Majewski, M. L. (1998). Optical properties of metallic films for vertical-cavity optoelectronic devices. *Appl. Optics*, **37**, 5271.

Rohrlich, F. (2000). The self-force and radiation reaction. *American Journal of Physics*, **68**, 1109.

Rybicki, G. B. and Lightman, A.P. (1979). *Radiative Processes in Astrophysics*. Wiley and Sons, Cambridge.

Tong, D. (2015). *Lectures on Electromagnetism*. Cambridge University Press, Cambridge.

Weinberg, S. (1972). *Gravitation and Cosmology: Principles and Applications of the General Theory of Relativity*. John Wiley and Sons, New York.

Zangwill, A. (2013). *Modern Electrodynamics*. Cambridge University Press, Cambridge.

Zee, A. (2016). *Group Theory in a Nutshell for Physicists*. Princeton University Press, Princeton, USA.

Index